Planning, Construction, and Statistical Analysis of Comparative Experiments

A complete list of the titles in this series appears at the end of this volume.

Planning, Construction, and Statistical Analysis of Comparative Experiments

FRANCIS G. GIESBRECHT
MARCIA L. GUMPERTZ

WILEY-
INTERSCIENCE

A JOHN WILEY & SONS, INC., PUBLICATION

Published by John Wiley & Sons, Inc., Hoboken, New Jersey.
Published simultaneously in Canada.

Library of Congress Cataloging-in-Publication Data:

Giesbrecht, Francis G., 1935–
Planning, construction, and statistical analysis of comparative experiments / Francis G. Giesbrecht, Marcia L. Gumpertz.
p. cm.—(Wiley series in probability and statistics)
Includes bibliographical references and index.
ISBN 0-471-21395-0 (acid-free paper)
1. Experimental design. I. Gumpertz, Marcia L., 1952– II. Title. III. Series.

QA279.G52 2004
519.5—dc21

2003053838

Printed in the United States of America.

10 9 8 7 6 5 4 3 2 1

Contents

Preface

This is a book based on years of teaching statistics at the graduate level and many more years of consulting with individuals conducting experiments in many disciplines. The object of the book is twofold. It is designed to be used as a textbook for students who have already had a solid course in statistical methods. It is also designed to be a handbook and reference for active researchers and statisticians who serve as consultants in the design of experiments.

We encountered very early in our work that the subject is not linear in the sense that there is a unique order in which topics must be presented—that topics A, B, and C must be presented in that order. All too often, we found that in one sense, A should be before B, and in another, B should be before A. Our aim is to give sufficient detail to show interrelationships, that is, how to reach destinations, yet not hide important points with too much clutter.

We assume that the reader of this book is comfortable with simple sampling, multiple regression, t-tests, confidence intervals, and the analysis of variance at the level found in standard statistical methods textbooks. The demands for mathematics are limited. Simple algebra and matrix manipulation are sufficient.

In our own teaching, we have found that we pick and choose material from various chapters. Chapters 1 to 5 and Chapter 7 provide a review of methods of statistical analysis and establish notation for the remainder of the book. Material selected from the remaining chapters then provides the backbone of an applied course in design of experiments. There is more material than can be covered in one semester. For a graduate level course in design of experiments targeted at a mix of statistics and non-statistics majors, one possible course outline covers Chapter 1, Sections 3.3 and 3.4, Chapter 4, Section 5.5, and Chapters 6 and 8 to 15. If the students have a very strong preparation in statistical methods, we suggest covering selections from Chapters 6 to 16 and 19 to 24, with just the briefest review of Chapters 1 to 5. We have also used material from Chapters 10 to 14, 16, and 19 to 24 for a more specialized course on factorial experiments for students interested in industrial quality control.

In a sense, it would have been much easier to write a book that contained a series of chapters that would serve as a straightforward text. However, our aim is also to provide a handbook for the investigator planning a research program.

As part of this, we have included many detailed plans for experiments. Data and SAS® code for the examples and exercises are provided on the companion Web site reached at one of the following URLs.

```
http://www4.stat.ncsu.edu/~gumpertz/ggdata
http://www.wiley.com/WileyCDA/Section/id-101289.html
```

We have also found it useful to separate the concepts of experimental design and treatment design, although we realize that in practice, the two concepts become intertwined. We try to reserve the term *experimental design* to situations where the investigator is manipulating experimental material to form blocks or replicates, and *treatment design* where the nature of the treatments themselves is manipulated.

We wish to thank our colleagues, students, research collaborators, and friends for providing the problems that define the foundation of this book. Questions raised by students are the basis for many of the exercises scattered throughout the text. Numeric problems come from consulting sessions with students and researchers. Many of the topics and tables included in this book are a direct outgrowth of some of the more challenging concerns raised during design consulting sessions. We hope that having this material available in one location and in a consistent format will provide consultants and investigators with access to a range of design possibilities for planning their experiments.

F. GIESBRECHT
M. GUMPERTZ

Raleigh, North Carolina

CHAPTER 1

Introduction

1.1 ROLE OF STATISTICS IN EXPERIMENT DESIGN

The basic premise in this book is that people learn by experience. We paraphrase Box and Meyer (1985), who point out that it is in our ability to learn that we differ most remarkably from animals. Although it is clear that technological learning has taken place throughout the history of humankind, change occurred very slowly until the last 300 or 400 years. One of the reasons for this was that for learning to take place, two rare events needed to coincide. An informative event had to occur and, in addition, a person able to draw logical conclusions and act upon them had to be aware of that informative event.

Passive surveillance of our surroundings is not enough. Careful attention and measurement as exemplified by quality charting methods practiced in many industries is a way of increasing the probability that rare, informative events will be noticed. Active intervention by experimentation increases the probability of such events actually occurring. Well-designed experiments conducted by qualified observers can dramatically increase the probability of learning because both the probability of an informative event occurring and the probability of the event being noticed are greatly increased.

Our aim in this book is to provide tools to construct better experiments and consequently, make the process of learning more efficient. We divide the process of conducting an experiment into a planning phase, an execution phase, and an analysis or interpretation phase. It is in the first phase that the investigator thinks about hypotheses to be tested (i.e., answers questions such as "Why do the study?", "What material do I want to use?", and "What do I want to learn?"). At this stage, the investigator thinks about conditions under which the study is to be performed, factors that need to be considered, and factors that can safely be ignored. Ethical considerations, as well as financial and time constraints, must be considered. It is the second stage that the layperson typically thinks of as conducting an experiment. It is at this stage where the actual treatments are

Planning, Construction, and Statistical Analysis of Comparative Experiments,
by Francis G. Giesbrecht and Marcia L. Gumpertz
ISBN 0-471-21395-0

applied, when the actual work is being done in the laboratory. The final stage consists of some sort of statistical analysis. This may be done formally, according to strict rules and methods as found in numerous statistics books, or very informally. All too often, it is thought that this is the stage where statistics can make its major contribution. Although proper examination of the results of an experiment is important, there is no way that a clever analysis can make up for a poorly designed study, a study that leaves out key factors, or inadvertently confounds and/or masks relevant factors. It is our view that the analysis of data from a well-designed and well-executed study is usually very simple. Often, proper interpretation of the results requires little more than computing a few averages.

Statistics can have the greatest impact on the quality of an experimental study while in the planning stage. The feasibility of any experiment is constrained by limitations on time, equipment, funds, and ethical considerations. Ethical constraints range from such mundane considerations as too large an experiment, which wastes resources, to too small an experiment, which also wastes resources because it fails to give answers, to more serious considerations, such as causing pain and suffering to living organisms or even to fellow human beings, or harming the environment. The goals of statistical experimental design are to provide tools that help to ensure that the experiment will be capable of providing answers to the research questions posed and that it will provide the information as efficiently as possible using the available resources and respecting all ethical constraints. An experimental plan that does not respect ethical constraints is not a good plan.

The statistician is rarely the lead investigator in scientific investigations, but statistics has some valuable insights and tools to offer to the process of designing experiments. In this book, emphasis is on such questions as:

1. How many experimental units should be examined? How much replication should there be?
2. How does one construct the experiment to provide as broad a basis for inference as possible?
3. How does one take into account the natural variability among experimental units?
4. How should treatments be allocated to the entities being studied?
5. How is the statistical analysis to be performed?

There are four key principles of design of experiments that need to be emphasized: (1) *representativeness*, (2) *randomization*, (3) *replication*, and (4) *error control or blocking*. These ideas are fundamental to making an experiment useful for providing answers to the research questions. Representativeness, proper randomization, and sufficient replication and error control are the solid bases that allow inferences to be made without bias and with enough precision to be meaningful. Randomization provides insurance that unanticipated factors do not accidentally introduce unwanted biases into experimental results. Replication and error control are necessary to ensure that results are generally true and not just the product of the special circumstances of a particular run. Each of these ideas is discussed in more detail in later sections of this chapter.

1.2 ORGANIZATION OF THIS BOOK

The emphasis in this book is on designs for controlled comparative experiments, as opposed to observational studies or sample surveys. The distinguishing feature of the controlled experiments is that the experimental material is manipulated by the investigator. It has been argued by Sprott and Farewell (1993) that the purpose of controlled experimentation is to elucidate causation, (i.e., to establish that treatment T applied under condition C will produce a predictable outcome O or, more generally, the likelihood of specific outcome $O(T)$. This is in contrast to the observational study or survey, where the investigator takes observations on the world as it exists, calculates summary statistics, and draws conclusions. Controlled experiments are typically characterized by relatively few observations, simple analyses, and relatively high information content per experimental unit. For example, a study of the effect of atmospheric ozone on wheat yield might involve four plots of wheat at each of two ozone levels, repeated for three years. We often see experiments with as few as 15 or 30 observations, but which provide much information. The safety of many food additives is typically established by experiments that involve no more than several hundred rodents and very simple statistical analyses. On the other hand, observational studies frequently involve thousands of observations and complex statistical analyses. An extreme example that comes to mind here is the problem of establishing a link between smoking and lung cancer or obesity and cancer in humans, cases where controlled experimentation is really not possible, yet thousands of observations are readily available. The emphasis in this book is on techniques for designing the relatively small studies which yield data that have a high information content.

Many types of designs have been developed to satisfy the particular constraints and objectives of different fields of research. Randomized complete block and split-plot designs were initially developed for agricultural trials. These simple designs are applicable in almost every field of research and are among the most widely used designs anywhere. Latin squares were also developed for agricultural applications and today are staples in medical and pharmaceutical clinical trials, in animal science, in marketing studies, and in well-run elections. Incomplete block designs were developed for agricultural trials comparing large numbers of varieties of crops and are also appropriate for comparisons of brands of consumer items. Fractional factorials, Plackett–Burman plans, and orthogonal arrays, all of which are really arrangements of treatments rather than experimental designs, were developed in agricultural research but are now frequently used for screening many possible factors in industrial processes, when cost and time are limiting factors. Response surface designs provide techniques for finding the best combination of settings to maximize yield of a product, minimize the number of defects, minimize cost, or maximize performance of a system in industrial experimentation.

The twin emphases throughout this book are on planning or designing experiments and on the corresponding statistical analyses. A key point to keep in mind is that the analysis and interpretation of a well-planned and well-executed experiment are typically quite simple and are based on a few reasonable and simple assumptions. Poorly planned experiments often lead to quite involved and computationally intensive statistical analyses, and even then, interpretations of the results often require complex and unverifiable assumptions. For the most part, we present statistical analyses based on the linear model and the analysis of variance. In our view the linear model provides a unified framework for evaluating proposed designs and for analyzing the results of completed experiments. However, other statistical analyses are often possible. For some of the simpler designs, we point out alternative methods of analysis.

The process of designing an experiment has three parts. First, it is necessary to understand the objectives and constraints clearly, state what quantities are to be estimated, and identify what hypotheses are to be tested. The comment that a problem is too complex, or involves too many ethical or financial constraints for a properly designed experiment, reflects poor understanding of the experimental process. The second task is to consider possible designs that are tailored to the aims of the experiment and yet satisfy all ethical, physical, and financial constraints. The final part of the process is to evaluate the proposed design to establish that the objectives can be achieved, that the estimates are likely to have sufficient precision, and that any proposed tests of hypotheses have adequate power. Outlining the statistical analysis before the experiment is actually conducted is always a good idea. It allows the investigator to see, clearly, what can be estimated and what hypotheses can be tested before any data are collected. It is all too common to find an experiment that appears reasonable and yet on closer examination fails to provide estimates of important quantities or does not allow testing the hypotheses of interest because of insufficient replication or flawed allocation of treatments! Outlining the statistical analysis during the planning stage helps prevent such mistakes.

1.3 REPRESENTATIVENESS AND EXPERIMENTAL UNITS

We use the terms *experimental units* or *plots* to refer to the basic elements in a comparative experiment. They are the items to which the treatments are assigned. Alternatively, it is also correct to say that experimental units are the units that are assigned to the individual treatments. The term *plots* reflects the agricultural origin of many of the concepts that will be used. The uninitiated occasionally make the mistake of thinking of the items that lead to individual observations as the experimental units. For example, the animals in a pen or the students in a class function as a unit, since the entire unit is assigned to the treatment in question. Similarly, the individual widgets produced under set conditions in a manufacturing study function as a unit, and the entire lot produced constitute the experimental unit. This remains true even though observations or numbers are collected on individual animals in a pen, or individual students in the class, or individual widgets. The same problem exists in the laboratory experiment, where repeated measurements or analyses of triplicate samples from the same experimental unit are often mistakenly thought of as coming from independent units. Proper identification of the experimental units in an experiment is of utmost importance, since differences among experimental units or plots provide the proper measure of experimental error to be used to judge whether observed differences can be attributed to chance or not. We must, however, point out that although the concept of experimental unit is so natural and obvious in most areas of research, there are situations where units are difficult to identify and the concept appears somewhat artificial. The concept still remains valuable. We discuss some cases in later chapters. We also note in passing that in some areas of research, the term *assemblies* is used in place of experimental units.

By representativeness we mean the extent to which the experimental material is representative of the units or material for which inferences are to be made. Are we safe in making inferences to the large population of animals in an entire species after conducting a nutrition experiment with a highly inbred strain of animals? A recent example where representativeness can be inadvertently sacrificed in an attempt to make an experiment more powerful is demonstrated by Felton and Gaylor (1989). They argue that experiments using single strains of animals provide more powerful tests of toxicity of chemical substances than multistrain experiments. There is a similar question of whether we are justified in recommending a variety of grain after testing it only on the research plots that have been used for scientific work for many years. Can we safely recommend for general use, a procedure that works well for a small well-trained cadre of technicians? Are the soil samples used by a researcher in civil engineering studying highway construction methods representative of the soils that will be encountered in practice? Are studies conducted in freshman or sophomore psychology classes of much value when we are trying to make inferences about students or humans in general? Are studies conducted with children from middle-class suburban homes adequate for identifying teaching techniques to be used with children from poorer sections of the inner city? How safe is it to infer dose requirements for medicines for elderly females when tests have been conducted on young adult males? This may well be the most important point in the validity of an experiment.

We realize full well that there are cases where it is not possible to conduct experiments with units that are truly representative of the target population. However, it must be kept in mind that in such cases, the final inferences are based on something beyond the experimental results at hand. In this category we include toxicity studies with rodents and eventual inferences to humans. However, it is generally true that if the material used in the experiment is not representative of the material to which the conclusions are to be applied, the experiment is somewhat flawed, regardless of the reason and regardless of the experiment's other characteristics.

Notice that we refer to two types of inferences involved in research work. First there are the statistical inferences, based on statistical analyses of the results of experiments. The statistical inferences are limited to the population of units from which the sample of experimental units actually used in the experiment was drawn. Then there is another type of inference that relies on knowledge of the subject matter. An "expert" may have good reason to argue that if something works with units drawn from one population, it will also work with units drawn from another population. For example, if an experiment is conducted using animals of type a as subjects, statistical inferences based on data from that experiment are limited to type a animals. However, an expert may claim that other animals deemed similar will react similarly. Statistics is not part of this inferential process. Similarly, statistical inferences about the performance of machine tools is limited to tools from one manufacturing plant if the selection of units for experimental testing was limited to tools from that plant. Inferences to tools from another plant depend on someone "knowing" that all the units from the two plants are alike.

A common mistake in scientific research is to restrict the pool of candidate experimental units too severely in an effort to make them as uniform as possible, thereby overly restricting the scope or the basis for the statistical inference from the experiment. Statistical inferences have the advantage of being based on sound probability theory and can be evaluated objectively. Expert knowledge inferences are only as good as the expert. They cannot be evaluated objectively. Festing and Lovell (1996) cite a dramatic example of the inference problem in the arena of testing effects of the carcinogens diethylstilbestrol (DES) and neutron irradiation in rats. In many animal studies of carcinogens (up to 1996) one particular inbred strain was used. This had the advantage that the animals were genetically identical, keeping the random error very small. The assumption made in interpreting results from these studies is an enormous one: that the carcinogen will have similar effects in other strains. Table 1.1 illustrates the problem. The ACI strain of rats developed mammary adenocarcinomas (MACs) under radiation in the presence of DES, but not otherwise, whereas the Sprague–Dawley strain seemed protected by the DES. Too often, unfortunately, expert knowledge misses this sort of thing.

Table 1.1 Occurrence of Mammary Adenocarcinomas and Fibroadenomas in Two Strains of Rat

	ACI Strain			Sprague–Dawley Strain		
Treatment	No.	MAC (%)	MFA (%)	No.	MAC (%)	MFA (%)
0 rad	13	0	0	33	3	3
0.4 rad	35	0	0	46	2	11
1.3 rad	24	0	4	35	3	14
4.0 rad	23	0	4	34	15	32
DES + 0 rad	23	52	0	33	0	0
DES + 0.4 rad	33	67	0	44	0	2
DES + 1.3 rad	23	83	0	35	0	0
DES + 4.0 rad	23	91	0	33	3	3

Source: Festing and Lovell (1996); abstracted from Shellabarger et al. (1978).

1.4 REPLICATION AND HANDLING UNEXPLAINED VARIABILITY

The precision (or inversely, the variability) of estimates and the power of hypothesis tests depend on the amount of replication and the magnitude of the unexplained variation among experimental units. In general, the more replication, the more precise the estimates. The amount of replication and the pattern of how many units are assigned to each treatment are things that the investigator can control, and much of this book is devoted to these topics. If an experiment is not replicated properly, it may not be possible to estimate the precision at all. It is important to know how precise estimates are in order to have some idea of the likelihood or assurance that an estimate falls within reasonable distance of the true (but unknown) value.

The precision of estimates also depends on the magnitude of unexplained variation in the experimental material. Consider a simple experiment to determine the effect of atmospheric ozone from air pollution on total per plant leaf area of wheat plants. The experimental units are plots of wheat and the measurement of interest is the average total leaf area for n randomly selected plants. Any variation among the plots that cannot be associated with ozone treatments is called *unexplained variation*. Differences among plots due to different soil conditions, microclimate variations, or genetic variation among plants all contribute to the unexplained variation. In addition, error in the leaf area measurements can contribute to the unexplained variation. The experimenter can reduce the unexplained variation by improving experimental techniques such as the depth to which seeds are planted, uniformity of plot preparation, and measurement techniques for determining leaf area and to an extent by increasing n, the number of plants measured.

In some cases it is possible to reduce unexplained variation by using the technique of *blocking*. This does not reduce the unexplained variation per se, but it assigns treatments to experimental units in such a way that the unexplained variation is screened out before making comparisons among the treatments. In our ozone example the plots can be grouped by soil condition and/or genetic type of plants. Blocking is possibly the most useful idea in all of statistical experimental design. It permits the investigator to allow for a larger variety of experimental units and thereby broadens the basis for statistical inference without risk of masking treatment effects. This topic is discussed at length in Chapter 5.

Finally, the unexplained variation can often be reduced by improving the model. This can be done by building in additional explanatory variables, also called *covariates*, or by changing the form of the model. The decision to measure additional variables should be made in the planning stages before the experiment is executed. For example, the amount of nitrogen in the soil affects the leaf area of wheat. If nitrogen concentration is to be used as a covariate, it should be measured in the soil at the beginning of the season, before the ozone exposures begin, so that which effects are due to the ozone and which due to the preexisting nitrogen level is unambiguous.

It should be noted that the common technique of adding covariates to the statistical model posits a linear model with various assumptions. Simple linear models without covariates permit analyses and tests of hypotheses based solely on the principles of randomization. The usual methods of analysis and testing hypotheses utilizing extended linear models rest on assumptions such as uncorrelated errors, constant variances, and normal distributions. A detailed discussion of the points involved can be found in Cox (1956). Incorporating additional explanatory variables into the analysis can, however, provide a large gain in the ability to differentiate effects due to the treatments from background variation. A cost is a set of somewhat more stringent assumptions.

Of the three methods described: (1) improving experimental technique, (2) blocking, and (3) improving the model, the first is usually up to the subject-matter scientist, and the latter two are statistical devices. The goal of all three is to reduce the unexplained variation so that the effects of treatments stand out from the background noise and can be identified more easily.

1.5 RANDOMIZATION: WHY AND HOW

The original notion of randomization is usually credited to R. A. Fisher and is accepted by most experimenters. Unfortunately, not all practice it. Too often, haphazard assignment of treatments to experimental units is the practice used instead of true randomization. The two are not the same. Another common error is to assume that any randomization scheme that assigns each treatment to every experimental unit equally frequently is adequate. This is not true. For proper randomization, all possible assignments of treatments to experimental units must be equally likely to appear. An alternative but fully equivalent requirement, which is often easier to verify, is that each pair of treatments (control can be thought of as a treatment) has exactly the same chance of being assigned to every pair of experimental units.

Consider an experiment that consists of two treatments, each assigned to five of 10 experimental units. A total of $\binom{10}{5}$ or 252 assignments are possible. A good randomization procedure will guarantee that each one of these is equally likely to happen. Simply dividing the experimental units into two groups of five (five good and five poor!) and then using a random coin toss to assign the treatment to one of the groups of five is not adequate. Why not? Also, it is not adequate to use an unbiased coin to assign treatments until enough have been assigned to the one treatment and then simply allocating the remainder to the other treatment. An easy-to-read reference in the *American Statistician* is Hader (1973).

Why not just assign the treatments in some haphazard fashion, or better yet, in some order that is convenient to the person conducting the experiment? After all, randomly selecting the order in which things are done can often lead to some inconvenience. It may mean that a certain piece of equipment has to be recalibrated repeatedly. The root question is: Why randomize? There are really two fundamental and very different reasons to randomize. Cox (1958, p. 85) states the positive advantages of randomization as assurances that:

(a) "In a large experiment it is very unlikely that the estimated treatment effects will be appreciably in error."

(b) "The random error of the estimated treatment effects can be measured and their level of statistical significance examined, taking into account all possible forms of uncontrolled variation."

With respect to the first of these points, it is important to be assured that the fact that experimental units on one treatment do better than those on an alternative treatment can only be attributed to the different treatments and not to anything else. If a nonrandomized design is used, it must be argued that the method of treatment assignment is very unlikely to be responsible for the treatment effect. There is no way to measure the uncertainty associated with this sort of statement. Randomization is the means to establish that the treatment caused the difference observed. The purpose of deliberate randomization in controlled scientific experiments is to separate causation from correlation. Farewell and Sprott (1992) refer to the two points as "the causal or scientific reason for randomization" and "the distributional or statistical reason." They refer to these as *causal inference* and *statistical inference*, respectively. They claim that "The statistical inference is a statement about the magnitude of the effect, while the causal inference is about what causes the effect described by the statistical inference." The causal inference is "the sole reason for which the experiment was performed." This requires deliberate randomization.

Sprott (2000, Chap. 10) illustrates the need for randomization with the hypothetical example of a trial in which 50 subjects are assigned to treatment and 50 are assigned to control. All 50 in the treatment group respond and none of the control group respond. Clearly, no statistical analysis is needed to support the inference that there is a difference between the group receiving treatment and the control group. Yet to infer that the treatment, and not some other outside factor, caused the difference requires some assurance that the subjects were assigned at random and not carefully selected, or possibly even allowed to select themselves.

It should be pointed out that the sharp distinction between these two inferences in the controlled experiment is blurred in the observational study or survey. In an observational study, factors are observed as nature presents them. In a controlled experiment, factors are manipulated at will. The primary purpose of a controlled experiment is to elucidate causation. The role of deliberate randomization in an experiment is to separate causation from correlation (Barnard 1986). There is no counterpart in an observational study. Whether and under what assumptions causal inferences may be made from nonrandomized studies is the subject of ongoing study and controversy. Dawid (2000) and the accompanying discussion articles give an idea of the complexity of the arguments.

D. R. Cox's second point has to do with the foundations of inference from an experiment. Classical tests of significance using Student's t-tests or Snedecor's F-tests, for example, rely on the assumption that the observations are randomly sampled from a theoretical distribution, such as the normal distribution. In a designed experiment, however, we assign treatments to experimental material that is available. Frequently, it is difficult to see how the material can be viewed as being selected from any particular theoretical distribution. Tests of significance can be justified, regardless of the underlying distribution of the experimental material, using an argument based on randomizing the allocation of treatments to the experimental units. A simple example of a randomization type of argument used to derive a test of significance uses the simple experimental setup described previously that consists of two treatments, each randomly assigned to five of 10 experimental units. We saw that 252 assignments are possible and that a good randomization procedure ensures that each assignment is equally likely. The experimenter will observe one of the assignments. Now assume that all the experimental units that received the treatment yielded more than any of the control units, (i.e., units that received no treatment). If the treatment had absolutely no effect, this would still be expected to happen, purely by chance, once in every 252 randomizations. One would expect it $1/252 \times 100$ or .4% of the time. If this happens, the experimenter can either claim that the treatments had no effect and that something very unusual (something that should happen only .4% of the time) has happened, or else the treatment had an effect on the units. The level of significance for this test is at the .004 or .4% level. This test is completely valid without making any distributional assumptions. The only requirement is that treatments be assigned at random. This is completely under the control of the investigator. Here randomization is used as a device to obtain a statistical test of significance. A classical discussion of the principles of randomization and constructing a test based purely on the randomization can be found in Chapter II of *The Design of Experiments* by Fisher (1935).

Unfortunately, there is no universal agreement in the statistical literature on the need for randomization. Generally speaking, however, when we find the claim that randomization is not required, we find that the causal inference component is being slighted in favor of the statistical inference. It is possible to construct a very coherent argument for statistical inference in which randomization is totally unnecessary and irrelevant. In our view, the problem is that for science, this is not the real question. The real question is the causal inference. Barnard (1986) summarizes the situation very succinctly when he writes:

> In the experimental verification of a supposed causal relation, in the common case when the perfectly uniform experimental units are not available, it is necessary to randomize the allocation of units between the various values or states of the input variable. It is sometimes thought that the main purpose of randomization is to provide an objective probability distribution for a test of significance. Its crucial function is in fact, to assure, with high probability, that differences in the output variables associated with changes in the input variables really are due to these changes and not to other factors.

1.6 ETHICAL CONSIDERATIONS

Ethical considerations arise in all studies and become especially prominent when animals and humans are involved. Festing and Lovell (1996) state that over 50 million animals are used in biomedical and environmental safety studies every year. They claim that responsible application of experimental design concepts can dramatically reduce the number of animals needed for these studies. Fractional factorial and saturated designs were developed for screening many factors in industrial settings where budget and facilities put severe limitations on the number of runs that can be performed. The same design principles can be used to reduce the number of animals used in experimentation.

Festing and Lovell point out that testing several strains of an animal species rather than one strain generally increases the power for detecting carcinogenic effects and certainly broadens the base for inference. If this base is broad enough, the need for additional tests is reduced.

Festing and Lovell also point out an additional problem that must be kept in mind. For example, if a screening test based on animals that has become accepted as standard is unreliable, then it is very difficult to validate any new procedures that do not rely on animals. The problem is that any new test will not correlate well with the standard test because of the variability in the standard. If the animal test is effectively not repeatable, then considerable disagreement will be found between it and any possible competitor. In particular, it will be difficult to promote an *in vitro* test to replace the established *in vivo* test. Consequently, any *in vitro* test is unlikely to be considered a valid alternative. Thus, making the screening tests more reliable could lead to validation of testing procedures that do not require animals at all. Animals used in testing the safety of chemicals, e.g., testing irritation and damage to eyes from household chemicals and cosmetics, are a case in point. In some of these tests, subjective scoring is used to measure irritation in animals' eyes. These scores can be highly variable from one rater to another. Development of better measures of irritation could reduce the variance and hence the number of animals needed for these tests. In fact, a reduction in this variance may make it possible to develop an acceptable *in vitro* test.

The ethical considerations are even more complicated and difficult in studies involving humans. In the clinical trials setting, and in fact in any experimental situation that deals with human subjects, the problem of whether to randomize treatment allocation tends to be confounded with the question of ethics. Is it ethical for a doctor to let some random device decide whether a patient is to receive a specific treatment? In research in education we frequently hear questions about the ethics (often phrased in terms of legality) of using a new teaching technique on a randomly selected group of students. Alternatively, we ask the question: Is it ethical to apply a new teaching technique to all students before a properly designed (randomized) study has been conducted to establish that it is as good as or better than the technique being replaced?

It appears to us that if the investigator has any information that one treatment regimen is better than another, then there is a real question of ethics, and probably the entire experiment should not be performed. Placing a restriction on how treatments are assigned, i.e., not randomizing, does not obviate the problem of providing a suboptimal treatment to some fellow creature. In fact, it appears to us that assigning treatments in a nonrandom manner is much worse than randomizing. Not only is there a chance that one selects something other than the best treatment (after all, it is not *known* which treatment is best) and consequently is guilty of doing something to another that is not so good, but the action fails to have the slight redeeming quality of providing solid scientific information. The same argument applies in a medical as well as an education experimental setting. The question of whether to randomize or not applies only to situations where to the best of our knowledge there is no good reason to select one treatment in preference to the other.

REVIEW EXERCISES

1.1 Write a short description of an experiment that interests you. Include information about the objectives of the experiment and the experimental design, paying specific attention to the issues of representativeness, replication, and randomization.

1.2 In April 1999, Dow Agrosciences Ltd. conducted an experiment in which it paid student volunteers to ingest small amounts of pesticides (Warrick 2000). Because of public protests against this type of experimentation, the Environmental Protection Agency convened a special panel on the use of human subjects in pesticide studies in July 1998. The panel's final recommendations are in a report available online at www.epa.gov/sab/pdf/ec0017.pdf. Can statistical experimental design considerations contribute anything to this debate? Write a short essay supporting your point of view.

1.3 The SCALE-UP program *(NCSU Bulletin 2001)* is a proposed method of teaching introductory science classes that is intended to make the learning experience in large classes more similar to that of small classes. In the SCALE-UP program students work in small teams on problems posed by a roving instructor, rather than listening to 1-hour lectures and attending a separate 3-hour lab each week. It is claimed that the failure rate of women students taking SCALE-UP classes is one-third that of women taking conventional classes in the same subject. An experiment was done using two SCALE-UP sections and two conventional sections of introductory physics, with 100 students, half female, in each section.

(a) How many experimental units are there in this experiment? Do you think the precision for comparison of failure rates will be high or low?

(b) Discuss the practical and ethical issues of assigning students to classes and teachers to sections.

CHAPTER 2

Completely Randomized Design

2.1 INTRODUCTION

The completely randomized design (CRD) is the simplest possible experimental design. We assume that the reader is already familiar with the basic statistical analysis of data from this design. Consequently, we focus on the design aspects: the assumptions, the concept of randomization, and assignment of treatments to the units. In Chapters 3 and 4 we review statistical analyses. We use the CRD in this chapter as a basis to discuss a number of important issues that will recur with other designs in future chapters.

2.2 COMPLETELY RANDOMIZED DESIGN

Although the CRD is the simplest, most commonly used design, in a sense it requires the most severe set of assumptions. It is the default design used by investigators if they do not stop and really think about the design problems. The assumptions are most severe in the sense that it is the optimal choice of design only in the case when all experimental units are completely exchangeable, i.e., are all equivalent and there is no a priori reason to believe that it is possible to select subsets of experimental units that are more alike than others. If it is possible to organize the experimental units into more homogeneous subsets, other, more efficient designs can be constructed. These possibilities are investigated in later chapters. A major advantage of the CRD is that the statistical analysis is not disturbed by unequal replication of the various treatments. This design permits simple statistical analysis and suffers the least when some experimental units die or become lost.

The CRD is often used when there is a serious risk that a significant number of experimental units will be lost during the experiment. The statistical analysis

Planning, Construction, and Statistical Analysis of Comparative Experiments,
by Francis G. Giesbrecht and Marcia L. Gumpertz
ISBN 0-471-21395-0

and subsequent interpretation remain relatively simple, even when large sections of the study are lost. The cost is some loss in *efficiency*, a concept that we discuss later when examining more complex designs.

2.2.1 Randomization

Recall that there are two distinct purposes to randomization. The first is to provide a basis for arguing cause and effect, and the second is to provide a basis for estimating experimental error. The aim in a proper randomization scheme for a completely randomized design is to ensure that every possible arrangement or assignment of treatments to experimental units has an equal chance of being selected. A direct way to achieve this is to make up pieces of paper, one for each possible random assignment of treatments to units, put all in a box, and then draw one at random. This is not really feasible for a large experiment.

Two simple procedures for t treatments and r_i replications of the ith treatment and a total of $\sum_i r_i = n$ units are as follows:

(a) Label the units $1, \ldots, n$
Label pieces of paper $1, \ldots, n$
Draw r_1 pieces from the box. Assign them to treatment 1.
Draw r_2 pieces for the next treatment, etc.
The last r_t are assigned to treatment t.
There are simplifications if all $\{r_i\}$ are equal.

(b) Using a calculator or computer, generate n uniform random numbers.
Assign these numbers to the n experimental units.
Sort the random numbers in increasing order.
Assign the units attached to the first r_1 random numbers to treatment 1.
The next r_2 random numbers are assigned to treatment 2. ... The last r_t units are assigned to treatment t.

Interactive software for design of experiments is available that makes it very easy to randomize the assignment of treatments to runs in a design.

Interestingly enough, it is *not* sufficient to have a randomization scheme in which every treatment has an equal chance of being assigned to every experimental unit. Exercise 2.1 demonstrates this with an example.

Exercise 2.1: Invalid Randomization. Consider an experiment in which the six experimental units to be used are permanently fixed in a row and two treatments are to be randomly assigned to the units. (One can think of fruit trees or a sequence of runs on a piece of laboratory equipment as the experimental units.) A scientist proposes using an unbiased coin to assign the treatments. He comes to the first unit, tosses the coin, and if it shows a head, assigns treatment a and otherwise, treatment b. This is repeated at the second unit, and so on, until three units have been assigned to one of the treatments. Then the remaining units are assigned to the other treatment. Show that this is not a valid randomization scheme. □

Dot notation is used extensively throughout the book. In this notation a dot in a subscript indicates either summation or averaging over the missing subscript. Generally, no bar over the variable indicates summation, and a bar indicates averaging. Examples of our usage:

$$\mu_{i\cdot} = \sum_{j=1}^{n} \mu_{ij}, \qquad \overline{\mu}_{i\cdot} = \sum_{j=1}^{n} \mu_{ij}/n,$$
$$\mu_{\cdot j} = \sum_{i=1}^{u} \mu_{ij}, \qquad \overline{\mu}_{\cdot j} = \sum_{i=1}^{u} \mu_{ij}/u.$$

The big exception to this use is $\mu = \sum_{i=1}^{u} \sum_{j=1}^{n} /un$. Since μ appears in numerous expressions, we simply write μ in place of $\overline{\mu}_{\cdot\cdot}$.

2.3 ASSUMPTION OF ADDITIVITY

The concept of additivity can be explained most easily by the following *thought experiment*. Assume that the investigator has u experimental units and is contemplating t not necessarily distinct treatments. Conceptually, it is possible for each treatment to be applied to every unit, giving a total of ut distinct responses. In reality, only a subset of these is possible. However, we illustrate the hypothetical responses in the array in Table 2.1. Let the symbol μ represent the average of all un responses, $\overline{\mu}_{i\cdot}$, the average in row i, and $\overline{\mu}_{\cdot j}$, the average in column j. Now when an investigator thinks about the response due to one treatment vs. another treatment, say treatment 1 compared to treatment 2, the reference is really to the difference between $\overline{\mu}_{\cdot 1}$ and $\overline{\mu}_{\cdot 2}$, with no thought about the experimental units, or possibly to the belief that all of the differences, μ_{11} vs. μ_{12}, μ_{21} vs. $\mu_{22}, \ldots, \mu_{u1}$ vs. μ_{u2}, are equal. The latter is the assumption of additivity of units and treatments.

More formally, we can define

$$\overline{\mu}_{\cdot 1} - \mu$$

Table 2.1 Array of Possible Responses

	Treatments				
	μ_{11}	μ_{12}	μ_{13}	$\cdots$	μ_{1t}
	μ_{21}	μ_{22}			
Units	μ_{31}				
	$\vdots$				$\vdots$
	μ_{u1}		$\cdots$		μ_{ut}

as the response to or *effect* of treatment 1, where μ is defined to be the mean of all $\{\overline{\mu}_{i\cdot}\}$, or equivalently, the mean of all $\{\overline{\mu}_{\cdot j}\}$, and $\overline{\mu}_{\cdot 1} - \overline{\mu}_{\cdot 2}$ as the difference between response to treatment 1 and treatment 2. Since we think of the control, no treatment applied, as one of the treatments, it is never possible to observe the response of an experimental unit without the addition (either positive or negative) of a treatment. Consequently, $\overline{\mu}_{i\cdot} - \mu$ is defined as the response of unit i. Under perfect additivity, the response of unit i to treatment j would be the sum of the overall average plus the unit effect plus the treatment effect. This can be written as

$$\mu + (\overline{\mu}_{i\cdot} - \mu) + (\overline{\mu}_{\cdot j} - \mu).$$

Deviations from perfect additivity can be summarized by the array of quantities

$$\mu_{ij} - [\mu + (\overline{\mu}_{i\cdot} - \mu) + (\overline{\mu}_{\cdot j} - \mu)] = \mu_{ij} - \overline{\mu}_{i\cdot} - \overline{\mu}_{\cdot j} + \mu$$

for $i = 1, \ldots, u$ and $j = 1, \ldots, t$. Under perfect additivity, all of these would be zero. The converse of additivity, i.e., nonadditivity, is also referred to as *interaction*. If treatments and experimental units do not interact, there is additivity, and if additivity does not hold, i.e., if there is nonadditivity, the two interact.

Since it is not possible to apply all treatments to each unit, neither the treatment responses nor the unit responses are actually observable. However, these responses are important. Inferences about the experiment involve these unobservable quantities. The investigator wishes to make inferences about the effect of a treatment if it were applied to any unit. In fact, the real aim is to make inferences about a population of similar units, not only just the sample that happened to be available for the study. The investigator must be satisfied with a sample from the array of responses.

In addition, it may well be impossible to observe the sample of true responses. There may also be (usually is) a random observational error with variance σ_ϵ^2 associated with each response. However, this does not really change the inference process. It can be shown that if additivity holds and there are no observational errors, the error mean square in the analysis of variance is an unbiased estimate of variance among experimental units, i.e., $\sigma_u^2 = \sum_i (\overline{\mu}_{i\cdot} - \mu)^2/(u-1)$. If there are independent observational errors, the error mean square estimates $\sigma_u^2 + \sigma_\epsilon^2$. Common practice is to ignore this level of detail and just use σ_ϵ^2 as the true residual error.

It is also informative to think of the array of responses in Table 2.1 in terms of the graph shown in Figure 2.1. Since the numbering of the experimental units is arbitrary, the order of the points along the abscissa is arbitrary. However, the key elements to note are the lines corresponding to the treatments. If there is perfect additivity of units and treatments, the space between any two lines remains constant across the graph. The lines are "parallel." Deviations from this would indicate

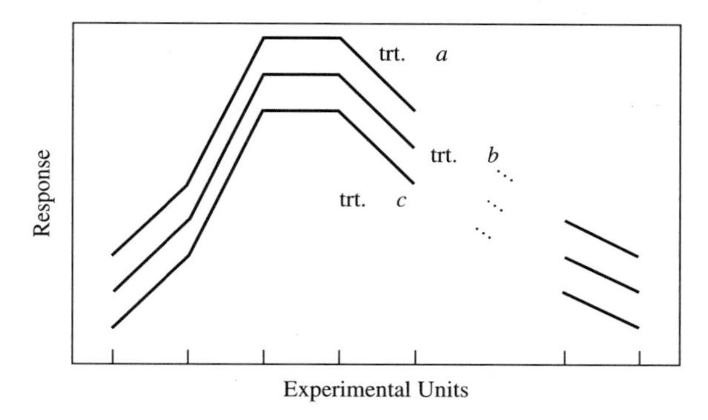

Figure 2.1 Additivity.

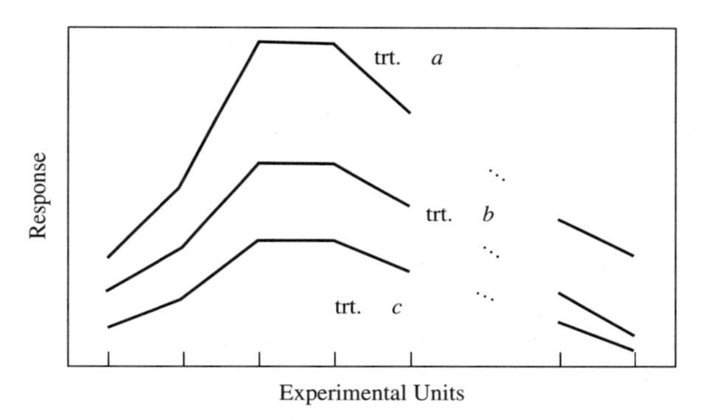

Figure 2.2 Removable nonadditivity.

nonadditivity. This may or may not be serious. A situation such as that illustrated in Figure 2.2 implies nonadditivity that can probably be corrected by transforming the data. For example, it may be that treatment a causes a 10% increase in size relative to the control treatment. In this case, larger units would have greater response than smaller units, i.e., nonadditivity. However, a log transformation will resolve this problem. Other cases may require different transformations.

There are statistical techniques such as examination of the residuals and the Box–Cox procedure (see, e.g., Steel et al. 1997) for finding possible transformations. In addition, experience shows that specific transformations stabilize the variances of certain types of data: logarithms for growth data, logarithms or square roots for count data, and the angular transformation for proportions (Snedecor and Cochran 1989). Recently, generalized linear models have been developed that extend regression and analysis of variance–type analysis to binary, count, and proportions data (Dobson 1990). For purposes of this book we assume that familiarity with the subject matter, careful thought, and examination of the

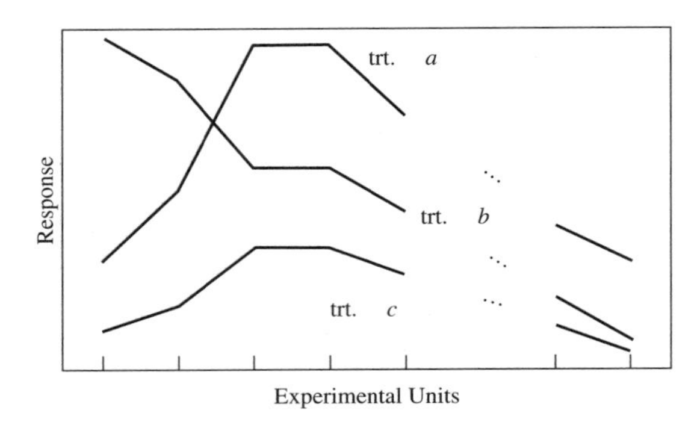

Figure 2.3 Nonremovable nonadditivity.

residuals allow the investigator to decide whether a transformation is needed and if so, to select one that is appropriate. For example, in work with microorganisms, it is quite often reasonable to assume that growth rate will be exponential and that treatments will affect the rate. Consequently, logarithms of counts are frequently analyzed. Alternatively, if the response is number of defects in a batch, then it is reasonable to use an arcsine transformation or to employ a generalized linear models analysis with a logit or a log as a link function.

A possible situation such as that illustrated in Figure 2.3 is much more serious. Now, no suitable transformation is possible. In fact, the graph suggests that one treatment has a positive effect relative to another treatment on some units and a negative effect on others. This kind of situation requires serious thought by the investigator before proceeding with the study. Some way to develop a more homogeneous set of experimental units must be found.

The concept of additivity under a slightly different guise will appear again when we look at more complex designs and factorial arrangements of treatments. The major difference will be that then we will be dealing with quantities that can be measured or estimated. Also, especially in the case of factorial experiments, the case of nonremovable nonadditivity will not be a hindrance but will be very informative to the investigator. The term *interaction* is normally used when there is nonadditivity in factorial experiments.

2.4 FACTORIAL TREATMENT COMBINATIONS

We begin by emphasizing a matter of terminology. Strictly speaking, a factorial is not a design, in the sense that, for example, the randomized block or Latin square are designs. It is really just a method of organizing the treatments. A factorial set of treatments can be placed into any sort of experimental design, including a completely randomized design. It is useful to refer to factorials as treatment designs to distinguish them from experimental designs. One then has a treatment design in an experimental design.

It has been pointed out before that while the principles of designing experiments are relevant to the wide range of human activity, the notion of factorial experiments had its roots in agriculture and related areas. For example, a quick scan of the titles of references given in Federer and Balaam (1972) for papers prior to 1940 that deal with factorial experiments reveals that of 25 papers, 18 deal with agriculture and related areas of research, two deal with education and psychology research methods and five appear to be more general and discuss principles of factorial experiments and apply to general research methodology. A consequence of this agricultural background is that many classical textbook examples tend to be drawn from that area. In current practice factorial experimentation is used extensively in industry, and factorial treatment designs are central to off-line quality control.

Two names that stand out in the early literature on factorial experiments are R. A. Fisher and F. Yates. The earliest paper that we are aware of that clearly and strongly advocates factorial experiments (referred to as *complex experiments* at that time) is by Fisher (1926) in the *Journal of the Ministry of Agriculture.* In this paper he writes:

> No aphorism is more frequently repeated in connection with field trials, than that we must ask Nature few questions, or, ideally, one question at a time. The writer is convinced that this view is wholly mistaken. Nature, he suggests, will best respond to a logical and carefully thought out questionnaire; indeed, if we ask her a single question, she will often refuse to answer until some other topic has been discussed.

Several years later in *The Design of Experiments* (1935), Fisher wrote the following about factorial experiments:

> In expositions of the scientific use of experimentation it is frequent to find an excessive stress laid on the importance of varying the essential conditions only one at a time. The experimenter interested in the causes which contribute to a certain effect is supposed, by a process of abstraction, to isolate these causes into a number of elementary ingredients, or factors, and it is often supposed, at least for purposes of exposition, that to establish controlled conditions in which all of these factors except one can be held constant, and then to study the effects of this single factor, is the essentially scientific approach to an experimental investigation. This ideal doctrine seems to be more nearly related to expositions of elementary physical theory than to laboratory practice in any branch of research. In the experiments merely designed to illustrate or demonstrate simple laws, connecting cause and effect, the relationships of which with the laws relating to other causes are already known, it provides a means by which the student may apprehend the relationship, with which to familiarize himself, in as simple a manner as possible. By contrast, in the state of knowledge or ignorance in which genuine research, intended to advance knowledge, has to be carried on, this simple formula is not very helpful. We are usually ignorant which, out of innumerable possible factors, may prove ultimately to be the most important, though we may have strong presuppositions that some few of them are particularly worthy of study. We have usually no knowledge that one factor will exert its effects independently of all others that can be varied, or that its effects are particularly simply related to variations in these other factors. On the contrary,

when factors are chosen for investigation, it is not because we anticipate that the laws of nature can be expressed with any particular simplicity in terms of these variables, but because they are variables which can be controlled or measured with comparative ease. If the investigator, in these circumstances, confines his attention to any single factor, we may infer either that he is the unfortunate victim of a doctrinaire theory as to how the experimentation should proceed, or that the time, material or equipment at his disposal is too limited to allow him to give attention to more than one narrow aspect of his problem.

The modifications possible to any complicated apparatus, machine or industrial process must always be considered as potentially interacting with one another, and must be judged by the probable effects of such interactions. If they have to be tested one at a time this is not because to do so is an ideal scientific procedure ..., but because to test them simultaneously would sometimes be too troublesome, or too costly. In many instances, as will be shown, the belief that this is so has little foundation. Indeed, in a wide class of cases an experimental investigation, at the same time as it is made more comprehensive, may also be made more efficient if by more efficient we mean that more knowledge and a higher degree of precision are obtainable by the same number of observations.

In a long paper, "Complex Experiments," in the *Journal of the Royal Statistical Society*, Yates (1935) credits R. A. Fisher with introduction of the terms *factorial design* and *factorial experiments*. Yates expands on the importance of testing factorial treatment combinations and examining interactions among factors:

Suppose that a new plant of agricultural importance is introduced into a country. What is the right way to set about determining the best varieties and the appropriate manurings and cultivations?

One procedure, extensively practiced, is to divide the problem into a large number of smaller problems and attack each one separately. One set of experiments will be started to determine the best variety, a second set to determine the best manuring, a third the best cultivations. Nor need, nor does, the subdivision stop there. Responses to the three standard fertilizers, nitrogen, phosphate and potash, for instance, may be relegated to separate experiments.

This procedure has on the face of it a deceptive appearance of simplicity. The questions formulated are themselves simple:—Is one variety better than another? Is the yield increased by the application of nitrogen? Their answers can be obtained with an apparently high precision. But there is one very cogent objection. Clearly the experimenter on fertilizers, who, we will imagine, decides to confine his enquiries at the start to response to nitrogen, must choose some variety on which to experiment. He will probably choose what he considers is the best variety. After three years of experiment the experimenter on varieties may announce that some other variety is markedly superior. Are all the experiments on fertilizers now worthless, in that they apply to a variety that will no longer be grown? The experimenter on fertilizers will probably answer that the response of the two varieties is not likely to be widely different, and that his conclusions therefore still hold. But he has no experimental proof of this, only his experience of other crops, and would not have even this last if he and all other experimenters on fertilizers had persistently only experimented on one variety.

If the experimenter on varieties is so rash as to criticize his results on these grounds, however, and has himself laid down some standard of manuring for his varietal trials, the experimenter on fertilizers can effectively turn the tables by pointing out that the varietal trials are also of little value, being carried out at a level of manuring different from what he proposes to recommend.

Had the enquiries been combined into one system of experiments, so that all varieties were tested in conjunction with all levels of nitrogen, this imaginary controversy could not have arisen; for definite information would be obtained on whether all varieties did, in fact, respond equally to nitrogen. Moreover, if it was found that they did, a considerable gain in efficiency on the primary questions would result (provided that the experimental errors per plot were not greatly increased), since each plot would enter into both primary comparisons and would thus be used twice over. If, on the other hand, differential response was demonstrated, then although the response of the chosen variety would be known with less accuracy than if the whole experiment had been carried out with that variety, yet the experimenter might count himself lucky in that the possibility of false conclusions due to using another variety in his fertilizer trials had been avoided. Moreover, the conclusion as to the best variety might also require modification in the light of the differences in response to fertilizer.

When new varieties are being selected this greatly understates the advantage, for the essential and valuable difference of one variety over another may lie just in its ability to respond to heavy dressings of fertilizer; at the customary levels of manuring, it may be, the yields are about the same. Nor need this response be direct. In the case of wheat, for instance, the limit of nitrogenous manuring is determined less by what the plant can make use of than by what it can stand up to without lodging.

In practice it will seldom be possible to include every variety it is desired to test in the fertilizer trials. This, however, is no argument against including a representative selection of varieties. If no substantial differences in fertilizer response are discovered with such a selection we may then, reasoning inductively, conclude that it is improbable that substantial differences exist for any variety.

It is not unusual to encounter now some of the same arguments and objections to factorial experiments faced by Fisher and Yates some 60 to 80 years ago [see Czitrom (1999) for a recent article]. We trust that readers of this book will become convinced that factorial experiments or, as some prefer to call them, factorial arrangements of treatments, represent one of the most useful tools that statistics has to offer.

2.4.1 Basic Concepts in Factorial Structure

The concepts of *main effects* and *interactions* are fundamental to the sections that follow. We begin with a list of hypothetical investigations each involving two factors, A and B.

- Levels of protein and energy in a feed ration affecting growth rate of animals

- Horsepower produced by an engine as affected by type of gasoline and nature of fuel injector used
- Yield of usable chips as a response to temperature and time adjustments in the fabricating process

Assume that these are at n_A levels or forms of factor A and n_B levels or forms of factor B. Refer to the levels or forms of A as a_i, with $i = 1, \ldots, n_A$ and levels or forms of B as b_j, with $j = 1, \ldots, n_B$. The model for the experiment consists of all possible $n_A n_B$ combinations of the two. Corresponding to each of the $n_A n_B$ combinations there is a true response, which we denote by μ_{ij}. We can think of these responses as an $n_A \times n_B$ array of values:

$$\begin{array}{cccc} \mu_{11} & \mu_{12} & \cdots & \mu_{1n_B} \\ \mu_{21} & \mu_{22} & \cdots & \mu_{2n_B} \\ \vdots & \vdots & \ddots & \vdots \\ \mu_{n_A1} & \mu_{n_A2} & \cdots & \mu_{n_An_B}. \end{array}$$

The next logical step is to compute marginal means, row means $\overline{\mu}_{1\cdot}, \overline{\mu}_{2\cdot}, \ldots, \overline{\mu}_{n_A\cdot}$ and column means $\overline{\mu}_{\cdot 1}, \overline{\mu}_{\cdot 2}, \ldots, \overline{\mu}_{\cdot n_B}$, and subtract the overall mean μ from each. These deviations from the overall mean as well as differences and/or contrasts among them are known as *main effects*, those based on row means are *A effects*, and those based on column means are *B effects*.

Now it may well be that the two-way array is adequately described by the grand mean, row means, and column means, or in symbols that

$$\begin{aligned} \mu_{ij} &\approxeq \mu + (\overline{\mu}_{i\cdot} - \mu) + (\overline{\mu}_{\cdot j} - \mu) \\ &= \overline{\mu}_{i\cdot} + \overline{\mu}_{\cdot j} - \mu \end{aligned}$$

for all values of i and j where we use "$\approxeq$" to mean "approximately equal." When exact equality holds, we say that there is no interaction, i.e., no $A \times B$ interaction in the table. Alternatively, some or even all of the quantities $\{\mu_{ij} - \overline{\mu}_{i\cdot} - \overline{\mu}_{\cdot j} + \mu\}$ may be appreciable. When this happens, we say that there is an $A \times B$ interaction, or that the two factors A and B *interact*. Note that it may be informative to think of the quantities that we have defined as

$$\mu_{ij} - \overline{\mu}_{i\cdot} - \overline{\mu}_{\cdot j} + \mu = (\mu_{ij} - \mu) - (\overline{\mu}_{i\cdot} - \mu) - (\overline{\mu}_{\cdot j} - \mu).$$

Any contrasts among these quantities are known as *interaction contrasts*. Notice that the concept of interaction is a "throwaway" concept in the sense that if there is no interaction, the two-way array can be described adequately by the simpler model, whereas the presence of interaction says that this is not so. Finding that there is an interaction means that one must seek more elaborate explanations for the model underlying the data.

We also see that the concept of interaction is closely related to the concept of nonadditivity discussed in Section 2.3. The major difference between the responses discussed in this section and the purely hypothetical experimental unit × treatment responses discussed in Section 2.3 is that, at least conceptually and possibly with some error, it is possible to observe all of the responses. In interaction among treatments, as in the type of nonadditivity of Section 2.3, there are also the two cases: interaction that can be removed by expressing the responses on an appropriate scale (a transformation), and interactions that cannot be removed. However, the problem is not nearly as serious as the problem of nonremovable experimental unit × treatment nonadditivity.

Except for very special cases, we never get to know the true $\{\mu_{ij}\}$ values. If we did, there would be no need for the experiments as we know them. We assume that observations deviate from the expected response, that we observe $\{y_{ij}\}$ where

$$y_{ij} = \mu_{ij} + \epsilon_{ij}.$$

In the general linear hypothesis theory, the $\{\epsilon_{ij}\}$ represent realizations of random errors from some distribution.

Replacing the $\{\mu_{ij}\}$ with $\{y_{ij}\}$ does not change any of the concepts, such as main effects and interactions, although we must realize that we are dealing with estimates rather than true parameters. There is an $n_A \times n_B$ array of $\{y_{ij}\}$ values, just as there was an array of $\{\mu_{ij}\}$ values. The estimates of the A main effects are the corresponding contrasts among the row means, $\overline{y}_{1\cdot}, \overline{y}_{2\cdot}, \ldots, \overline{y}_{n_A\cdot}$, and the estimates of the B main effects are the corresponding contrasts among the column means $\overline{y}_{\cdot 1}, \overline{y}_{\cdot 2}, \ldots, \overline{y}_{\cdot n_B}$. Interaction estimates are contrasts among the $\{y_{ij} - \overline{y}_{i\cdot} - \overline{y}_{\cdot j} + \overline{y}_{\cdot\cdot}\}$.

The final point to notice is that things extend directly to more than just two factors. Existence of a three-factor interaction, say $A \times B \times C$, simply means that there are two different settings of the A factor such that at least one of the $B \times C$ tables shows interaction and the two tables do not show the *same* magnitude of interaction. In the interpretation of the $A \times B \times C$ interaction, the factors can be permuted, i.e., the presence of a three-factor interaction indicates that the magnitude of the $A \times C$ interaction depends on the level of factor B and the $B \times C$ interaction depends on the level of factor A. The definition of three-way interactions extends directly to k-factor interactions. One interprets a k-factor interaction by looking at the $(k-1)$-factor interactions with one factor held constant.

Almost invariably, whenever an experimenter thinks about a problem realistically, it turns out that there is a long list of potential factors, each at two or more levels, that may have an influence on the response of interest. Many of the methods in the second half of this book, which are especially useful for industrial off-line quality control, were developed to handle many factors in a small number of experimental runs, with the aim of keeping the time and expense manageable.

2.5 NESTED FACTORS

Sometimes the levels of one factor depend on the levels of another. For example, the model number of a refrigerator depends on the brand. In this type of situation it would be impossible to build a factorial treatment structure with factor A as brand and factor B as model number. Besides having different models, the different brands might have different numbers of models. If we wanted to study these two factors, we would have to include a different set of model numbers for each brand. This type of treatment structure is called a *nested treatment structure*. In this example, factor B, model number, is nested within factor A, brand, and written as $B(A)$.

We list some other examples of nested factors in various settings.

- Testing for differences in success rates between hospitals and surgeons within hospitals. Surgeons are nested within hospitals.
- Comparing teaching methods in several schools and several classes at each school. Classes are nested within schools.
- Comparing two dry cleaning processes. Each process uses different solvents. When comparing three amounts of solvent for each process, the amounts of solvent differ between the two processes. The amount of solvent is nested within the process.

2.5.1 Basic Concepts and Notation for Nested Structure

There are n_A levels of factor A, but the number of levels of nested factor B may vary depending on the level of factor A. For the ith level of factor A, there are n_{Bi} levels of factor B. The true response is labeled μ_{ij}, where $i = 1, \ldots, n_A$ and $j = 1, \ldots, n_{Bi}$.

In nested structures it may or may not make sense to average over the levels of factor B for each level of factor A. Such marginal means of factor A are labeled $\overline{\mu}_{1\cdot}, \overline{\mu}_{2\cdot}, \ldots, \overline{\mu}_{n_A\cdot}$. The second example of Section 2.5, in which classes are nested within schools, provides an example where the marginal means are of primary interest. These are the means for each school, averaged over the classes at that school. The refrigerator example presents a different kind of situation. Each marginal A mean would be the brand mean averaged over models. Suppose that brand X has models T101, T201, and T203 and brand Y has models V3303 and AV3304. If the objective is to estimate the energy efficiency of the line of refrigerators, it might be of interest to compare the marginal mean of brand X with brand Y. In this scenario, weighted averages, where the weights are proportional to the sales volumes of the models, might be more appropriate for comparing the energy efficiency of the brands. On the other hand, if interest centers on brand comparisons of the length of time that a freezer stays cold when the power is out, it may not be at all reasonable to compare the means of two brands that have very different numbers and types of models. In this case, interest will probably center on comparisons of particular models or sets of models.

For nested factors it does not often (ever?) make sense to average over the levels of the nesting factor, i.e., to take a marginal mean. What we have in mind can easily be seen in the refrigerator example. It is impossible to average model T101 across brands. The same idea applies to classes within schools. Suppose that there are three sections (i.e., classes) of the same course at each school, labeled 1, 2, and 3. In this scenario the section numbers are arbitrary, so section 1 at one school does not correspond in any way to section 1 at another school. Hence, even though the numbering is the same, it still is not of any interest to average the section 1 results over schools.

For a nested factor B, the focus will often be on comparisons among the levels of B within a particular level of factor A. The true mean can be decomposed into three parts,

$$\mu_{ij} = \mu + (\overline{\mu}_{i\cdot} - \mu) + (\mu_{ij} - \overline{\mu}_{i\cdot}).$$

In this decomposition, the effect of the nesting factor A is written as the deviation of the A mean from the grand mean μ, but the effect of the nested factor B is specific to the level of A. The effect of B within a specific level of A is written as $B(A)$ and is the deviation between μ_{ij} and the mean of just the relevant (the ith) level of A.

These ideas extend to more than two factors and can be combined with factorial treatment structures. We illustrate some possibilities with two dry cleaning scenarios.

1. Suppose that the objective is to compare the effectiveness of factor A, two dry cleaning processes, on factor B, two types of dirt: oil based and milk based. Each dry cleaning process uses a different solvent. The first process uses 10 to 30 units of its solvent and the second process uses 1 to 5 units of its solvent. Factor C is the amount of solvent. In this scenario factor C is nested within factor A, and factor B is crossed with the other two factors. The decomposition of the mean includes main effects for A and B and the $A \times B$ interaction, the nested $C(A)$ effect, and the interaction of factors B and C for each process, $B \times C(A)$. The latter effect quantifies the difference in the effect of the amount of solvent on different types of dirt within each process. The decomposition of the mean is

$$\begin{aligned}\mu_{ijk} &= \mu + (\overline{\mu}_{i\cdot\cdot} - \mu) + (\overline{\mu}_{\cdot j\cdot} - \mu) + (\overline{\mu}_{ij\cdot} - \overline{\mu}_{i\cdot\cdot} - \overline{\mu}_{\cdot j\cdot} + \mu) \\ &\quad + (\overline{\mu}_{i\cdot k} - \overline{\mu}_{i\cdot\cdot}) + (\mu_{ijk} - \overline{\mu}_{ij\cdot} - \overline{\mu}_{i\cdot k} + \overline{\mu}_{i\cdot\cdot}). \qquad (2.1)\end{aligned}$$

2. Now take a different set of factors. Suppose that A again represents two dry cleaning processes. Now suppose that the two processes take different lengths of time to clean, where the first process takes 1 to 2 hours and the second process is a much speedier process that takes 10 to 20 minutes. Factor B is the length of time for cleaning: 1, 1.5, or 2 hours for process 1 and 10 or 20 minutes for process 2. Factor C is the amount of solvent, nested within each process as in scenario 1. In this scenario both factors B and C and their interaction are nested within factor A. The decomposition includes a main effect for factor A plus nested effects $B(A)$ for time to

clean within process, $C(A)$ for the amount of solvent within the process, and $B \times C(A)$ for the time by the amount of solvent interaction within each process. Algebraically, the decomposition is

$$\mu_{ijk} = \mu + (\overline{\mu}_{i..} - \mu) + (\overline{\mu}_{ij.} - \overline{\mu}_{i..}) + (\overline{\mu}_{i\cdot k} - \overline{\mu}_{i..}) + (\mu_{ijk} - \overline{\mu}_{ij.} - \overline{\mu}_{i\cdot k} + \overline{\mu}_{i..}). \qquad (2.2)$$

REVIEW EXERCISES

2.1 (L. Duffie, personal communication) Intercropping, i.e., planting alternating rows of two different crops in the same field, is used as a management tool to control early leaf spot in peanut. The following experiment to compare four different intercropping treatments (monoculture, monoculture with fungicide, alternate rows of peanut and corn, and alternating four-row strips of peanut and corn) is proposed. (1) Randomly assign five plots to each intercropping treatment. (2) Place fungus inoculum in the center of each of the 20 plots. (3) At 6 weeks after planting, measure disease at three locations in each plot.

(a) What is the experimental unit, and how many units are there?

(b) Randomly assign the four treatments to the 20 plots.

2.2 A forester is interested in finding the concentration of auxin that maximizes the rooting of cuttings. The treatments under consideration are four concentrations of auxin plus a control with no auxin. There will be four control cuttings and four cuttings treated with auxin at each concentration. Randomly assign the auxin concentrations to the cuttings, and randomly assign the cuttings to positions in the greenhouse.

2.3 (Based on an experiment by J. Trussell, personal communication) An engineering professor wants to know whether students learn better if homework is graded than if it is assigned but not required to be turned in. He wants to test each method on two sections each of three electrical engineering courses and three math courses. The response is the section's mean score on a standardized exam.

(a) What factors are nested, and what factors are crossed in this experiment, if any? What interactions can be defined for this experiment?

(b) Describe how to randomize this experiment in detail. What is the experimental unit, and how many units are there?

(c) Write out an appropriate decomposition of the response. Explain the meaning of each term in the decomposition.

2.4 A university library wants to compare ease of use of four different bibliography software programs for three different populations: students, faculty, and staff. Each participant is given a user manual and as much time as they like to learn to use one of the programs. When they feel ready, they are given a test of their proficiency and a questionnaire asking how long it took to learn to use the software and rating the difficulty of learning the software. The response of interest is their proficiency score.

(a) What factors are nested, and what factors are crossed in this experiment, if any? What interactions can be defined for this experiment?

(b) Suppose that it is desired to use 20 persons from each of the three populations. Describe how to randomize this experiment in detail, including how to select the individuals. What is the experimental unit in your experiment, and how many units are there?

(c) Write out an appropriate decomposition of the response. Draw a graph showing the meaning of each term in the decomposition.

CHAPTER 3

Linear Models for Designed Experiments

3.1 INTRODUCTION

The process of evaluating competing designs for an experiment depends on understanding the statistical methods that will be used to analyze the data resulting from the experiment. The purpose of the typical experiment is to test research hypotheses and/or estimate unknown parameters. The goal of experimental design is to increase the precision of estimates and the power of hypothesis tests. An examination of the statistical analysis provides a guide to the choice of design. It is most helpful to walk through the statistical analysis before an experiment is conducted. In this walk-through the investigator discovers whether the experiment provides estimates of important parameters, the expected precision of parameter estimates, whether the research hypotheses are testable, and the power of the tests under the proposed design. These provide a basis for evaluating competing experimental designs. For these reasons we go through the ideas and notation of linear statistical models in some depth.

The most widely used statistical analyses of designed experiments are based on linear models. In this chapter we review linear models, least squares estimation of parameters, and discuss various parameterizations of analysis of variance models. It is assumed that the reader is already familiar with analysis of variance, multiple regression, and ordinary least squares estimation. In Section 3.2 we highlight some important features of linear models and present the notation and terminology that will be used throughout the book. Matrix notation proves to be very convenient when dealing with linear models. The models take on a simple but general form regardless of number of factors or regressor variables present. Many different models fall out as special cases. Understanding the material of this chapter is basic to analyzing data from the designs in this book. The

Planning, Construction, and Statistical Analysis of Comparative Experiments,
by Francis G. Giesbrecht and Marcia L. Gumpertz
ISBN 0-471-21395-0

reader may find it helpful to use this chapter and its appendices for reference while working through subsequent chapters.

3.2 LINEAR MODEL

3.2.1 Simple Linear Regression Model

In introductory books we see discussions of linear regression models and analysis of variance models. Both are special cases of the general linear model. In a general linear model the response is modeled as a linear combination of the unknown parameters and a random error term. The simplest is the simple linear (straight-line) regression model

$$y_i = \beta_0 + \beta_1 x_i + \epsilon_i, \qquad i = 1, \ldots, n, \tag{3.1}$$

where the $\{\epsilon_i\}$ are independent, identically distributed (iid) $(0, \sigma_\epsilon^2)$ random variables. The $\{\epsilon_i\}$ represent the noise or variation in the response that is not otherwise explained by the model. Standard practice is to refer to these as *random errors.* For example, suppose that a soil scientist applies magnesium (Mg) to the soil and then measures the amount of Mg taken up into the stem tissue of corn plants with the aim of regressing the concentration of Mg in the stem tissue on the Mg concentration in the soil. Some potential sources of random error in the tissue concentrations are genetic variation, microclimate variation, variability in soil moisture, and variability in other soil nutrients.

The set of equations given in expression (3.1) can be displayed in more detail as

$$\begin{aligned} y_1 &= \beta_0 + \beta_1 x_1 + \epsilon_1 \\ y_2 &= \beta_0 + \beta_1 x_2 + \epsilon_2 \\ &\vdots \\ y_n &= \beta_0 + \beta_1 x_n + \epsilon_n. \end{aligned}$$

Any standard linear model with ***iid*** errors can be written in matrix notation as

$$\boldsymbol{y} = \boldsymbol{X\beta} + \boldsymbol{\epsilon}, \tag{3.2}$$

where $\boldsymbol{\epsilon}$ is distributed with mean vector $\boldsymbol{0}$ and covariance matrix $\sigma_\epsilon^2 \boldsymbol{I}$. In the simple linear regression case, the matrices are

$$\boldsymbol{y} = \begin{bmatrix} y_1 \\ y_2 \\ y_3 \\ \vdots \\ y_n \end{bmatrix}, \quad \boldsymbol{X} = \begin{bmatrix} 1 & x_1 \\ 1 & x_2 \\ 1 & x_3 \\ \vdots & \\ 1 & x_n \end{bmatrix}, \quad \boldsymbol{\beta} = \begin{bmatrix} \beta_0 \\ \beta_1 \end{bmatrix}, \quad \text{and} \quad \boldsymbol{\epsilon} = \begin{bmatrix} \epsilon_1 \\ \epsilon_2 \\ \epsilon_3 \\ \vdots \\ \epsilon_n \end{bmatrix}.$$

The matrix $\boldsymbol{X}$ is known as the *design* or *model matrix*. The term *design matrix* is used more often in the analysis of variance context, and the term *model matrix* in the regression context. The $\boldsymbol{\beta}$ vector contains all of the *location parameters* in the model. The word *location* indicates that these parameters determine the expected value of $\boldsymbol{y}$ rather than the variance. (Variance parameters are frequently called *scale parameters*.) The design matrix contains a separate column for each parameter in $\boldsymbol{\beta}$.

3.2.2 Simple Cell Means Model

Another example of a linear model is the one-factor analysis of variance model. The simplest form of this model is

$$y_{ij} = \mu_i + \epsilon_{ij}, \tag{3.3}$$

where the $\epsilon_{ij} \sim \text{iid}(0, \sigma_\epsilon^2)$, for $i = 1, \ldots, t$ treatment groups, and $j = 1, \ldots, r_i$ observations in the ith group. This form of the model is known as the *cell means model*, because there is a separate parameter μ_i for the mean of each treatment group or cell. The model in matrix notation, for the case of $t = 3$ treatment groups and $r_1 = 2, r_2 = 3,$ and $r_3 = 2$, is

$$\begin{bmatrix} y_{11} \\ y_{12} \\ y_{21} \\ y_{22} \\ y_{23} \\ y_{31} \\ y_{32} \end{bmatrix} = \begin{bmatrix} 1 & 0 & 0 \\ 1 & 0 & 0 \\ 0 & 1 & 0 \\ 0 & 1 & 0 \\ 0 & 1 & 0 \\ 0 & 0 & 1 \\ 0 & 0 & 1 \end{bmatrix} \begin{bmatrix} \mu_1 \\ \mu_2 \\ \mu_3 \end{bmatrix} + \begin{bmatrix} \epsilon_{11} \\ \epsilon_{12} \\ \epsilon_{21} \\ \epsilon_{22} \\ \epsilon_{23} \\ \epsilon_{31} \\ \epsilon_{32} \end{bmatrix}. \tag{3.4}$$

3.3 PRINCIPLE OF LEAST SQUARES

3.3.1 Least Squares and the Normal Equations

Least squares is one of the unifying principles that provides a basis for fitting models to data. In the case of linear models, least squares generally leads to very tractable solutions. When the errors are uncorrelated and have common variance, *ordinary least squares* is the method of choice for fitting models. By *fitting a model* we mean finding values, say $\widetilde{\beta}_0$ and $\widetilde{\beta}_1$ for β_0 and β_1 in equation (3.1) [or $\widetilde{\boldsymbol{\beta}}$ for $\boldsymbol{\beta}$ in equation (3.2)] such that when these values are inserted for β_0 and β_1 in the model to compute $\{\widetilde{y}_i\}$, these values are in some sense close to the $\{y_i\}$. The least squares principle says to minimize the sum of squared deviations,

$$Q = \sum_{i=1}^{n} (y_i - \widetilde{y}_i)^2.$$

In matrix notation, this is written more compactly as

$$Q = (\mathbf{y} - \widetilde{\mathbf{y}})^t(\mathbf{y} - \widetilde{\mathbf{y}}) = (\boldsymbol{y} - \boldsymbol{X}\widetilde{\boldsymbol{\beta}})^t(\boldsymbol{y} - \boldsymbol{X}\widetilde{\boldsymbol{\beta}}).$$

The ordinary least squares (OLS) estimators denoted by $\widehat{\beta}_0$ and $\widehat{\beta}_1$ are the values for $\widetilde{\beta}_0$ and $\widetilde{\beta}_1$ that achieve this minimum. These values are obtained most easily by solving the *normal equations*, written in matrix notation as

$$\boldsymbol{X}^t\boldsymbol{X}\widehat{\boldsymbol{\beta}} = \boldsymbol{X}^t\boldsymbol{y}. \tag{3.5}$$

These estimators are linear in the $\{y_i\}$ and can be shown to be unbiased and have the smallest variance in the class of all possible unbiased linear estimators. They are called the *best linear unbiased estimator* (BLUE). Note that the term BLUE is a technical statistical term. We also call

$$(\boldsymbol{y} - \boldsymbol{X}\widehat{\boldsymbol{\beta}})^t(\boldsymbol{y} - \boldsymbol{X}\widehat{\boldsymbol{\beta}}) \tag{3.6}$$

the *residual* or *error sum of squares* (SSE).

For the example with three treatments displayed in model (3.4), the normal equations are

$$\begin{bmatrix} 2 & 0 & 0 \\ 0 & 3 & 0 \\ 0 & 0 & 2 \end{bmatrix} \begin{bmatrix} \widehat{\mu}_1 \\ \widehat{\mu}_2 \\ \widehat{\mu}_3 \end{bmatrix} = \begin{bmatrix} y_{11} + y_{12} \\ y_{21} + y_{22} + y_{23} \\ y_{31} + y_{32} \end{bmatrix}.$$

The solution to this system is easy to obtain. The least squares estimates of the parameters are the treatment means,

$$\begin{bmatrix} \widehat{\mu}_1 \\ \widehat{\mu}_2 \\ \widehat{\mu}_3 \end{bmatrix} = \begin{bmatrix} (y_{11} + y_{12})/2 \\ (y_{21} + y_{22} + y_{23})/3 \\ (y_{31} + y_{32})/2 \end{bmatrix}.$$

3.3.2 Simple Analysis of Variance Model for Treatments in a CRD

Analysis of variance models are more commonly written with a term for the overall mean plus terms for treatment effects, as, for example,

$$y_{ij} = \mu + \tau_i + \epsilon_{ij} \qquad \text{for } j = 1, \ldots, r_i \text{ and } i = 1, \ldots, t. \tag{3.7}$$

In this formulation, the $\{\tau_i\}$ parameters represent treatment effects, where the word *effect* has a technical definition. The *effect* of the ith treatment is the difference between the true (but unknown) ith treatment mean μ_i and the common parameter μ:

$$\text{Treatment effect:} \quad \tau_i = \mu_i - \mu.$$

When the number of replications for each treatment, $\{r_i\}$, are all equal, μ is generally defined as the average of all treatment means. This is the most usual case, and the definition that is most often of interest. Notice, however, that treatment means can be expressed with model (3.7) even if μ is not defined, because the expected value $\mathrm{E}[y_{i\cdot}] = \mu + \tau_i$. In experiments where the numbers of observations are different for different treatments, the symbol μ is sometimes defined as a weighted average of the treatment means. Different weight values can be useful in some situations.

In this formulation the example given in equation (3.4) becomes

$$\begin{bmatrix} y_{11} \\ y_{12} \\ y_{21} \\ y_{22} \\ y_{23} \\ y_{31} \\ y_{32} \end{bmatrix} = \begin{bmatrix} 1 & 1 & 0 & 0 \\ 1 & 1 & 0 & 0 \\ 1 & 0 & 1 & 0 \\ 1 & 0 & 1 & 0 \\ 1 & 0 & 1 & 0 \\ 1 & 0 & 0 & 1 \\ 1 & 0 & 0 & 1 \end{bmatrix} \begin{bmatrix} \mu \\ \tau_1 \\ \tau_2 \\ \tau_3 \end{bmatrix} + \begin{bmatrix} \epsilon_{11} \\ \epsilon_{12} \\ \epsilon_{21} \\ \epsilon_{22} \\ \epsilon_{23} \\ \epsilon_{31} \\ \epsilon_{32} \end{bmatrix}. \tag{3.8}$$

3.3.3 Normal Equations for the Analysis of Variance Model

The normal equations for model (3.7) are

$$\begin{bmatrix} \sum_i r_i & r_1 & r_2 & \cdots & r_t \\ r_1 & r_1 & 0 & \cdots & 0 \\ r_2 & 0 & r_2 & \cdots & 0 \\ \vdots & \vdots & \vdots & \ddots & \vdots \\ r_t & 0 & 0 & \cdots & r_t \end{bmatrix} \begin{bmatrix} \widehat{\mu} \\ \widehat{\tau}_1 \\ \vdots \\ \widehat{\tau}_t \end{bmatrix} = \begin{bmatrix} \sum_i \sum_j y_{ij} \\ \sum_j y_{1j} \\ \vdots \\ \sum_j y_{tj} \end{bmatrix}. \tag{3.9}$$

The normal equations for the special case, model (3.8), are

$$\begin{bmatrix} 7 & 2 & 3 & 2 \\ 2 & 2 & 0 & 0 \\ 3 & 0 & 3 & 0 \\ 2 & 0 & 0 & 2 \end{bmatrix} \begin{bmatrix} \widehat{\mu} \\ \widehat{\tau}_1 \\ \widehat{\tau}_2 \\ \widehat{\tau}_3 \end{bmatrix} = \begin{bmatrix} y_{11} + y_{12} + y_{21} + y_{22} + y_{23} + y_{31} + y_{32} \\ y_{11} + y_{12} \\ y_{21} + y_{22} + y_{23} \\ y_{31} + y_{32} \end{bmatrix}. \tag{3.10}$$

Now we encounter some difficulties in solving the normal equations. We notice that the first equation is the sum of the remaining equations. There are really only t independent equations, but $t + 1$ unknown parameters. The $\boldsymbol{X}$ matrix as illustrated in the example in equation (3.8) is not full-column rank; i.e., column 1 is the sum of the remaining columns. Consequently, $\boldsymbol{X}^t\boldsymbol{X}$ is also non-full rank and cannot be inverted. The system of equations can be shown to be consistent; i.e., there are solutions, but there is no unique solution, no unique estimator for μ, τ_1, τ_2, and τ_3. Basically, there are just too many parameters in this model. One of them is redundant. In this special case, the problem is relatively easy to recognize.

Exercise 3.1: Solving Non-Full-Rank Systems. Working with the system of equations (3.10), show that you can set $\widehat{\mu} = 0$ and obtain a solution to the system. Next set $\widehat{\tau}_1 = 0$ and obtain a solution. Show that the two solutions are different. Show that for the two solutions, you get the same value for $\widehat{\tau}_2 - \widehat{\tau}_3$. □

There are two approaches to circumventing this non-unique solution problem. The first, called the *non-full-rank* approach, is to work with the overparameterized form of the model and focus on *estimable* quantities rather than on the model parameters per se. If $\boldsymbol{l}^t\boldsymbol{\beta}$ is a linear combination of the parameters in $\boldsymbol{\beta}$, it is said to be *estimable* if and only if there exists a linear combination of the elements of $\boldsymbol{y}$ with expected value equal to $\boldsymbol{l}^t\boldsymbol{\beta}$. Saying that $\boldsymbol{l}^t\boldsymbol{\beta}$ is estimable is just another way of saying that it is possible to find a quantity $\boldsymbol{a}^t\boldsymbol{y}$ that is an unbiased estimator for $\boldsymbol{l}^t\boldsymbol{\beta}$. The second approach is to put sufficient restrictions on the model parameters to make each parameter represent a separate unambiguous quantity. In essence this reduces the number of parameters and defines away the problem. This will be called the *full-rank approach*.

Non-Full-Rank Approach

Any observation provides an unbiased estimate of its expected value. Since a treatment mean is just a linear function of observations, it follows that the expected value of every treatment mean is estimable. For instance, in the model (3.7), μ is undefined and is, in fact, not estimable, but $\mu + \tau_i = \mu_i$ represents the ith treatment mean and is estimable. The analysis of variance for the non-full-rank model (3.7) focuses on estimating and testing hypotheses about estimable quantities such as $\mu + \tau_i$. This approach relies exclusively on the fact that linear combinations of estimable quantities are again estimable. Since the $\{\mu + \tau_i\}$ are all estimable, it follows that contrasts such as $\sum_i c_i\tau_i$ are also estimable, provided that $\sum_i c_i = 0$. The difference between two treatment means is an example of such a contrast.

This approach uses generalized inverses to solve the normal equations (3.5). The generalized inverse of $\boldsymbol{X}^t\boldsymbol{X}$ is denoted $(\boldsymbol{X}^t\boldsymbol{X})^-$. It is not unique, so the resulting solution to the normal equations, $(\boldsymbol{X}^t\boldsymbol{X})^-\boldsymbol{X}^t\boldsymbol{y}$, is not unique either; however, $\widehat{\boldsymbol{y}}$ and the estimates of all *estimable* functions of the parameters are unique, regardless of the choice of generalized inverse. This approach has many computational advantages and is used by all-purpose statistical software such as SAS® `proc GLM`. Linear models textbooks by Graybill (1961), Searle (1971), and Christensen (1996b) provide in-depth coverage of non-full rank models. In this book we concentrate mainly on the full-rank approach, with enough discussion of the non-full-rank approach to understand the methods used in statistical software.

Full-Rank Approach

In the full-rank method, the approach is to reduce the number of parameters, or to constrain them sufficiently, to give a unique solution to the normal equations. There are two general routes that can be taken to achieve this: (1) delete unnecessary parameters from the model, or (2) define the parameters with restrictions so that the normal equations have unique solutions.

The cell means model (3.3) takes route (1). This model is obtained from model (3.7) by eliminating μ. The special case example, model (3.4), has three parameters to describe three treatment means, and the columns of the design matrix are all linearly independent. Hence there is one unique solution to the normal equations (3.5), namely,

$$\widehat{\boldsymbol{\beta}} = (\boldsymbol{X}^t\boldsymbol{X})^{-1}\boldsymbol{X}^t\boldsymbol{y}. \tag{3.11}$$

The cell means approach can be very useful in dealing with missing data.

The approach of defining parameters with restrictions consists of explicitly defining all the parameters, or equivalently adding "side conditions" to the model. In the simple case represented by model (3.7) there is only one undefined parameter (μ). If we define μ to be the mean of the treatment means, we get the side condition

$$\sum_{i=1}^{t} \tau_i = 0 \qquad \text{for all } i$$

as a consequence. Now there are two alternatives. One can solve the system of equations, i.e., equation (3.9), by adding an extra equation to the system. The alternative is simply to substitute the restriction to get rid of one parameter, i.e., to let $\tau_t = -\sum_{i=1}^{t-1} \tau_i$ in the model. To illustrate, model (3.8) now becomes

$$\begin{aligned}
y_{11} &= \mu + \tau_1 \phantom{{}-\tau_2} + \epsilon_{11}\\
y_{12} &= \mu + \tau_1 \phantom{{}-\tau_2} + \epsilon_{12}\\
y_{21} &= \mu \phantom{{}+\tau_1} + \tau_2 + \epsilon_{21}\\
y_{22} &= \mu \phantom{{}+\tau_1} + \tau_2 + \epsilon_{22}\\
y_{23} &= \mu \phantom{{}+\tau_1} + \tau_2 + \epsilon_{23}\\
y_{31} &= \mu - \tau_1 - \tau_2 + \epsilon_{31}\\
y_{32} &= \mu - \tau_1 - \tau_2 + \epsilon_{32}.
\end{aligned}$$

The parameter τ_3 is not needed and we now have the full column rank design matrix:

$$\boldsymbol{X} = \begin{bmatrix} 1 & 1 & 0\\ 1 & 1 & 0\\ 1 & 0 & 1\\ 1 & 0 & 1\\ 1 & 0 & 1\\ 1 & -1 & -1\\ 1 & -1 & -1 \end{bmatrix} \quad \text{and} \quad \boldsymbol{\beta} = \begin{bmatrix} \mu\\ \tau_1\\ \tau_2 \end{bmatrix}.$$

This technique is referred to as *effect coding*. In effect coding, the first column of the design matrix contains ones, corresponding to the constant term μ. The next $t-1$ columns contain coding for treatment effects. The ith of these columns is coded 1, -1, or 0 for experimental units in treatment i, experimental units in treatment t, and all other experimental units, respectively. No column is needed for treatment t because $\tau_t = -\sum_{i=1}^{t-1} \tau_i$.

The normal equations now become

$$\begin{bmatrix} 7 & 0 & 1 \\ 0 & 4 & 2 \\ 1 & 2 & 5 \end{bmatrix} \begin{bmatrix} \widehat{\mu} \\ \widehat{\tau_1} \\ \widehat{\tau_2} \end{bmatrix} = \begin{bmatrix} y_{11} + y_{12} + y_{21} + y_{22} + y_{23} + y_{31} + y_{32} \\ (y_{11} + y_{12}) - (y_{31} + y_{32}) \\ (y_{21} + y_{22} + y_{23}) - (y_{31} + y_{32}) \end{bmatrix}. \tag{3.12}$$

This system can be solved easily. The inverse of the $\boldsymbol{X}^t\boldsymbol{X}$ matrix is

$$1/54 \begin{bmatrix} 8 & 1 & -2 \\ 1 & 17 & -7 \\ -2 & -7 & 14 \end{bmatrix},$$

giving

$$\widehat{\mu} = 1/54\Big[8\big(y_{11} + \cdots + y_{32}\big) + \big((y_{11} + y_{12}) - (y_{31} + y_{32})\big) - 2\big((y_{21} + y_{22} + y_{23}) - (y_{31} + y_{32})\big)\Big], \tag{3.13}$$

$$\widehat{\tau_1} = 1/54\Big[\big(y_{11} + \cdots + y_{32}\big) + 17\big((y_{11} + y_{12}) - (y_{31} + y_{32})\big) - 7\big((y_{21} + y_{22} + y_{23}) - (y_{31} + y_{32})\big)\Big], \tag{3.14}$$

$$\widehat{\tau_2} = 1/54\Big[-2\big(y_{11} + \cdots + y_{32}\big) - 7\big((y_{11} + y_{12}) - (y_{31} + y_{32})\big) + 14\big((y_{21} + y_{22} + y_{23}) - (y_{31} + y_{32})\big)\Big], \tag{3.15}$$

and from the restriction

$$\widehat{\tau_3} = -\widehat{\tau_1} - \widehat{\tau_2}.$$

Exercise 3.2: Verify Expected Value. Verify by substituting the expected values for the $\{y_{ij}\}$ that the estimators given in (3.13), (3.14), and (3.15) are really unbiased for the parameters as advertised. □

Exercise 3.3: Verify Simplification. Demonstrate that the effect coding technique illustrated leads to much nicer equations when each treatment has exactly r repetitions. □

Exercise 3.4: Estimate Contrasts. Demonstrate that using the estimators given in (3.13), (3.14), and (3.15) to estimate the contrast $\tau_3 - \tau_1$ leads to the same answer as setting $\mu = 0$ and solving the simpler system of equations. □

3.3.4 Computing Sums of Squares

In equation (3.6) we indicated that the residual sum of squares or error sum of squares is $(\boldsymbol{y} - \boldsymbol{X}\widehat{\boldsymbol{\beta}})^t(\boldsymbol{y} - \boldsymbol{X}\widehat{\boldsymbol{\beta}})$. We can write

$$(\boldsymbol{y} - \boldsymbol{X}\widehat{\boldsymbol{\beta}})^t(\boldsymbol{y} - \boldsymbol{X}\widehat{\boldsymbol{\beta}}) = \boldsymbol{y}^t\boldsymbol{y} - \widehat{\boldsymbol{\beta}}^t\boldsymbol{X}^t\boldsymbol{y} - \boldsymbol{y}^t\boldsymbol{X}\widehat{\boldsymbol{\beta}} + \widehat{\boldsymbol{\beta}}^t\boldsymbol{X}^t\boldsymbol{X}\widehat{\boldsymbol{\beta}}.$$

Now using the fact that $\widehat{\boldsymbol{\beta}}^t\boldsymbol{X}^t\boldsymbol{y}$ and $\boldsymbol{y}^t\boldsymbol{X}\widehat{\boldsymbol{\beta}}$ are equal (both are scalars), and equation (3.5), which tells us that $\boldsymbol{X}^t\boldsymbol{X}\widehat{\boldsymbol{\beta}} = \boldsymbol{X}^t\boldsymbol{y}$, it follows that

$$\begin{aligned} \text{SSE} &= (\boldsymbol{y} - \boldsymbol{X}\widehat{\boldsymbol{\beta}})^t(\boldsymbol{y} - \boldsymbol{X}\widehat{\boldsymbol{\beta}}) \\ &= \boldsymbol{y}^t\boldsymbol{y} - \widehat{\boldsymbol{\beta}}^t\boldsymbol{X}^t\boldsymbol{y}. \end{aligned}$$

The error sum of squares is equal to the total sum of squares, $\boldsymbol{y}^t\boldsymbol{y}$ minus the piece, $\widehat{\boldsymbol{\beta}}^t\boldsymbol{X}^t\boldsymbol{y}$, attributable to the model being fitted. The latter piece is called the (uncorrected) *sum of squares due to regression*. It is typically denoted by $\mathrm{R}(\boldsymbol{\beta})$.

In the non-full-rank model case, care must be exercised to make sure that if some parameters are deleted, or side conditions applied to reduce the number of free parameters, the reduced $\boldsymbol{X}$ matrix and the reduced set of parameters are used when calculating $\mathrm{R}(\boldsymbol{\beta})$.

Since there are n independent observations, the total sum of squares, $\boldsymbol{y}^t\boldsymbol{y}$, has n degrees of freedom. To obtain the degrees of freedom for $\mathrm{R}(\boldsymbol{\beta})$, we count the number of independent parameters in the $\boldsymbol{\beta}$ vector. Let this be p. It follows that the residual sum of squares has $n - p$ degrees of freedom. It can be shown that $\text{MSE} = \text{SSE}/(n - p)$ is an unbiased estimate of σ_ϵ^2, defined in Section 3.3.2. It can also be shown that the sum of squares due to regression is unchanged regardless of the trick—setting a parameter equal to zero or applying a side condition—used in fitting the non-full rank models. This is demonstrated in Exercise 3.5.

Exercise 3.5: Computing R(β)

(a) Show that $\widehat{\mu}$, $\widehat{\tau}_1$, and $\widehat{\tau}_2$ obtained from equations (3.13), (3.14), and (3.15), together with the right-hand sides of equation (3.12), give $\mathrm{R}(\boldsymbol{\beta})$, or what in this case would more likely be called $\mathrm{R}(\mu, \tau_1, \tau_2, \tau_3)$, i.e., show that

$$\begin{aligned} &\widehat{\mu} \times (y_{11} + \cdots + y_{32}) + \widehat{\tau}_1 \times \big((y_{11} + y_{12}) - (y_{31} + y_{32})\big) \\ &\qquad + \widehat{\tau}_2 \times \big((y_{21} + y_{22} + y_{23}) - (y_{31} + y_{32})\big) \\ &= (y_{11} + y_{12})^2/2 + (y_{21} + y_{22} + y_{23})^2/3 + (y_{31} + y_{32})^2/2. \end{aligned}$$

(b) Show that the same result can be obtained by setting $\mu = 0$ in the model (3.8).

(c) Use the estimates obtained under the two different procedures to evaluate the $\{\widehat{y}_i\}$. You should get the same answers for both methods. □

3.4 PARAMETERIZATIONS FOR ROW–COLUMN MODELS

We now proceed to look at the analysis of linear models for the row–column classification. We encounter row–column models first in randomized block designs and later in our discussions of factorial experiments. Whereas the basic principles underlying the randomized block design and the factorial treatment structure are very different, the linear models used in the statistical analyses are very similar. Unfortunately, the similarity of the numeric computations involved may lead to confusion. We attempt to circumvent this by discussing a general row–column structure and then specialize to the randomized block design and the factorial in later chapters.

3.4.1 Non-Full-Rank Model

To demonstrate some ideas about the meanings of the parameters and estimation of the parameters, we write out the model for the row–column classification with interaction,

$$y_{ijk} = \mu + \rho_i + \gamma_j + (\rho\gamma)_{ij} + \epsilon_{ijk}, \tag{3.16}$$

where $\epsilon_{ijk} \sim \text{iid}(0, \sigma_\epsilon^2)$ for $i = 1, \ldots, R$, $j = 1, \ldots, C$, and $k = 1, \ldots, n$. In this model $\{\rho_i\}$ and $\{\gamma_j\}$ represent parts of the response attributable to the row effect and the column effect, respectively, and $(\rho\gamma)_{ij}$ the contribution unique to the row–column combination. The $\{\epsilon_{ijk}\}$ represent errors unique to the individual responses.

A detailed example of model (3.16) with three rows, two columns, and two replications is

$$\boldsymbol{y} = \begin{bmatrix} y_{111} \\ y_{112} \\ y_{121} \\ y_{122} \\ y_{211} \\ y_{212} \\ y_{221} \\ y_{222} \\ y_{311} \\ y_{312} \\ y_{321} \\ y_{322} \end{bmatrix}, \quad \boldsymbol{X} = \begin{bmatrix} 1 & 1 & 0 & 0 & 1 & 0 & 1 & 0 & 0 & 0 & 0 & 0 \\ 1 & 1 & 0 & 0 & 1 & 0 & 1 & 0 & 0 & 0 & 0 & 0 \\ 1 & 1 & 0 & 0 & 0 & 1 & 0 & 1 & 0 & 0 & 0 & 0 \\ 1 & 1 & 0 & 0 & 0 & 1 & 0 & 1 & 0 & 0 & 0 & 0 \\ 1 & 0 & 1 & 0 & 1 & 0 & 0 & 0 & 1 & 0 & 0 & 0 \\ 1 & 0 & 1 & 0 & 1 & 0 & 0 & 0 & 1 & 0 & 0 & 0 \\ 1 & 0 & 1 & 0 & 0 & 1 & 0 & 0 & 0 & 1 & 0 & 0 \\ 1 & 0 & 1 & 0 & 0 & 1 & 0 & 0 & 0 & 1 & 0 & 0 \\ 1 & 0 & 0 & 1 & 1 & 0 & 0 & 0 & 0 & 0 & 1 & 0 \\ 1 & 0 & 0 & 1 & 1 & 0 & 0 & 0 & 0 & 0 & 1 & 0 \\ 1 & 0 & 0 & 1 & 0 & 1 & 0 & 0 & 0 & 0 & 0 & 1 \\ 1 & 0 & 0 & 1 & 0 & 1 & 0 & 0 & 0 & 0 & 0 & 1 \end{bmatrix},$$

$$\boldsymbol{\beta} = \begin{bmatrix} \mu \\ \rho_1 \\ \rho_2 \\ \rho_3 \\ \gamma_1 \\ \gamma_2 \\ (\gamma\rho)_{11} \\ (\gamma\rho)_{12} \\ (\gamma\rho)_{21} \\ (\gamma\rho)_{22} \\ (\gamma\rho)_{31} \\ (\gamma\rho)_{32} \end{bmatrix}, \quad \boldsymbol{\epsilon} = \begin{bmatrix} \epsilon_{111} \\ \epsilon_{112} \\ \epsilon_{121} \\ \epsilon_{122} \\ \epsilon_{211} \\ \epsilon_{212} \\ \epsilon_{221} \\ \epsilon_{222} \\ \epsilon_{311} \\ \epsilon_{312} \\ \epsilon_{321} \\ \epsilon_{322} \end{bmatrix}. \tag{3.17}$$

3.4.2 Cell Means Model

As it is written, model (3.16) is overparameterized. Notice that the last six columns of the $\boldsymbol{X}$ matrix in Section 3.4.1 span the column space of $\boldsymbol{X}$; i.e., all other columns can be constructed as linear combinations of these six columns. In a sense, all other columns are unneeded. We could write the model as

$$y_{ijk} = \mu_{ij} + \epsilon_{ijk}, \tag{3.18}$$

where $\epsilon_{ijk} \sim \text{iid}(0, \sigma_\epsilon^2)$ for $i = 1, \ldots, R$, $j = 1, \ldots, C$, and $k = 1, \ldots, n$.

Our special example then becomes

$$\boldsymbol{y} = \begin{bmatrix} 1 & 0 & 0 & 0 & 0 & 0 \\ 1 & 0 & 0 & 0 & 0 & 0 \\ 0 & 1 & 0 & 0 & 0 & 0 \\ 0 & 1 & 0 & 0 & 0 & 0 \\ 0 & 0 & 1 & 0 & 0 & 0 \\ 0 & 0 & 1 & 0 & 0 & 0 \\ 0 & 0 & 0 & 1 & 0 & 0 \\ 0 & 0 & 0 & 1 & 0 & 0 \\ 0 & 0 & 0 & 0 & 1 & 0 \\ 0 & 0 & 0 & 0 & 1 & 0 \\ 0 & 0 & 0 & 0 & 0 & 1 \\ 0 & 0 & 0 & 0 & 0 & 1 \end{bmatrix} \begin{bmatrix} \mu_{11} \\ \mu_{12} \\ \mu_{21} \\ \mu_{22} \\ \mu_{31} \\ \mu_{32} \end{bmatrix} + \begin{bmatrix} \epsilon_{111} \\ \epsilon_{112} \\ \epsilon_{121} \\ \epsilon_{122} \\ \epsilon_{211} \\ \epsilon_{212} \\ \epsilon_{221} \\ \epsilon_{222} \\ \epsilon_{311} \\ \epsilon_{312} \\ \epsilon_{321} \\ \epsilon_{322} \end{bmatrix}.$$

This is the cell means representation of the model for the two-way array with interaction. It just has a separate parameter for each cell mean.

3.4.3 Effect Coding

By writing the cell means model as equation (3.18) we have tacitly defined μ_{ij} to be the mean response for cell ij, i.e., for the row i, column j combination. However, we have not properly defined the μ, $\{\rho_i\}$, $\{\gamma_j\}$, and $\{(\rho\gamma)_{ij}\}$ in model

(3.16). When we restrict ourselves to the balanced case, the case where there are n responses in each cell, matters are straightforward. We write

$$\begin{aligned}
\overline{\mu}_{i\cdot} &= \frac{1}{C}\sum_{j}^{C}\mu_{ij} \\
\overline{\mu}_{\cdot j} &= \frac{1}{R}\sum_{i}^{R}\mu_{ij} \\
\mu &= \frac{1}{RC}\sum_{i}^{R}\sum_{j}^{C}\mu_{ij} \\
\rho_i &= \overline{\mu}_{i\cdot} - \mu \\
\gamma_j &= \overline{\mu}_{\cdot j} - \mu \\
(\rho\gamma)_{ij} &= \mu_{ij} - \overline{\mu}_{i\cdot} - \overline{\mu}_{\cdot j} + \mu.
\end{aligned}$$

These definitions imply that $\sum_i \rho_i = 0$, $\sum_j \gamma_j = 0$, $\sum_i (\rho\gamma)_{ij} = 0$ for each j, and $\sum_j (\rho\gamma)_{ij} = 0$ for each i.

We apply these constraints by replacing ρ_R by $-\sum_i^{R-1}\rho_i$, γ_C by $-\sum_j^{C-1}\gamma_j$, $(\rho\gamma)_{Rj}$ by $-\sum_i^{R-1}(\rho\gamma)_{ij}$ for $j = 1, \ldots, C-1$, $(\rho\gamma)_{iC}$ by $-\sum_j^{C-1}(\rho\gamma)_{ij}$ for $i = 1, \ldots, R-1$, and $(\rho\gamma)_{RC}$ by $\sum_i^{R-1}\sum_j^{C-1}(\rho\gamma)_{ij}$. The design matrix contains a column of ones for μ, $R-1$ columns for row effects, $C-1$ columns for column effects, and $(R-1)(C-1)$ columns for interactions. The ith indicator variable for row effects is coded 1, -1, and 0 for observations from the ith row, the Rth row, and any other row, respectively. Similarly, the jth indicator variable for the column effects is coded 1, -1, and 0 for observations from the jth, the Cth, and other columns. The columns of the design matrix for the row–column interactions may be obtained by multiplying the row and column indicator variables together. Our special three-row, two-column example then becomes

$$\boldsymbol{X} = \begin{bmatrix}
1 & 1 & 0 & 1 & 1 & 0 \\
1 & 1 & 0 & 1 & 1 & 0 \\
1 & 1 & 0 & -1 & -1 & 0 \\
1 & 1 & 0 & -1 & -1 & 0 \\
1 & 0 & 1 & 1 & 0 & 1 \\
1 & 0 & 1 & 1 & 0 & 1 \\
1 & 0 & 1 & -1 & 0 & -1 \\
1 & 0 & 1 & -1 & 0 & -1 \\
1 & -1 & -1 & 1 & -1 & -1 \\
1 & -1 & -1 & 1 & -1 & -1 \\
1 & -1 & -1 & -1 & 1 & 1 \\
1 & -1 & -1 & -1 & 1 & 1
\end{bmatrix} \quad \text{and} \quad \boldsymbol{\beta} = \begin{bmatrix}
\mu \\ \rho_1 \\ \rho_2 \\ \gamma_1 \\ (\rho\gamma)_{11} \\ (\rho\gamma)_{21}
\end{bmatrix}.$$

The $\boldsymbol{X}$ matrix has full column rank; i.e., it is not possible to express any column as a linear function of the remaining columns. It follows that $\boldsymbol{X}^t\boldsymbol{X}$ is nonsingular and the normal equations have a unique solution. This is true not only for our little example, but for the $R \times C$ row–column case in general. Once the normal equations have been solved, one can use the restriction equations to obtain estimates of the parameters that were removed from the system.

3.4.4 Reference Cell Coding

We mentioned in Section 3.3.1 that the normal equations (3.5) could be solved in different ways. We have considered three methods of making $X^t X$ nonsingular (1) using the cell means model, (2) apply effects coding restrictions to the parameter definitions to reduce their actual number, and (3) reduce the model by eliminating some parameters. This last method, although very powerful, must be used with care since it appears to redefine parameters in the model. The term *reference cell coding* is often used when referring to this method.

In reference cell coding we simply delete ρ_R, γ_C, and $(\rho\gamma)_{ij}$ when $i = R$ or $j = C$ from the vector of parameters (and the corresponding columns from the X matrix). For the three row, two column example the result of deleting these columns from the X matrix (3.17) is

$$\boldsymbol{y} = \begin{bmatrix} y_{111} \\ y_{112} \\ y_{121} \\ y_{122} \\ y_{211} \\ y_{212} \\ y_{221} \\ y_{222} \\ y_{311} \\ y_{312} \\ y_{321} \\ y_{322} \end{bmatrix}, \quad \boldsymbol{X} = \begin{bmatrix} 1 & 1 & 0 & 1 & 1 & 0 \\ 1 & 1 & 0 & 1 & 1 & 0 \\ 1 & 1 & 0 & 0 & 0 & 0 \\ 1 & 1 & 0 & 0 & 0 & 0 \\ 1 & 0 & 1 & 1 & 0 & 1 \\ 1 & 0 & 1 & 1 & 0 & 1 \\ 1 & 0 & 1 & 0 & 0 & 0 \\ 1 & 0 & 1 & 0 & 0 & 0 \\ 1 & 0 & 0 & 1 & 0 & 0 \\ 1 & 0 & 0 & 1 & 0 & 0 \\ 1 & 0 & 0 & 0 & 0 & 0 \\ 1 & 0 & 0 & 0 & 0 & 0 \end{bmatrix}, \quad \boldsymbol{\beta} = \begin{bmatrix} \mu \\ \rho_1 \\ \rho_2 \\ \gamma_1 \\ (\rho\gamma)_{11} \\ (\rho\gamma)_{21} \end{bmatrix}.$$

This X matrix has full column rank. It follows that the matrix $X^t X$ is nonsingular and the normal equations have a unique solution. Again, this is true, not only for our little example, but for the $R \times C$ row–column case in general. In general, there are $(R - 1)$ columns for row effects, $(C - 1)$ columns for column effects, and $(R - 1)(C - 1)$ columns for the interactions.

The names of the parameters in this method are the same as for effect coding method, but the meanings are very different. However, one can translate back to the original set of parameters or to the treatment means. In the three-row, two-column example, the treatment means can be read from the rows of $X\boldsymbol{\beta}$. Thus, $\mathrm{E}[y_{11\cdot}] = \mu + \rho_1 + \gamma_1 + (\rho\gamma)_{11}$. But notice that $\mathrm{E}[y_{12\cdot}] = \mu + \rho_1$ and $\mathrm{E}[y_{32\cdot}] = \mu$. Hence μ represents the mean of the $(3, 2)$ treatment, which is the "reference cell" in this parameterization; it does *not* represent the mean of all treatments. The ρ_i parameter represents $\mathrm{E}[y_{i2\cdot}] - \mathrm{E}[y_{32\cdot}]$. The interaction terms also represent deviations from reference cells and not from factor means under this parameterization. For instance, $(\rho\gamma)_{11}$ represents $\left(E[y_{11\cdot}] - E[y_{12\cdot}]\right) - \left(E[y_{31\cdot}] - E[y_{32\cdot}]\right)$. Notice that all contrasts of the form $\mu_{ij} - \mu_{i'j'}$ are estimable. The point is that these are "estimable quantities" and consequently, are available to us regardless of the trick used to solve the normal equations.

REVIEW EXERCISES

3.1 Conduct an experiment to see whether pennies spin longer than nickels or quarters. For each size coin, spin 10 separate coins, for a total of 30 coins. Are the assumptions necessary for fitting model (3.7) satisfied? To answer this question, do the following:

(a) Describe your randomization scheme and discuss why randomization might be necessary. Discuss whether there might be any better way of selecting run order than complete randomization.

(b) Construct a box plot of the spinning times for each type of coin to assess visually whether the variances are homogeneous for the three types of coins. Label each axis clearly.

(c) Keep track of the sequence of coins that you spin and plot the response against sequence number (let j = sequence number and y_{ij} be the jth response for the ith type of coin), with a separate symbol for each type of coin.

(d) Compute the mean $\overline{y}_{i\cdot}$ and standard deviation s_i for each type of coin, where i refers to type of coin. Compute standardized responses $z_{ij} = (y_{ij} - \overline{y}_{i\cdot})/s_i$ by subtracting off the appropriate mean and dividing by the standard deviation for each coin. Then plot the standardized response z_{ij} against the standardized response for the coin thrown previously. This type of graph displays temporal autocorrelation.

(e) Summarize your findings as to whether the assumptions of model (3.7) are satisfied.

3.2 Using your favorite statistical software, generate five sets of 10 observations that follow the regression model $y_i = 1 + 2x_i + \epsilon_i$, where the ϵ_i are normally distributed with mean zero and variance $\sigma^2 = 4$ and $x_i = i/2$. (See Appendix 3B.)

(a) Plot y vs. x for each of the five data sets.

(b) Estimate the regression parameters and standard errors for each of the five data sets.

(c) Explain with words and formulas what quantities the computed standard errors estimate.

(d) Compute the mean and standard deviation of the five estimates of β_0 and β_1.

3.3 A manufacturer is studying the drying times of various types of adhesives. She wants to compare three types of adhesives on three different surfaces.

The surfaces under consideration are (1) smooth plastic, (2) a pebbly plastic surface, and (3) a scored plastic surface. She will do five replications of each treatment combination.

(a) Write a paragraph instructing the manufacturer how to conduct this experiment. Include specification of the number of pieces of plastic to use, how to randomize the treatments, and any other considerations of importance.

(b) Write down the model for analysis of variance of this experiment. Include the assumptions about the random errors in your specification of the model. Include ranges of the subscripts.

(c) Write out the elements of the design matrix for one replication, and the elements of $\boldsymbol{\beta}$ using effects coding.

(d) Write down the formulas for the least squares estimates and standard errors of the following quantities: (1) the main effect of adhesive 1; (2) the mean of adhesive 1; (3) the difference between the effect of adhesive 1 and 2 on a pebbly surface and the effect on a scored plastic surface. (See Appendix 3A.)

(e) What is the variance of the mean difference between smooth plastic and the average of the two rough plastic surfaces?

3.4 Cohen (2001) reported on a test of a gasoline additive that is proposed to make gasoline burn more completely and reduce NOx emissions. Three fueling regimes were tested: (1) R: reference fuel used for every tank fill, (2) T: test fuel used for every tank fill, and (3) A: alternately filling tank with test fuel and reference fuel. Seven types of vehicles were used, each falling into a different class of fuel emission standards. Four vehicles of each type were tested: one each for fueling regimes T and A, and two vehicles for fueling regime R. All vehicles were driven 1000 miles using the reference fuel and NOx emissions were measured. Then each vehicle was driven 8000 miles under its assigned testing regime. At the end of 8000 miles, NOx emissions were measured again. The mean changes in log(NOx) emissions were .13, .09, and .05 for fueling regimes R, T, and A, respectively. Suppose that the model (3.16) (two-factor analysis of variance) was used to analyze these data. The exact MSE is not given in the report, but assuming that MSE $= .10$, answer the following questions. (See Appendix 3A.)

(a) What is the standard error of the mean difference in emissions between the regimes that use the test additive (the average of regimes T and A) and the reference fuel regime (R)?

(b) Compute the standard error of the mean difference between regimes T and A.

(c) Write out a matrix $\boldsymbol{L}$ and a vector of parameters $\boldsymbol{\beta}$ that express the contrasts of parts (**a**) and (**b**). Are these two contrasts orthogonal?

APPENDIX 3A: LINEAR COMBINATIONS OF RANDOM VARIABLES

Linear combinations of random variables and their standard errors take simple forms in matrix notation. Let $\boldsymbol{y} = \begin{bmatrix} y_1 \\ \vdots \\ y_n \end{bmatrix}$ be a vector of random variables, $\boldsymbol{\mu} = \begin{bmatrix} \mu_1 \\ \vdots \\ \mu_n \end{bmatrix}$ the vector of expected values, and $\boldsymbol{a} = \begin{bmatrix} a_1 \\ \vdots \\ a_n \end{bmatrix}$ and $\boldsymbol{c} = \begin{bmatrix} c_1 \\ \vdots \\ c_n \end{bmatrix}$ two vectors of constants. The linear function $z = \sum_i a_i y_i$ can then be written compactly as $\boldsymbol{a}^t \boldsymbol{y}$. Also, $\text{E}[z] = \boldsymbol{a}^t \boldsymbol{\mu}$. If the variances and covariances are written in the form of an $n \times n$ matrix

$$\boldsymbol{V} = \begin{bmatrix} \sigma_1^2 & \cdots & \sigma_{1n} \\ \vdots & \ddots & \vdots \\ \sigma_{n1} & \cdots & \sigma_n^2 \end{bmatrix},$$

then

$$\text{Var}\left[\boldsymbol{a}^t \boldsymbol{y}\right] = \boldsymbol{a}^t \boldsymbol{V} \boldsymbol{a}$$

$$\text{Cov}[\boldsymbol{a}^t \boldsymbol{y}, \boldsymbol{c}^t \boldsymbol{y}] = \boldsymbol{a}^t \boldsymbol{V} \boldsymbol{c}.$$

We can put all this together to compute the variance–covariance matrix of sets of linear combinations of random variables. Suppose that we want to compute $\boldsymbol{a}^t \boldsymbol{y}$ and $\boldsymbol{c}^t \boldsymbol{y}$. We can construct a matrix containing all of the $\{a_i\}$ and $\{c_i\}$ coefficients and compute the variance–covariance matrix at one time. In general, if $\boldsymbol{L}$ is a $k \times n$ matrix of coefficients, the variance-covariance matrix for $\boldsymbol{L}\boldsymbol{y}$ is

$$\text{Var}\left[\boldsymbol{L}\boldsymbol{y}\right] = \boldsymbol{L}\boldsymbol{V}\boldsymbol{L}^t.$$

In the case of computing the variance–covariance matrix of $\boldsymbol{a}^t \boldsymbol{y}$ and $\boldsymbol{c}^t \boldsymbol{y}$, the $\boldsymbol{L}$ matrix would be

$$\boldsymbol{L} = \begin{bmatrix} a_1 & a_2 & \cdots & a_n \\ c_1 & c_2 & \cdots & c_n \end{bmatrix},$$

and the variance–covariance matrix would be

$$\boldsymbol{LVL}^t = \begin{bmatrix} \text{Var}\left[\boldsymbol{a}^t \boldsymbol{y}\right] & \text{Cov}[\boldsymbol{a}^t \boldsymbol{y}, \boldsymbol{c}^t \boldsymbol{y}] \\ \text{Cov}[\boldsymbol{a}^t \boldsymbol{y}, \boldsymbol{c}^t \boldsymbol{y}] & \text{Var}\left[\boldsymbol{c}^t \boldsymbol{y}\right] \end{bmatrix}.$$

The ordinary least squares estimator of $\boldsymbol{\beta}$ is a linear combination of elements of $\boldsymbol{y}$; it is $\widehat{\boldsymbol{\beta}} = (\boldsymbol{X}^t \boldsymbol{X})^{-1} \boldsymbol{X}^t \boldsymbol{y}$. Ordinary least squares is used when the errors all have the same variance and are uncorrelated, so that $\boldsymbol{V} = \sigma_\epsilon^2 \boldsymbol{I}$. The variance–covariance matrix of $\widehat{\boldsymbol{\beta}}$ is then

$$\text{Var}[\widehat{\boldsymbol{\beta}}] = (\boldsymbol{X}^t \boldsymbol{X})^{-1} \boldsymbol{X}^t \boldsymbol{V} \boldsymbol{X} (\boldsymbol{X}^t \boldsymbol{X})^{-1} = \sigma_\epsilon^2 (\boldsymbol{X}^t \boldsymbol{X})^{-1}.$$

We can find the variances and covariances of linear combinations of the regression coefficients and contrasts of treatment effects and interactions in a similar way:

$$\begin{aligned}\mathrm{Var}\big[\boldsymbol{L}\widehat{\boldsymbol{\beta}}\,\big] &= \boldsymbol{L}\mathrm{Var}\big[\widehat{\boldsymbol{\beta}}\big]\boldsymbol{L}^t \\ &= \sigma_\epsilon^2 \boldsymbol{L}(\boldsymbol{X}^t\boldsymbol{X})^{-1}\boldsymbol{L}^t.\end{aligned}$$

In actual applications, the variance σ_ϵ^2 is unknown and the estimate, $\widehat{\sigma}_\epsilon^2 = \mathrm{MSE}$, is substituted. The estimated variance of a linear combination of parameter estimates is

$$\widehat{\mathrm{Var}}\big[\boldsymbol{L}\widehat{\boldsymbol{\beta}}\big] = \widehat{\sigma}_\epsilon^2 \boldsymbol{L}(\boldsymbol{X}^t\boldsymbol{X})^{-1}\boldsymbol{L}^t.$$

The estimated variances of linear combinations of parameter estimates obtained in this way are unbiased because $\widehat{\sigma}_\epsilon^2$ is an unbiased estimate of σ_ϵ^2; i.e.,

$$\mathrm{E}\big[\widehat{\sigma}_\epsilon^2 \boldsymbol{L}(\boldsymbol{X}^t\boldsymbol{X})^{-1}\boldsymbol{L}^t\big] = \sigma_\epsilon^2 \boldsymbol{L}(\boldsymbol{X}^t\boldsymbol{X})^{-1}\boldsymbol{L}^t.$$

APPENDIX 3B: SIMULATING RANDOM SAMPLES

For statistical work it is often useful to simulate random phenomena and then study the properties of different statistical estimators. Statistical software and many calculators include functions for generating (pseudo) random numbers from different probability distributions. We use both the normal distribution and the uniform distribution extensively in this text. Typically, random number functions generate numbers from a standard distribution, such as the standard normal distribution with mean 0 and variance 1, written as $N(0, 1)$, which can then be transformed to the distribution of interest. For example, the **RANNOR** function in SAS® generates numbers from a $N(0, 1)$ distribution. Typically, the user specifies a seed, which is some arbitrary number, to start off the process of generating random numbers. To obtain a set of uncorrelated random numbers from a normal distribution with mean 10 and variance 25, the procedure is to generate a set of standard normal variates, multiply each generated value by the standard deviation, 5, and add the mean, 10. Why does this work?

The SAS® code to generate 10 $N(10, 25)$ observations is

```
data one;
   seed=4085321;
   do i=1 to 10;
   y=10+5*rannor(seed);
   output;
   end;
```

CHAPTER 4

Testing Hypotheses and Determining Sample Size

4.1 INTRODUCTION

The material in this chapter can be divided into three distinct parts. First we present three methods of testing hypotheses. Next we present three techniques for determining appropriate and effective sample sizes. In the last part we discuss an approach that is appropriate for those cases where the object is to select the best from a set of treatments. The classical example of this is the selection of the best in a variety trial in agriculture.

In Sections 4.2.1 and 4.2.2, tests derived under the linear model with independent normally distributed errors are discussed. This is a review of material usually covered in a first course on analysis of variance. The initial objective here is to develop the tools necessary to evaluate the large class of hypotheses available in the context of the linear model. Comparisons of treatment means in analysis of variance problems as well as tests of significance of terms in multiple regression models fall into this framework. The ultimate aim is to construct the machinery to compute power functions for the linear hypothesis tests. In Section 4.3 we present a nonparametric test procedure for comparing several treatment means. Although this procedure does not have the full flexibility of the general linear hypothesis tests, it has the advantage that it does not require the errors to be normally distributed. Then in Section 4.4, randomization tests are discussed. These tests are not only nonparametric but also free of any distributional assumptions about the random errors. The validity of the randomization tests depends solely on the randomization employed in execution of the experiment. The validity is fully under the control of the experimenter.

In Sections 4.5 to 4.7 we present details for determining appropriate and effective sample sizes. The material in Section 4.5 builds on normal theory and the

Planning, Construction, and Statistical Analysis of Comparative Experiments,
by Francis G. Giesbrecht and Marcia L. Gumpertz
ISBN 0-471-21395-0

linear hypothesis tests discussed in Section 4.2. Methods for calculating power for any test that can be cast as a linear hypothesis test are illustrated. The underlying concept is to calculate the probability of rejecting a null hypothesis when it is not true. In Section 4.6 the techniques are extended to samples for proportions. In Section 4.7 we present an alternative approach to calculating required sample size based on the notion of desired confidence interval widths.

4.2 TESTING HYPOTHESES IN LINEAR MODELS WITH NORMALLY DISTRIBUTED ERRORS

One general and widely applicable method of testing hypotheses about parameters is to compare the fit of a *full model* without restrictions on the parameters to the fit of a *reduced model* where some of the parameters are restricted to equal hypothesized values. If the error sum of squares [eq. (3.6)] for the reduced model is much larger than the error sum of squares for the full model, or equivalently, the sum of squares due to regression $R(\beta)$ is much larger for the full model than for the reduced model, we conclude that the parameters do not equal the hypothesized values. Recall from Section 3.3.4 that

$$\boldsymbol{y}^t\boldsymbol{y} = \text{SSE} + \widehat{\boldsymbol{\beta}}^t\boldsymbol{X}^t\boldsymbol{y}. \tag{4.1}$$

Tests of hypotheses can be obtained conveniently from either the error sum of squares or the sum of squares due to regression.

Interesting tests of hypotheses can typically be set up in terms of subsets of the parameters in the model. Suppose that the full model has $k + 1$ parameters,

$$y_i = \beta_0 + \beta_1 x_{i,1} + \cdots + \beta_k x_{i,k} + \epsilon_i. \tag{4.2}$$

The k terms in addition to the intercept could be either regression parameters or terms in an analysis of variance model (or some of each). Suppose further that we wish to test the hypothesis that the last m of the parameters are really zero,

$$\text{H}_0 : \beta_{k-m+1} = \cdots = \beta_k = 0. \tag{4.3}$$

A reasonable strategy is to fit the model containing all $k + 1$ parameters and then fit a reduced model containing only the first $k - m + 1$ parameters and see whether the full model accounts for an appreciably larger portion of the variation among the y values than does the reduced model. The reduced model would be

$$y_i = \beta_0 + \beta_1 x_{i,1} + \cdots + \beta_{k-m} x_{i,k-m} + \epsilon_i. \tag{4.4}$$

The difference between the error sum of squares for a full model and for a reduced model is called the *extra sum of squares* or the *reduction in sum of squares* due to fitting the extra parameters and is denoted $\mathrm{R}(\cdot|\cdot)$,

$$\begin{aligned}\mathrm{R}(\beta_{k-m+1},\ldots,\beta_k|\beta_0,\beta_1,\ldots,\beta_{k-m})\\ &= \mathrm{R}(\beta_0,\ldots,\beta_k) - \mathrm{R}(\beta_0,\ldots,\beta_{k-m})\\ &= \mathrm{SSE(model\ 4.4)} - \mathrm{SSE(model\ 4.2)}. \end{aligned} \quad (4.5)$$

The statistic for testing the hypothesis shown in equation (4.3) is

$$F = \frac{\mathrm{R}(\beta_{k-m+1},\ldots,\beta_k|\beta_0,\beta_1,\ldots,\beta_{k-m})/m}{\mathrm{MSE(model\ 4.2)}}.$$

This test statistic has numerator degrees of freedom $\mathrm{df}_1 = m$, which is the difference between the number of parameters in the full and reduced models. The denominator degrees of freedom are $\mathrm{df}_2 = n-k-1$, the degrees of freedom for MSE in the full model. The test criterion is:

$$\text{Reject } \mathrm{H}_0 \text{ if } F > F(1-\alpha, \mathrm{df}_1, \mathrm{df}_2),$$

where α is the desired significance level and $F(1-\alpha, \mathrm{df}_1, \mathrm{df}_2)$ denotes the $1-\alpha$ quantile of the F-distribution with df_1 and df_2 degrees of freedom. Under the assumptions of the standard linear model, i.e., independent normally distributed errors with constant variance, this is the *likelihood ratio test* of $\mathrm{H}_0 : \beta_{k-m+1} = \cdots = \beta_k = 0$.

In either regression or analysis of variance it is sometimes of interest to test whether fitting the model is any better than just fitting an overall mean to the data. This is a special case of the hypothesis shown in equation (4.3). The test of whether the full model fits better than the simple mean of the data consists of comparing the full model to the reduced model,

$$y_i = \beta_0 + \epsilon_i, \quad (4.6)$$

where $i = 1,\ldots,n$. The error sum of squares for a model with just an overall mean is known as the *corrected total sum of squares*,

$$\mathrm{SSTO} = \sum_{i=1}^{n}(y_i - \overline{y}_.)^2 = \boldsymbol{y}^t\boldsymbol{y} - \frac{(y_.)^2}{n}, \quad (4.7)$$

where n is the total number of observations in the sample. For testing whether all terms except the intercept in a regression model with $k+1$ parameters are zero, the reduction in sum of squares is

$$\begin{aligned}\mathrm{R}(\beta_1,\ldots,\beta_k|\beta_0) &= \mathrm{R}(\beta_0,\beta_1,\ldots,\beta_k) - \mathrm{R}(\beta_0)\\ &= \widehat{\boldsymbol{\beta}}^t\boldsymbol{X}^t\boldsymbol{y} - \frac{(y_.)^2}{n}.\\ &= \mathrm{SSE(model\ 4.6)} - \mathrm{SSE(model\ 4.2)} \end{aligned} \quad (4.8)$$

Equation (4.8) is also known as the *regression sum of squares* (SSR) or *model sum of squares*.

4.2.1 General Linear Hypothesis Tests

Any linear hypothesis can also be written in matrix notation as $\text{H}_0 : \boldsymbol{L\beta} = \boldsymbol{m}_0$, where the elements of $\boldsymbol{L}$ select linear combinations of the parameters to test and $\boldsymbol{m}_0$ is a vector of constants. We demonstrate for a one-factor analysis of variance with $t = 4$ treatment levels and $r_i = 3$ replicates for each treatment. The design matrix and parameter vector, using effects coding, are

$$\boldsymbol{y} = \begin{bmatrix} y_{11} \\ y_{12} \\ y_{13} \\ y_{21} \\ y_{22} \\ y_{23} \\ y_{31} \\ y_{32} \\ y_{33} \\ y_{41} \\ y_{42} \\ y_{43} \end{bmatrix}, \quad \boldsymbol{X} = \begin{bmatrix} 1 & 1 & 0 & 0 \\ 1 & 1 & 0 & 0 \\ 1 & 1 & 0 & 0 \\ 1 & 0 & 1 & 0 \\ 1 & 0 & 1 & 0 \\ 1 & 0 & 1 & 0 \\ 1 & 0 & 0 & 1 \\ 1 & 0 & 0 & 1 \\ 1 & 0 & 0 & 1 \\ 1 & -1 & -1 & -1 \\ 1 & -1 & -1 & -1 \\ 1 & -1 & -1 & -1 \end{bmatrix}, \quad \boldsymbol{\beta} = \begin{bmatrix} \mu \\ \tau_1 \\ \tau_2 \\ \tau_3 \end{bmatrix}, \quad \boldsymbol{\epsilon} = \begin{bmatrix} \epsilon_{11} \\ \epsilon_{12} \\ \epsilon_{13} \\ \epsilon_{21} \\ \epsilon_{22} \\ \epsilon_{23} \\ \epsilon_{31} \\ \epsilon_{32} \\ \epsilon_{33} \\ \epsilon_{41} \\ \epsilon_{42} \\ \epsilon_{43} \end{bmatrix}.$$

First test the treatment effects: $\text{H}_0 : \tau_1 = \tau_2 = \tau_3 = 0$. The $\boldsymbol{L}$-matrix depends critically on the parameterization used. In this example, using effects coding, the $\boldsymbol{L}$-matrix and $\boldsymbol{m}_0$ vector for H_0 are

$$\boldsymbol{L} = \begin{bmatrix} 0 & 1 & 0 & 0 \\ 0 & 0 & 1 & 0 \\ 0 & 0 & 0 & 1 \end{bmatrix} \quad \text{and} \quad \boldsymbol{m}_0 = \begin{bmatrix} 0 \\ 0 \\ 0 \end{bmatrix}.$$

The hypothesis being tested is then

$$\text{H}_0 : \begin{bmatrix} \tau_1 \\ \tau_2 \\ \tau_3 \end{bmatrix} = \begin{bmatrix} 0 \\ 0 \\ 0 \end{bmatrix}.$$

Any linear hypothesis can be written in the same matrix form, just by changing the $\boldsymbol{L}$-matrix. For instance, to test the hypothesis that the effect of treatment 2 is twice that of treatment 1, restate the hypothesis as $\text{H}_0 : 2\tau_1 - \tau_2 = 0$. In matrix notation this is $\text{H}_0 : \boldsymbol{L\beta} = m_0$, where $\boldsymbol{L} = \begin{bmatrix} 0 & 2 & -1 & 0 \end{bmatrix}$ and $m_0 = 0$. This hypothesis involves a single linear combination of parameters, so $\boldsymbol{L}$ has only one row and m_0 is a scalar. The matrix formulation is especially useful for more complicated models, because the same formulas hold for all linear models and all linear hypotheses.

Exercise 4.1: Set Up L-Matrices of Contrast Coefficients Using Effects Coding. For a one-factor analysis of variance model using effects coding for four treatment levels, write out the $\boldsymbol{L}$-matrix and the degrees of freedom (numerator and denominator) for testing the following hypotheses: (a) $\text{H}_0 : \tau_2 = 0$, (b) $\text{H}_0 : \tau_4 = 0$, and (c) $\text{H}_0 : \tau_1 = \tau_4$. □

Exercise 4.2: Set Up L-Matrices of Contrast Coefficients Using the Non-Full-Rank Parameterization. For a one-factor analysis of variance model for four treatments using a non-full-rank model, write out the $\boldsymbol{L}$-matrix and the degrees of freedom for testing the hypotheses (a) $\text{H}_0 : \tau_1 = \tau_2 = \tau_3 = \tau_4$, and (b) $\text{H}_0 : \tau_1 = \tau_4$. □

The sum of squares for any linear hypothesis H_0 about the linear model, where $\boldsymbol{X}$ and $\boldsymbol{L}$ are both full rank, can be written as a quadratic form,

$$\text{SS}(\text{H}_0) = (\boldsymbol{L}\widehat{\boldsymbol{\beta}} - \boldsymbol{m_0})^t(\boldsymbol{L}(\boldsymbol{X}^t\boldsymbol{X})^{-1}\boldsymbol{L}^t)^{-1}(\boldsymbol{L}\widehat{\boldsymbol{\beta}} - \boldsymbol{m_0}). \tag{4.9}$$

[See Neter et al. (1990) for a derivation.] This quadratic form involves a set of contrasts and the inverse of its variance:

$$\text{SS}(\text{H}_0) = (\boldsymbol{L}\widehat{\boldsymbol{\beta}} - \boldsymbol{m_0})^t(\text{Var}[\boldsymbol{L}\widehat{\boldsymbol{\beta}}]/\sigma_\epsilon^2)^{-1}(\boldsymbol{L}\widehat{\boldsymbol{\beta}} - \boldsymbol{m_0}). \tag{4.10}$$

If the hypothesis involves only one linear combination of parameters, so that $\boldsymbol{L}$ has only one row and $\boldsymbol{L}\boldsymbol{\beta}$ is a scalar, the quadratic form [eq. (4.9)] reduces to $(\boldsymbol{L}\widehat{\boldsymbol{\beta}} - \boldsymbol{m_0})^2/(\text{Var}[\boldsymbol{L}\widehat{\boldsymbol{\beta}}]/\sigma_\epsilon^2)$. The hypothesis is tested by comparing the test statistic $F = \text{MS}(\text{H}_0)/\text{MSE}$ to quantiles of an F-distribution with df_H and df_E degrees of freedom, where $\text{MS}(\text{H}_0) = \text{SS}(\text{H}_0)/\text{df}_H$. In general, the degrees of freedom for the hypothesis equal the number of rows in $\boldsymbol{L}$, i.e., $\text{df}_H = \text{rank}(\boldsymbol{L})$. This is the number of components being tested in the hypothesis. The degrees of freedom for error always equal the total number of observations minus the number of terms in the model: i.e., $\text{df}_E = n -$ rank($\boldsymbol{X}$).

For the standard linear model discussed in this chapter, which has independent normally distributed errors with constant variances, the likelihood ratio F-statistic coincides with the test statistic based on the quadratic form in equation (4.9). However, this is not true if the data are not normally distributed or the errors are not independent with constant variance, or if the model is not linear in the parameters (Myers 1990).

4.2.2 Distribution of the Test Statistic

It is important to understand that quantities such as $\text{MS}(\text{H}_0)$ and MSE are realizations of random variables, and that $F = \text{MS}(\text{H}_0)/\text{MSE}$ is also a realization of a random variable. All three of these random variables have distributions with means, variances, and so on. When H_0 is true, the F-statistic has one distribution, and when H_0 is false, it has a different distribution. The situation is shown in Figure 4.1.

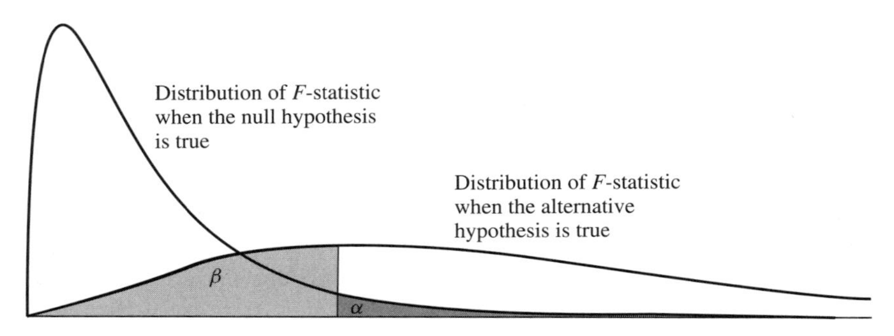

Figure 4.1 Analysis of variance F-test.

The F-test is an example of a common general test procedure, which goes as follows. Select a method of analysis, that is, a test statistic such that the distribution of the underlying random variable (F in this case) is known, or at least is known when H_0 is true. Then conduct the experiment and obtain a realization of the test statistic. Reject H_0 only if the realized value of the test statistic is extreme. To gauge whether the realized value is extreme, compare it to the distribution of the underlying random variable. In the F-test, compare the realized value with the $1 - \alpha$ quantile of the F-distribution, where α is the desired significance level. Reject the null hypothesis if, under H_0, the probability of obtaining a value as extreme or more extreme than the realized value is less than α. The extreme $100\alpha\%$ of the distribution of the random variable is called the *critical region*. In Figure 4.1 the vertical line indicates a *critical value* and the critical region consists of all values greater than the critical value. Under the null hypothesis, the area of the critical region is α and is dark gray. If the null hypothesis is false and the alternative distribution really describes the distribution of the statistic, the probability of rejecting H_0 includes all the area to the right of the critical value (white and gray) under the curve describing the alternative distribution. This probability is called the *power* of the test. The probability of rejecting H_0 is larger than α if H_0 is not true, which of course is what makes the test procedure useful. Another thing to notice in Figure 4.1 is the light gray area labeled β. This region shows the probability of not rejecting H_0 if the alternative is really the true distribution. This is the probability of a type II error, also known as $\beta = 1 - \text{power}$.

A common, and in many respects better, way to report the results of a statistical analysis is to report the p-value as a summary statistic or measure of the weight of evidence against H_0. The *p-value* is the quantile of the null distribution that corresponds to the realized value of the test statistic. It is the smallest significance level that would result in rejecting the null hypothesis, given the realized value of the test statistic.

Turning to the F-ratio specifically, this particular form of test statistic is based on the assumptions of normality, independence, and variance homogeneity of the random errors in the model. Under normality, the sums of squares divided by the corresponding expected mean squares follow a χ^2-distribution. A ratio of independent χ^2 random variables, each divided by its degrees of freedom, has an F-distribution:

$$\frac{\chi_1^2/\mathrm{df}_1}{\chi_2^2/\mathrm{df}_2} \sim F \qquad \text{if } \chi_1^2 \text{ and } \chi_2^2 \text{ are independent.}$$

When the null hypothesis is true, the F-ratio follows what is called a *central F-distribution*, and when an alternative is true it follows a *noncentral F-distribution*. Examples of these two types of F-distribution are depicted in Figure 4.1.

4.2.3 One-Factor Analysis of Variance

In the one-factor analysis of variance for t treatments, based on the model

$$y_{ij} = \mu + \tau_i + \epsilon_{ij}, \qquad j = 1, \ldots, r_i, \quad i = 1, \ldots, t, \text{ and } \sum r_i = n, \tag{4.11}$$

one question that is often tested is whether there are any nonzero treatment effects, $\mathrm{H}_0 : \tau_i = 0$, for all $i = 1, \ldots, t$. The *treatment sum of squares* (SSTR) is just the model sum of squares, i.e., the difference in error sum of squares between a model that includes treatment effects [model (4.11)] and a model that has only a term for the grand mean:

$$\mathrm{SSTR} = \mathrm{R}(\tau_1, \ldots, \tau_t|\mu) = \widehat{\boldsymbol{\beta}}^t \boldsymbol{X}^t \boldsymbol{y} - \frac{y_{..}^2}{n}.$$

The latter term, $y_{..}^2/n$, is often called the *correction term* (CT). Another way to compute the sum of squares for treatment effects is to write a quadratic form as shown in Section 4.2.1. There SSTR was computed for a one-factor model with four treatments, using effects coding. The treatment sum of squares is SSTR = SS(H_0) from equation (4.10), where $\boldsymbol{L}\widehat{\boldsymbol{\beta}}$ is

$$\boldsymbol{L}\widehat{\boldsymbol{\beta}} = \begin{bmatrix} 0 & 1 & 0 & 0 \\ 0 & 0 & 1 & 0 \\ 0 & 0 & 0 & 1 \end{bmatrix} \begin{bmatrix} \widehat{\mu} \\ \widehat{\tau}_1 \\ \widehat{\tau}_2 \\ \widehat{\tau}_3 \end{bmatrix}.$$

Under this parameterization,

$$\begin{bmatrix} \widehat{\mu} \\ \widehat{\tau}_1 \\ \widehat{\tau}_2 \\ \widehat{\tau}_3 \end{bmatrix} = \begin{bmatrix} \frac{1}{t}\sum_i^t \overline{y}_{i.} \\ \overline{y}_{1.} - \widehat{\mu} \\ \overline{y}_{2.} - \widehat{\mu} \\ \overline{y}_{3.} - \widehat{\mu} \end{bmatrix}.$$

Table 4.1 One-Way Analysis of Variance Sums of Squares and Expected Mean Squares[a]

Source	df	Regression Form	SS	E[MS]
Mean	1	$R(\mu)$	$y_{..}^2/n$	
Treatment	$t-1$	$R(\tau_1, \ldots, \tau_t \mid \mu)$	$\sum_{i=1}^t r_i(\overline{y}_{i.} - \overline{y}_{..})^2$	$\sigma_\epsilon^2 + \frac{\sum r_i(\tau_i - \overline{\tau})^2}{t-1}$
Error	$n-t$		$\sum_i^t \sum_j^{r_i} (y_{ij} - \overline{y}_{i.})^2$	σ_ϵ^2
Total (uncorrected)	n		$\sum_i^t \sum_j^{r_i} y_{ij}^2$	

[a] There are t treatments, r_i observations in the ith treatment, a total of n observations, and $\overline{\tau} = \sum r_i \tau_i / n$.

The *analysis of variance table* is a convenient way to display the sums of squares for testing standard hypotheses. The regression notation form of the sums of squares gives an explicit way of representing the hypothesis being tested, and the form labeled "SS" in Table 4.1 gives an intuitive understanding of what the quantities measure. The expectations of the treatment and error mean squares for a one-way analysis of variance are also given in Table 4.1.

The test of the hypothesis $H_0 : \tau_1 = \ldots = \tau_t = 0$ against $H_a : \tau_i$ not all equal to zero uses an F-statistic, comparing the ratio $F = \text{MSTR}/\text{MSE}$ with the appropriate quantile from the central F-distribution. If the data are normally distributed [i.e., $\epsilon_{ij} \sim \text{NID}(0, \sigma_\epsilon^2)$], the F-ratio follows a central F-distribution when the null hypothesis is true. The rejection rule is:

$$\text{Reject } H_0 \text{ if } F > F(1-\alpha, \text{df}_1, \text{df}_2),$$

where $F(1-\alpha, \text{df}_1, \text{df}_2)$ indicates the $1-\alpha$ quantile of the F-distribution with degrees of freedom df_1 and df_2. In this case the degrees of freedom are $\text{df}_1 = t-1$ and $\text{df}_2 = n-t$ for the numerator and denominator, respectively. Under H_0, the distribution of the F-ratio follows from the assumption that the errors are uncorrelated, normally distributed, and have homogeneous variances ($\text{NID}(0, \sigma_\epsilon^2)$). Under those conditions SSTR and SSE are independent and

$$\frac{\text{SSTR}}{\text{E[MSTR]}} \sim \chi^2(\text{df}_1)$$

$$\frac{\text{SSE}}{\text{E[MSE]}} \sim \chi^2(\text{df}_2)$$

$$F = \frac{\text{SSTR}}{\text{df}_1\text{E[MSTR]}} \div \frac{\text{SSE}}{\text{df}_2\text{E[MSE]}}.$$

If E[MSTR] and E[MSE] are the same, they cancel each other out. In Table 4.1 we see that $\text{E[MSTR]} = \text{E[MSE]} = \sigma_\epsilon^2$ when H_0 is true, but not otherwise.

Under H_0, then,

$$F = \frac{\text{MSTR}}{\text{MSE}} \sim F(\text{df}_1, \text{df}_2).$$

4.3 KRUSKAL–WALLIS TEST

What if the errors do not have homogeneous variances or are not normally distributed? If the r_i are close to being equal, the variance homogeneity and normality assumptions are not very crucial. A test statistic or estimation procedure that is not very sensitive to departures from the assumptions is said to be *robust*. The F-test becomes more and more sensitive, i.e., less and less robust, to heteroscedasticity and nonnormality as matters become more unbalanced. Transforming the data to stabilize the variances is often good practice, but it is not always possible to find such a transformation.

If we are not willing to assume normality, the Kruskal–Wallis test (called the *Wilcoxon test* when there are two samples) is a good nonparametric alternative (Hollander and Wolf 1973). Recall that there are $n = \Sigma r_i$ observations. Rank all of the observations from 1 to n. Let s_{ij} be the rank of y_{ij} in the complete set. Now compute the test statistic

$$H = \frac{12}{n(n+1)} \times \sum_{i=1}^{t} \frac{\left(\sum_{j=1}^{r_i} s_{ij}\right)^2}{r_i} - 3(n+1).$$

If there are ties, assign average rank to all those tied. Let there be m distinct values and let the number that take the first distinct value be d_1, the second distinct value be $d_2, \ldots$. Then compute

$$H = \frac{\frac{12}{n(n+1)} \sum_{i=1}^{t} \left(\frac{\left(\sum_{j=1}^{r_i} s_{ij}\right)^2}{r_i} \right) - 3(n+1)}{1 - \frac{\sum_{h=1}^{m}(d_h^3 - d_h)}{n^3 - n}}.$$

Note: If there are no ties, all $d_h = 1$ and $m = n$.

Reject $H_0 : \tau_1 = \cdots = \tau_t$ if $H \geq \chi^2_{(t-1)}$ at the $1 - \alpha$ level. This rule is appropriate for large samples. For small samples the distribution of the test statistic may not be closely approximated by the χ^2-distribution. In this case, an exact p-value can be computed using statistical software such as SAS® `Proc Npar1way`. One can construct pictures similar to Figure 4.1 when using the Kruskal–Wallis or Wilcoxon test or any other nonparametric test.

***Exercise 4.3: Tomato Starch—Nonparametric Analysis of Variance*.** An experiment was conducted to compare the starch content of tomato plants grown in sandy soil supplemented by one of three different nutrients: N_0, N_1, or N_2. Fifteen tomato seedlings of one particular variety were selected for the study,

Table 4.2 Stem Starch Content

Nutrient N_0	Nutrient N_1	Nutrient N_2
22	12	7
20	14	9
21	15	7
18	10	6
16	9	5

with five assigned to each nutrient group. All seedlings were planted in a sand culture and maintained under a controlled environment. The seedlings assigned to nutrient N_0 served as the control group; they received distilled water only. Plants assigned to nutrient N_1 were fed a weak concentration of Hoagland nutrient. Those assigned to nutrient N_2 were fed Hoagland nutrient at the full, usually recommended strength. The weak concentration was exactly half the standard full strength. The stem starch contents determined 25 days after planting are recorded in Table 4.2, in micrograms per milligram. Use the Kruskal–Wallis or Wilcoxon statistic to test the null hypothesis that the levels of Hoagland solution had no effect. What are your conclusions? □

4.4 RANDOMIZATION TESTS

Comparison of the F-ratio to quantiles of the F-distribution is based on the assumption that the responses are randomly selected from a normal distribution. The robustness of the F-distribution to nonnormality can be justified by results derivable from consequences of proper randomization. Many years of practical experience confirm the robustness of the F-test. Comparison of the Kruskal–Wallis statistic to quantiles of the χ^2-distribution is also an approximation. For either of these statistics, *exact* tests of hypotheses requiring no distributional assumptions may be obtained by comparing the test statistic to a reference distribution created by randomization rather than comparing the statistic to the F- or χ^2-distribution. This kind of test is called a *randomization test*. The rationale for this type of test depends solely on the manner in which the treatments are assigned to the experimental units. In a randomization test the experimental units and their responses are considered fixed and eventually known, and the assignment of treatments to experimental units is what is random. In this framework it is not necessary to think of experimental units as having been selected from any particular distribution. If the experimental units have not been randomly selected, however, making inferences from the observed units to the wider population is still not justified; the principle of representativeness of the experimental units is just as important as when doing F-tests or Kruskal–Wallis tests.

The idea behind a randomization test is this: if treatments really have no effect, it shouldn't matter which treatments are assigned to which units. The response for any particular experimental unit would be unchanged no matter which treatment was assigned to it. This is exactly the situation under the null hypothesis of no treatment effect. The act of randomization has simply split the experimental units into groups identified with the treatments. This grouping has also defined a value for the F-statistic if one chooses to compute it. However, a different F-statistic value would have resulted if a different random assignment had been realized. The point is that the randomization dictates the F-statistic. Under H_0, the treatments have no effect. Now assume for a moment that the F-statistic actually realized turned out to be very large. This could be due to one of two things. Just by chance, all the plots with large responses could have been assigned to one treatment and all plots with a small response to another treatment, a very unusual event; the other possibility is that H_0 was wrong and treatments really had an effect. The practice is to reject such a rare event as being due to chance and conclude that H_0 is not true.

The mechanics of this test are relatively simple. After the data have been collected, one performs the usual analysis of variance, computes the F-statistic, and records it. Then one computes the F-statistics for all (or at least a large sample) of the other possible random assignments of treatments. The empirical distribution of these F-statistics define what is called a *reference distribution*. Then if the original F-statistic is unusual, e.g., larger than say 95% of the F-statistics in the reference distribution, one rejects H_0. The important thing to notice in the rerandomizations of the treatment labels is that *the response value stays with the experimental unit*, regardless of what treatment label is assigned to the unit. The null hypothesis says that treatments do not affect the responses. In hypothesis testing the probabilities of outcomes are computed under the assumption that the null hypothesis is true, and that hypothesis is rejected if the probability of observing the realized outcome is very small under H_0. The beauty of the randomization test procedure is that the only underlying assumption is that randomization was performed properly. The careful investigator can easily ensure that this is the case. No external distributional assumptions are needed.

We illustrate with an example. Consider an experiment with three treatments to be applied to six experimental units, two units to each treatment. For convenience think of the six units as six animals named A, B, C, D, E, and F. The null hypothesis is that the treatments, injections with three different growth hormones, have no effect whatsoever on growth over the next few weeks. The treatments are assigned at random or, equivalently, the experimental units are assigned at random to the treatments, the treatments applied, and the weight gains ascertained. The data are now analyzed using analysis of variance to obtain an F-value. Note that this F-statistic represents a realization of the F-random variable. Now the randomization really amounted to nothing more than selecting at random one from the $\binom{6}{2}\binom{4}{2}\binom{2}{2} = 90$ sequences of possible random assignments. Sixteen of the possible assignments are shown in Table 4.3. SAS® code for producing them is available on the companion Web site.

For each of the 90 possible assignments, we can compute an F-statistic. In these computations, remember that the animal's response stays with the animal. For example, suppose that animals A and C were given treatment 1, B and E

Table 4.3 Sixteen Possible Assignments of Six Experimental Units to Three Treatments

Trt. I	Trt. II	Trt. III
A,B	C,D	E,F
A,B	C,E	D,F
A,B	C,F	D,E
A,C	B,D	E,F
A,C	B,E	D,F
A,C	B,F	D,E
A,D	B,C	E,F
A,D	B,E	C,F
A,D	B,F	C,E
A,E	B,C	D,F
A,E	B,D	C,F
A,E	B,F	C,D
A,F	B,C	D,E
A,F	B,D	C,E
A,F	B,E	C,D
C,D	A,B	E,F

were given treatment 2, and animals D and F were given treatment 3. Suppose that the observed responses for the six animals were $y_A = 12$, $y_B = 10$, $y_C = 11$, $y_D = 15$, $y_E = 14$, and $y_F = 13$. The treatment sum of squares would then be

$$\text{SSTR} = \frac{(12+11)^2 + (10+14)^2 + (15+13)^2}{2} - \frac{(12+10+11+15+14+13)^2}{6}.$$

The SSTR for the first possible assignment of treatments in Table 4.3 (trt. 1: A,B; trt. 2: C,D; trt. 3: E,F) would be

$$\text{SSTR} = \frac{(12+10)^2 + (11+15)^2 + (14+13)^2}{2} - \frac{(12+10+11+15+14+13)^2}{6} = 7.$$

Notice that SSTR is the same for the first and the sixteenth treatment assignments listed; in both cases SSTR = 7. This is because in both arrangements units A and B are in the same treatment, C and D are in another treatment, and E and F are in a third treatment. In fact, for each of the first 15 treatment assignments in Table 4.3 there are $3! = 6$ possible assignments that result in the same value of SSTR. The reference distribution is constructed by computing the 15 possible unique SSTR and F-values.

Since under the null hypothesis, treatments have no effect and we randomly selected one from these assignments, we really are claiming that any one of these 15 possible F-values would have had an equal chance of appearing if the null hypothesis were the true state of affairs. Now if, for example, our realized F-value were the third largest in the set and we reject the null hypothesis that treatment had no effect because the F-statistic was too large, we are making a test of significance at the $\alpha = \frac{3}{15} = .20$ level of significance; that is, there is a 20% chance of obtaining an F-statistic as large as, or larger then, that obtained purely due to chance when the null hypothesis was really true. Similarly, if we reject the null hypothesis only if our F-value realized from the experiment is the largest one in the set, we are performing the test at the $\alpha = \frac{1}{15} = .067$ level of significance. There is one chance in 15 of observing such an F-value under the null hypothesis. Here we have estimated the p-value as d/b, where d is the number of F-values in the reference distribution that are greater than or equal to the sample F-value observed in the experiment, and b is the number of possible F-values in the reference distribution. Note that the p-value is also often estimated as $(d+1)/(b+1)$. If $(b+1)\alpha$ is an integer, where α is the significance level, then either method of estimating the p-value gives an exact α-level test.

Randomization theory can be used to give confidence that F-tests give reasonable results even when the data are not normally distributed. It is possible to derive the expected mean squares for a randomized complete block design under the randomization model (see Mead 1988, Chap. 9). Under this model the expected mean squares are similar to the expected mean squares under the infinite population model; furthermore, the F-ratio approximately follows an F-distribution.

If you do not want to rely on the robustness of the F-test, the randomization method of analysis itself can be applied. In fact, one need not even use the F-statistic for the basic analysis. Any reasonable statistic can be used. The statistic must be chosen with some care to ensure that it really is sensitive to the types of departure from the null hypothesis that are important to the experimenter. The standard analysis based on normal theory is usually a good guide for selecting a test statistic.

In the development of randomization tests given here, we have only discussed testing the basic null hypothesis that there are no treatment effects. Randomization tests can also be extended to test other, more restricted null hypotheses, although these tests are not as straightforward. For example, in the two-factor analysis of variance with factors A and B, the test of the A main effect is a restricted null hypothesis, meaning that the reference distribution cannot be obtained by permuting all possible assignments of treatments to experimental units. This is so because even if H_0 is true, there may be an effect of the other factor, B. Good (2000) describes several test statistics that can be used to test main effects and interactions in factorial experiments. Edgington (1986) discusses the actual hypotheses being tested by randomization tests in a precise way.

Exercise 4.4: Randomization Test—Small Example. For the example of Table 4.3, with observed responses $y_A = 12, y_B = 10, y_C = 11, y_D = 15, y_E = 14, y_F = 13$ test the hypothesis that there are no treatment effects using a randomization test. State the p-value and your conclusions. The data are available on the companion Web site. □

Exercise 4.5: Randomization Test Comparing Two Treatments. A scientist has six animals that are to be used in an experiment to test two different diets. The plan is to put three animals on the first diet and the other three on the second diet. The scientist will take initial and final weights and use these to compute weight gains. Let the six animals be identified as Jo, Pete, Stu, Pokey, Zip, and Mo. List the 20 possible ways that the six animals could be assigned to the two treatments. Assume that the differences between diets are completely irrelevant to weight gains, i.e., H_0 of no treatment effects is true, and that the weight gains for the animals are 4 lb, 5 lb, 6 lb, 7 lb, 8 lb, and 9 lb, respectively, regardless of treatment received. List the F-values that would have resulted from each of the 20 possible randomizations. □

4.5 POWER AND SAMPLE SIZE

An important question that a researcher faces in designing an experiment is "How many observations are needed?" Suppose that the researcher wishes to test $H_0 : \boldsymbol{L\beta} = \boldsymbol{m_0}$ against the alternative $H_A : \boldsymbol{L\beta} \neq \boldsymbol{m_0}$. For this, the idea of *statistical power* is important. We would like the sample to be sufficiently large that we have a reasonable chance of rejecting $H_0 : \boldsymbol{L\beta} = \boldsymbol{m_0}$ when in fact $\boldsymbol{L\beta} = \boldsymbol{m_A}$ and $\boldsymbol{m_A} \neq \boldsymbol{m_0}$. The probability of rejecting the null hypothesis when it is false is called the *power* of the test.

More specifically, we compute the probability of detecting that $\boldsymbol{L\beta} \neq \boldsymbol{m_0}$ when the vector of differences is really $\boldsymbol{\delta} = \boldsymbol{m_A} - \boldsymbol{m_0}$. The vector of true values is some vector $\boldsymbol{L\beta}$. This vector can be $\boldsymbol{m_A}$, or $\boldsymbol{m_0}$, or some other value. We hypothesize that it is $\boldsymbol{m_0}$. We want to be warned if it differs from $\boldsymbol{m_0}$ by too much. Note that the vector of differences we are trying to detect, $\boldsymbol{m_A} - \boldsymbol{m_0}$ (i.e., $\boldsymbol{\delta}$), is a vector of numbers that we are free to choose, usually based on knowledge of the subject matter. The subject-matter scientist usually has some idea of what magnitude of differences, $\boldsymbol{\delta}$, would be important to detect in the scientific process under study.

The following example demonstrates how power depends on the magnitude of the difference to be detected δ, the sample size, and the variance σ_ϵ^2.

Example 4.1: Virginia Pine Christmas Trees. In a study to optimize the production of Virginia pine Christmas trees by rooted cuttings (Rosier 2002), researchers compared shoot production of trees that had been "stumped" to that of intact trees. The stumped trees were cut back to leave one basal whorl of branches and the control trees were left intact. The trees were stumped in March when the trees were dormant; then later in the year the number of 4-inch-long shoots were counted for each tree in the study. The researchers found that the number of shoots decreased when trees were stumped, but rooting ability increased. In planning a follow-up experiment, one possible experimental layout would assign four trees to each plot and assign each plot to one of the stumping treatments: control or stumped. Intact trees produce about 250 shoots per tree. A difference in the average number of shoots per tree of 75 or more, in either direction, would be an important outcome for the researchers developing new tree production methods. The standard deviation of the plot means, from preliminary experiments where each plot has four trees, is expected to be about 50 (C. L. Rosier, personal communication). How many plots should be left intact, and how many should be cut back, to determine whether there is a difference in the number of shoots? The following discussion demonstrates how to use power curves to answer this question. The issue of how to compute the power curves is taken up in Sections 4.5.1 and 4.5.2.

In general, power increases as sample size increases. Figure 4.2*a* shows the power of the test that $\mu_1 = \mu_2$ against the alternative that $\mu_1 \neq \mu_2$ as a function of the sample size for three different values of the error variance. Each of the three power curves in the left panel represents a different value of the error variance. In this example μ_1 is the control mean and μ_2 is the mean for stumped trees. Suppose that the standard deviation is $\sigma_\epsilon = 50$. To determine how large a sample would be needed to detect, with 80% power, a difference of $|\delta| = |\mu_1 - \mu_2| = 75$, first draw a horizontal line on the graph at power = .8 until it intersects the solid power curve. Then drop a vertical line from this intersection to the x-axis. The intersection occurs at $n = 8$. We conclude that if $\sigma_\epsilon = 50$, $n = 8$ plots per treatment would be a large enough sample to have 80% probability of detecting a mean difference of 75 shoots per tree in a two-sided test at the $\alpha = .05$ significance level.

The power of a test also depends on the variance among plots. Specifically, power increases as the variance σ_ϵ^2 decreases. If the variance is smaller than the guessed value of $\sigma_\epsilon = 50$, say $\sigma_\epsilon = 30$, the dashed curve in Figure 4.2*a* indicates that four plots would be sufficient to achieve 80% power. On the other hand, if the variance is larger than expected, say $\sigma_\epsilon = 70$, the dash/dot curve indicates that more than 10 plots would be needed per treatment to achieve the same power. It is often a good idea to plot power curves for different values of the variance to see what sample size would be needed if the variance turns out to be larger than expected (or hoped).

Figure 4.2*b* demonstrates how the power of the test depends on the size of the difference to be detected, δ. In general, power increases as δ increases. This figure shows that for a sample size of eight plots and $\sigma_\epsilon = 50$, the power for detecting a difference of 50 is only about 45%, whereas the power for detecting a difference of 100 between the two treatment means is over 95%. Two important features to notice in Figure 4.2 are that (1) when the null hypothesis is true (when $\mu_1 = \mu_2$ in this example) the power equals the significance level α, and (2) power, which is a probability, approaches 1.0 as δ increases.

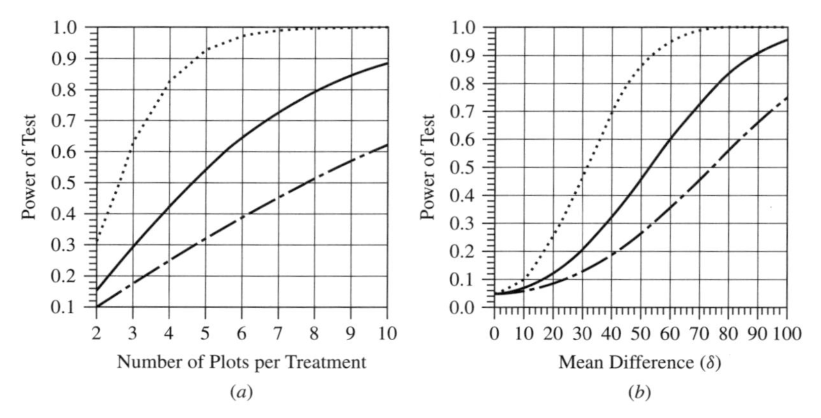

Figure 4.2 Power of a two-sided hypothesis test as a function of (*a*) the number of experimental units per treatment r when $\delta = \mu_1 - \mu_2 = 75$, and (*b*) the difference between the two treatment means δ when $r = 8$. A separate curve is shown for each of three standard deviations: $\sigma_\epsilon = 30$ (dashed line), 50 (solid line), or 70 (dash/dot line). The significance level is $\alpha = .05$.

The probability of detecting a difference represents the power of the test. In Figure 4.1 this is the region to the right of the critical value under the alternative distribution, whereas the significance level, α, is the region to the right of the critical value under the hypothesized distribution. Power is also equal to $1 - \beta$, where β is the probability of a type II error. A good reference for this material is the small book *Sample Size Methodology* by Desu and Raghavarao (1990).

The general definition of power of a hypothesis test is

$$\text{power} = \Pr(\text{reject H}_0 \mid \text{H}_0 \text{ is false}).$$

A hypothesis test may test one linear combination of the parameters or contrast, as in the Christmas tree stumping example, or it may test several linear combinations at the same time. The definition of power given here is stated in terms of vectors for the case where several contrasts or linear combinations are tested at once. For testing a specific set of linear combinations of the parameters $\boldsymbol{L\beta}$, the power of the test is the probability of rejection given that the difference between the hypothesized vector $\boldsymbol{m_0}$ and the true vector $\boldsymbol{L\beta}$ is $\boldsymbol{\delta}$,

$$\text{power} = \Pr(\text{reject H}_0 : \boldsymbol{L\beta} = \boldsymbol{m_0} \mid \boldsymbol{L\beta} = \boldsymbol{m_A}).$$

As we have seen in the Christmas tree example, the power of a hypothesis test depends on several things: (1) the magnitudes of differences we want to detect, $\boldsymbol{\delta} = \boldsymbol{m_A} - \boldsymbol{m_0}$; (2) sample size, or more generally, the design of the experiment; (3) the variance of y, σ_ϵ^2; and (4) the significance level α.

If the hypothesis consists of only one linear combination, the vectors reduce to scalar quantities as in the Christmas tree example. On the other hand, in an experiment with t treatments, the test of the treatment effect in a one-factor analysis of variance consists of $t - 1$ contrasts among the treatments, i.e., $t - 1$

linear combinations of treatment means. To compute the power for detecting differences among t means, it is necessary to specify the $t-1$ elements of $\boldsymbol{m}_A$. Lenth (2001) provides a good discussion of methods of eliciting reasonable values for the sizes of the effects that the researcher would like to detect. Often, researchers are interested in detecting a treatment effect whenever the difference between any pair of treatments exceeds a certain value, d. A good technique for ensuring enough power is to set one μ_i to $-d/2$, another to $d/2$, and all the rest of the treatment means to zero. This gives the smallest possible power for any case where there is a difference of d or larger between treatment means.

In the same article, Lenth lists several programs and Web resources for computing power. The Web site *http://www.stat.uiowa.edu/ rlenth/Power/* provides a very easy way to compute power and sample size for the most common experimental designs and allows the user to specify the model and hypotheses to be tested.

4.5.1 Probability of Rejecting the Null Hypothesis

The distribution of the F-ratio is known both when $\text{H}_0 : \boldsymbol{L\beta} = \boldsymbol{m}_0$ is true and when it is not true, assuming that the data are normally distributed. In the first instance this distribution is known as the central F-distribution denoted by $F(\text{df}_1,\ \text{df}_2)$ and in the latter case it is known as the noncentral F-distribution commonly denoted by $F'(\text{df}_1,\ \text{df}_2,\ \lambda)$, where λ is the noncentrality parameter. The noncentrality parameter λ measures the "distance" between the hypothesized vector $\boldsymbol{m}_0$ and the alternative $\boldsymbol{m}_A$, standardized by $\text{Var}[\widehat{\boldsymbol{L\beta}}] = \sigma_\epsilon^2 \boldsymbol{L}(\boldsymbol{X}^t\boldsymbol{X})^{-1}\boldsymbol{L}^t$. In this book λ is defined as

$$\lambda = (\boldsymbol{m}_A - \boldsymbol{m}_0)^t(\text{Var}[\widehat{\boldsymbol{L\beta}}])^{-1}(\boldsymbol{m}_A - \boldsymbol{m}_0). \tag{4.12}$$

This has the same form as $\text{SS}(\text{H}_0)/\sigma_\epsilon^2$ [eq. (4.9)] except that the alternative value $\boldsymbol{m}_A$ appears rather than the estimated value $\widehat{\boldsymbol{L\beta}}$. In the one-factor analysis of variance test for treatment effects, this works out to be

$$\lambda = \frac{\sum r_i(\tau_i - \overline{\tau}.)^2}{\sigma_\epsilon^2}, \qquad \text{where} \quad \overline{\tau}. = \sum_{i=1}^{t} r_i\tau_i \Big/ \sum r_i.$$

Note that usage in the literature is not at all consistent. Some writers divide by 2, defining the noncentrality parameter to be $\sum r_i(\tau_i - \overline{\tau}.)^2/2\sigma_\epsilon^2$, and others define the noncentrality parameter as the square root, $\sqrt{\sum r_i(\tau_i - \overline{\tau}.)^2/2\sigma_\epsilon^2}$. Users must be cautious.

The power for testing the hypothesis $\text{H}_0 : \boldsymbol{L\beta} = \boldsymbol{m}_0$ with significance level α when $\boldsymbol{L\beta} = \boldsymbol{m}_A$ is

$$\text{power} = \Pr\left\{ \frac{\text{SS}(\text{H}_0)/\text{df}_H}{\text{MSE}} > F(1-\alpha, \text{df}_H, \text{df}_E) \,\Big|\, \lambda \right\}.$$

Probabilities from the central and noncentral F-distribution can be computed easily using statistical software. In general, the *survival function* is the probability of exceeding a specified value, whereas the *cumulative distribution function* (cdf) is the probability of being less than or equal to a specified value. In SAS® the probability that $F > c$, where the F-statistic follows a noncentral $F(\text{df}_1, \text{df}_2, \lambda)$-distribution and c is any constant, may be computed using the survival function with syntax `sdf('F',c,df₁, df₂,λ)`.

Testing a Single-Degree-of-Freedom Hypothesis

A t-test may be used for testing a single-degree-of-freedom hypothesis. In this case, write the linear combination being tested as $\boldsymbol{l}^t\boldsymbol{\beta}$, where $\boldsymbol{l}^t$ is a row vector of coefficients. Then the test statistic

$$t = \frac{\boldsymbol{l}^t\widehat{\boldsymbol{\beta}} - \boldsymbol{m}_0}{\sqrt{\text{MSE}(\boldsymbol{l}^t\left(\boldsymbol{X}^t\boldsymbol{X}\right)^{-1}\boldsymbol{l})}} \sim t'(\text{df}_E, \phi)$$

follows a noncentral t-distribution with noncentrality parameter

$$\phi = \frac{\boldsymbol{m}_A - \boldsymbol{m}_0}{\sqrt{\text{Var}[\boldsymbol{l}^t\widehat{\boldsymbol{\beta}}]}},$$

where $\text{Var}[\boldsymbol{l}^t\widehat{\boldsymbol{\beta}}] = \sigma_\epsilon^2\boldsymbol{l}^t\left(\boldsymbol{X}^t\boldsymbol{X}\right)^{-1}\boldsymbol{l}$. Student's t-statistics can be used to test either one- or two-sided hypotheses, whereas F-statistics can only be used to test two-sided hypotheses.

How to Obtain σ_ϵ^2

For the purpose of computing power we have assumed up to this point in the discussion that σ_ϵ^2 is known. In practice, this is rarely true. Now what does one do? There are three possibilities. The first is to use existing knowledge or experience to make the best guess for σ_ϵ^2 that one possibly can, and proceed. The second alternative is to posit a range of plausible values of σ_ϵ^2 and compute sample size for several different values of σ_ϵ^2. This will give an idea of the sample sizes needed to obtain a specified power under best- and worst-case scenarios for the variance. The final alternative, which is sometimes feasible, is to take a preliminary sample of size n_1 and use it to perform an analysis and compute s^2, the unbiased estimate of σ_ϵ^2. If the calculations yield an n value less than n_1, one has sufficient data. If $n > n_1$, collect more data to make up the shortfall.

4.5.2 Power Calculations

Suppose that an experimenter plans to study the relative responses to six different treatments and there are sufficient resources to allow eight replicates of each treatment. The experimenter plans a one-factor analysis of variance using model (4.11) with the normality assumption, $y_{ij} = \mu + \tau_i + \epsilon_{ij}$, where $\epsilon_{ij} \sim \text{NID}(0, \sigma_\epsilon^2)$, for $i = 1, \ldots, 6$ and $j = 1, \ldots, 8$. The null hypothesis is $\text{H}_0 : \tau_1 = \tau_2 = \cdots = \tau_6$ vs. $\text{H}_A : \tau_i$ not all equal.

The question from the experimenter is: What are my chances of declaring a significant treatment effect when testing at the 5% level of significance?; i.e., what is the probability that the F-ratio exceeds the critical value $F(.95, 5, 42) = 2.44$? If the chances are good that it will be possible to discern treatment differences with this number of replicates, the researcher will carry out the experiment. If the power is too low, it may be wiser not to conduct the experiment as currently conceived. The power depends on the true but unknown values of $\tau_1, \tau_2, \ldots, \tau_6$.

Suppose that the experimenter knows from experience and/or reading the literature that the mean is approximately $\mu = 75$ and that the coefficient of variation should be approximately equal to 20%. (Recall that $\text{CV} = 100 \times \sigma_\epsilon/\mu$.) From this we compute that $\sigma_\epsilon^2 = 225$ is reasonable ($\sigma_\epsilon = .2\mu = 15$). The experimenter also expects that the mean for treatment 1 will be about 65, the mean for treatment 2 will be about 85, and the remaining four will be 70, 75, 75, and 80, respectively. The corresponding treatment effects are $\tau_1 = -10$, $\tau_2 = 10$, $\tau_3 = -5$, $\tau_4 = 0$, $\tau_5 = 0$, and $\tau_6 = 5$.

The general procedure for computing power is first to compute the noncentrality parameter and then look up the probabilities of the noncentral F-distribution. We first compute the noncentrality parameter λ,

$$\begin{aligned}\lambda &= \frac{8\sum(\tau_i - \overline{\tau}_{..})^2}{\sigma_\epsilon^2} \\ &= \frac{8 \times [(-10)^2 + (10)^2 + (-5)^2 + (0)^2 + (0)^2 + (5)^2]}{225} \\ &= 8.89.\end{aligned}$$

The power can then be computed using the SAS® statements in Display 4.1, where $c = 2.44$ is the critical value for the test, lambda is the noncentrality parameter λ, $\text{df}_1 = 5$, and $\text{df}_2 = 42$.

The result is that power = .55. We would conclude that a sample size of eight observations per treatment is not large enough to give good power for the F-test of treatment effects.

This method is very flexible and can be used for assessing power of testing any linear hypothesis about the treatment means. Suppose that in the scenario above, treatments 3 and 4 have one active ingredient, and treatments 5 and 6 have a different active ingredient. We can compute what the power for testing the hypothesis $H_0 : \mu_3 + \mu_4 = \mu_5 + \mu_6$ (written alternatively as $H_0 : \mu_3 + \mu_4 - \mu_5 - \mu_6 = 0$) would be for the researcher's guessed values

```
data one;
 df1=5;  df2=42;  c=2.44;
 lambda=8.89;
 power=sdf('F',c,df1,df2,lambda);
```

Display 4.1 SAS® code for computing power of an F-test.

```
data one;
   dfe=42;  r=8;  c=2.018;  lbeta=-10;
   stderrlbeta=sqrt(4*225/8);
   phi=lbeta/stderrlbeta;
   power=cdf('t',-c,dfe,phi) + sdf('t',c,dfe,phi);
```

Display 4.2 SAS® power computations for a two-sided t-test.

of the parameters. In this case $\boldsymbol{L\beta_A} = 70 + 75 - 75 - 80 = -10$. Also notice that for eight replications $\text{Var}[\boldsymbol{L}\widehat{\boldsymbol{\beta}}] = 4\sigma_\epsilon^2/8$. The noncentrality parameter for an F-test is $\boldsymbol{L\beta_A}(\text{Var}[\boldsymbol{L}\widehat{\boldsymbol{\beta}}])^{-1}(\boldsymbol{L\beta}_A)'$ (note that $\boldsymbol{L\beta_0} = 0$ here), which reduces to $\lambda = \left(\boldsymbol{L\beta}_A\right)^2/\text{Var}[\boldsymbol{L}\widehat{\boldsymbol{\beta}}]$ for a single-degree-of-freedom hypothesis. The noncentrality parameter works out to be $\lambda = (-10)^2/(4 \cdot 225/8)$ for the test that $\mu_3 + \mu_4 - \mu_5 - \mu_6 = 0$ using the values supplied by the researcher. The results for this test are $\lambda = .89$ and power $= 15\%$.

Since this is a single-degree-of-freedom test, a Student's t-test could be used just as well. A *two-sided* t-test coincides with the F-test given above. Care must be exercised to include the area in both tails of the noncentral t-distribution when computing the power of a two-sided test. The SAS® code for the t-test is shown in Display 4.2.

The general notation given in expression (4.12) applies no matter how complicated the linear combinations. We demonstrate use of the general form for this simple example. Written in terms of the treatment effects, and using the constraint that $\sum \tau_i = 0$ (effects coding), the hypothesis is $\text{H}_0 : \tau_3 + \tau_4 = -\tau_1 - \tau_2 - \tau_3 - \tau_4$. Putting all the parameters on the left side of the equation gives $\text{H}_0 : \tau_1 + \tau_2 + 2\tau_3 + 2\tau_4 = 0$. This hypothesis, written in matrix notation, is $\text{H}_0 : \boldsymbol{L\beta} = 0$, where

$$\boldsymbol{\beta} = \begin{bmatrix} \mu \\ \tau_1 \\ \tau_2 \\ \tau_3 \\ \tau_4 \\ \tau_5 \end{bmatrix} \quad \text{and} \quad \boldsymbol{X}^t\boldsymbol{X} = 8\begin{bmatrix} 6 & 0 & 0 & 0 & 0 & 0 \\ 0 & 2 & 1 & 1 & 1 & 1 \\ 0 & 1 & 2 & 1 & 1 & 1 \\ 0 & 1 & 1 & 2 & 1 & 1 \\ 0 & 1 & 1 & 1 & 2 & 1 \\ 0 & 1 & 1 & 1 & 1 & 2 \end{bmatrix},$$

$$\boldsymbol{L} = \begin{bmatrix} 0 & 1 & 1 & 2 & 2 & 0 \end{bmatrix}, \quad m_0 = 0, \quad \text{and} \quad m_A = -10.$$

```
proc iml;
 sigmasq=225;  df1=1;  df2=42;  r=8;
 L={0 1 1 2 2 0};
 x1={1,1,1,1,1,1}||designf({1,2,3,4,5,6});
 xpx=r*x1'*x1;  mA= -10;  m0=0;
 lambda=(mA-m0)'*inv(L*inv(xpx)*L')*(mA-m0)/sigmasq;
 c=4.07;
 power=sdf('F',c,df1,df2,lambda);
```

Display 4.3 Example power computations for testing a general linear hypothesis in SAS/IML®.

The noncentrality parameter [eq. (4.12)] and power for this test can be computed in SAS/IML using the code in Display 4.3.

To summarize the results of the power analysis, we tell the experimenter that with eight replicates of six treatments and using the 5% significance level, there is roughly a 50:50 chance of obtaining statistical significance when testing whether there are any differences among the six treatment means. If the experimenter's interest is primarily in comparing the average of treatments 3 and 4 with treatments 5 and 6, the situation is much worse; there is only a 15% chance of declaring a significant difference for this comparison.

4.5.3 Experiment with a Small Twist: Sample Size for a Study of Computer–Human Interactions

An experiment comparing how well children aged 9 to 13 interact with a computer using a drag-and-drop (D) versus a point-and-click (P) mouse interaction style reported that children made fewer errors and took less time to complete tasks using the point-and-click style (Inkpen 2001). In this experiment several factors were studied, including mouse interaction style and the gender of the child, but no attempt was made to look at changes in ability with age. Each child performed a sequence of tasks using one style first, then another sequence of tasks using the second style, with the order of styles being randomized for each child. The mean difference in time to complete a movement between the D and P styles was 81 milliseconds (ms) with a standard error of 26 ms.

Suppose that you are to plan a new experiment to determine whether there are any differences among ages and whether such differences depend on gender. How many boys and girls should be in each age group?

First there is the question of how many age groups to use. If it is known beforehand that time to complete a movement decreases linearly with age between the ages of 9 and 13, the most efficient design to estimate the slope would put half the children at age 9 and half at age 13. If a quadratic function describes the relationship, at least three ages would be required to estimate the parameters of the curve, and using equally spaced levels provides good precision of both the slope and curvature estimates. Similarly, if it is not known whether the relationship is linear or curved, and if we want to be able to test whether the relationship is linear, at least three ages are required. If it is suspected that ability might improve up to a certain age and then level off, but the age that this happens is unknown and might vary among children, it would be wisest to include children of every age from 9 to 13. D. R. Cox in the book *Planning of Experiments* (1958) gives a good discussion of choosing treatment levels.

Exercise 4.6: Three Levels of x Versus Two Levels for Simple Linear Regression. For a simple experiment studying the effect of age on time to complete a task, show that the slope of a straight-line regression is estimated more precisely if three children aged 9 and three children aged 13 are tested than if two children at each of the ages 9, 11, and 13 are tested. □

Let us suppose that the time to complete a task decreases up to a certain age but then flattens out before age 13. To get good estimates of the slope in the age range where ability is changing, we must have sufficient numbers of observations in the early ages, but since we don't know the age at which the curve flattens out, and since there are only five years in the range of interest, prudence suggests testing equal numbers of children at each of the five ages.

Now we are at a point where we can compute power of hypothesis tests to decide how many boys and how many girls to study at each of the five ages from 9 to 13. The experiment will have just two factors: gender and age, with equal numbers of girls and boys at each age. The twist is this: The response measured on each child is the *difference* in time to complete a set of tasks using the D style versus using the P style of mouse interaction. The model will include a term for gender, G_i, a term for age, A_j, and the interaction: $y_{ijk} = \mu + G_i + A_j + (G \times A)_{ij} + \epsilon_{ijk}$, where $\epsilon_{ijk} \sim \text{NID}(0, \sigma_\epsilon^2)$, for $i = 1, 2$ genders, $j = 1, \ldots, 5$ ages, and $k = 1, \ldots, n$ children of each age and gender. Since y is the difference between the two mouse styles for an individual child, the test of $\text{H}_0 : \mu = 0$ tests whether there is a difference in mean response time for the D style of interaction vs. the P style of interaction averaged over both genders and all ages, under the model with sum-to-zero constraints.

To compute the power of a hypothesis test, it is necessary to specify the magnitude of differences that we would like to detect. We can make up any scenario we like at this point. Just for demonstration purposes, suppose that we want to be able to detect a difference between the two mouse interaction styles if the mean difference between drag-and-drop and point-and-click times is 50 ms at any age and the time for each mouse style decreases about 25 ms per year. The mean difference between the two mouse styles is constant over time in this scenario, and we assume that the difference for boys is the same as for girls. We must also specify the variance, and for this we will first assume that the variance is $\sigma_\epsilon^2 = 45{,}000$, which is roughly based on the results in Inkpen (2001).

Under this scenario, $m_0 = 0$ and $m_A = 50$, $\text{Var}[\widehat{\mu}] = \sigma_\epsilon^2/10n$, so the noncentrality parameter for a t-test is $\phi = (m_A - m_0)/\sqrt{\text{Var}[\widehat{\mu}]} = 50\sqrt{10n}/\sqrt{45{,}000} = .75\sqrt{n}$. The power of the two-sided test $\text{H}_0 : \mu = 0$ vs. the alternative $\text{H}_A : \mu \neq 0$ when the significance level is $\alpha = .05$ is

$$\Pr\left\{ \frac{\widehat{\mu}}{\sqrt{\text{Var}[\widehat{\mu}]}} < t(.025, \text{df}e) \middle| \phi \right\} + \Pr\left\{ \frac{\widehat{\mu}}{\sqrt{\text{Var}[\widehat{\mu}]}} > t(.975, \text{df}e) \middle| \phi \right\}.$$

The results in Table 4.4 indicate that 15 children of each gender at each age would be sufficient to detect a difference of 50 ms with 80% power, using a two-sided test at the 5% significance level. The same computations could be run positing a larger or smaller value for the variance to see how large the sample size would need to be if the variance is really larger or smaller.

Table 4.4 Power of the Test That $\mu = 0$ in the Computer–Human Interaction Study

No. Children of Each Gender	10	11	12	13	14	15	16	17	18	19	20
Power	.65	.69	.73	.75	.79	.82	.84	.86	.88	.90	.91

Now consider another scenario. Suppose that we are interested in the power to detect an interaction between age and mouse interaction style. This interaction is estimated by the age effects A_j since the response variable is the difference between the two mouse interaction styles. Hence the hypothesis to test is that the means for the five ages are all equal. This test has four degrees of freedom, so requires an F-test rather than a t-test.

For this scenario, suppose, as before, that the mean difference between style D and style P is 50 ms and that children's times decrease at a rate of 25 ms per year for the D style, but that they decrease under the P style only until age 11, after which ability stays about the same into adulthood. This could be achieved with the set of means shown in Table 4.5. The mean of y is $\mu = 50$ and the age effects of the difference between the two mouse interaction styles are $A_1 = 15$, $A_2 = 15$, $A_3 = 15$, $A_4 = -10$, and $A_5 = -35$. The noncentrality parameter for the F-test is

$$\lambda = \frac{2n\sum_j {A_j}^2}{\sigma_\epsilon^2} = \frac{2n[15^2 + \cdots + (-35)^2]}{45{,}000} = \frac{4000n}{45{,}000} = .089n$$

when $\sigma_\epsilon^2 = 45{,}000$. The power for different values of n are given in Table 4.6. From these calculations it is evident that no sample size in the range $n = 30$

Table 4.5 Hypothesized Mean Time to Complete Tasks

	Age				
Style	9	10	11	12	13
D	1392	1367	1342	1317	1292
P	1327	1302	1277	1277	1277
Difference $\mu_{.j}$	65	65	65	40	15

Table 4.6 Power for Detecting Interaction between Age and Mouse Interaction Style

No. Children of Each Gender	30	31	32	33	34	35	36	37	38	39	40
Power	.22	.22	.23	.24	.24	.25	.26	.26	.27	.28	.28

to 40 will allow detection of an interaction between age and mouse interaction style that is this small.

4.6 SAMPLE SIZE FOR BINOMIAL PROPORTIONS

In many applications the response variable is *binary* rather than continuous, meaning that the response can take on only one of two values (e.g., success or failure, defective or not defective, alive or dead). The mean for a particular treatment is then a proportion, which typically follows a binomial distribution rather than a normal distribution. For large sample sizes (larger than 100, say), the binomial distribution is similar to a normal distribution, so the standard techniques for normally distributed responses developed in the earlier sections of this chapter apply.

As mentioned in Section 4.3, the analysis of variance is robust to nonnormality, but transformation of the response variable to make the variances more homogeneous is often good practice. The variance of a binomial random variable is not constant but is a function of the probability of success p and the sample size n. Specifically, the variance of the proportion of successes, $\widehat{p}$, is $\text{Var}(\widehat{p}) = p(1-p)/n$. If all of the proportions for different treatments are in the range $\widehat{p} = .2$ to $.8$, the variances of the $\widehat{p}$ range only from $.16/n$ to $.25/n$, so they are fairly homogeneous. However, if the probability of success is $p = .05$, then $\text{Var}(\widehat{p}) = .0475/n$, which is a fifth as large as for $p = .50$. The variance of $\widehat{p}$ may be stabilized by use of the arcsine transformation, or *angular transformation*. If $y = n\widehat{p} \sim \text{Binomial}(n, p)$, the transformed variable

$$z = \arcsin\sqrt{\widehat{p}}$$

is approximately normally distributed with mean $\arcsin\sqrt{p}$ and variance $1/4n$. We then use a χ^2-approximation for the sum of squares of contrasts. In a linear model of the transformed proportions $z = X\boldsymbol{\beta} + \boldsymbol{\epsilon}$, the sum of squares for the linear hypothesis $\boldsymbol{L\beta} - \boldsymbol{m_0} = \boldsymbol{0}$ [eq. (4.10)], divided by the variance of z, approximately follows a χ^2-distribution. The degrees of freedom df_H equal the number of parameters being tested. When the null hypothesis is true, the quadratic form follows a central χ^2-distribution, and when the null hypothesis is false, it follows a noncentral χ^2-distribution:

$$\begin{aligned}
\text{Under H}_0: &\quad (\boldsymbol{L}\widehat{\boldsymbol{\beta}} - \boldsymbol{m_0})^t(\text{Var}[\boldsymbol{L}\widehat{\boldsymbol{\beta}}])^{-1}(\boldsymbol{L}\widehat{\boldsymbol{\beta}} - \boldsymbol{m_0}) \dot{\sim} \chi^2(\text{df}_H), \\
\text{If } \boldsymbol{L\beta} = \boldsymbol{m_A}: &\quad (\boldsymbol{L}\widehat{\boldsymbol{\beta}} - \boldsymbol{m_0})^t(\text{Var}[\boldsymbol{L}\widehat{\boldsymbol{\beta}}])^{-1}(\boldsymbol{L}\widehat{\boldsymbol{\beta}} - \boldsymbol{m_0}) \dot{\sim} \chi^{2\prime}(\text{df}_H, \lambda),
\end{aligned}$$

where $\text{df}_H = \text{rank}(\boldsymbol{L})$ and the noncentrality parameter is [eq. (4.12)]:

$$\lambda = (\boldsymbol{m_A} - \boldsymbol{m_0})^t(\text{Var}[\boldsymbol{L}\widehat{\boldsymbol{\beta}}])^{-1}(\boldsymbol{m_A} - \boldsymbol{m_0}).$$

In the one-factor analysis of variance that tests the equality of t proportions $p_1, \ldots, p_i, \ldots, p_t$, with n_i observations for the ith treatment, suppose that the alternative proportions that we want to detect are $p_{A1}, \ldots, p_{At}$. The noncentrality parameter for the test of no treatment effect simplifies to

$$\lambda = \sum_{i=1}^{t} 4n_i(\arcsin\sqrt{p_{Ai}} - \overline{\zeta}.)^2,$$

where $\overline{\zeta}. = \sum_{i=1}^{t} n_i \arcsin\sqrt{p_{Ai}} \,/ \sum n_i$.

Example 4.2: Manufacture of Windshield Molding. Martin et al. (1987) described an experiment conducted to improve a stamping process in manufacturing windshield molding. During the stamping process debris can cause dents in the product. The controllable factors studied in the experiment were (A) polyfilm thickness (.0025, .00175), (B) oil mixture (1 : 20, 1 : 10), (C) glove fabric (cotton, nylon), and (D) underside of metal blanks (dry, oily). The response variable was the number of good parts produced out of 1000. We use this example for two purposes: (1) to demonstrate the sample size calculations for arcsine-transformed variables, recognizing that because of the large sample size the transformation is really not necessary; and (2) to discuss batch-to-batch variation, which is as important an experimental design issue for binary and proportion data as for normally distributed data.

This experiment was a fractional factorial (this type of design will be addressed in Chapter 14) conducted in such a way that information about the D main effect and the AC interaction were not both estimable. In the original analysis, factors A, C, and D were found to have large effects. Suppose that the plant engineers want to conduct a follow-up experiment to determine more conclusively whether or not factor D, oil on the underside of the metal blanks, really does have any effect on the proportion of good parts produced, and to try to estimate the effect of D separate from the AC interaction. Since factor B does not appear to have a large effect in either analysis, the follow-up experiment will focus on factors A, C, and D. The follow-up experiment will test all eight combinations of factors A, C, and D. The question is: How many parts should be produced under each of the settings?

First, the approach to answering this question will differ depending on how the experiment is run. Two possibilities come to mind: (1) each part is made in a separate run and the treatments are run in random order, or (2) some number of parts assigned to one treatment are produced together in a batch and batches are run in random order. We present the sample size calculations for the first experimental scenario first. In scenario 2, the errors within a batch are binomial, but there is usually additional variation between batches.

In scenario 1, which assumes that the parts will be run individually in random order, the number of runs equals the number of parts. To determine how many parts should be made and tested under each treatment setting, the engineers must specify the magnitude of difference they wish to detect, the significance level α, and the desired power. The engineers might like to know how large a sample is needed if factor D causes a change of anywhere from .10 to .30 in the probability of producing a good part. We demonstrate the calculations using a 5% significance level and power of 85%, assuming that the same number of parts will be tested for all treatment combinations. In the previous experiment the overall proportion of good parts was about 40%, so let us take that as the probability of producing a good part, p. We will label the proportion of good parts in level 1 of factor D as $\widehat{p}_{D_1}$. The transformed estimated proportion of good parts in level 1 of factor D has variance $\text{Var}[\arcsin\sqrt{\widehat{p}_{D_1}}] = 1/(4 \cdot 2 \cdot 2r)$, where r is the number of parts tested in each $A \times B \times D$ treatment combination, and each factor has two levels. For a mean difference of .10 between oily and dry metal blank undersides, in an experiment with three factors each at two levels, the noncentrality parameter is about [note that the expression is not exact because $(\arcsin\sqrt{.45} + \arcsin\sqrt{.35})/2$ does not exactly equal $\arcsin\sqrt{.40}$]

$$\lambda = 16r\left[\left(\arcsin\sqrt{.45} - \arcsin\sqrt{.40}\right)^2 + \left(\arcsin\sqrt{.35} - \arcsin\sqrt{.40}\right)^2\right]$$
$$= .084r.$$

The power of the test is obtained by referring to the noncentral χ^2 distribution. The sample sizes needed to obtain 85% power for detecting differences of .10 to .30 are shown in Table 4.7, and the SAS® code that produced these computations is shown in Display 4.4. The conclusion is that if the engineers think that it is important to be able to detect a difference of .10 in the proportion of parts that are good for the two levels of factor D, about 110 parts should be produced for each of the eight treatment combinations.

The second issue to highlight in this example is the question of whether the windshield moldings were produced and tested in batches. In the discussion so far we have assumed that the parts are produced in random order and that the proportion of good parts shows no trend or correlation over time. If in reality the parts are produced and tested in batches, there could well be variation among the batches in addition to the binomial variation used in the sample size calculations of Example 4.2. This phenomenon is called *extrabinomial variation* or

Table 4.7 Number of Observations Needed to Obtain 85% Power in Each Treatment Combination of a 2 x 2 x 2 Factorial Where the Response Variable Is Binary and the Mean Proportion Is .40

Difference $p_{\cdot\cdot 1} - p_{\cdot\cdot 2}$	Sample Size r
0.10	106
0.20	27
0.30	12

```
data samplesize;
p=.40;  c95=cinv(.95,1);
do d=.05 to .15 by .05;
do r=5 to 110;
lambda=4*4*r*((arsin(sqrt(p+d))−arsin(sqrt(p)))**2
      +      (arsin(sqrt(p−d))−arsin(sqrt(p)))**2);
power=sdf('chisquared',c95,1,lambda);
output;  end;  end;
```

Display 4.4 SAS® code to compute sample sizes of Table 4.7.

overdispersion. The variation among batches is often much larger than the within-batch variation. Hamada and Nelder (1997) reanalyzed the windshield molding data using a generalized linear model, which explicitly models the binomial distribution, and more important, they incorporated extrabinomial variation in their model. In contrast to the original analysis, Hamada and Nelder concluded that factor D was not important.

When objects are produced and/or tested in batches and there is a possibility of variation among batches, the appropriate variance for testing treatment effects is larger than the formulas for binomial variance would indicate. Tests of significance will not be correct if the batch-to-batch variation is ignored. Replicate batches of each treatment are the best way to obtain an estimate of the batch-to-batch variation. The batch-to-batch variation can be estimated from the residual sum of squares of the batch means, and the sample size is computed by the methods of Section 4.5.2.

Example 4.2 (continued): Windshield Molding—Parts Tested in Batches. In the windshield slugging experiment, 1000 parts were run under each treatment combination. If these were actually run in batches of 1000 parts, the batch-to-batch variation would probably be larger than the within-batch binomial variation. We may be able to infer something about the magnitude of the extrabinomial variation from the analysis of variance of the arcsine square-root transformed proportions. In this example it is possible to get only a rough estimate of the variance among batches because there were no replications of batches treated alike. Several ways of estimating a variance from unreplicated experiments are discussed in Chapter 12. One quick way, although not necessarily the best, is to assume that the two-factor interactions are negligible and to pool their sums of squares together to estimate the error variance. The mean squared error obtained this way is MSE = .0073. Note that this is much larger than what the variance would be if the observations were independent binomial random variables, in which case the residual variance would be $1/4r = .00025$.

Under the assumption that the batch effects are random and normally distributed, the treatment effects are tested using the usual F-tests from the analysis of variance of the batch means. Using the method of Section 4.5.2 and rounding the batch mean variance to .01, the noncentrality parameter for detecting a difference of $p_{\cdot\cdot 1} - p_{\cdot\cdot 2} = .10$ when the hypothesized mean is $p = .40$ is

$$\lambda = (2)(2)(b)\left(\arcsin\sqrt{.45} - \arcsin\sqrt{.35}\right)^2 / [(2)(.01)]$$
$$= 2.09b,$$

where b is the number of batches to be run for each treatment. Five batches would give power $= .88$ to detect a difference this large with a 5% significance level. This sample size (five batches of 1000 parts each for each treatment combination) is several times larger than the sample size computed under the assumption of binomial variance, which was just 110 parts per treatment combination. This demonstrates how crucial it can be to recognize and account for batch-to-batch variation.

4.7 CONFIDENCE INTERVAL WIDTH AND SAMPLE SIZE

An alternative approach to the problem of sample size determination that has much appeal is given by Beal (1989). The approach is based on the notion of controlling the confidence interval width, rather than controlling the power of a test of a null hypothesis. In part at least, the appeal of this method rests on the fact that confidence intervals are frequently more informative and therefore more appropriate than tests of significance.

Beal's argument requires a concept of "power" relevant to confidence intervals. He defines power $\phi(h)$ as the probability that a double-sided confidence interval has half-width less than or equal to h, i.e., width no greater than $2h$. The half-width of a confidence interval is also sometimes called the *margin of error*. The user specifies h as a measure of the desired precision, and chooses a sample size to give a high value of the probability ϕ that this precision will be attained. The probability ϕ can be decomposed into two parts: ϕ_{I}, the probability that the half-width is no greater than h given that the true parameter value is in the interval; and ϕ_{E}, the probability that the half-width is no greater than h given that the true value is not in the interval. It follows that $\phi(h) = \alpha\phi_{\mathrm{E}}(h) + (1 - \alpha)\phi_{\mathrm{I}}(h)$. The argument now is that one typically will be more interested in controlling $\phi_{\mathrm{I}}(h)$ via sample size rather than $\phi(h)$.

Beal has constructed tables that control $\phi_{\mathrm{I}}(h)$ for two-sided confidence intervals on the mean and on differences between two means, which are reproduced in part in Tables 4.8 and 4.9. He also shows how the tables can be used to control $\phi(h)$. The arguments for the tables are: (1) the confidence level, $1 - \alpha$, (2) the desired "power," and (3) the desired upper bound to h, the half-length of the interval, expressed as standard deviation units, i.e., as h/σ.

Example 4.2 (continued): Windshield Molding—Sample Size to Control the Confidence Interval Width. As an example of the use of these tables, consider the windshield slugging experiment scenario in Example 4.2 in which

Table 4.8 Number of Observations Required for Confidence Interval on Mean

	$\alpha = .01$			$\alpha = .05$			$\alpha = .10$		
	Power ϕ_I			Power ϕ_I			Power ϕ_I		
h/σ	.90	.80	.50	.90	.80	.50	.90	.80	.50
.1	713	698	668	422	410	388	303	293	275
.2	192	185	170	116	110	99	84	80	71
.5	39	36	30	24	22	18	18	17	13
1.0	14	13	10	9	8	6	7	7	5
2.0	7	6	5	5	5	4	4	4	3
4.0	5	4	4	3	3	3	3	3	2

Source: Excerpted from Beal (1989). Published with the permission of the International Biometric Society.

Table 4.9 Number of Observations Required for Confidence Interval on Differences between Means

	$\alpha = .01$			$\alpha = .05$			$\alpha = .10$		
	Power ϕ_I			Power ϕ_I			Power ϕ_I		
h/σ	.90	.80	.50	.90	.80	.50	.90	.80	.50
.2	357	349	334	211	206	194	152	147	138
.5	64	61	55	39	37	32	29	27	23
1.0	20	18	15	13	11	9	10	9	7
2.0	8	7	6	5	5	4	4	4	3
4.0	4	4	3	3	3	3	3	3	3

Source: Excerpted from Beal (1989). Published with the permission of the International Biometric Society.

the parts were produced in batches. The response variable is $z = \arcsin\sqrt{\widehat{p}}$, where $\widehat{p}$ is the proportion of parts that are good in a batch. Suppose that 40% of the parts are good under one particular treatment ($\widehat{p} = .40$ corresponds to $z = .68$), and the standard deviation of z is $\sigma = .10$. It is desired to construct a confidence interval for E[z] that has half-width no wider than $h = .10$. For a double-sided confidence interval, confidence level of $1 - \alpha = .95$, and "power" equal to $\phi_I = .80$, a sample size of $b = 8$ batches for each treatment would be required. If the objective is to construct a confidence interval of the difference in E[z] between two treatments with a half-width no wider than $h = .10$, then $b = 11$ batches for each treatment would be required.

One-sided confidence intervals are handled similarly except that α is replaced by $\alpha/2$ in the confidence statement. The tables given by Beal are much more extensive than those given here. Also, it is explained in the original article that the numbers given here really represent upper bounds in many cases where h/σ is less than 1.

4.8 ALTERNATIVE ANALYSIS: SELECTING AND SCREENING

In this section we introduce a completely different approach to statistical analysis of data from an experiment and, to some extent, the designing of experiments. The approach traces back to work in statistics over the last 40 plus years. Rather than attempt even a brief review of the developments, we refer interested readers to two general books, *Selecting and Ordering Populations: A New Statistical Methodology* by Gibbons et al. (1977) and *Design and Analysis of Experiments for Statistical Selection, Screening, and Multiple Comparisons* by Bechhofer et al. (1995). In addition, readers with particular interest in plant breeding are urged to consult two dissertations, "On Statistical Selection in Plant Breeding" by Dourleijn (1993) and "Statistical Selection: Multiple Comparison Approach" by Driessen (1992).

The classical statistical analyses such as described in Chapter 3 and in the preceding sections of this chapter are most often concerned with testing homogeneity of means, as in the analysis of variance. However, there are cases where such tests of homogeneity are not relevant to the experimenter. The fundamental idea is that the scientist or researcher is not really interested in testing hypotheses, because the experimenter knows or is willing to assume that there are differences. Instead, the object is to select the "best," where "best" has to be defined. It may be defined in terms of the mean of the distribution of responses, the variance of the distribution of responses, or possibly the probability of success or failure on repeated trials. The experimenter will look at the results of the experiment and no doubt select what appears to be the best. This selection may be either correct or incorrect. The overriding strategy will be to design experiments that guarantee, in some sense, the experimenter's odds of making the correct selection. The object may be to select the single best treatment, or it may be to select a subset of treatments containing the best one.

At this point it is probably fair to say that the approach has been most frequently adopted by people in operations research and industrial engineering. However, the methodology has application in many areas. An agronomist may wish to select the best of several varieties, or may be content to select a subset of five varieties for further study, with reasonable assurance that the best is in the set selected. A medical researcher may wish to select the best drug, or select a subset of alternative drugs, before going to a more elaborate, more expensive experiment to identify the best. Similar examples exist in manufacturing process studies, in marketing studies, and so on.

4.8.1 Basic Formulation: Selecting the Best

We briefly outline a formulation for this problem that is usually referred to as the *indifference zone approach* to selecting the best. The analysis of the data from the experiment will be trivial. The experimenter simply picks the winner and declares that it is the best treatment. The difficulty is that this decision could be incorrect. The object is to plan an experiment that leads to a correct selection with a suitably specified probability. We follow usual convention and denote this by Pr(CS).

In the simplest case the experimenter is faced with $t \geq 2$ treatments or populations that are denoted by $\mathbb{P}_1, \mathbb{P}_2, \ldots, \mathbb{P}_t$. Population $\mathbb{P}_i$ is assumed to have mean μ_i and variance σ_i^2. The ordered means are denoted by $\mu_{[1]} \leq \mu_{[2]} \leq \cdots \leq \mu_{[t]}$. Obviously, this true ordering is not known. Formally, the pairing between the $\{\mu_{[i]}\}$ and the $\{\mathbb{P}_i\}$ is unknown. In particular it is not known which of the $\{\mathbb{P}_i\}$ has mean $\mu_{[t]}$. If it were known, there would be no problem.

In order to present the following discussion reasonably succinctly, we use the following notation:

$\mu_1, \mu_2, \ldots, \mu_t$	These are the actual unknown means.
$\mu_{[1]}, \mu_{[2]}, \ldots, \mu_{[t]}$	These are the ordered μ's. We wish to know which is $\mu_{[t]}$.
$\widehat{\mu}_1, \widehat{\mu}_2, \ldots, \widehat{\mu}_t$	These are the estimates of the μ's. We know these.
$\widehat{\mu}_{[1]}, \widehat{\mu}_{[2]}, \ldots, \widehat{\mu}_{[t]}$	These are the ordered $\widehat{\mu}$'s. We know these.
$\widehat{\mu_{[1]}}, \widehat{\mu_{[2]}}, \ldots, \widehat{\mu_{[t]}}$	These would be estimates of the $\mu_{[i]}$. We wish we had these or at least could identify $\widehat{\mu_{[t]}}$ for a correct decision.

The experimenter usually will claim that $\widehat{\mu}_{[t]} = \widehat{\mu_{[t]}}$, i.e., the largest of the observed means $\widehat{\mu}_{[t]}$, represents the estimate of the largest mean, $\widehat{\mu_{[t]}}$. If this is true, this represents a correct decision.

The experiment yields estimates $\widehat{\mu}_1, \widehat{\mu}_2, \ldots, \widehat{\mu}_t$ and by sorting, $\widehat{\mu}_{[1]}, \widehat{\mu}_{[2]}, \ldots, \widehat{\mu}_{[t]}$. The largest, $\widehat{\mu}_{[t]}$, is selected to identify the population $\mathbb{P}_i$ that has mean $\mu_{[t]}$. This decision by the experimenter is either correct or incorrect. The idea behind the indifference zone approach is that it is important to select the correct population whenever

$$\mu_{[t]} - \mu_{[t-1]} \geq \delta^*,$$

where $0 \leq \delta^* \leq \infty$. The constant δ^* can be interpreted as the smallest difference worth detecting. This concept gives rise to the term *indifference zone*. If $\mu_{[t]} - \mu_{[t-1]} < \delta^*$, the experimenter does not care. Selecting δ^* is a subject-matter decision, to be made by the scientist, not by the statistician.

Since it is possible to make a correct selection by using a random choice and not even looking at the data, $1/t$ provides a lower bound for Pr(CS), the probability of a correct decision. The actual probability of a correct decision when looking at data depends on a large number of things. First, it depends on the distribution involved. A reasonable first assumption is normal distributions with common variance, σ_ϵ^2, and equal-sized independent samples. Additionally, there is a question about the relative sizes of the other means, $\mu_{[1]}, \mu_{[2]}, \ldots, \mu_{[t-1]}$. The standard approach is to assume that all means are the same except for $\mu_{[t]}$; i.e., $\mu_{[1]} = \mu_{[2]} = \cdots = \mu_{[t-2]} = \mu_{[t-1]} = \mu_{[t]} - \delta^*$. The scientist is trying to identify population $\mathbb{P}_i$ that has mean $\mu_{[t]}$. This can easily be shown to be the "worst" or "least favorable" configuration, because if $\mu_{[t]} - \mu_{[t-1]} > \delta^*$, it should be even easier to identify population $\mathbb{P}_i$ that has mean $\mu_{[t]}$. Also if, say, $\mu_{[1]}$is smaller than $\mu_{[t-1]}$, it should become easier to eliminate the population that has mean $\mu_{[1]}$ from consideration and improve the odds of selecting population $\mathbb{P}_i$ that has mean $\mu_{[t]}$, i.e., make a correct decision. The same comment can be made about all other candidate populations or means. Under the constraints given, if the means are not equal to $\mu_{[t-1]}$, the populations in a sense remove themselves from competition and make it easier to identify the population $\mathbb{P}_i$ that has mean $\mu_{[t]}$.

Now if σ_ϵ^2 and δ^* are assumed known, and the experiment provides independent normally distributed $\widehat{\mu}_1, \widehat{\mu}_2, \ldots, \widehat{\mu}_t$, each with variance σ_ϵ^2/n, the rule of selecting population $\mathbb{P}_i$ to have $\mu_{[t]}$ determines Pr(CS), the probability of a correct selection. For the purpose of designing an experiment, we invert this relationship. The experimenter selects a value P^* between $1/t$ and 1 as an upper bound for Pr(CS), the probability of a correct selection, and then calculates the required n. The actual calculations (which involve evaluating multivariate normal probability integrals) are difficult. However, extensive tables have been constructed and can be found in both Gibbons et al. (1977) and Bechhofer et al. (1995).

A convenient and very adequate approximation in the range $2 \leq t \leq 11$ and $.75 \leq P^* \leq .99$ is obtained by using the formula

$$n = \lceil 2(\sigma_\epsilon Z_{t-1,.5}^{(1-P^*)}/\delta^*)^2 \rceil, \tag{4.13}$$

where the notation $\lceil x \rceil$ indicates the smallest integer larger than x and $Z_{t-1,.5}^{(1-P^*)}$ is computed as

$$\begin{aligned}
z = {} & .472209894 - 3.613322738(1 - P^*) - .340997641(t - 1) \\
& + 6.433011943(1 - P^*)^2 + .025107585(t - 1)^2 \\
& + 1.251176599 \ln(t - 1) - .367905796 ln(1 - P^*) \\
& + 2.042239477(1 - P^*) \ln(t - 1) + .011083934(t - 1) \ln(1 - P^*) \\
& - .000835002(t - 1)^3 - 7.938821059(1 - P^*)^3 \\
& - .525345772(1 - P^*)(t - 1) + .260225704(1 - P^*)^2(t - 1) \\
& + .015895664(1 - P^*)(t - 1)^2 - 1.662668596(1 - P^*)^2 \ln(t - 1) \\
& - .000493189(t - 1)^2 \ln(1 - P^*).
\end{aligned}$$

In practice, the user will know t, the number of treatments; will need σ_ϵ^2 from prior experience; and will need to select δ^*, the smallest difference worth detecting, and P^*, the lower bound for the probability of a correct decision. The formula then provides the number of replications needed.

To illustrate the computations, consider a hypothetical example with $t = 8$ treatments. Assume that $\sigma_\epsilon^2 = 100$ and that the investigator feels that a difference of 6 is important. Using formula (4.13) and the approximation for $Z_{t-1,.5}^{(1-P^*)}$ we obtain n values 13, 23, and 31 for $P^* = .75, .90$, and .95. The experimenter needs 13 replications on each treatment to be assured of at least a 75% chance of correctly identifying the best treatment if it differs from the next best by 6 or more.

Bechhofer et al. (1995) give the Fortran code for a program called MULTZ to calculate $Z_{t-1,\ .5}^{(1-P^*)}$. In addition, there is also a public-domain Fortran program MVNPRD written by Dunnett (1989,1993), available via ftp from the statlib archive maintained at Carnegie Mellon University at the Internet address *lib.stat.cmu.edu* or at the URL *http://lib.stat.cmu.edu/apstat/*, which can be used to calculate $Z_{t-1,\ .5}^{(1-P^*)}$.

There is also the question of the robustness of the method to lack of normality and lack of homogeneity of the variances. Simulations using the uniform distribution (an extremely short tail distribution) and using the three-degree-of-freedom t-distribution (an extremely long tail distribution) have shown that lack of normality is not serious, provided that the distributions are symmetrical. However, as in most statistical procedures, nonhomogeneous errors present a much more serious problem. Lack of independence can also be a problem, but this problem should be unlikely in the experimental situation.

4.8.2 Subset Selection Formulation

In a sense this topic relates more to statistical analysis than to design of experiments; however, there is a design aspect. A common scenario is where a large number of treatments or varieties must be compared to select the best, and a preliminary experiment with a few replications of each treatment has been performed. Now the task is to select a promising subset for further evaluation. The initial experiment does not have sufficient replication to identify the best treatment conclusively. A larger subset increases the odds of including the best but makes the second stage more difficult. We can only indicate the general approach to the problem here. The interested reader is referred to two books (Gibbons et al. 1977) and (Bechhofer et al. 1995) and the dissertation by (Dourleijn 1993).

The basic setup is just as before. The experimenter has n observations from each of $t \geq 2$ populations $\mathbb{P}_1, \mathbb{P}_2, \ldots, \mathbb{P}_t$. Population $\mathbb{P}_i$ has unknown mean μ_i and variance σ_ϵ^2. It will be assumed that σ_ϵ^2 is known. The goal, a correct selection (CS), is to select a (random-size) subset of the $\{\mathbb{P}_i\}$ that contains the population with mean $\mu_{[t]}$. The requirement is that the experimenter wants

$$\Pr(\text{CS}\big|\{\mu_i\}, \sigma_\epsilon^2) \geq P^*,$$

where $1/t \leq P^* < 1$ holds for all $\{\mu_i\}, \sigma^2$.

The key point in this formulation is that the size of the subset selected is not specified. It is random. Notice that one can make the subset consist of all $\{\mathbb{P}_i\}$. This guarantees a correct selection according to our criterion. However, this decision is uninformative. An alternative is to insist that the subset contain only one population. Now the subset is maximally informative, in the sense that it contains no extraneous populations. However, the probability of making this correct selection is much lower. In fact, the lower bound on P^* clearly comes from the case where σ_ϵ^2 is known and a subset of size 1 is selected. This can be achieved without any data.

Note that this sort of procedure has considerable appeal for a preliminary screening step, a step in which the object is to screen out or eliminate the inferior populations before proceeding on to a more elaborate study to identify the best.

Gupta(1956,1965) has proposed the following rule: Include $\mathbb{P}_i$ in the subset selected if and only if

$$\overline{y}_i \geq \overline{y}_{[t]} - Z_{t-1,.5}^{(1-P^*)} \sigma_\epsilon \sqrt{2/n}.$$

If the differences $\mu_{[t]} - \mu_i$ for $1 \leq i < t$ are small, the subset selected will be larger than if the differences $\mu_{[t]} - \mu_i$ are large. If the differences are large, $\mu_{[t]}$ is easier to locate. Also, a large σ_ϵ^2 will lead to a large selected subset. It is possible for the procedure to fail, i.e., to return the uninformative complete set.

REVIEW EXERCISES

4.1 Revisit the tomato starch experiment of Exercise 4.3.

(a) If you were to test the hypothesis of no treatment effect using a randomization test with the F-ratio as the test statistic, how many possible values of F would there be in the reference distribution?

(b) Carry out the test of part (a) using a reference distribution of 999 random treatment assignments. State the p-value and your conclusion.

4.2 In the example of Section 4.5.2, suppose that the experimenter has enough resources to consider using more than eight replicates. Compute the power for testing treatment effects when $\mu_1 = 65$, $\mu_2 = 85$, $\mu_3 = 70$, $\mu_4 = 75$, $\mu_5 = 75$, $\mu_6 = 80$, and $\sigma_\epsilon = 15$ using $n = 8, 9, \ldots, 16$ replicates and plot power against number of replicates. How many replicates would you recommend using to obtain a power of 80%?

4.3 Suppose that in a new experiment to compare the six treatments of Section 4.5.2 the experimenter is using improved experimental techniques that reduce the amount of unexplained variation. The treatment means are again assumed to be $\boldsymbol{\mu} = [65, 85, 70, 75, 75, 80]^t$. The coefficient of variation using the improved experimental techniques is expected to be about 15%.

(a) How many replicates would you recommend using to obtain a power of 80% for testing the hypothesis that there are no treatment effects?

(b) Suppose that treatment 1 is an untreated control, treatment 2 is a new experimental treatment, and treatments 3 through 6 are various established treatments that have been in use for some time. The hypothesis of interest is that treatment 2, the new treatment, gives a significantly different response than the average of the older treatments, treatments 3 through 6. How many replicates would you recommend studying to obtain 80% power for testing this hypothesis?

4.4 An agronomist is studying the effects of elevated CO_2 on rice yields. Two treatments will be compared: ambient CO_2 and elevated CO_2.

(a) For each sample size from $n = 4$ to $n = 10$ replicates, compute the power for testing at the 5% significance level that the CO_2 has no effect on mean yield, for values of $(\mu_1 - \mu_2)/\sigma_\epsilon$ ranging from .5 to 3. Plot power versus $(\mu_1 - \mu_2)/\sigma_\epsilon$ for each sample size. Write one or two paragraphs explaining to the agronomist how to use these power curves.

(b) Elevated atmospheric CO_2 causes increased plant growth in many crops. The agronomist wants to test whether this is the case with rice. Plot power curves for the one-sided test of $H_0 : \mu_1 = \mu_2$ vs. $H_a : \mu_1 < \mu_2$.

4.5 A crop scientist is planning an experiment to compare germination of seeds from ten different lines of corn. In each flat 20 seeds are planted, all from the same line. In a preliminary experiment the standard deviation between flats was about .01. Suppose that approximately 90% of seeds are expected to germinate and the researcher would like to detect a difference of 3% between the lines with the highest and lowest germination rates. How many flats should be planted in each line to be able to detect such a difference with 80% power?

4.6 A statistician has proposed a new test statistic for use in regression when some of the explanatory variables are missing for some of the experimental units. She is planning a simulation experiment to evaluate the statistical properties of the new test statistic for two different sample sizes ($n = 10$ and $n = 30$) and three different patterns of missing covariates (x_1 missing 10%, x_1 missing 20%, x_1 and x_2 each missing 10% of observations). She is interested in testing whether the rejection rate using the new test statistic is significantly different from the nominal level of 5% and would like to be warned if the true rejection rate is greater than 6% or less than 4%. In this simulation experiment the statistician will generate r samples for each of the six combinations of sample size and missing covariate patterns. All of the samples will be generated from the true model under the null hypothesis. The protocol is to fit the regression model and compute the test statistic for each sample and then to count the number of samples in which the null hypothesis is rejected. The proportion of times the null hypothesis is rejected is the empirical size of the test. How many samples should be generated for each "treatment" combination?

4.7 A manufacturer of light-emitting diode (LED) displays for grocery store checkout counters might design an experiment to test a component of the display under three different environmental conditions, where the response variable is whether or not the component switches on. The response is coded 0 if the component does not switch on and 1 if it does. The mean for each environmental condition is the proportion of times the component switches on. Suppose that the component fails to switch on about 3% of the time in a standard environment and the manufacturer wants to know under what conditions the component fails more often than in the standard environment. He wants to be warned if there is an environmental condition for which the proportion failing is more than .02 above the standard environment. How many samples would be required to test that the difference in the proportion of failures exceeds .02? Use a one-sided test with a 10% significance level, and compute the sample size to obtain a power of 70%.

CHAPTER 5

Methods of Reducing Unexplained Variation

5.1 RANDOMIZED COMPLETE BLOCK DESIGN

In Chapter 2 we introduced the simplest of all designs, the CRD. We also pointed out that this design required the most severe assumptions about the experimental units, that the experimental units are exchangeable. In this chapter we relax these assumptions somewhat. We allow for a priori differences among experimental units but require that it is possible for the experimenter to organize experimental units into subsets called *blocks* which are much more uniform than the whole population. An auxiliary requirement in this chapter will be that the subsets of experimental units, i.e., blocks, are large enough to accommodate a *complete* replicate of the full set of all treatments: consequently, the name *randomized complete block designs* (RCBDs). This requirement on block size will be lifted in a subsequent chapter.

5.2 BLOCKING

Blocking, the process of organizing experimental units into blocks, is one of the most fundamental techniques of experimental design. The basic idea is simple: to remove unwanted variability by dividing the collection of available experimental units into homogeneous subsets. Treatments are then applied to these homogeneous subsets. When the investigator then makes comparisons among the treatments, the treatment effects show up more clearly because extraneous sources of variation among the units have been removed.

5.2.1 Terminology

The term *blocking* traces back to the agricultural heritage of the entire field of design of experiments. The technique was first applied in field experiments

Planning, Construction, and Statistical Analysis of Comparative Experiments,
by Francis G. Giesbrecht and Marcia L. Gumpertz
ISBN 0-471-21395-0

where the experimental units were plots of land. The plots were part of larger blocks of land. A field typically consists of a number of blocks. Since there is often considerable variability in the nature and fertility of the soil in a field, the technique developed was to identify blocks that were relatively uniform and then partition them into the actual plots. The term *blocks* is still used, even when experimental units consist of animals, machines, machine operators, stores in a marketing study, and so on. In some of the more complex designs that we will look at, the terms *rows* and *columns* are used in place of blocks. The terms *stratum* and *strata* are occasionally used in place of block and blocks.

We also point out that in some cases, experimenters refer to blocks as replications. We prefer not to do this since the term *replications*, or more succinctly, "reps," is a more general term that also applies to experimental units in a CRD or any other design. However, the reader should be aware that this usage exists.

5.2.2 Purpose of Blocking

Blocking is a very clever, yet very simple technique that allows an investigator to reconcile two somewhat opposing aims of experimental design. The first aim, as mentioned above, is to minimize unwanted variability among experimental units as much as possible, so that subtle effects of treatments can be detected. A second, equally important aim is to make inferences relevant to a population of interest. This requires that the treatments be tested under the full range of conditions inherent in the population of interest. To accomplish this it is important to test treatments under as diverse conditions as possible (selected within reason, of course) in order to provide as broad a basis for inference as possible. Testing treatments under diverse conditions eliminates the need to extrapolate from one set of conditions to another. The obvious strategy here is to select widely varying experimental units. This is in direct conflict with the notion of standardizing experimental units and conditions. The problem with an experiment conducted under very uniform conditions is that statistical inferences are limited to those very uniform conditions.

As an example we can consider a hypothetical experiment involving a chemical reaction. The question of interest may be the relative speeds of a reaction in the presence of one of three different forms of a catalyst. It is suspected that the amount of water present in the mixture as well as external environmental conditions, e.g., atmospheric pressure (difficult to control) and temperature (easy to control), may affect the speed of the reaction. One option would be to tightly control the amount of water and temperature and conduct the study only on days when the temperature and barometric pressure are at some chosen level. This sort of experiment will answer the question of relative rates of the reactions in the presence of the catalysts under specific conditions. However, this is of limited value. One would really like to make statements without having to specify that the barometric pressure and temperature must be at specific levels. A better plan would be to perform tests on different days and at different temperatures. Temperatures can and must be selected judiciously to reflect a reasonable range of interest. Also, one would not test one catalyst at one temperature and a second at a higher temperature. Days could be considered as blocks and all catalysts tested on each day. The amount of water present in the mixture may be a factor with a slightly different character. It may be a factor in the reaction that is of interest in and of itself. This would be a factorial experiment. These will be dealt with later. However, it may be an extraneous variable that is difficult to control in practice but can be controlled for purposes of the experiment. Now it is a factor to utilize in blocking, just as are temperature and atmospheric pressure.

In general, we note that blocking can be used to control for well-understood effects in experimental material, such as differences between genders in biological experiments or between old and new pieces of equipment. Blocking may also be used to eliminate haphazard unpredictable variations, such as day-to-day inconsistencies or possible batch-to-batch differences in chemical reagents in laboratory experiments.

In an interesting article in *The Scientist* (Clemmitt 1991), the author shows that it is not uncommon for researchers inadvertently to design experiments that do not allow inferences to the population of interest by insisting on uniformity of the experimental subjects. This can have serious consequences. The article, addressed to medical researchers, is about ethical considerations in research on human subjects. The author points out that most clinical trials are conducted with relatively young white men as subjects. The reasons for doing this are (1) to work with a relatively homogeneous population, and (2) the ready availability of white male subjects. The problem is that inferences and recommendations will be made for the entire human population, not just for white males. As examples of the inherent risks, the article points out that black and white persons respond differently to diuretics. Also, there appears to be more and more evidence that the anti-AIDS drug AZT acts differently in white men than in black men. The key point is that if one restricts the experimental material to a very uniform subset, future inferences are restricted to that subset. Any extrapolation to other subjects must be based on extraexperimental evidence, e.g., expert opinion.

The trick to reconciling the two contradictory aims of minimizing unexplained variability and making inferences to a broad population is to select a large variety of experimental units and then group them into blocks that are as homogeneous as possible. The treatments are then evaluated by making comparisons within blocks, but evaluating the comparison in the different blocks, i.e., under varying conditions. Proper statistical analysis of the data removes the differences among blocks, yet retains the robustness of the inference process that comes from making the comparisons under diverse conditions.

5.2.3 Nature of Blocks

We list a number of experimental situations and describe several different types of blocking criteria.

- In a laboratory experiment that uses substantial quantities of some reagent, it is often a good idea to block on bottles or shipments of reagent. There may be subtle shipment-to-shipment differences that affect the outcome. Blocking on time is often advantageous, since reagent bottles may change over time because of evaporation or degradation.

- In an industrial experiment, the parts coming from one machine may differ in subtle ways from parts coming from another machine. Again, blocking will not be costly, but can lead to important insights.
- In the classical field experiment in agronomy, experimental units are generally organized into blocks for one of two reasons. First, there is the question of homogeneity of plots. Second, there is a matter of convenience. In many of the tasks related to the plots, such as harvesting, it may not always be certain that all of the units can be harvested at the same time. The blocking scheme conveniently splits the units into manageable subsets.
- In animal research, we have already alluded to gender and variety differences. Frequently, an investigator can set up the experiment with some blocks using animals of variety A, other blocks with animals of variety B, and so on, with very little additional cost. The advantage is protection against inadvertent severe treatment $\times$ variety interaction such as that cited by Shellabarger et al. (1978).
- In animal research, experimental error can often be reduced considerably by using litters of animals to form blocks. The fact that littermates have experienced a common prenatal environment tends to make them more alike than are animals selected randomly.
- Burrows et al. (1984) examined enzyme-linked immunosorbent assay (ELISA) data from a number of studies and concluded that the 96 well plates displayed definite row and column effects and that using a blocking structure that removed these effects led to significantly more efficient experiments.
- Technicians may be given identical instructions, yet differ in subtle ways in their execution of some tasks in an experiment.

5.2.4 Mechanics of Blocking

The mechanics of constructing a randomized complete block design are simple. One sorts the experimental units into subsets that are as homogeneous as possible. Then randomly assign treatments to units within blocks using an independent randomization for each block.

One can view the sorting step as either trying to make the differences among blocks as large as possible, or as making the differences within blocks as small as possible. Randomization is restricted to units within blocks, whereas in the CRD there is no restriction on the randomization.

5.2.5 Disadvantages of Blocking

The disadvantages of blocking are not severe, but they do exist. Introducing blocking into an experiment makes both the organization and the statistical analysis of the experiment somewhat more complex. In some more complex schemes that we will look at, the randomization process becomes quite complex. A complex blocking scheme does increase the possibility of error and consequently, requires more care. Although the statistical analysis becomes more complex, the ready availability of statistical software has mitigated the problem of more complex computations. Inadvertent loss of observations or incorrect treatment assignment can make computations and interpretation of results from an experiment with a complex blocking structure more difficult.

Blocking used incorrectly can present a problem. For example, it is not uncommon to see blocks in field experiments designed to make fieldwork convenient and not honor fertility gradients. Similarly, it may be convenient to use the four tires on an automobile as a basis for blocking and neglect the fact that there are major differences between front and rear tires. The most important consideration in establishing blocks is to make sure that the experimental units within the blocks are homogeneous. Although convenience is always an important consideration in any experimental plan, careful consideration is warranted before one lets convenience compromise efficiency. In a future chapter we discuss what to do if homogeneous blocks are not large enough to hold all the treatments.

If an experiment has a small number of treatments and few replications, a randomized block design can have substantially fewer degrees of freedom for estimating the error variance than a completely randomized design. This can be a serious consideration for small experiments comparing two or three treatments. For example, an experiment with two treatments and four replicates conducted as a CRD would have $t(r-1) = 6$ error degrees of freedom, whereas an RCBD with two treatments in four blocks has just $(t-1)(b-1) = 3$ degrees of freedom for error. With more treatments or more replications the error degrees of freedom become large in either design and there is not much practical difference. The power computations of Section 5.3.4 are useful for choosing between an RCBD and a CRD in the planning stages of an experiment.

Another error that is encountered far too frequently is that an experiment is conducted as an RCBD but analyzed as a CRD. The result is that block-to-block differences are merged into the error term, resulting in considerable loss of power.

5.2.6 Blocks, Fixed or Random

One occasionally hears the question: Are blocks fixed or are blocks random? Factors with randomly selected levels are called *random effects*, whereas factors with specifically selected levels are called *fixed effects*. In some cases an experimenter will use blocks to account for known factors in experimental material. Examples of this would be sex in a biology experiment or blocks based on a fertility gradient in a field. In these cases block effects would be considered fixed. In other cases the experimenter uses blocks to control for unknown or unexpected effects, such as differences between days, between locations in an agronomy experiment, or between observers. Now blocks represent a random sample from some universe, and block effects are said to be *random*. Thus, the answer to the question of whether block effects are fixed or random depends on the criteria used to construct blocks. If all types are represented in the experiment, blocks represent a fixed effect. If block effects are fixed, future applications will be on the same (equivalent) blocks, e.g., on the same two genders. If block effects are random, future applications of treatments will be to future experimental units, presumably from different randomly selected blocks.

We take the pragmatic view that blocks consist of collections of experimental units. The object of the experiment is to make inferences, predictions of what will happen if the treatments are applied to new units from the population. In this sense the experimenter wants blocks to be random. Good practice is to select as wide a range of blocks as possible.

The choice as to blocks random or fixed depends on the experimenter's judgment as to whether the blocks can reasonably be considered to represent a random selection from the population of interest. This choice then affects the scope of the statistical inferences that can be made from the results of the experiment. If the blocks are considered to be random, statistical inferences about treatment effects apply to the entire population. On the other hand, if the blocks are considered to be fixed, statistical inferences really apply only to the blocks used in the experiment. Fortunately, the question of whether block effects are fixed or random does not affect the statistical analysis for the standard randomized complete block design. In these standard RCBD experiments it is a question of the scope of the statistical inferences.

Researchers are often forced to extend inferences from a fixed set of blocks to a much larger population. Such extensions are not statistical and do not carry with them protections such as confidence limits and standard errors. They are based entirely on the expertise and knowledge of the person making the inferences. It must be realized that such extensions are beyond the experiments. This is not to say that they are invalid. The striking example that comes to mind is inferences from experiments conducted with one species of animals to another species, such as rodents to humans.

5.2.7 Block x Treatment Interaction

Standard assumptions for an RCBD generally include the statement that there are no block $\times$ treatment interactions in the experiment. This is often referred to as block-treatment additivity. Recall that in a good experiment one wants blocks to be as different as possible, but we have to include a caveat here. If the blocks are very different from each other, the treatments promote one response in some blocks and a different response in other blocks, i.e., a block $\times$ treatment interaction. Now we face the question: What happens when block $\times$ treatment interactions occur? We look at three distinct cases.

We consider first the case when blocks are fixed and clearly defined in the experiment and can be repeated precisely. Here we have in mind blocking factors such as gender or variety in biological experiments, model of machine in industrial experiments, source of chemical reagent for a laboratory experiment, and so on. In these cases a block × treatment interaction will inflate (bias) the error term, making it more difficult to identify the treatment effects. If each of the block species is represented several times, i.e., several blocks of males and several blocks of females, it is possible to treat gender as a factor in the analysis of variance and identify the interaction. In a sense, this is not really a block × treatment interaction, but rather, a type of two-factor interaction. If the blocks have been constructed by sorting experimental units on, say, a fertility gradient, or time in a laboratory experiment where reagents change with time, the techniques outlined in Section 4.8 in Scheffé (1956) and in Section 15.8 in Steel et al. (1997) can be applied to identify a possible interaction. These types of interactions are usually easily interpreted and tend to provide useful information.

In the second case we consider blocks random and have only one complete replicate of treatments in each block. In this case the error term in the conventional analysis of variance is appropriate, although the interaction does dilute the power. The interaction inflates both the error and the treatment mean square and weakens the test of significance. This loss of power can become serious if the interaction is large. Again, more detailed statistical analyses (Scheffé 1956) can be used to tease out the interaction.

The final case we consider is the one where each block contains two or more complete replications of all treatments. Now the experimenter is in a position to do a conventional statistical analysis that properly identifies the interaction.

We also point out that identifying an interaction or nonadditivity is not something to be feared. It represents an opportunity for better understanding of what is going on. We emphasized earlier that there is an advantage to making blocks as different as possible. The wider range serves to give a broader basis for inference. When blocks cover a very broad range, it is more likely that large block × treatment interactions will be encountered. However, this is not necessarily a bad thing. In fact, such a discovery often leads to very important insights. The bottom line is that small block × treatment interactions have very negligible effects, and large block × treatment interactions are usually easy to notice and lead to important insights. A very nice design is obtained by making blocks large enough to accommodate each treatment two times. This permits a very sensitive test for block × treatment interactions. Some of the points raised in this section are examined in more detail in Section 5.4.

5.3 FORMAL STATISTICAL ANALYSIS FOR THE RCBD

5.3.1 Classical Model

The classical model for an RCBD design is

$$y_{ij} = \mu + \beta_i + \tau_j + \epsilon_{ij}, \tag{5.1}$$

where $i = 1, \ldots, b$ blocks and $j = 1, \ldots, t$ treatments. The $\{\beta_i\}$ represent block effects, the $\{\tau_j\}$ represent treatment effects, and the $\{\epsilon_{ij}\}$ represent random errors associated with individual plots. In addition, it is usually assumed that the $\{\epsilon_{ij}\}$ are iid$(0, \sigma_\epsilon^2)$ or iid $N(0, \sigma_\epsilon^2)$. Further assumptions and/or restrictions on this model are discussed in Section 5.4.

Example 5.1: Effect of Elevated CO_2 on Rice Grain Yield. This experiment from Kobayashi et al. (1999) and Okada et al. (2001) provides a good vehicle to demonstrate the process of deciding how to assign experimental units to blocks. The central objective of this experiment was to compare rice growth and yield under two CO_2 regimes: ambient and elevated. The experiment was conducted in rice paddy fields rented from local farmers at a site in northern Japan. In the elevated CO_2 plots a ring of tubing emitting CO_2 into the paddy field was installed around a 12-m-diameter plot. The level of CO_2 emitted was dynamically controlled to maintain a specified level above the ambient CO_2 concentration. In the ambient plots no CO_2 was emitted.

Each farmer owned several paddy fields, and in the years prior to the experiment each farmer managed his own paddy fields. Note, however, that for the duration of the actual experiment, all fields were managed by experiment station personnel using a common protocol. The layout of the field site in Table 5.1 lists the owner of each field and the yield in 1997, the year before this experiment began. It is evident that there were some substantial differences in yield among the farmers. Each farmer used different fertilizer regimes and management practices that affected the fertility of the fields. The means (and standard deviations in parentheses) for each of the farmers that owned four or more paddy fields were 724(25), 678(26), 745(22), and 801(27) g/m^2 for farmers 1, 5, 7, and 9, respectively.

For practical reasons only four replicates could be used. First, the researchers could afford to build and supply only four elevated CO_2 rings, and second, it was necessary to leave two rice paddies between every pair of treatment rings in each row to ensure that CO_2 from one ring would not accidentally contaminate

Table 5.1 Spatial Layout of Rice Paddies: Rice Yields in 1997, the Year Before the Rice FACE Experiments Began

Farmer			1*	1	1	1*	2	2	3	3
Yield (g/m^2)			714	762	708	712		814	737	657
Farmer			4	4	5	5+	5	5	5+	
Yield (g/m^2)			596	608	672	657	650	702	709	
Farmer	6	6	7	7*	7	7	7*	7	8	
Yield (g/m^2)		744	741	743	747	772	759	706	693	
Farmer		9	9+	9	9	9+	10	10	10	
Yield (g/m^2)		767	815	828	778	817	719	736	679	

Source: Data courtesy of K. Kobayashi, personal communication.
*Indicates a ring at top end of field was used in experiment.
+Indicates a ring at bottom end of field was used in experiment.

another ring. Based on the 1997 yields, it appeared that blocking by farmer would be an effective way to obtain homogeneous sets of rice paddies. Each of the farmers who owned at least four paddy fields (farmers 1, 5, 7, and 9) was specified as one block. Two paddy fields at least three fields apart were selected from each block. If there were more than four fields in a block, these two fields were selected randomly. The elevated CO_2 treatment was randomly assigned to one of the fields and the ambient CO_2 treatment to the other. At the end of the season, the grain yield of rice was measured at three locations, i.e., subsamples, in each of the eight experimental plots. Table 5.2 shows the ln(grain yield) value for 1998 for each subsample and the mean for each plot.

A good way to analyze experiments with subsampling, where there are an equal number of subsamples in each experimental unit, is to compute the means of the subsamples in each unit and then do further analysis on these means. Subsampling is discussed at more length in Section 5.4.7. All further analyses of the rice-CO_2 data will be done on the means of the three locations for each ring.

We return to this example throughout the chapter to demonstrate details of the statistical analysis. The data are reproduced on the companion Web site.

5.3.2 Analysis of Variance Table

A common statistical analysis of data from the randomized complete block is the analysis of variance based on model (5.1). The form of this analysis is shown in Table 5.3.

The analysis of variance table typically has an additional column for mean square values. The entries in this column are obtained by dividing the sum of

Table 5.2 Effect of Elevated CO_2 on ln(Grain Yield) of Rice

	Ambient CO_2		Elevated CO_2	
Block	Subsamples	Ring Mean	Subsamples	Ring Mean
1	6.184, 6.225, 6.220	6.21	6.406, 6.422, 6.406	6.41
2	6.302, 6.218, 6.241	6.25	6.352, 6.456, 6.451	6.42
3	6.077, 6.144, 6.065	6.10	6.198, 6.246, 6.323	6.26
4	6.086, 6.201, 6.132	6.14	6.422, 6.206, 6.287	6.30

Table 5.3 Analysis of Variance for RCBD

Source	df	SS
Mean	1	$y_{..}^2/bt$
Blocks	$b-1$	$t\sum_i^b(\overline{y}_{i\cdot} - \overline{y}_{..})^2$
Treatment	$t-1$	$b\sum_j^t(\overline{y}_{\cdot j} - \overline{y}_{..})^2$
Error	$(b-1)(t-1)$	$\sum_{i,j}(y_{ij} - \overline{y}_{i\cdot} - \overline{y}_{\cdot j} + \overline{y}_{..})^2$
Total	bt	$\sum_{i,j} y_{ij}^2$

squares entries by the degrees of freedom entries. Common abbreviations are df for degrees of freedom and MSB, MST, and MSE for the mean square column entries for block, treatment, and error lines, respectively. The MSE is an unbiased estimate of σ_ϵ^2. Treatment effects are tested by comparing the mean square from the treatments line with the mean square from the error line, i.e., MST with MSE. The hypothesis is $\text{H}_0 : \tau_j = 0$ for $j = 1, \ldots, t$ vs. $\text{H}_a :$ not all $\tau_j = 0$, and the criterion for rejection is:

$$\text{Reject H}_0 \text{ if } F = \frac{\text{MST}}{\text{MSE}} > F(1 - \alpha, t - 1, (b - 1)(t - 1)).$$

5.3.3 Treatment Means, Contrasts, and Standard Errors

The estimates of the treatment means are simply the averages for each treatment across the blocks,

$$\widehat{\mu}_{\cdot j} = \overline{y}_{\cdot j}.$$

The standard error of a treatment mean depends on whether the block effects are random or fixed. If block effects are fixed, the standard error of a treatment mean is

$$s(\widehat{\mu}_{\cdot j}) = \sqrt{\frac{\text{MSE}}{b}}.$$

In the case of random block effects, the standard error must include variation among blocks, and so is larger. This case is discussed in detail in Section 5.4.5.

If we let $C = \sum_j c_j \tau_j$ represent a contrast among treatment effects and C_0 an hypothesized value for the contrast, we can test the hypothesis that $C = C_0$ by computing $(\widehat{C} - C_0)/s(\widehat{C})$, where $s^2(\widehat{C})$ is the estimate of the variance of $\widehat{C}$, and comparing with a t value with $(b - 1)(t - 1)$ degrees of freedom. Note that $\widehat{C}$ is simply the contrast among means, $\sum_j c_j \overline{y}_{\cdot j}$. The variance of a contrast of treatment means is

$$\text{Var}\Big[\sum_j c_j \overline{y}_{\cdot j}\Big] = \sum c_j^2 \times \sigma_\epsilon^2 / b$$

(because the errors are assumed to be uncorrelated), which is estimated by

$$s^2(\widehat{C}) = \sum c_j^2 \times \text{MSE}/b.$$

The difference between two treatment means is a contrast (with coefficients $c_1 = 1$ and $c_2 = -1$), so the variance of a difference between two treatment means is estimated as

$$\text{Var}[\overline{y}_{\cdot j} - \overline{y}_{\cdot j'}] = \frac{2\text{MSE}}{b}.$$

The t-test for testing the difference between two means thus involves computing the ratio $(\overline{y}_{\cdot j} - \overline{y}_{\cdot j'})/\sqrt{2\text{MSE}/b}$ and comparing with the value from the t-table.

Example 5.1 (continued): Effect of Elevated CO_2—Analysis of Variance. The analysis of variance computed for the ring means data shown in Table 5.2 is given in Table 5.4.

The two means are $\widehat{\mu}_{\cdot 1} = 6.18$ and $\widehat{\mu}_{\cdot 2} = 6.35$. The rice exposed to elevated CO_2 shows average ln(grain yield) .17 unit higher than rice grown in ambient air. The variance of this difference is

$$\text{Var}[\widehat{\mu}_{\cdot 2} - \widehat{\mu}_{\cdot 1}] = \frac{2\sigma_\epsilon^2}{b}.$$

The variance among the $\{\epsilon_{ij}\}$, σ_ϵ^2 is estimated by MSE in the analysis of variance of Table 5.4. Substituting MSE for σ_ϵ^2, the standard error of the difference between treatment means is

$$s(\widehat{\mu}_{\cdot 2} - \widehat{\mu}_{\cdot 1}) = \sqrt{\frac{2\text{MSE}}{b}} = .0095.$$

Our findings about the effect of CO_2 can be summarized nicely in a confidence interval. The 95% confidence interval for the mean difference is

$$\widehat{\mu}_{\cdot 2} - \widehat{\mu}_{\cdot 1} \pm t(.975, 3)s(\widehat{\mu}_{\cdot 2} - \widehat{\mu}_{\cdot 1}),$$

where $t(.975, 3) = 3.182$ is the 97.5th percentile of Student's t-distribution. This gives an interval for the mean difference of .14 to .20.

Alternatively, since it is expected that increasing CO_2 will make plants grow faster and taller, because they take up CO_2 in photosynthesis, a one-sided test of $H_0 : \mu_{\cdot 1} = \mu_{\cdot 2}$ vs. $H_a : \mu_{\cdot 2} > \mu_{\cdot 1}$ might be desired. The test statistic is

$$t = \frac{\widehat{\mu}_{\cdot 2} - \widehat{\mu}_{\cdot 1}}{s(\widehat{\mu}_{\cdot 2} - \widehat{\mu}_{\cdot 1})} = 17.9$$

but is compared to the 95th percentile of Student's t-distribution. This gives a p-value of .0002, so we conclude that mean ln(grain yield) of rice grown in elevated CO_2 is significantly higher than for rice grown in ambient air.

Table 5.4 Effect of Elevated CO_2 on ln(Yield) of Rice: Analysis of Variance of Ring Means of Table 5.2 and Expected Mean Squares

Source	df	SS	MS	E[MS]
Block	3	.03408	.01136	$\sigma_\epsilon^2 + 2\sigma_b^2$
CO_2	1	.06015	.06015	$\sigma_\epsilon^2 + 4\sum_{j=1}^t \tau_j^2/(t-1)$
Error	3	.0005444	.0001815	σ_ϵ^2

5.3.4 Power of Hypothesis Tests

The noncentrality parameter [equation (4.12), Section 4.5.1] for testing the hypothesis of no treatment effects in a randomized complete block design has a simple form. It works out to be

$$\lambda = b \sum_j (\tau_j - \overline{\tau}_.)^2 / \sigma_\epsilon^2. \tag{5.2}$$

The power of the test is

$$\text{power} = \Pr\left\{ \frac{\text{MST}}{\text{MSE}} > F(1-\alpha, t-1, (b-1)(t-1)) \middle| \lambda \right\}.$$

***Exercise 5.1: Noncentrality Parameter*.** For $t = 3$ treatment levels and $b = 4$ blocks derive the expression $\lambda = b \sum_j (\tau_j - \overline{\tau})^2 / \sigma_\epsilon^2$ from the general quadratic form for the noncentrality parameter, equation (4.12). □

5.3.5 Nonparametric Friedman–Kendall Test

As an alternative to analysis of variance one can do a distribution-free analysis. The recommended procedure is the Friedman–Kendall test. Let the observations be denoted by y_{ij}, and let the ranks within the blocks be denoted by R_{ij}. Compute

$$S = \frac{12 \sum_j^t \left(\sum_i R_{ij} - b(t+1)/2 \right)^2}{t(t+1)b}.$$

This statistic is then compared with the critical value $\chi^2(1-\alpha, t-1)$. Reject H_0 that there are no treatment effects if S is larger than the critical value. If there are ties in the data, assign average ranks to the R_{ij} values. Then subtract from the denominator of S

$$\frac{1}{t-1} \sum_{i=1}^{b} \left[\sum_{k=1} (d_{ik}^3 - d_{ik}) \right],$$

where there are d_{i1} tied observations in the first group of tied observations in block i, d_{i2} are in the second group of tied observations in block i, and so on.

5.4 MODELS AND ASSUMPTIONS: DETAILED EXAMINATION

In Section 5.3 we presented a basic linear model and discussed the analysis of variance for the RCBD. The discussion there did not include any mention of possible block×treatment interactions. However, the topic had appeared several times in Section 5.2. In this section we proceed to examine some of the assumptions (models) underlying the analysis and some of the assertions in the discussions in more detail. In particular, we look at the consequences of interactions between blocks and treatments.

To provide a systematic treatment we define a somewhat more general model for the RCBD which assumes a block × treatment interaction and also that each treatment is represented m times in each block. The model is written as

$$y_{ijk} = \mu + \beta_i + \tau_j + (\beta \times \tau)_{ij} + \epsilon_{ijk}, \tag{5.3}$$

where $i = 1, \ldots, b$, $j = 1, \ldots, t$, and $k = 1, \ldots, m$ indicate the block, treatment, and experimental unit, respectively. The k subscript is omitted in the common case when $m = 1$; i.e., there is exactly one experimental unit per treatment per block.

5.4.1 Expected Value of the Error Mean Square

We begin with only the assumption that ϵ_{ijk} are iid$(0, \sigma_\epsilon^2)$. We will add more assumptions as we find that they are required. We begin with the mean square for residual:

$$\text{E[MSE]} = \text{E}\left[\frac{1}{bt(m-1)} \sum_{i,j,k} (y_{ijk} - \overline{y}_{ij\cdot})^2\right] = \sigma_\epsilon^2. \tag{5.4}$$

This quantity is available only if $m > 1$. The term *pure error* is often used because there is no contamination from possible block × treatment interactions.

5.4.2 Expected Value of the Block x Treatment Mean Square

Next consider the mean square for block × treatment interaction:

$$\begin{aligned}
\text{E[MSBT]} &= \text{E}\left[\frac{m}{(b-1)(t-1)} \sum_{i,j} (\overline{y}_{ij\cdot} - \overline{y}_{i\cdot\cdot} - \overline{y}_{\cdot j\cdot} + \overline{y}_{\cdots})^2\right] \\
&= \text{E}\left[\frac{m}{(b-1)(t-1)} \left[\sum_{i,j} \Big((\beta \times \tau)_{ij} - \overline{(\beta \times \tau)}_{i\cdot}\right.\right. \\
&\qquad \left.\left. - \overline{(\beta \times \tau)}_{\cdot j} + \overline{(\beta \times \tau)}_{\cdot\cdot}\Big)^2 + \sum_{i,j} \left(\overline{\epsilon}_{ij\cdot} - \overline{\epsilon}_{i\cdot\cdot} - \overline{\epsilon}_{\cdot j\cdot} + \overline{\epsilon}_{\cdots}\right)^2\right]\right] \\
&= \frac{m}{(b-1)(t-1)} \sum_{i,j} \text{E}\Big[\Big((\beta \times \tau)_{ij} - \overline{(\beta \times \tau)}_{i\cdot} \\
&\qquad - \overline{(\beta \times \tau)}_{\cdot j} + \overline{(\beta \times \tau)}_{\cdot\cdot}\Big)^2\Big] + \sigma_\epsilon^2.
\end{aligned} \tag{5.5}$$

The best test of the hypothesis of no block $\times$ treatment interaction is available by computing MSBT/MSE when $m > 1$. If there are no block $\times$ treatment interactions, equation (5.5) reduces to

$$\mathrm{E[MSBT]} = \sigma_\epsilon^2. \tag{5.6}$$

When $m = 1$, then MSE in equation (5.4) is not available and we are back to the classical model in Section 5.3. An uncontaminated estimate of σ_ϵ^2 is available only if there is no block $\times$ treatment interaction. In that case it is also common to refer to this mean square as MSE rather than MSBT.

5.4.3 Fixed Block Effects

In a sense this provides the simplest case. The experimenter is faced with a set of treatments to be used in future specific conditions. The treatments are to be tested under all of these specific conditions. Since all test conditions are examined, it is reasonable to define μ and the $\{\beta_i\}$ as in Section 3.4.3 so that $\sum \beta_i = 0$ and $\sum_i (\beta \times \tau)_{ij} = 0$ for all j. It is also reasonable to define the $\{\tau_j\}$ so that $\sum \tau_j = 0$ and $\sum_j (\beta \times \tau)_{ij} = 0$ for all i. The expected mean square in equation (5.5) then reduces to

$$\mathrm{E[MSBT]} = \frac{m}{(b-1)(t-1)} \sum_{i,j} (\beta \times \tau)_{ij}^2 + \sigma_\epsilon^2. \tag{5.7}$$

Analogous calculations show that under these assumptions the expected value of the treatment mean square is

$$\begin{aligned} \mathrm{E[MST]} &= \mathrm{E}\left[\frac{mb}{t-1} \sum_j (\overline{y}_{\cdot j \cdot} - \overline{y}_{\cdots})^2\right] \\ &= mb\left[\frac{1}{t-1} \sum_i \tau_j^2\right] + \sigma_\epsilon^2. \end{aligned} \tag{5.8}$$

When $m > 1$ and a pure error exists, there is a valid F-test for treatments. If $m = 1$, we have the situation discussed in Section 5.2.7.

Exercise 5.2: Verify E[MST] Formula. Verify that if we had not assumed that $\sum \tau_j = 0$, (but keep the assumption that $\sum_i (\beta \times \tau)_{ij} = \sum_j (\beta \times \tau)_{ij} = 0$) we would have obtained

$$\mathrm{E[MST]} = mb\left[\frac{1}{t-1} \sum_i \left(\tau_j - \overline{\tau}_{\cdot}\right)^2\right] + \sigma_\epsilon^2$$

in equation (5.8). □

5.4.4 Random Block Effects

Now consider the case where the blocks used represent a random sample from a large population of blocks. In the elevated CO_2 example, the blocks (farmers) could be interpreted as a random sample from the population of rice growers in the region. In many cases this may be a more realistic framework since often the object of an experiment is to test treatments and provide guidance for future applications. The analysis of variance arithmetic for an RCBD remains constant regardless of whether block effects are fixed or random, provided that treatments appear equally frequently in all blocks and no observations are lost. The RCBD model with random block effects is a special case of a *mixed model*, which is the term used for models that contain both fixed and random effects. Section 5.5.2 gives methods for handling mixed models for more general cases, including analysis of variance with missing data, regression, and analysis of covariance.

Considering the blocks as a random sample implies that the $\{\beta_i\}$ are random variables with some distribution. This is written more explicitly as $\beta_i \sim \text{iid}(0,\sigma_b^2)$. With this assumption there are questions about the nature of a reasonable parameterization of the block $\times$ treatment interaction. Since the blocks are a random sample from some population, it follows naturally that the $\{(\beta \times \tau)_{ij}\}$ associated with any treatment represents a random sample from some distribution. This is conveniently written as $(\beta \times \tau)_{ij} \sim \text{iid}(0,\sigma_{bt}^2)$ for any fixed j. If the experiment involves the full set of treatments being considered and all t treatments are represented equally frequently in each block, the set $\{(\beta\times\tau)_{i1}, \ldots, (\beta\times\tau)_{it}\}$ represents the full population for specific block i and it follows without loss of generality that one can assume that $\sum_j (\beta \times \tau)_{ij} = 0$ for all i. More elaboration of this point is given in Appendix 5A.

The expected error mean square in equation (5.5) now reduces to

$$\text{E[MSBT]} = m\sigma_{bt}^2 + \sigma_\epsilon^2. \tag{5.9}$$

We also compute

$$\begin{aligned}
\text{E[MSB]} &= \text{E}\left[\frac{mt}{b-1}\sum_i(\overline{y}_{i\cdot\cdot} - \overline{y}_{\cdot\cdot\cdot})^2\right] \\
&= mt\text{E}\left[\frac{1}{b-1}\left[\sum_i(\beta_i - \overline{\beta}_\cdot)^2 + \sum_i\left(\overline{(\beta \times \tau)}_{i\cdot} - \overline{(\beta \times \tau)}_{\cdot\cdot}\right)^2\right.\right. \\
&\qquad\qquad \left.\left. + \sum_i(\overline{\epsilon}_{i\cdot\cdot} - \overline{\epsilon}_{\cdot\cdot\cdot})^2\right]\right] \\
&= mt\sigma_b^2 + \sigma_\epsilon^2
\end{aligned} \tag{5.10}$$

and

$$\begin{aligned}
\mathrm{E[MST]} &= \mathrm{E}\left[\frac{mb}{t-1}\sum_j(\overline{y}_{.j.} - \overline{y}_{...})^2\right] \\
&= mb\mathrm{E}\left[\frac{1}{t-1}\left[\sum_i\left(\tau_j - \overline{\tau}_.\right)^2 + \sum_j\left(\overline{(\beta\times\tau)}_{.j} - \overline{(\beta\times\tau)}_{..}\right)^2\right.\right. \\
&\qquad\left.\left. + \sum_j\left(\overline{\epsilon}_{.j.} - \overline{\epsilon}_{...}\right)^2\right]\right] \\
&= mb\left[\frac{1}{t-1}\sum_i\tau_j^2\right] + m\sigma_{bt}^2 + \sigma_\epsilon^2.
\end{aligned} \tag{5.11}$$

We now see a valid F-test for treatment effects even if $m = 1$, as discussed in Section 5.2.7. The block $\times$ treatment interaction dilutes the power of the F-test but does not invalidate it.

A simple algorithm for writing down the expected mean squares for a factorial model with fixed and random effects is given in many statistical methods texts, such as Snedecor and Cochran (1989).

5.4.5 Standard Errors, Tests of Hypotheses, and Estimation of Variance Components

We summarize the situation for the important case where $m = 1$. If there are no block $\times$ treatment interactions, the question of blocks fixed or random is irrelevant. The error mean square is used for testing treatment effects. A common requirement is to test hypotheses about treatment differences, or more generally about treatment contrasts. If $\{c_j\}$ are constants with $\sum c_j = 0$, the treatment contrast $\sum c_j\tau_j$ is estimated by $\sum c_j\overline{y}_{.j}$. This contrast has variance $\sum c_j^2\sigma_\epsilon^2/b$, and σ_ϵ^2 is estimated by the error mean square. Since the error sum of squares has $(b-1)(t-1)$ degrees of freedom, one can construct confidence intervals directly. Notice also that if we compute

$$\begin{aligned}
\mathrm{E}\left[\sum_j c_j\overline{y}_{.j}\right] &= \mathrm{E}\left[\sum_j c_j(\mu + \overline{\beta}_. + \tau_j)\right] \\
&= \sum c_j\tau_j,
\end{aligned}$$

the block parameters drop out, regardless of whether blocks are fixed or random.

Individual treatment means can be reported. If blocks are fixed, $\overline{y}_{.j}$ can be interpreted as estimating $\mu + \overline{\beta}_. + \tau_j$ and has variance σ_ϵ^2/b. If blocks are a random sample from some population, then $\overline{y}_{.j}$ is an unbiased estimate of $\mu + \tau_j$ with variance

$$(\sigma_b^2 + \sigma_\epsilon^2)/b.$$

This requires an estimate of σ_b^2. From equations (5.6) and (5.10) it follows that

$$\widehat{\sigma}_b^2 = \frac{\text{MSB} - \text{MSE}}{t}.$$

If blocks are random, equations (5.9) and (5.11) indicate that the tests of hypotheses are still valid even if there are block $\times$ treatment interactions. However, the power is reduced. If blocks are fixed, and there are block $\times$ treatment interactions, equations (5.7) and (5.8) tell us that there is no valid test of the hypothesis of no treatment effects. In a sense this is purely academic, since if there are block $\times$ treatment interactions, treatments do have an effect, but the effects also depend on the blocks. To make progress, the experimenter needs to apply treatments to several plots within each block or be willing to make some further restrictive assumptions such as required for Tukey's test for nonadditivity. (Scheffé 1956) and Christensen (1996a, Sec. 10.4) also discuss Tukey's test in some detail.

Example 5.1 (continued): Effect of Elevated CO_2—Variance Components and Standard Errors. In the rice/CO_2 example (Table 5.4), the variance component for blocks is estimated to be $\widehat{\sigma}_b^2 = (.01136 - .0001815)/2 = .0056$. Standard errors of treatment means and differences are estimated by substituting estimates of the variance components into the expression for the variance of the desired quantity. The treatment means, standard error of a mean, and the standard error of a difference of two means for the rice/CO_2 study are shown in Table 5.5.

In this experiment the blocks represent a sample of farmers growing rice in this area of Japan and a sample of fertility levels. We are not interested in comparing these particular farmers but in estimating the mean CO_2 effect regardless of who owns and farms the fields, and we would like our estimates to apply to other farmers beyond those in this sample. Therefore, blocks should be considered as random effects, and the correct standard errors are given in the column labeled "Blocks Random" in Table 5.5.

The column labeled "Blocks Fixed" is not appropriate for this example, but is shown for comparison. What we see is that the estimates of treatment means are

Table 5.5 Effect of Elevated CO_2 on Rice: Means and Standard Errors of Data in Table 5.2 under the Assumption That Block Effects Are Either Fixed or Random

	Blocks Fixed	Blocks Random
Ambient CO_2 mean	$\overline{y}_{\cdot 1} = 6.18$	$\overline{y}_{\cdot 1} = 6.18$
Elevated CO_2 mean	$\overline{y}_{\cdot 2} = 6.35$	$\overline{y}_{\cdot 2} = 6.35$
$s(\overline{y}_{\cdot j})$	$\sqrt{\dfrac{\text{MSE}}{b}} = .0067$	$\sqrt{\dfrac{\widehat{\sigma}_b^2 + \text{MSE}}{b}} = .038$
$s(\overline{y}_{\cdot 1} - \overline{y}_{\cdot 2})$	$\sqrt{\dfrac{2 \cdot \text{MSE}}{b}} = .0095$	$\sqrt{\dfrac{2 \cdot \text{MSE}}{b}} = .0095$

the same whether blocks are treated as fixed or random. Similarly, the standard error of the difference between two treatment means, $s(\overline{y}_{.1} - \overline{y}_{.2})$, is the same whether blocks are treated as fixed or random. This is the standard error that is usually of most interest in a comparative experiment. On the other hand, the standard error of a single treatment mean is much larger if blocks are random than if blocks are fixed. This is because the variance of the treatment mean is based on the variability both among farmers and among the fields owned by the same farmer. All of this variability has to be taken into account when we make confidence intervals for the treatment means if we want to capture the variability likely to be seen among farmers.

5.4.6 Comments on Multiple Units per Treatment per Block

The classical RCBD has only one experimental unit per block–treatment combination, and the corresponding model (5.1) assumes that there is no block $\times$ treatment interaction (i.e., that the effect of CO_2 is similar in all blocks). More information can be obtained about block $\times$ treatment interactions if the experiment includes more than one experimental unit per treatment per block. In the rice/CO_2 experiment (Example 5.1) this would be accomplished by having more than one rice paddy per treatment in each block, and model (5.3) would be appropriate. If there were multiple paddies for each treatment in each block, it would be possible to measure the variability in the response to CO_2 among blocks and compare this variability to the variability among rice paddies treated alike within the same block.

The design in which each treatment is replicated the same number of times in each block is a good design to use if it is reasonable to think that there may be interaction between block and treatment effects and it is important to be able to estimate the interaction separately from the residual within-block variance. However, in practice it is usually not advantageous to use multiple units of each treatment within a block. The reason is that if blocking is being used to provide replication and the aim is to make inferences to the broader population from which the blocks are sampled, it is better to increase the number of blocks than to increase the number of units per treatment within a block. The blocks have been constructed so as to maximize homogeneity among units within a block and at the same time (by random sampling) to represent the entire range of variability in the population. The goal is to estimate the treatment effects under the entire range of conditions from which the blocks are sampled. To obtain precise estimates of these treatment effects and powerful tests of them, it is necessary to have a sufficiently large sample of blocks.

Exercise 5.3: How to Allocate Plots to Blocks. Suppose that you are planning an agricultural experiment in which 24 plots are available and the objective is to compare two treatments. The plots could be arranged in (1) four blocks each containing three plots of each of the two treatments, or (2) 12 blocks each containing one plot of each of the two treatments.

(a) Compute the variance of the difference in the two treatment means for each of the two design alternatives, assuming that the variance in a block of six plots is the same as the variance in a block of two plots and $\sigma_{bt}^2 = \sigma_\epsilon^2 = 1$.
(b) Comment on which design you recommend. □

5.4.7 Subsampling

Consider the following example. A food processor is trying to improve the quality of a product from an ultrahigh-temperature processing plant. It is known that there is some leeway in combinations of time and temperature required to produce adequate destruction of microorganisms. A study was designed to test four different time–temperature combinations. The procedure was to process the food under the conditions selected and then store for a specific period of time in a commercial warehouse before testing for quality of product. Since conditions vary considerably from warehouse to warehouse and throughout the year, it was decided to use warehouses as blocks. A batch of product was prepared and sublots of product processed under the required conditions and placed in a warehouse. Some time later the process was repeated and the product placed in another warehouse. This was repeated a total of six times over a period of several months. Fifty days after processing, samples of the food product were retrieved and evaluated. For each warehouse and each process condition, five evaluations scores were obtained, for a total of 120 observations.

This example is interesting because there are 24 experimental units (six warehouses times four treatments) in the experiment. What looks like 120 observations are really subsamples or repeated observations on the same sublots of food product. Only 24 distinct lots of food product were processed. The experiment has four treatments in six blocks. We warn the reader that it is not unusual to see data from this sort of experiment analyzed incorrectly as 20 experimental units in each block with each treatment repeated five times.

The appropriate formal statistical model for this situation is a modification of model (5.3). In general, it can be written as

$$y_{ijk} = \mu + \beta_i + \tau_j + (\beta \times \tau)_{ij} + \epsilon_{ij} + \eta_{k(ij)},$$

where the $\{\eta_{k(ij)}\}$ represent additional error terms associated with individual sublots. They are assumed to be independently, identically distributed with mean zero and variance σ_η^2. We assume here that $k = 1, \ldots, m$ with $m \geq 1$.

Under this model, the expected value of MSE [see equation (5.4)] is σ_η^2. The expected value of MSBT [see equation (5.5)] is

$$\frac{m}{(b-1)(t-1)} \sum_{i,j} \mathrm{E}\Big[\Big((\beta \times \tau)_{ij} - \overline{(\beta \times \tau)}_{i.} - \overline{(\beta \times \tau)}_{.j} + \overline{(\beta \times \tau)}_{..}\Big)^2\Big]$$
$$+ m\sigma_\epsilon^2 + \sigma_\eta^2.$$

We notice immediately that there is no valid F-test for the null hypothesis of no interaction between blocks and treatments.

Similar manipulations under various assumptions always leave us with $m\sigma_\epsilon^2 + \sigma_\eta^2$, where we had σ_ϵ^2 alone before. The MSE with $bt(m-1)$ degrees of freedom is really of limited value unless one can assume that $\sigma_\epsilon^2 = 0$. It is very unlikely that such an assumption would be warranted in practice.

Example 5.1 (continued): Effect of Elevated CO_2—Analysis of Subsamples. Here we reanalyze the rice/CO_2 data of Table 5.2 using the subsamples instead of the ring means. The CO_2 was applied to whole rings, where there were two rings, i.e., two treatments, in each of four blocks. Each ring was 12 m in diameter and rice yield was measured on three different subplots within each ring. The subplots constitute subsamples in this experiment. The analysis of variance of the subplot log(yield) is presented in Table 5.6 under the assumption that there is no block $\times$ CO_2 interaction. Compare this with the analysis of ring means in Table 5.4.

The sums of squares are the same as in Table 5.4 except for a multiple of three, because sums of squares are computed on a per-measurement basis. The symbol ϕ_t stands for $\sum_j \tau_j^2/(t-1)$. The F-ratios and degrees of freedom for block and CO_2 effects are the same in both tables. The additional piece of information that is gained from an analysis of the subsamples is a measure of the variability within plots. In this experiment, surprisingly, the variability within the exposure plots is larger than the variability among plots. The SAS® code for this example is available on the companion Web site.

Exercise 5.4: Rice Experiment—Analysis of Variance with Subsampling. Show by writing out the algebraic forms that the sums of squares in Table 5.6 are exactly three times those of Table 5.4. □

Exercise 5.5: Rice Experiment—Variance Components. Estimate the variance components for blocks, rings, and subplots within rings. Then estimate the variance of ln(yield). If any variance component estimate is negative, replace it with zero. Finally, express the variance components as proportions of the total variance and discuss which sources of variation contribute the most to the total. □

Table 5.6 Analysis of Variance of Subsamples of Table 5.2

Source	df	SS	MS	E[MS]	F
Block	3	.1023	.0341	$\sigma_\eta^2 + 3\sigma_\epsilon^2 + 6\sigma_b^2$	62.6
CO_2	1	.1804	.1804	$\sigma_\eta^2 + 3\sigma_\epsilon^2 + 12\phi_t$	331.6
Experimental error	3	.001632	.0005441	$\sigma_\eta^2 + 3\sigma_\epsilon^2$	
Subsampling error	16	.05371	.003357	σ_η^2	

5.5 STATISTICAL ANALYSIS WHEN DATA ARE MISSING IN AN RCBD

5.5.1 Blocks Fixed

If blocks are fixed and data are missing, the general linear hypothesis techniques introduced in Section 4.2 are appropriate. As a simple example of the type of analyses required, we consider the case of three blocks and three treatments, missing treatment 1 in block 1. The design matrix and parameter vector, using effects coding, are

$$\boldsymbol{y} = \begin{bmatrix} y_{12} \\ y_{13} \\ y_{21} \\ y_{22} \\ y_{23} \\ y_{31} \\ y_{32} \\ y_{33} \end{bmatrix}, \quad \boldsymbol{X} = \begin{bmatrix} 1 & 1 & 0 & 0 & 1 \\ 1 & 1 & 0 & -1 & -1 \\ 1 & 0 & 1 & 1 & 0 \\ 1 & 0 & 1 & 0 & 1 \\ 1 & 0 & 1 & -1 & -1 \\ 1 & -1 & -1 & 1 & 0 \\ 1 & -1 & -1 & 0 & 1 \\ 1 & -1 & -1 & -1 & -1 \end{bmatrix}, \quad \boldsymbol{\beta} = \begin{bmatrix} \mu \\ \beta_1 \\ \beta_2 \\ \tau_1 \\ \tau_2 \end{bmatrix}.$$

The matrix $\boldsymbol{L}$ for testing $\begin{bmatrix} \tau_1 \\ \tau_2 \end{bmatrix} = \boldsymbol{0}$ is $\boldsymbol{L} = \begin{bmatrix} 0 & 0 & 0 & 1 & 0 \\ 0 & 0 & 0 & 0 & 1 \end{bmatrix}$. The sum of squares for treatment effects is then

$$\mathrm{SS}(L) = (\boldsymbol{L}\widehat{\boldsymbol{\beta}})^t(\boldsymbol{L}(\boldsymbol{X}^t\boldsymbol{X})^{-1}\boldsymbol{L}^t)^{-1}\boldsymbol{L}\widehat{\boldsymbol{\beta}}.$$

Another way of looking at sums of squares in the analysis of variance is as *extra sums of squares* (Section 4.2). The two types of sums of squares most commonly computed are called *sequential* and *partial* sums of squares. Partial sums of squares are the extra sums of squares for adding a factor given that all other effects are in the model. Usually, in analysis of variance when data are missing, partial sums of squares should be used. In the model $y_{ij} = \mu + \beta_i + \tau_j + \epsilon_{ij}$ the partial sum of squares for blocks is $\mathrm{R}(\beta_1, \ldots, \beta_b|\mu, \tau_1, \ldots, \tau_t)$ and for treatments is $\mathrm{R}(\tau_1, \ldots, \tau_t|\mu, \beta_1, \ldots, \beta_b)$. The partial sum of squares for treatment effects tests the hypothesis that all treatment means are equal; i.e., that $\sum_i \mu_{i1}/b = \cdots = \sum_i \mu_{it}/b$ for $j = 1, \ldots, t$. The partial sums of squares can be found either by subtracting error sums of squares from full and reduced models, as the extra sum of squares

$$\begin{aligned} &\mathrm{SSE}(\mu, \beta_1, \ldots, \beta_b) - \mathrm{SSE}(\mu, \beta_1, \ldots, \beta_b, \tau_1, \ldots, \tau_t), \\ &= \mathrm{R}(\mu, \beta_1, \ldots, \beta_b, \tau_1, \ldots, \tau_t) - \mathrm{R}(\mu, \beta_1, \ldots, \beta_b), \end{aligned}$$

or by computing the quadratic form for the general linear hypothesis test. For linear models these two methods of computing sums of squares coincide as long as care is taken to parameterize the model so that the same hypothesis is tested in each case.

Sequential sums of squares (called *Type I sums of squares* in SAS® `proc GLM`) depend on the order that terms are entered into the model. They are the extra sums of squares for adding a factor given that *preceding* effects are already in the model. These are most useful for polynomial regression models. In the model $y_{ij} = \mu + \beta_i + \tau_j + \epsilon_{ij}$ the sequential sum of squares for blocks is $\mathrm{R}(\beta_1, \ldots, \beta_b|\mu)$ and the sequential sum of squares for treatments is $\mathrm{R}(\tau_1, \ldots, \tau_t|\mu, \beta_1, \ldots, \beta_b)$. If the model had been written with the treatment term before the block term, as $y_{ij} = \mu + \tau_j + \beta_i + \epsilon_{ij}$, the sequential sum of squares for treatment effects would be $\mathrm{R}(\tau_1, \ldots, \tau_b|\mu)$. Sequential sums of squares are often useful for hand computations and for deriving the distributions of statistics, but when data are missing in analysis of variance, they do not usually test the hypotheses of interest.

Missing Value Formulas—A Neat Trick

A neat strategy to use when block–treatment combinations have one or at most a few values missing is to replace missing observations with some value, that is, to *impute* values and then carry out the analysis of variance on the new "complete" data set, including the imputed value. This method uses the computing formulas for balanced data, so can require much less computing power than doing a full least squares analysis on unbalanced data. In the past, this was the most commonly used method of handling missing data, and it is still interesting to study the idea of imputation.

The criterion for imputing is to insert a value that will contribute nothing to the error sum of squares. If y_{ij} is missing and we insert the least squares value $\widehat{y}_{ij}$, the error sum of squares is unchanged. It is possible to find $\widehat{y}_{ij}$ without extensive arithmetic.

Without loss of generality we can proceed as though treatment 1 in block 1 is missing. Our aim is to substitute a value z for y_{11} so that we can use the standard computing formulas. We want to find the value of z such that

$$z = \widehat{y}_{11} = \frac{y_{1\cdot} + z}{t} + \frac{y_{\cdot 1} + z}{b} - \frac{y_{\cdot\cdot} + z}{bt}. \tag{5.12}$$

Solving for z leads to the equation

$$z = \frac{by_{1\cdot} + ty_{\cdot 1} - y_{\cdot\cdot}}{(b-1)(t-1)}. \tag{5.13}$$

Using this estimated value, we can compute treatment contrasts such as

$$\widehat{\tau}_1 - \widehat{\tau}_2 = \frac{y_{\cdot 1} + z}{b} - \overline{y}_{\cdot 2}.$$

This technique extends directly to more than one missing value. If, say, N values are missing, N equations like (5.12) must be set up and solved simultaneously. Imputing these values into the data set leads to a complete set and the usual simple numerical calculations. Error degrees of freedom are reduced for each value imputed.

5.5.2 Blocks Random

Developing the Model

The RCBD model with random block effects is a special case of a *mixed model*, which is the term used for models that contain both fixed and random effects. For purposes of this discussion we use matrix notation. We confine ourselves to the model used in equations (5.9), (5.10), and (5.11) but assume, $m = 1$, no interactions and initially no missing data. We order the data so that $\boldsymbol{y} = \begin{bmatrix} y_{11} & y_{12} & \cdots & y_{1t} & y_{21} & \cdots & y_{bt} \end{bmatrix}^t$ represents the data. The model is then written as

$$\boldsymbol{y} = \mathbf{1}\mu + \boldsymbol{Z\beta} + \boldsymbol{X\tau} + \boldsymbol{\epsilon}, \tag{5.14}$$

where $\boldsymbol{Z} = \boldsymbol{I}_b \otimes \mathbf{1}_t$, $\boldsymbol{X} = \mathbf{1}_b \otimes \boldsymbol{I}_t$, $\boldsymbol{\beta}$ is the vector of block effects, $\boldsymbol{\tau}$ is the vector of treatment effects, $\boldsymbol{\epsilon}$ is the vector of random errors, and where $\boldsymbol{J}_n$, $\boldsymbol{I}_n$, and $\mathbf{1}_n$ are the $n \times n$ matrix with all elements equal to 1, the $n \times n$ identity matrix, and the column of n ones, respectively. We drop the subscript when the size is clear from the context. The variance–covariance matrix of $\boldsymbol{y}$ is block diagonal and can be written as $\text{Var}[\boldsymbol{y}] = \boldsymbol{I}_b \otimes \boldsymbol{V}$, where $\otimes$ denotes the Kronecker product and $\boldsymbol{V}$ is a $t \times t$ matrix that takes the form $\sigma_b^2 \boldsymbol{J}_t + \sigma_\epsilon^2 \boldsymbol{I}_t$.

Notice that $\boldsymbol{ZZ}^t = \boldsymbol{I}_b \otimes \boldsymbol{J}_t$ and that one can write

$$\text{Var}[\boldsymbol{y}] = \sigma_b^2 \boldsymbol{ZZ}^t + \sigma_\epsilon^2 \boldsymbol{I}_b \otimes \boldsymbol{I}_t. \tag{5.15}$$

The latter form is convenient because it illustrates the modifications required to accommodate the missing observations case. The modification is to delete rows from model (5.14). We use (5.15) as the definition of the variance–covariance matrix of $\boldsymbol{y}$.

Estimation for Mixed Models

The type of estimation presented up to now is called *ordinary least squares* (OLS). When working with correlated data, more efficient parameter estimates may be obtained by incorporating the correlation pattern into the model. The concept of minimizing the error sum of squares can be generalized to handle correlated data. In ordinary least squares the objective is to find parameter values that minimize the error sum of squares

$$(\boldsymbol{y} - \boldsymbol{X\tau})^t(\boldsymbol{y} - \boldsymbol{X\tau}).$$

Under the usual assumptions of homogeneous variances and uncorrelated observations, this method provides the most efficient estimators possible. Under these assumptions, the OLS estimators are the *best linear unbiased estimators*, meaning that they are the ones with the smallest possible standard errors that are unbiased and are linear combinations of the data $\boldsymbol{y}$. When the data are correlated, *generalized least squares* (GLS) provides the best linear unbiased estimator. GLS, as the name implies, is a generalization of ordinary least squares.

Generalized least squares minimizes a weighted sum of squares where the weight matrices involve the correlations. Specifically, the GLS estimator minimizes the quadratic form

$$(\boldsymbol{y} - \boldsymbol{X\tau})^t \boldsymbol{V}^{-1} (\boldsymbol{y} - \boldsymbol{X\tau}). \tag{5.16}$$

In this equation observations are differentially weighted according to which other observations also provide information. The idea is that if two observations are highly correlated with each other (because they are subsamples of the same experimental unit, say), they do not contain as much information as two independent observations would. The weighting matrix in the GLS estimating equations is the inverse of the covariance matrix. If the variances are homogeneous and observations are in fact uncorrelated, so that $\boldsymbol{V} = \sigma_\epsilon^2 \boldsymbol{I}$, the GLS objective function (5.16) reduces to the ordinary error sum of squares $(\boldsymbol{y} - \boldsymbol{X\tau})^t (\boldsymbol{y} - \boldsymbol{X\tau})$.

Notice that there is a minor change in notation. The $\boldsymbol{X}$ in (5.16) is really $\boldsymbol{1} \| \boldsymbol{X}$ in (5.14). However, we can simply ignore this fact since we know that the model is not full rank and that at least one restriction will be required on the fixed-effects part of the model. We can conveniently use $\mu = 0$ as one of the restrictions and simply eliminate it from the model.

The GLS objective function (5.16) is minimized by taking partial derivatives with respect to $\boldsymbol{\tau}$ and setting them equal to zero. This process yields the normal equations $\boldsymbol{X}^t \boldsymbol{V}^{-1} \boldsymbol{X} \widehat{\boldsymbol{\tau}} = \boldsymbol{X}^t \boldsymbol{V}^{-1} \boldsymbol{y}$. The GLS estimator is the solution to these normal equations,

$$\widehat{\boldsymbol{\tau}}_{\text{GLS}} = (\boldsymbol{X}^t \boldsymbol{V}^{-1} \boldsymbol{X})^{-1} \boldsymbol{X}^t \boldsymbol{V}^{-1} \boldsymbol{y}, \tag{5.17}$$

provided that $\boldsymbol{X}$ is full-column rank. The standard errors of $\widehat{\boldsymbol{\tau}}_{\text{GLS}}$ are obtained from the diagonal elements of $\text{Var}[\widehat{\boldsymbol{\tau}}_{\text{GLS}}]$, where

$$\begin{aligned}\text{Var}[\widehat{\boldsymbol{\tau}}_{\text{GLS}}] &= (\boldsymbol{X}^t \boldsymbol{V}^{-1} \boldsymbol{X})^{-1} \boldsymbol{X}^t \boldsymbol{V}^{-1} \ \text{Var}[\boldsymbol{y}] \boldsymbol{V}^{-1} \boldsymbol{X} (\boldsymbol{X}^t \boldsymbol{V}^{-1} \boldsymbol{X})^{-1} \\ &= (\boldsymbol{X}^t \boldsymbol{V}^{-1} \boldsymbol{X})^{-1}.\end{aligned}$$

In practice, the covariance matrix $\boldsymbol{V}$ is unknown and must be estimated. If an estimated covariance matrix $\widehat{\boldsymbol{V}}$ is substituted for $\boldsymbol{V}$ in equation (5.17), the resulting estimator is called the *estimated generalized least squares* (EGLS) estimator. Estimated generalized least squares will not necessarily be more efficient than ordinary least squares, because of the variability introduced by estimated $\boldsymbol{V}$. For small samples ordinary least squares may be preferable, provided that the standard errors of treatment effects are computed correctly. For large samples, estimated generalized least squares is more efficient and more powerful than ordinary least squares estimation. There are several algorithms for computing EGLS estimates. SAS® `Proc Mixed` provides maximum likelihood (ML), restricted maximum likelihood (REML), and minimum norm quadratic estimation (MINQUE), with REML as the default. Maximum likelihood and REML are iterative algorithms; they begin with some starting values for the variance components and update the estimates at each step until no further improvement can be obtained. For the designs in this book, if no data are missing, the algorithms should converge in one or two steps. Maximum likelihood estimates of variance components are slightly biased; in some simple cases it is known that REML adjusts for this bias. Minimum norm quadratic estimation is not iterative, and it can be used as the first step in REML estimation. MINQUE provides unbiased estimates of the variance components and the algorithm is described in detail below.

Exercise 5.6: Standard Errors When Errors Are Not Independent and Identically Distributed. If Var[$\boldsymbol{y}$] is not $\sigma_\epsilon^2\boldsymbol{I}$, what is the variance–covariance matrix of the ordinary least squares estimator $\widehat{\boldsymbol{\tau}}_{\mathrm{OLS}}$? What is printed out for standard errors of the parameter estimates in ordinary least squares software such as SAS® `proc GLM`? *Note*: the design matrix for proc GLM contains both x and z. □

Estimated Generalized Least Squares

We develop minimum norm quadratic estimation (MINQUE) in this subsection. It is not really necessary that $\boldsymbol{y}$ be full column rank, so we present the more general form. It is convenient to redefine $\boldsymbol{V}_r = (\sigma_b^2/\sigma_\epsilon^2)\boldsymbol{Z}\boldsymbol{Z}^t + \boldsymbol{I}$. If σ_b^2 and σ_ϵ^2 or at least $\sigma_b^2/\sigma_\epsilon^2$ are known, the best linear unbiased estimate of τ, the generalized least squares estimate, is available as

$$\widehat{\tau} = (\boldsymbol{X}^t\boldsymbol{V}_r^{-1}\boldsymbol{X})^-\boldsymbol{X}^t\boldsymbol{V}_r^{-1}\boldsymbol{y}.$$

Also,

$$\mathrm{Var}[\widehat{\tau}] = \sigma_\epsilon^2(\boldsymbol{X}^t\boldsymbol{V}_r^{-1}\boldsymbol{X})^-.$$

If only the ratio $\sigma_b^2/\sigma_\epsilon^2$ is known, the estimate of σ_ϵ^2 is obtained by

$$\boldsymbol{y}^t\boldsymbol{V}_r^{-1}\big(\boldsymbol{V}_r - \boldsymbol{X}(\boldsymbol{X}^t\boldsymbol{V}_r^{-1}\boldsymbol{X})^-\boldsymbol{X}^t\big)\boldsymbol{V}_r^{-1}\boldsymbol{y}\big/(n - \mathrm{rank}(X)),$$

where n is the number of observations.

If neither σ_b^2 and σ_ϵ^2 or even $\sigma_b^2/\sigma_\epsilon^2$ are known, the first step is to estimate σ_b^2 and σ_ϵ^2. This is followed by estimated generalized least squares. To estimate σ_b^2 and σ_ϵ^2, by MINQUE define

$$\boldsymbol{V}_0 = \boldsymbol{Z}\boldsymbol{Z}^t + \boldsymbol{I}$$

$$\boldsymbol{Q}_0 = \boldsymbol{V}_0^{-1} - \boldsymbol{V}_0^{-1}\boldsymbol{X}(\boldsymbol{X}^t\boldsymbol{V}_0^{-1}\boldsymbol{X})^-\boldsymbol{X}^t\boldsymbol{V}_0^{-1}$$

and then solve the two-equation system

$$\begin{bmatrix} \mathrm{tr}(\boldsymbol{Q}_0\boldsymbol{Q}_0) & \mathrm{tr}(\boldsymbol{Q}_0\boldsymbol{Z}\boldsymbol{Z}^t\boldsymbol{Q}_0) \\ \mathrm{tr}(\boldsymbol{Q}_0\boldsymbol{Z}\boldsymbol{Z}^t\boldsymbol{Q}_0) & \mathrm{tr}(\boldsymbol{Z}^t\boldsymbol{Q}_0\boldsymbol{Z}\boldsymbol{Z}^t\boldsymbol{Q}_0\boldsymbol{Z}) \end{bmatrix} \begin{bmatrix} \widehat{\sigma_\epsilon^2} \\ \widehat{\sigma_b^2} \end{bmatrix} = \begin{bmatrix} \boldsymbol{y}^t\boldsymbol{Q}_0\boldsymbol{Q}_0\boldsymbol{y} \\ \boldsymbol{y}^t\boldsymbol{Q}_0\boldsymbol{Z}\boldsymbol{Z}^t\boldsymbol{Q}_0\boldsymbol{y} \end{bmatrix}$$

for $\widehat{\sigma}_b^2$ and $\widehat{\sigma}_\epsilon^2$. These estimates are both unbiased. If the $\{\beta_i\}$ and $\{\epsilon_{ij}\}$ have symmetrical distributions, an unbiased, estimated generalized least squares estimate of τ is

$$\widetilde{\tau} = (\boldsymbol{X}^t \widehat{\boldsymbol{V}}_r^{-1} \boldsymbol{X})^- \boldsymbol{X}^t \widehat{\boldsymbol{V}}_r^{-1} \boldsymbol{y},$$

where

$$\widehat{\boldsymbol{V}}_r = (\widehat{\sigma}_b^2 / \widehat{\sigma}_\epsilon^2) \boldsymbol{Z}\boldsymbol{Z}^t + \boldsymbol{I}.$$

An estimate of the variance–covariance matrix of $\widetilde{\tau}$ is $\widehat{\sigma}_\epsilon^2 (\boldsymbol{X}^t \widehat{\boldsymbol{V}}_r^{-1} \boldsymbol{X})^-$.

We can obtain closed form expressions for $\widehat{\tau}_1$ and $\widehat{\tau}_2$ by partitioning the fixed part of the model, i.e., $\boldsymbol{X}\boldsymbol{\tau} = \boldsymbol{X}_1\boldsymbol{\tau}_1 + \boldsymbol{X}_2\boldsymbol{\tau}_2$ and estimating $\boldsymbol{\tau}_2$ after $\boldsymbol{\tau}_1$. The object of this partitioning is to test hypotheses about $\widehat{\tau}_2$ after allowing for $\widehat{\tau}_1$. Set up the normal equations

$$\begin{bmatrix} \boldsymbol{X}_1^t \widehat{\boldsymbol{V}}_r^{-1} \boldsymbol{X}_1 & \boldsymbol{X}_1^t \widehat{\boldsymbol{V}}_r^{-1} \boldsymbol{X}_2 \\ \boldsymbol{X}_2^t \widehat{\boldsymbol{V}}_r^{-1} \boldsymbol{X}_1 & \boldsymbol{X}_2^t \widehat{\boldsymbol{V}}_r^{-1} \boldsymbol{X}_2 \end{bmatrix} \begin{bmatrix} \widetilde{\tau}_1 \\ \widetilde{\tau}_2 \end{bmatrix} = \begin{bmatrix} \boldsymbol{X}_1^t \widehat{\boldsymbol{V}}_r^{-1} \boldsymbol{y} \\ \boldsymbol{X}_2^t \widehat{\boldsymbol{V}}_r^{-1} \boldsymbol{y} \end{bmatrix}.$$

Solve first for $\widetilde{\tau}_1$.

$$(\boldsymbol{X}_1^t \widehat{\boldsymbol{V}}_r^{-1} \boldsymbol{X}_1)\widetilde{\tau}_1 = \boldsymbol{X}_1^t \widehat{\boldsymbol{V}}_r^{-1} \boldsymbol{y} - \boldsymbol{X}_1^t \widehat{\boldsymbol{V}}_r^{-1} \boldsymbol{X}_2 \widetilde{\tau}_2$$

and

$$\widetilde{\tau}_1 = (\boldsymbol{X}_1^t \widehat{\boldsymbol{V}}_r^{-1} \boldsymbol{X}_1)^- \boldsymbol{X}_1^t \widehat{\boldsymbol{V}}_r^{-1} \boldsymbol{y} - (\boldsymbol{X}_1^t \widehat{\boldsymbol{V}}_r^{-1} \boldsymbol{X}_1)^- \boldsymbol{X}_1^t \widehat{\boldsymbol{V}}_r^{-1} \boldsymbol{X}_2 \widetilde{\tau}_2. \tag{5.18}$$

Also obtain

$$(\boldsymbol{X}_2^t \widehat{\boldsymbol{V}}_r^{-1} \boldsymbol{X}_2)\widetilde{\tau}_2 = \boldsymbol{X}_2^t \widehat{\boldsymbol{V}}_r^{-1} \boldsymbol{y} - \boldsymbol{X}_2^t \widehat{\boldsymbol{V}}_r^{-1} \boldsymbol{X}_1 \widetilde{\tau}_1. \tag{5.19}$$

Substituting equation (5.18) into (5.19) gives

$$\begin{aligned}(\boldsymbol{X}_2^t \widehat{\boldsymbol{V}}_r^{-1} \boldsymbol{X}_2)\widetilde{\tau}_2 = \boldsymbol{X}_2^t \widehat{\boldsymbol{V}}_r^{-1} \boldsymbol{y} &- (\boldsymbol{X}_2^t \widehat{\boldsymbol{V}}_r^{-1} \boldsymbol{X}_1)(\boldsymbol{X}_1^t \widehat{\boldsymbol{V}}_r^{-1} \boldsymbol{X}_1)^- \boldsymbol{X}_1^t \widehat{\boldsymbol{V}}_r^{-1} \boldsymbol{y} \\ &+ (\boldsymbol{X}_2^t \widehat{\boldsymbol{V}}_r^{-1} \boldsymbol{X}_1)(\boldsymbol{X}_1^t \widehat{\boldsymbol{V}}_r^{-1} \boldsymbol{X}_1)^- \boldsymbol{X}_1^t \widehat{\boldsymbol{V}}_r^{-1} \boldsymbol{X}_2 \widetilde{\tau}_2,\end{aligned}$$

from which we obtain

$$\begin{aligned}\widetilde{\tau}_2 = &\left[\boldsymbol{X}_2^t \widehat{\boldsymbol{V}}_r^{-1} \boldsymbol{X}_2 - (\boldsymbol{X}_2^t \widehat{\boldsymbol{V}}_r^{-1} \boldsymbol{X}_1)(\boldsymbol{X}_1^t \widehat{\boldsymbol{V}}_r^{-1} \boldsymbol{X}_1)^- \boldsymbol{X}_1^t \widehat{\boldsymbol{V}}_r^{-1} \boldsymbol{X}_2\right]^- \\ &\left[\boldsymbol{X}_2^t \widehat{\boldsymbol{V}}_r^{-1} - (\boldsymbol{X}_2^t \widehat{\boldsymbol{V}}_r^{-1} \boldsymbol{X}_1)(\boldsymbol{X}_1^t \widehat{\boldsymbol{V}}_r^{-1} \boldsymbol{X}_1)^- \boldsymbol{X}_1^t \widehat{\boldsymbol{V}}_r^{-1}\right]\boldsymbol{y}\end{aligned}$$

and

$$\text{Var}\left[\widetilde{\tau}_2\right] = \sigma_\epsilon^2 \left[\boldsymbol{X}_2^t \widehat{\boldsymbol{V}}_r^{-1} \boldsymbol{X}_2 - (\boldsymbol{X}_2^t \widehat{\boldsymbol{V}}_r^{-1} \boldsymbol{X}_1)(\boldsymbol{X}_1^t \widehat{\boldsymbol{V}}_r^{-1} \boldsymbol{X}_1)^- \boldsymbol{X}_1^t \widehat{\boldsymbol{V}}_r^{-1} \boldsymbol{X}_2\right]^-.$$

Testing Hypotheses in Mixed Models

For a randomized complete block design with no missing data the F-tests of Section 5.4.5 provide exact tests and should be used. When some data are missing or the design is unbalanced in some other way, hypotheses about the fixed effects $\boldsymbol{\tau}$ can be tested using either a Wald or a likelihood ratio statistic. The likelihood ratio statistic for testing a fixed effect is twice the difference in the fitted log likelihood value between a full and a reduced model, where the two models are fitted by maximum likelihood. For experiments in which the number of levels of the random effects is large, the likelihood ratio statistic follows a χ-square distribution with degrees of freedom equal to the difference in the number of parameters in the full and reduced models. In the case of a randomized block design with random block effects, the determining factor in whether a sample can be considered large or not is the number of blocks. The likelihood ratio test can be quite liberal, i.e., have a much larger type I error rate than the nominal level, if the sample size is not large (Welham and Thompson 1997). Methods of adjusting the likelihood ratio test for small samples have been developed recently by Welham and Thompson (1997) and Zucker et al. (2000).

The Wald statistic is computed similar to the computation method for the general linear test statistic in Section 4.2.1:

$$\begin{aligned} W &= (\boldsymbol{L}\widetilde{\boldsymbol{\tau}} - \boldsymbol{m_0})^t (\widehat{\text{Var}}[\boldsymbol{L}\widetilde{\boldsymbol{\tau}}])^{-1} (\boldsymbol{L}\widetilde{\boldsymbol{\tau}} - \boldsymbol{m_0}) \\ &= (\boldsymbol{L}\widetilde{\boldsymbol{\tau}} - \boldsymbol{m_0})^t (\boldsymbol{L}(\boldsymbol{X}^t \widehat{\boldsymbol{V}}^{-1} \boldsymbol{X})^- \boldsymbol{L}^t)^{-1} (\boldsymbol{L}\widetilde{\boldsymbol{\tau}} - \boldsymbol{m_0}). \end{aligned}$$

As the number of blocks becomes large, this statistic tends in distribution to a χ-square with degrees of freedom equal to the number of linearly independent rows of $\boldsymbol{L}$. The exact distribution of this statistic is unknown. The χ-square approximation is appropriate if the number of blocks is large.

For small samples, the following F-approximation produces type I error rates that are closer to the nominal level:

$$\frac{W}{\text{rank}(\boldsymbol{L})} \mathrel{\dot\sim} F(\text{df}_1, \text{df}_2),$$

where

$$\begin{aligned} \text{df}_1 &= \text{rank}(L) \\ \text{df}_2 &= \text{degrees of freedom for estimating } \boldsymbol{L}(\boldsymbol{X}^t \boldsymbol{V}^{-1} \boldsymbol{X})^- \boldsymbol{L}^t. \end{aligned}$$

This approximation is obtained by analogy with the ordinary least squares case, in which

$$F = \frac{(\boldsymbol{L}\widehat{\boldsymbol{\tau}})^t (\boldsymbol{L}(\boldsymbol{X}^t \boldsymbol{X})^- \boldsymbol{L}^t)^{-1} (\boldsymbol{L}\widehat{\boldsymbol{\tau}})/\text{df}_1}{\text{MSE}}.$$

To see the correspondence, notice that if the $\boldsymbol{y}$ values are uncorrelated, $\widehat{\boldsymbol{V}} = \text{MSE}\,\boldsymbol{I}$, so

$$(\boldsymbol{L}(\boldsymbol{X}^t\widehat{\boldsymbol{V}}^{-1}\boldsymbol{X})^-\boldsymbol{L}^t)^{-1} = \frac{(\boldsymbol{L}(\boldsymbol{X}^t\boldsymbol{X})^-\boldsymbol{L}^t)^{-1}}{\text{MSE}}$$

and $F = W/\text{df}_1$.

The denominator degrees of freedom df_2 are found in various ad hoc ways. At this time there is no theoretically correct answer for the denominator degrees of freedom because the statistic does not exactly follow an F-distribution, even asymptotically. From simulation studies it has been found that the probability of a type I error is closer to α using the F-approximation than using the Wald statistic with a chi-square approximation. The simplest choice for denominator degrees of freedom is to determine what the appropriate error term would have been if no data were missing, and compute the degrees of freedom as if all terms in the model were fixed effects. Two more sophisticated methods of determining the denominator degrees of freedom, Satterthwaite's approximation (Giesbrecht and Burns 1985; McLean and Sanders 1988; Fai and Cornelius 1993) and a method proposed by Kenward and Roger (1997), produce type I error rates close to the nominal levels. Kenward and Roger's method also adjusts the standard errors of parameter estimates to reflect the fact that EGLS uses an estimated covariance matrix rather than a known covariance matrix.

To demonstrate the Wald-type tests, we analyze a small simulated data set (given in Table 5.7). The data have been simulated from model (5.14) with treatment means as given in Table 5.8 and $\sigma_b^2 = 1$, and $\sigma_\epsilon^2 = 1$. The data set is missing observations for treatment 1 in blocks 1 and 2. The REML estimated variance components are $\widehat{\sigma}_b^2 = .89$ and $\widehat{\sigma}_\epsilon^2 = 1.32$, and the estimated covariance matrix of $\widetilde{\boldsymbol{\mu}} = [\widehat{\mu}_{\cdot 1} \cdots \widehat{\mu}_{\cdot 6}]^t$ is

$$\widehat{\text{Var}}[\widetilde{\boldsymbol{\mu}}] = \begin{bmatrix} .96 & .22 & .22 & .22 & .22 & .22 \\ .22 & .55 & .22 & .22 & .22 & .22 \\ .22 & .22 & .55 & .22 & .22 & .22 \\ .22 & .22 & .22 & .55 & .22 & .22 \\ .22 & .22 & .22 & .22 & .55 & .22 \\ .22 & .22 & .22 & .22 & .22 & .55 \end{bmatrix}.$$

Table 5.7 Simulated Data for RCBD with Four Blocks, Six Treatments, and Two Missing Observations

Block	Trt. 1	Trt. 2	Trt. 3	Trt. 4	Trt. 5	Trt. 6
1	—	2.66	2.80	3.43	5.25	6.47
2	—	6.41	7.04	6.09	5.69	8.43
3	4.93	6.07	6.19	3.26	4.99	5.95
4	3.13	4.48	5.22	6.71	6.34	7.35

Table 5.8 Simulated Data for an RCBD with Six Treatments in Four Blocks, One Observation per Cell, and Two Missing Observations[a]

Treatment	$\mu_{\cdot j}$	$\widetilde{\mu}_{\cdot j}$	$s(\widehat{\mu}_{\cdot j})$
1	4.5	3.9	.98
2	4.7	4.9	.74
3	4.9	5.3	.74
4	5.1	4.9	.74
5	5.3	5.6	.74
6	5.5	7.1	.74

[a]True and estimated treatment means and standard errors.

We demonstrate testing the hypothesis that there is no significant linear trend. The null hypothesis is $H_0 : \boldsymbol{L\mu} = 0$, where $\boldsymbol{L}$ contains the orthogonal polynomial coefficients $\begin{bmatrix} -5 & -3 & -1 & 1 & 3 & 5 \end{bmatrix}$ against the alternative that there is a nonzero linear trend. The value of the Wald statistic is

$$W = \frac{(\boldsymbol{L}\widetilde{\boldsymbol{\mu}})^2}{\widehat{\text{Var}}[\boldsymbol{L}\widetilde{\boldsymbol{\mu}}]} = \left(\frac{17.09}{5.76}\right)^2 = 8.80.$$

The Wald statistic is compared to percentiles of the chi-square distribution with one degree of freedom, yielding a p-value of .0027. Since the hypothesis has only one component ($\boldsymbol{L}$ has one row), the F-statistic is the same as W. However, in the F-test the statistic is compared to quantiles of the $F(1, 13)$ distribution, which gives a p-value of .01. The simple method of computing denominator degrees of freedom gives $\text{df}_2 = 13$, which is the residual degrees of freedom for an RCBD minus two missing values, $(b-1)(t-1) - 2$. For this simulated data set it happens that Satterthwaite's approximation, computed in `proc MIXED` in SAS®, gives very similar denominator degrees of freedom, $\text{df}_2 = 13.2$. In general, the denominator degrees of freedom obtained by the two methods will not necessarily be close to each other. Notice that the Wald test using the chi-square distribution is much more liberal than the F-test. This is because the chi-square approximation assumes that the sample size is infinitely large, which is equivalent to having infinitely large denominator degrees of freedom in an F-test.

5.6 ANALYSIS OF COVARIANCE

The aim of analysis of covariance is the same as that of blocking: to reduce unexplained variability in the responses. Blocking does this by physically placing experimental units into homogeneous sets. Analysis of covariance attempts to reduce the error variance mathematically by including more explanatory variables in the model. This approach works well if there are some auxiliary variables that are useful predictors of the response associated with each experimental unit. We refer to such variables as *covariables* or as *covariates*. Some examples in different types of experiments are: (1) in field experiments, the previous year's crop yield; (2) in animal experiments, the initial body weight or age; (3) in animal experiments, an anatomical condition that is not affected by the treatment but can be determined only by sacrificing the animal; (4) in human experiments, performance on a pretest; and (5) in laboratory experiments, the ambient temperature, humidity, or atmospheric pressure.

The analysis of covariance allows the experimenter to incorporate auxiliary information into the analysis of the experiment. It uses simple linear regression along with the analysis of variance to adjust for differences among experimental units. For a CRD with t treatments and n_i observations on treatment i, the model, a generalization of model (3.7), is written as

$$y_{ij} = \mu + \tau_i + \beta x_{ij} + \epsilon_{ij} \tag{5.20}$$

where y_{ij} and x_{ij} denote the observations and covariates, respectively, for $j = 1, \ldots, n_i$ and $i = 1, \ldots, t$, and $\epsilon_{ij} \sim N \text{ iid}(0, \sigma_\epsilon^2)$.

From the perspective of planning experiments, the most important feature of analysis of covariance is that the covariate must not be affected by the treatments. The usual setting is that the concomitant variable is available prior to the experiment. Strictly speaking, this need not always be the case, but the assumption that the treatment does not affect the covariate is required. The alternative setting, where the covariate is another random variable that may be influenced by the treatments, falls into the general class of multivariate analysis. This is beyond the scope of this book. The crucial point at the planning stage of an experiment is to consider possible covariates and to be sure to measure, *before* the treatments are applied, any covariates that might be affected by the treatments.

The researcher may have the choice of using the covariable to construct blocks or of running a completely randomized design using analysis of covariance. Covariables are obvious candidates for constructing blocks. If it happens that for some reason or other it is not possible to use the covariate to construct suitably homogeneous blocks, analysis of covariance is indicated. This would be the case, for example, if the covariate, although not a function of the treatments, only becomes available some time after the treatments have been assigned. If the potential covariate values are nearly constant for groups of experimental units, one should use the covariate as a basis for blocking. If such a simple blocking is not obvious, the general recommendation (Cox 1957) is as follows: If the expected correlation between x and y is less than .3, ignore the covariate; if the correlation is expected to lie between .3 and .6, use the covariate to construct blocks as best you can; and if the correlation is expected to be larger than .6, use analysis of covariance. The reduction in variance of estimates of treatment differences (Cochran 1957) obtained by using the covariance analysis is of the order $(1 - \rho^2)[1 + 1/(\text{df error} - 2)]$.

5.6.1 Covariance Model and Adjusted Treatment Means

In general, the analysis of covariance can be adapted to any design and can be modified to take several covariates into account. The parameters are estimated by ordinary least squares as in Chapter 3, and standard linear models software such as SAS® `Proc GLM` can be used to fit the model and obtain estimates of the treatment effects and standard errors and tests of hypotheses.

In analysis of covariance the means of the covariates vary among the treatments. The unadjusted treatment means of the responses, $\overline{y}_{i\cdot} = \widehat{\mu} + \widehat{\tau}_i + \widehat{\beta}\overline{x}_{i\cdot}$, incorporate these covariate means. The idea of analysis of covariance is that more accurate comparisons among treatments may be made if the treatment means are adjusted to a common value of the covariate. The *adjusted treatment mean* is the fitted value of the response evaluated at the overall mean value of the covariate, $\widehat{\mu} + \widehat{\tau}_i + \widehat{\beta}\overline{x}_{\cdot\cdot}$, rather than at the treatment mean of the covariate. A small numerical example showing unadjusted and adjusted treatment means is given in Table 5.9. The standard error of a treatment mean is reduced from 6.31 in the analysis of variance to about 4.4 in the analysis of covariance. Verification of these results is left to the reader.

Exercise 5.7: Small Numerical Example of Analysis of Covariance. Plot y vs. x of Table 5.9 using a separate color for each treatment. Fit model (5.20) to the data using your favorite linear models software. Verify that if analysis of variance had been done, completely ignoring the covariate, the standard error of a treatment mean would be 6.31. Verify the adjusted treatment means and standard errors in Table 5.9. □

To see how analysis of covariance reduces the bias and variance of estimated treatment effects, it is convenient to rewrite the model as

$$y_{ij} = \mu + \beta\overline{x}_{\cdot\cdot} + \tau_i + \beta(\overline{x}_{i\cdot} - \overline{x}_{\cdot\cdot}) + \beta(x_{ij} - \overline{x}_{i\cdot}) + \epsilon_{ij}. \tag{5.21}$$

Table 5.9 Data and Treatment Means for Analysis of Covariance Example

Trt.	Y	X	Trt.	Y	X
1	101	9	3	114	1
	79	0		122	6
	95	6		109	2
2	105	9	4	92	3
	72	4		103	7
	91	6		96	10

Trt.	$\overline{y}_{i\cdot}$	$\overline{x}_{i\cdot}$	Adj. Trt. Mean	Std. Error
1	91.67	5.00	92.29	4.26
2	89.33	6.33	86.64	4.34
3	115.00	3.00	120.58	4.59
4	97.00	6.67	93.48	4.39

This keeps the relationship with model (3.7) clear. Next we write $\mu^* = \mu + \beta\overline{x}_{\cdot\cdot}$ and $\tau_i^* = \tau_i + \beta(\overline{x}_{i\cdot} - \overline{x}_{\cdot\cdot})$ and the transformed model as

$$y_{ij} = \mu^* + \tau_i^* + \beta(x_{ij} - \overline{x}_{i\cdot}) + \epsilon_{ij}. \tag{5.22}$$

Notice that the $\beta(\overline{x}_{i\cdot} - \overline{x}_{\cdot\cdot})$ piece contributes to the apparent treatment effect (bias), and the $\beta(x_{ij} - \overline{x}_{i\cdot})$ piece appears as part of the error term and contributes to variance (precision). The point to the analysis of covariance is to improve both the accuracy and the precision of contrasts by removing both of these. The reason for the transformation from model (5.21) to (5.22) is to emphasize the relationship to the analysis of variance for the CRD and the RCBD.

The derivation of the sums of squares and algebraic formulas for the analysis of covariance may be found in Hinkelmann and Kempthorne (1994), and Steel and Torrie (1980) and Christensen (1996a) demonstrate the hand computations for the analysis of covariance in the CRD and RCBD. An old but very good discussion of the analysis of covariance can be found in the special covariance issue of *Biometrics* in 1957.

Finally, a warning: The analysis of covariance is not very robust against failures of the analysis of variance (normality) assumptions. One must have complete faith in the model. Alternatively, the analysis of variance is quite robust. It can be shown that the analysis of variance tends to give the correct answers even when most of the assumptions fail. This cannot be demonstrated for the analysis of covariance (Cox 1957).

REVIEW EXERCISES

5.1 Hald (1952), in his book *Statistical Theory with Engineering Applications*, discusses an example of the study of permeability (to air?) of sheets of building material determined on three different machines on nine different days. He reports the data in units of seconds, implying that the work has been done using a standard volume. The object of the study was to test for differences among machines. Since the day-to-day environmental differences affect the tests, tests were conducted on nine days. Days are blocks. On each day they tested three randomly selected sheets of building material, one on each machine. (They actually cut each sheet into eight pieces, tested all eight on one machine on that day, and averaged the results.) The data are given in the accompanying table. Hald took logarithms of the data before performing the analysis of variance because he had reason to believe that logarithms of the data were more nearly normally distributed than the original observations. Compute the Friedman–Kendall test of the null hypothesis of no differences among machines. Compare with the conclusion you would arrive at if you used the analysis of variance.

Day	Machine 1	Machine 2	Machine 3
1	25.35	20.23	85.50
2	28.00	17.41	26.67
3	82.04	32.06	24.10
4	77.09	47.10	52.60
5	59.16	16.87	20.89
6	46.24	25.35	42.95
7	82.79	16.94	21.28
8	70.00	38.28	48.86
9	34.67	43.25	50.47

5.2 Test the hypothesis that there is no difference between the means of elevated and ambient plots in the rice experiment of Example 5.1. Do either a two-sided t-test or an F-test. State the null and alternative hypotheses and write out the test statistic, the rejection criterion, and your conclusion. Do the t- and F-tests coincide? Prove your answer.

5.3 In the rice experiment described in Example 5.1, assume that $\sigma^2 = 0.0002$. The researchers are interested in testing that elevated CO_2 has no effect on yield. Using a 5% significance level, what is the smallest difference in treatment means that could be detected with 80% power?

5.4 The experiment described in Example 5.1 is to be repeated next year. Give your recommendation for the number of blocks that should be used in the coming year. Write out your assumptions and show your calculations.

5.5 An engineer compared the strength of six different adhesives. In the experiment three technicians each tested all six adhesives. The analysis of variance table follows.

Source	df	SS
Technician	2	29.99
Adhesive	5	3.98
Error	10	24.46

(a) Estimate the proportion of the total variance attributable to differences among technicians.

(b) If you were to run another experiment comparing the strength of adhesives, would you recommend doing an RCBD or a completely randomized design (CRD)? Draw the layout of each design and discuss how many technicians would be required in each case.

5.6 For each of the following scenarios, write out an appropriate statistical model and the usual assumptions. Define all terms and the ranges of subscripts.

(a) Six farms were selected at random in eastern North Carolina for the purpose of comparing effects of four fertilizers on copper concentration in the soil. The four fertilizers were assigned randomly to four plots on each farm. Six soil samples were taken from each plot and analyzed for copper concentration.

(b) Indoor air pollution was measured in 36 office buildings. Three factors were of interest: age of building (pre-1950, 1950–1975, 1976–present); height of building (one, two, or three floors); and type of windows (fixed, operable). Two buildings were selected in each category, and measurements were taken in two rooms on the first floor of each building.

(c) If a computer has a math coprocessor, programs can be written to either use it or not to use it. In this experiment 10 computers were used. Each computer was equipped with a math coprocessor and had two Fortran compilers. On each computer the programmer compiled a test program using each of the two compilers, both with and without the math coprocessor.

5.7 *Experiment.* You are to conduct an experiment to test the effect of butter vs. margarine on the diameter of cookies. In the preliminary and design stages of this process you will have to do several things, including (1) discuss what extraneous factors might affect the outcome; (2) think about possible blocking factors; (3) consider cost, time, feasibility; (4) decide on an experimental design. After you have done the experiment, write up the experimental design, the statistical analysis, and the results. Describe the experimental design in enough detail that another person could replicate it. In your statistical analysis, include a table of means and standard errors of means and/or differences. In the results section, state your conclusions and evaluate whether the experiment answered the initial question. Give recommendations for improving the design if this experiment were redone in the future.

5.8 The following experiment was run to test whether there was any difference in run time of four different versions of SAS. The tests were run between 9:00 and 10:00 A.M. on three different days selected at random. On each day four versions of SAS were compared: versions 6.12, 7.0, 8.0, and 8.1. For each version the time to generate 1000 numbers from a standard normal distribution was recorded. The time (in seconds) it took to generate the data sets is given in the table.

	SAS Version			
Day	6.12	7.0	8.0	8.1
1	.25	.13	.18	.11
2	.26	.13	.18	.46
3	.26	.15	.16	.11

(a) Write down an appropriate model (including assumptions) for this experiment.

(b) Test whether there is a significant difference in mean run time among versions.

(c) Estimate the mean run time and standard error for version 8.1.

(d) Estimate the mean difference in run time between version 6.12 and version 8.1.

(e) The value for day 2 for version 8.1 looks rather high. What would you recommend doing about this?

5.9 An experiment studying the effects of CO_2 on rice and wheat was initiated in Anzhen, Jiangsu Province, China in 2001. The field layout listing the plot number and rice yield in the year 2000 are shown in the accompanying diagram. Wheat was grown in fields 21 to 48 in 2000 but not in fields 7 to 20. Fields 37 and 38 belonged to a different farmer than the other fields in 2000, and had mustard planted in them in 2000. There is a small road along the top edge of the diagram and a large road along the left edge.

41	10.1				
42	10.7	22	10.4		
43	9.9	23	10.8		
44	10.0	24	10.0		
45	9.8	25	9.8		
46	10.0	26	10.6		
47	10.7	27	10.4	7	10.2
48	8.2	28	10.8	8	10.5
		29	10.4	9	10.3
		30	10.1	10	10.2
		31	9.2	11	11.0
		32	10.8	12	9.9
				13	10.0
				14	10.6
				15	11.2
		36	9.1	16	10.0
		37	8.4		
		38	9.5		

Source: Data courtesy of Yong Han and Zhu Jiangguo of the Laboratory of Material Cycling in Pedosphere, Institute of Soil Science, Chinese Academy of Sciences, Nanjing.

Suppose that it is February 2001 and it is your job to provide statistical advice on the design of the experiment for the summer of 2001. Two treatments are to be compared in 2001: elevated CO_2 and ambient CO_2.

(a) Would you recommend using an RCBD or a CRD? Discuss the issues involved in making this decision and justify your recommendation. Consider power in your discussion, and show your calculations.

(b) For a design with five replicates, explain how to randomize your design. Draw a diagram of the proposed design. If your design is an RCBD, show which plots are in each block.

(c) The CO_2 is very expensive and researchers can afford to build just a few replicate exposure rings for each treatment. What is the minimum number of replicates that would allow detection of an increase of 1.5 kg under elevated CO_2? State your assumptions.

APPENDIX 5A: INTERACTION OF A RANDOM BLOCK EFFECT AND A FIXED TREATMENT EFFECT

When block effects are random, it seems to us that the most natural definition of the ith block mean is the average response, in block i, of the t treatments $\overline{\mu}_{i\cdot}$. For a given block, the fixed (treatment) effects are averaged over the levels included in the experiment, because these levels constitute the entire population of treatment effects.

On the other hand, random effects are drawn from some population with an infinite number of possible values. Averaging (i.e., taking the expectation) is done over the entire population of possible values, so the relevant mean of the jth treatment is its expectation over all possible blocks, $\mathrm{E}[\mu_{ij}]$ (which cannot now be written as $\overline{\mu}_{\cdot j}$).

Using these definitions of block and treatment means, the main effects and two-factor interactions are defined as

$$\beta_i = \overline{\mu}_{i\cdot} - \mu$$
$$\tau_j = \mathrm{E}[\mu_{ij}] - \mu$$
$$(\beta \times \tau)_{ij} = \mu_{ij} - \overline{\mu}_{i\cdot} - \mathrm{E}[\mu_{ij}] + \mu.$$

The parameter μ is usually defined to be the grand mean. Taking expectations over both block and treatment levels gives the grand mean,

$$\mu = \mathrm{E}\left[\frac{1}{t}\sum_{j=1}^{t}\mu_{ij}\right] = \mathrm{E}[\overline{\mu}_{i\cdot}] = \frac{1}{t}\sum_{j=1}^{t}\mathrm{E}[\mu_{ij}].$$

These definitions taken together imply that the main effects and interactions sum to zero when added over the levels of the fixed effect. Thus, when summing over the fixed treatment levels, $\sum_{j=1}^{t}\tau_j = 0$ and $\sum_{j=1}^{t}(\beta \times \tau)_{ij} = 0$ for each block i. The sum-to-zero constraint for the fixed effects comes directly from the definitions of treatment effects and μ, $\sum_{j=1}^{t}\tau_j = \sum_{j=1}^{t}\mathrm{E}[\mu_{ij}] - t\mu = 0$. On the other hand, the random effects do not sum to zero because the mean of the blocks in the sample is not the same as the expectation of the block means. $\sum_{i=1}^{b}\beta_i = \sum_{i=1}^{b}\overline{\mu}_{i\cdot} - b\mathrm{E}[\overline{\mu}_{i\cdot}]$. Similarly, the interactions do not sum to zero when summed over the random blocks. The expected mean squares given in this text are based on the definitions given here and these sum-to-zero constraints. Note that there is not universal agreement on these points. SAS® `Procs GLM and Mixed` use a different set of definitions that do not lead to the sum-to-zero constraints. Dean and Voss (1999) present this alternative parameterization.

CHAPTER 6

Latin Squares

6.1 INTRODUCTION

Sometimes there are two blocking factors that can be used to reduce extraneous variability rather than just one. Blocking on both factors may increase precision. For example, suppose that five kinds of furniture glue are being compared. The strength of the glue bond is affected by the temperature and humidity at the time the pieces are glued together and by the moisture content of the wood. Temperature and humidity vary across days, so one possible blocking factor could be the day that the pieces are glued together. The moisture content of the wood might vary among batches of lumber, so batch could be a second blocking factor. A Latin square for five treatments $T_1, \ldots, T_5$ blocked on day and batch of lumber is shown in Table 6.1.

All five types of glue are tried each day on each batch of lumber. Days and batches can both be considered blocks. The two types of blocking are superimposed on each other. This is a clever way of controlling both kinds of unwanted variability simultaneously.

The Latin square design requires that the number of treatments equal the number of levels of both blocking criteria. An $n \times n$ Latin square allows for control of variability, i.e., blocking in two directions, and provides n replications for each of the n levels of a treatment factor. The big advantage of Latin squares is that this is accomplished with only n^2 experimental units. Latin squares have the potential of large gains in efficiency without having to use excessive experimental material.

One of the earliest and possibly still the most common applications of the Latin square is in field experiments in agriculture. Two blocking factors in this case might be distance from a river and a gradient of soil type changes. Latin square designs are useful in industrial engineering, as in our opening example or where the two blocking factors might be machine operator and day of the week or in marketing studies where stores may form the columns, seasons (or

Planning, Construction, and Statistical Analysis of Comparative Experiments,
by Francis G. Giesbrecht and Marcia L. Gumpertz
ISBN 0-471-21395-0

Table 6.1 Latin Square for Five Treatments

Batch	Day 1	Day 2	Day 3	Day 4	Day 5
1	T_1	T_3	T_4	T_5	T_2
2	T_4	T_2	T_1	T_3	T_5
3	T_2	T_5	T_3	T_4	T_1
4	T_3	T_1	T_5	T_2	T_4
5	T_5	T_4	T_2	T_1	T_3

weeks) the rows, and treatments are marketing programs. They are important in social science, medical, and animal science applications where a subject (human or animal) forms one blocking factor and time of application forms the other blocking factor. Finally, an important application is the design of fair election ballots, ballots which account for the fact that the candidate whose name is on the top line of a ballot has a distinct advantage over other candidates.

Latin square designs are the most common representatives of a large, relatively complex class of designs known as row–column designs. Formally, Latin squares can also be interpreted as examples of fractions of factorial experiments (Chapter 14). Common practice is to treat them separately, as special cases of row–column designs, because they represent a logical extension of the notion of blocking to more dimensions. Mandl (1985) gives an interesting example that reflects the dual nature of the Latin square and fractional factorial. The problem considered arises in compiler testing. He discusses an example of a declaration such as `static int ABCDE;` in the programing language C. The rules say that in C, `static` can be replaced by any member from a set of four and `int` by any member from another set of four. The compiler must correctly allocate storage, initialize, and so on, correctly for all combinations. In addition; the operation must work with the four binary arithmetic operators, $\{'+', '-', '/', '*'\}$. In the compiler testing experiment, only $4 \times 4 = 16$ combinations are run. This example can be interpreted as a one-fourth replicate of the $4 \times 4 \times 4$ factorial, although Mandl chose to think of it as a Latin square.

6.2 FORMAL STRUCTURE OF LATIN SQUARES

To a mathematician the term *Latin square* refers to a structure, an $n \times n$ array of n symbols, such that each symbol appears exactly once in each row and in each column. The nature of the symbols is immaterial. To a statistician or experimental scientist, the Latin square is a design or plan for an experiment, based on such a structure.

If we let the integers $0, 1, \ldots, n-1$ be the symbols, we can always construct an $n \times n$ Latin square with the element in row i and column j:

$$l_{ij} = i + j - 2 \text{ (modulo } n) \qquad \text{for } i = 1, 2, \ldots, n \quad \text{and} \quad j = 1, 2, \ldots, n. \tag{6.1}$$

Rearranging or permuting rows does not disturb the structure as long as whenever a row is moved *the entire row is moved at once*. Similarly, permuting columns does not disrupt the Latin square structure. If the symbols represent treatments, the treatments are grouped into complete replicates in two different ways, by rows and by columns. The study of Latin squares is important, not only because of their direct application, but because they also form the basis for a large collection of tools and techniques to construct more complex designs. Many important results in advanced design theory and coding theory derive directly from Latin squares.

A *standard Latin square*, or equivalently, a Latin square in standard or reduced form, is one in which the columns have been permuted so that the treatment names in the first row are in ascending order and the rows have been permuted so that the names of the treatments in the first column are in ascending order. The formula for generating Latin squares [eq. (6.1)] produces standard Latin squares. Sometimes one also hears the term *semistandard*. This refers to a Latin square in which either only the rows or only the columns are in order. The concepts of standard and semistandard become important when we consider establishing when two Latin squares are really different. Two Latin squares are said to be *different* if they fail to agree exactly in one or more cells, after they have both been put in standard form. It can be shown that all 3×3 Latin squares are equal or equivalent if they are put into standard form. There are four different 4×4 standard Latin squares. Any 4×4 square falls into one of these groups. It has been established by enumeration that there are 56 standard 5×5 Latin squares and 9408 standard 6×6 standard Latin squares.

The 9408 standard 6×6 Latin squares can be grouped into 22 *transformation sets*. Transformation sets of reduced Latin squares are sets of squares with the property that any member of the set can be obtained from any other member of the set by appropriately permuting letters and then rearranging rows and columns to again obtain reduced form. The transformation sets are not all of equal size. Yates (1933) lists all 4×4 Latin squares and one square from each of the 5×5 and 6×6 transformation sets.

6.2.1 Randomization

The requirement for a valid randomization is that every possible random assignment of treatments to experimental units, subject to the constraints implied by the design, has an equal probability of appearing. Given a standard $t \times t$ Latin square, there are $t(t-1)$ possible permutations. For the 3×3 case, there is only one standard square. To perform a valid randomization, either randomly permute two of the rows and all three columns or randomly permute two columns and all three rows. It does not hurt to permute all rows and all columns randomly, but this is not necessary to produce all possible squares. For the 4×4 squares, randomly select one of the standard squares from Appendix 6A and then randomly permute rows $2, \ldots, 4$ and all columns (or columns $2, \ldots, 4$ and all rows).

```
* Select a Transformation Set;
* Transformation Set I: probability=50/56'
* Transformation Set II: probability=6/56';
data transfset;
   p1=50/56; p2=6/56;
   seed=28187; x=rantbl(seed,p1,p2);
   put 'Transformation Set' x;  run;
```

Display 6.1 SAS® code for randomly selecting a 5×5 transformation set.

For the 5×5 and 6×6 squares, one representative of each transformation set is listed in Appendix 6A along with the number of different standard squares, s, in each transformation set. The first step in a proper randomization is to randomly select one of the standard squares, appropriately weighting by the probability of selecting each transformation set ($s/56$ for the 5×5 case and $s/9408$ for the 6×6 case). Random selection of a transformation set is demonstrated in Display 6.1. Then randomly assign treatments to symbols, randomly permute $t - 1$ of the rows, and randomly permute all the columns. For the 7×7 case, Appendix 6A gives a set of 21 standard squares, which have the property that none can be obtained from any one of the remaining 20 by permuting letters and rearranging in reduced form. These are representatives of transformation sets. Unfortunately, the number of reduced squares in each transformation set is unknown. Selecting one of the 21 squares at random and then permuting all letters, all rows, and all columns appears to be an adequate randomization. For larger squares, it is safe to take any square and permute rows, columns, and treatments. Example SAS® code for randomizing a 5×5 Latin square is given on the companion Web site.

Exercise 6.1: Standard Latin Squares. How many transformation sets are there for 4×4 Latin squares, and how many standard squares are there in each set? □

6.2.2 Analysis of Variance

Some assumptions are made in employing a Latin square design. Recall that in the RCBD the usual assumption is that there is no block $\times$ treatments interaction. Similarly, in the Latin square design, we assume that there are no interactions between treatments and either of the blocking factors. Specifically, we assume that treatment $\times$ row, treatment $\times$ column, and row $\times$ column interactions are zero. If only one square is used, this is a necessary assumption because there are not enough observations to estimate both the interactions and residual error. These assumptions may be tested using Tukey's test for additivity. Hinkelmann and Kempthorne (1994) present such statistics for testing various possible types of nonadditivity.

The analysis of variance of data from a Latin square is based on the model

$$y_{ij} = \mu + r_i + c_j + \tau_{d(i,j)} + \epsilon_{ij}, \tag{6.2}$$

where $\epsilon_{ij} \sim \text{NID}(0,\sigma_\epsilon^2)$ and

$$\left.\begin{array}{l} i = 1, \ldots, t \\ j = 1, \ldots, t \\ d(i,j) = 1, \ldots, t \end{array}\right\} \quad \text{only } t^2 \text{ of the } t^3 \text{ combinations appear.}$$

The symbol $\tau_{d(i,j)}$ indicates the effect of the treatment assigned to the (i, j)th cell. The $d(i, j)$ subscripts depend on the specific Latin square selected. In a sense, they define the design or are defined by the design. Treatment means are denoted by $\{\overline{y}_d\}$.

The analysis of variance is obtained via least squares. Table 6.2 gives the analysis of variance table in two forms: (1) in symbolic representation of the difference in error sums of squares between a reduced and a full model, and (2) as sums of squares of differences between treatment means and overall means.

Notice that in the analysis of variance table $\text{R}(r_1, \ldots, r_t|\mu) = \text{R}(r_1, \ldots, r_t |c_1, \ldots, c_t, \tau_1, \ldots, \tau_t, \mu)$. Since the rows, columns, and treatments are all orthogonal to each other, the sum of squares for each term can be computed ignoring the other two. The practical meaning of orthogonality of two factors A and B is that every level of factor B is tested in combination with each level of factor A equally often. Similarly, $\text{R}(c_1, \ldots, c_t|\mu) = \text{R}(c_1, \ldots, c_t|r_1, \ldots, r_t, \tau_1, \ldots, \tau_t, \mu)$

Table 6.2 Sums of Squares in the Analysis of Variance for a Latin Square Design

Source	Symbolic Representation	df	SS
Mean	$\text{R}(\mu) = CT$	1	$\overline{y}_{..}^2/t^2$
Row	$\text{R}(r_1, \ldots, r_t\|\mu)$ $= \text{R}(r_1, \ldots, r_t$ $\|c_1, \ldots, c_t, \tau_1, \ldots, \tau_t, \mu)$	$t-1$	$t\sum_i(\overline{y}_{i.} - \overline{y}_{..})^2$
Column	$\text{R}(c_1, \ldots, c_t\|\mu)$ $= \text{R}(c_1, \ldots, c_t$ $\|r_1, \ldots, r_t, \tau_1, \ldots, \tau_t, \mu)$	$t-1$	$t\sum_j(\overline{y}_{.j} - \overline{y}_{..})^2$
Treatment	$\text{R}(\tau_1, \ldots, \tau_t\|\mu)$ $= \text{R}(\tau_1, \ldots, \tau_t$ $\|r_1, \ldots, r_t, c_1, \ldots, c_t, \mu)$	$t-1$	$t\sum_d(\overline{y}_d - \overline{y}_{..})^2$
Error		$(t-2)(t-1)$	$\sum_i\sum_j\sum_{d(i,j)}(y_{ij}$ $-\overline{y}_{i.} - \overline{y}_{.j} - \overline{y}_d + 2\overline{y}_{..})^2$
Total		t^2	$\sum_i\sum_j y_{ij}^2$

and $R(\tau_1, \ldots, \tau_t|\mu) = R(\tau_1, \ldots, \tau_t|r_1, \ldots, r_t, c_1, \ldots, c_t, \mu)$. These nice relationships hold only as long as no values are missing and no treatment assignments are mixed up.

6.2.3 Treatment Means and Standard Errors

In a Latin square row, column, and treatment, contrasts are all orthogonal to each other. Every treatment appears once in every row and in every column, so the row and column effects balance out in each treatment mean. Hence the ordinary least squares estimates of treatment means are just the sample means of the treatments, $\widehat{y}_d = \overline{y}_d$. Also, all pairs of treatments appear in each row and each column, so all treatment contrasts can be made within a row and within a column. Treatment contrasts are estimated by computing the corresponding contrast of the sample means. Since the $\{\overline{y}_d\}$ are means of t independent observations, the standard error of a treatment contrast is, as in a randomized complete block design or a completely randomized design,

$$s\left(\sum_d c_d\overline{y}_d\right) = \sqrt{\frac{\text{MSE}}{t}\sum_d c_d^2},$$

where c_d are contrast coefficients that sum to zero.

Exercise 6.2: Treatment Contrasts. Show that row and column effects both drop out of treatment contrasts. □

6.2.4 Missing Values

If some data are missing or if treatments are accidentally assigned to the wrong row–column cell of the square, the analysis of variance can be conducted by ordinary least squares using effects coding for the model. The partial sums of squares can then be computed by subtraction as SS rows $= R(r_1, \ldots, r_t|c_1, \ldots, c_t, \tau_1, \ldots, \tau_t, \mu)$, SS columns $= R(c_1, \ldots, c_t|r_1, \ldots, r_t, \tau_1, \ldots, \tau_t, \mu)$, and SS treatment $= R(\tau_1, \ldots, \tau_t\ |r_1, \ldots, r_t, c_1, \ldots, c_t, \mu)$. The degrees of freedom for an effect equals the difference in the number of parameters between the full and reduced models.

Alternatively, the sums of squares can be computed after fitting the full model by expressing the hypothesis as $H_0 : \boldsymbol{L\beta} = \mathbf{0}$, and writing the sum of squares in the quadratic form $SS(L) = (\widehat{\boldsymbol{L\beta}})^t(\boldsymbol{L}(\boldsymbol{X}^t\boldsymbol{X})^-\boldsymbol{L}^t)^{-1}\widehat{\boldsymbol{L\beta}}$. The degrees of freedom for the effect being tested is the number of linearly independent rows of $\boldsymbol{L}$. Statistical software for general linear models such as SAS® `Proc GLM` typically handles analysis of variance this way, whether or not data are missing.

Sometimes both rows and columns are considered random effects. If no data are missing, the analysis of variance is exactly the same as if all effects are fixed. If many observations are missing, one possibility is to proceed as though rows and columns were both fixed and perform an ordinary least squares analysis. Logically, one has problems making inferences to future settings that involve other rows and columns. The alternative is to use estimated generalized least squares, where the variance components for rows and columns are estimated using maximum likelihood, restricted maximum likelihood, (or MINQUE see Section 5.5.2). The EGLS estimates do not, in general, coincide with the ordinary least squares estimates when data are missing or the design is not balanced. Hypotheses about the fixed effects are tested using either Wald-type tests (with the F-approximation for small samples), or using a likelihood ratio test.

Imputing Missing Values

Sometimes it is useful to know how to impute a missing value in a Latin square under model (6.2). The derivation is similar to that of the missing value formula for a randomized complete block design (Section 5.5.1). Assume that the observation on treatment $d = d(i, j)$ in row i and column j is missing. The object is to find a value to substitute for the missing y_{ij} that if inserted into the data set, would not change the error sum of squares calculated as in Table 6.2. Note that this is the value where $d(i, j) = d$ in the design. This value is the same as the predicted value, $\widehat{y}_{ij}$, obtained from ordinary least squares.

The value $z = \widehat{y}_{ij}$ must then satisfy

$$z = \frac{y_{i\cdot} + z}{t} + \frac{y_{\cdot j} + z}{t} + \frac{y_d + z}{t} - 2\frac{y_{\cdot\cdot} + z}{t^2}.$$

Solving for z gives the formula for the missing value:

$$z = \frac{t(y_{i\cdot} + y_{\cdot j} + y_d) - 2y_{\cdot\cdot}}{(t-2)(t-1)}. \tag{6.3}$$

An approximate test of the treatment effect can be obtained by doing the following computations:

1. Replace the missing value with z computed according to equation (6.3). Compute SSE for the full model using the usual calculations for balanced data (from Table 6.2) performed on the augmented data set which contains the original data and z. Remember that z contributes nothing to the error sum of squares.
2. Now ignore the treatment classification and treat the data as though it were from a two-way classification with just rows and columns as the factors. Compute the missing value under this reduced model using equation (5.13) and augment the original data with this estimate. Compute SSE for the reduced model using the equation in Table 5.3; call this SSE (reduced).
3. Compute the treatment sum of squares by subtraction,

$$\mathrm{R}(\tau_1, \ldots, \tau_t | \mu, r_1, \ldots, r_t, c_1, \ldots, c_t) = \mathrm{SSE(reduced)} - \mathrm{SSE(full)}.$$

Exercise 6.3: Missing Data in a Latin Square. For a design in which $d(1, 1) = 1$ is missing, derive the variance of $\widehat{\tau}_1 - \widehat{\tau}_2$. *Hints:* (1) Show that $\text{Var}[y_1 + z] = t\sigma^2 + t^2\sigma^2/(t-1)(t-2)$; and (2) show that $\text{Cov}[y_1 + z, y_2] = 0$, where y_1 and y_2 are totals for treatments 1 and 2. □

6.2.5 Advantages and Disadvantages of Latin Squares

The prime strength of the Latin square is that it provides better control of the experimental material by allowing the experimenter to control two sources of variation. This results in very efficient use of resources. Two early studies looked at the efficiencies of Latin squares in agricultural field experiments. Cochran (1938) looked at a number of experiments and concluded that the Latin square was about 1.37 times as efficient as the RCBD and about 2.22 times as efficient as the CRD. This can be interpreted as meaning that one plot in a Latin square provides as much information as 1.37 plots in a RCBD or 2.22 plots in a CRD. In another study, Ma and Harrington (1948) concluded that the Latin square was about 1.27 times as efficient as the RCBD, i.e., that one plot in a Latin square design was equivalent to 1.27 plots in a RCBD. Savings of this order are important when budgets are tight.

On the other hand, for agricultural work the fact that the number of replications must equal the number of treatments tends to be restrictive. If the number of treatments is large, it may not be possible or desirable to use as much replication as required by a Latin square design and if the number of treatments is small ($t = 3$, in particular), the Latin square provides too few degrees of freedom for error. There is a trade-off between the reduction in random noise and the decrease in degrees of freedom due to adding the second blocking factor. We can judge whether the reduction in error variance is large enough to offset the loss of degrees of freedom by computing the power of hypothesis tests under the two designs.

An additional consideration for use of the Latin square design is that the experimenter must be confident that the effects of treatments are additive with respect to both row and column differences. In other words, the scientist must feel confident that one row does not favor one of the treatments while another row favors a different treatment. This is basically the same consideration as additivity of blocks and treatments discussed in Section 5.2.7. We sometimes hear the admonition that the rows and columns must not interact; i.e., row and column effects must be additive. Violation of this stricture of interaction of rows and columns is not as serious, in the sense that it does not introduce biases among treatment differences, provided that the randomization was carried out properly. However, row × column interaction does reduce the power of the test of the null hypothesis; i.e., it causes a loss of sensitivity.

Latin squares tend to work very well in laboratory settings and in industrial experiments where several factors are often important, and the experimenter has good control over how things are done. They are also ideal for experimentation on human or animal subjects, where more than one treatment can be applied to each subject sequentially over time. In these situations a Latin square may be the only reasonable design.

6.3 COMBINING LATIN SQUARES

It is often a good idea to use multiple Latin squares, especially for experiments with a small number of treatments, because a single Latin square does not provide much replication. Squares can be combined in many different ways.

Case 1: There are s separate squares with a separate randomization for each square.

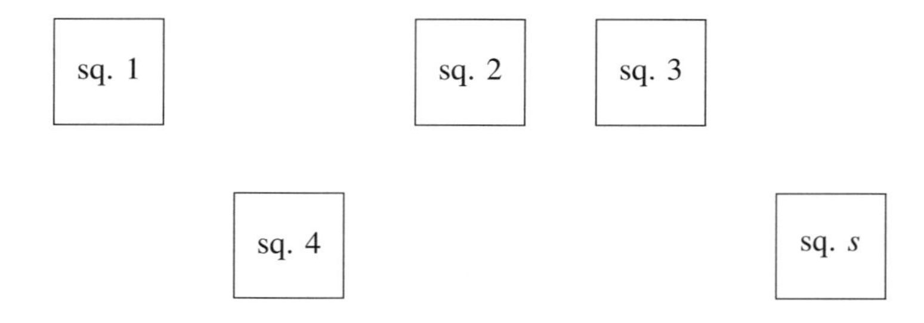

The statistical model for this case, in which rows and columns of one square do not bear any relationship to rows and columns of the other squares, is

$$y_{hij} = \mu + s_h + r(s)_{i(h)} + c(s)_{j(h)} + \tau_{d(h,i,j)} + (s \times \tau)_{hd(h,i,j)} + \epsilon_{hij},$$

where $h = 1, \ldots, s$ squares, $i = 1, \ldots, t$ rows within each square, $j = 1, \ldots, t$ columns within each square, and $d(h, i, j) = 1, \ldots, t$ treatments. The $d(h, i, j)$ are the treatment subscripts defined by the designs (randomization) and depend on the square and the row and column within squares. Note here that parentheses are used two different ways. The notation $d(h, i, j)$ indicates that d is a function of $h, i,$ and j; whereas $r(s)$ indicates that rows are nested within squares. The meaning of the parentheses should be clear from the context. There are a total of st^2 observations. The squares, serving as replication, are typically treated as random effects.

The analysis of variance table corresponding to this model is shown in Table 6.3. If the squares represent random effects, the correct F-ratio for testing treatment effects is $F =$ MS treatment/(MS treatment $\times$ square) with $t - 1$ and $(s - 1)(t - 1)$ degrees of freedom. If the squares represent fixed effects, the correct F-ratio is $F =$ MS treatment/MSE with $t - 1$ and $s(t - 1)(t - 2)$ degrees of freedom.

The residual SSE is computed from interactions between treatments and rows within squares and columns within squares. The assumption is that these interactions are negligible, so this error estimates the variability among units in a square after accounting for all the terms in the model. Sometimes, SS square $\times$ treatment

Table 6.3 Analysis of Variance for Multiple Latin Squares Where the Rows and Columns of Different Squares Do Not Correspond with Each Other

Source	df	
Mean	1	
Square	$s-1$	
Treatment	$t-1$	
Row (square)	$s(t-1)$	
Column (square)	$s(t-1)$	
Treatment × square	$(s-1)(t-1)$	If squares are fixed effects, these
Error	$s(t-1)(t-2)$	are often pooled for error
Total	st^2	

and SSE are pooled together to obtain the error sum of squares. One rule of thumb that preserves the type I error rate fairly well is to pool a term with error if the p-value for that effect exceeds .25 (Bozivich et al. 1956).

Case 2: There are s squares with rows common across squares. Again, each square has a separate randomization. An example of this design would be a study involving a questionnaire with t questions given to st subjects, where each question could appear in t different positions on the questionnaire. The appropriate statistical model is

$$y_{hij} = \mu + s_h + r_i + c(s)_{j(h)} + (s \times r)_{hi} + \tau_{d(h,i,j)} + (s \times \tau)_{hd(h,i,j)} + \epsilon_{hij}.$$

The design can be illustrated as follows:

pos. 1 – – pos. t	sq. 1	sq. 2	sq. 3	– –	– –	– –	sq. s

This plan is often referred to as the *Latin rectangle.* Using this plan with four or five treatments, the same number of rows, and twice as many columns makes a nice design that relieves the problem of small Latin squares having too few degrees of freedom for error.

A common application of this design is in questionnaire design for sample surveys. The Latin square is used to address the problem that the position of a question may influence the response. If there are t questions, a $t \times t$ Latin square can be used to construct t forms of the questionnaire, such that each question is in each position (row) exactly once. The columns then correspond to subjects. Since there will be many subjects, several squares could be used, with columns assigned to new subjects. Note that in this example, subject effects could be considered either fixed or random, depending on how the subjects were selected and on the objectives of the study. The identical problem occurs in the

Table 6.4 Analysis of Variance for Latin Rectangle with Columns Arranged in Separate Squares

Source	df	
Mean	1	
Square	$s - 1$	
Column (square)	$s(t - 1)$	
Row	$t - 1$	
Treatment	$t - 1$	
Square × row	$(s - 1)(t - 1)$	These three are
Square × treatment	$(s - 1)(t - 1)$	often pooled to
Error	$s(t - 1)(t - 2)$	form error
Total	st^2	

design of election ballots. Position on the ballot is very important. The name at the top of the ballot has a distinct advantage. A Latin square can be used to design a set of ballots such that each name appears in each position equally frequently. Additional issues that arise in connection with the order of treatments, when the columns of a Latin square represent a single subject (human, animal, or otherwise) are addressed in detail in Chapter 9.

The analysis of variance for a Latin rectangle with distinct squares is shown in Table 6.4. If squares represent random effects, the correct F-statistic for testing treatment effects is F = MS treatment/(MS square × treatment). Similarly, if squares are random effects, fixed row effects are tested using F = MS row/(square × MS row). This would be useful in testing whether position affected the responses on a questionnaire. If squares represent fixed effects, the F-ratios for testing all other effects have MSE in the denominator.

Breaking out sums of squares for square, for square × row, and square × treatment makes sense only if the columns are really grouped into squares. For example, the columns could be students and the squares classes, or the columns classes and the squares schools. If there are no actual "squares," the sums of squares for squares and columns (squares) are pooled together to make

Table 6.5 Analysis of Variance for Latin Rectangle Where There Are No Distinct Squares

Source	df
Mean	1
Column	$st - 1$
Row	$t - 1$
Treatment	$t - 1$
Error	$(t - 1)(st - 2)$
Total	st^2

Monitor Type	Lighting Environment				
	1	2	3	4	5
1 2 3 4 5	5 Pull-down menu designs				

Figure 6.1 Multiple Latin squares with the same rows and columns: pull-down menu experiment.

SS columns with $st - 1$ degrees of freedom. The analysis of variance table reduces to the same form as for a single Latin square, but the degrees of freedom are slightly different. The sources and degrees of freedom for a Latin rectangle with no distinct squares are shown in Table 6.5. The statistical model simplifies considerably, to

$$y_{ij} = \mu + r_i + c_j + \tau_{d(ij)} + \epsilon_{ij}.$$

Case 3: The same square is repeated s times with common rows and columns. As an example, consider a software designer who wants to know which of five possible pull-down menu designs is easiest to use. Five menu designs on five different monitor types in five different lighting environments are to be compared. Each of these squares is repeated across s groups of 25 people (Figure 6.1). The statistical model for this design is

$$y_{hij} = \mu + s_h + r_i + c_j + (s \times r)_{hi} + (s \times c)_{hj} + \tau_{d(ij)} + (s \times \tau)_{hd(ij)} + \epsilon_{hij}.$$

The classical analysis of variance for this design is given in Table 6.6. The Latin square type of error is computed from interactions among the rows, columns, and treatments. If a different randomization is used for each square, there may not be any reason to partition the last two sums of squares in Table 6.6.

In the example of testing pull-down menu designs, each square might be a different facility, with 25 people tested at each facility. If squares represent random effects and rows, columns, and treatments are all fixed effects, the interaction with squares is used to test each of the fixed effects. In particular, the treatment effect is tested using the statistic $F =$ MS treatment/(MS square $\times$ treatment).

If the units are not really grouped into squares, the sums of squares for squares and interactions with squares would not be broken out in the analysis of variance. In the pull-down menu example, where the blocking factors are lighting environments and monitor types, if 50 people were tested but not in groups of 25, and if the randomization of rows or columns of the design is not confined to permutations within a square, the sums of squares for squares and interactions with squares should be omitted.

Exercise 6.4: Randomization of Multiple Squares. Discuss the practical aspects of running the pull-down menu experiment, such as how many monitors are needed, how many rooms or cubicles, and how to randomize the experiment. □

Table 6.6 Analysis of Variance for Multiple Latin Squares Where All the Squares Share Common Rows and Columns

Source	df	
Mean	1	
Square	$s-1$	
Column	$t-1$	
Square × column	$(s-1)(t-1)$	
Row	$t-1$	
Square × row	$(s-1)(t-1)$	
Treatment (≡ menu design)	$t-1$	
Square × treatment	$(s-1)(t-1)$	
Latin square type of error	$(t-1)(t-2)$	Often pooled to
Latin square type of error × square	$(s-1)(t-1)(t-2)$	form error
Total	st^2	

Exercise 6.5: Latin Square Design for Testing Pull-Down Menu Styles. For the pull-down menu experiment, do the following.

(a) Write down the statistical model corresponding to this experimental design and corresponding to Table 6.6. Define all symbols and subscripts. State what interactions are assumed not to exist.

(b) From your experience using computers, what assumptions of the Latin square model do you think are reasonable? Are there any assumptions that you think might be violated? □

Exercise 6.6: Random Square Effects. In the pull-down menu experiment, suppose that there are two squares and the same randomization is used for each square, and use the model corresponding to Table 6.6.

(a) Suppose that squares are considered random effects, but all other factors are considered fixed effects. Write down the test statistic and rejection criterion for testing whether there are differences among menu designs.

(b) Again considering squares to be a random effect, write down the variance of a difference between the means for two menu designs. □

Example 6.1. These data come from a uniformity trial of dairy cows (Lucas 1948). In a *uniformity trial* no actual treatments are imposed; the experiment is done to study the patterns of variability in the population. For the purpose of this example, however, the analysis is performed as though there were three treatments, A, B, and C. The response variable is a measure of milk yield, adjusted for fat content. The experiment was constructed as a set of four 3×3 squares. The columns consist of individual dairy cows, and rows consist of time periods. It is not clear from the discussion in the report whether the rows are nested within squares or go across all squares. For demonstration purposes, we analyze the data both ways. The data are as follows:

Cow no.:	Square 1: 1		2		3		Square 2: 1		2		3	
Row 1	*C*	35.3	*A*	46.1	*B*	42.7	*A*	64.2	*C*	29.8	*B*	29.8
Row 2	*A*	53.5	*B*	28.6	*C*	33.9	*B*	27.9	*A*	26.7	*C*	30.8
Row 3	*B*	24.9	*C*	38.5	*A*	35.5	*C*	38.0	*B*	37.7	*A*	36.4
Total		113.7		113.2		112.1		130.1		94.2		97.0

Cow no.:	Square 3: 1		2		3		Square 4: 1		2		3	
Row 1	*A*	26.8	*C*	37.2	*B*	29.8	*C*	61.5	*A*	26.7	*B*	38.6
Row 2	*C*	24.1	*B*	36.4	*A*	26.4	*B*	46.4	*C*	25.0	*A*	29.6
Row 3	*B*	50.5	*A*	23.4	*C*	24.5	*A*	24.9	*B*	20.6	*C*	29.2
Total		101.4		97.0		80.7		132.0		72.3		97.4

The row totals for the squares are given in Table 6.7. The treatment totals follow.

A 420.2 *B* 413.9 *C* 407.8

For case 1, rows within squares, we demonstrate use of some hand-computing forms of sums of squares. These forms make hand computation very quick.

$$\text{CT} = \frac{y_{\cdots}^2}{st^2} = \frac{1241.9^2}{36} = 42,842.10$$

$$\text{Total SS} = \sum_{h,i,j} y_{hij}^2$$

$$= 35.3^2 + 46.1^2 + \cdots + 20.6^2 + 29.2^2 = 46,741.57$$

$$\text{SS squares} = \sum_h \frac{y_{h\cdot\cdot}^2}{t^2} - \text{CT}$$

$$= \frac{339.0^2 + 321.3^2 + 279.1^2 + 302.5^2}{9} - \text{CT}$$

$$= 219.87$$

Table 6.7 Row Totals

Row	Square 1	Square 2	Square 3	Square 4	Total
1	124.1	123.8	93.8	126.8	468.5
2	116.0	85.4	86.9	101.0	389.3
3	98.9	112.1	98.4	74.7	384.1
Total	339.0	321.3	279.1	302.5	1241.9

$$\text{SS row (square)} = \frac{\sum_h \sum_i y_{hi\cdot}^2}{t} - \frac{\sum_h y_{h\cdot\cdot}^2}{t^2}$$

$$= \frac{124.1^2 + 116.0^2 + \cdots + 74.7^2}{3}$$

$$- \frac{339.0^2 + 321.3^2 + 279.1^2 + 302.5^2}{9}$$

$$= 843.35$$

$$\text{SS column (square)} = \frac{\sum_h \sum_j y_{h\cdot j}^2}{t} - \frac{\sum_h y_{h\cdot\cdot}^2}{t^2}$$

$$= \frac{113.7^2 + 130.1^2 + \cdots + 97.4^2}{3}$$

$$- \frac{339.0^2 + 321.3^2 + 279.1^2 + 302.5^2}{9}$$

$$= 961.47$$

$$\text{SS treatments} = \frac{\sum_d y_{\cdot d}^2}{st} - \text{CT}$$

$$= \frac{420.2^2 + 413.9^2 + 407.8^2}{12} - \text{CT}$$

$$= 6.41$$

$$\text{SS square} \times \text{ treatment} = \frac{\sum_h \sum_d y_{hd}^2}{t} - \frac{\sum_h y_{h\cdot\cdot}^2}{t^2} - \frac{\sum_d y_{\cdot d}^2}{st} + \text{CT}$$

$$= \frac{(53.5 + 46.1 + 35.5)^2 + \cdots + (61.5 + 25.0 + 29.2)^2}{3}$$

$$- \text{CT} - 219.87 - 6.41$$

$$= 969.46$$

If the rows are different for each square, the analysis of variance table is as shown in Table 6.8.

If squares represent random effects, the test statistic for treatment effects is F = MS treatment/(MS square $\times$ treatment). This would be the case, for example, if squares represent different barns and the four different barns are used for replication. Since rows represent time, this model would be appropriate if each barn were sampled at a different set of times. If the squares represent fixed effects, such as ages of cows, the test statistic is F = MS treatment/MSE.

The treatment means are estimated by the sample treatment means, since all treatments appear in every row, column, and square. If the squares are random effects, the standard error of a difference between two treatment means is, e.g., $s(\overline{y}_A - \overline{y}_B) = \sqrt{2\text{MS square} \times \text{treatment}/st} = \sqrt{(2)(161.6)/12} = 5.29$. If the squares represent fixed effects, the standard error of the difference of two

Table 6.8 Analysis of Variance: Dairy Cow Example

Source	SS	df	
Mean	42,842.10	1	
Among squares	219.87	3	
Among treatments	6.41	2	
Among rows (squares)	843.35	8	
Among columns (squares)	961.47	8	
Squares × treatments	969.46	6	Pool
Error	898.91	8	these?
Total	46,741.57	36	

treatment means is computed from MSE; it is $s(\overline{y}_A - \overline{y}_B) = \sqrt{2\text{MSE}/st} = \sqrt{(2)(112.4)/12} = 4.33$.

If rows extended across all squares, as would probably be logical if rows were time, then in place of the sum of squares due to rows within squares, we compute a sum of squares for rows. Algebraically, the previous sum of squares due to rows within squares decomposes into two sums of squares: one for rows and one for the row by square interaction. These sums of squares are computed as

$$\begin{aligned}
\text{SS row} &= \frac{\sum_i y_{\cdot i \cdot}^2}{st} - \text{CT} \\
&= \frac{468.5^2 + 368.3^2 + 384.1^2}{12} - \text{CT} = 372.86 \\
\text{SS square} \times \text{row} &= \frac{\sum_h \sum_i y_{hi\cdot}^2}{t} - \frac{\sum_h y_{h\cdot\cdot}^2}{t^2} - \frac{\sum_i y_{\cdot i \cdot}^2}{st} + \text{CT} \\
&= \frac{124.1^2 + 116.0^2 + \cdots + 74.7^2}{3} - \text{CT} - 372.86 - 219.87 \\
&= 470.49.
\end{aligned}$$

The analysis of variance table now takes the form shown in Table 6.9.

Table 6.9 Analysis of Variance: Dairy Cow Example

Source	SS	df	
Mean	42,842.10	1	
Square	219.87	3	
Treatment	6.41	2	
Row	372.86	2	
Column (square)	961.47	8	
Square × row	470.49	6	Pool
Square × treatment	969.46	6	these?
Error	898.91	8	
Total	46,741.57	36	

Exercise 6.7: Decomposition of SS rows(Squares). Show algebraically that SS rows (squares) decomposes into SS rows and SS rows × squares. □

Exercise 6.8: Sums of Squares. In case (b) of the dairy cow example, where rows extend across all squares, write out the definitional formulas for the sums of squares for the entire analysis of variance table. □

6.4 GRAECO–LATIN SQUARES AND ORTHOGONAL LATIN SQUARES

Graeco–Latin squares extend the idea of blocking to three blocking factors. The following industrial example shows four factors in a 4×4 Graeco–Latin square. The aim of the experiment is to compare the effect of four sources of raw material on a finished product. The investigator wants to control for day of the week, operator, and machine. With a Graeco–Latin square design it is possible to control all three blocking factors plus the treatment factor using only $4 \times 4 = 16$ runs, whereas a full factorial would require $4^4 = 256$ runs. The setup might look as shown in Table 6.10. Notice that each machine is paired with each material exactly once, and every material is tried on each machine. Thus one can consider machine type a blocking factor, just as operator and day of the week.

This design has the same types of restrictive assumptions as the Latin square design. The model is

$$Y_{ij} = \mu + r_i + c_j + l_{dg(i,j)} + \tau_{dl(i,j)} + \epsilon_{ij},$$

where

$$r_i = \text{row effect}$$
$$c_j = \text{column effect}$$
$$l_{dg(i,j)} = \text{effect associated with Greek alphabet}$$
$$\tau_{dl(i,j)} = \text{effect associated with Latin alphabet.}$$

Notice that the two treatment subscripts, $dg(i, j)$ and $dl(i, j)$, both run from 1 to t and are defined by, or define, the design.

Table 6.10 4 x 4 Graeco–Latin Square

	Monday	Tuesday	Wednesday	Thursday
Jo	machine A material α	machine B material β	machine C material γ	machine D material δ
Harry	machine B material δ	machine A material γ	machine D material β	machine C material α
Tom	machine C material β	machine D material α	machine A material δ	machine B material γ
Pete	machine D material γ	machine C material δ	machine B material α	machine A material β

This model has no interactions; that is, all interactions are assumed zero. There are also far fewer possible squares. For the case of four treatments, there are only three possible Graeco–Latin squares. The randomization procedure is simply to pick one of the three squares randomly and randomly assign treatments to the letters A, B, C, D and layers to the Greek letters $\alpha, \beta, \gamma, \delta$.

Graeco–Latin squares are constructed by superimposing two *mutually orthogonal Latin squares* (MOLS). Mutually orthogonal Latin squares are squares with the property that if any two of the squares are superimposed, every possible ordered treatment pair occurs once. The 4×4 mutually orthogonal Latin squares are

$$\begin{array}{cccc} 1 & 2 & 3 & 4 \\ 3 & 4 & 1 & 2 \\ 4 & 3 & 2 & 1 \\ 2 & 1 & 4 & 3 \end{array} \qquad \begin{array}{cccc} 1 & 2 & 3 & 4 \\ 2 & 1 & 4 & 3 \\ 3 & 4 & 1 & 2 \\ 4 & 3 & 2 & 1 \end{array} \qquad \begin{array}{cccc} 1 & 2 & 3 & 4 \\ 4 & 3 & 2 & 1 \\ 2 & 1 & 4 & 3 \\ 3 & 4 & 1 & 2 \end{array}.$$

If the number of treatments is a prime number or a power of a prime, there are $t - 1$ mutually orthogonal Latin squares. In some cases there exist several complete sets of mutually orthogonal Latin squares, $(t - 1)$ squares per set. When t is not prime or a power of a prime, the maximum number of mutually orthogonal Latin squares that can be constructed is an open question. It is known that it is no larger than $t - 1$. For $t = 6$ it is not possible to construct even one pair of mutually orthogonal Latin squares. For values of $t > 2$ and $t \neq 6$ there exist at least one pair of mutually orthogonal squares, and for each $t > 10$ there exists at least one set of three mutually orthogonal squares. Appendix 6B lists sets of mutually orthogonal Latin squares for $t = 3, 4, 5$, and 7, and gives a recipe for constructing sets of mutually orthogonal Latin squares when t is a prime, a power of a prime, or a product of two primes. A good reference for those interested is the book *Combinatorial Designs* (Wallis 1988).

When it comes to constructing Graeco–Latin squares for an experiment, the possibilities for randomization are very limited. Randomization is limited to randomly assigning treatments to letters and then randomizing rows and columns after squares have been superimposed on one another. We also caution the user that in general it is not true that a given Latin square has an orthogonal mate, though, there are many cases where there are several orthogonal mates.

Mutually orthogonal Latin squares are very important building blocks in the construction of more complex designs, especially the lattice designs and crossover designs balanced for carryover effects.

Exercise 6.9: Orthogonal 3 x 3 Latin Squares. Since 3 is a prime number, there must exist two 3×3 orthogonal Latin squares. Construct these two squares. □

Exercise 6.10: Construction of Latin Squares Where t Is Prime. It was pointed out earlier [eq. (6.1)] that a $t \times t$ Latin square could be constructed by writing down the array with element in row i and column j equal to

$$d(i, j) = (i - 1) + (j - 1) \text{ (modulo } t) \qquad \text{for } i = 1, 2, \ldots, t$$
$$\text{and } j = 1, 2, \ldots, t.$$

Demonstrate for $t = 5$ that the three Latin squares with elements $\{d(i, j)^{(2)}\}$, $\{d(i, j)^{(3)}\}$, and $\{d(i, j)^{(4)}\}$, respectively, are pairwise orthogonal and orthogonal to the original, where

$$\begin{aligned} d(i, j)^{(2)} &= 2(i - 1) + (j - 1) \text{ (modulo } t) \\ d(i, j)^{(3)} &= 3(i - 1) + (j - 1) \text{ (modulo } t) \\ d(i, j)^{(4)} &= 4(i - 1) + (j - 1) \text{ (modulo } t) \\ &\text{for } i = 1, 2, \ldots, t \quad \text{and} \quad j = 1, 2, \ldots, t. \end{aligned}$$

This scheme can be used for any prime number $t > 2$. For t equal to the power of a prime number, an analogous scheme can be used, based on elements of a finite field. □

***Exercise 6.11: Orthogonal Mate for* 4 x 4 *Latin Square*.** Show that it is not possible to construct an orthogonal mate for

$$\begin{array}{cccc} \text{A} & \text{B} & \text{C} & \text{D} \\ \text{B} & \text{C} & \text{D} & \text{A} \\ \text{C} & \text{D} & \text{A} & \text{B} \\ \text{D} & \text{A} & \text{B} & \text{C} \end{array}$$

□

6.4.1 Special 6 x 6 Case

As mentioned earlier, it is not possible to find a pair of 6×6 orthogonal Latin squares, i.e., there are no 6×6 Graeco–Latin square designs. This has been known for a long time; however, it is possible to find two 6×6 Latin squares that are very close to being orthogonal. Street and Street (1987) show the following two squares.

$$\begin{array}{cccccc} 5 & 6 & 3 & 4 & 1 & 2 \\ 2 & 1 & 6 & 5 & 3 & 4 \\ 6 & 5 & 1 & 2 & 4 & 3 \\ 4 & 3 & 5 & 6 & 2 & 1 \\ 1 & 4 & 2 & 3 & \underline{5} & \underline{6} \\ 3 & 2 & 4 & 1 & \underline{6} & \underline{5} \end{array} \quad \text{and} \quad \begin{array}{cccccc} 1 & 2 & 5 & 6 & 3 & 4 \\ 6 & 5 & 1 & 2 & 4 & 3 \\ 4 & 3 & 6 & 5 & 1 & 2 \\ 5 & 6 & 4 & 3 & 2 & 1 \\ 2 & 4 & 3 & 1 & \underline{5} & \underline{6} \\ 3 & 1 & 2 & 4 & \underline{6} & \underline{5} \end{array}.$$

Comparing these two squares, things seem to be in order down to the four marked elements in the lower right corner. Interchanging 5 and 6 in one of the squares does not help. Although these two are not orthogonal, if one is in a bind and needs a Graeco–Latin square, (where 5×5 is too small and 7×7 is too large), this pair can be used. Contrasts among the first four treatments assigned via the first square are orthogonal to contrasts among the first four treatments assigned via the second square. There is some, not too strong correlation between contrasts involving treatments 5 and 6 from either or both sets.

6.5 SOME SPECIAL LATIN SQUARES AND VARIATIONS ON LATIN SQUARES

6.5.1 Systematic Latin Squares

A simple construction for an $n \times n$ Latin square is to number the rows and columns $0, 1, \ldots, (n-1)$ and then write the element in row i and column j as $i + j$ modulo n. This produces a very systematic square. A modification is to write the element in row i and column j as $i + kj$ modulo n, where k is some integer $0 < k < n$, and prime to n. This recipe yields the *Knut Vik square*.

A couple of cautions are in order regarding Knut Vik squares. First, Knut Vik squares are to be avoided in experiments designed to balance out spatial correlation, because in these squares each treatment has the same neighbors in every row and every column. Other Latin squares are very useful in designing experiments to control for spatial position in two directions, especially the complete Latin squares to be discussed in the next section. Second, since Knut Vik squares are so easy to construct, they tend to be used more often than other squares. The experimenter is warned not to restrict the randomization to just one or other of these squares. The problem is not with the square, but with the randomization. If the legitimate randomization throws up one of these squares, use it, but do not restrict the randomization to these squares.

6.5.2 Complete Latin Squares

A *row complete* Latin square is a Latin square that has every distinct pair of treatments in adjacent cells in a row exactly once. Similarly, a *column complete* square is one with every distinct pair of treatments in adjacent cells in a column exactly once. We encounter these squares in Chapter 9 as designs balanced for carryover effects. A *complete* Latin square is both row complete and column complete. Complete Latin squares are ideal for comparing treatments when spatial correlation is a concern. These designs control how often each treatment is in a neighboring plot to each other treatment.

Row complete $n \times n$ squares are known to exist for all even n and a few odd values. For even n, a row complete square can be developed by writing the first row as

0	n	1	$n-1$	2	·	·	·	$n/2-1$	$(n-1)$	$n/2$

Subsequent rows are created by adding one to each element and reducing modulo n when necessary, then adding two to each element, and so on, to complete the n rows. A complete square is obtained by rearranging rows so that the first column has the same structure as the displayed row.

A *quasi-complete* Latin square is defined as a Latin square with pairs of elements adjacent to each other in either order, exactly twice in the columns and exactly twice in the rows. This is much less restrictive than the ordering required for complete squares. Quasi-complete $n \times n$ Latin squares with n odd exist. Complete Latin squares are useful as polycross designs in agricultural work.

This is a very specialized area, and we refer the interested reader to papers by Freeman (1979a,b, 1981) and the references in those papers.

6.5.3 Semi-Latin Squares

The semi-Latin square design has been used in agricultural research work and is recommended occasionally, despite the fact that there are some questions about the validity of the usual error term in the analysis of variance. These plans appear to be particularly suitable for greenhouse experiments where plants are arrayed in rows and columns on benches. Bailey (1992) defines the semi-Latin square as a design consisting of n^2k experimental units divided into n rows and n columns in such a way that the intersection of each row and each column contains k experimental units, and nk treatments assigned in such a way that each row and each column contains a complete replicate. This definition is not really complete in the sense that it does not clearly define the randomization to be used. The question of randomization then affects the subsequent statistical analysis of the data. It also appears that there may be some confusion in terminology. We begin our discussion by quoting Yates (1935), where he writes: "In this arrangement there are twice (or more generally k times) as many rows as columns, and every treatment occurs once in each column and once in each pair of (or each k) rows which are treated as a unit." Some have restricted the definition to $k = 2$, which gave twice as many rows as columns and hence the name. We consider k more general.

Partitioning the nk treatments into n sets of k and then assigning the sets to the $n \times n$ Latin square yields a split-plot experiment. The actual assignment of treatments to units is random within each set, in addition to the Latin square randomization. Proper statistical analysis of data from such experiments is discussed in Chapter 7. There are no questions about the nature or validity of the statistical analysis of data from such an experiment. However, one of the implications of such an analysis is that some treatment comparisons are more precise than others.

Rojas and White (1957) define the semi-Latin square somewhat differently. In their definition, the nk treatments are assigned in sets of k to the $n \times n$ Latin square with the sole restriction that each treatment appear in every row and every column exactly once. They give the semi-Latin square with 12 treatments and three rows and three columns shown in Table 6.11. This is not a split-plot experiment and it is not clear how the randomization was or should be performed. Rojas and White (1957) show that the error mean square from the analysis of variance of data from this design is biased because the model does not account for all the correlations among the errors, and that in general it is not possible to predict the direction of the bias. They also present the results of an analysis using corn yield uniformity data showing relative biases in the error term as large as 20%, but usually less than 10%.

Darby and Gilbert (1958) define the semi-Latin square in terms of k mutually orthogonal Latin squares. They partition the treatments into k sets of n. Each set is then randomized on a Latin square. The squares are then superimposed

Table 6.11 Semi-Latin Square from Rojas and White

	Column											
Row	1				2				3			
1	2	6	3	5	1	11	10	8	4	9	7	12
2	11	7	12	8	2	4	5	9	3	1	10	6
3	1	9	4	10	3	7	12	6	2	8	11	5

on one another. Finally, the subsets of k at the row–column intersections are randomized. This still has some of the flavor of a split-plot experiment, although it is not exactly a split-plot. Their analysis of variance includes rows with $n - 1$ degrees of freedom, columns with $n - 1$ degrees of freedom, treatments with $nk - 1$ degrees of freedom, whole-plots (adjusted for treatments) with $(n - 1)^2$ degrees of freedom and a split-plot error with $(n - 1)(nk - n - 1)$ degrees of freedom by subtraction. They derive the expected values of the corresponding mean squares, estimate variance components, and compute appropriate standard errors for treatment contrasts. The reader is urged to consult the paper for details. An example of the (Darby and Gilbert) semi-Latin square (without proper randomization) for 12 treatments in four rows and columns constructed via the three orthogonal Latin squares:

```
1 2 3 4    1 2 3 4    1 2 3 4
2 1 4 3    3 4 1 2    4 3 2 1
3 4 1 2    4 3 2 1    2 1 4 3
4 3 2 1    2 1 4 3    3 4 1 2
```

is shown in Table 6.12. One can obtain estimates of σ_u^2 and σ_ϵ^2 by using the following SAS® code:

```
proc glm ;
class row col wp trt ;
model y = row col trt wp  ;
random wp ; run ;
```

Alternatively, one can use `proc MIXED` to obtain standard errors for all treatment differences using the following code:

```
proc mixed ;
class row col wp trt ;
model y = row col trt ;
random wp ;
lsmeans trt/diff = all ; run ;
```

The estimates of the contrasts and the estimates of the variances of the contrasts are unbiased. However, the statistical tests and confidence intervals are approximate, because variance components are used to construct the variances and degrees of freedom must be approximated. Our suggestion is to consider using Youden squares in place of the semi-Latin square, if at all possible.

Table 6.12 Semi-Latin Square Given by Darby and Gilbert

	Column											
Row	1			2			3			4		
1	1	5	9	2	6	10	3	7	11	4	8	12
2	2	7	12	1	8	11	4	5	10	3	6	9
3	3	8	10	4	7	9	1	6	12	2	5	11
4	4	6	11	3	5	12	2	8	9	1	7	10

6.6 FREQUENCY SQUARES

Frequency squares, or *F-squares* as they have come to be known, represent a class of designs, variations of the Latin squares, that deserve much more consideration than they are normally given. They were formally introduced into the statistical literature by Addelman (1967), although a number of papers dealing with topics that are clearly related (Finney 1945, 1946a, b; Freeman 1966) had appeared earlier. The concept is most easily introduced via the following 6×6 Latin square:

$$\begin{array}{cccccc} C & A & E & B & F & D \\ E & B & C & D & A & F \\ A & F & B & E & D & C \\ B & D & F & A & C & E \\ D & C & A & F & E & B \\ F & E & D & C & B & A \end{array}.$$

Now transform this square by replacing D with A and E with B and F with C to obtain the F-square:

$$\begin{array}{cccccc} C & A & B & B & C & A \\ B & B & C & A & A & C \\ A & C & B & B & A & C \\ B & A & C & A & C & B \\ A & C & A & C & B & B \\ C & B & A & C & B & A \end{array}.$$

The result is a six-row, six-column design for three treatments, each replicated 12 times. We will use the notation $F(6; 2, 2, 2)$ to indicate that two treatment symbols, two treatment symbols, and two treatment symbols have been collapsed to one symbol each. Alternatively, we can also take this to mean that of the $6 \times 6 = 36$ units in the square, 6×2 receive the first treatment, 6×2 the second treatment, and 6×2 the third treatment. Similarly, we write $F(6; 2, 2, 1, 1)$ for the design obtained by replacing D with A and E with B, and leaving F and C. Here we have 6×2 units assigned to the first and second treatments and 6×1 to each of the other treatments. In this notation $F(6; 1, 1, 1, 1, 1, 1)$ would represent a 6×6 Latin square. Other possible F-squares are $F(6; 3, 3)$ and $F(6; 4, 2)$. We see immediately that there is a large

Table 6.13 Analysis of Variance Table

Source	df
Mean	1
Rows	5
Columns	5
Treatments	2
Error	23
Total	36

variety of F-squares. The derivation from the Latin squares also makes it clear that each treatment will appear equally frequently in each column and in each row of the square. It follows that treatment contrasts are orthogonal to both row and column effects.

Notice that $F(6; 4, 2)$ and $F(6; 2, 2, 1, 1)$ both have at least one treatment with more replication than another. However, frequencies are proportional and the simple analysis of variance formulas hold. Notice that our notation does not uniquely identify an individual F-square. For example, there are several different $F(6; 2, 2, 2)$ squares. Also, this definition is not consistent with some of the notation used in the literature.

The model for the analysis of variance is related to the model for the Latin square. We write the model for the $n \times n$ F-square with k treatments as

$$y_{ij} = \mu + r_i + c_j + \tau_{d(i,j)} + \epsilon_{ij},$$

where $i = 1, \ldots, n$, $j = 1, \ldots, n$, and $d(i, j) = 1, \ldots, k$. Notice that as is usual with the Latin square, not all combinations of subscripts appear. The analysis of variance for $F(6; 2, 2, 2)$ takes the form shown in Table 6.13. Since each treatment is replicated 12 times, the variance of an estimated treatment difference is $\sigma_\epsilon^2/6$.

In an F-square it is also possible to extract information about the row by treatment and column by treatment interactions. If these interactions are included in the model, the simple hand computation formulas are not appropriate, but the general linear tests of Section 4.2.1 can be used. General linear model software such as SAS® `Proc GLM` can be used to produce these test statistics in the analysis of variance table. If row or column effects are random, the usual F-ratio is not appropriate, and generalized least squares should be used.

Exercise 6.12: Independence Assumption in Analysis of F-Square. Show that any contrast, say the difference between two row means, is uncorrelated with any contrast among treatment means. From this argue that the row sum of squares and the treatment sum of squares are independent, provided that the $\{\epsilon_{ij}\}$ are independent, identically distributed normal variables. □

6.6.1 Applications

The immediate potential application that comes to mind is experimental material that has distinct row and column effects, but is not suitable for a Latin square, because of the restriction that number of rows and columns must equal the number of treatments or because of the sparsity of error degrees of freedom for small squares. For example, a 6×6 array requires six treatments. An investigator may find it convenient to use only three treatments, i.e., an $F(6; 3, 2, 1)$ design or an $F(6; 2, 2, 2)$ design. Similarly, a very appealing set of possible $F(4; 2, 2)$ designs provides for two treatments in a 4×4 square. Burrows et al. (1984) observed that there are important row and column effects among wells in microtiter plates used in enzyme-linked immunosorbent assay (ELISA) techniques. Standard microtiter plates have an 8×12 array of wells. F-squares provide a convenient tool for constructing appropriate designs with these restrictions.

In field experiments it is not unusual to find that an experiment is to be conducted on plots that have been used previously and contain residual effects from prior treatments. If the prior experiment was a 6×6 Latin square, this presents difficulties. However, 6×6 F-squares orthogonal to the Latin square provide useful options.

The orthogonal F-squares provide a very flexible tool in cases where one has a third blocking factor, but the constraints imposed by a Graeco–Latin square are too severe. Finally, there are cases where an ongoing procedure has been based on using standard Latin square designs, but suddenly for some reason the number of treatments is reduced. The F-square is the obvious design to substitute.

6.6.2 Orthogonal F-Squares

The concept of orthogonal squares extends directly from the Latin square to the F-square. In both cases the definition hinges on the frequency of pairings when one square is superimposed on another. Recall that for two Latin squares to be orthogonal, we required that all pairs of symbols, one from the first square and one from the second square, appear together when one square was superimposed on the other. Consider two F-squares, $F(n; a_1, a_2, \ldots, a_k)$ and $F(n; b_1, b_2, \ldots, b_m)$. In the first, treatment i appears in na_i cells and in the second treatment, j appears in nb_j cells. The two squares are said to be *orthogonal* if when the two squares are superimposed on one another, the combination, treatment i from the first and treatment j from the second will appear $a_i b_j$ times. For example, the two $F(4; 2, 2)$ squares

$$
\begin{array}{cccc}
1 & 0 & 1 & 0 \\
1 & 0 & 1 & 0 \\
0 & 1 & 0 & 1 \\
0 & 1 & 0 & 1
\end{array}
\qquad
\begin{array}{cccc}
1 & 1 & 0 & 0 \\
0 & 0 & 1 & 1 \\
0 & 0 & 1 & 1 \\
1 & 1 & 0 & 0
\end{array}
$$

are orthogonal. These two squares were obtained from

$$\begin{matrix} A & B & C & D \\ C & D & A & B \\ D & C & B & A \\ B & A & D & C \end{matrix}$$

by first replacing A and C with 1 and B and D with 0 and then replacing A and B with 1 and C and D with 0.

Exercise 6.13: Orthogonality of F-Squares. Since an ordinary Latin square design is a special case of an F-square, it makes perfectly good sense to talk about a Latin square and an F-square being orthogonal. Verify that the two 6×6 squares

$$\begin{matrix} 1 & 2 & 3 & 4 & 5 & 6 \\ 2 & 1 & 6 & 5 & 4 & 3 \\ 3 & 4 & 1 & 2 & 6 & 5 \\ 4 & 6 & 5 & 1 & 3 & 2 \\ 5 & 3 & 2 & 6 & 1 & 4 \\ 6 & 5 & 4 & 3 & 2 & 1 \end{matrix} \quad \text{and} \quad \begin{matrix} 1 & 2 & 3 & 4 & 2 & 5 \\ 4 & 5 & 2 & 3 & 1 & 2 \\ 2 & 3 & 2 & 5 & 4 & 1 \\ 2 & 1 & 4 & 2 & 5 & 3 \\ 5 & 4 & 1 & 2 & 3 & 2 \\ 3 & 2 & 5 & 1 & 2 & 4 \end{matrix}$$

are orthogonal. □

6.6.3 Constructing Orthogonal F-Squares

We begin by pointing out that even for $N > 2$ and $N \neq 6$, it is not always possible to find a Latin square that is orthogonal to an existing Latin square. For example, it is not difficult to show that one cannot find a Latin square orthogonal to

$$\begin{matrix} A & B & C & D \\ D & A & B & C \\ C & D & A & B \\ B & C & D & A \end{matrix}.$$

To show this, one begins by trying to locate *transversals* of this square. A transversal (often called a *directrix*) of an $n \times n$ Latin square is a set of n cells, one from each row, one from each column, and containing one of each of the symbols. To illustrate, for the 4×4 square given, try finding a transversal by selecting A from row 1 and B from row 2. This forces C from column two, row four, and it is impossible to find D. A little trial and error quickly reveals that no transversal exists. The treatments in one square of a pair of orthogonal Latin squares form transversals of the other square.

By showing that the 4×4 square given does not have a transversal, we have verified that it is not possible to construct an orthogonal Latin square. It is also impossible to construct an $F(4{:}3,1)$ that is orthogonal to the 4×4. However, it is interesting to note that an $F(4{:}2,2)$ orthogonal to the 4×4 exists. Begin by selecting A and B from the first row of the Latin square and match these with zeroes in the F-square. Then select any pair of letters from the second row. Then examine pairs from the third row and see if the F-square can be completed. Three different solutions are

0	0	1	1		0	1	0	1		1	0	0	1
0	1	1	0		0	1	0	1		0	1	1	0
1	1	0	0		1	0	1	0		1	0	0	1.
1	0	0	1		1	0	1	0		0	1	1	0

We are aware of no simple general method for finding orthogonal F-squares. Finding transversals (or establishing that one does not exist) is simple. For an $n \times n$ Latin square, there are n choices from the first row, $n - 1$ from the second row, and so on. Once a transversal has been found, the process can be repeated to find the second, and so on. If there are no transversals, one works with pairs of cells, from the first row, pairs from the second, and so on. In general, large F-squares can be obtained by using a Kronecker product, like construction using smaller F-squares. If A_1 and A_2 are F-squares with k_1 and k_2 treatments, respectively, then $A_1 \circledast A_2$ is an F-square with $k_1 k_2$ treatments. If B_1 and B_2 are F-squares with k_1 and k_2 treatments, and A_1 and B_1 are orthogonal, and A_2 and B_2 are orthogonal, then $A_1 \circledast A_2$ and $B_1 \circledast B_2$ are orthogonal. This construction is explained in more detail in Appendix 6B.

6.6.4 Special 6 × 6 Squares

Since the 6×6 square is such a thorn in practice, we include a series of examples. Recall that in Section 6.4.1 we had two 6×6 squares that were close to being orthogonal. If we convert one of them to a $F(6; 2, 1, 1, 1, 1)$ by replacing levels 5 and 6 by a new level, we have a pair of orthogonal squares.

There is a set of four mutually orthogonal $F(6; 2, 2, 2)$ squares:

0	0	1	1	2	2		0	0	2	2	1	1
0	0	1	1	2	2		1	2	0	1	2	0
1	1	2	2	0	0		1	1	0	0	2	2
1	1	2	2	0	0		2	0	1	2	0	1
2	2	0	0	1	1		2	2	1	1	0	0
2	2	0	0	1	1		0	1	2	0	1	2

0	2	1	0	1	2		0	1	2	1	0	2
2	1	2	1	0	0		2	2	1	0	1	0
1	0	2	1	2	0		1	2	0	2	1	0
0	2	0	2	1	1		0	0	2	1	2	1.
2	1	0	2	0	1		2	0	1	0	2	1
1	0	1	0	2	2		1	1	0	2	0	2

It is not known whether or not these are part of a larger set. As far as we know, the theory of maximal possible sizes of sets of mutually orthogonal F-squares of various types and sizes is still open. The interested reader is referred to *Discrete Mathematics Using Latin Squares* by Laywine and Mullen (1998).

The next example is not an F-square but is similar. It is a plan for nine treatments with four replications in a 6×6 square:

$$\begin{array}{cccccc}
A & B & D & E & G & H \\
B & C & E & F & H & I \\
C & A & F & D & I & G \\
G & H & A & B & D & E \\
H & I & B & C & E & F \\
I & G & C & A & F & D
\end{array}.$$

The analysis of variance requires a general linear models program. Pairs of treatments fall into three classes, two of which are equivalent. Some pairs lie in a common row twice and a common column twice. An example is the A and E pair. Some pairs lie in a common row twice and a common column four times. An example is A and B. An equivalent set consists of those in a common row four times and a common column twice. An example of this is the B and E pair. Variances of estimated treatment differences are $.56\sigma^2$ or $.61\sigma^2$, compared to $.5\sigma^2$ if this were a completely randomized design. (Keep in mind that σ^2 in a completely randomized design would be larger.) In a completely randomized design, correlations between treatment differences would be either .0 or $\pm.5$. In this design, with the removal of row and column effects, the correlations deviate by less than .05 from the .0 or $\pm.5$ values.

The final example is a plan for four treatments with nine replications in a 6×6 square:

$$\begin{array}{cccccc}
A & B & C & D & A & C \\
D & A & B & C & B & D \\
C & D & A & B & D & A \\
B & C & D & A & C & B \\
A & B & D & C & A & B \\
B & C & A & D & C & D
\end{array}.$$

The analysis of variance again requires a general linear models program. All estimated treatment differences have variance $.24\sigma^2$, which compares to $2\sigma^2/9$ for the completely randomized design. The correlations between treatment differences adjusted for row and column effects are equal to .0 or $\pm.5$.

6.7 YOUDEN SQUARE

Youden proposed these plans as a solution to a problem that he repeatedly encountered in his research work conducted in a greenhouse. The problem was that the plants were in pots on long, narrow benches along the wall. The light coming in from the side had different effects on the various rows of pots. There was also a gradient going in the other direction from one end of the greenhouse to the other. This would dictate a Latin square design; however, the bench was too narrow for a full Latin square. Youden's proposal was to use a subset of the rows in a Latin square.

Rather than begin with a general definition, consider the first three rows of the 7 × 7 Latin square:

$$\begin{array}{ccccccc} A & B & C & D & E & F & G \\ B & C & D & E & F & G & A \\ D & E & F & G & A & B & C \\ C & D & E & F & G & A & B \\ E & F & G & A & B & C & D \\ F & G & A & B & C & D & E \\ G & A & B & C & D & E & F \end{array}.$$

This is a *Youden square*. The special features of this design are that each letter appears exactly once in each row. Also, each letter appears either once or not at all in a column. In addition, when we examine pairs of letters, for example (A, B) or (E, F), we find that all of the $\binom{7}{2} = 21$ possible (unordered) pairs of treatments appear exactly once in a column. The implication is that every treatment difference can be measured within a column exactly once. This suggests "nice" statistical properties.

It is informative to consider the combinatorics of the situation for a moment. There are seven treatments. This implies $\binom{7}{2} = 21$ unordered pairs of treatments. Since each column has three places, there is room for exactly $\binom{3}{2} = 3$ treatment pairs in each column, and since there are seven columns, $3 \times 7 = 21$ treatment pairs in all. This suggests that there is a possibility that one can find an arrangement of seven treatments each replicated three times in a 3 × 7 array, such that every pair appears exactly once in a column. The first three rows of the given square represent such an arrangement. This is a Youden square (even though it is a rectangle rather than a square).

Alternatively, we can consider a 4 × 7 array. Now there is room for $\binom{4}{2} \times 7 =$ 42 pairs or 21 pairs, each two times. The bottom four rows of the given square do contain each of the 21 unordered pairs exactly two times—hence a 4 × 7 Youden square for seven treatments. It is also clear that we cannot set up such a nice pattern using five rows. In this case there is room for 10 pairs in a column, for a total of 70 pairs. We cannot fit in 21 pairs, each an equal number of times.

6.7.1 Possible Applications

One of the first applications of the Youden square design was for greenhouse experiments. The scenario consists of seven treatments: two gradients (along and across the bench), and room for only three or four rows, but room for all seven columns on a bench. The standard 8 × 12 microtiter plates used in ELISA work have room for two 5 × 4 and two 7 × 4 Youden square designs.

Another very useful application of this design is in the construction of questionnaires for interviewing people. Here the scenario might consist of seven questions and seven subjects, but each subject can only tolerate three or four questions, because they become tired (or irritated). If this happens, the subjects might tend to respond differently to a question depending on whether it is for the first, second, or third question on the questionnaire. The Youden square allows for the removal of possible position effects.

6.7.2 Statistical Analysis

The Youden square is really a special case of a balanced incomplete block design (BIB). In Chapter 8 we discuss these designs in detail. It is informative to proceed in an intuitive manner, although we will eventually give a formal approach to the statistical analysis. Specifically, we consider the design with three rows and seven columns discussed above. First of all, label the observations:

$$\begin{array}{ccccccc} y_{11A} & y_{12B} & y_{13C} & y_{14D} & y_{15E} & y_{16F} & y_{17G} \\ y_{21B} & y_{22C} & y_{23D} & y_{24E} & y_{25F} & y_{26G} & y_{27A} \\ y_{31D} & y_{32E} & y_{33F} & y_{34G} & y_{35A} & y_{36B} & y_{37C} \end{array}.$$

The standard analysis uses the same model as for a Latin square:

$$y_{ijd(i,j)} = \mu + r_i + c_j + \tau_{d(i,j)} + \epsilon_{ijd(i,j)},$$

where $i = 1, 2$ or 3, $j = 1, 2, \ldots, 7$ and $d(i, j) = A, B, \ldots, G$.

The estimate of the treatment A effect is estimated from within the columns that include that treatment and is based on

$$\frac{2y_{11A} - y_{21B} - y_{31D} + 2y_{35A} - y_{15E} - y_{25F} + 2y_{27A} - y_{17G} - y_{37C}}{7}.$$

To see what this estimates, substitute the model (without the $\{\epsilon_{ijd(i,j)}\}$) and simplify. This term estimates

$$\begin{aligned} \frac{6\tau_A - \tau_B - \tau_C - \tau_D - \tau_E - \tau_F - \tau_G}{7} \\ &= \frac{7\tau_A - (\tau_A + \tau_B + \tau_C + \tau_D + \tau_E + \tau_F + \tau_G)}{7} \\ &= \tau_A - \overline{\tau}_{\cdot\cdot} \end{aligned}$$

This is free of row and column effects. We can do the same thing for $\tau_B - \overline{\tau}_{\cdot}$, $\tau_C - \overline{\tau}_{\cdot}$, and so on. Also, these can be combined to form estimates of $\tau_A - \tau_B$.

Exercise 6.14: Treatment Difference in Youden Square. It has been shown that the linear function

$$\frac{2y_{11A} - y_{21B} - y_{31D} + 2y_{35A} - y_{15E} - y_{25F} + 2y_{27A} - y_{17G} - y_{37C}}{7}$$

is an unbiased estimate of $\tau_A - \overline{\tau}_{\cdot\cdot}$ Display the corresponding estimate of $\tau_B - \overline{\tau}_{\cdot\cdot}$ Combine the two into one linear function and then verify that $\text{Var}(\widehat{\tau}_A - \widehat{\tau}_B) = 6\sigma^2/7$. □

These computations are a special case of the computations for a balanced incomplete block design. The general forms for balanced incomplete block designs use the following quantities:

$$kQ_i = kT_i - B(t)_i \qquad \text{for } i = 1, \ \ldots, t, \tag{6.4}$$

where

T_i = sum of observations on treatment i

$B(t)_i$ = sum of block totals for those blocks (columns in our case) that contain treatment i

k = number of observations per block

r = number of replicates (in this case, $r = k$).

In general $\widehat{\tau_i} - \widehat{\overline{\tau}}. = rQ_i/\lambda t$, where t is the number of treatments and λ is the number of times that two treatments appear together in the same column. In this example, $k = r = 3$, $t = 7$, and $\lambda = 1$. Also, $\text{Var}\,[\widehat{\tau_i} - \widehat{\tau_{i'}}] = 2k\sigma^2/\lambda t$. This result is general, not obvious, but will be verified in Chapter 8.

For any balanced incomplete block design, the sum of squares for treatments eliminating columns (or blocks) is equal to $k/\lambda t \sum_i Q_i^2$.

Example 6.2. This experiment (from Natrella (1963) involves a study of seven thermometers designated by the symbols $A, B, C, \ldots, G$. These thermometers were placed in a water bath in order to compare their readings. However, the temperature in the bath could not be kept exactly constant. Three readings were to be made on each thermometer. There was reason to suspect that the order in which the thermometers in a bath were read would be a factor. The plan of the study and the data are shown in Table 6.14.

The thermometers had scale divisions of .1 of a degree and were read to the third place using a magnifying glass. Only the last two places were reported in Natrella (1963) and are given in Table 6.14. Each thermometer appeared once in each position, i.e., first once, second once, and third once. Hence, the order sum of squares is orthogonal to (free of) the differences between thermometers. One might initially have been tempted to ignore that effect in the design. Interestingly,

Table 6.14 Study Plan and Data for Thermometer Example

	Order of Readings within Set:					
Set	1		2		3	
1	*A*	56	*B*	31	*D*	35
2	*E*	16	*F*	41	*A*	58
3	*B*	41	*C*	53	*E*	24
4	*F*	46	*G*	32	*B*	46
5	*C*	54	*D*	43	*F*	50
6	*G*	34	*A*	68	*C*	60
7	*D*	50	*E*	32	*G*	38

Table 6.15 Analysis of Variance of Thermometer Readings

Source	df	SS	MS
Mean	1	39, 260.190	
Order	2	15.524	7.762
Sets	6	497.333	82.889
Thermometers	6	2640.000	440.000
Error	6	15.143	2.524
Total	21		

the position effect turned out to be appreciable. The complete analysis of variance is given in Table 6.15. There also are big differences between sets, which were expected. The sum of squares due to sets is not free of the effects due to the different thermometers.

The estimated means for thermometers A and B were $\overline{y}_A = 64.2$ and $\overline{y}_B = 41.8$, respectively, both with standard error $s(\overline{y}_d) = 1.02$. The mean difference between these two thermometers is estimated to be $\overline{y}_A - \overline{y}_B = 22.4$ with standard error 1.47.

6.7.3 Construction of Youden Squares

The construction of Youden squares is typically not completely trivial. For example, a different 3×7 Youden square can be obtained from the 7×7 Latin square given at the start by selecting rows 1, 4, and 7. There are collections of Youden square plans in Chapter 13 of Cochran and Cox (1957) and in Natrella (1963). Alternatively, Federer (1955), in *Experimental Design, Theory and Application*, suggests using a method outlined in Smith and Hartley (1948) to convert a BIB with the number of blocks equal to the number of treatments to a Youden square.

The recommended randomization procedure for a Youden square is:

1. Randomly assign the treatments to the symbols.
2. Randomly rearrange the columns.
3. Randomly rearrange the rows.

If extra replication is necessary (i.e., it is necessary to replicate the plan), steps (2) and (3) are repeated to provide a new randomization. However, generally it is preferable to use complete Latin squares rather than replicated Youden squares. Cochran and Cox (1957) present a series of designs obtained by adding a Youden square to a Latin square. These plans tend to be useful in cases where the number of treatments is relatively small, yet a fairly large number of replicates is required. Many of these plans were developed for use with experimentation with fruit trees and other perennial plants. They generally fall into the large class of designs commonly referred to as row–column designs.

Exercise 6.15: Construction of a Youden Square. Show that a Youden square can always be obtained by deleting one row of a Latin square. □

REVIEW EXERCISES

6.1 Suppose that a potter in Sandhills, North Carolina wants to test methods of applying glaze and formulations of glaze. He wants to compare three formulations and two methods of applying each formulation, for a total of six treatment combinations. This potter knows that temperature varies somewhat from front to back in the kiln and may vary from top to bottom as well. There are six shelves in the kiln.

(a) Propose a suitable design for this experiment. Randomly assign treatments and units to positions in the kiln. Draw a diagram of your proposed design.

(b) Write out the model that matches your proposed design.

(c) Write out the sources and degrees of freedom for the analysis of variance.

6.2 In a study of the relationship between forage quality and age of switchgrass at harvest, five steers were randomly assigned to five ages of switchgrass

Table 6.16 Switchgrass Data

Period	Treatment	Animal ID	Dry Matter Intake	Dry Matter Digestion	Average Daily Wt. Gain
1	56	34	1.37	50.48	−.09
	28	751	1.34	52.02	−.57
	0	93	1.83	60.60	.95
	14	69	1.54	53.54	.09
	42	68	1.34	46.58	.38
2	56	93	1.09	47.67	−.61
	28	69	1.54	50.79	−.25
	0	751	2.35	61.64	.21
	14	68	1.62	49.41	−.46
	42	34	1.63	54.21	.00
3	56	69	1.32	46.99	−.20
	28	34	1.68	49.41	−.25
	0	68	1.91	62.34	.79
	14	93	1.50	53.17	.07
	42	751	1.27	51.59	−.60
4	56	751	1.30	46.77	.13
	28	68	1.47	49.83	−.53
	0	69	2.11	63.82	2.00
	14	34	1.79	56.09	.73
	42	93	1.26	46.96	.00
5	56	68	1.44	49.90	.81
	28	93	1.54	52.77	.06
	0	34	2.36	63.82	1.06
	14	751	1.94	58.06	.81
	42	69	1.37	54.19	−.63

(harvest day = 0, 14, 28, 42, or 56) in a Latin square designed by Burns et al. (1997). There were five intake periods for each steer, each lasting 21 days. The amount of dry matter intake (kg/100 kg body weight) and dry matter digestion (%) for the last 14 days of the intake period was recorded for each steer, as well as the steer's average daily weight gain (kg) (Table 6.16).

(a) Draw a diagram of the experimental layout using the actual steer ID numbers and treatment numbers.

(b) Write out the analysis of variance table for dry matter intake. Are there significant differences among ages of switchgrass? State the rejection criterion and your conclusion.

(c) Compute the standard error of the mean difference between two treatments.

6.3 A plant pathologist wants to compare the rate of photosynthesis in five varieties of bean that are being exposed to elevated atmospheric ozone (a major component of smog). The plants are exposed to ozone in one large controlled exposure chamber that can hold up to 30 plants. This chamber looks something like a refrigerator with a large door in the front wall. The lights are located in the center of the ceiling of the chamber, as is the fan for circulating the ozone. Plants are grown in pots in the chamber from the time seeds are planted until harvest.

(a) Propose a design for this experiment. Draw a diagram of the proposed design.

(b) Propose a second design for this experiment and draw its diagram.

(c) What design would you recommend? If you need more information before making a recommendation, what specific information do you need, and what computations would you do with that information in hand?

(d) This experiment could be replicated over time. Would you recommend doing so? What are the issues to consider?

6.4 Lattemai et al. (1996) studied the effect on milk yield of adding molasses to the silage-fed dairy cows. They used a 3×3 Latin square design, with three cows fed three levels of molasses: 0, 40, or 100 liters per ton of fresh red clover, in three 10-day recording periods. The means of two of the response variables that they studied are presented here: milk yield and milk fat, both in kilograms per cow per day. The least significant difference ($\text{LSD} = t_{.975}(\text{df})\sqrt{s(\overline{y}_d - \overline{y}_{d'})}$) for comparing two treatment means (treatment d vs. d') at the $\alpha = .05$ significance level is also included in the accompanying table.

Molasses	Milk Yield	Milk Fat
0	19.8	39
40	20.0	40
100	20.4	39
LSD	.4	.8

(a) Write out an appropriate statistical model for this experiment. Include definitions of all the terms and statements of all necessary assumptions.

(b) Compute the sums of squares for treatments and for error for milk yield, and test whether there is a significant treatment effect. State the null and alternative hypotheses, the test statistic, the p-value, and your conclusion.

6.5 Use the milk fat data of Review-Exercise 6.4 to answer the following questions.

(a) Compute as much of the analysis of variance table as is possible from the data given.

(b) Various authors have reported finding an increase in milk fat concentration with increased consumption of forage. In this study, consumption was higher when molasses was added to the feed, yet the amount of milk fat produced did not appear to be affected by adding molasses. If the true milk-fat means were 39, 40, and 41.5 for 0, 40, and 100 liters of molasses per ton, respectively (i.e., if milk fat actually increases 2.5 kg per cow per day per 100 liters of molasses per ton of fresh red clover), how many cows would be needed to have a power of at least 80% for testing that there is no treatment effect at a 5% significance level?

6.6 In an agricultural experiment comparing eight varieties in an 8×8 Latin square, where the plots are laid out in rows and columns, suppose that you are also interested in testing the effect of two levels of pesticide. The pesticide can be applied to the individual plots, but you can use no more than the 64 plots.

(a) Propose a design for this experiment. Include a drawing of the design. Explain how to randomize the design.

(b) Write out the model for this design. Include all assumptions.

(c) List the desirable properties of the design proposed.

6.7 A sports clothing manufacturer is comparing five different materials for lightweight jackets. They want to test the materials under five different types of exercise. The response variable is a measure of moisture that accumulates inside the jacket when the wearer performs the different types of exercise. This could be affected by the humidity in the outside air.

(a) The manufacturer wants to use five people for this experiment. Would you choose a Latin square or a Graeco–Latin square with humidity as a third blocking factor for this experiment? For each design discuss the logistical issues and scope of inference.

(b) Draw a diagram of your chosen design.

(c) Explain how to randomize the design.

(d) Write down the model for the design, including all assumptions.

6.8 *Experiment.* Compare five methods of computing the corrected total sum of squares for a single sample in SAS®. The five methods are (1) Use the `CSS` function in `Proc Means`; (2) Use a do loop in `Proc IML` and compute $(y_i - \overline{y}_.)^2$ at each iteration and sum these up over all iterations; (3) Compute the vector $\boldsymbol{y} - \overline{y}_.$ in `Proc IML` then use the `ssq` function to compute the corrected sum of squares; (4) Compute the vector $\boldsymbol{r} = \boldsymbol{y} - \overline{y}_.$ in `Proc IML` then compute css= $\boldsymbol{r}^t\boldsymbol{r}$; and (5) Compute the vector $\boldsymbol{r} = \boldsymbol{y} - \overline{y}_.$ in `Proc IML` then use the ## subscript reduction operator to compute the sum of squares of $\boldsymbol{r}$.

The response of interest is the length of time that it takes to compute the sum of squares using each of the five methods. Some blocking factors to consider are (1) the SAS® session, and (2) the order in the session. First generate 10,000 normal random numbers and store them in a data set. Then use this data set for testing the five methods.

APPENDIX 6A: SOME STANDARD LATIN SQUARES

There is only one 3 × 3 reduced Latin square:

```
A B C
B C A.
C A B
```

There are four 4 × 4 reduced Latin squares:

```
A B C D    A B C D    A B C D    A B C D
B A D C    B C D A    B D A C    B A D C
C D B A    C D A B    C A D B    C D A B
D C A B    D A B C    D C B A    D C B A
```

There are two transformation sets of 5 × 5 Latin squares. One representative square from each set, along with the number of standard squares in each transformation set (s) is given below.

```
ABCDE  ABCDE
BADEC  BCDEA
CEABD  CDEAB
DCEAB  DEABC
EDBCA  EABCD
 s=50   s=6
```

There are 22 transformation sets of 6 × 6 Latin squares. One representative square from each set, along with the number of reduced squares in each set (s), is given below.

```
ABCDEF ABCDEF ABCDEF ABCDEF ABCDEF ABCDEF ABCDEF ABCDEF
BCFADE BCFEAD BCFAED BAFECD BAEFCD BAECFD BAFEDC BAFECD
CFBEAD CFBADE CFBADE CFBADE CFBADE CFBADE CEBFAD CFBADE
DEABFC DAEBFC DEABFC DCEBFA DEABFC DEFBCA DCABFE DEABFC
EADFCB EDAFCB EADFCB EDAFBC EDFCBA EDAFBC EFDCBA ECDFBA
FDECBA FEDCBA FDECBA FEDCAB FCDEAB FCDEAB FDEACB FDECAB
s=1080 s=1080 s=1080 s=1080 s=1080 s=54 s=540 s=540

ABCDEF ABCDEF ABCDEF ABCDEF ABCDEF ABCDEF ABCDEF ABCDEF
BCDEFA BAEFCD BAFCDE BAEFCD BCFADE BCAFDE BCAFDE BCAEFD
CEAFBD CFAEDB CEABFD CFABDE CFBEAD CABEFD CABEFD CABFDE
DFBACE DCBAFE DFEACB DEBAFC DAEBFC DFEBAC DFEBCA DEFBAC
EDFBAC EDFCBA ECDFBA EDFCBA EDAFCB EDFCBA EDFABC EFDACB
FAECDB FEDBAC FDBEAC FCDEAB FEDCBA FEDACB FEDCAB FDECBA
s=540 s=360 s=360 s=360 s=180 s=120 s=120 s=120

ABCDEF ABCDEF ABCDEF ABCDEF ABCDEF ABCDEF
BCAFDE BCAEFD BAFEDC BADFCE BAECFD BCAFDE
CABEFD CABFDE CDABFE CFAEBD CEAFDB CABEFD
DFEBAC DFEBAC DFEACB DEBAFC DCFABE DEFABC
EDFACB EDFCBA ECBFAD EDFCAB EFDBAC EFDCAB
FEDCBA FEDACB FEDCBA FCEBDA FDBACE FDEBCA
s=60 s=40 s=36 s=36 s=36 s=20
```

We give a set of 21 reduced 7×7 Latin squares.

```
ABCDEFG ABCDEFG ABCDEFG ABCDEFG ABCDEFG ABCDEFG ABCDEFG
BADCFGE BADCFGE BADCFGE BADCFGE BADCFGE BADCFGE BADCFGE
CDABGEF CDABGEF CDABGEF CDABGEF CDABGEF CDABGEF CDABGEF
DEFGABC DEFGABC DEFGABC DEFGABC DEFGABC DEFGABC DEFGABC
ECGFBAD ECGFBAD ECGFBAD ECGFBAD ECGFBAD ECGFBDA ECGFDAB
FGBECDA FGBEDCA FGEACDB FGEACDB FGBECAD FGEADCB FGBECDA
GFEADCB GFEACDB GFBEDCA GFBEDCA GFEADCB GFBECAD GFEABCD

ABCDEFG ABCDEFG ABCDEFG ABCDEFG ABCDEFG ABCDEFG ABCDEFG
BADCFGE BADCFGE BADCFGE BADCFGE BADCFGE BADCFGE BADCFGE
CDABGEF CDABGEF CDABGEF CDABGEF CDABGEF CDABGEF CDABGEF
DEFGABC DEFGABC DEFGABC DEFGABC DEFGABC DEFGABC DEFGABC
ECGFDAB EFGABCD EFGACDB EFGADCB EFGADCB EGBFCAD EGBFCDA
FGEABCD FGBECDA FGBEDCA FGBECAD FGBECDA FCGEBDA FCGEBAD
GFBECDA GCEFDAB GCEFBAD GCEFBDA GCEFBAD GFEADCB GFEADCB

ABCDEFG ABCDEFG ABCDEFG ABCDEFG ABCDEFG ABCDEFG ABCDEFG
BADCFGE BADCFGE BADCFGE BADCFGE BADCFGE BADCFGE BADCFGE
CDABGEF CDABGEF CDABGEF CDABGEF CDABGEF CDABGEF CDABGEF
DEFGABC DEFGABC DEFGACB DEFGACB DEFGACB DEFGACB DEFGACB
EGBFCDA EGBFDCA ECGFBDA ECGFBDA ECGFBDA ECGFBDA EFGABDC
FCGEDAB FCGEBAD FGBECAD FGBEDAC FGEACBD FGEADBC FGBECAD
GFEABCD GFEACDB GFEADBC GFEACBD GFBEDAC GFBECAD GCEFDBA
```

APPENDIX 6B: MUTUALLY ORTHOGONAL LATIN SQUARES

There are two mutually orthogonal 3×3 Latin squares:

```
A B C          A B C
B C A   and    C A B.
C A B          B C A
```

There are three mutually orthogonal 4×4 Latin squares:

```
A B C D   A B C D   A B C D
B A D C   C D A B   D C B A
C D A B   D C B A   B A D C.
D C B A   B A D C   C D A B
```

In general, if n is a prime number or a power of a prime, it can be shown that there exists at least one (exactly one for $n = 3, 4, 5, 7$, and 8) set of mutually orthogonal Latin squares. If n is not a prime number, the number of mutually orthogonal Latin squares is greatly reduced, and in general the number is unknown. It is not possible to find any 6×6 Latin square that is orthogonal to another; i.e., there is no 6×6 Graeco–Latin square.

If n is a prime number or the power of a prime, there exists a finite field with n elements. Denote these elements by $a_0, a_1, \ldots, a_{n-1}$. Using these elements, a set of $n - 1$ mutually orthogonal Latin squares, $S_1, S_2, \ldots, S_{n-1}$, can be constructed. If we number the rows $0, \ldots, n - 1$ and the columns $0, \ldots, n - 1$, the element in row i, column j of the kth square is given by

$$a_i a_k + a_j.$$

We illustrate first for $n = 5$ and $a_0, \ldots, a_4$ equal to 0, 1, 2, 3, 4. The four mutually orthogonal squares (digits have been replaced by letters) are

```
A B C D E     A B C D E     A B C D E     A B C D E
B C D E A     C D E A B     D E A B C     E A B C D
C D E A B,    E A B C D,    B C D E A,    D E A B C.
D E A B C     B C D E A     E A B C D     C D E A B
E A B C D     D E A B C     C D E A B     B C D E A
```

Seven is a prime number and we can follow the recipe for the 5×5 squares exactly. Eight is the power of a prime. Consequently, we follow the recipe again but must use the elements and arithmetic from the finite field of eight elements. The seven mutually orthogonal 8×8 squares are

```
A B C D E F G H    A B C D E F G H    A B C D E F G H
B A D C F E H G    C D A B G H E F    D C B A H G F E
C D A B G H E F    E F G H A B C D    G H E F C D A B
D C B A H G F E    G H E F C D A B    F E H G B A D C
E F G H A B C D'   F E H G B A D C'   B A D C F E H G'
F E H G B A D C    H G F E D C B A    C D A B G H E F
G H E F C D A B    B A D C F E H G    H G F E D C B A
H G F E D C B A    D C B A H G F E    E F G H A B C D
```

A	B	C	D	E	F	G	H
E	F	G	H	A	B	C	D
F	E	H	G	B	A	D	C
B	A	D	C	F	E	H	G
H	G	F	E	D	C	B	A’
D	C	B	A	H	G	F	E
C	D	A	B	G	H	E	F
G	H	E	F	C	D	A	B

A	B	C	D	E	F	G	H
F	E	H	G	B	A	D	C
H	G	F	E	D	C	B	A
C	D	A	B	G	H	E	F
D	C	B	A	H	G	F	E’
G	H	E	F	C	D	A	B
E	F	G	H	A	B	C	D
B	A	D	C	F	E	H	G

A	B	C	D	E	F	G	H
G	H	E	F	C	D	A	B
B	A	D	C	F	E	H	G
H	G	F	E	D	C	B	A
C	D	A	B	G	H	E	F’
E	F	G	H	A	B	C	D
D	C	B	A	H	G	F	E
F	E	H	G	B	A	D	C

A	B	C	D	E	F	G	H
H	G	F	E	D	C	B	A
D	C	B	A	H	G	F	E
E	F	G	H	A	B	C	D
G	H	E	F	C	D	A	B˙
B	A	D	C	F	E	H	G
F	E	H	G	B	A	D	C
C	D	A	B	G	H	E	F

We complete this appendix by pointing out that when $n = pq$ and p and q are both prime or powers of primes (both larger than 2), one can construct a pair of mutually orthogonal Latin squares using a Kronecker product–like construction. If A_1 and A_2 are mutually orthogonal Latin squares with elements $\{a(1)_{i,j}\}$ and $\{a(2)_{i,j}\}$, respectively, and B_1 and B_2 another pair with elements $\{b(1)_{i,j}\}$ and $\{b(2)_{i,j}\}$, respectively, the two squares $A_1 \circledast B_1$ and $A_2 \circledast B_2$ are mutually orthogonal. By $A \circledast B$ we mean the product

$$\begin{bmatrix} a_{11}b_{11} & \ldots & a_{11}b_{1q} & \ldots & \ldots & a_{1p}b_{11} & \ldots & a_{1p}b_{1q} \\ \vdots & & \vdots & \ldots & \ldots & \vdots & & \vdots \\ a_{11}b_{q1} & \ldots & a_{11}b_{qq} & \ldots & \ldots & a_{1p}b_{q1} & \ldots & a_{1p}b_{qq} \\ \vdots & \vdots & \vdots & \vdots & \vdots & \vdots & \vdots & \vdots \\ \vdots & \vdots & \vdots & \vdots & \vdots & \vdots & \vdots & \vdots \\ a_{p1}b_{11} & \ldots & a_{p1}b_{1q} & \ldots & \ldots & a_{pp}b_{11} & \ldots & a_{pp}b_{1q} \\ \vdots & & \vdots & \ldots & \ldots & \vdots & & \vdots \\ a_{p1}b_{q1} & \ldots & a_{p1}b_{qq} & \ldots & \ldots & a_{pp}b_{q1} & \ldots & a_{pp}b_{qq} \end{bmatrix}.$$

This is a simple exercise to construct a pair of mutually orthogonal 9×9 Latin squares from the two mutually orthogonal 3×3 Latin squares.

Leonhard Euler, who introduced Latin squares in 1783, conjectured that there were no orthogonal Latin squares of order 6, 10, 14 or any order of the form $4n + 2$.

A famous pair of orthogonal 10×10 Latin squares (with rows and columns rearranged to have the first in reduced form) was constructed by Bose et al. in 1960, which established the falsity of the Euler conjecture:

```
A B C D E F G H I J
B F G A H I J C D E
C H I J F E D B A G
D G A B I J C E F H
E I J H A G F D C B
F A B C J D E G H I
G J H I C B A F E D
H C D E G A B I J F
I D E F B C H J G A
J E F G D H I A B C
```

and

```
A C E G J I H F D B
J G B D C A F I H E
B H I J E D C A G F
I A C E B G J H F D
C J H I F E D B A G
H B D F A J I G E C
D I J H G F E C B A
G D F A I H B E C J
F E G B H C A D J I
E F A C D B G J I H
```

.

A very readable discussion of the construction of mutually orthogonal Latin squares, i.e., Graeco-Latin squares, can be found in Laywine and Mullen (1998).

APPENDIX 6C: POSSIBLE YOUDEN SQUARES

The following is a list of possible Youden squares with fewer than 100 treatments and r not more than 10 rows or replications.

Possible Youden squares are:

$t = b$	$r = k$	λ	$t = b$	$r = k$	λ
3	2	1	15	7	3
4	3	2		8	4
5	4	3	16	6	2
6	5	4		10	6
7	3	1	19	9	4
	4	2		10	5
	6	5	21	5	1
8	7	6	25	9	3
9	8	7	31	6	1
10	9	8		10	3
11	5	2	37	9	2
	6	3	57	8	1
	10	9	73	9	1
13	4	1	91	10	1
	9	6			

where t is the number of treatments and b is the number of blocks (columns). For the Youden square $t = b$. Also r is the number of times each treatment appears in the experiment and k is the number of observations in each block. For the Youden square $r = k$. Finally, λ, is the number of times that two distinct treatments will appear together in a common block (column) (this is constant for all pairs of treatments in an experiment).

CHAPTER 7

Split-Plot and Related Designs

7.1 INTRODUCTION

There are two chapters in this book that deal with split-plot and related designs. In this, the first, we consider the case of two (or more) factors, but without a formal factorial treatment structure. The emphasis will be on reasons for selecting the design, the appropriate randomization, the advantages and disadvantages of the design, the error structure, and the subsequent statistical analysis of the data. We will see that the error structure, which is a consequence of the randomization, determines the form of the analysis of variance and also the standard errors of contrasts among various treatment means. In Chapter 22 we superimpose a formal factorial treatment structure on the design and examine both the possibilities and the ramifications.

7.2 BACKGROUND MATERIAL

The early development of the split-plot designs was in agriculture research in response to a very common problem in field experiments. An example of the problem encountered can be described as follows: A scientist wishes to study two factors, say methods of cultivation and varieties of grain. The first factor, with t levels, requires the use of large complex equipment and consequently, demands relatively large plots of land. This and cost immediately place a restriction on the number of plots. Natural field variability also implies relatively large plot-to-plot variance. Because of the nature of the equipment used for planting the grain, the second factor, which consists of s varieties, can be accommodated in much smaller plots. The large plots are split into smaller plots at the planting stage. This in turn has two implications: (1) there is less variability among these smaller plots because they are closer together, and (2) there are many more of

Planning, Construction, and Statistical Analysis of Comparative Experiments,
by Francis G. Giesbrecht and Marcia L. Gumpertz
ISBN 0-471-21395-0

them. More plots and less variability among plots means smaller standard errors for contrasts, i.e., more information.

The standard solution gives an experiment with two strata. The whole-plot stratum, the base experiment, consists of large plots organized as either a CRD, RCBD, or Latin square with room for the t levels of the first factor. The split-plot stratum consists of the split-plots, smaller plots obtained by splitting each of the large plots into s parts. The treatments assigned to the large whole-plots are replicated r times, and treatments assigned to the split-plots are replicated rt times. There is much more information on the split-plot factor, not only because of the extra replication, but also a smaller split-plot-to-split-plot variance. The interaction contrasts (between whole- and split-plot treatment) also fall into the split-plot stratum and benefit from the smaller variance.

Since split-plot experiments represent an inherently sensible way to organize many experiments, they occur frequently in research work. One occasionally hears the comment that an experimenter should assign the most important factor to the split-plot stratum and the less important factor to the whole-plot stratum. This is true. However, in practice, external constraints often dictate the assignment of factors.

The distinguishing feature of all split-plot (or split-unit) experiments is that there are two distinct randomizations. The first, in stratum 1, takes place when the levels of the whole-plot treatment are randomly assigned to the whole-plots. The second, in stratum 2, takes place as the levels in the split-plot treatment are assigned in the split-plots. When one is examining the results of an experiment designed by someone else and has difficulty establishing the exact nature of the randomization employed, it is a good idea to suspect a split-plot experiment. Also, one should be aware of the possibility of strip-plot designs. These are discussed later in the chapter. Many split-plot plans can easily be modified to become strip-plot experiments. These have advantages and disadvantages.

7.3 EXAMPLES OF SITUATIONS THAT LEAD TO SPLIT-PLOTS

7.3.1 Agriculture Field Experiments

It is clear from the discussion in Section 7.2 that variations of the split-plot or split-unit design occur frequently in agricultural research, and many examples can be found in statistics textbooks. We look at two such examples.

The first example, which has become a classic, is the alfalfa cutting experiment that first appeared in Snedecor (1946) and in numerous books since then. In the experiment, three varieties of alfalfa were assigned to plots in an RCBD with six blocks. In 1943 the complete plots were harvested twice before being split into four split-plots per whole-plot. In each whole-plot, three split-plots were harvested, on Sept. 1, Sept. 20, and Oct. 7, respectively, and a fourth was not harvested. The measured response was the yield on the split-plots in 1944. The object of the experiment was to see if late harvesting in one year would have a deleterious effect on the subsequent year's yield.

The merit of the example is that it clearly illustrates the nature of the split-plot experiment. The varieties were planted in relatively large plots. This required one randomization. Subsequently, the whole-plots were partitioned into split-plots and the second treatment assigned. This required a second randomization. Each variety was assigned to six whole-plots, while each split-plot treatment (date of cutting) was assigned to 18 split-plots. The experiment provides more information (smaller standard errors for treatment differences) for the split-plot treatment. The nature of the experiment dictated that varieties be assigned to whole-plots. Of course, in practice this sort of experiment would also be replicated across several years. We also note in passing that one can identify three strata: (1) blocks defined within total area; (2) whole-plots defined within blocks; and (3) split-plots defined within whole-plots. Treatments have been assigned to only the last two strata.

The second example that illustrates principles involved in the split-plot experiment is the oat seed treating experiment discussed in Section 16.3 of Steel et al. (1997). In this experiment lots of oat seeds obtained from four different sources constitute one factor and different fungicide treatments of the seed constitute the second factor. The actual treatment consists of physically mixing measured amounts of fungicide powder with specified amounts of seed. There were three fungicides and a control with no fungicide. The 16 seed source $\times$ fungicide treatment combinations must be planted separately. We infer from Steel et al.'s description of the experiment that within each block, each seed source was randomly assigned to a group of four plots and then fungicide treatments randomly assigned within groups. Responses were obtained on the 16 plots within each block. Steel et al.'s analysis of variance of a set of data from the experiment indicates that the variance for testing differences among seed sources is about 3.5 times as large as the variance for testing fungicide differences. The experimenter has taken advantage of the fact that one can sort 16 experimental units into groups of four that are more nearly alike than units in the complete set. The groups of four constitute whole-plots.

We notice, however that other options would have been possible. First, it would have been possible to perform the experiment as a 4×4 factorial in an RCBD with four blocks of 16. In this case there would be only one randomization, and it would not have been a split-plot experiment. This experiment would not restrict plots with the same seed source to be physically close together. Plots with seeds from different sources would be just as likely to be close together. In this case, the variance for testing variety differences would be equal to the variance for testing fungicide treatment differences. It is reasonable to assume that the variance for these tests would be a weighted average of the two error variances observed, or roughly 1.5 times as large as the variance for testing the split-plot factor.

A third alternative would have been to assign fungicide treatments to whole-plots and seed sources to split-plots. Notice that all three options involve roughly the same amount of work. In this last case it is reasonable to assume that the variance for testing fungicide treatment differences would have been three to four times as large as the variance for testing variety differences. The latter should be of the same order as the variance for testing fungicide treatment differences in the experiment actually performed.

The key feature in this example is that the experimenter had options when assigning treatments to strata. The experimenters chose to obtain better information on fungicide treatment differences and sacrifice some information on seed source differences. In many cases the nature of the treatments dictate the assignment to whole-plots or split-plots.

Exercise 7.1: Variations on a Split-Plot Experiment

(a) In the second example in Section 7.3.1, we discussed two possible variations of a split-plot experiment. Draw a diagram for each variation, illustrating a potential randomization outcome.

(b) In the same section we also mentioned that it would have been possible to conduct the experiment in an RCBD. Draw a diagram illustrating how a typical treatment assignment would appear in this case and how it differs from the split-plot. □

7.3.2 Environmental Engineering Experiments

Air stripping is an efficient, relatively low-cost procedure to remove volatile organic compounds from contaminated groundwater. The procedure is to let the contaminated water flow down through a packed column while air is injected into the bottom of the column. In a research study several different packing materials as well as combinations of water and airflow are to be tested. It is relatively simple to shift the flow rates of water and air. One simply changes the rate of the appropriate pump. Changing packing material is much more difficult. The column must be taken apart and then reassembled. The nature of the factors dictates a split-unit design. A reasonable strategy is to select a packing material randomly, build a column, test the flow rate combinations in random order, select a second packing material, rebuild the column, and so on. Packing materials are assigned to a whole-unit stratum, and water and airflow rates are assigned to a split-unit stratum. Complete randomization of the order of treatments, i.e., a CRD, would lead to an exorbitant increase in amount of work because the column would have to be repacked many more times.

7.3.3 The Food Industry

Split-plot experiments have a natural place in the food production industry. Consider a hypothetical cake baking study. Assume that r recipes and c baking conditions are to be studied. There are a number of possibilities. A simple split-plot experiment with recipes as a whole-plot factor and baking condition as a split-plot factor can be set up if cake batters are made up using recipes in random order. Each batch of batter is then split into c portions. The portions are then baked under the c conditions. A new random baking order is selected for each batter. Replication is provided by repeating recipes.

An alternative scheme would be to make up enough batter to make one cake from each of the r recipes. All r are then baked at one time in an oven at one of the c conditions. Now we have an experiment with baking conditions as a whole-plot factor and recipe as a split-plot factor.

A third alternative that has some appeal is a strip-plot design. Now the experimenter would make up batches of each of the batters large enough, to partition each into c cakes and then bake the cakes in sets, with one cake of each recipe in each set. One can think of this as a row and column construction. The rows represent recipes and the columns, baking conditions. The advantage here is that in the absence of replication, only r batches of batter need be mixed and the oven need only be set up c times. Strip-plot designs are discussed in detail in Section 7.6.

7.3.4 Industrial Quality Research

Box and Jones (1992) describe an example of a split-plot experiment as part of a robust product design project. They credit the example to John Schuerman and Todd Algrim, engineers with DRG Medical Packaging Company in Madison, Wisconsin. The object of the project was to develop a package material that would give a better seal under the wide range of possible sealing process conditions used by potential customers. The package manufacturer identified a number of factors that were believed to affect the quality of the seal. In the whole-plot part of the experiment, sample lots of eight different packaging materials were produced. These lots of material were then sent to a customer's plant, where they were each subdivided into six sublots. The sublots were used in six different sealing processes. This constitutes the split-plot part of the study.

7.4 STATISTICAL ANALYSIS OF SPLIT-PLOT EXPERIMENTS

7.4.1 Split-Plot Experiment with Whole-Plots in a CRD

Statistical Model

The statistical model for a split-plot mirrors the design of the experiment. The two randomization steps in the split-plot experiment, one in each stratum, dictate a model with two error terms. We consider an experiment with whole-plots arranged in a CRD. If we let W represent the whole-plot treatment and S the split-plot treatment, the linear statistical model is written as

$$y_{ijk} = \mu + w_i + \epsilon(1)_{ij} + s_k + (w \times s)_{ik} + \epsilon(2)_{ijk}, \qquad (7.1)$$

where

$$\left.\begin{aligned} \epsilon(1)_{ij} &\sim \text{iid}(0, \sigma_1^2) \\ \epsilon(2)_{ijk} &\sim \text{iid}(0, \sigma_2^2) \end{aligned}\right\} \text{ mutually independent}$$

for $i = 1, \ldots, t$, $j = 1, \ldots, r$ and $k = 1, \ldots, s$.

The whole-plot stratum of the model contains the whole-plot treatment effects, $\{w_i\}$ and the whole-plot error terms, $\{\epsilon(1)_{ij}\}$. If we include the mean μ, this part of the model is analogous to the model for the CRD given in Section 3.3.2. The split-plot stratum contains the split-plot treatment effects $\{s_k\}$, the whole-plot treatment $\times$ split-plot treatment interaction effects $\{(w \times s)_{ik}\}$, and $\{\epsilon(2)_{ijk}\}$ the experimental errors associated with individual split-plots. All nonrandom terms on the right-hand side of the model (except μ) are assumed to be deviations from their respective means.

Analysis of Variance

The analysis of variance for the split-plot experiment in the CRD is an extension of the analysis for the CRD. There are in a sense, separate analyses with separate error terms for the two strata. The formal analysis of variance is given in Table 7.1.

The W $\times$ S sum of squares is conveniently computed as $\sum_i^t \sum_k^s r\overline{y}_{i\cdot k}^2 - rst\overline{y}_{\dots}^2 -$ SSW$-$SSS, where SSW is the whole-plot treatment sum of squares and SSS is the split-plot treatment sum of squares. The error(2) sum of squares is conveniently obtained by subtraction. Entries for a fourth column in the table (not shown) titled "Mean Squares" is obtained by dividing sum of squares entries by degrees of freedom.

Expected Mean Squares

The expected values of the mean squares in Table 7.1 are given in Table 7.2. These expected values are derived using the methods of Section 5.4. The quantities ϕ_w, ϕ_s, and $\phi_{w\times s}$ represent quadratic forms, $\sum_j w_j^2/(t-1)$, $\sum_k s_k^2/(s-1)$, and $\sum_{jk}(w \times s)_{jk}^2/((t-1)(s-1))$, respectively. These quadratic forms will be zero under the appropriate null hypotheses. From the expected mean squares it follows directly that error(1) is used to test the hypothesis of no whole-plot treatment effect and error(2) used to test hypotheses of no interaction or split-plot treatment effects. The test for interaction is performed first, since significance at that level establishes the effects of treatments and makes other tests moot.

Table 7.1 Degrees of Freedom and Sums of Squares for a Split-Plot Experiment with Whole-Plots Arranged in a CRD

Source	df	SS
W	$t-1$	$rs\sum_i^t(\overline{y}_{i\cdot\cdot} - \overline{y}_{\dots})^2$
Error(1)	$t(r-1)$	$s\sum_i^t\sum_j^r(\overline{y}_{ij\cdot} - \overline{y}_{i\cdot\cdot})^2$
S	$s-1$	$rt\sum_k^s(\overline{y}_{\cdot\cdot k} - \overline{y}_{\dots})^2$
W $\times$ S	$(t-1)(s-1)$	$r\sum_i^t\sum_k^s(\overline{y}_{i\cdot k} - \overline{y}_{i\cdot\cdot} - \overline{y}_{\cdot\cdot k} + \overline{y}_{\dots})^2$
Error(2)	$t(r-1)(s-1)$	$\sum_i^t\sum_j^r\sum_k^s(y_{ijk} - \overline{y}_{ij\cdot} - \overline{y}_{i\cdot k} + \overline{y}_{i\cdot\cdot})^2$
Total (corrected)	$rts-1$	$\sum_i^t\sum_j^r\sum_k^s(y_{ijk} - \overline{y}_{\dots})^2$

Table 7.2 Expected Mean Squares and F-Ratios for a Split-Plot Experiment with Whole-Plots Arranged in a CRD

Source	E[MS]	F-ratio
W	$\sigma_2^2 + s\sigma_1^2 + rs\phi_w$	MSW/MSE(1)
Error(1)	$\sigma_2^2 + s\sigma_1^2$	
S	$\sigma_2^2 + rt\phi_s$	MSS/MSE(2)
W × S	$\sigma_2^2 + r\phi_{w\times s}$	MSW × S/MSE(2)
Error(2)	σ_2^2	
Total		

The analyses in the whole-plot stratum and the split-plot stratum are orthogonal since all levels of each factor are tested in combination with every level of the other factor. Estimates of interactions between whole- and split-unit factors are contrasts that are orthogonal to both whole-plot and split-plot treatment contrasts.

Exercise 7.2: Orthogonality of Contrasts. Show that the whole-plot contrast $\sum_i c_i \overline{y}_{i\cdot\cdot}$ is orthogonal to the split-plot contrast $\sum_k c_k^* \overline{y}_{\cdot\cdot k}$. □

Example 7.1. Since there is considerable algebra involved in the discussion in the next section, we include a small artificial example for reference purposes. Let A be the whole-plot factor at three levels and with two replicates and B the split-plot factor at four levels. The data are given in Table 7.3. The analysis of variance is given in Table 7.4.

Standard Errors of Main-Effect Contrasts

The standard errors in the split-plot are somewhat more complex than in the designs encountered up to this point. Consider first contrasts among levels of the split-plot treatment. We write the general form of a split-plot contrast as $\sum_k c_k \overline{y}_{\cdot\cdot k}$. A contrast is a linear combination of treatment means in which

Table 7.3 Simulated Data for a Small Split-Plot Experiment

		Level of B			
Level of A	Replication	1	2	3	4
1	1	11.04	8.05	9.89	9.47
	2	15.59	11.16	11.42	12.34
2	1	10.89	6.86	7.32	11.04
	2	7.43	3.55	6.44	7.71
3	1	8.21	7.69	8.99	11.74
	2	10.34	10.35	7.65	14.68

Table 7.4 Analysis of Variance Table for Data in Table 7.3

Source	df	MS	F-Value
A	2	24.88	1.95
Error(1)	3	12.78	
B	3	14.23	12.16
$A \times B$	6	4.15	3.55
Error(2)	9	1.17	
Total	23		

$\sum_k c_k = 0$. We compute the expected value by writing the contrast in terms of the model,

$$\mathrm{E}\Big[\sum_k c_k \overline{y}_{\cdot\cdot k}\Big] = \sum_k c_k \mathrm{E}\big[\overline{y}_{\cdot\cdot k}\big]$$

$$= \sum_k c_k s_k. \tag{7.2}$$

Since only random effects contribute to the variance, and the whole-plot errors cancel out we write

$$\mathrm{Var}[\sum_k c_k \overline{y}_{\cdot\cdot k}] = \mathrm{Var}\Big[\sum_k c_k\Big(\sum_i \sum_j \epsilon(2)_{ijk}/rt\Big)\Big]$$

$$= \sum_k c_k^2 \sigma_2^2/rt. \tag{7.3}$$

Since MSE(2) is an unbiased estimate of σ_2^2, it follows that the estimated standard error of a split-plot treatment contrast is of the form

$$s\Big(\sum_k c_k \overline{y}_{\cdot\cdot k}\Big) = \sqrt{\sum_k c_k^2 \mathrm{MSE}(2)/rt}. \tag{7.4}$$

A special but important case is the difference, $\overline{y}_{\cdot\cdot k} - \overline{y}_{\cdot\cdot k'}$ where $k \neq k'$ is obtained by setting $c_k = 1$ and $c_{k'} = -1$ and all other coefficients equal to zero. This has expected value $s_k - s_{k'}$ and variance $2\sigma_2^2/rt$. It follows that

$$s(\overline{y}_{\cdot\cdot k} - \overline{y}_{\cdot\cdot k'}) = \sqrt{2\mathrm{MSE}(2)/rt}. \tag{7.5}$$

For example, in the small artificial example, one can look at the difference between the first two levels of B. The $\{c_k\}$ are $\{1, -1, 0, 0\}$. The contrast has value $10.58 - 7.94 = 2.64$. The estimated standard error is $\sqrt{[1^2 + (-1)^2 + 0^2 + 0^2]1.17/6} = .62$. Confidence intervals computed for contrasts and are based on $t(r-1)(s-1)$ degrees of freedom.

The general form of a whole-plot treatment contrast takes the form $\sum_i c_i \overline{y}_{i\cdot\cdot}$. This has expected value

$$\mathrm{E}\Big[\sum_i c_i \overline{y}_{i\cdot\cdot}\Big] = \sum_i c_i \mathrm{E}[\overline{y}_{i\cdot\cdot}]$$
$$= \sum_i c_i w_i. \tag{7.6}$$

In this case the whole-plot errors come into the variance

$$\mathrm{Var}\Big[\sum_i c_i \overline{y}_{i\cdot\cdot}\Big] = \mathrm{Var}\Big[\sum_i c_i \Big(\sum_j \epsilon(1)_{ij}/r\Big) + \sum_i c_i \Big(\sum_j \sum_k \epsilon(2)_{ijk}/rs\Big)\Big]$$
$$= \sum_i c_i^2 (\sigma_2^2 + s\sigma_1^2)/rs. \tag{7.7}$$

Since MSE(1) is an unbiased estimate of $\sigma_2^2 + s\sigma_1^2$, it follows that the estimated standard error of a whole-plot treatment contrast of the form is

$$s\Big(\sum_i c_i \overline{y}_{i\cdot\cdot}\Big) = \sqrt{\sum_i c_i^2 \mathrm{MSE}(1)/rs}. \tag{7.8}$$

For whole-plot treatment differences we have

$$s(\overline{y}_{i\cdot\cdot} - \overline{y}_{i'\cdot\cdot}) = \sqrt{2\mathrm{MSE}(1)/rs}. \tag{7.9}$$

For example, one can look at the difference between the first and third levels of A in the small artificial example. The constants $\{c_i\}$ are $\{1, 0, -1\}$. The contrast has value $11.12 - 9.96 = 1.16$. The estimated standard error of the contrast is $\sqrt{(1^2 + 0^2 + (-1)^2)12.78/8} = 1.79$. Confidence intervals for contrasts will be based on $t(r-1)$ degrees of freedom.

Standard Errors of Interaction Contrasts

If there is significant whole-plot treatment by split-plot treatment interaction, matters are more complex. First consider standard errors for contrasts among the split-plot treatment levels at a given whole-plot treatment level. Such contrasts can be written as $\sum_k c_k \overline{y}_{i\cdot k}$ given any specific choice for i. An example of this sort of contrast would be the quadratic effect of split-plot factor B with whole-plot factor A at a specific level. It follows that expected value

$$\mathrm{E}\Big[\sum_k c_k \overline{y}_{i\cdot k}\Big] = \sum_k c_k \mathrm{E}\Big[\overline{y}_{i\cdot k}\Big]$$
$$= \sum_k c_k \big(s_k + (w \times s)_{ik}\big). \tag{7.10}$$

Also,

$$\begin{aligned}\operatorname{Var}\Big[\sum_k c_k \overline{y}_{i\cdot k}\Big] &= \operatorname{Var}\Big[\sum_k c_k\Big(\sum_j \epsilon(2)_{ijk}/r\Big)\Big]\\ &= \sum_k c_k^2\sigma_2^2/r. \qquad (7.11)\end{aligned}$$

It follows that in general

$$s\Big(\sum_k c_k \overline{y}_{i\cdot k}\Big) = \sqrt{\sum_k c_k^2 \text{MSE}(2)/rt} \qquad (7.12)$$

and in particular for the difference between two split-plot treatment means at a given whole-plot treatment level,

$$s(\overline{y}_{i\cdot k} - \overline{y}_{i\cdot k'}) = \sqrt{2\text{MSE}(2)/rt}. \qquad (7.13)$$

Alternatively, it may be reasonable to inquire about contrasts among whole-plot treatment levels, either at the same split-plot treatment level or at different split-plot treatment levels. In general, these contrasts are of the form $\sum_i \sum_k c_{ik}\overline{y}_{i\cdot k}$ where $\sum_i \sum_k c_{ik} = 0$. A simple example of such a contrast could be the quadratic effect of A, the whole-plot factor at a specific level of B, the split-plot factor. A more complex contrast would involve a comparison between the quadratic effects of the whole-plot factor at two different split-plot factor levels. Again, it follows that expected value

$$\text{E}\Big[\sum_i\sum_k c_{ik}\overline{y}_{i\cdot k}\Big] = \sum_i\sum_k c_{i\cdot k}w_k + \sum_i\sum_k c_{ik}(w \times s)_{ik}.$$

This contrast has variance

$$\begin{aligned}\operatorname{Var}\Big[\sum_i\sum_k c_{ik}\overline{y}_{i\cdot k}\Big] &= \operatorname{Var}\Big[\sum_i\sum_k c_{ik}\Big(\sum_j \epsilon(1)_{ij}/r\Big)\\ &\quad + \sum_i\sum_k c_{ik}\Big(\sum_j \epsilon(2)_{ijk}/r\Big)\Big]\\ &= \sum_i\sum_k c_{ik}^2\sigma_1^2/r + \sum_i\sum_k c_{ik}^2\sigma_2^2/r\\ &= \sum_i\sum_k c_{ik}^2(\sigma_1^2+\sigma_2^2)/r. \qquad (7.14)\end{aligned}$$

At this point we encounter a difficulty. There is no error mean square with expected value $\sigma_1^2 + \sigma_2^2$ in the analysis of variance (Table 7.2).

However,

$$\mathrm{E}\left[\frac{(s-1)\mathrm{MSE}(2)+\mathrm{MSE}(1)}{s}\right]=\sigma_1^2+\sigma_2^2.$$

Notice that this can be interpreted as the weighted average of the two error mean squares. It follows that

$$\widehat{\mathrm{Var}}\Big[\sum_i\sum_k c_{ik}\overline{y}_{i\cdot k}\Big]=\sum_i\sum_k\frac{c_{ik}^2}{r}\left[\frac{(s-1)\mathrm{MSE}(2)+\mathrm{MSE}(1)}{s}\right]. \qquad (7.15)$$

The corresponding standard error is the positive square root.

Two special cases of this that are frequently of interest are differences between two whole-plot treatment means at either the same split-plot treatment level or at two different split-plot treatment levels. The estimate of the standard error for these comparisons is

$$\sqrt{\frac{2}{r}\left[\frac{(s-1)\mathrm{MSE}(2)+\mathrm{MSE}(1)}{s}\right]}. \qquad (7.16)$$

In our example, this can be used to look at differences such as $\overline{y}_{1\cdot 1}-\overline{y}_{2\cdot 2}$.

To test hypotheses or to construct confidence intervals, the degrees of freedom associated with the standard errors are required. In this case two error mean squares have been combined and there is no exact count for degrees of freedom. We use the formula given by Satterthwaite (1946) to give an approximate value. First, note that the expression displayed in (7.15) is a special case of the form

$$\sum_i^m a_i\mathrm{MSE}(i). \qquad (7.17)$$

This is approximately distributed as a χ^2 random variable with

$$\Big[\sum_i^m a_i\mathrm{MSE}(i)\Big]^2\Big/\sum_i^m\frac{\left[a_i\mathrm{MSE}(i)\right]^2}{\mathrm{df}_i} \qquad (7.18)$$

degrees of freedom, where MSE(i) has df_i degrees of freedom. For the variance in equation (7.15) we have $m=2$, $a_1=(s-1)$, $a_2=1$, $\mathrm{df}_1=(r-1)(t-1)-1$, and $\mathrm{df}_2=t(r-1)(s-1)-1$.

7.4.2 Variance Components, Variance of y, and Variance of Treatment Means

The variance among whole-plots and the variance among split-plots within a whole-plot both contribute to the variance of y. $\mathrm{Var}[y_{ijk}]=\sigma_1^2+\sigma_2^2$. We find an unbiased estimator for the variance among whole-plots, σ_1^2, by comparing the expectations of MSE(1) and MSE(2). Hence,

$$\widehat{\sigma_1^2} = \frac{\text{MSE}(1) - \text{MSE}(2)}{s}.$$

In the small artificial example, $\widehat{\sigma_1^2} = (12.78 - 1.17)/4 = 2.90$. The total variance of y (unexplained by factors A and B) is estimated to be $\widehat{\text{Var}}\left[y_{ijk}\right] = 2.90 + 1.17 = 4.07$, and the variance among whole-plots accounts for $2.90/4.07 = 71\%$ of the total.

The variability of whole-plots also contributes to the variance of a treatment mean,

$$\text{Var}[\overline{y}_{i \cdot k}] = \frac{\sigma_1^2 + \sigma_2^2}{r}.$$

In the small artificial example, $r = 2$ and the standard error of the mean of one combination of the levels of factors A and B is $s(\overline{y}_{i \cdot k}) = \sqrt{4.07/2} = 1.43$.

7.4.3 Error(2) Bigger Than Error(1)

We can see from the model given in equation (7.1) and even more clearly from the expected mean squares for the analysis of variance shown in Table 7.2 that error(1) is greater than or equal to error(2). The entire structure of split-plot experiments is built on this premise. Yet in practice it sometimes occurs that the estimate MSE(2) of error(2) is larger than MSE(1), the estimate of error(1). The question then is: What does one do in the statistical analysis? At this point there is no universally agreed-upon answer. Some replace MSE(1) with MSE(2). Others ignore the whole-plots all together and reanalyze the data as a factorial experiment. In effect, this pools the sums of squares for the two error terms. Both of these strategies entail a shift in the model. One model provided the basis for the construction of the plan and the randomizations and a different model is used for the final statistical analysis. We take the view that one should stick with the original model, the one dictated by the randomization and proceed with the analysis. Accept the fact that estimated standard errors of split-plot treatment differences are larger than standard errors of whole-plot treatment differences. This agrees with the recommendation by Janky (2000). His review of the literature and a simulation study indicate that the anticipated gain in power, the usual motivation for pooling, is typically not realized. The cost incurred by a rule to pool error terms is an increase in the size of the test (the Type I error rate) and if one attempts to adjust for this, there is an additional risk of power loss.

We must remember that we are dealing with realizations of random variables and that the problem of estimates of split-plot errors being too large can happen due to chance under any circumstance. It will happen occasionally when the investigator is using the correct model and all assumptions are valid. However, if the problem occurs too frequently in an area of research, there is reason for concern. It suggests that some assumption is not valid, that there are negative correlations within whole-plots, or possibly that there is something wrong in the randomizations. There may be some interaction among experimental units that induces some unanticipated correlation structure.

7.4.4 Split-Plot Experiment with Whole-Plots in an RCBD

Statistical Model

This design differs from the preceding one in that the whole-plots are organized into r blocks. The experiment still involves two strata, two randomizations. The model for the whole-plot stratum now includes a term for the block effects:

$$y_{hik} = \mu + b_h + w_i + \epsilon(1)_{hi} + s_k + (w \times s)_{ik} + \epsilon(2)_{hik}, \tag{7.19}$$

where $\{b_h\}$ is the part of the response attributed to differences among blocks. As discussed in Chapter 5, the block effects may be either fixed or random, depending on the way the blocks were selected. If the block effects are assumed random, the model includes the additional assumption that

$$b_h \sim \text{iid}(0, \sigma_b^2),$$

where all three random effects are uncorrelated with each other; i.e., $b_h, \epsilon(1)_{hi}$, and $\epsilon(2)_{hik}$ are all mutually uncorrelated. In our discussion we assume that the blocks are random.

Analysis of Variance

The structure of the analysis of variance again follows naturally after the pattern established in Section 7.4.1 and Chapter 5. The key again is that the experiment consists of two strata, two different types of experimental units. The analysis of variance table is shown in Table 7.5. Notice also that the error(1) sum of squares is usually calculated as

$$s \sum_h^r \sum_i^t \overline{y}_{hi\cdot}^2 - rts\,\overline{y}_{\cdots}^2 - \text{SS blocks} - \text{SSW}$$

and error(2) sum of squares is calculated by subtraction.

Formally, the SSE(l) or whole-plot error sum of squares can be computed and thought of as the interaction of whole-plot treatments and blocks. The error is measured as the failure of the treatments to perform identically across blocks. Also, the split-plot error can be computed and/or thought of as two pieces, the interaction of split-plot treatments and blocks with $(s-1)(r-1)$ degrees of freedom and the three-way interaction of whole-plots, split-plots, and blocks with $(t-1)(s-1)(r-1)$ degrees of freedom. There is rarely any reason to keep these separate, although one could use the first for error to test split-plot

Table 7.5 Analysis of Variance for a Split-Plot Experiment with Whole-Plots in an RCBD

Source	df	SS
Blocks	$r-1$	$st\sum_h^r(\overline{y}_{h\cdot\cdot} - \overline{y}_{\cdot\cdot\cdot})^2$
W	$t-1$	$rs\sum_i^t(\overline{y}_{\cdot i\cdot} - \overline{y}_{\cdot\cdot\cdot})^2$
Error(1)	$(r-1)(t-1)$	$s\sum_h^r\sum_i^t(\overline{y}_{hi\cdot} - \overline{y}_{h\cdot\cdot} - \overline{y}_{\cdot i\cdot} + \overline{y}_{\cdot\cdot\cdot})^2$
S	$s-1$	$rt\sum_k^s(\overline{y}_{\cdot\cdot k} - \overline{y}_{\cdot\cdot\cdot})^2$
W × S	$(t-1)(s-1)$	$r\sum_i^t\sum_k^s(\overline{y}_{\cdot ik} - \overline{y}_{\cdot i\cdot} - \overline{y}_{\cdot\cdot k} + \overline{y}_{\cdot\cdot\cdot})^2$
Error(2)	$(r-1)t(s-1)$	$\sum_h^r\sum_i^t\sum_k^s(y_{hik} - \overline{y}_{hi\cdot} - \overline{y}_{h\cdot k} + \overline{y}_{h\cdot\cdot})^2$
Total (corrected)	$rts-1$	$\sum_h^r\sum_i^t\sum_k^s(y_{hik} - \overline{y}_{\cdot\cdot\cdot})^2$

Table 7.6 Expected Mean Squares and F-Ratios for a Split-Plot Experiment with Whole-Plots Arranged in an RCBD and Random Block Effects

Source	E[MS]	F-Ratio
Blocks	$\sigma_2^2 + s\sigma_1^2 + st\sigma_b^2$	
W	$\sigma_2^2 + s\sigma_1^2 + rs\phi_w$	MSW/MSE(1)
Error(1)	$\sigma_2^2 + s\sigma_1^2$	
S	$\sigma_2^2 + rt\phi_s$	MSS/MSE(2)
W × S	$\sigma_2^2 + r\phi_{w\times s}$	MSW × S/MSE(2)
Error(2)	σ_2^2	
Total		

treatment effects and the second to test for interaction. Mean squares are obtained by dividing sums of squares by degrees of freedom.

Expected Mean Squares

The expected mean squares are derived directly from the model and are shown in Table 7.6. The whole-plot factor is tested by comparing MSW with the whole-plot error, MSE(1). Also, split-plot treatments are applied to the split-plot units, and differences among split-plot treatment means are evaluated using the MSE(2) computed from differences among split-plots.

Standard Errors of Contrasts

The standard errors for treatment and interaction contrasts are the same as for the CRD. Details are given in Section 7.4.1.

Example 7.2. The Japanese study of the effects of elevated atmospheric CO_2 on rice (Example 5.1) was actually a split-plot experiment. There were four blocks of rice paddies (Kim et al. 2001), each corresponding to a different farmer. In each block two paddies were assigned randomly to one of two levels of CO_2, elevated (E) or ambient (A). Relatively large exposure rings were used so that the effects of elevated CO_2 could be studied under actual farming conditions. The exposure ring centers were located at least 90 m apart to minimize the possibility of CO_2 traveling from an elevated CO_2 plot to contaminate an ambient plot. A second factor, amount of nitrogen fertilizer, was applied to split-plots within each rice paddy. Split-plots with different nitrogen levels were separated from each other by a PVC barrier sunk into the ground to prevent paddy water mixing between split-plots. Table 7.7 gives the mean fertility (percent of spikelets that were fertile) for each level of nitrogen in each ring in 1999.

The mean percent of spikelets that are fertile decreased as the level of nitrogen increased and there may have been a slight increase in the number of fertile spikelets as CO_2 concentration increased (Table 7.8). In the tables, the response is labeled y_{hij}, where $h = 1, \ldots, 4$ blocks, $i = 1, 2$ CO_2 levels, and $j = 1, 2, 3$ nitrogen levels. The estimated standard error of the difference between means at two different CO_2 levels is denoted $s(\overline{y}_{\cdot i \cdot} - \overline{y}_{\cdot i' \cdot})$. Reading from Table 7.8, the standard error for the difference between mean fertility under elevated and ambient CO_2 is .55. The effect of elevated CO_2 on the percent of spikelets that are fertile was not large, estimated by a 95% confidence interval to lie somewhere between a decrease of .79 and an increase of 2.71. This 95% confidence interval for the mean difference is computed as $\overline{y}_{\cdot i \cdot} - \overline{y}_{\cdot i' \cdot} \pm t(3, .975)s(\overline{y}_{\cdot i \cdot} - \overline{y}_{\cdot i' \cdot}) = 94.49 - 93.53 \pm (3.182)(.55)$. On the other hand, the effect of nitrogen on fertility was clearly negative. For example, the 95% confidence intervals for the decrease in mean fertility from low nitrogen to high nitrogen under ambient CO_2 is [2.33, 6.27], and under elevated CO_2 it is similar: [3.06, 6.99].

The standard error for a treatment mean is based on the sum of all the variance components in the model: in this case, blocks, whole-plot error, and split-plot error. $\text{Var}[\overline{y}_{\cdot ik}] = (\sigma_b^2 + \sigma_1^2 + \sigma_2^2)/r$. The variance components are estimated by setting the mean squares equal to their expected values and solving for the

Table 7.7 Effects of Elevated Atmospheric CO_2 on Rice Yield: 1999 Plot Means for Each Combination of CO_2 and Nitrogen Level

		Fertile Spikelets (%)			
CO_2	Nitrogen	Block 1	Block 2	Block 3	Block 4
A	L	95.8	95.0	95.5	95.1
A	M	95.2	93.3	95.4	92.9
A	H	92.0	92.0	92.5	87.7
E	L	97.1	95.7	96.3	96.0
E	M	96.2	94.9	96.5	96.3
E	H	93.6	92.6	89.1	89.7

Source: K. Kobayashi, personal communication.

Table 7.8 Mean Fertility (Percent of Spikelets That Were Fertile) and Standard Errors in 1999 Study of Effects of Elevated CO_2 on Rice

Treatment Means		
Nitrogen	Ambient CO_2	Elevated CO_2
Low	95.35	96.28
Medium	94.19	95.94
High	91.05	91.25

Standard Error	Notes
$s(\overline{y}_{\cdot ik}) = .72$	
$s(\overline{y}_{\cdot 1\cdot} - \overline{y}_{\cdot 2\cdot}) = .55$	1 = ambient, 2 = elevated CO_2
$s(\overline{y}_{\cdot\cdot k} - \overline{y}_{\cdot\cdot k'}) = .64$	k, k' indicate different N levels
$s(\overline{y}_{\cdot ik} - \overline{y}_{\cdot ik'}) = .90$	

Table 7.9 Analysis of Variance for 1999 Study of Effects of Elevated CO_2 and Nitrogen Level on Fertility (% of Spikelets That Are Fertile) of Rice

Source	df	SS	MS	F-Ratio	F-Value	p-Value
Block	3	12.53	4.17			
CO_2	1	5.51	5.51	$\frac{\text{MSCO}_2}{\text{MSE(1)}}$	3.06	.18
Error(1)	3	5.39	1.80			
N	2	100.29	50.14	$\frac{\text{MSN}}{\text{MSE(2)}}$	30.80	$< .0001$
$CO_2 \times$ N	2	2.41	1.20	$\frac{\text{MSCO}_2\times\text{N}}{\text{MSE(2)}}$	.74	.50
Error(2)	12	19.54	1.63			

variance components. For this experiment $\widehat{\sigma}_b^2 = (\text{MS blocks} - \text{MSE}(1))/6 = .40$, $\widehat{\sigma}_1^2 = (\text{MSE}(1) - \text{MSE}(2))/3 = .06$, and $\widehat{\sigma}_2^2 = \text{MSE}(2) = 1.63$.

The test of CO_2 ($F = 3.06$, p-value $= .18$, Table 7.9) is not powerful enough to detect any difference in mean fertility between the paddies with elevated CO_2 and those in ambient conditions. Notice that if we had incorrectly constructed the F-ratio using the split-plot error MSE(2) in the denominator, the result would have looked more significant, with an apparent (incorrect) p-value of .09. CO_2 fumigation was performed on an entire ring; hence CO_2 effects are most appropriately compared to variation among rings (i.e., to variation among rice paddies). Nitrogen fertilizer, on the other hand, was applied to split-plots within each exposure ring, so the nitrogen effect is a split-plot factor and should be compared to the within-ring variance. The test statistics for the nitrogen main effect and the interaction of CO_2 and nitrogen both contain MSE(2) in the denominator. Comparing these F-ratios with quantiles of the $F(2, 12)$ distribution, we find that there is a highly significant effect of nitrogen and that the effect of nitrogen is similar for both CO_2 levels. The SAS® code and output for this example, along with data for three response variables, are available on the companion Web site.

7.4.5 Split-Plot Model in the Mixed Model Framework

The process of assigning whole-plot factors to plots and split-plot factors to split-plots within the whole-plots induces a correlated error structure similar to that for randomized block experiments with random block effects (Section 5.5.2). For a split-plot experiment with whole-plots arranged in a CRD [model (7.1)], the covariance for two split-plot observations within the same whole-plot is

$$\begin{aligned} \text{Cov}\left[y_{ijk}, y_{ijk'}\right] &= \text{Cov}\left[\epsilon(1)_{ij} + \epsilon(2)_{ijk}, \epsilon(1)_{ij} + \epsilon(2)_{ijk'}\right] \\ &= \sigma_1^2, \end{aligned}$$

where $k \neq k'$ indicates two different split-plots. On the other hand, the covariance between two split-plot observations in different whole-plots is zero. Since the variance of an observation is $\sigma_1^2 + \sigma_2^2$, it follows that the correlation between two split-plot observations within the same whole-plot is

$$\text{Corr}\left[y_{ijk}, y_{ijk'}\right] = \frac{\sigma_1^2}{\sigma_1^2 + \sigma_2^2}.$$

It is this correlation structure that underlies the form for the analysis of variance.

If the whole-plots are arranged in a randomized complete block design with random block effects, the blocks also contribute to the correlation structure. In this case, observations in different blocks are uncorrelated, those in the same block are correlated with each other, and observations in the same whole-plot are more highly correlated yet. The covariances are

$$\text{Cov}\left[y_{hik}, y_{h'i'k'}\right] = \begin{cases} 0 & \text{if } h \neq h' \\ \sigma_b^2 & \text{if } h = h' \text{ and } i \neq i' \\ \sigma_b^2 + \sigma_1^2 & \text{if } h = h' \text{ and } i = i'. \end{cases}$$

When no data are missing in a split-plot design and when there are no covariates, the expected mean squares have the simple algebraic forms we have seen in previous sections, which lead to F-tests for the whole-plot and split-plot factors. If some data are missing, the expected mean squares no longer have simple forms. In these cases it is often preferable to switch to generalized least squares to estimate contrasts and standard errors (see Section 5.5.2). Note that in the event that no data are missing and there are no covariates, generalized least squares estimators coincide with the estimators presented in Sections 7.4.1 and 7.4.4.

Example 7.2 (continued). The split-plot model for the rice/CO_2/nitrogen experiment may be fit using REML with software such as SAS® `Proc Mixed`. Since the design is a standard split-plot and no data are missing, the generalized least squares estimates have closed-form solutions and the REML estimates of the variance components coincide with the method of moments estimates obtained by setting the mean squares equal to their expectations in Section 7.4.4. The Wald-type F-tests coincide with those of Table 7.9. Note that the estimated generalized least squares estimates of the treatment means coincide with the ordinary treatment means and the ordinary least squares estimates because this is a full factorial with equal sample sizes for all treatments. Using mixed model software (e.g., `Proc Mixed`) does have the advantage that the standard errors so computed take the correlations induced by the split-plot structure into account. Mixed model software produces the same standard errors as developed in Section 7.4.4. Not all of these standard errors can be obtained from ordinary least squares software such as `Proc GLM`. The `Proc Mixed` code and output are available on the companion Web site.

7.4.6 Split-Plot Experiment with Whole-Plots in a Latin Square

Statistical Model

This design differs from previous ones only in that the whole-plots are now organized into a Latin square with t rows and t columns. The experiment still involves two strata, two randomizations. The model for the whole-plot stratum now includes terms for row and column effects. We take model (6.2) from Chapter 6 and incorporate the split-plot feature.

$$y_{ijk} = \mu + r_i + c_j + \tau_{d(i,j)} + \epsilon(1)_{ij} + s_k + (\tau \times s)_{d(i,j)k} + \epsilon(2)_{ijk}, \qquad (7.20)$$

where $i = 1, \ldots, t$, $j = 1, \ldots, t$, $d(i, j) = 1, \ldots, t$, and $k = 1, \ldots, s$. Also, the $\{r_i\}$ are the row effects, $\{c_j\}$ are the column effects, $\{\tau_{d(i,j)}\}$ are the whole-plot treatment effects, $\{s_k\}$ are the split-plot treatment effects, and $\{(\tau \times s)_{d(i,j)k}\}$ are the whole-plot treatment × split-plot treatment interaction effects. In addition, the $\{\epsilon(1)_{ij}\}$ are the whole-plot errors and $\{\epsilon(2)_{ijk}\}$ the split-plot errors. The analysis of variance and standard errors for treatment comparisons follow the pattern established. The basic analysis of variance form is shown in Table 7.10.

7.5 SPLIT-SPLIT-PLOT EXPERIMENTS

All the designs discussed in this chapter can be extended in the sense of further splitting of the experimental units, i.e., splitting the split-plot units into split-split-plots, which are assigned to levels of a third factor. The model for the split-split-plot design with whole-plots in a CRD is

$$\begin{aligned} y_{ijk\ell} = \mu + w_i + \epsilon(1)_{ij} + s_k + (w \times s)_{ik} + \epsilon(2)_{ijk} + ss_\ell, \\ + (w \times ss)_{i\ell} + (s \times ss)_{k\ell} + (w \times s \times ss)_{ik\ell} + \epsilon(3)_{ijk\ell}. \end{aligned} \qquad (7.21)$$

Table 7.10 Expected Mean Squares and F-Ratios for a Split-Plot Experiment with Whole-Plots Arranged in a Latin square

Source	df	E[MS]	F-ratio
Rows	$(t-1)$		
Columns	$(t-1)$		
W	$(t-1)$	$\sigma_2^2 + s\sigma_1^2 + st\phi_w$	MSW/MSE(1)
Error(1)	$(t-1)(t-2)$	$\sigma_2^2 + s\sigma_1^2$	
S	$(s-1)$	$\sigma_2^2 + t^2\phi_s$	MSS/MSE(2)
W × S	$(t-1)(s-1)$	$\sigma_2^2 + t\phi_{w\times s}$	MSW × S/MSE(2)
Error(2)	$t(t-1)(s-1)$	σ_2^2	
Total			

The analysis of variance is a direct extension of the analysis shown in Table 7.1. Standard errors can be derived by extending the formulas in Section 7.4.1. Standard errors for main-effect contrasts are straightforward. Standard errors for interaction contrasts need a bit more care.

7.6 STRIP-PLOT EXPERIMENTS

The strip-plot experiment, also known as a strip-block or split-block experiment, represents a variation of the split-plot experiment that has much to offer in appropriate situations. We introduce the strip-block by way of two examples. In the first, consider the classical field experiment in agriculture. A piece of land available for experimentation is in the form of a relatively large rectangle. The two treatments planned are s modes of seedbed preparation with a large piece of equipment replicated r_c times and seeding v varieties of some crop with a large mechanical planter replicated r_r times. A convenient way to perform the experiment is to divide the field into $r_c s$ strips in one direction and randomly assign modes of seedbed preparation to the strips (Figure 7.1). Each seedbed preparation

Variety	Seedbed Preparation							
	3	1	4	3	1	2	4	2
3								
2								
2								
1								
3								
1								

Figure 7.1 Schematic of a strip-plot experiment with two replicates of four seedbed preparations and two replicates of three varieties.

mode is assigned to r_c strips. For convenience, we call these columns. This is one randomization. It defines one stratum in the experiment.

Next, the field is divided into vr_r strips at right angles to the original (Figure 7.1). The varieties of the crop are randomly assigned to these strips, which we shall call rows, with r_r rows assigned at random to each variety. This establishes a stratum for variety. In addition, it establishes a third stratum for the interaction of seedbed preparation and varieties. To see this, we note that both randomizations affect the assignment in this stratum. We also note that if fertility gradients are suspected, strips (either one or both) can be grouped into sets, i.e., blocking factors introduced in one or both directions.

This is an example of a strip-plot or strip-block experiment where the stripping is dictated by the nature of the experimental treatments. It is a convenient way to organize things if one needs to use large pieces of equipment. Note that eventually the experimenter harvests the subplots defined by the intersection of the row and column strips. A feature of this design is that it provides most information on the interaction.

The second example, a cake baking study, involves a slightly different experimental procedure, but the design principles are the same. Assume that a food product developer wants to develop a new cake recipe. A set of a similar recipes are to be tested. In addition, b different baking regimes are to be tested. A convenient way to organize the study is first to mix cake batters, one batch of each recipe. Make each batch large enough to provide b cakes. Assign one cake from each batch to each of the ovens and each oven to a baking regime. We assume that individual ovens are large enough to hold a cakes. The recipes form rows and the baking regimes form columns, two strata. In addition, there is the third stratum, the interaction. Observations are on individual cakes. This entire procedure is repeated, i.e., replicated, r times.

The big advantage in this experimental design is the reduction in amount of work. The alternative of mixing and baking ab individual cakes for one complete replicate is much more work. Again, the best information is at the interaction level. In many cases this is a good thing since interactions are often very important. This experiment allows the developer to check the robustness of the recipes to variations in baking routine.

7.6.1 Statistical Model for the Strip-Plot

We present the linear statistical model appropriate for the cake baking example:

$$y_{hij} = \mu + \text{rep}_h + a_i + \epsilon(r)_{hi} + b_j + \epsilon(c)_{hj} + (a \times b)_{ij} + \epsilon_{hij} \qquad (7.22)$$

where

$$\left.\begin{aligned} \epsilon(r)_{hi} &\sim \text{iid}(0, \sigma_r^2) \\ \epsilon(c)_{hj} &\sim \text{iid}(0, \sigma_c^2) \\ \epsilon_{hij} &\sim \text{iid}(0, \sigma_\epsilon^2) \end{aligned}\right\} \text{ mutually independent}$$

for $h = 1, \ldots, r$, $i = 1, \ldots, a$, and $j = 1, \ldots, c$. We also assume that the replicate effects $\{\text{rep}_h\}$, recipe effects $\{a_i\}$, and baking regimen effects $\{b_j\}$ are all deviations from a mean and that the interaction effects are defined to sum to zero in both directions. Replicate effects could be assumed to be random effects as well.

Exercise 7.3: Strip-Plot Model. Write out the statistical model for the strip-plot field experiment discussed in Section 7.6 and compare and contrast with the model shown in (7.22). □

7.6.2 Analysis of Variance and Standard Errors

The analysis of variance based on model (7.22) is given in Table 7.11. The expected values of the mean squares for the analysis are given in Table 7.12. The standard errors for row (treatment A) and column (treatment B) comparisons are

$$\sqrt{\sum_i c_i^2[\mathrm{MSE}(r)]/rb}$$

Table 7.11 Analysis of Variance Table for a Strip-Plot Experiment

Source	df	SS
Replications	$r-1$	$ab\sum_h^r(\overline{y}_{h\cdot\cdot} - \overline{y}_{\cdots})^2$
A	$a-1$	$rb\sum_i^a(\overline{y}_{\cdot i\cdot} - \overline{y}_{\cdots})^2$
MSE(r)	$(r-1)(a-1)$	$b\sum_h^r\sum_i^a(\overline{y}_{hi\cdot} - \overline{y}_{h\cdot\cdot} - \overline{y}_{\cdot i\cdot} + \overline{y}_{\cdots})^2$
B	$b-1$	$ra\sum_j^b(\overline{y}_{\cdot\cdot j} - \overline{y}_{\cdots})^2$
MSE(c)	$(r-1)(b-1)$	$a\sum_h^r\sum_j^b(\overline{y}_{h\cdot j} - \overline{y}_{h\cdot\cdot} - \overline{y}_{\cdot\cdot j} + \overline{y}_{\cdots})^2$
A × B	$(a-1)(b-1)$	$r\sum_i^a\sum_j^b(\overline{y}_{\cdot i\cdot} - \overline{y}_{\cdot i\cdot} - \overline{y}_{\cdot\cdot j} + \overline{y}_{\cdots})^2$
MSE(ϵ)	$(r-1)(a-1)(b-1)$	by subtraction
Total (corrected)	$rab-1$	$\sum_h^r\sum_i^a\sum_j^b(y_{hij} - \overline{y}_{\cdots})^2$

Table 7.12 Expected Mean Squares for the Analysis of Variance in Table 7.11

MS	df	Expected Values
MS rep.	$r-1$	—
MS A	$a-1$	$\sigma_\epsilon^2 + b\sigma_r^2 + rb\phi_a$
MSE(r)	$(r-1)(a-1)$	$\sigma_\epsilon^2 + b\sigma_r^2$
MS B	$b-1$	$\sigma_\epsilon^2 + a\sigma_c^2 + ra\phi_b$
MSE(c)	$(r-1)(b-1)$	$\sigma_\epsilon^2 + a\sigma_c^2$
MS AB	$(a-1)(b-1)$	$\sigma_\epsilon^2 + r\phi_{ab}$
MSE(ϵ)	$(r-1)(a-1)(b-1)$	σ_ϵ^2
Total		

and

$$\sqrt{\sum_j c_j^2[\text{MSE}(c)]/ra},$$

respectively, where $\{c_i\}$ and $\{c_j\}$ represent sets of arbitrary constants that sum to zero. Standard errors for interaction contrasts are somewhat more complex. The general form of these contrasts is $\sum_i \sum_j c_{ij}\overline{y}_{\cdot ij}$, where $\{c_{ij}\}$ are arbitrary constant coefficients that sum to zero. For example, one can select:

1. $c_{ij} = 1$, $c_{ij'} = -1$, for some i and some $j \neq j'$ and all other coefficients equal to zero to compare two B treatments at a given A level.
2. $c_{ij} = 1$, $c_{i'j} = -1$, for some $i \neq i'$ and a specific j and all other coefficients equal to zero to compare two A treatments at a given B level.
3. $c_{ij} = 1$, $c_{i'j'} = -1$, for specific $i \neq i'$, and $j \neq j'$ and all other coefficients equal to zero to compare two means for different A and B levels.

The variance of the general contrast form is

$$\begin{aligned}\text{Var}\Big[\sum_i\sum_j c_{ij}\overline{y}_{\cdot ij}\Big] &= \text{Var}\Big[\sum_i\sum_j c_{ij}\sum_h \epsilon(r)_{hi}/r\Big]\\ &\quad + \text{Var}\Big[\sum_i\sum_j c_{ij}\sum_h \epsilon(c)_{hj}/r\Big]\\ &\quad + \text{Var}\Big[\sum_i\sum_j c_{ij}\sum_h \epsilon_{hij}/r\Big] \qquad (7.23)\end{aligned}$$

by the independence assumption. Now we must look at special cases. In case 1 (same level of A, different levels of B) we have $\sum_j c_{ij} = 0$, leaving us with

$$\begin{aligned}\text{Var}\Big[\sum_i\sum_j c_{ij}\overline{y}_{\cdot ij}\Big] &= \text{Var}\Big[\sum_i\sum_j c_{ij}\sum_h \epsilon(c)_{hj}/r\Big]\\ &\quad + \text{Var}\Big[\sum_i\sum_j c_{ij}\sum_h \epsilon_{hij}/r\Big].\end{aligned}$$

This evaluates to

$$\sum_i\sum_j c_{ij}^2[\sigma_c^2 + \sigma_\epsilon^2]/r.$$

An unbiased estimate of $\sigma_c^2 + \sigma_\epsilon^2$ is given by $\big[\text{MSE}(c) + (a-1)\text{MSE}(\epsilon)\big]/a$. There is also the question of degrees of freedom for conducting tests and constructing confidence intervals. We use Satterthwaite's formula, given in expressions (7.17) and (7.18), to compute an approximation. In this case we have $m = 2$, $a_1 = 1$, MSE(1) = MSE(c), $\mathrm{df}_1 = (r-1)(b-1)$, $a_2 = a - 1$, MSE(2) = MSE(ϵ), and $\mathrm{df}_2 = (r-1)(a-1)(b-1)$.

In case 2 we have

$$\mathrm{Var}\Big[\sum_i \sum_j c_{ij}\overline{y}_{\cdot ij}\Big] = \sum_i \sum_j c_{ij}^2\big[\sigma_r^2 + \sigma_\epsilon^2\big]/r.$$

An unbiased estimate of $\sigma_r^2 + \sigma_\epsilon^2$ is given by $\big[\mathrm{MSE}(r) + (b-1)\mathrm{MSE}(\epsilon)\big]/b$. The approximate degrees of freedom are again obtained using expressions (7.17) and (7.18). In this case we have $m = 2$, $a_1 = 1$, MSE(1) = MSE(r), $\mathrm{df}_1 = (r-1)(a-1)$, $a_2 = b - 1$, MSE(2) = MSE(ϵ), and $\mathrm{df}_2 = (r-1)(a-1)(b-1)$.

In case 3 we obtain

$$\mathrm{Var}\Big[\sum_i \sum_j c_{ij}\overline{y}_{\cdot ij}\Big] = \sum_i \sum_j c_{ij}^2\big[\sigma_r^2 + \sigma_c^2 + \sigma_\epsilon^2\big]/r.$$

An unbiased estimate of $\sigma_r^2 + \sigma_c^2 + \sigma_\epsilon^2$ is given by

$$\big[a\mathrm{MSE}(r) + b\mathrm{MSE}(c) + (ab - a - b)\mathrm{MSE}(\epsilon)\big]/ab.$$

The Satterthwaite's approximation now involves $m = 3$, $a_1 = a$, MSE(1) = MSE(r), $\mathrm{df}_1 = (r-1)(a-1)$, $a_2 = b$, MSE(2) = MSE(c), $\mathrm{df}_2 = (r-1)(b-1)$, $a_3 = (ab - a - b)$, MSE(3) = MSE(ϵ), and $\mathrm{df}_3 = (r-1)(a-1)(b-1)$.

Exercise 7.4: Strip-Plot Analysis of Variance. Construct the analysis of variance table corresponding to the field experiment discussed in Section 7.6 and modeled in Exercise 7.3. Compare and contrast with the analysis of variance shown in Table 7.11. □

7.7 COMMENTS ON FURTHER VARIATIONS

The possible extensions to the techniques illustrated in this chapter are limitless. For example, one can proceed to any number of levels of splitting plots. Similarly, the strip-plot design can be extended by splitting strips within strips or columns within columns. It is also possible to extend the design to three or more dimensions. In addition, splits can be incorporated into the interaction cells. A major difficulty is encountered if one finds that the interactions are significant and needs to evaluate standard errors for interaction contrasts. Standard errors for contrasts within individual strata are simple. Matters become more complex when comparisons involve, say, split-split-plot treatment means at different split-plot or whole-plot treatments. In principle, appropriate formulas are derived as in the split-plot examples.

7.8 ANALYSIS OF COVARIANCE IN SPLIT-PLOTS

7.8.1 Analysis of Covariance with One Split-Plot Covariate

The analysis of covariance for the split-plot experiment involves many complications, due to the excessive number of possibilities. For example, it is possible to have a covariate for the whole-plots and not for the split-plots, or for the split-plots and not for the whole-plots. Alternatively, it is possible to have different covariates for the whole and split-plots. Adjustments to the treatment means can be messy. The large number of possible models means that the experimenter must think things through very carefully. There is no simple "one way" to use and adjust for covariates in split-plot experiments. We develop the analysis of covariance model and derive the sums of squares and parameter estimates in detail in this section.

In our development we assume that covariates are part of the experimental units rather than responses to the treatments applied. This implies that treatments do not affect the covariates. Generally, although not always, this means that the covariates are available to the experimenter when the experiment is being planned or executed. Thus, it is important to decide what covariates to measure during the planning stage and take the appropriate measurements before the experiment has begun. In the model, the covariates are observable constants. Our view is that applying the covariance arithmetic to cases where covariates are affected by treatments or represent realizations of random variables is a multivariate analysis technique and is outside the scope of this book.

Developing the Model for One Covariate at the Split-Plot Level

To illustrate some of the problems, we consider a very basic model for a split-plot experiment with whole-plots arranged in an RCBD and one covariate that is associated with the split-plot experimental units,

$$y_{hij} = \mu + r_h + w_i + \epsilon(1)_{hi} + s_j + (w \times s)_{ij} + \beta x_{hij} + \epsilon(2)_{hij}, \qquad (7.24)$$

where $h = 1, \ldots, r$, $i = 1, \ldots, t$ and $j = 1, \ldots, s$ and the covariate is x_{hij}. We will assume that both the whole- and split-plot treatments represent fixed effects, implying that $\overline{w}_{.} = \overline{s}_{.} = \overline{(w \times s)}_{.j} = \overline{(w \times s)}_{i.} = 0$. The $\{x_{hij}\}$ are observed constants. We further assume that

$$\left.\begin{array}{l} \epsilon(1)_{hi} \sim \text{iid } N(0, \sigma_1^2) \\ \epsilon(2)_{hij} \sim \text{iid } N(0, \sigma_2^2) \end{array}\right\} \text{ mutually independent.}$$

We follow the general method of analysis first outlined in Section 5.6.1 and rewrite the model to isolate the covariate's contribution to bias and variance. Since there are two types of experimental units, two sources of error, the first step is to split x_{hij} into two pieces, $\overline{x}_{hi\cdot}$, corresponding to the whole-plots and $(x_{hij} - \overline{x}_{hi\cdot})$ to go with the split-plots. These represent the x_w and x_s that will be introduced. This split makes the algebra for developing the test statistics tractable.

We now generalize the model slightly to allow different regression coefficients, β^w for the whole-plot part of the analysis and β^s for the split-plot part. The model becomes

$$\begin{aligned} y_{hij} &= \mu + r_h + w_i + \beta^w \overline{x}_{hi\cdot} + \epsilon(1)_{hi} + s_j + (w \times s)_{ij} \\ &\quad + \beta^s(x_{hij} - \overline{x}_{hi\cdot}) + \epsilon(2)_{hij}. \end{aligned} \tag{7.25}$$

We return to this point later. What we want to do next is to write the model in a form that explicitly shows how the covariate contributes to the bias of the estimated whole- and split-plot factor effects and the variance components. For this we use the identities

$$\overline{x}_{hi\cdot} = \overline{x}_{\cdots} + (\overline{x}_{h\cdot\cdot} - \overline{x}_{\cdots}) + (\overline{x}_{\cdot i\cdot} - \overline{x}_{\cdots}) + (\overline{x}_{hi\cdot} - \overline{x}_{h\cdot\cdot} - \overline{x}_{\cdot i\cdot} + \overline{x}_{\cdots})$$

and

$$\begin{aligned} (x_{hij} - \overline{x}_{hi\cdot}) &= (\overline{x}_{\cdot\cdot j} - \overline{x}_{\cdots}) + (\overline{x}_{\cdot ij} - \overline{x}_{\cdot i\cdot} - \overline{x}_{\cdot\cdot j} + \overline{x}_{\cdots}) \\ &\quad + (x_{hij} - \overline{x}_{hi\cdot} - \overline{x}_{\cdot ij} + \overline{x}_{\cdot i\cdot}) \end{aligned}$$

to rewrite the model as

$$\begin{aligned} y_{hij} &= \mu + \beta^w \overline{x}_{\cdots} + r_h + \beta^w(\overline{x}_{h\cdot\cdot} - \overline{x}_{\cdots}) + w_i + \beta^w(\overline{x}_{\cdot i\cdot} - \overline{x}_{\cdots}) \\ &\quad + \beta^w(\overline{x}_{hi\cdot} - \overline{x}_{h\cdot\cdot} - \overline{x}_{\cdot i\cdot} + \overline{x}_{\cdots}) + \epsilon(1)_{hi} + s_j + \beta^s(\overline{x}_{\cdot\cdot j} - \overline{x}_{\cdots}) + (w \times s)_{ij} \\ &\quad + \beta^s(\overline{x}_{\cdot ij} - \overline{x}_{\cdot i\cdot} - \overline{x}_{\cdot\cdot j} + \overline{x}_{\cdots}) + \beta^s(x_{hij} - \overline{x}_{hi\cdot} - \overline{x}_{\cdot ij} + \overline{x}_{\cdot i\cdot}) + \epsilon(2)_{hij} \\ &= \mu^* + r_h^* + w_i^* + \beta^w(\overline{x}_{hi\cdot} - \overline{x}_{h\cdot\cdot} - \overline{x}_{\cdot i\cdot} + \overline{x}_{\cdots}) + \epsilon(1)_{hi} + s_j^* + (w \times s)_{ij}^* \\ &\quad + \beta^s(x_{hij} - \overline{x}_{hi\cdot} - \overline{x}_{\cdot ij} + \overline{x}_{\cdot i\cdot}) + \epsilon(2)_{hij}, \end{aligned} \tag{7.26}$$

where

$$\begin{aligned} \mu^* &= \mu + \beta^w \overline{x}_{\cdots} \\ r_h^* &= r_h + \beta^w(\overline{x}_{h\cdot\cdot} - \overline{x}_{\cdots}) \\ w_i^* &= w_i + \beta^w(\overline{x}_{\cdot i\cdot} - \overline{x}_{\cdots}) \\ s_j^* &= s_j + \beta^s(\overline{x}_{\cdot\cdot j} - \overline{x}_{\cdots}) \end{aligned}$$

and

$$(w \times s)_{jk}^* = (w \times s)_{ij} + \beta^s(\overline{x}_{\cdot ij} - \overline{x}_{\cdot i\cdot} - \overline{x}_{\cdot\cdot j} + \overline{x}_{\cdots}).$$

The extra pieces in μ^*, r_h^*, w_i^*, s_j^*, and $(w \times s)_{ij}^*$ represent the contributions to bias from the experimental units via the covariate and $\beta^w(\overline{x}_{hi\cdot} - \overline{x}_{h\cdot\cdot} - \overline{x}_{\cdot i\cdot} + \overline{x}_{\cdots})$ and $\beta^s(x_{hij} - \overline{x}_{hi\cdot} - \overline{x}_{\cdot ij} + \overline{x}_{\cdot i\cdot})$ the contributions to variance. The analysis of covariance provides adjustments to remove all of these.

Considerations When Using Statistical Software

When calculations are performed using regression software or a matrix programming language, one needs to put whole-plot means of the covariate, $\overline{x}_{hi\cdot\cdot}$, into the model as one covariate term and deviations from the whole-plot means, $x_{ijk} - \overline{x}_{hi\cdot}$ as another covariate term. In SAS® `Proc GLM`, the tests of the covariate on both the split- and whole-plot basis are obtained from the type I sums of squares. The remaining tests for treatments and interaction are obtained from the type II or type III sums of squares. The *least squares means* in `Proc GLM` are the fitted values of y evaluated at $\overline{x}_{\cdots\cdot}$. For example, the least squares means of the whole-plot factor are $\widehat{y}_{\cdot i\cdot,adj} = \widehat{\mu} + \widehat{w}_i + \widehat{\beta^w}\overline{x}_{\cdots\cdot}$. These are also known as *adjusted treatment means*. The algebraic forms of the sums of squares and contrasts among treatment means are derived in the following pages.

Analysis of Covariance Details and Derivation of F-Tests

Once model (7.26) has been constructed this way, the whole-plot part of the design matrix is orthogonal to the split-plot part and it is possible to do the analysis of covariance and estimate all the parameters by hand. The first step is to construct a compact analysis of covariance table (Table 7.13) corresponding to the analysis in Table 7.5. The quantities in the column labeled "y-Variable" are the usual analysis of variance sums of squares. The second of the two columns, under the heading "Whole-Plot Covariate," contains sums of squares computed using the whole-plot covariate. The other column contains sums of cross-products involving the y-variable and the whole-plot covariate. Similarly, the two columns under the "Split-Plot Covariate" heading contain sums of squares and cross-products involving the split-plot covariate. The x_w and x_s subscripts identify terms computed using the whole-plot and split-plot covariates, respectively. For instance,

$$\mathrm{E}(1)_{x_w x_w} = s \sum_h \sum_i (\overline{x}_{hi\cdot} - \overline{x}_{\cdot i\cdot} - \overline{x}_{h\cdot\cdot} + \overline{x}_{\cdots})^2$$

and

$$\mathrm{E}(2)_{x_s x_s} = \sum_h \sum_i \sum_j (x_{hij} - \overline{x}_{hi\cdot} - \overline{x}_{\cdot ij} + \overline{x}_{\cdot i\cdot})^2.$$

Table 7.13 Analysis of Covariance Table

Source	y-Variable	Whole-Plot Covariate		Split-Plot Covariate	
Mean	M_{yy}	$\mathrm{M}_{x_w y}$	$\mathrm{M}_{x_w x_w}$		
Blocks	B_{yy}	$\mathrm{B}_{x_w y}$	$\mathrm{B}_{x_w x_w}$		
W	W_{yy}	$\mathrm{W}_{x_w y}$	$\mathrm{W}_{x_w x_w}$		
Error (1)	$\mathrm{E}(1)_{yy}$	$\mathrm{E}(1)_{x_w y}$	$\mathrm{E}(1)_{x_w x_w}$		
S	S_{yy}			$\mathrm{S}_{x_s y}$	$\mathrm{S}_{x_s x_s}$
W × S	$\mathrm{W} \times \mathrm{S}_{yy}$			$\mathrm{W} \times \mathrm{S}_{x_s y}$	$\mathrm{W} \times \mathrm{S}_{x_s x_s}$
Error (2)	$\mathrm{E}(2)_{yy}$			$\mathrm{E}(2)_{x_s y}$	$\mathrm{E}(2)_{x_s x_s}$
Total	T_{yy}				

For convenience we list the following six expected values:

$$\mathrm{E}\big[\mathrm{E}(1)_{yy}\big] = (r-1)(t-1)(\sigma_2^2 + s\sigma_1^2) + s(\beta^w)^2 \sum_h \sum_i (\overline{x}_{hi\cdot} - \overline{x}_{\cdot i\cdot} - \overline{x}_{h\cdot\cdot} + \overline{x}_{\cdots})^2$$

$$\mathrm{E}\big[\mathrm{E}(1)_{x_w y}\big] = s\beta^w \sum_h \sum_i (\overline{x}_{hi\cdot} - \overline{x}_{\cdot i\cdot} - \overline{x}_{h\cdot\cdot} + \overline{x}_{\cdots})^2$$

$$\mathrm{E}\big[\big(\mathrm{E}(1)_{x_w y}\big)^2\big] = s^2(\beta^w)^2\Big[\sum_h \sum_i (\overline{x}_{hi\cdot} - \overline{x}_{\cdot i\cdot} - \overline{x}_{h\cdot\cdot} + \overline{x}_{\cdots})^2\Big]^2 + s(\sigma_2^2 + s\sigma_1^2)\sum_h \sum_i (\overline{x}_{hi\cdot} - \overline{x}_{\cdot i\cdot} - \overline{x}_{h\cdot\cdot} + \overline{x}_{\cdots})^2$$

$$\mathrm{E}\big[\mathrm{E}(2)_{yy}\big] = (r-1)t(s-1)\sigma_2^2 + (\beta^s)^2 \sum_h \sum_i \sum_j (x_{hij} - \overline{x}_{hi\cdot} - \overline{x}_{\cdot ij} + \overline{x}_{\cdot i\cdot})^2$$

$$\mathrm{E}\big[\mathrm{E}(2)_{x_s y}\big] = \beta^s \sum_h \sum_i \sum_j (x_{hij} - \overline{x}_{hi\cdot} - \overline{x}_{\cdot ij} + \overline{x}_{\cdot i\cdot})^2$$

$$\mathrm{E}\big[\big(\mathrm{E}(2)_{x_s y}\big)^2\big] = (\beta^s)^2\Big[\sum_h \sum_i \sum_j (x_{hij} - \overline{x}_{hi\cdot} - \overline{x}_{\cdot ij} + \overline{x}_{\cdot i\cdot})^2\Big]^2 + \sigma_2^2 \sum_h \sum_i \sum_j (x_{hij} - \overline{x}_{hi\cdot} - \overline{x}_{\cdot ij} + \overline{x}_{\cdot i\cdot})^2$$

Using these pieces, it follows that the ratio

$$\widehat{\beta^w} = \mathrm{E}(1)_{x_w y}/\mathrm{E}(1)_{x_w x_w}$$

is an unbiased estimate of β^w and

$$\widehat{\beta^s} = \mathrm{E}(2)_{x_s y}/\mathrm{E}(2)_{x_s x_s}$$

is an unbiased estimate of β^s. Also,

$$\begin{aligned}\mathrm{Var}[\widehat{\beta}_w] &= (\sigma_2^2 + s\sigma_1^2)\big/ s\sum_h \sum_i (\overline{x}_{hi\cdot} - \overline{x}_{\cdot i\cdot} - \overline{x}_{h\cdot\cdot} + \overline{x}_{\cdots})^2 \\ &= (\sigma_2^2 + s\sigma_1^2)\big/\mathrm{E}(1)_{x_w x_w}\end{aligned}$$

and

$$\text{Var}[\widehat{\beta}_s] = \sigma_2^2 / \sum_h \sum_i \sum_j (x_{hij} - \overline{x}_{hi\cdot} - \overline{x}_{h\cdot j} + \overline{x}_{h\cdot\cdot})^2$$
$$= \sigma_2^2 / \text{E}(2)_{x_s x_s}.$$

Notice that if we had not generalized the model in (7.25) to allow different covariates in the whole-plot and split-plot strata, the best linear unbiased estimate of the regression coefficient would have been considerably more complex. Allowing two regression coefficients even though only one covariate is observed wastes one degree of freedom in the statistical analysis, but simplifies the theory. If the model includes a separate regression coefficient for each stratum, there are exact tests of all the hypotheses in the analysis of covariance, whereas if the model includes a single regression coefficient in the split-plot stratum, the tests are approximate. Note that there is no wasted degree of freedom if there really is one covariate in the whole-plot stratum and another in the split-plot stratum and the two effects are additive in the whole-plot stratum.

Adjusting the whole-plot error for the covariate gives us

$$\text{MSE}(1)^a = \left(\text{E}(1)_{yy} - \frac{\text{E}(1)^2_{x_w y}}{\text{E}(1)_{x_w x_w}} \right) \Big/ [(r-1)(t-1)-1] \tag{7.27}$$

with $(r-1)(t-1)-1$ degrees of freedom. This is an unbiased estimate of $\sigma_2^2 + s\sigma_1^2$. Adjusting the split-plot error gives us

$$\text{MSE}(2)^a = \left(\text{E}(2)_{yy} - \frac{\text{E}(2)^2_{x_s y}}{\text{E}(2)_{x_s x_s}} \right) \Big/ [(r-1)t(s-1)-1] \tag{7.28}$$

with $(r-1)t(s-1)-1$ degrees of freedom. This is an unbiased estimate of σ_2^2. Notice the superscript "a", indicating a mean square adjusted for the covariate.

To provide proper tests of hypotheses, the treatment and interaction sums of squares must also be adjusted for the covariates. In the whole-plot stratum, the adjusted whole-plot treatment sum of squares with $(t-1)$ degrees of freedom is

$$\text{MSW}^a = \left(\text{W}_{yy} - \frac{\left[\text{W}_{x_w y} + \text{E}(1)_{x_w y}\right]^2}{\left[\text{W}_{x_w x_w} + \text{E}(1)_{x_w x_w}\right]} + \frac{\text{E}(1)^2_{x_w y}}{\text{E}(1)_{x_w x_w}} \right) \Big/ (t-1).$$

Similarly, in the split-plot stratum,

$$\text{MSS}^a = \left(\text{S}_{yy} - \frac{\left[\text{S}_{x_s y} + \text{E}(2)_{x_s y}\right]^2}{\text{S}_{x_s x_s} + \text{E}(2)_{x_s x_s}} + \frac{\text{E}(2)^2_{x_s y}}{\text{E}(2)_{x_s x_s}} \right) \Big/ (s-1)$$

and

$$\text{MSW} \times \text{S}^a = \left(\text{W} \times \text{S}_{yy} - \frac{\left[\text{W} \times \text{S}_{x_s y} + \text{E}(2)_{x_s y}\right]^2}{\text{W} \times \text{S}_{x_s x_s} + \text{E}(2)_{x_s x_s}} + \frac{\text{E}(2)^2_{x_s y}}{\text{E}(2)_{x_s x_s}} \right) \Big/ (t-1)(s-1),$$

respectively.

Tests of hypotheses are performed using the adjusted mean squares. For whole-plot treatments we have

$$F = \text{MSW}^a / \text{MSE}(1)^a$$

with $t - 1$ and $(r - 1)(t - 1) - 1$ degrees of freedom. In the split-plot stratum we test the split-plot treatment using

$$F = \text{MSS}^a / \text{MSE}(2)^a$$

with $s - 1$ and $(r - 1)t(s - 1) - 1$ degrees of freedom and the interaction of whole-plot and split-plot treatments using

$$F = \text{MSW} \times \text{S}^a / \text{MSE}(2)^a$$

with $(t - 1)(s - 1)$ and $(r - 1)t(s - 1) - 1$ degrees of freedom.

Treatment Contrasts: Main Effects

The purpose of introducing the covariate into the model is to provide more accurate as well as more precise estimates of treatment contrasts. We now look at adjusting for the covariate to remove biases attributable to differences among experimental units. We examined above the reduction in residual variance provided by the analysis of covariance. In this section we reduce the bias by adjusting treatment means and contrasts.

We begin with the whole-plot part of the design. From model (7.26) we have unbiased estimates of μ^* and $\{w_i^*\}$, i.e.,

$$\overline{y}_{\cdot i \cdot} = \widehat{\mu}^* + \widehat{w}_i^*$$

and

$$\sum_i c_i \overline{y}_{\cdot i \cdot} = \sum_i c_i \widehat{w}_i^*$$

with $\sum_i c_i = 0$. However, these are not the estimates we want. We want unbiased estimates of contrasts of the form $\sum_i c_i w_i$. Using the definitions implied in equation (7.26), we have unbiased estimates

$$\overline{y}_{\cdot i \cdot} = \widehat{\mu} + \widehat{\beta}^w \overline{x}_{\ldots} + \widehat{w}_i + \widehat{\beta}^w (\overline{x}_{\cdot i \cdot} - \overline{x}_{\ldots})$$

and

$$\sum_i c_i \overline{y}_{\cdot i \cdot} = \sum_i c_i \widehat{w}_i + \sum_i c_i \widehat{\beta}^w \overline{x}_{\cdot i \cdot}.$$

Since the $\{x_{hij}\}$ are observed constants, we can rewrite these expressions as

$$\widehat{\mu} + \widehat{w}_i = \overline{y}_{\cdot i \cdot} - \widehat{\beta}^w \overline{x}_{\cdot i \cdot}$$

and

$$\sum_i c_i \widehat{w}_i = \sum_i c_i \overline{y}_{\cdot i \cdot} - \sum_i c_i \widehat{\beta}^w \overline{x}_{\cdot i \cdot}.$$

These are the contrasts we want, the adjusted main effect contrasts. Also,

$$\mathrm{Var}\Big[\sum_i c_i \widehat{w}_i\Big] = \Big[\sum_i c_i^2/r + \frac{(\sum_i c_i \overline{x}_{\cdot i \cdot})^2}{\mathrm{E}(1)_{x_w x_w}}\Big](\sigma_2^2 + s\sigma_1^2)/s.$$

$\mathrm{MSE}(1)^a$ provides the unbiased estimate of $\sigma_2^2 + s\sigma_1^2$ with $(r-1)(t-1)-1$ degrees of freedom.

Comparisons among the split-plot treatment levels are similar. We have

$$\overline{y}_{\cdot\cdot j} = \widehat{\mu}^* + \widehat{s}_j^*$$

and

$$\sum_j c_j \overline{y}_{\cdot\cdot j} = \sum_j c_j \widehat{s}_j^*.$$

As before, we want unbiased estimates of contrasts of the form $\sum_j c_j s_j$. Using the definitions in (7.26) we have unbiased estimates

$$\overline{y}_{\cdot\cdot j} = \widehat{\mu} + \widehat{\beta}^w \overline{x}_{\dots} + \widehat{s}_j + \widehat{\beta}^s(\overline{x}_{\cdot\cdot j} - \overline{x}_{\dots})$$

and

$$\sum_j c_j \overline{y}_{\cdot\cdot j} = \sum_j c_j \widehat{s}_j + \sum_j c_j \widehat{\beta}^s \overline{x}_{\cdot\cdot j}.$$

We rewrite these expressions as

$$\widehat{\mu} + \widehat{s}_j = \overline{y}_{\cdot\cdot j} - (\widehat{\beta}^s \overline{x}_{\cdot\cdot j} - \overline{x}_{\dots}) - \widehat{\beta}^w \widehat{x}_{\dots}$$

and

$$\sum_j c_j \widehat{s}_j = \sum_j c_j \overline{y}_{\cdot\cdot j} - \sum_j c_j \widehat{\beta}^s \overline{x}_{\cdot\cdot j}.$$

Also,

$$\text{Var}\Big[\sum_j c_j\widehat{s}_j\Big] = \Big[\sum_j c_j^2/rt + \frac{(\sum_j c_j\overline{x}_{\cdot\cdot j})^2}{\text{E}(2)_{x_s x_s}}\Big]\sigma_2^2.$$

MSE(2)[a] provides the unbiased estimate of σ_2^2 with $(r-1)t(s-1)-1$ degrees of freedom.

Interaction Contrasts

Interaction contrasts have the form $\sum_{ij} c_{ij}\overline{y}_{\cdot ij}$ with $\sum_{ij} c_{ij} = 0$. We again encounter the complications first illustrated in Section 7.4.1. The standard errors depend on the nature of the contrasts.

We begin with contrasts among split-plot levels for a fixed whole-plot level. These are of the form $\sum_j c_{ij}\overline{y}_{\cdot ij}$ with $\sum_j c_{ij} = 0$ with a specific i value. We have

$$\begin{aligned}
\text{E}\Big[\sum_j c_{ij}\overline{y}_{\cdot ij}\Big] &= \sum_j c_{ij}\big(s_j^* + (w\times s)_{ij}^*\big) \\
&= \sum_j c_{ij}\big(s_j + (w\times s)_{ij} + \beta^s(\overline{x}_{\cdot ij} - \overline{x}_{\cdot i\cdot})\big).
\end{aligned}$$

The adjusted contrast is

$$\sum_j c_{ij}\big(\widehat{s}_j + \widehat{(w\times s)}_{ij}\big) = \sum_j c_{ij}\overline{y}_{\cdot ij} - \widehat{\beta}^s(\overline{x}_{\cdot ij} - \overline{x}_{\cdot i\cdot}).$$

This contrast has variance

$$\sum_j c_{ij}^2\left[\frac{1}{r} - \frac{(\overline{x}_{\cdot ij} - \overline{x}_{\cdot i\cdot})^2}{\text{E}(2)_{x_s x_s}}\right]\sigma_2^2.$$

Next we consider contrasts among whole-plot treatment levels, either at the same split-plot treatment level or at different split-plot treatment levels. These contrasts take the general form $\sum_i\sum_j c_{ij}\overline{y}_{\cdot ij}$, where $\sum_i\sum_j c_{ij} = 0$. We have

$$\begin{aligned}
\text{E}\Big[\sum_i\sum_j c_{ij}\overline{y}_{\cdot ij}\Big] &= \sum_i\sum_j c_{ij}\big(w_i^* + s_j^* + (w\times s)_{ij}^*\big) \\
&= \sum_i\sum_j c_{ij}\big(w_i + \beta^w(\overline{x}_{\cdot i\cdot} - \overline{x}_{\cdots}) \\
&\qquad + s_j + (w\times s)_{ij} + \beta^s(\overline{x}_{\cdot ij} - \overline{x}_{\cdot i\cdot})\big).
\end{aligned}$$

The adjusted contrast is

$$\begin{aligned}
\sum_i\sum_j c_{ij}(\widehat{w}_i + \widehat{s}_j + \widehat{(w\times s)}_{ij}) &= \sum_i\sum_j c_{ij}(\overline{y}_{\cdot ij} - \widehat{\beta}^w(\overline{x}_{\cdot i\cdot} - \overline{x}_{\cdots}) \\
&\qquad - \widehat{\beta}^s(\overline{x}_{\cdot ij} - \overline{x}_{\cdot i\cdot})\big)
\end{aligned}$$

and has variance

$$\sum_i \sum_j c_{ij}^2 \left[(\sigma_1^2 + \sigma_2^2)/r + (\sigma_1^2 + s\sigma_2^2) \frac{(\overline{x}_{\cdot i \cdot} - \overline{x}_{\dots})^2}{\text{E}(1)_{x_w x_w}} + \sigma_2^2 \frac{(\overline{x}_{\cdot ij} - \overline{x}_{\cdot i \cdot})^2}{\text{E}(2)_{x_s x_s}} \right].$$

Unfortunately, there is no "nice" estimate of this variance. We can obtain a moderate simplification by splitting this into two cases. Consider first whole-plot contrasts at one split-plot level. This implies that the $\{c_{ij}\}$ are zero for all $j \neq j'$. The general adjusted contrast then becomes

$$\sum_i c_{ij'}(\widehat{w}_i + \widehat{s}_{j'} + \widehat{(w \times s)}_{ij'}) = \sum_i c_{ij'}(\overline{y}_{\cdot ij'} - \widehat{\beta}^s (\overline{x}_{\cdot ij'} - \overline{x}_{\cdot i \cdot})),$$

with variance

$$\sum_i c_{ij'}^2 \left[(\sigma_1^2 + \sigma_2^2)/r + \sigma_2^2 \frac{(\overline{x}_{\cdot ij'} - \overline{x}_{\cdot i \cdot})^2}{\text{E}(2)_{x_s x_s}} \right]$$

$$= \sum_i c_{ij'}^2 \left((\sigma_1^2 + \sigma_2^2)/r \right) + \frac{\sum_i c_{ij'}^2 (\overline{x}_{\cdot ij'} - \overline{x}_{\cdot i \cdot})^2}{\text{E}(2)_{x_s x_s}} \sigma_2^2.$$

An unbiased estimate of this variance is given by

$$\sum_i c_{ij'}^2 \left[\frac{s-1}{sr} + \frac{(\overline{x}_{\cdot ij'} - \overline{x}_{\cdot i \cdot})^2}{\text{E}(2)_{x_s x_s}} \right] \text{MSE}(2)^a + \sum_i c_{ij'}^2 \left(\frac{1}{sr} \right) \text{MSE}(1)^a.$$

If this variance is to be used for testing or constructing a confidence interval, an approximation for degrees of freedom must be used. One approach is to use Satterthwaite's formula given in equation (7.18).

For the more general case we have the unbiased estimate of the variance

$$\sum_i \sum_j c_{ij}^2 \left[\frac{s-1}{sr} + \frac{s^2-1}{s} \frac{(\overline{x}_{\cdot i \cdot} - \overline{x}_{\dots})^2}{\text{E}(1)_{x_w x_w}} + \frac{(\overline{x}_{\cdot ij} - \overline{x}_{\cdot i \cdot})^2}{\text{E}(2)_{x_s x_s}} \right] \text{MSE}(2)^a$$

$$+ \sum_i \sum_j c_{ij}^2 \left[\frac{1}{sr} + \frac{(\overline{x}_{\cdot i \cdot} - \overline{x}_{\dots})^2}{s\text{E}(1)_{x_w x_w}} \right] \text{MSE}(1)^a.$$

Degrees of freedom again require Satterthwaite's approximation given in equation (7.18).

General Comments

A key point to recognize is that a covariate defined for the split-plots also affects the whole-plots (unless the definition is such that the covariates always sum to a common constant for each whole-plot). Alternatively, it is possible to define covariates only at the whole-plot level. These will have no effect on the split-plot stratum. The formulas given adapt directly to this situation.

Similar results carry over to the strip-plot experiment. Here one can define a covariate in the bottom stratum, the row × column stratum, and have it either affect or not affect the row and column strata. The condition is whether or not covariates sum to zero. It is also possible to define covariates that apply only to the row stratum or only to the column stratum.

7.8.2 Analysis of Covariance with One Whole-Plot Covariate

As a hypothetical example of a whole-plot-level covariate, we can consider an herbal cream that is to be applied to patches of skin of animals. Several formulations of cream are to be tested. To account for biological variability, several replications of each formulation are made up. All replications of all formulations are applied to skin patches on r males and r females to see if responses differ by gender. Age of the subjects provides a covariate.

In this section we illustrate the analysis of covariance for the split-plot experiment with whole-plots in an RCBD and one covariate associated with the whole-plots. One can picture an animal experiment where animals with somewhat different ages are used for whole-plot experimental units, and then skin patches are used for split-plot units.

Developing the Model

We begin with the model very similar to the one displayed in (7.24),

$$y_{hij} = \mu + r_h + w_i + \beta x_{hi} + \epsilon(1)_{hi} + s_j + (w \times s)_{ij} + \epsilon(2)_{hij},$$

where $h = 1, \ldots, r$, $i = 1, \ldots, t$, and $j = 1, \ldots, s$, and the covariate is x_{hi}. We again assume that both the whole- and split-plot treatments represent fixed effects, implying that $\overline{w}. = \overline{s}. = \overline{(w \times s)}_{.j} = \overline{(w \times s)}_{i.} = 0$. The $\{x_{hi}\}$ are observed constants. We also make the usual assumptions about $\{\epsilon(1)_{hi}\}$ and $\{\epsilon(2)_{hij}\}$.

The method of analysis again follows that outlined in Section 5.6.1. We rewrite the model to isolate the covariate's contribution to bias and variance. However, now we only need

$$x_{hi\cdot} = \overline{x}_{\cdot\cdot} + (\overline{x}_{h\cdot} - \overline{x}_{\cdot\cdot}) + (\overline{x}_{\cdot i} - \overline{x}_{\cdot\cdot}) + (x_{hi} - \overline{x}_{h\cdot} - \overline{x}_{\cdot i} + \overline{x}_{\cdot\cdot}).$$

The model is rewritten as

$$\begin{aligned} y_{hij} &= \mu + \beta\overline{x}_{\cdot\cdot} + r_h + \beta(\overline{x}_{h\cdot} - \overline{x}_{\cdot\cdot}) + w_i + \beta(\overline{x}_{\cdot i} - \overline{x}_{\cdot\cdot}) \\ &\quad + \beta(x_{hi} - \overline{x}_{h\cdot} - \overline{x}_{\cdot i} + \overline{x}_{\cdot\cdot}) + \epsilon(1)_{hi} + s_j + (w \times s)_{ij} + \epsilon(2)_{hij} \\ &= \mu^* + r_h^* + w_i^* + \beta(x_{hi} - \overline{x}_{h\cdot} - \overline{x}_{\cdot i} + \overline{x}_{\cdot\cdot}) + \epsilon(1)_{hi} + s_j + (w \times s)_{ij} \\ &\quad + \epsilon(2)_{hij}, \end{aligned}$$

where

$$\mu^* = \mu + \beta\overline{x}_{..},$$
$$r_h^* = r_h + \beta(\overline{x}_{h.} - \overline{x}_{..}),$$
$$w_i^* = w_i + \beta(\overline{x}_{.i} - \overline{x}_{..}).$$

The pieces in μ^*, r_h^*, and w_i^* represent the contributions to bias from the experimental units via the covariate, and $\beta(x_{hi} - \overline{x}_{h.} - \overline{x}_{.i} + \overline{x}_{..})$ represents the contribution to variance. The analysis of covariance provides adjustments to remove all of these. This model can be fitted directly in SAS® `Proc GLM`. The algebraic forms for the sums of squares and estimated treatment contrasts are derived in the next section.

Hand Computations in the Analysis of Covariance

Once the model has been constructed, the next step is to construct the compact analysis of covariance table (Table 7.14) corresponding to the analysis in Table 7.5.

There are no covariate adjustments in the split-plot stratum. In the whole-plot stratum we have

$$\widehat{\beta} = \mathrm{E}(1)_{xy}/\mathrm{E}(1)_{xx}$$
$$\mathrm{Var}\,\widehat{\beta} = (\sigma_2^2 + s\sigma_1^2)/\mathrm{E}(1)_{xx}$$
$$\mathrm{MSE}(1)^a = \left(\mathrm{E}(1)_{yy} - \frac{\mathrm{E}(1)_{xy}^2}{\mathrm{E}(1)_{xx}}\right) \Big/ [(r-1)(t-1)-1]$$
$$\mathrm{MSW}^a = \left(\mathrm{W}_{yy} - \frac{[\mathrm{W}_{xy} + \mathrm{E}(1)_{xy}]^2}{\mathrm{W}_{xx} + \mathrm{E}(1)_{xx}} + \frac{\mathrm{E}(1)_{xy}^2}{\mathrm{E}(1)_{xx}}\right) \Big/ (t-1).$$

Contrasts

Since there is only a whole-plot covariate, only the whole-plot treatment contrasts are adjusted. As in the preceding example, we begin with the contrast

$$\sum_i c_i\overline{y}_{.i.} = \sum_i c_i\widehat{w}_i + \widehat{\beta}\sum_i c_i(\overline{x}_{.i} - \overline{x}_{..})$$

with $\sum_i c_i = 0$. We rewrite this as

$$\sum_i c_i\widehat{w}_i = \sum_i c_i\overline{y}_{.i.} - \widehat{\beta}\sum_i c_i\overline{x}_{.i}.$$

Table 7.14 Analysis of Covariance Table for the Split-Plot with Only a Whole-Plot Covariate

Source	Y-Variable	Covariate	
Mean	M_{yy}	M_{xy}	M_{xx}
Blocks	B_{yy}	B_{xy}	B_{xx}
W	W_{yy}	W_{xy}	W_{xx}
Error (1)	$E(1)_{yy}$	$E(1)_{xy}$	$E(1)_{xx}$
S	S_{yy}		
W × S	$W \times S_{yy}$		
Error (2)	$E(2)_{yy}$		
Total	T_{yy}	T_{xy}	T_{xx}

This contrast has variance

$$\left(\sum_i c_i^2 \frac{1}{rs} + \frac{\left(\sum_i c_i \overline{x}_{\cdot i}\right)^2}{E(1)_{xx}}\right)(\sigma_2^2 + s\sigma_1^2).$$

The estimate of $\sigma_2^2 + s\sigma_1^2$ with $(r-1)(t-1)-1$ degrees of freedom is given by $MSE(1)^a$.

7.8.3 Whole-Plot and Split-Plot Covariates

In some cases it is reasonable to consider two different covariates, one for the whole-plots and one for the split-plots. We return to our animal example with individual animals as whole-plots and age as the covariate and patches of skin as split-plot units. Now assume that the animals must be sacrificed before the skin patches can be harvested and that changes begin to take place very rapidly in the skin once the animal is dead. The time from sacrifice until the harvested patch is observed or at least stabilized can serve as a very useful split-plot covariate. As in our discussion in Section 7.8.1, the split-plot covariate carries over into the whole-plot stratum. This leads to a very reasonable but relatively complex model with two covariates in the whole-plot stratum and one covariate in the split-plot stratum. We will not pursue the details.

7.8.4 Missing Values

Missing plots in a split-plot experiment with covariates are a problem. The usual recommendation is that if a split-plot value is missing, use the RCBD missing value formula for the missing split-plot and then continue on to the whole-plot analysis as though nothing were missing. In a sense, a missing whole-plot is easier to handle.

Federer and Meredith (1992) present an easy-to-read article that spells out the details of the analysis of covariance for the split-plot, the split-split-plot, and the split-block experiment. They illustrate the appropriate models, the analysis of variance, the formulas for the adjusted means, and the appropriate standard errors.

7.9 REPEATED MEASURES

Repeated measures experiments are closely related to split-plot and strip-plot experiments. In repeated measures experiments, also sometimes called *longitudinal studies*, subjects are assigned to treatment groups and observations are taken on each subject over time or some other dimension. The terminology comes from biomedical applications where the subjects are human volunteers or animals. In a pharmacokinetic study, for example, blood samples might be taken from each volunteer at several time points after a drug is administered. In a laboratory or manufacturing experiment the "subject" might be a test tube or a piece of equipment that is sampled over time. In agricultural experiments, repeated measurements could be taken at several depths in the soil on a given plot of land or at several distances from an irrigation ditch. In these cases the "subjects" might be plots of land, and the repeated factor is depth or distance rather than time.

Example 7.3. A study of the effects of melatonin on aging in the reproductive system of female rats (Diaz et al. 2000) provides an example. Melatonin has been found to inhibit some aging processes. It is suspected that melatonin may affect pituitary responsiveness to the reproductive hormone GnRH. In this study 40 3-month old and 38 23-month-old female rats were randomly assigned to either the melatonin or a placebo treatment. After the animals had received melatonin or placebo injections for a month, blood samples were taken and the levels of several reproductive hormones were measured. Then each animal was injected with GnRH and blood samples were taken 15, 30, 60, and 90 minutes after the GnRH injection.

The first difference between repeated measures and split-plot experiments is that in split-plots the levels of the split-plot factor are randomly assigned to subunits within each whole-plot, whereas a repeated measures factor such as points in time or depths cannot be randomly assigned to the subunits. Since time or depth (or other repeated measures factor) cannot be randomized, some protection against inadvertent biases provided by randomization is lost.

Taking repeated measurements over time or space (lack of randomization) also has an impact on the appropriateness of the usual assumptions about correlations between subplots in the model. The split-plot model assumes that all pairs of split-plots are equally correlated and that this correlation is positive. In the repeated measures case it is more reasonable to assume that correlations between two observations close together in space or time are strong and that the correlations diminish as the gap increases. Hence, repeated measures analyses usually incorporate a time-series correlation structure for the within-subject errors. If nothing is known about the correlation pattern, an unstructured correlation matrix with $s(s-1)/2$ correlation parameters becomes appropriate. We see that the split-plot model, which has one correlation parameter for all pairs of split-unit observations, puts many more constraints on the correlation structure than does an unstructured repeated measures model.

Repeated measures studies also differ from the typical time-series studies in emphasis and in design. Repeated measures experiments involve several subjects, each measured over a relatively short series of points. The subjects are randomly assigned to treatments according to some experimental design, and the objective may be to study the effects of treatments or model the effects over time. Time-series analysis focuses more on very long series of observations, usually just one or two series, with the objective being to forecast into the future.

We give this brief introduction to repeated measures in this chapter because this type of experiment is extremely common and often mistaken for a split-plot. In some special cases the split-plot analysis is appropriate for repeated measures data. There are two main cases where this is true: (1) the subunit errors within a subject are all equally correlated; or (2) there are only two repeated measurements on each subject, and these are equally spaced in time for all subjects. If all subjects are measured at, for example, the same two time points, there is only one correlation parameter among time points, so any time series or unstructured correlation matrix coincides with the equicorrelation structure.

Example 7.4. A study of the effects of silvicultural techniques on forest productivity in the Croatan National Forest provides an example of an experiment with a repeated measures factor. The response variables were the levels of nitrate (NO_3^-) and ammonium (NH_4^+) measured at different depths in the soil. The repeated measures factor is depth in the soil (N. Duarte, personal communication). The study focuses on the long-term effects of tree removal practices on soil health as indicated by the two compounds. The treatments in the experiment consisted of six combinations of methods of organic matter (tree) removal and levels of soil compaction. The study area was divided into three blocks, based on soil type and drainage class. Block 1 was better drained, and block 3 was more poorly drained than the others. Each block contained six plots. The plots were established and trees were removed 10 years prior to the soil sampling described here. Soil samples were taken at two depths at each of four locations in each plot. Since spatial variability within a plot was not of interest, the four locations were averaged for each depth before any further statistical analysis, resulting in a total of 36 (3 blocks $\times$ 6 treatments $\times$ 2 depths) observations. Since there are only two depths, this experiment may be analyzed as a split-plot by the methods of Section 7.4.

Example 7.5. DeMates (1990) describes an example of an experiment that illustrates a number of points covered in this chapter. The object of the study was to reduce part-weight variation over a range of targets for an injection molding process. The whole-plot stratum of the experiment involved several factors, but for purposes of this discussion can be treated as one with eight levels. The split-plot stratum again involved several factors that were compounded into one factor at two levels. In addition, there was a third stratum with one factor (injection pressure) at eight levels. The experiment was performed by first selecting a whole-plot treatment level, selecting one level of the split-plot factor and then systematically going through the eight injection pressures. Four observations were taken at each injection pressure.

The experiment can be looked at as a split-plot with eight whole-plots, two split-plots within each whole-plot, and then eight repeated measures at the split-split-plot level. In addition, four subsamples were taken at each injection pressure. If one ignores the possibility of serial correlation introduced by the lack of randomization, one can analyze the data as a split-split-split-plot experiment (no split-split-split-plot stratum treatments) or as a split-split-plot if one first summarizes the sets of four observations. Alternatively, since there is a sequence of injection pressures, a good analysis would be to estimate a linear (or quadratic) trend for each of the 8×2 split-plot combinations and then analyze the estimates of trend as a split-plot experiment.

REVIEW EXERCISES

7.1 Methane is produced as a by-product of digestion of animal feed; hence methane production is used as a measure of forage quality and digestibility. Eun et al. (2003) tested the effects of forage/concentrate ratio and saliva flow rates on methane production in ruminant digestion. An experiment to study three saliva flow rates and three forage ratios was done using three artificial fermenters over six periods of time. In each period all three fermenters were set at the same saliva flow rate (dilution rate), and the three forage ratios were assigned at random to the three fermenters. A separate randomization of forage ratios was done in each period.

(a) Draw a diagram of the experimental layout.

(b) Write an appropriate statistical model for this experiment. Include definitions of all terms, ranges of subscripts, and assumptions about random effects.

7.2 Suppose that your task is to design an experiment for a manufacturer of plastic pellets. The raw material is fed through an extruder, which keeps the raw materials at a specified temperature and has screws that mix the materials. The factors under study are screw style (two types) and raw material (four types).

(a) Two experimental designs are under consideration: (1) a completely randomized design with two replicates and (2) a split-plot design with screw style as the whole-plot factor and raw materials as the split-plot factor. In the second design, the order of the screw styles is completely randomized, with two replicates. Draw the experimental layout for each design and show how to randomize each.

(b) Suppose that it takes 30 minutes to change the screws in the extruder and 5 minutes to change the raw materials. Suppose also that the time to reset both factors is just the sum of the individual times to reset each factor. How many replicates of the split-plot design would it be possible to run in the same time that it would take to run two replicates of the CRD?

(c) Suppose that $\sigma_1^2 = 1$ and $\sigma_2^2 = 1$, where σ_1^2 represents the among-runs variance component, where the extruder has been disassembled between the runs and σ_2^2 represents variability among runs within the same assembling of the extruder. Suppose also that the main objective of the experiment is to determine the best screw style. How many replicates of the split-plot design would be necessary to obtain a standard error for the mean difference between two screw styles as small or smaller than in a completely randomized design with two replicates?

7.3 Test whether the average decrease in fertility in Table 7.8 from the low to the medium level of nitrogen is the same magnitude as the decrease from the medium to the high level of nitrogen. State the null and alternative hypotheses, the criterion for rejection, and your conclusion.

7.4 The data for the silvicultural experiment described in Example 7.4 are given in Table 7.15. Consider the four combinations of organic matter and soil compaction as four treatments.

(a) Estimate the between-plot variance component σ_1^2 and the within-plot variance component σ_2^2.

(b) Compute a 95% confidence interval for the mean of compaction level 0 vs. 2 at the soil surface.

(c) Estimate the standard error for the mean of compaction level 0 and check it against the standard error reported by your software.

Table 7.15 Mean Ammonium (NH_4) Concentration at the Surface and at 5 cm Depth in Experimental Plots in the Croatan National Forest[a]

Factor		Block 1 Depth		Block 2 Depth		Block 3 Depth	
O	C	Surface	5 cm	Surface	5 cm	Surface	5 cm
0	0	5.9249	4.3280	7.2951	5.5796	3.1194	5.6948
0	2	5.7004	4.9441	7.8114	3.9977	6.7300	9.6583
1	0	3.3863	3.4636	3.8361	2.8936	4.0993	2.8230
1	2	10.9314	9.5983	3.2403	3.8276	4.7151	5.3373
2	0	5.5351	3.7102	3.1920	4.1234	2.5725	2.8753
2	2	8.5418	6.5739	3.8662	4.6499	3.3117	5.6909

Source: Data courtesy of Natasha Duarte, personal communication.

[a]Forest plots in three organic matter treatments (O) and two levels of soil compaction (C) were arranged in four randomized complete blocks.

CHAPTER 8

Incomplete Block Designs

8.1 INTRODUCTION

Blocking has been introduced as a tool to increase precision by grouping experimental units into homogeneous sets. In Chapter 5 the randomized complete block was introduced. The criterion there was to sort units into sets or blocks of size t so that each block could accommodate the full set of t treatments.

In this chapter we confront the problem where the number of experimental units per block is dictated by some external constraint and is smaller than the number of treatments in the experiment. The reason may be an excessive number of treatments or it may be a severe limit on block size. Regardless, the result is *incomplete block* designs, that is, designs in which each block, by design, contains only a subset of the treatments. These cover the full range from experiments from using identical twins, i.e., blocks of two, to field experiments where blocks may be as large as 30 or 40 and the number of varieties (treatments) to be compared may exceed 100. We will also consider cases where the number of treatments is not a multiple of the block size, necessitating the use of blocks of several sizes in one experiment.

8.1.1 Series of Examples

Incomplete block designs are used in a wide range of disciplines. Examples include:

1. The classical nutrition and physiology studies that use identical twins to minimize differences between experimental units in a block.
2. The study in veterinary medicine in which four different ointments can be tested on individual sled dogs' feet.

Planning, Construction, and Statistical Analysis of Comparative Experiments,
by Francis G. Giesbrecht and Marcia L. Gumpertz
ISBN 0-471-21395-0

3. A food technology study in which the taste of 10 different formulations of a food product are to be compared by members of a taste panel. An individual panel member can only evaluate five or six items in one session before taste fatigue sets in.
4. An agricultural scientist is studying 15 different cultural treatments of strawberry plants in field patches. It is known that harvesting more than 10 patches in one day may be a problem. Convenience dictates blocks of size 10.
5. A national education assessment project requires that students of a specific age be examined on 18 different items. However, one student cannot be examined on more than five or six items.
6. A civil engineering study to evaluate a series of road-building techniques. The nature of the road-building equipment dictates that trial sections of roadway exceed some minimal size. Testing all techniques in one sequence is not feasible because of changes in soil and anticipated traffic conditions.

8.1.2 Models, Definitions, and Statistical Analysis

At this point it is sufficient to define incomplete block designs as block designs in which the number of treatments, t, exceeds the block size. Let b be the number of blocks. The basic statistical model that will be used to analyze data from such an experiment takes the familiar form

$$y_{ij} = \mu + \beta_i + \tau_j + \epsilon_{ij} \tag{8.1}$$

for $i = 1, \ldots, b$ and $j = 1, \ldots, t$. The $\{\beta_i\}$ represent block effects, the $\{\tau_j\}$ represent treatment effects, and the $\{\epsilon_{ij}\}$ represent random errors. The incomplete block nature of these designs dictates that not all possible combinations of i and j occur. As in the case of randomized complete block designs, blocks may be specifically chosen or a random sample, i.e., fixed or random. We are also making the assumption that no treatment is allowed to occur more than once in a block. Incomplete block designs for which this condition holds are called *binary designs*. It is also convenient to define the $t \times b$ *incidence matrix* $\boldsymbol{N}$ for the design. Formally, $\boldsymbol{N}^t$ has elements $\{n_{ij}\}$, with $n_{ij} = 1$ if block i contains treatment j and zero otherwise.

In matrix notation the model is written as

$$\boldsymbol{y} = \mathbf{1}\mu + \boldsymbol{Z\beta} + \boldsymbol{X\tau} + \boldsymbol{\epsilon}, \tag{8.2}$$

where $\mathbf{1}$ represents a vector of ones, $\boldsymbol{Z}$ is the design matrix for the blocks, and $\boldsymbol{X}$ is the design matrix for the treatment effects. The vectors $\boldsymbol{\beta}$ and $\boldsymbol{\tau}$ contain the block effects and treatment effects, respectively. With this notation it is possible to write $\boldsymbol{N} = \boldsymbol{X}^t\boldsymbol{Z}$.

8.1.3 Statistical Analysis, Intrablock Information

Traditional Least Squares Analysis

All-purpose linear models software such as SAS® `Proc GLM` produces the traditional least squares analysis. Example 8.2 provides a demonstration statistical analysis. In this section we provide the rationale and derivations for estimators and sums of squares. Working from the model shown in (8.1) and assuming that $\{\beta_i\}$ and $\{\tau_j\}$ are constants, we can apply least squares and obtain the normal equations

$$
\begin{array}{rcrcrcll}
\sum_i r_i\widehat{\mu} & + & \sum_i k_i\widehat{\beta}_i & + & \sum_j r_j\widehat{\tau}_j & = & y_{..} & \mu \text{ equation} \\
k_i\widehat{\mu} & + & k_i\widehat{\beta}_i & + & \sum_j n_{ij}\widehat{\tau}_j & = & y_{i.} & \text{block equations} \\
r_j\widehat{\mu} & + & \sum_i n_{ij}\widehat{\beta}_i & + & r_j\widehat{\tau}_j & = & y_{.j} & \text{treatment equations.}
\end{array}
$$

Exercise 8.1: Incomplete Block Normal Equations. Write out the design matrix for the incomplete block design using the non-full-rank model in (8.1) and verify the normal equations given. □

These equations are to be solved for estimates of treatment effects eliminating block effects. Begin by solving the block equations for $\{\widehat{\beta}_i\}$. Rearranging the second line of the normal equations gives the estimated block effects (as if treatments were not in the model):

$$
\begin{aligned}
\widehat{\beta}_i &= \frac{1}{k_i}\left(y_{i.} - \sum_j n_{ij}\widehat{\tau}_j - k_i\widehat{\mu}\right) \\
&= \frac{1}{k_i}y_{i.} - \frac{1}{k_i}\sum_j n_{ij}\widehat{\tau}_j - \widehat{\mu}
\end{aligned}
$$

for $i = 1, \ldots, b$. Now substitute these solutions into the treatment equations (third line) to obtain the treatment effects adjusted for the block effects. This substitution gives

$$
r_j\widehat{\mu} + \sum_i n_{ij}\left(\frac{1}{k_i}y_{i.} - \frac{1}{k_i}\sum_j n_{ij}\widehat{\tau}_j - \widehat{\mu}\right) + r_j\widehat{\tau}_j = y_{.j}
$$

or equivalently,

$$
r_j\widehat{\mu} + \sum_i \frac{n_{ij}}{k_i}\left(y_{i.} - \sum_h n_{ih}\widehat{\tau}_h - k_i\widehat{\mu}\right) + r_j\widehat{\tau}_j = y_{.j}.
$$

Notice that $\sum_i n_{ij}\widehat{\mu} = r_j\widehat{\mu}$, because n_{ij} indicates when treatment j is in block i and that treatment j appears in r_j blocks. Making this substitution and rearrangement gives

$$r_j\widehat{\tau}_j - \sum_i \sum_h \frac{n_{ij}n_{ih}}{k_i}\widehat{\tau}_h = y_{\cdot j} - \sum_i \frac{n_{ij}}{k_i} y_{i\cdot}$$
$$= Q_j. \tag{8.3}$$

The *adjusted treatment total*, Q_j, represents the deviation of the jth treatment total from the total of the block means of all blocks in which treatment j appears.

To make any more progress with the algebra, some additional conditions are required. In general, this is a messy system of equations and a computer is needed to obtain a solution. In a later section we examine the special case, the balanced incomplete block design.

To get an intuitive feeling for what is happening, we consider a small example with four blocks of size 3 and four treatments. The design layout is as follows:

block 1	y_{11}	y_{12}	y_{13}		(no treatment 4)
block 2	y_{21}	y_{22}		y_{24}	(no treatment 3)
block 3	y_{31}		y_{33}	y_{34}	(no treatment 2)
block 4		y_{42}	y_{43}	y_{44}	(no treatment 1).

The model is as shown in (8.1). There are three independent intrablock estimates of $\tau_1 - \tau_2$. From blocks 1 and 2 we have $y_{11} - y_{12}$ and $y_{21} - y_{22}$. These are clear. However, blocks 3 and 4 together provide a third estimate that must not be ignored. This estimate is

$$\left(y_{31} - \frac{y_{33} - y_{34}}{2}\right) - \left(y_{42} - \frac{y_{43} - y_{44}}{2}\right).$$

Similar intrablock estimates exist for the other treatment differences. The least squares analysis outlined combines all of these pieces of information properly.

Least Squares Analysis Using Matrix Notation

It is informative to look at the least squares analysis using matrix notation. We begin with the model as given in equation (8.2) and with both treatments and blocks fixed. We want estimates of $\boldsymbol{\tau}$ free of other parameters. We modify the model slightly by dropping μ. We can justify this by realizing that we have an overparameterized model, can shift μ into the $\boldsymbol{\beta}$ vector, then reparameterizing either with $\mu = 0$ or, equivalently, replacing each β_i by $\mu + \beta_i$. Either way, the $\boldsymbol{Z}$ matrix remains. The model can be written in partitioned matrix form as

$$\boldsymbol{y} = \begin{bmatrix} \boldsymbol{X} & \boldsymbol{Z} \end{bmatrix} \begin{bmatrix} \boldsymbol{\tau} \\ \boldsymbol{\beta} \end{bmatrix} + \boldsymbol{\epsilon}.$$

The normal equations were written out algebraically in the beginning of this section. In partitioned matrix form they are

$$\begin{bmatrix} \boldsymbol{X}^t\boldsymbol{X} & \boldsymbol{X}^t\boldsymbol{Z} \\ \boldsymbol{Z}^t\boldsymbol{X} & \boldsymbol{Z}^t\boldsymbol{Z} \end{bmatrix} \begin{bmatrix} \widehat{\boldsymbol{\tau}} \\ \widehat{\boldsymbol{\beta}} \end{bmatrix} = \begin{bmatrix} \boldsymbol{X}^t \\ \boldsymbol{Z}^t \end{bmatrix} \boldsymbol{y}.$$

From the lower half of this system, we have

$$\widehat{\boldsymbol{\beta}} = (\boldsymbol{Z}^t\boldsymbol{Z})^{-1}\boldsymbol{Z}^t\boldsymbol{y} - (\boldsymbol{Z}^t\boldsymbol{Z})^{-1}\boldsymbol{Z}^t\boldsymbol{X}\widehat{\boldsymbol{\tau}}. \tag{8.4}$$

Substituting $\widehat{\boldsymbol{\beta}}$ into the upper half of the system and rearranging gives the *reduced normal equations* for estimating the treatment effects:

$$\boldsymbol{X}^t(\boldsymbol{I} - \boldsymbol{Z}(\boldsymbol{Z}^t\boldsymbol{Z})^{-1}\boldsymbol{Z}^t)\boldsymbol{X}\widehat{\boldsymbol{\tau}} = \boldsymbol{X}^t(\boldsymbol{I} - \boldsymbol{Z}(\boldsymbol{Z}^t\boldsymbol{Z})^{-1}\boldsymbol{Z}^t)\boldsymbol{y}. \tag{8.5}$$

The right-hand side is the vector of the adjusted treatment totals,

$$\boldsymbol{Q} = \boldsymbol{X}^t(\boldsymbol{I} - \boldsymbol{Z}(\boldsymbol{Z}^t\boldsymbol{Z})^{-1}\boldsymbol{Z}^t)\boldsymbol{y} \tag{8.6}$$

and the estimated treatment effects are

$$\widehat{\boldsymbol{\tau}} = \left(\boldsymbol{X}^t(\boldsymbol{I} - \boldsymbol{Z}(\boldsymbol{Z}^t\boldsymbol{Z})^{-1}\boldsymbol{Z}^t)\boldsymbol{X}\right)^{-} \boldsymbol{Q}, \tag{8.7}$$

where $(\cdot)^-$ denotes a generalized inverse.

The elements of the vector $\boldsymbol{Q}$ correspond exactly to the $\{Q_j\}$ in equation (8.3). The variances and covariances of the adjusted treatment effects are obtained from the generalized inverse of the *information matrix*:

$$\boldsymbol{A} = \boldsymbol{X}^t(\boldsymbol{I} - \boldsymbol{Z}(\boldsymbol{Z}^t\boldsymbol{Z})^{-1}\boldsymbol{Z}^t)\boldsymbol{X}. \tag{8.8}$$

Direct substitution of equation (8.7) into equation (8.4) gives

$$\widehat{\boldsymbol{\beta}} = (\boldsymbol{Z}^t\boldsymbol{Z})^{-1}\boldsymbol{Z}^t\boldsymbol{y} - (\boldsymbol{Z}^t\boldsymbol{Z})^{-1}\boldsymbol{Z}^t\boldsymbol{X}\boldsymbol{A}^-\boldsymbol{Q}. \tag{8.9}$$

From general least squares theory and using equations (8.7) and (8.9), we obtain

$$\mathrm{R}(\tau, \beta) = \boldsymbol{y}^t\boldsymbol{X}\boldsymbol{A}^-\boldsymbol{Q} - \boldsymbol{y}^t\boldsymbol{Z}(\boldsymbol{Z}^t\boldsymbol{Z})^{-1}\boldsymbol{Z}^t\boldsymbol{y} - \boldsymbol{y}^t\boldsymbol{Z}(\boldsymbol{Z}^t\boldsymbol{Z})^{-1}\boldsymbol{Z}^t\boldsymbol{X}\boldsymbol{A}^-\boldsymbol{Q}.$$

Since

$$\mathrm{R}(\beta) = \boldsymbol{y}^t\boldsymbol{Z}(\boldsymbol{Z}^t\boldsymbol{Z})^{-1}\boldsymbol{Z}^t\boldsymbol{y},$$

it follows that

$$\begin{aligned} \mathrm{R}(\tau|\beta) &= \boldsymbol{y}^t\boldsymbol{X}\boldsymbol{A}^-\boldsymbol{Q} - \boldsymbol{y}^t\boldsymbol{Z}(\boldsymbol{Z}^t\boldsymbol{Z})^{-1}\boldsymbol{Z}^t\boldsymbol{X}\boldsymbol{A}^-\boldsymbol{Q} \\ &= \boldsymbol{y}^t(\boldsymbol{I} - \boldsymbol{Z}(\boldsymbol{Z}^t\boldsymbol{Z})^{-1}\boldsymbol{Z}^t)\boldsymbol{X}\boldsymbol{A}^-\boldsymbol{Q}, \end{aligned}$$

which we can also write as

$$\widehat{\boldsymbol{\tau}}^t \boldsymbol{Q}.$$

Recall that we shifted μ into $\boldsymbol{\beta}$, and consequently, $\mathrm{R}(\tau|\beta)$ is really $\mathrm{R}(\tau_1, \ldots, \tau_t| \mu, \beta_1, \ldots, \beta_b)$.

8.1.4 Analysis of Variance

For the analysis of variance for the incomplete block designs, we return to the concepts of extra sums of squares discussed in Sections 4.2 and 5.5. The analysis of variance is based on the linear model given in two different notation systems in equations (8.1) and (8.2).

The basic form of the analysis of variance table corresponding to our model used for testing treatment effects is illustrated in Table 8.1. Note that the sums of squares of Table 8.1 are partial sums of squares (Section 5.5) and are called *type III sums of squares* in SAS® `Proc GLM`.

The sum of squares $\mathrm{R}(\tau_1, \ldots, \tau_t|\mu, \beta_1, \ldots, \beta_b)$ is computed as $\widehat{\boldsymbol{\tau}}^t \boldsymbol{Q}$. In general, computing this sum of squares manually is a daunting task. We will see that relatively simple computing formulas exist for some special cases. In practice, general linear models software makes computing the analysis of variance simple. An example is shown in Section 8.4.1.

Here we describe the mechanics that go into computing the sums of squares. The treatment sum of squares is computed from only the intrablock information. There are two sums of squares that are not of interest in themselves but that are useful in computing some other sums of squares. The block sum of squares, $\mathrm{R}(\beta_1, \ldots, \beta_b|\mu)$ is easy to compute by hand. It is the corrected block sum of squares with treatments ignored completely. Using a matrix computing package, it would be computed as $\boldsymbol{y}^t \mathbf{Z}(\mathbf{Z}^t \mathbf{Z})^{-1} \mathbf{Z}^t \boldsymbol{y} - \boldsymbol{y}^t \mathbf{1} \mathbf{1}^t \boldsymbol{y}/kb$ in the notation established by equation (8.2). Similarly, $\mathrm{R}(\tau_1, \ldots, \tau_t|\mu)$ is easily computed manually as the corrected treatment sum of squares with no regard to blocks. Now we can compute all the remaining sums of squares by hand. SSTO $= \mathrm{R}(\beta_1, \ldots, \beta_b|\mu) + \mathrm{R}(\tau_1, \ldots, \tau_t|\mu, \beta_1, \ldots, \beta_b) + \mathrm{SSE}$, so SSE may be computed by subtraction. Similarly, $\mathrm{R}(\beta_1, \ldots, \beta_b|\mu, \tau_1, \ldots, \tau_t)$ does not need to be computed directly. It can be obtained by subtraction as SSTO $-$ SSE $-$ $\mathrm{R}(\tau_1, \ldots, \tau_t|\mu)$.

Table 8.1 Analysis of Variance for an Incomplete Block Design

Source	df	Partial SS
Blocks	$b-1$	$\mathrm{R}(\beta_1, \ldots, \beta_b\|\mu, \tau_1, \ldots, \tau_t)$
Treatment	$t-1$	$\mathrm{R}(\tau_1, \ldots, \tau_t\|\mu, \beta_1, \ldots, \beta_b)$
Error	$rt-b-t+1$	SSE
Total	$rt-1$	

A different way of looking at the treatment sum of squares uses the quadratic form for testing a linear hypothesis. The hypothesis is $H_0 : \boldsymbol{L\tau} = \mathbf{0}$, where $\boldsymbol{L}$ contains coefficients that select the parameters to be tested. The sum of squares is a quadratic form measuring the statistical distance between the estimated vector $\boldsymbol{L\tau}$ and the hypothesized vector $\mathbf{0}$:

$$\text{SS(L)} = (\boldsymbol{L\widehat{\tau}})^t(\boldsymbol{L}(\boldsymbol{A}^-\boldsymbol{L}^t)^{-1}\boldsymbol{L\widehat{\tau}}.$$

8.1.5 Interblock Information

Up to this point our discussion of the statistical analysis of incomplete block designs has dealt exclusively with intrablock information. This is analogous to the randomized complete block situation where every treatment is represented equally frequently in every block. Differences between blocks are not a function of treatments if every block contains the same complement of treatments.

In incomplete block designs, things are different. If we return for a moment to the simple example with four treatments and four blocks of three on page 200 we notice that the totals $(y_{11} + y_{12} + y_{13})$ and $(y_{21} + y_{22} + y_{24})$ would be expected to differ not only because of random error, or because of differences between blocks, but also because the first set contains treatment 3 and not treatment 4, whereas the second set contains treatment 4 and not 3. The remaining two treatments are common to the two blocks. We find similar differences between other pairs of block totals. To an extent, differences among block totals reflect treatment differences. Blocks containing "good" treatments will have higher totals than blocks with "poorer" treatments. The object of interblock statistical analysis is to recover this information.

First, a key point. If blocks are fixed, i.e., do not represent a random sample from some population, it is not possible to recover interblock information. For example, if our example dealt with growth rates of dogs in response to diets and the investigator had purposefully selected three northern sled dogs for block 1, three young racing hounds for block 2, three working sheep dogs for block 3, and three toy poodles for block 4, it would not be possible to recover any interblock information. This is not to say that this could not be a good experiment. One of the strengths of the wide selection of blocks is that it provides a very broad basis for inference. If a treatment difference shows up with such different blocks, the investigator will probably feel fairly secure in making rather broad inferences to other types of dogs.

If however, blocks represent a random sample from some population, it is straightforward to recover interblock information. Let us assume that the blocks in the earlier example represent four randomly selected animals of some species and the experiment consists of treatments applied to patches of skin. There are four treatments, but for some reason only three can be applied per animal. Assume further that the block effects, the $\{\beta_i\}$, are random variables. Now consider the four block totals, $y_{1\cdot}$, $y_{2\cdot}$, $y_{3\cdot}$, and $y_{4\cdot}$. These four totals represent realizations of four uncorrelated random variables with expected values $3\mu + \tau_1 + \tau_2 + \tau_3$, $3\mu +$

$\tau_1 + \tau_2 + \tau_4$, $3\mu + \tau_1 + \tau_3 + \tau_4$, and $3\mu + \tau_2 + \tau_3 + \tau_4$, respectively, and common variance $9\sigma_b^2 + 3\sigma_\epsilon^2$. We can apply least squares and obtain estimates.

Traditional Hand Calculations

In general, for an incomplete block design with b randomly selected blocks, we can form the b block totals and apply least squares. The totals have expected values that depend on the treatments in the blocks and variance $\{k_i^2\sigma_b^2 + k_i\sigma_\epsilon^2\}$. If we work from the model given in equation (8.1) with t treatments, we obtain

$$y_{i\cdot} = k_i\mu + \sum_j n_{ij}\tau_j + \epsilon_i^*$$

for $i = 1, \ldots, b$. Notice that ϵ_i^* consists of k_i copies of β_i and the sum of k_i independent $\{\epsilon_{ij}\}$ values. Hence, the $\{\epsilon_i^*\}$ have mean zero and variance $k_i^2\sigma_b^2 + k_i\sigma_\epsilon^2$. After some algebra, ordinary least squares based on this model yields the normal equations

$$\sum_j r_j\widetilde{\mu} + \sum_j r_j\widetilde{\tau}_j = y_{\cdot\cdot}$$

and

$$\sum_i k_i n_{ij}\widetilde{\mu} + \sum_i\sum_h n_{ij}n_{ih}\widetilde{\tau}_h = \sum_i n_{ij}y_{i\cdot} \tag{8.10}$$

for $j = 1, \ldots, t$. In the general case we still have a messy system of equations to solve, even with the side conditions $\sum_j \widetilde{\tau}_j = 0$. In a later section we return to these equations and solve after applying additional conditions.

Calculations in Matrix Notation

We begin with the model as given in equation (8.2) but with the added condition that the $\{\beta_i\}$ are iid$(0, \sigma_b^2)$ random variables. One can then rewrite the model as

$$\boldsymbol{y} = \boldsymbol{1}\mu + \boldsymbol{X\tau} + \boldsymbol{\epsilon}^*, \tag{8.11}$$

where $\boldsymbol{\epsilon}^*$ has variance $\boldsymbol{V}^* = \sigma_b^2\boldsymbol{ZZ}^t + \sigma_\epsilon^2\boldsymbol{I}$. Now if σ_b^2 and σ_ϵ^2, or at least the ratio $\sigma_b^2/\sigma_\epsilon^2$, were known and one could compute a matrix proportional to $\boldsymbol{V}^*$, generalized least squares applied to model (8.11) would be the analysis of choice.

The traditional hand calculations outlined above can be interpreted as replacing $\boldsymbol{X\tau}$ in model (8.2) with

$$(\boldsymbol{I} - \boldsymbol{Z}(\boldsymbol{Z}^t\boldsymbol{Z})^{-1}\boldsymbol{Z}^t)\boldsymbol{X\tau} + \boldsymbol{Z}(\boldsymbol{Z}^t\boldsymbol{Z})^{-1}\boldsymbol{Z}^t\boldsymbol{X\tau}$$

and then applying ordinary least squares to two models:

$$\boldsymbol{y} = \boldsymbol{1}\mu + (\boldsymbol{I} - \boldsymbol{Z}(\boldsymbol{Z}^t\boldsymbol{Z})^{-1}\boldsymbol{Z}^t)\boldsymbol{X\tau} + \boldsymbol{\epsilon}$$

$$\boldsymbol{y} = \boldsymbol{1}\mu + \boldsymbol{Z}(\boldsymbol{Z}^t\boldsymbol{Z})^{-1}\boldsymbol{Z}^t\boldsymbol{X\tau} + \boldsymbol{\epsilon}^*.$$

Even if one assumes that σ_b^2 and σ_ϵ^2 were known, in general the generalized least squares estimates will differ from a weighted combination of the two separate estimates.

Since in practice, σ_b^2 and σ_ϵ^2 are not known, there is a question about what to do. This means that there is a variance component estimation problem associated with the problem of estimating and testing hypotheses about the $\{\tau_j\}$. In addition, some form of estimated generalized least squares seems warranted. Proper examination of this topic is beyond the scope of this book. However, we make some comments. First, the user should bear in mind that it is not possible to show that estimated generalized least squares always provides better estimates (smaller variance) or more powerful tests than the intrablock analysis alone. Then there is the additional compounding problem that occasionally, by chance, the among-block variance observed is smaller than the within-block variance.

Patterson and Thompson (1971) proposed a modified maximum likelihood (MML) procedure for estimating σ_b^2 and σ_ϵ^2 for the recovery of interblock information in incomplete block designs. The procedure consists of dividing the contrasts among observed yields into two sets: (1) contrasts between treatment totals, and (2) contrasts with zero expectation. The method then maximizes the joint likelihood of all the contrasts in the second set. The maximization involves a nonlinear system of equations and iterative calculations. Once $\widehat{\sigma}_b^2$ and $\widehat{\sigma}_\epsilon^2$ are obtained, estimated generalized least squares estimates are obtained by solving

$$(\boldsymbol{X}^t \widehat{\boldsymbol{V}}^{-1} \boldsymbol{X})\widehat{\boldsymbol{\tau}} = \boldsymbol{X}^t \widehat{\boldsymbol{V}}^{-1} \boldsymbol{y}, \qquad \text{where} \quad \widehat{\boldsymbol{V}} = \widehat{\sigma}_b^2 \boldsymbol{Z}\boldsymbol{Z}^t + \widehat{\sigma}_\epsilon^2 \boldsymbol{I}. \tag{8.12}$$

Note that there are problems if the maximization leads to a negative value for $\widehat{\sigma}_b^2$. Also note that neither the $\{\widehat{\tau}_j\}$ nor the estimates of the variance components are unbiased. These estimates are often referred to as restricted maximum likelihood (REML) estimates.

An alternative approach is to use the (Rao 1971) minimum norm quadratic unbiased estimates (MINQUE) of σ_b^2 and σ_ϵ^2 discussed in Section 5.5.2. In this procedure the variance components are estimated by solving the system

$$\begin{bmatrix} \text{tr}(\boldsymbol{Q}_0 \boldsymbol{Q}_0) & \text{tr}(\boldsymbol{Q}_0 \boldsymbol{Z}\boldsymbol{Z}^t \boldsymbol{Q}_0) \\ \text{tr}(\boldsymbol{Q}_0 \boldsymbol{Z}\boldsymbol{Z}^t \boldsymbol{Q}_0) & \text{tr}(\boldsymbol{Z}^t \boldsymbol{Q}_0 \boldsymbol{Z}\boldsymbol{Z}^t \boldsymbol{Q}_0 \boldsymbol{Z}) \end{bmatrix} \begin{bmatrix} \widehat{\sigma}_\epsilon^2 \\ \widehat{\sigma}_b^2 \end{bmatrix} = \begin{bmatrix} \boldsymbol{y}^t \boldsymbol{Q}_0 \boldsymbol{Q}_0 \boldsymbol{y} \\ \boldsymbol{y}^t \boldsymbol{Q}_0 \boldsymbol{Z}\boldsymbol{Z}^t \boldsymbol{Q}_0 \boldsymbol{y}, \end{bmatrix}$$

where $\boldsymbol{Q}_0 = \boldsymbol{V}_0^{-1} - \boldsymbol{V}_0^{-1}\boldsymbol{X}(\boldsymbol{X}^t \boldsymbol{V}_0^{-1} \boldsymbol{X})^- \boldsymbol{X}^t \boldsymbol{V}_0^{-1}$, $\boldsymbol{V}_0 = \boldsymbol{I} + \boldsymbol{Z}\boldsymbol{Z}^t$, and $\text{tr}(\cdot)$ is the sum of the elements on the main diagonal of the matrix. As advertised, these estimates are unbiased. Also, since these estimates are quadratic, under assumed normality, the estimated generalized least squares estimates of the $\{\widehat{\tau}_j\}$ obtained via (8.12) are unbiased.

If very good estimates of σ_b^2 and σ_ϵ^2, say, σ_b^{2*} and σ_ϵ^{2*} are available from prior work, minimum variance quadratic unbiased estimates (MIVQUE) of the σ_b^2 and σ_ϵ^2 can be obtained by using $\boldsymbol{V}^* = \sigma_\epsilon^{2*}\boldsymbol{I} + \sigma_b^{2*}\boldsymbol{Z}\boldsymbol{Z}^t$ in place of $\boldsymbol{V}_0$ in the formula to compute $\widehat{\sigma}_b^2$ and $\widehat{\sigma}_\epsilon^2$. This procedure has much to offer in a continuing research program that utilizes incomplete block designs repeatedly under reasonably constant conditions.

A third approach is to apply maximum likelihood (ML) methods, assuming that the data are normally distributed. This method has little to recommend it. The ML estimates of σ_b^2 and σ_ϵ^2 are neither quadratic nor unbiased. Consequently, it is also not possible to establish unbiasedness for the $\{\widehat{\tau}_j\}$.

REML, MINQUE, MIVQUE, and ML are all available in `Proc Mixed` in SAS®. Example 8.2 demonstrates the syntax. REML and ML tend to converge rapidly, and in fact for many incomplete block designs, convergence is achieved on the first cycle; i.e., iteration is not necessary. In these cases, REML, MINQUE, and MIVQUE are all equivalent. Actually, MINQUE and MIVQUE do not require normality, whereas REML and ML do. However, in practice, the arithmetic computations are very similar. SAS® also has a method called MIVQUE(0), which sets $\boldsymbol{V} = \boldsymbol{I}$. This method is not recommended.

8.1.6 Randomization

The proper randomization for an incomplete block design depends on several factors. To begin with, it is a block design and treatments must be assigned at random to the experimental units within the blocks. However, there are further considerations. Not all treatments are able to appear in each block. The nature of the incomplete block design selected, a topic that we examine in considerable detail later in this chapter, determines the sets of treatments that appear together in blocks. Consequently, there is a need for randomization in the assignment of sets to blocks. For example, a nutrition experiment with three treatments using identical twins for experimental units will have treatments A and B assigned to one-third of the sets of twins, treatments A and C assigned to one-third, and treatments B and C assigned to the last third. This is in addition to random assignment of treatments to the individuals within pairs.

An incomplete block design is said to be *resolvable* if it is possible to organize the blocks into sets such that within each set each treatment appears an equal number of times. Not all incomplete block designs are resolvable. An example of a resolvable design is illustrated in Table 8.2, in which each treatment appears twice in each set. Later, in Tables 8.5 and 8.6 we give two examples of incomplete block designs that are not resolvable. In these it is not possible to organize the blocks into sets forming complete replicates.

Resolvability opens a number of possibilities. It may be possible to group the blocks so that a major portion of variability is removed by differences among-sets. This allows for random block assignment within each set. Since blocks are nested within sets and treatments appear with equal frequencies in each set, an among-sets sum of squares, orthogonal to treatments, can be removed from the block sums of squares in the analysis of variance displayed in Table 8.1. A point in favor of such designs is that if matters go badly amiss with the experiment, there is still the option of ignoring the blocking (with some loss of power) and analyzing as a randomized complete block design with sets of blocks as the new

Table 8.2 Example of a Resolvable Incomplete Block Design

Set	Block	Treatments	Set	Block	Treatments
1	1	0 1 2 3 4 5	3	7	0 1 4 5 6 8
	2	3 4 5 6 7 8		8	0 2 3 4 7 8
	3	0 1 2 6 7 8		9	1 2 3 5 6 7
2	4	0 1 3 4 6 7	4	10	1 2 3 4 6 8
	5	1 2 4 5 7 8		11	0 2 4 5 6 7
	6	0 2 3 5 6 8		12	0 1 3 5 7 8

blocking factor. With proper organization and randomization, this design still removes a large part of the extraneous variation.

In a resolvable incomplete block design one can also view each of the sets as a separate incomplete block design. If an experiment must be carried out in stages and there is a risk that work may need to be terminated before completion of the entire plan of study, the sets provide reasonable breaking points. This type of consideration is often important in agricultural work, where weather may cause an interruption in work.

8.2 EFFICIENCY OF INCOMPLETE BLOCK DESIGNS

This section is somewhat more technical than the remainder of the chapter. It may well be skipped on first reading. This section is inserted to provide a background for some of the statements and claims that will be made in subsequent sections. Also, it provides a background for the potential user who has a problem where experimentation is very costly and constraints very exacting. There are cases where it may be worthwhile to expend considerable effort to select the most efficient design.

To start we emphasize that efficiency as used in this section is not to be confused with potential gains from blocking to reduce variability among experimental units. We have in mind an experiment in which it has already been decided that the experimental units will be organized into blocks but the problem dictates that it is not possible to put all treatments into a block; i.e., an incomplete block design is required. The only choices remaining involve selecting subsets of treatments that are to appear together. In this section we assume that blocks are fixed, or at least that no attempt will be made to recover interblock information. We will assume t treatments and b blocks. Equal block size and equal replication will often be relaxed. In fact, the restriction that a treatment appears at most once in a block will also be relaxed in some cases.

Variances of contrasts among the elements of $\{\widehat{\tau}_j\}$ are of key importance when examining the results of an experiment. Since the $\{\widehat{\tau}_j\}$ are obtained as solutions to the system of equations (8.5), it follows that the information matrix $\boldsymbol{A} = \boldsymbol{X}^t(\boldsymbol{I} - \boldsymbol{Z}(\boldsymbol{Z}^t\boldsymbol{Z})^{-1}\boldsymbol{Z}^t)\boldsymbol{X}$, or equivalently, its generalized inverse, are key to evaluating designs. In general, if there are t treatments, the $t \times t$ matrix $\boldsymbol{A}$ has rank $\leq (t-1)$. If $\boldsymbol{W}$ is a generalized inverse of $\boldsymbol{A}$, estimates of variances and covariances of estimable contrasts among the elements of $\{\widehat{\tau}_j\}$ can be obtained from $\boldsymbol{W}\widehat{\sigma}_\epsilon^2$. For example, $\widehat{\text{Var}}[\widehat{\tau}_j - \widehat{\tau}_{j'}] = (w_{jj} - 2w_{jj'} + w_{j'j'})\widehat{\sigma}_\epsilon^2$. The efficiency factor for this estimate is $1/(w_{jj} - 2w_{jj'} + w_{j'j'})$. Pairwise efficiency factors can be defined for all estimable pairs. Often, a reasonable strategy is to look at averages of the pairwise efficiency factors of subsets of differences.

In a given experiment some differences may be more important than others. With current computing facilities it is not difficult to search for treatment assignments that favor contrasts of particular interest. As an example, we consider a flavor study of 20 peanut lines. Since the taste panel could evaluate only five samples in one session, an incomplete block design was necessary. However, in the set of lines being tested, several had been obtained by inserting a gene that was expected to improve storage properties. A question in the study was whether this gene also affected flavor. The investigator wanted to make all pairwise comparisons but comparisons between lines that differed only in the presence or absence of the one gene were of particular importance and required more replication. Functions of pairwise efficiency factors were the appropriate values to consider in the process of evaluating competing designs for the study.

For a second example, consider a biology experiment with 10 treatments using animals, say male rats, as experimental units. Assume further that the animals are relatively expensive, litters provide unequal block sizes, and that litters arrive in a sequence spread over time. At any given stage the experimenter can check the variances of treatment contrasts in the design up to that point. The assignment for the next block can then be adjusted to improve the homogeneity among the variances of treatment contrasts, i.e., the balance in the experiment. Notice that the unequal block sizes do not compromise the assumption of common within-block variance required of all block designs.

We note in passing that the term *balance* is used in different senses in the incomplete block design literature. In some cases the term is used to mean that all the estimated treatment differences have a common variance. In other cases the term is used to mean that all distinct pairs of treatments appear together in a block an equal number of times. This means that all the off-diagonal elements in the incidence matrix $\boldsymbol{N}$ defined on page 198 are equal. If an incomplete block design has all blocks size k and all treatments replicated r times, balance in the sense that all distinct pairs of treatments appear together in a block λ times implies balance in the sense that all estimated treatment differences have a common variance.

A more comprehensive or omnibus criterion for measuring efficiency can be obtained by considering the eigenvalues of the information matrix $\boldsymbol{A}$. The nonzero eigenvalues of $\boldsymbol{A}$ are the inverses of the variances of the *canonical variables*, where the canonical variables (also called *principal components*) are linear combinations of estimated treatment effects chosen to be independent of each other. The first canonical variable is that contrast of estimated treatment effects that has the smallest variance of any such contrast, and the last canonical variable is the contrast with the largest possible variance. In a sense, basing a measure of efficiency on the eigenvalues of $\boldsymbol{A}$ considers the set of all possible treatment contrast estimates. Since $\boldsymbol{A}$ is singular, at least one eigenvalue will always be zero. In the important class of designs called *connected designs*, all remaining eigenvalues are > 0. A design is said to be *connected* if it is not possible to split the set of treatments into two disjoint subsets such that no treatment in the first set appears in a block with any treatment in the second subset. In a connected design all differences of the form $\tau_j - \tau_{j'}$ for $j \neq j'$ can be estimated using only intrablock information.

A frequent measure of a design's efficiency is the harmonic mean of the nonzero eigenvalues. Using the harmonic mean of the eigenvalues as a criterion is equivalent to looking at the average of all $\binom{t}{2}$ pairwise efficiency factors. Designs that maximize this harmonic mean are said to be *A-optimal*. Other criteria are *E-optimality* (maximizes the smallest nonzero eigenvalue) and *D-optimality* (maximizes the determinant of the information matrix $\boldsymbol{A}$). The significance of E-optimality is that it minimizes the variance of the contrast with the largest variance, thereby ensuring that the largest variance of any contrast among treatment effects is as small or smaller than in any competing design. D-optimality maximizes the geometric mean of the nonzero eigenvalues. Other optimality criteria have been proposed as well. Raghavarao (1971), Dey (1986), and John and Williams (1995) provide more thorough discussions of the concepts of balance and efficiency.

8.3 DISTRIBUTION-FREE ANALYSIS FOR INCOMPLETE BLOCK DESIGNS

Skillings and Mack (1981) give a distribution-free analysis procedure for data from the general class of incomplete block designs. We assume model (8.1). The errors, $\{\epsilon_{ij}\}$, are assumed to be independent, identically distributed with some continuous distribution function. The hypothesis of interest is $\text{H}_0 : \tau_1 = \tau_2 = \cdots = \tau_t$.

The steps in the analysis are as follows:

1. Within each block, rank the observations from 1 to k_i, where k_i is the number of treatments in block i. If ties occur, assign average ranks.
2. Let r_{ij} be the rank assigned to y_{ij} if the observation is present; otherwise, let $r_{ij} = (k_i + 1)/2$.
3. Compute an adjusted treatment effect for treatment j,

$$a_j = \sum_{i=1}^{b} \left(\frac{12}{k_i + 1} \right)^{1/2} \left(r_{ij} - \frac{k_i + 1}{2} \right).$$

Note that this is a weighted deviation of $\{r_{ij}\}$ from the mean under H_0 with the coefficient $\sqrt{12/(k_i + 1)}$ providing slightly more weight to those blocks with fewer observations and presumably more accurate rankings. Also, the fact that we are setting $r_{ij} = (k_i + 1)/2$ for those cases where y_{ij} is missing really means that we are ignoring those values since the term that goes into our computations is $r_{ij} - (k_i + 1)/2$.

Under the null hypothesis and assumptions listed earlier, the $\{r_{ij}\}$ amount to nothing more than the random assignment of the integers $1, 2, \ldots, k_i$ to k_i cells. It follows that $\mathrm{E}[r_{ij}] = (k_i + 1)/2$, $\mathrm{Var}[r_{ij}] = (k_i + 1)(k_i - 1)/12$, and $\mathrm{Cov}[r_{ij}, r_{ih}] = -(k_i + 1)/12$ for $h \neq j$. Also, r_{ij} and r_{lh} are independent if $i \neq l$. It follows that under these conditions, the variances and covariances among the $\{a_j\}$ are given by

$$\mathrm{Var}(a_j) = \sum_{i=1}^{b} (k_i - 1) n_{ij} \qquad \text{for } j = 1, 2, \ldots, t$$

$$\mathrm{Cov}(a_j, a_h) = -\sum_{i=1}^{b} n_{ij} n_{ih} \qquad \text{for } 1 \leq j \neq h \leq t.$$

To allow sufficient generality we define λ_{jh} to be the count of the number of times that treatment j and treatment h appear together in a block. It follows that

$$\mathrm{Cov}(a_j, a_h) = -\lambda_{jh} \qquad \text{for } 1 \leq j \neq h \leq t$$

$$\mathrm{Var}(a_j) = \sum_{h=1}^{t} \lambda_{jh} - \lambda_{jj} \qquad \text{for } j = 1, \ldots, t.$$

Note that in a balanced incomplete block design (to be discussed in Section 8.4), $\lambda_{jh} = \lambda$ for all $j \neq h$ and $\lambda_{jj} = r$ the number of replications for each treatment.

To construct the test of $\mathrm{H}_0 : \tau_1 = \tau_2 = \cdots = \tau_t$, let $\boldsymbol{a}$ be the column of adjusted treatment sums $(a_1, a_2, \ldots, a_t)$ and $\boldsymbol{G}$ the $t \times t$ variance–covariance matrix consisting of the variances and covariances among the $\{a_j\}$ defined above:

$$\boldsymbol{G} = \begin{bmatrix} \mathrm{Var}(a_1) & \mathrm{Cov}(a_1, a_2) & \cdots & \mathrm{Cov}(a_1, a_t) \\ \mathrm{Cov}(a_2, a_1) & \mathrm{Var}(a_2) & \cdots & \mathrm{Cov}(a_2, a_t) \\ \vdots & \vdots & \vdots & \vdots \\ \mathrm{Cov}(a_t, a_1) & \mathrm{Cov}(a_t, a_2) & \cdots & \mathrm{Var}(a_t) \end{bmatrix}.$$

The test statistic is $\boldsymbol{T} = \boldsymbol{a}^t \boldsymbol{G}^- \boldsymbol{a}$, where $\boldsymbol{G}^-$ is any generalized inverse of $\boldsymbol{G}$. In most cases $\boldsymbol{G}$ will have rank $t-1$ and one can compute the T statistic by deleting a_t (the last element) from $\boldsymbol{a}$ and the last row and column of $\boldsymbol{G}$ and then using the regular matrix inverse. H_0 is rejected for large values of the statistic. For large

b (number of blocks) the statistical test is performed by comparing the statistic computed with the critical values for a χ^2 with $t-1$ degrees of freedom. Skillings and Mack present the results of a simulation study of the adequacy of the use of the χ^2 approximation for the test of significance for small b. It appears that b should really be of the order of 20 or more to be really safe. If the unbalance is very severe, the rank of $\boldsymbol{G}$ may be less than $t-1$. However, for this to happen, at least several of the $\{\lambda_{jh}\}$ defined previously must be zero. In that case, one must not only use the generalized inverse of $\boldsymbol{G}$ to compute the test statistic, but one must use something like an eigenvalue routine to determine the rank of $\boldsymbol{G}$ to use for the degrees of freedom for the χ^2-test.

Note that the proposed test is based wholly upon the within or intrablock information. Also, the procedure as described above is rather general. In the case of a balanced incomplete block design with no missing values, the formulas simplify considerably.

Example 8.1. Consider the hypothetical data from an unbalanced block design shown in Table 8.3. The observations $\{y_{ij}\}$ are given together with the ranks $\{r_{ij}\}$ in parentheses.

$$
\begin{aligned}
a_1 &= \left(\frac{12}{5}\right)^{1/2}(1-2.5) + \left(\frac{12}{5}\right)^{1/2}(3-2.5) + \left(\frac{12}{4}\right)^{1/2}(1-2) \\
&\quad + \left(\frac{12}{6}\right)^{1/2}(2-3) + \left(\frac{12}{4}\right)^{1/2}(3-2) + \left(\frac{12}{5}\right)^{1/2}(2-2.5) \\
&\quad + \left(\frac{12}{6}\right)^{1/2}(2-3) + \left(\frac{12}{5}\right)^{1/2}(2-2.5) \\
&= -5.927 \qquad \text{etc.}
\end{aligned}
$$

$$\boldsymbol{a}^t = (-5.927, 14.312, 5.792, -4.969, -9.208)$$

$$\text{Var}(A_1) = 24$$

$$\text{Var}(A_2) = 23$$

$$\text{Cov}(A_1, A_2) = -5 \qquad \text{etc.}$$

Table 8.3 Simulated Data for an Unbalanced Block Design

	Block									
Trt.	1	2	3	4	5	6	7	8	9	10
1	28(1)	43(3)	23(1)	76(2)	62(3)			54(2)	15(2)	39(2)
2	86(4)	54(4)		86(4)		68(4)	83(3)	93(4)	76(5)	
3	38(2)	38(2)	38(2)	97(5)	58(2)	36(2)	97(4)	75(3)	35(4)	58(3)
4	77(3)		51(3)	78(3)		14(1)	27(1)	28(1)	12(1)	68(4)
5		22(1)		19(1)	13(1)	54(3)	44(2)		32(3)	17(1)

$$\boldsymbol{G} = \begin{bmatrix} 24 & -5 & -8 & -6 & -5 \\ -5 & 23 & -7 & -6 & -5 \\ -8 & -7 & 30 & -8 & -7 \\ -6 & -6 & -8 & 25 & -5 \\ -5 & -5 & -7 & -5 & 22 \end{bmatrix}.$$

Since none of the $\{\lambda_{jh}\}$ are zero, we know that the rank of this matrix is 4. If we were in doubt, one could use the statement `D=EIGVAL(G);` in `PROC IML` of `SAS`® and count the nonzero eigenvalues found in the $\boldsymbol{G}$ matrix. The desired test statistic is computed as `T = a'*GINV(G)*a;`. For this example, $T = 13.13$, which exceeds the 5% critical value $\chi^2 = 9.49$. Although we know that $\boldsymbol{G}$ has rank 4, a check using `EIGVAL` reveals the eigenvalues 37.64, 30.85, 28.41, 27.10, and 0.00.

The fact that there are only 10 blocks casts some doubt on the adequacy of the χ^2 assumption for the test statistic. However, since the test statistic exceeds the critical value by such a wide margin, it is probably safe to accept the conclusion from the test.

8.4 BALANCED INCOMPLETE BLOCK DESIGNS

A very large and possibly the most important class of incomplete block designs consists of balanced incomplete block (BIB) designs. A BIB is a binary incomplete block design for t treatments in b blocks, which satisfies the following conditions: (1) every treatment has equal replication r; (2) every block contains the same number of experimental units, k; and (3) every pair of treatments occurs together in a block λ times. The condition that the design is binary simply means that a treatment occurs at most once in a block. We actually used BIBs as special examples at some points in previous sections.

The conditions above are not really independent. For example, t treatments each replicated r times means rt observations. Similarly, b blocks of size k means bk observations. Hence, $rt = bk$. In addition, b blocks of size k provides room for $b\binom{k}{2} = bk(k-1)/2$ pairs of treatments. However, there are $\binom{t}{2}$ possible pairs of treatments. Each pair occurs λ times for a total of $\lambda t(t-1)/2$ pairs in the experiment. Consequently, $bk(k-1) = \lambda t(t-1)$ and $rt(k-1) = \lambda t(t-1)$, which is commonly written as

$$\lambda = \frac{r(k-1)}{t-1}.$$

The constants t, b, r, k, and λ are called *parameters* of the design. Since λ must be an integer, this condition puts a severe restriction on possible parameter values, on possible sizes of BIBs. An additional constraint on the parameters, which is somewhat more difficult to establish and which we will simply accept, is that $t \leq b$.

8.4.1 Statistical Analysis of the BIB

Intrablock Analysis

In Section 8.1.3 we discussed the statistical analysis of intrablock information for the general incomplete block design. We continue with that development but with the extra constraints from the BIB. We started with the normal equations, solved for the $\{\widehat{\beta}_i\}$, substituted back into the treatment equations, and eventually obtained

$$r_j\widehat{\tau}_j - \sum_i \sum_h \frac{n_{ij}n_{ih}}{k_i}\widehat{\tau}_h = y_{\cdot j} - \sum_i \frac{n_{ij}}{k_i} y_{i\cdot\cdot}$$

At that stage we noted that we would have to impose some conditions to make further progress. We now specialize to the BIB.

Notice first that $n_{ij}n_{ih} = 1$ if the j and h treatment are both in block i and zero otherwise. It follows that $\sum_i n_{ij}n_{ih}$ is just the number of blocks containing both treatments j and h. Consequently,

$$\sum_i n_{ij}n_{ih} = \begin{cases} \lambda & \text{when } h \neq j \\ r & \text{when } h = j. \end{cases}$$

Substituting leads to

$$\begin{aligned} rk\widehat{\tau}_j - r\widehat{\tau}_j - \lambda \sum_{h \neq j} \widehat{\tau}_h &= ky_{\cdot j} - \sum_i n_{ij} y_{i\cdot} \\ &= kQ_j. \end{aligned} \tag{8.13}$$

Now impose the side condition $\sum_h \widehat{\tau}_h = 0$, which implies that $\sum_{h \neq j} \widehat{\tau}_h = -\widehat{\tau}_j$. Note that there has been no need to impose any side conditions on the $\{\beta_i\}$ because they have been eliminated from the equations. We started as though they were fixed effects to justify the normal equations, but then they simply dropped out. There was no need for any extra assumptions.

Finally, obtain

$$rk\widehat{\tau}_j - r\widehat{\tau}_j + \lambda\widehat{\tau}_j = kQ_j. \tag{8.14}$$

The estimator $\widehat{\tau}_j$ follows directly as

$$\begin{aligned} \widehat{\tau}_j &= \frac{k}{r(k-1)+\lambda} Q_j \\ &= \frac{k}{t\lambda} Q_j \end{aligned} \tag{8.15}$$

for $j = 1, \ldots, t$. These estimates are linear functions of the observations, so we have

$$\mathrm{Var}\left[\widehat{\tau}_j\right] = \frac{k(t-1)}{\lambda t^2}\sigma_\epsilon^2.$$

Since, $\widehat{\mu} = \overline{y}_{\cdot\cdot}$ we can write the estimate of the jth adjusted treatment mean as

$$\widehat{\mu}_{\cdot j} = \overline{y}_{\cdot\cdot} + \widehat{\tau}_j \tag{8.16}$$

with variance

$$\mathrm{Var}\left[\widehat{\mu}_{\cdot j}\right] = \left(\frac{1}{rt} + \frac{k(t-1)}{\lambda t^2}\right)\sigma_\epsilon^2, \tag{8.17}$$

provided that blocks are fixed.

Exercise 8.2: Variance of a Treatment Mean in a BIB. Show that expression (8.17) is correct by showing that the covariance of $y_{\cdot\cdot}$ and Q_j equals zero. Verification of $\mathrm{Var}\left[\widehat{\tau}_j\right]$ is left to Exercise 8.3. □

It follows directly from our derivations that the intrablock estimated difference between two treatments is

$$\widehat{\tau}_j - \widehat{\tau}_{j'} = \frac{k}{t\lambda}(Q_j - Q_{j'}).$$

It also follows that

$$\mathrm{Var}\left[\widehat{\tau}_j - \widehat{\tau}_{j'}\right] = \frac{2k}{t\lambda}\sigma_\epsilon^2.$$

Exercise 8.3: Variance of $\widehat{\tau}_j$. The variance of a treatment effect $\widehat{\tau}_j$ and of the difference between two treatment means $\widehat{\tau}_j - \widehat{\tau}_{j'}$ can be obtained as variance of linear combinations of several terms. The variances and covariances needed are

$$\mathrm{Var}\left[y_{\cdot j}\right] = r\sigma_\epsilon^2, \qquad \mathrm{Cov}\left[y_{\cdot j}, y_{\cdot j'}\right] = 0$$

$$\mathrm{Var}\left[\textstyle\sum_i n_{ij}\overline{y}_{i\cdot}\right] = \frac{r\sigma_\epsilon^2}{k}, \qquad \mathrm{Cov}\left[y_{\cdot j}, \textstyle\sum_i n_{ij}\overline{y}_{i\cdot}\right] = \frac{r\sigma_\epsilon^2}{k}$$

$$\mathrm{Cov}\left[y_{\cdot j}, \textstyle\sum_i n_{ij'}\overline{y}_{i\cdot}\right] = \frac{\lambda\sigma_\epsilon^2}{k} \qquad \text{for } j \neq j',$$

$$\mathrm{Cov}\left[\textstyle\sum_i n_{ij}\overline{y}_{i\cdot}, \sum_v n_{vj'}\overline{y}_{v\cdot}\right] = \frac{\lambda\sigma_\epsilon^2}{k} \qquad \text{for } j \neq j'.$$

(a) Verify each of the statements given.
(b) Derive the variance of $\widehat{\tau}_j$.
(c) Derive the variance of $\widehat{\tau}_j - \widehat{\tau}_{j'}$. □

It was shown in Section 8.1.3 that, in general, $\mathrm{R}(\tau_1, \ldots, \tau_t | \mu, \beta_1, \ldots, \beta_b)$ in the analysis of variance given in Table 8.1 can be computed as

$$\sum_j \widehat{\tau}_j Q_j.$$

For the BIB design these computations are relatively straightforward. This completes the intrablock analysis of variance for the BIB design.

To relate these computations back to he general matrix notation material in Section 8.1.3, we note that for the BIB design, $\boldsymbol{X}^t\boldsymbol{X} = r\boldsymbol{I}$, $\boldsymbol{Z}^t\boldsymbol{Z} = k\boldsymbol{I}$, and $\boldsymbol{X}^t\boldsymbol{Z} = \boldsymbol{N}$. Also, $\boldsymbol{A}$ simplifies to $r\boldsymbol{I} - (1/k)\boldsymbol{N}\boldsymbol{N}^t$.

Exercise 8.4: Elements of Key Matrix in the BIB. For a BIB design:

(a) Derive the elements of $\boldsymbol{Q} = \boldsymbol{X}^t(\boldsymbol{I} - \boldsymbol{Z}(\boldsymbol{Z}^t\boldsymbol{Z})^{-1}\boldsymbol{Z}^t)\boldsymbol{y}$.
(b) Show that the information matrix $\boldsymbol{A}$ has diagonal elements $r(k-1)/k$ and off-diagonal elements $-\lambda/k$.
(c) Show the correspondence between the reduced normal equations in (8.3), (8.5), and (8.14). □

Example 8.2. We demonstrate the intrablock analysis of variance for a resolvable balanced incomplete block design on a simulated data set using SAS® `Proc GLM` and the analysis treating blocks as random effects using `Proc Mixed`. Consider a design with six treatments. Imagine that you are testing six types of eyedrops on dogs, and one type of eyedrop is tried in each eye. The dogs serve as blocks in this experiment, so each block has two units, i.e., two eyes. Now imagine that the dogs are grouped into five replicate groups of three dogs each, each group tested by a different technician. Thus there are $b = 15$ blocks, $r = 5$ replicates, $t = 6$ treatments, $k = 2$ observations per block, and $\lambda = 1$.

This design is resolvable, and each technician measures a complete set of six treatments. Model (8.18) is appropriate.

$$y_{ijk} = \mu + \rho_i + \beta_{j(i)} + \tau_k + \epsilon_{ijk}, \tag{8.18}$$

where ρ_i represents the ith replicate effect and $i = 1, \ldots, 5$ technicians, $\beta_{j(i)}$ represents the effect of the jth block within the ith replicate where $j = 1, \ldots, 3$ dogs within each replicate, τ_k represents the kth treatment effect ($t = 1$ or 2), and $\{\epsilon_{ijk}\}$ represents the within-block error. In this example we assume that replicate effects are random, so there are three random effects terms in this model: technicians, dogs within technicians, and within-dog errors. Note that if

```
proc glm data=sim;
   class rep block trt;
   model y=rep block(rep) trt;
   lsmeans trt;
   estimate 'trt 1 vs 4' trt 1 0 0 $-$1 0 0;
```

Display 8.1 SAS® code for the intrablock analysis of the simulated eyedrop data.

dogs were grouped into replicates based on age or on severity of eye condition or some other factor with a finite number of levels of interest, it might make sense to model replicates as a fixed effect. The assumptions about the random effects are $\rho_i \sim$ iid $N(0, \sigma_r^2)$, $\beta_{j(i)} \sim$ iid $N(0, \sigma_b^2)$, $\epsilon_{ijk} \sim$ iid $N(0, \sigma_\epsilon^2)$, and all the random effects are independent of each other. The data (available on the companion Web site) were generated from model (8.18) with parameter values $\mu = 2, \tau_1 = 0, \tau_2 = .5, \tau_3 = -.5, \tau_4 = -.6, \tau_5 = -.4, \tau_6 = 1.0, \sigma_r^2 = 1, \sigma_b^2 = .25$, and $\sigma_\epsilon^2 = .50$.

The SAS® code for the intrablock analysis is given in Display 8.1. There are several things to be aware of in interpreting the results. First, sequential and partial sums of squares do not coincide for this type of design. The partial sums of squares are the ones that are of most interest. Second, the least squares means and the simple means for the treatments do not coincide. The simple means involve unwanted block effects and should not be used. The true means μ_k from which the data were simulated for this example, the simple treatment means $\overline{y}_{..k}$, the ordinary least squares means $\widehat{\mu}_{k,\text{OLS}}$ from `Proc GLM`, and the estimated generalized least squares means $\widehat{\mu}_{k,\text{EGLS}}$ from `Proc Mixed` are shown in Table 8.4. The standard error for a least squares mean reported by `Proc GLM` is not correct if the replicate and block(replicate) effects are random. However, the standard errors for treatment contrasts are correct, because the block and replicate effects cancel out of the estimated contrasts. Note that `Proc Mixed` does estimate appropriate standard errors for treatment means, which `Proc GLM` cannot do. The `Proc Mixed` standard errors for treatment means are underestimated to an

Table 8.4 Simulated Eyedrop Data[a]

Treatment	True Mean μ_k	$\overline{y}_{..k}$	$\widehat{\mu}_{k,\text{OLS}}$	$\widehat{\mu}_{k,\text{EGLS}}$
1	2.0	1.76	1.89	1.82
2	2.5	2.14	2.34	2.24
3	1.5	2.06	1.77	1.91
4	1.4	1.42	1.55	1.49
5	1.6	2.06	1.74	1.90
6	3.0	2.78	2.94	2.86
	$s(\widehat{\mu}_k - \widehat{\mu}_{k'})$		.44	.39

[a]The true treatment means, the simple treatment averages, the ordinary least squares means, and the estimated generalized least squares means, along with the standard error of a difference of two means.

```
proc mixed data=sim;
   class rep block trt;
   model y=trt;
   random rep block(rep);
   estimate 'trt 1 vs 4' trt 1 0 0 $-$1 0 0;
   lsmeans trt;
   title2 'Analysis of variance with
           recovery of interblock information';
   title3 'Reps and blocks(rep) are random';
```

Display 8.2 Estimated generalized least squares analysis of the simulated eyedrop data.

unknown degree, however, because they do not take into account the variability due to estimating the covariance matrix.

The SAS® `Proc Mixed` code for analysis incorporating interblock information is shown in Display 8.2. The variance components for replicate and blocks (replicate) are estimated by REML, and the treatment means are estimated by EGLS. Estimated generalized least squares is an iterative procedure. It is important to check that the procedure does converge, which it should do quickly for a balanced incomplete block design.

In this example the variance components are estimated to be $\widehat{\sigma}_r^2 = .77, \widehat{\sigma}_b^2 = .24$, and $\widehat{\sigma}_\epsilon^2 = .29$. Using these values we would estimate that the variability among replicates (which represent technicians in the example scenario) accounts for 59% of the variance of y, while the variance among dogs within a technician accounts for 18%, and the variance between eyes within a dog accounts for the remaining 22%.

The EGLS means differ somewhat from the means from the intrablock analysis, and the standard error of a difference between two treatment means is a bit smaller. Hence the combined intra- and interblock EGLS estimation appears to be more efficient than the intrablock analysis. Estimates of the variance components, however, are not particularly close to their true values, $\sigma_r^2 = 1.0, \sigma_b^2 = .25$, and $\sigma_\epsilon^2 = .50$, so it is hard to know which analysis to prefer. In general, the combined analysis will provide more efficient estimates than the intrablock analysis if the number of blocks is large. A Monte Carlo simulation could be done to resolve the question, and in this case a small simulation verifies that the EGLS estimates of differences in treatment means have slightly lower standard errors than the intrablock estimates.

Exercise 8.5: OLS or EGLS? Run a Monte Carlo simulation to confirm whether the intrablock analysis or the combined intra- and interblock analysis (EGLS) produces more efficient estimated treatment differences (small standard errors) for the resolvable incomplete block design of Example 8.4. A Monte Carlo simulation is a computer experiment designed to study the properties of proposed statistical methods. To carry out this simulation, generate a sample of 30 values from model (8.18) with parameter values as in Table 8.2 and $\sigma_r^2 = 1.0, \sigma_b^2 = .25$ and $\sigma_\epsilon^2 = .50$. Fit the model using both the intrablock analysis (SAS® `Proc GLM`) and the EGLS analysis (SAS® `Proc Mixed`). Compute the standard error of a treatment difference under each estimation method and store these two standard errors. Repeat this procedure 500 times (i.e. generate 500 samples). Report the mean standard error for each method of estimation. The method (intrablock analysis or EGLS) that gives the smaller average standard error is more efficient. □

Interblock Analysis

We return to the recovery of interblock information for the general incomplete block design in Section 8.1.5 and examine the special case of the BIB, which has equal block sizes, equal replication, and equal λ for all treatment pairs. It is illuminating to study the traditional Yates method of combining intra- and interblock information. With the side condition $\sum_j \widetilde{\tau}_j = 0$, equation (8.10) can be solved to give

$$\widetilde{\tau}_j = \frac{\sum_i n_{ij} y_{i\cdot} - rk\widetilde{\mu}}{r - \lambda}$$

for $j = 1, \ldots, t$. It follows that

$$\widetilde{\tau}_j - \widetilde{\tau}_{j'} = \frac{\sum_i n_{ij} y_{i\cdot} - \sum_i n_{ij'} y_{i\cdot}}{r - \lambda}$$

$$\text{Var}\left[\widetilde{\tau}_j - \widetilde{\tau}_{j'}\right] = \frac{2k}{r - \lambda}(k\sigma_b^2 + \sigma_\epsilon^2).$$

The intrablock estimates $\widehat{\tau}_j - \widehat{\tau}_{j'}$ and the interblock estimates $\widetilde{\tau}_j - \widetilde{\tau}_{j'}$ can then be combined with weights equal to the reciprocals of their estimated variances. This is the method of analysis found in many of the older statistics methods books.

8.4.2 Efficiency Gains in the BIB from Blocking

In this section we look at the gains in efficiency provided by BIB designs as a consequence of the blocking pattern and using an intrablock analysis. The approach is to compare the BIB design with another design with the same number of treatments and replicates. Consider first a RCBD as the alternative. In a BIB design, the variance of the estimate of the difference between two treatment means each replicated r times in blocks of size k is $\text{Var}\left[\widehat{\tau}_h - \widehat{\tau}_j\right] = 2k\sigma^2_{\text{BIB}}/\lambda t$, where the BIB subscript has been added to σ^2 to show that this is the residual variance for the BIB design. If in place of this a RCBD with r blocks had been performed, the variance of the estimate of a treatment difference would be $\text{Var}\left[\widehat{\tau}_h - \widehat{\tau}_j\right] = 2\sigma^2_{\text{RCBD}}/r$, where the RCBD subscript on the σ^2 indicates that this is the residual variance in the RCBD. Keep in mind that σ^2_{BIB} and σ^2_{RCBD} represent different variances, because the two experiments have different block sizes. The relative efficiency of the BIB compared to the RCBD is the ratio of these two variances of treatment differences,

$$\text{RE} = \frac{2\sigma^2_{\text{RCBD}}}{r} \div \frac{2k\sigma^2_{\text{BIB}}}{\lambda t} = \frac{\lambda t}{rk} \times \frac{\sigma^2_{\text{RCBD}}}{\sigma^2_{\text{BIB}}}.$$

The quantity $\sigma^2_{\text{RCBD}}/\sigma^2_{\text{BIB}}$ is expected to be greater than 1. The BIB design was set up with smaller and consequently, more homogeneous blocks than the RCBD. The term

$$E = \frac{\lambda t}{rk},$$

called the *efficiency factor*, is always less than 1. The truth of this last statement can be seen from the condition $\lambda(t-1) = r(k-1) \Rightarrow \lambda(t-1)/r(k-1) = 1$ along with $r > \lambda$ and $t > k$. The efficiency factor is frequently used as a basis for comparing BIB designs. The λ in the numerator of the expression for E reflects the fact that one gets more information the more treatment pairs occur together within blocks.

Experimenters are really more interested in the relative efficiency than the efficiency factor, because that tells them how much real gain in efficiency to expect from use of the more complex incomplete block design. Unfortunately, there is no way to estimate σ^2_{RCBD} from a single BIB experiment to get an estimate of gain in efficiency. One must rely on experience or data from uniformity trials. Uniformity trials are sham experiments conducted as experiments but with no treatments, whose sole purpose is to obtain estimates of variance among experimental units. Except in field research in agriculture, this type of information is difficult to obtain.

Agricultural researchers working with various crops report relative efficiencies up to over 1.5 for BIB and α-designs compared to randomized complete block designs (Robinson et al. 1988; Kempton et al. 1994; Yau 1997; Sarker et al. 2001). Burrows et al. (1984) found that in enzyme-linked immunosorbent assay (ELISA) work, large variability among wells on a single plate can be a problem. In their application detecting viral infections in plants, they observed relative efficiencies in the range 1.2 to 2.7 for balanced lattice designs compared to RCB designs.

8.4.3 Power Calculations

The power of hypothesis tests is a good criterion for choosing the number of blocks. It is possible to relate power and sample size calculations for the BIB design back to the power calculations for the RCBD in Section 4.5. The expected value of the treatment sum of squares in the analysis of variance in Table 8.1 for the BIB design is

$$\frac{t\lambda}{k}\sum_j(\tau_j - \overline{\tau}.)^2 + (t-1)\sigma^2_\epsilon.$$

It follows that the noncentrality parameter λ^* corresponding to equation (5.2) is

$$\lambda^* = \frac{t\lambda}{k}\sum_j(\tau_j - \overline{\tau}.)^2/\sigma^2_\epsilon.$$

To see if a proposed BIB design with parameters t, b, λ, and r has sufficient power to reject an hypothesized complement of treatment effects, we compute λ^* and evaluate the probability that a noncentral F with parameters $t-1$, $rt-b-t+1$, and λ^* exceeds the selected critical value for a central F with $t-1$ and $rt-b-t+1$ degrees of freedom. If this probability (power) is sufficient, the design is acceptable. Otherwise, a larger design with more replication is required.

Note the unfortunate choice in conventional notation. There is the λ design parameter in the BIB literature and λ the noncentrality parameter in F-, t-, and χ^2-distributions. We are using λ^* for the noncentrality parameter in this chapter.

A small example shows the benefit of doing a power computation in the planning stage. Consider a small experiment with three treatments in blocks of size 2 consisting of the blocks (1,2), (1,3), and (2,3). Suppose that you would like to be able to detect a treatment effect if the treatment means were $\mu_1 = 1, \mu_2 = 3$, and $\mu_3 = 5$ and the residual variance was $\sigma_\epsilon^2 = .5$. If only the three blocks are used, $\text{df}_E = 1$ and the critical value for rejection is huge: $F(.95, 2, 1) = 199.5$. Consequently, the power of the test is very low. The noncentrality parameter is $\lambda^* = 24$ and the power to detect treatment effects is a dismal .20. If we replicate the design, doubling the number of blocks, $\text{df}_E = 4$ and the critical value is $F(.95, 2, 4) = 6.9$, so the power should be much better. In this case the noncentrality parameter doubles to $\lambda^* = 48$ and the power is greatly improved to power $= .98$.

The matrix form of power computations developed in Section 4.5 is completely general and applies here. The noncentrality parameter for testing the hypothesis $\text{H}_0 : \boldsymbol{L\tau} = \boldsymbol{m}_0$, reproduced from equation (4.12), is

$$\lambda^* = (\boldsymbol{m}_A - \boldsymbol{m}_0)^t \left(\text{Var}\left[\boldsymbol{L}\widehat{\boldsymbol{\tau}}\right]\right)^{-1} (\boldsymbol{m}_A - \boldsymbol{m}_0),$$

where $\text{Var}\left[\boldsymbol{L}\widehat{\boldsymbol{\tau}}\right] = \sigma_\epsilon^2 \boldsymbol{L}\boldsymbol{A}^- \boldsymbol{L}^t$. The numerator degrees of freedom for testing this hypothesis are the number of linearly independent rows of $\boldsymbol{L}$, and the error degrees of freedom are $rt-b-t+1$. Thus, the power for testing any linear hypothesis of interest may be computed simply by setting up appropriate $\boldsymbol{L}$ and $\boldsymbol{m}_A$ matrices.

8.4.4 Surveys with Sensitive Questions

When a sample survey questionnaire includes questions that a respondent may not like to answer truthfully because the questions ask for sensitive medical information or probe embarrassing or criminal behavior, incomplete block designs provide a method of protecting respondents' privacy. This should increase the chances of obtaining honest and unbiased responses. In these designs each respondent is given a questionnaire containing a subset of the possible questions; thus each respondent plays the role of a "block" in an incomplete block design. The method outlined here applies to questionnaires in which all blocks have at least one quantitative question.

Instead of recording the answer to every question, only the total is recorded for each respondent. From the respondent totals it is possible to estimate the mean response for each question, but not any particular individual's response to a particular question. The statistical analysis is just an interblock analysis in which the block totals are fitted to a model that includes a term for each question on the questionnaire. The design matrix is the incidence matrix with a row for each block and a column for each question and 0/1 entries indicating which questions are in each block. If there is only one sensitive question, a supplemented block design which is similar to a reinforced design but has unequal block sizes, can be very efficient. Raghavarao and Federer (1979) provide a detailed discussion of the use of balanced incomplete block designs and supplemented block designs for surveys with sensitive questions.

8.4.5 Constructing BIB Designs

The general topic of constructing incomplete block designs or even BIB designs is broad and well beyond the scope of this book. We give a brief catalog of incomplete block designs at the end of this chapter. However, to make the catalog manageable, we need some of the simpler design construction techniques. In part, the object is to help the reader become familiar with tools needed to use and possibly extend the catalog. We begin with types of BIB designs.

A Trivial Construction

Possibly the simplest construction of a BIB design for t treatments is to begin with a RCBD with t blocks and eliminate a different treatment from each block. The result is a BIB for t treatments in $b = t$ blocks of size $k = t - 1$ with $r = t - 1$ and $\lambda = k - 1$. A BIB with $t = b$ is called a symmetric design.

The Irreducible Design

For a given number of treatments t and block size k, it is always possible to construct the *irreducible* BIB design by taking all t treatments in sets of k at a time. While this design is always available, it is typically not practical because it requires too many blocks. In fact, the irreducible design requires $b = \binom{t}{k}$ blocks. These designs have $r = \binom{t-1}{k-1}$ and $\lambda = \binom{t-2}{k-2}$.

Youden Squares

Youden squares, introduced in Section 6.7, represent an important class of BIBs. Youden squares are constructed by taking appropriate subsets of k rows from $t \times t$ Latin squares. The rows form complete blocks and the columns incomplete blocks of size k. There are t treatments in t columns and k rows. A catalog of possible Youden squares is given in Appendix 6C. It can be shown (Smith and Hartley 1948) that any BIB in which the number of treatments is equal to the number of blocks can be set up as a Youden square. Youden squares are attractive designs because they allow for the removal of two sources of variation in the sense of the Latin square. The major problem with Youden squares is that few sizes are available. Since Youden square designs for t treatments and blocks of size k are obtained by selecting k rows from a Latin square, a quick check on availability is to see if the proposed size of plan has enough room. For t treatments there must be room for an exact multiple of $\binom{t}{2}$ pairs. A plan with t blocks of size k has room for $t\binom{k}{2}$ pairs. These must be compatible. This requirement is both necessary and sufficient for the design to exist.

Some Other Constructions of BIBs

In this section we look at some of the simpler techniques of constructing BIBs from other existing BIBs. The techniques are not meant to be an exhaustive coverage of the field but an aid in the use of the catalog given in the appendix.

Complements of Existing Designs

A simple technique for constructing one BIB from another is to replace the treatments in each block by the *complementary set* of treatments, i.e., the set of treatments that are *not* in the block. If the initial design is a BIB t treatments in b blocks of size k, the complementary design will have t treatments in b blocks of size $t - k$. An illustration is given in Tables 8.5 and 8.6.

Residual Designs

A *symmetrical BIB* is one with $t = b$. An important characteristic of symmetrical BIBs is that any two distinct blocks have exactly λ treatments in common. Recall that λ is defined as the number of times that two distinct treatments appear together in a block. The *residual* of a symmetrical BIB design is obtained by

Table 8.5 Balanced Incomplete Block Design

Block										
1	2	3	4	5	6	7	8	9	10	$t = 5$ treatments
										$b = 10$ blocks
0	0	0	1	2	0	0	0	1	1	$k = 3$ obs. per block
1	1	3	2	3	1	2	2	2	3	$r = 6$ replicates
2	4	4	3	4	3	3	4	4	4	$\lambda = 3$

Table 8.6 Complementary Design Obtained from Design in Table 8.5

Block										
										$t^* = t = 5$ treatments
										$b^* = b = 10$ blocks
1	2	3	4	5	6	7	8	9	10	$k^* = t - k = 2$ obs. per block
										$r^* = b - r = 4$ replicates
3	2	1	0	0	2	1	1	0	0	$\lambda^* = \dfrac{r^*(k^* - 1)}{t - 1} = 1$
4	3	2	4	1	4	4	3	3	2	

Table 8.7 Example of a Symmetrical BIB Design with Residual and Derived Designs

Original Design

Block 1	2	3	4	5	6	7
3	1	1	1	2	1	2
5	4	2	2	3	3	4
6	6	5	3	4	4	5
7	7	7	6	7	5	6

$t = 7$
$b = 7$
$k = 4$
$r = 4$
$\lambda = 2$

Residual Design, Omit Treatments from Block 7

Block 1	2	3	4	5	6
3	1	1	1	3	1
7	7	7	3	7	3

$t^* = t - k = 3$
$b^* = b - 1 = 6$
$k^* = k - \lambda = 2$
$r^* = k = 4$
$\lambda^* = \lambda = 2$

Derived Design, Retain Only Treatments from Block 7

Block 1	2	3	4	5	6
5	4	2	2	2	4
6	6	5	6	4	5

$t^* = k = 4$
$b^* = b - 1 = 6$
$k^* = \lambda = 2$
$r^* = k - 1 = 3$
$\lambda^* = \lambda - 1 = 1$

deleting from the design all treatments found in one selected block. If the original design has parameters t, $b = t$, k, r, and λ, the residual will have parameters $t^* = t - k$, $b^* = b - 1$, $k^* = k - \lambda$, $r^* = k$, and $\lambda^* = \lambda$. Table 8.7 illustrates a symmetrical BIB design.

Derived Designs

Derived designs are constructed in a similar fashion. In this case the procedure is to select one block from a symmetrical BIB design, call it S, and then in all of the other blocks retain only the treatments in S. This results in a BIB design with k treatments in $b - 1$ blocks of size λ, as demonstrated in Table 8.7.

8.5 LATTICE DESIGNS

Lattice designs, also called *quasi-factorials*, represent a large class of incomplete block designs that can be constructed by making an association between treatments and the treatment combinations in a factorial experiment. Historically, this association was important because it also provided some very convenient algorithms for the manual statistical analysis of the data. With modern computing facilities these computational aspects are largely irrelevant. Some but not all lattice designs are BIB designs. Lattice designs have a place in experimental work because they are relatively simple to construct, are resolvable, and have good statistical properties.

Since lattice designs rely on underlying factorial schemes, they are rather limited in treatment (t) and block size (k) configurations. For example, if the number of treatments is a power of 2, the block size must be a power of 2. More generally, if $t = p^n$, where p is a prime number, blocks must be of size $k = p^m$, where $m < n$. The terms *simple, triple, quadruple, ... lattice designs* are applied to designs obtained by using $2, 3, 4, \ldots$ replicates.

As a special case of these designs, we consider the BIB designs with $t = p^2$ and p prime or a power of a prime created via orthogonal Latin squares. We know from Chapter 6 that for this case there exists at least one set of $(p-1)$ $p \times p$ mutually orthogonal Latin squares. Let these be labeled $\text{SQ}_1, \ldots, \text{SQ}_{p-1}$. In a base square called T, label the individual cells $1, \ldots, t$. For replicates 1 and 2, define blocks using the columns of T and the rows of T, respectively. For replicate 3 define blocks using the "treatments" of SQ_1. For replicates $4, \ldots, p+1$, define blocks using "treatments" in $\text{SQ}_2, \ldots, \text{SQ}_{p-1}$. We illustrate with $t = 9$. The two orthogonal 3×3 Latin squares are

$$\text{SQ}_1 = \begin{matrix} A & B & C \\ C & A & B \\ B & C & A \end{matrix} \qquad \text{SQ}_2 = \begin{matrix} \alpha & \beta & \gamma \\ \beta & \gamma & \alpha \\ \gamma & \alpha & \beta \end{matrix}.$$

Label the cells in the base square

$$T = \begin{matrix} 1 & 2 & 3 \\ 4 & 5 & 6 \\ 7 & 8 & 9 \end{matrix}.$$

The complete design is shown in Table 8.8. In general, this BIB design has $t = p^2, b = p(p+1), k = p, r = p+1$, and $\lambda = 1$.

Rectangular lattices for $t = p(p-\ell)$ for $\ell < p$ have a very similar construction. However, we must consider two cases, p prime or power of a prime, and otherwise. In the first case there are $p-1$ mutually orthogonal Latin squares. Label the squares $\text{SQ}_1, \ldots, \text{SQ}_{p-1}$. In SQ_1 eliminate cells corresponding to ℓ "treatments." Label the remaining cells $1, \ldots, p(p-\ell)$. From rows and columns of the first square and from "treatments" in the remaining squares, construct p replicates, each consisting of p blocks of $p-\ell$ units. In total we have an incomplete block design with $t = p(p-\ell)$, $b = p^2$, $k = p-\ell$, and $r = p$. Note

Table 8.8 BIB Design Constructed from Two Orthogonal 3 x 3 Latin Squares

Block											
Rows of T			Columns of T			"Treatments" SQ_1			"Treatments" SQ_2		
1	2	3	4	5	6	7	8	9	10	11	12
1	4	7	1	2	3	1	2	3	1	2	3
2	5	8	4	5	6	5	6	4	6	4	5
3	6	9	7	8	9	9	7	8	8	9	7

that this is not a BIB design and hence there is no λ value. If $p = 6$, there is only one Latin square. Otherwise, there are at least two orthogonal Latin squares.

Occasionally, we encounter the terms two-dimensional, three-dimensional, four-dimensional, etc. lattice design. These are lattice designs for p^2, p^3, p^4, etc. treatments in blocks of size $k = p$ for p a prime or a power of a prime number. Notice that the number of blocks rises very quickly. The construction is straightforward but not pursued here.

A complete treatment of lattice designs relies heavily on a thorough understanding of concepts to be developed in Chapters 10, 11, 13, and 17. For those readers already familiar with the concept of confounding in 2^N factorials, consider an experiment with eight treatments to be performed in blocks of four. Think of the eight treatments as the eight treatment combinations in a 2^3 factorial. In the 2^3 there are three main effects and four interactions. We can think of an experiment with seven replicates, each consisting of two blocks of four obtained by confounding once each of the main effects and interactions. In terms of the original eight treatments, this is a BIB design with $t = 8$, $b = 14$, $k = 4$, $r = 7$, and $\lambda = 3$. In general, if all effects and interactions in the underlying factorial scheme are present equally frequently in the underlying confounding schemes, the result is a balanced lattice, i.e., a BIB design.

Exercise 8.6: Simple Lattice Design. (This exercise requires concepts from Chapter 13.) Display the 14 blocks in the BIB design obtained by working from a 2^3 factorial by blocking on each effect and interaction in turn. Verify by counting that $r = 7$ and $\lambda = 3$. □

A lattice construction can also be used to generate designs for two-way elimination of heterogeneity for p^2 treatments with p prime. This is particularly useful in field experiments. In the p^2 factorial there are exactly $p + 1$ effects and $(p - 1)$-degree of freedom components of the interactions. If p is odd, organize these into $(p + 1)/2$ pairs. If p is even, organize into $p/2$ pairs. From each pair, construct a $p \times p$ array by confounding one with rows and the other with columns. The arrays form replicates. Although this is not exactly an incomplete block design, a simple variation of the general incomplete block analysis of variance can be used to analyze the data. The major difference is that both rows and columns (they are orthogonal) within arrays sums of squares are removed from the analysis.

8.6 CYCLIC DESIGNS

We begin our discussion of cyclic designs with an example of a design with t treatments in blocks of size k. Let the treatments be labeled 0 to $t - 1$. Set up an initial block of k treatments. Generate successive blocks by adding 1 (reducing modulo t if necessary) to all the treatments in the block. This is repeated until the initial block reappears. For example, a design for $t = 8$ treatments in blocks of size $k = 4$ constructed from the initial block (0, 1, 2, 5) would consist of the initial block plus new blocks (1, 2, 3, 6), (2, 3, 4, 7), (3, 4, 5, 0), (4, 5, 6, 1), (5, 6, 7, 2), (6, 7, 0, 3), and (7, 0, 1, 4). In this case there are $b = t = 8$ blocks and $r = 4$ replicates.

A number of observations follow directly. First, cyclic designs are easy to generate. All that is needed is an initial block. Second, cyclic designs exist for all combinations of t and k. A set of t distinct blocks obtained via the cyclic development of an initial block is called a *full set*. Any one of the blocks in a set can serve as an initial block to generate that set. A full set defines an incomplete block design with t treatments in $b = t$ blocks of size k. Each treatment is replicated $r = k$ times. In addition, each treatment appears once in each of the k positions. This means that the design can be used as a row–column design with blocks defining rows and position columns, in the sense of Youden squares. This has meaning if the experimental units have fixed positions in space or time, but does not have meaning if experimental units are fully interchangeable and move around.

If t and k are not relatively prime and the position of the treatments in the block is not maintained, the cyclic development applied to particular initial blocks can yield partial sets consisting of t/d blocks, where d is any common divisor of t and k. This happens for certain choices of initial block. These are called *partial sets*. More extensive incomplete block designs can be constructed by combining full and partial sets.

Cyclic designs with reasonable balance are easy to generate, but finding designs with optimal properties is more difficult. The cyclic generation of the designs per se has little to do with the efficiency of the designs generated. The nature of the set of blocks generated depends on the initial block selected. Statistical properties of cyclic designs depend on concurrences, the number of times that treatment pairs occur together in a block. Some cyclic designs are BIB designs, but most are not. The number of times that treatment j and treatment j' with $j \neq j'$ appear together in a block can be determined from the initial block. If two treatments each appear r times in a cyclic design, the two will have the same total number of concurrences with other treatments. It follows that average number of concurrences is not a useful criterion to use to rank designs. However, a design with all pairs having more uniform concurrences will be a better design. A design with all pairwise concurrences equal is a BIB design. Efficient computer searches in the class of cyclic designs for designs with good statistical properties are possible. Initial blocks to generate the sets provide sufficient input information. Notice that the incomplete block design generated in the example is not a BIB design.

Exercise 8.7: Cyclic Example Is Not BIB. Verify that the example of a cyclic design generated to start Section 8.6 is not a BIB design. □

Exercise 8.8: Generating a Partial Set. Show that the cyclic development of (0, 2, 4, 6) in place of the initial block used at the start of Section 8.6 yields a partial set. □

Exercise 8.9: Comparison of Designs. Use the initial block (0, 1, 2, 4, 7) to generate an incomplete block design with $t = 8$ and $k = 5$. Verify that 24 treatment pairs appear together in a block three times and four pairs appear together in a block two times. We say that the design has $\lambda_1 = 3$ and $\lambda_2 = 2$. Compare this design with one generated from the initial block (0, 1, 2, 3, 4). Note that both designs have each treatment replicated $r = 5$ times. Which design would you recommend? Why? □

An easy-to-use catalog of efficient cyclic designs in the range $t = 4$ to 15 and $r \leq 10$ is given in Appendix 8A.3. A catalog for $t = 10$ to 60 and $r = k = 3$ to 10 can be found in Lamacraft and Hall (1982). The efficiencies are based on intrablock information using the designs as incomplete block designs and not row–column designs. The efficiency criterion is the harmonic mean of the nonzero eigenvalues of $(1/r)\boldsymbol{A}$, where $\boldsymbol{A}$ is defined in equation (8.8). Since we have equal block sizes and equal treatment replication here, $\boldsymbol{A}$ simplifies to $r\boldsymbol{I} - (1/k)\boldsymbol{N}\boldsymbol{N}^t$. A few of the designs in the catalog turn out to be BIB designs, and as such are A-optimal.

8.6.1 Dicyclic and Tricyclic (n-Cyclic) Designs

While the cyclic generation of incomplete block designs is quite general, there is still a need for more flexibility. One simple approach to adding such flexibility is to label treatments as products of factors and then cycle on the individual factors. For example, if the number of treatments t can be written as a product $t = m_1 m_2$, the treatment labels can be written as pairs of numbers, 0 to $m_1 - 1$ and 0 to $m_2 - 1$. Dicyclic designs are then developed by selecting an initial block and cyclically incrementing both numbers. To illustrate the method, consider 16 treatments. Label them using pairs of factors, each from 0 to 3, i.e., labels are 00, 01, 02, 03, 20, ..., 33. Using (00, 01, 02, 10, 23, 30) as initial block leads to the design

$$
\begin{array}{ll}
(00,\ 01,\ 02,\ 10,\ 23,\ 30) & (01,\ 02,\ 03,\ 11,\ 20,\ 31) \\
(02,\ 03,\ 00,\ 12,\ 21,\ 32) & (03,\ 00,\ 01,\ 13,\ 22,\ 33) \\
(10,\ 11,\ 12,\ 20,\ 33,\ 00) & (11,\ 12,\ 13,\ 21,\ 30,\ 01) \\
(12,\ 13,\ 10,\ 22,\ 31,\ 02) & (13,\ 10,\ 11,\ 23,\ 32,\ 03) \\
(20,\ 21,\ 22,\ 30,\ 03,\ 10) & (21,\ 22,\ 23,\ 31,\ 00,\ 11) \\
(22,\ 23,\ 20,\ 32,\ 01,\ 12) & (23,\ 20,\ 21,\ 33,\ 02,\ 13) \\
(30,\ 31,\ 32,\ 00,\ 13,\ 20) & (31,\ 32,\ 33,\ 01,\ 10,\ 21) \\
(32,\ 33,\ 30,\ 02,\ 11,\ 22) & (33,\ 30,\ 31,\ 03,\ 12,\ 23)
\end{array}
\tag{8.19}
$$

It is left as an exercise to show that this is a BIB design with $t = 16$, $b = 16$, $k = 6$, $r = 6$, and $\lambda = 2$. A number of the designs in the catalog of BIBs given in Section 8A.2 are obtained in this way.

The method can be extended to tricyclic designs and beyond by writing t as the product of more factors. For example, a tricyclic $3 \times 4 \times 5$ scheme for 60 treatments labels the treatments 000, 001, ..., 004, 100, ..., 234.

Exercise 8.10: Dicyclic Design Is a BIB Design. Show that the dicyclic design (8.19) is BIB design with parameters $t = 16$, $b = 16$, $k = 6$, $r = 6$, and $\lambda = 2$. □

8.6.2 Generalized Cyclic Designs

Another generalization in the development of cyclic designs uses an incrementing number larger than 1 in the cyclic development of the designs. This again requires that it be possible to write t as the product of two numbers. The idea is that with $t = mn$, it is possible to divide the treatments into m sets of n. It follows from this that blocks may be developed by cyclic addition of m (and reduction modulo t if necessary) to an initial block to produce designs with treatments replicated equally. An initial block leads to n blocks, unless position within blocks is not maintained. In that case blocks can repeat after some fraction of n blocks is reached. Standard practice is to retain only unique blocks. Generally, several initial blocks are required and the sets of generated blocks combined.

We illustrate the construction with an example. We begin with $t = 30 = mn$ in $b = 15 = n$ blocks of size $k = 4$, beginning with initial block (0, 2, 3, 25) and incrementing by $m = 2$. The design is

$$
\begin{array}{lll}
(0,\ 2,\ 3,\ 25) & (10,\ 12,\ 13,\ 5) & (20,\ 22,\ 23,\ 15) \\
(2,\ 4,\ 5,\ 27) & (12,\ 14,\ 15,\ 7) & (22,\ 24,\ 25,\ 17) \\
(4,\ 6,\ 7,\ 29) & (14,\ 16,\ 17,\ 9) & (24,\ 26,\ 27,\ 19). \\
(6,\ 8,\ 9,\ 1) & (16,\ 18,\ 19,\ 11) & (26,\ 28,\ 29,\ 21) \\
(8,\ 10,\ 11,\ 3) & (18,\ 20,\ 21,\ 13) & (28,\ 0,\ 1,\ 23)
\end{array}
$$

Hall and Jarrett (1981) present a catalog of efficient generalized cyclic designs for $10 \leq t \leq 60$, $r \leq k$, and for $(r, k) = (2,4), (2,6), (2,8), (2,10), (3,6), (3,9), (4,6), (4,8), (4,10)$, and $(5,10)$. In the full generality, generalized cyclic designs do not require that all treatments be replicated equally or that all block sizes be equal.

8.6.3 Rotational Designs

Rotational designs are another variation on the cyclic design theme. Several of the BIB designs in Section 8A.2 are rotational designs. In a rotational construction for t treatments, numbered $0, 1, \ldots, t-1$, modulo $(t-1)$ arithmetic is used rather than modulo t arithmetic. All but $t-1$ of the blocks are constructed by cycling a starting block as in cyclic designs. The remaining $t-1$ blocks include treatment t in every block and are formed by cycling the remaining $k-1$ treatments. The last treatment is labeled ∞ to indicate that it should not be cycled in constructing the design.

We give an example of a BIB design for $t = 8$ treatments in blocks of size $k = 4$. The starting blocks are (0, 1, 3, 6) and (∞, 0, 1, 3). The resulting blocks, after cycling them each modulo 7, are

$$\begin{array}{ll}
(0,\ 1,\ 3,\ 6) & (\infty,\ 0,\ 1,\ 3) \\
(1,\ 2,\ 4,\ 0) & (\infty,\ 1,\ 2,\ 4) \\
(2,\ 3,\ 5,\ 1) & (\infty,\ 2,\ 3,\ 5) \\
(3,\ 4,\ 6,\ 2) & (\infty,\ 3,\ 4,\ 6). \\
(4,\ 5,\ 0,\ 3) & (\infty,\ 4,\ 5,\ 0) \\
(5,\ 6,\ 1,\ 4) & (\infty,\ 5,\ 6,\ 1) \\
(6,\ 0,\ 2,\ 5) & (\infty,\ 6,\ 0,\ 2)
\end{array}$$

The symbol ∞ represents the eighth treatment in this design.

8.7 α-DESIGNS

The α-designs developed by Patterson and Williams (1976) and Patterson et al. (1978) present a very general solution to the problem of generating resolvable incomplete block designs with good statistical properties for any number of treatments in any reasonable block size and amount of replication. Clearly, with a set of (basically inconsistent) conditions, something had to be sacrificed. When it is not possible to have t a multiple of block size, they choose to sacrifice the equal-block-size constraint. These designs can be thought of as resolvable generalized cyclic incomplete block designs with incrementing constant equal to k. The designs are generated by cyclically developing suitable generating arrays. These generating arrays were obtained by a combination of heuristics and computing to produce designs with acceptable efficiency. Recall that it can be shown that BIB designs achieve optimal efficiency and consequently are recommended when they exist. The α-design generating procedure that we give will not in general give the most efficient design. It is recommended that the potential user always check to make sure that no BIB design exists before selecting an α-design. The generating arrays provided have been examined to give designs with near maximal, if not maximal, efficiency for cases where BIB designs do not exist. Generating arrays are given in Section 8A.4 for a wide range of t and k values.

8.7.1 Construction of α-Designs

Section 8A.4 contains a series of tables comprising an extensive set of generating arrays. These arrays are the basis for a broad array of efficient designs for t in the range from 20 to 100 and k from 4 to 10. For the actual construction we consider first the case where $t = ks$. This leads to s blocks of size k in each replicate. We then provide some alternatives when t is not a multiple of k, and consequently the equal block size restriction must be sacrificed.

Number of Treatments an Exact Multiple of Block Size

Our presentation is in two steps. We begin with an α-array and illustrate the construction of a design and then explain how the α-arrays are extracted from the tables in Section 8A.4. For our example, we generate a design for $t = 30$ treatments in blocks of size $k = 5$. For this design we have $s = 6$ blocks per replicate. The α-array for this design with $r = 3$ replicates is the $r \times k$ array

$$\begin{array}{ccccc} 0 & 0 & 0 & 0 & 0 \\ 0 & 1 & 3 & 2 & 4. \\ 0 & 5 & 2 & 3 & 1 \end{array}$$

There are several alternative algorithms that can be used to get from this array to the actual design. A particularly convenient one, which can be implemented in SAS® `Proc Plan`, is as follows:

1. Relabel the elements α_{ij} of the generating array as $a_{ij} = k\alpha_{ij} + (j-1)$. We call the relabeled generating array the A-array. For $t = 30, k = 5, s = 6$, and $r = 3$, the A-array is

$$\begin{array}{rrrrr} 0 & 1 & 2 & 3 & 4 \\ 0 & 6 & 17 & 13 & 24. \\ 0 & 26 & 12 & 18 & 9 \end{array}$$

2. The ith row of the A-array is the starting block for the ith replicate. The elements of this row are incremented by k modulo t to form the remaining blocks of the ith replicate. The first replicate of the design is shown in Table 8.9.
3. The next set of blocks is generated by incrementing row 2 (by $k = 5$) of the A-array. The second set of blocks is shown in Table 8.10.
4. The third set of blocks is obtained by incrementing the third row of the A-array shown in Table 8.11.

Once the structure of the rs blocks has been determined, the symbols within blocks are randomized, the block assignment within replicates is randomized and the actual treatments randomly assigned to the treatment symbols.

Generating arrays are given in compact form in Section 8A.4. To obtain a specific generating array, first use t and k to fix s. Use s to select the proper table. Select the first r rows and k columns as the appropriate α-array.

Table 8.9 First Replicate of α-Design

Block	Treatments
1	(0, 1, 2, 3, 4)
2	(5, 6, 7, 8, 9)
3	(10, 11, 12, 13, 14)
4	(15, 16, 17, 18, 19)
5	(20, 17, 18, 19, 20)
6	(25, 26, 27, 28, 29)

Table 8.10 Second Replicate of α-Design

Block	Treatments
7	(0, 6, 17, 13, 24)
8	(5, 11, 22, 18, 29)
9	(10, 16, 27, 23, 4)
10	(15, 21, 2, 28, 9)
11	(20, 26, 7, 3, 14)
12	(25, 1, 12, 8, 19)

Table 8.11 Third Replicate of α-Design

Block	Treatments
13	(0, 26, 12, 18, 9)
14	(5, 1, 17, 23, 14)
15	(10, 6, 22, 28, 19)
16	(15, 11, 27, 3, 24)
17	(20, 16, 2, 8, 29)
18	(25, 21, 7, 13, 4)

Number of Treatments Not a Multiple of Block Size

If t is a prime number or at least not a multiple of some reasonable block size, unequal block sizes are unavoidable. A reasonable recommendation is to generate an α-design with the smallest possible $ks > t$ with reasonable k and reduce $ks - t$ blocks to block size $k - 1$ in each replicate by eliminating $ks - t$ symbols. Consider, for example, a design for 27 treatments in blocks of five. We developed above a design for 30 treatments in blocks of five. Now if three treatments, one from each of three different blocks in one column are eliminated, the result is a design with nine blocks of four and nine blocks of five.

8.8 OTHER INCOMPLETE BLOCK DESIGNS

There is an extensive and, as far as experimenters are concerned, poorly organized body of literature dealing with incomplete block designs. Some of the designs, such as the *partially balanced incomplete block* designs which we have ignored, were developed with emphasis on simple statistical analysis of data rather than statistical efficiency. This was important before the advent of cheap and extensive computing. Then there is the mathematics literature, where the emphasis is on mathematical structures rather than statistical properties. In addition, there is a large body of literature discussing cyclic designs which is beyond the scope of this book. A good reference for details is John and Williams (1995). There is also a large body of literature dealing with experimental designs that eliminate heterogeneity in two directions, and much recent work on designs that attempt to take advantage of the observation that experimental units located close together tend to be more alike than units farther apart. There is an ever-increasing number of commercially available computer programs for generating statistical designs, and in particular, incomplete block designs.

8.8.1 Reinforced Designs

Occasionally, one encounters situations where there is one clearly established standard treatment in which other treatments are to be compared to that standard. An example would be a variety testing program where a large number of new varieties are to be compared to an old standard variety. A common strategy is to set up an incomplete block design for the "new" treatments and then add control to each block. This leads to designs generally referred to as *reinforced designs*. The introduction of control provides considerable interblock information for comparisons among new treatments.

REVIEW EXERCISES

8.1 Construct a BIB for 16 treatments in 20 blocks of four by starting with a set of three mutually orthogonal 4×4 Latin squares.

8.2 An animal scientist wants to compare six fertility treatments for hogs. She is going to use animals from several different farms. At each farm it is feasible to test only two or three different treatments.

(a) Draw a BIB design (unrandomized) for three animals per farm.

(b) Suppose that the within-farm variance is 1.50 regardless of the number of treatments tested on the farm. Compute the variance of the difference of two treatment means under the design using three animals per farm, treating farms as fixed block effects.

(c) Suppose that the between-farm variance component is also 1.50. Compute the variance of the difference of two treatment means under the design using three animals per farm, treating farms as random block effects. *Hint*: the variance of a GLS estimate is discussed in Section 5.5.2.

(d) Would you recommend using two animals per farm or three animals per farm, supposing that you use the same total number of animals? Does the answer differ depending on whether blocks are considered random or fixed?

8.3 A photo processor wants to compare 10 types of photographic paper. He can process six rolls of film in one morning.

(a) Propose two designs for this comparison. Draw the layout of each design. Explain how to randomize each design.

(b) Compute the variance (or variances) of a pairwise treatment contrast in each design using the intrablock analysis.

(c) What is the efficiency factor (or factors) for each design? Explain exactly what the efficiency factors measure; i.e., explain what two designs are being compared by each efficiency factor.

8.4 An electronics testing room has space for five TV sets to be viewed side by side. Design an experiment to compare 11 models, testing them in sets of five at a time.

(a) List the models in each set of your proposed design.

(b) Compute the efficiency factor of this design. Explain what this means in words and draw a diagram of the reference design.

8.5 A plant pathologist wants to test the susceptibility of peanuts to 40 different isolates of a fungus. To do these tests, sections of peanut stems are infected with the virus and put into a box to incubate for a period of time. Each box can hold five stems (each with a different isolate). Suggest a design for this experiment with two replications of each isolate.

8.6 Suggest a design for six treatments in blocks of size 4 with four replicates. What are the important features of your design?

8.7 (Data from A. Meshaw, personal communication) In a study of the effects on turkeys of *in ovo* exposure to gastrointestinal peptides and growth hormones, four treatments were studied: saline (S), thyroxine (T4), epidermal growth factor (EGF), and peptide YY (PYY). The turkeys were kept in cages on a rack with three shelves and four quadrants per shelf. The treatment assignments and the responses are shown below.

	Quadrant							
Level	1		2		3		4	
1	S	1776	T4	505	EGF	1174	PYY	2603
2	T4	519	EGF	1170	PYY	2601	S	1675
3	EGF	1173	PYY	2607	S	1671	T4	501

(a) What kind of experimental design is this? Write down an appropriate model and assumptions.

(b) Estimate the mean difference between the treatment S and the average of the other treatments, along with its standard error.

(c) Construct a 95% confidence interval for the difference between the S and EGF treatments.

8.8 Brownie et al. (1993) discuss a corn variety trial in which the corn varieties were arranged in a rectangular triple lattice. Thirty varieties of corn were tested, and there were three replicates of each variety. Each replicate consisted of six rows of plots, where each row was a block of five plots.

(a) Show how to construct such a design and draw a layout of the experiment.

(b) Compute the variance of the difference between the treatment 1 and 2 effects, assuming that block and replicate effects are fixed effects. Compute the same variance assuming that block and replicate effects are random effects and the variance components are $\sigma_r^2 = 1, \sigma_b^2 = 1$, and $\sigma_\epsilon^2 = 1$ for replicates, blocks, and within-block error, respectively.

(c) Compute the variance of the difference between the treatment 1 and treatment 6 means, assuming that block and replicate effects are fixed effects. Compute the same variance assuming that block and replicate effects are random effects and the variance components are $\sigma_r^2 = 1, \sigma_b^2 = 1$, and $\sigma_\epsilon^2 = 1$ for replicates, blocks, and within-block error, respectively.

8.9 Quantitative ELISA (enzyme-linked immunosorbent assay) can be used to detect viral infections in plants. For this type of bioassay, samples are put into wells on a plate, and typically there are 96 wells per plate, arranged in eight rows and 12 columns. It has been observed that there is spatial variation among wells on a plate and between plates (Burrows et al. 1984). In one experiment for 16 treatments, each plate is divided into six squares of 16 wells each, but only five of the squares are used. Each square has four 2×2 blocks of four wells.

(a) Give the name of a suitable design for this layout and draw a diagram of your design.

(b) Write down a model matching your design.

8.10 Is it possible to construct a triple lattice for 36 treatments in blocks of six? A quadruple lattice?

8.11 Consider cyclic designs for $t = 7$ treatments.

(a) Consider the simple cyclic design for $t = 7$ treatments and $b = 14$ constructed by combining the two sets obtained by cyclic development of (0, 1, 2) and (0, 1, 3). Point out why this is not a BIB.

(b) Verify that the design consisting of (0, 1, 3), (1, 2, 4), (2, 3, 5), (3, 4, 6), (4, 5, 0), (5, 6, 1), and (6, 0, 2) is a BIB and consequently, efficient.

(c) Which is a better design: the cyclic design with 14 blocks constructed from starting blocks (0, 1, 2) and (0, 1, 3) or the design generated from the two initial blocks (0, 2, 4) and (0, 3, 5)? Explain why you think so.

(d) Compute the average efficiency factors of the two designs in part(**c**) using either a matrix programming language or analysis of variance software.

APPENDIX 8A: CATALOG OF INCOMPLETE BLOCK DESIGNS

We provide a number of tables that should prove useful to a person planning an investigation. Section 8A.1 gives a guide to incomplete block designs with specified parameters. Since BIB designs can be shown to achieve maximal efficiency, this guide directs the user to the catalog of BIB designs in Section 8A.2 whenever possible. In this catalog we give the possible BIB designs for $4 \leq t \leq 100$ and $r \leq 15$. We note in passing that designs with more replication can be obtained by combining BIB designs. A more extensive listing of BIB designs and in particular BIB designs for larger numbers of replicates may be found in Colbourn and Dinitz (1996). When a suitable BIB does not exist, we direct the reader to the table of α-designs in Section 8A.4 and efficient cyclic designs in Section 8A.3.

8A.1 Some Available Incomplete Block Designs

Catalog of BIB, Lattice, Cyclic, and Alpha Designs[a]

t	k	$r=2$	$r=3$	$r=4$	$r=5$	$r=6$
4	2	DL	BL			
	3		YS			
5	2			BIB		
	3					BIB
	4			YS		
6	2	RL,C	RL	C	BIB	C
	3		C		BIB	C
	4			C		
	5				YS	
7	2	C		C		BIB
	3		YS			C
	4			YS		
	5				C	
	6					YS
8	2	RL,C	RL	RL,C		C
	3		C		C	
	4			C		
	5				C	
	6					C
9	2	C		C		C
	3	DL	TL,C	BL		C
	4			C		
	5				C	
	6					C
10	2	RL,C	RL	RL,C	RL	C
	3		C			C
	4			C		BIB
	5				C	
	6					C

(*continued*)

Catalog of BIB, Lattice, Cyclic, and Alpha Designs[a] (*continued*)

t	k	$r = 2$	$r = 3$	$r = 4$	$r = 5$	$r = 6$
11	2	C		C		C
	3		C			C
	4			C		
	5				BIB	
	6					BIB
12	2	RL,C	RL	RL,C	RL	RL,C
	3	RL	RL,C	RL		C
	4			C		
	5				C	
	6					C
13	2	C		C		C
	3		C		C	
	4			BIB		
	5				C	
	6					C
15	3	RL	RL	RL	RL	
16	4	DL	TL	QL	BL	
	6					BIB
18	3	RL	RL	RL	RL	RL
20	4	α	α	α	RL	
21	5				BIB	
24	4	α	α	α	RL	RL
25	5	DL	TL	QL	SL	BL
31	6					BIB

[a]BIB, balanced incomplete block; BL, balanced lattice; YS, Youden square; DL, TL, QL = double, triple, quadruple lattices; RL, rectangular lattice; C, cyclic design; α, alpha design.

8A.2 Catalog of BIBs

Brief Catalog of BIB Designs for $4 \leq t \leq 100$ and $r \leq 15$

t	b	r	k	λ	Comment
4	6	3	2	1	4 symbols, 2 at a time
	4	3	3	2	From 4 symbols, leave out 1 at a time
5	10	4	2	1	5 symbols, 2 at a time
	5	4	4	3	From 5 symbols, leave out 1 at a time
	10	6	3	3	5 symbols, 3 at a time
6	15	5	2	1	6 symbols, 2 at a time
	10	5	3	2	Residual of design (11, 11, 5, 5, 2)
	6	5	5	4	From 6 symbols, leave out 1 at a time
	15	10	4	6	6 symbols, 4 at a time
7	7	3	3	1	Cyclic. Starting block (0, 1, 3).
	7	4	4	2	Complement of design (7, 7, 3, 3, 1)
	21	6	2	1	7 symbols, 2 at a time
	7	6	6	5	From 7 symbols, leave out 1 at a time

Brief Catalog of BIB Designs for $4 \leq t \leq 100$ and $r \leq 15$ (*continued*)

t	b	r	k	λ	Comment
8	28	7	2	1	8 symbols, 2 at a time
	14	7	4	3	Rotational. Starting blocks (0, 1, 3, 6), (∞, 0, 1, 3)
	8	7	7	6	From 8 symbols, leave out 1 at a time
9	12	4	3	1	Balanced lattice
	36	8	2	1	9 symbols 2 at a time
	18	8	4	3	Cochran and Cox (1957, p. 474), Colbourn and Dinitz (1996, Table I.1.17)
	12	8	6	5	Complement of design (9, 12, 4, 3, 1)
	9	8	8	7	From 9 symbols, leave out 1 at a time
	18	10	5	5	Complement of design (9, 18, 8, 4, 3)
10	15	6	4	2	Residual of design (16, 16, 6, 6, 2)
	45	9	2	1	10 symbols, 2 at a time
	30	9	3	2	Rotational. Starting blocks (∞, 0, 1), (0, 1, 4), (0, 2, 4), partial (0, 3, 6)
	18	9	5	4	Residual of design (19, 19, 9, 9, 4)
	15	9	6	5	Complement of design (10, 15, 6, 4, 2)
	10	9	9	8	From 10 symbols, leave out 1 at a time
11	11	5	5	2	Cyclic. Starting block (1, 2, 3, 5, 8)
	11	6	6	3	Complement of design (11, 11, 5, 5, 2)
	55	10	2	1	11 symbols 2 at a time
	11	10	10	9	11 symbols, leave out 1 at a time
	55	15	3	3	Cyclic. Starting blocks (0, 1, 10), (0, 2, 9), (0, 4, 7), (0, 8, 3), (0, 5, 6)
12	44	11	3	2	Rotational. Starting blocks (0, 1, 3), (0, 1, 5), (0, 4, 6), (∞, 0, 3)
	33	11	4	3	Rotational. Starting blocks (0, 1, 3, 7), (0, 2, 7, 8), (∞, 0, 1, 3)
	22	11	6	5	Rotational. Starting blocks (0, 1, 3, 7, 8, 10), (∞, 0, 5, 6, 8, 10)
13	13	4	4	1	Cyclic. Starting block (0, 1, 3, 9)
	26	6	3	1	Colbourn and Dinitz (1996, Table I.1.20)
	13	9	9	6	Complement of design (13, 13, 4, 4, 1)
	26	12	6	5	Cyclic. Starting blocks (1, 3, 4, 9, 10, 12), (2, 5, 6, 7, 8, 11)
	39	15	5	5	Cyclic. Starting blocks (0, 1, 8, 12, 5), (0, 2, 3, 11, 10), (0, 4, 6, 9, 7)
14	26	13	7	6	Colbourn and Dinitz (1996)[a]
15	35	7	3	1	Cochran and Cox (1957, p. 478), Colbourn and Dinitz (1996, Table I.1.21)
	15	7	7	3	Cochran and Cox (1957, p. 524), Colbourn and Dinitz (1996, Table I.1.23)
	15	8	8	4	Complement of design (15, 15, 7, 7, 3)
	42	14	5	4	Rotational. Starting blocks (∞, 0, 1, 3, 8), (0, 1, 2, 5, 6), (0, 2, 4, 7, 10)
	35	14	6	5	Rotational. Starting blocks (∞, 0, 3, 5, 7, 13), (1, 6, 9, 10, 11, 12), (1, 2, 4, 8, 9, 11)

(*continued*)

Brief Catalog of BIB Designs for $4 \le t \le 100$ and $r \le 15$ (*continued*)

t	b	r	k	λ	Comment
16	20	5	4	1	Balanced lattice
	16	6	6	2	Dicyclic (00, 01, 02, 10, 23, 30)
	24	9	6	3	Cochran and Cox (1957, p. 478), Colbourn and Dinitz (1996, Table I.5.21)
	16	10	10	6	Complement of design (16, 16, 6, 6, 2)
	80	15	3	2	Cyclic. Starting blocks (0, 1, 2), (0, 2, 8), (0, 3, 7), (0, 4, 7), (0, 5, 10)
	48	15	5	4	Colbourn and Dinitz (1996)[a]
	40	15	6	5	Combine 16 blocks of (16, 16, 6, 6, 2) with 24 blocks of (16, 24, 9, 6, 3)
	30	15	8	7	Associate 16 symbols with 16 treatment combinations in a 2^4 factorial. One can put a 2^4 into 2 blocks of 8 by confounding any one of 15 interactions or main effects, giving a total of 30 different blocks. This is one of the early methods used by Yates to develop incomplete block designs.
19	57	9	3	1	Cyclic. Starting blocks (0, 1, 4), (0, 2, 9), (0, 5, 11)
	19	9	9	4	Cochran and Cox (1957, p. 526), Colbourn and Dinitz (1996, Table I.1.26)
	19	10	10	5	Complement of design (19, 19, 9, 9, 4)
	57	12	4	2	Cyclic. Starting blocks (0, 1, 7, 11), (0, 2, 3, 14), (0, 4, 6, 9)
21	21	5	5	1	Cyclic. Starting block (0, 1, 4, 14, 16)
	70	10	3	1	Cyclic. Starting blocks (0, 1, 3), (0, 4, 12), (0, 5, 11), partial (0, 7, 14)
	30	10	7	3	Cochran and Cox (1957, p. 480)
	42	12	6	3	Cyclic. Starting blocks (0, 2, 10, 15, 19, 20), (0, 3, 7, 9, 10, 16)
	35	15	9	6	Colbourn and Dinitz (1996)[a]
22	77	14	4	2	Cyclic. Starting blocks (0, 4, 16, 17), (0, 12, 14, 21), (0, 14, 16, 19), (0, 4, 11, 15)
	44	14	7	4	Cyclic. Starting blocks (0, 2, 6, 8, 9, 10, 13), (0, 3, 5, 6, 12, 13, 17)
23	23	11	11	5	Cyclic. Starting block (1, 13, 8, 12, 18, 4, 6, 9, 2, 3, 16)
25	30	6	5	1	Balanced lattice
	50	8	4	1	Dicyclic. Starting blocks (00, 01, 10, 44), (00, 02, 20, 33)
	25	9	9	3	Cochran and Cox (1957, p. 529)
	100	12	3	1	Cyclic. Starting blocks (0, 1, 3), (0, 4, 11), (0, 5, 13), (0, 6, 15)
26	65	15	6	3	Rotational dicyclic. Starting blocks (00, 01, 02, 11, 24, 40), (00, 02, 04, 22, 23, 30), (∞, 00, 10, 20, 30, 40), (∞, 00, 12, 24, 31, 43), (∞, 00, 14, 23, 32, 41)
27	117	13	3	1	Cyclic. Starting blocks (0, 1, 3), (0, 4, 11), (0, 5, 15), (0, 6, 14), partial (0, 9, 18)

Brief Catalog of BIB Designs for $4 \leq t \leq 100$ and $r \leq 15$ (*continued*)

t	b	r	k	λ	Comment
	39	13	9	4	Colbourn and Dinitz (1996)[a]
	27	13	13	6	Colbourn and Dinitz (1996)[a]
28	63	9	4	1	Cochran and Cox (1957, p. 481)
	36	9	7	2	Residual of design (37, 37, 9, 9, 2)
	42	15	10	5	Colbourn and Dinitz (1996)[a]
29	58	14	7	3	Cyclic. Starting blocks (1, 16, 24, 7, 25, 23, 20) and (2, 3, 19, 14, 21, 17, 11)
31	31	6	6	1	Cochran and Cox (1957, p. 530)
	31	10	10	3	Cochran and Cox (1957, p. 531)
	155	15	3	1	Cyclic. Starting blocks (0, 1, 12), (0, 2, 24), (0, 3, 8), (0, 4, 17), (0, 6, 16)
	93	15	5	2	Cyclic. Starting blocks (1, 2, 4, 8, 16), (3, 17, 24, 12, 6), (9, 20, 10, 5, 18)
	31	15	15	7	Colbourn and Dinitz (1996)[a]
33	44	12	9	3	Residual of design (45, 45, 12, 12, 3)
36	84	14	6	2	Rotational. Starting blocks partial (∞, 0, 7, 14, 21, 28), partial (∞, 0, 7, 14, 21, 28), (4, 15, 16, 20, 24, 26), (2, 3, 12, 15, 18, 20)
	36	15	15	6	Colbourn and Dinitz (1996)[a]
37	37	9	9	2	Cyclic. Starting block (1, 16, 34, 26, 9, 33, 10, 12, 7)
	111	12	4	1	Cyclic. Starting blocks (0, 1, 3, 24), (0, 4, 6, 32), (0, 7, 17, 25)
	130	13	4	1	Cyclic. Starting blocks (0, 1, 4, 13), (0, 2, 7, 24), (0, 6, 14, 25), partial (0, 10, 20, 30)
	40	13	13	4	Colbourn and Dinitz (1996)[a]
41	82	10	5	1	Cyclic. Starting blocks (1, 10, 18, 16, 37), (36, 32, 33, 2, 20)
43	86	14	7	2	Cyclic. Starting blocks (0, 1, 11, 19, 31, 38, 40), (0, 2, 10, 16, 25, 38, 42)
45	99	11	5	1	Tricyclic $3 \times 3 \times 5$. Starting blocks (010, 020, 102, 202, 001), (210, 120, 222, 112, 001), partial (000, 001, 002, 003, 004)
	55	11	9	2	Colbourn and Dinitz (1996)[a]
	45	12	12	3	Shrikhande and Singh (1962)
49	56	8	7	1	Balanced lattice
56	56	11	11	2	Colbourn and Dinitz (1996)[a]
	70	15	12	3	Colbourn and Dinitz (1996)[a]
57	57	8	8	1	Cochran and Cox (1957, p. 533)
61	183	15	5	1	Cyclic. Starting blocks (0, 1, 3, 13, 34), (0, 4, 9, 23, 45), (0, 6, 17, 24, 32)
64	72	9	8	1	Balanced lattice
66	143	13	6	1	Colbourn and Dinitz (1996, Table I.2.34)
	78	13	11	2	Colbourn and Dinitz (1996)[a]
71	71	15	15	3	Colbourn and Dinitz (1996)[a]
73	73	9	9	1	Cochran and Cox (1957, p. 534)

(*continued*)

Brief Catalog of BIB Designs for $4 \le t \le 100$ and $r \le 15$ (*continued*)

t	b	r	k	λ	Comment
76	190	15	6	1	Colbourn and Dinitz (1996)[a]
79	79	13	13	2	Colbourn and Dinitz (1996)[a]
81	90	10	9	1	Balanced lattice
91	91	10	10	1	Cyclic. Starting block (0, 1, 6, 10, 23, 26, 34, 41, 53, 55)
	195	15	7	1	Dicyclic 7 × 13. Starting block (∞, 1_1, 1_{12}, 2_4, 2_9, 4_3, 4_{10})

[a]In *CRC Handbook*, but not in a very accessible form.

8A.3 Efficient Cyclic Designs

Starting Blocks for $4 \le t \le 15$ and $r \le 10$[a]

		kth Treatment, t =											
r	First $k - 1$ Treatments	4	5	6	7	8	9	10	11	12	13	14	15
2	0	1	1	1	1	1	1	1	1	1	1	1	1
4	0	2	2	2	3	3	3	3	3	3	5	4	4
6	0	1	3	3	2	2	2	2	5	5	2	6	2
8	0	1	4	5	4	4	4	4	2	2	4	3	7
10	0	2	1	4	5	5	5	5	4	4	3	5	5
3	01	2	2	3	3	3	3	4	4	4	4	4	4
6	02	1	3	1	3	7	6	7	7	5	7	7	8
9	01	3	2	3	3	4	5	3	3	6	4	6	5
4	013	—	2	2	6	7	7	6	7	7	9	7	7
8	014	—	2	2	6	7	8	2	6	6	6	6	6
5	0124	—	—	5	5	7	7	7	7	7	7	9	10
10	0234	—	—	5	5	7	8	9	8				
10	0236									7	11	12	10
6	01236	—	—	—	5	5	5	5	10	10	10	10	10
7	012347	—	—	—	—	5	5	9	9	9	9	9	10
8	0123468	—	—	—	—	—	5	9	9	9	9	11	11
9	01234579	—	—	—	—	—	—	8	8	8	10	10	10
10	0123456910	—	—	—	—	—	—	—	7	7	7	12	12

Source: Reproduced with permission from John (1981).
[a]Starting blocks for efficient cyclic designs for $16 \le t \le 30$ can be found in John (1987).

8A.4 Generating Arrays for α-Designs for $t = 20$ to 100, $k = 4$ to 10, and $r = 2$ to 4*

In these generating arrays $t = sk$, where s is the number of blocks per replicate and k is the number of observations per block. Designs for $k = 4$ and higher are constructed by using the first k columns of a generating array. These arrays are

*These generating arrays are reproduced with permission from Patterson et al. (1978). Paterson and Patterson (1983) give an algorithm for choosing suitable generating arrays for larger experiments.

not recommended for designs with $k < 4$. Designs for $r = 2$ to 4 replicates are constructed by using the first r rows of the generating array.

$k = 4, t = 20$
or $k = 5, t = 25; s = 5$

$$\begin{bmatrix} 0 & 0 & 0 & 0 & 0 \\ 0 & 1 & 2 & 3 & 4 \\ 0 & 4 & 3 & 2 & 1 \\ 0 & 2 & 4 & 1 & 3 \end{bmatrix}$$

$k = 4, \ldots, 10; s = 10$
$t = 40, 50, \ldots, 100$

$$\begin{bmatrix} 0 & 0 & 0 & 0 & 0 & 0 & 0 & 0 & 0 & 0 \\ 0 & 1 & 3 & 5 & 4 & 6 & 7 & 8 & 9 & 2 \\ 0 & 9 & 6 & 7 & 5 & 3 & 2 & 4 & 8 & 6 \\ 0 & 5 & 9 & 2 & 6 & 1 & 4 & 7 & 2 & 3 \end{bmatrix}$$

$k = 4, t = 24; k = 5, t = 30$
$k = 6, t = 36; s = 6$

$$\begin{bmatrix} 0 & 0 & 0 & 0 & 0 & 0 \\ 0 & 1 & 3 & 2 & 4 & 5 \\ 0 & 5 & 2 & 3 & 1 & 1 \\ 0 & 4 & 5 & 1 & 2 & 3 \end{bmatrix}$$

$k = 4, \ldots, 9; s = 11$
$t = 44, 55, \ldots, 99$

$$\begin{bmatrix} 0 & 0 & 0 & 0 & 0 & 0 & 0 & 0 & 0 \\ 0 & 1 & 4 & 9 & 2 & 5 & 6 & 3 & 7 \\ 0 & 6 & 8 & 7 & 3 & 1 & 5 & 9 & 4 \\ 0 & 7 & 1 & 5 & 6 & 3 & 10 & 4 & 1 \end{bmatrix}$$

$k = 4, \ldots, 7; s = 7$
$t = 28, 35, \ldots, 49$

$$\begin{bmatrix} 0 & 0 & 0 & 0 & 0 & 0 & 0 \\ 0 & 1 & 2 & 4 & 3 & 5 & 6 \\ 0 & 3 & 6 & 5 & 2 & 1 & 4 \\ 0 & 2 & 4 & 1 & 6 & 3 & 5 \end{bmatrix}$$

$k = 4, \ldots, 8; s = 12$
$t = 48, 60, \ldots, 96$

$$\begin{bmatrix} 0 & 0 & 0 & 0 & 0 & 0 & 0 & 0 \\ 0 & 1 & 7 & 9 & 4 & 11 & 10 & 5 \\ 0 & 2 & 5 & 6 & 11 & 3 & 4 & 1 \\ 0 & 3 & 1 & 4 & 8 & 10 & 7 & 6 \end{bmatrix}$$

$k = 4, \ldots, 8; s = 8$
$t = 32, 40, \ldots, 64$

$$\begin{bmatrix} 0 & 0 & 0 & 0 & 0 & 0 & 0 & 0 \\ 0 & 1 & 3 & 5 & 2 & 4 & 6 & 7 \\ 0 & 2 & 7 & 3 & 5 & 1 & 0 & 6 \\ 0 & 6 & 1 & 4 & 3 & 6 & 2 & 5 \end{bmatrix}$$

$k = 4, \ldots, 7; s = 13$
$t = 52, 65, \ldots, 91$

$$\begin{bmatrix} 0 & 0 & 0 & 0 & 0 & 0 & 0 \\ 0 & 1 & 3 & 9 & 12 & 8 & 6 \\ 0 & 4 & 8 & 2 & 10 & 5 & 7 \\ 0 & 10 & 11 & 1 & 6 & 12 & 8 \end{bmatrix}$$

$k = 4, \ldots, 9; s = 9$
$t = 36, 45, \ldots, 81$

$$\begin{bmatrix} 0 & 0 & 0 & 0 & 0 & 0 & 0 & 0 & 0 \\ 0 & 1 & 3 & 7 & 2 & 4 & 5 & 6 & 8 \\ 0 & 8 & 6 & 2 & 3 & 1 & 7 & 5 & 4 \\ 0 & 7 & 4 & 3 & 5 & 6 & 2 & 1 & 7 \end{bmatrix}$$

$k = 4, \ldots, 7; s = 14$
$t = 56, 70, \ldots, 98$

$$\begin{bmatrix} 0 & 0 & 0 & 0 & 0 & 0 & 0 \\ 0 & 1 & 9 & 11 & 2 & 5 & 3 \\ 0 & 8 & 10 & 13 & 6 & 11 & 14 \\ 0 & 10 & 7 & 2 & 1 & 12 & 11 \end{bmatrix}$$

$k = 4, t = 60; k = 5, t = 75$
$k = 6, t = 90; s = 15$

$$\begin{bmatrix} 0 & 0 & 0 & 0 & 0 & 0 \\ 0 & 1 & 3 & 7 & 10 & 14 \\ 0 & 8 & 12 & 2 & 13 & 3 \\ 0 & 7 & 14 & 5 & 11 & 8 \end{bmatrix}$$

CHAPTER 9

Repeated Treatments Designs

9.1 INTRODUCTION

In general, we call any design that calls for more than one treatment to be applied to one or more experimental units a *repeated treatments* design. *Crossover designs* constitute a subset of these designs. The crossover terminology reflects that in these cases the order of treatments is shifted within experimental units or interchanged from one experimental unit to another (i.e., cross over). Occasionally, the terms *switch back* and *double reversal* are also encountered. Although these designs were first developed in agricultural research (Cochran 1939; Lucas 1948; Williams 1949; Patterson 1951), they have a definite place in bioassay procedures (Finney 1987), clinical trials (Huitson et al. 1982), psychological experiments (Keppel 1982), industrial research (Raghavarao 1990), and laboratory work (Daniel 1975).

9.2 REPEATED TREATMENTS DESIGN MODEL

9.2.1 General Model

Repeated treatments implies that experimental units will be used more than once (i.e., at two or more time periods). Consequently, any potential model will need to contain parameters for unit effects, period effects, and possible carryover effects. *Carryover effects* are modifications of the response in the current period that are due to the treatment in the preceding period. We will denote the general repeated treatments design by the abbreviation RTD(t, n, p), where t is the number of treatments, n the number of experimental units, and p the number of periods. The repeated treatments on individual units also imply that measurements taken may well be dependent. The lack of independence can manifest itself as either a

Planning, Construction, and Statistical Analysis of Comparative Experiments,
by Francis G. Giesbrecht and Marcia L. Gumpertz
ISBN 0-471-21395-0

correlated error structure, or as part of the location parameter structure, or in other words, as a unit effect, a period effect, and possibly an additional carryover effect. We will usually assume that the latter is the case, although there is considerable literature dealing with correlated error structures. A common assumption in the correlated error model is that the correlation between successive observations on an experimental unit is a decreasing function of the number of periods separating the two.

We write the general model for data from the repeat treatments design with t treatments applied in p periods to n experimental units as

$$y_{ij} = \mu + \pi_i + u_j + \tau_{d(i,j)} + \gamma_{d(i-1,j)} + \epsilon_{ij}, \tag{9.1}$$

where

$$\begin{aligned}
\pi_i &= \text{period effect, } i = 1, \ldots, p \\
u_j &= \text{unit effect, } j = 1, \ldots, n \text{ and } u_j \sim \text{iid}(0, \sigma_u^2) \\
\tau_{d(i,j)} &= \text{(direct) treatment } d(i, j) \text{ effect in the } i\text{th period for the } j\text{th unit} \\
&\quad\; d(i, j) = 1, \ldots, t \\
\gamma_{d(i-1,j)} &= \text{carryover of treatment effect from period } i-1 \text{ to period } i \text{ on unit } j \\
\epsilon_{ij} &= \text{ experimental error } \sim \text{iid}(0, \sigma_e^2).
\end{aligned}$$

We assume that the γ parameter does not exist for period 1. We will, however, look at the case where there is a two-period carryover effect which we can denote by γ^*. The $\{\gamma_d\}$ (and $\{\gamma_d^*\}$) are often referred to as *residuals* or *residual effects*. Additionally, there are other special cases of this model, such as no period effects, and so on.

In the various designs that we consider, *balance* will always be a key concern. We will always insist on experimental units and periods being *mutually balanced*. By this we mean that each unit will be used in each period. This balance will guarantee that the sums of squares attributed to these two will be orthogonal. Generally, we will also insist that each treatment appear in each period. However, in some cases, not all treatments will be applied to all units, and in some cases individual treatments will not necessarily follow every treatment. The latter becomes important when we consider carryover effects. We will use the term *minimal balanced* if the design requires units and periods balanced, treatments and periods balanced, and residuals balanced with treatments in the sense that each treatment follows every other treatment an equal number of times. We will use the term *strongly balanced* if the design insists that each treatment follows every treatment (including itself) an equal number of times, in addition to the unit and period and the treatment and period balance.

9.2.2 Some Potential Applications

One of the early applications for repeat treatments designs in agriculture was in research with dairy cows. The cost of cows limited the number available for experimentation. The object of the plans was to apply all, or at least a sequence of treatments to each cow. Comparing treatments on a within-animal basis eliminated the large variability that exists among cows. In a sense, each cow served as its own control. An additional problem with dairy cow experiments is the link to lactation cycles. If individual treatments each have a short duration, several can be applied in one lactation cycle. However, milk production depends on stage in the lactation cycle, i.e., a large period effect. Also, there is a definite risk of a carryover effect. If the individual treatment extended through a complete cycle, periods would be linked to age of the animal.

A similar series of problems arises with studies on fruit trees in orchards. Individual trees are the obvious choice for experimental unit. Years form the natural periods. The tree-to-tree variability makes it appealing to compare management or fertilizer practices on a within-tree basis. Again, there is a very real possibility of year-to-year carryover effects. Completely analogous problems exist for experiments involving field management practices. Often, the nature of the treatments is such that, realistically, they can only be applied to moderate-sized fields with large-scale equipment. Cost then limits the number of fields. Years define a major period effect. There is also the possibility of carryover effects. Crop rotation experiments fall into this category.

Denbow (1980) used a repeat treatments design to study the relations of cations to temperature regulation in chickens. The experimental procedure involved implanting a small tube into the brain of live birds and injecting cation solutions. Since the surgical procedure was difficult, the study was limited to four birds, used repeatedly.

Huitson et al. (1982) discuss the flaws and merits of crossover designs in research in the pharmaceutical industry. The more typical example here involves only two treatments, and in general, a sequence of treatments is applied to a group of patients. We will not go into details of the crossover designs, but refer the reader to Jones and Kenward (1989). The basic problem in this work is finding a sufficient number of similar patients to set up a nonrepeat treatments design. Unfortunately, these studies face some ethical constraints in addition to the problems mentioned in the previous examples. In many cases one would like a "washout" period between treatments to reduce the carryover effects. In medical research, sometimes this is not possible and sometimes it is. The question of period $\times$ treatment interactions also tends to appear in these studies. A thorough discussion of these topics is beyond the scope of this book.

Daniel (1975) discusses a repeat treatments example encountered when studying the calibration of an autoanalyzer, a machine that accepts batches of samples and analyzes them sequentially. Samples to be analyzed take the place of treatments. The potential problem is that if one sample is excessively high in some characteristic, the reading on the subsequent sample will be affected. Raghavarao (1990) discusses a hypothetical example of a company having four plants that is interested in studying production and accident rates for different time intervals between rotation of shifts. Weekly, biweekly, triweekly, and monthly shift changes were considered. Since it is not feasible to shut down the plants between shift rotations, the plan must allow for carryover effects.

9.3 CONSTRUCTION OF REPEATED TREATMENTS DESIGNS

9.3.1 Minimal Balanced RTDs with Equal Number of Treatments, Periods, and Experimental Units ($t = p = n$) and t Even

The RTD(t, t, t)s are special cases of Latin squares. Their construction depends on whether t is even or odd. A simple procedure when t is even is to use special generating columns to construct the designs. Begin by writing down the list of treatments numbered 0 to $t - 1$, ordered in such a way that the successive differences using modulo t arithmetic include all of the numbers 1 to $t - 1$. This can always be done when t is an even number. We illustrate for $t = 4$:

Generating Column	Differences
0	
	$1 - 0 = 1$
1	
	$3 - 1 = 2$
3	
	$6 - 3 = 3$ (note: $2 = 6$ modulo 4).
2	

This forms the first column of the square. Remaining columns are generated cyclically. Generate the second column by adding one (reducing modulo t when necessary) to each element. In like manner, generate column 3 from column 2, and so on, to complete the Latin square. In our example, we generate

$$
\begin{matrix} 0 \\ 1 \\ 3 \\ 2 \end{matrix} \longrightarrow \begin{matrix} 1 \\ 2 \\ 0 \\ 3 \end{matrix} \longrightarrow \begin{matrix} 2 \\ 3 \\ 1 \\ 0 \end{matrix} \longrightarrow \begin{matrix} 3 \\ 0 \\ 2 \\ 1. \end{matrix}
$$

***Exercise 9.1:* 4 x 4 *Latin Square Balanced for Carryover*.** For $t = 4$ there are two generating columns such that the successive differences (in modulo 4 arithmetic) include all the numbers 1 to 3. Find the other one and generate its Latin square. □

Different 6×6 squares balanced for first-order carryover effects can be obtained by selecting one of the following four columns and developing cyclically, that is, by adding one and reducing modulo 6 when necessary:

$$
\begin{array}{cccc}
0 & 0 & 0 & 0 \\
1 & 2 & 4 & 5 \\
5 & 1 & 5 & 1 \\
2 & 4 & 2 & 4 \\
4 & 5 & 1 & 2 \\
3 & 3 & 3 & 3
\end{array}.
$$

Twenty-four different 8×8 Latin squares balanced for carryover effects can be obtained by expanding columns selected from the following array:

$$
\begin{array}{cccccccccccccccccccccccc}
0 & 0 & 0 & 0 & 0 & 0 & 0 & 0 & 0 & 0 & 0 & 0 & 0 & 0 & 0 & 0 & 0 & 0 & 0 & 0 & 0 & 0 & 0 & 0 \\
1 & 1 & 1 & 1 & 2 & 2 & 2 & 2 & 3 & 3 & 3 & 3 & 5 & 5 & 5 & 5 & 6 & 6 & 6 & 6 & 7 & 7 & 7 & 7 \\
3 & 6 & 7 & 7 & 1 & 3 & 5 & 7 & 1 & 2 & 5 & 5 & 3 & 3 & 6 & 7 & 1 & 3 & 5 & 7 & 1 & 1 & 2 & 5 \\
6 & 5 & 2 & 3 & 5 & 6 & 1 & 6 & 2 & 7 & 1 & 6 & 2 & 7 & 1 & 6 & 2 & 7 & 2 & 3 & 5 & 6 & 3 & 2 \\
2 & 3 & 6 & 6 & 3 & 5 & 7 & 1 & 6 & 1 & 2 & 2 & 6 & 6 & 7 & 2 & 7 & 1 & 3 & 5 & 2 & 2 & 5 & 6 \\
7 & 7 & 3 & 5 & 6 & 1 & 6 & 5 & 5 & 5 & 7 & 1 & 7 & 1 & 3 & 3 & 3 & 2 & 7 & 2 & 3 & 5 & 1 & 1 \\
5 & 2 & 5 & 2 & 7 & 7 & 3 & 3 & 7 & 6 & 6 & 7 & 1 & 2 & 2 & 1 & 5 & 5 & 1 & 1 & 6 & 3 & 6 & 3 \\
4 & 4 & 4 & 4 & 4 & 4 & 4 & 4 & 4 & 4 & 4 & 4 & 4 & 4 & 4 & 4 & 4 & 4 & 4 & 4 & 4 & 4 & 4 & 4
\end{array}.
$$

A suitable randomization scheme is to select a square at random, permute the columns randomly, and then assign treatments randomly to the symbols. It is not possible to randomize the rows without breaking up the balance for carryover effects.

Plans for more than eight treatments balanced for carryover effects are rarely needed as experimental designs. However, since larger designs are sometimes required in questionnaire construction, we give a method to construct a generating column for large, even values of t. The method is to write the symbols $0, 1, 2, \ldots, t/2-1$ in the odd-numbered positions, i.e., $1, 3, 5, \ldots, t-1$ and then $t-1, t-2, \ldots, t/2$ in positions $2, 4, \ldots, t$. We illustrate part of the construction for $t = 20$:

$$
0\ 19\ 1\ 18\ 2\ 17\ 3\ 16\ 4\ 15\ 5\ 14\ 6\ 13\ 7\ 12\ 8\ 11\ 9\ 10.
$$

The square is then completed by transposing this row into a column and then generating the remaining 19 columns cyclically.

9.3.2 Minimal Balanced RTDs with $t = p = n$ and t Odd

It can be shown that there is no $t \times t$ Latin square balanced for carryover effects for $t = 3, 5$, or 7. Hedayat and Afsarinejad (1978) give squares for $t = 9$, $t = 15$, and $t = 27$. Beyond this little is known about the existence of other possible squares with t odd. In general, one must use $n = 2t$ experimental units.

9.3.3 Minimal Balanced RTDs with Number of Experimental Units a Multiple of Number of Treatments ($n = \lambda t$)

An important case here is $\lambda = 2$ with t odd. Generally, it appears that these squares exist. For example, for $t = 5$, it is possible to use the following two generating columns to produce two Latin squares, which together give a minimal balanced RTD(5, 10, 5). We illustrate:

Generating Column 1	Difference	Generating Column 2	Difference
0		0	
	1		2
1		2	
	2		4
3		1	
	1		3
4		4	
	3		4
2		3	

Generate a square from each of the generating columns by using the cyclic procedure. The design resulting from the pair of generating columns is

$$\begin{array}{ccccc} 0&1&2&3&4\\ 1&2&3&4&0\\ 3&4&0&1&2\\ 4&0&1&2&3\\ 2&3&4&0&1 \end{array} \quad \begin{array}{ccccc} 0&1&2&3&4\\ 2&3&4&0&1\\ 1&2&3&4&0\\ 4&0&1&2&3\\ 3&4&0&1&2 \end{array}.$$

It is simple to verify that in this design, every treatment follows every other treatment exactly twice.

There are seven other pairs of generators suitable for generating pairs of 5×5 squares. Three different pairs of squares can be generated from the first three pairs of columns and four different pairs of squares by selecting one column from each of the last two sets given below.

$$\begin{array}{c|c} 0&0\\1&2\\3&1\\4&4\\2&3 \end{array} \quad \begin{array}{c|c} 0&0\\1&4\\4&1\\2&3\\3&2 \end{array} \quad \begin{array}{c|c} 0&0\\3&2\\2&3\\1&4\\4&1 \end{array} \quad \begin{array}{cc|cc} 0&0&0&0\\2&1&3&4\\1&3&4&2\\3&2&2&3\\4&4&1&1 \end{array}.$$

For $t = 7$ the number of possible pairs of suitable cyclic squares increases dramatically. By selecting one generator from among the first eight columns of the following array and a second generator from the last eight columns, 64 different designs can be produced.

$$\begin{array}{cccccccc|cccccccc} 0&0&0&0&0&0&0&0&0&0&0&0&0&0&0&0\\ 1&3&3&5&6&6&6&6&2&2&2&2&4&4&5&5\\ 6&1&2&6&2&2&3&5&4&5&6&6&2&6&1&6\\ 3&2&1&3&1&3&1&2&5&6&1&4&5&1&2&2\\ 2&6&5&2&5&1&2&3&1&4&4&5&6&2&4&4\\ 5&5&6&1&3&5&5&1&6&1&5&1&1&5&6&1\\ 4&4&4&4&4&4&4&4&3&3&3&3&3&3&3&3 \end{array}.$$

Hedayat and Afsarinejad (1978) give examples of pairs of 9×9 and 15×15 squares and references to other odd-sized squares that are balanced for carryover effects.

The randomization scheme for designs balanced for carryover effects requires three steps.

1. Select a pair of generating columns at random (for $t = 5$ there are seven pairs; for $t = 7$ there are 64 pairs to choose from).
2. Randomly assign columns to subjects.
3. Randomly assign treatment codes to treatments.

Russell (1991) gives a general recipe for constructing pairs of $t \times t$ squares such that each treatment follows every other treatment an equal number of times. This can be used for $t > 7$. The steps are:

1. Define $m = (t + 1)/2$ and q the largest integer $\leq (m + 1)/2$.
2. Construct the first column of the first square with odd-numbered elements $0, m+1, m+2, \ldots, 2m-2, q$ and even-numbered elements $m, m-1, \ldots, q+1, q-1, \ldots, 1$
3. Generate the first square cyclically.
4. From a column in the square, construct the column of $t - 1$ differences between elements and immediate predecessors $[(a_i - a_{i-1})$ modulo $t]$. One of the 0 to $t - 1$ will be missing and one will be duplicated.
5. The columns of the second square must have differences that complement those from the first square in the sense that the union of the two sets of differences contains all 0 to $t - 1$ values twice.
6. Reorder these differences so that they correspond to a column of t elements. No systematic method for doing this is known. However, the condition that no sums of $k < t$ contiguous differences equals zero (modulo t) is sufficient and can be used to construct an easily programmed computer search.
7. From the reordered differences construct an initial column.
8. Generate the second square cyclically.

We illustrate with $t = 13$, $m = 7$, and $q = 4$. The odd-numbered elements in the first column of the first square are $0, 8, 9, 10, 11, 12, 4$, and the even-numbered elements are $7, 6, 5, 3, 2, 1$. The differences between elements and predecessor are $7, 1, 11, 3, 9, 5, 6, 8, 4, 10, 2, 3$. The differences for the second square must consist of $1, 2, 4, 5, 6, 7, 8, 9, 10, 11, 12, 12$. A suitable rearrangement is $1, 5, 11, 12, 7, 8, 6, 4, 10, 9, 12, 2$, and a suitable first column for the second square consists of $0, 1, 6, 4, 3, 10, 5, 11, 2, 12, 8, 7, 9$.

9.3.4 Strongly Balanced Designs with Number of Treatments Equal to Number of Experimental Units $t = n$ and an Extra Period $(p = t + 1)$

The carryover treatments in a Latin square can be made orthogonal to the direct treatments by adding an extra treatment period that is a repeat of the last row. An example of an extra-period design for four treatments is

A B C D
B C D A
D A B C.
C D A B
C D A B

In this design the carryover treatments are orthogonal to direct treatments, columns, and rows. However, the direct treatments are no longer orthogonal to columns.

Note that the number of observations in the extra-period design is t more than in the standard Latin square. For this reason the extra-period design increases the precision of the estimates of treatment effects slightly. The big advantage of the extra-period design, however, is that it increases the precision of estimates of carryover effects substantially (see Exercise 9.2). The model for the extra-period design is identical to the model defined in equation (9.1).

Exercise 9.2: Latin Square with Extra Period. For the four-treatment Latin square design balanced for carryover effects and the extra-period design given above, compute $\text{Var}[\widehat{\tau}_1 - \widehat{\tau}_2]$ and $\text{Var}[\widehat{\gamma}_1 - \widehat{\gamma}_2]$ up to the unknown σ_ϵ^2. □

Notice that the direct and carryover effects are orthogonal since each carryover effect appears with each direct effect equally frequently. Also, the order of the carryover and the direct treatment effects could be switched in a least squares analysis and the sequential sums of squares for the two would be unchanged. This is not true for the design without the extra period. Note also that now the period effect is not orthogonal to all other factors, so the corresponding sequential and partial sums of squares for these would not coincide.

9.3.5 Minimal Balanced RTDs with Fewer Periods Than Treatments $(p < t)$

Afsarinejad (1990) gives procedures for constructing a class of RTD(t, n, p)s which have every treatment following every other treatment equally often, in addition to having every treatment appear equally frequently in every period and every experimental unit used the same number of periods. Necessary and sufficient conditions for plans to exist in this class are that $p < t$, $n = \lambda t$, and $\lambda(p - 1) = t - 1$, where λ is any integer multiplier. Possible plans with $t \leq 20$ and $n \leq 50$ exist for the array of parameters shown in Table 9.1.

To construct the actual plan for the experiment, first notice that there always exists a k such that n can be written as $n = 4k, n = 4k + 1, n = 4k + 2$, or

Table 9.1 Parameter Values for Minimal Balanced RTDs

t	n	p	t	n	p
3	6	2	9	36	3
4	12	2	10	30	4
5	10	3	11	22	6
	20	2	13	26	7
6	30	2		39	5
7	14	4	15	30	8
	21	3	16	48	6
	42	2	17	34	9
9	18	5	19	38	10

as $n = 4k + 3$. The actual plan is then obtained by developing cyclically, in turn, reducing modulo t when necessary, each of the columns given by the $p \times \lambda$ array:

$$\begin{array}{ccccccc} c_1 & c_p & \cdots & c_{(i-1)p-(i-2)} & \cdots & c_{(\lambda-1)p-(\lambda-2)} \\ c_2 & c_{p+1} & \cdots & c_{(i-1)p-(i-3)} & \cdots & c_{(\lambda-1)p-(\lambda-3)} \\ \vdots & \vdots & & \vdots & & \vdots \\ c_p & c_{2p} & \cdots & c_{ip-(i-1)} & \cdots & c_t \end{array}$$

where $(c_1, c_2, \ldots, c_t)$ is

$$\begin{array}{ll} (0, t-1, 1, t-2, 2, \ldots, (t-2)/2, t/2) & \text{if } t \text{ is even} \\ (0, t-1, 2, t-3, 4, \ldots, (t-1)/2, (t+3)/2, \ldots, (t-1), 0) & \text{if } t = 4k+1 \\ (0, t-1, 2, t-3, 4, \ldots, (t-3)/2, (t+1)/2, \ldots, (t-1), 0) & \text{if } t = 4k+3. \end{array}$$

To illustrate, consider that $t = 4$ and $p = 2$. This implies that $n = 12$, $\lambda = 3$, and $(c_1, c_2, c_3, c_4) = (0\ 3\ 1\ 2)$. The $\lambda = 3$ columns to be developed cyclically (reduced modulo 4) are $\begin{smallmatrix}0\\3\end{smallmatrix}$, $\begin{smallmatrix}3\\1\end{smallmatrix}$, and $\begin{smallmatrix}1\\2\end{smallmatrix}$. The actual design is

$$\begin{array}{cccccccccccc} 0 & 1 & 2 & 3 & 3 & 0 & 1 & 2 & 1 & 2 & 3 & 0 \\ 3 & 0 & 1 & 2 & 1 & 2 & 3 & 0 & 2 & 3 & 0 & 1. \end{array}$$

Similarly, with $t = 9$ and $p = 5$, we have $n = 18$, $\lambda = 2$, and $(c_1, \ldots, c_9) = (0\ 8\ 2\ 6\ 4\ 6\ 2\ 8\ 0)$, giving the following design:

$$\begin{array}{cccccccccccccccccc} 0 & 1 & 2 & 3 & 4 & 5 & 6 & 7 & 8 & 4 & 5 & 6 & 7 & 8 & 0 & 1 & 2 & 3 \\ 8 & 0 & 1 & 2 & 3 & 4 & 5 & 6 & 7 & 6 & 7 & 8 & 0 & 1 & 2 & 3 & 4 & 5 \\ 2 & 3 & 4 & 5 & 6 & 7 & 8 & 0 & 1 & 2 & 3 & 4 & 5 & 6 & 7 & 8 & 0 & 1. \\ 6 & 7 & 8 & 0 & 1 & 2 & 3 & 4 & 5 & 8 & 0 & 1 & 2 & 3 & 4 & 5 & 6 & 7 \\ 4 & 5 & 6 & 7 & 8 & 0 & 1 & 2 & 3 & 0 & 1 & 2 & 3 & 4 & 5 & 6 & 7 & 8 \end{array}$$

Randomization for these designs is limited to randomly assigning units to columns and treatments to the symbols. Randomizing order is not possible, since that would destroy the carryover balance.

Exercise 9.3: Construction of Treatment Sequence. Verify that the treatment sequence (0 8 2 6 4 6 2 8 0) given for the $t = 9$, $p = 5$, and $n = 18$ design is correct. □

Exercise 9.4: Orthogonality of Periods and Carryover. Verify that periods are orthogonal to carryover effects in the RTD(9, 18, 5) illustrated. □

Exercise 9.5: Orthogonality of Sums of Squares. For the minimal balanced RTD(t, n, p), which pairs of sums of squares are equal? Explain in terms of orthogonality.

(a) $\mathrm{R}(\pi_1, \ldots, \pi_t | \mu, u_1, \ldots, u_n, \tau_1, \ldots, \tau_t, \gamma_1, \ldots, \gamma_t)$ and $\mathrm{R}(\pi_1, \ldots, \pi_t | \mu)$.

(b) $R(\tau_1, \ldots, \tau_t, | \mu, \pi_1, \ldots, \pi_t, u_1, \ldots, u_n, \gamma_1, \ldots, \gamma_t)$ and $R(\tau_1, \ldots, \tau_t | \mu, \pi_1, \ldots, \pi_t, u_1, \ldots, u_n)$. □

9.3.6 Strongly Balanced RTDs with Fewer Periods Than Treatments ($p < t$)

Afsarinejad (1990) gives procedures for constructing a class of RTD(t, n, p)s which have every treatment following every treatment (including itself) equally often, in addition to having every treatment appear equally frequently in every period and every experimental unit used the same number of periods. Necessary and sufficient conditions for plans to exist in this class are that $p < t$, $n = \lambda t$, and $\lambda(p-1) = t$. Possible plans with $t \leq 20$ and $n \leq 50$ exist for the array of parameters shown in Table 9.2.

The construction is similar to that for the minimal balanced RTDs, except that $(c_1, c_2, \ldots, c_t, c_{t+1})$ is

$$(0, t-1, 1, t-2, \ldots, (t-2)/2, t/2, t/2)$$
$$(0, t-1, 2, t-3, \ldots, (t+3)/2, (t-1)/2, (t-1)/2, \ldots, t-1, 0)$$
$$(0, t-1, 2, t-3, \ldots, (t-3)/2, (t+1)/2, (t+1)/2, \ldots, t-1, 0),$$

if t is even, if $t = 4k + 1$, or if $t = 4k + 3$, respectively. These values are then used to construct the λ columns of length p as before. These columns are

Table 9.2 Parameter Values for Strongly Balanced RTDs

t	n	p	t	n	p	t	n	p
3	9	2	7	49	2	12	36	5
4	8	3	8	16	5		48	4
	16	2		32	3	14	28	8
5	25	2	9	27	4	15	45	6
6	12	4	10	20	6	16	32	9
	18	3		50	3	18	36	10
	36	2	12	24	7	20	40	11

again expanded cyclically and reduced modulo t when necessary. Specifically, for $t = 10$ and $p = 6$, which implies that $n = 20$ and $\lambda = 2$, we have $(c_1, \ldots, c_{11}) =$ (0 9 1 8 2 7 3 6 4 5 5), giving the design

0 1 2 3 4 5 6 7 8 9 7 8 9 0 1 2 3 4 5 6
9 0 1 2 3 4 5 6 7 8 3 4 5 6 7 8 9 0 1 2
1 2 3 4 5 6 7 8 9 0 6 7 8 9 0 1 2 3 4 5
8 9 0 1 2 3 4 5 6 7 4 5 6 7 8 9 0 1 2 3
2 3 4 5 6 7 8 9 0 1 5 6 7 8 9 0 1 2 3 4
7 8 9 0 1 2 3 4 5 6 5 6 7 8 9 0 1 2 3 4

For $t = 9$ and $p = 4$, which implies that $n = 27$ and $\lambda = 3$, we have $(c_1, \ldots, c_{10}) =$ (0 8 2 6 4 4 6 2 8 0), giving the design

0 1 2 3 4 5 6 7 8 6 7 8 0 1 2 3 4 5 6 7 8 0 1 2 3 4 5
8 0 1 2 3 4 5 6 7 4 5 6 7 8 0 1 2 3 2 3 4 5 6 7 8 0 1
2 3 4 5 6 7 8 0 1 4 5 6 7 8 0 1 2 3 8 0 1 2 3 4 5 6 7
6 7 8 0 1 2 3 4 5 6 7 8 0 1 2 3 4 5 0 1 2 3 4 5 6 7 8.

9.3.7 Balanced RTDs with Fewer Units than Treatments, $n < t$

Occasionally, one encounters situations where the number of experimental units available is less than the number of treatments. An example would be the effect of safety programs on the number of accidents in factories or highway markings and signs on sections of highways. Sites are limited, but many periods with carryover are available. In general, the strategy is to select suitable columns from the plans developed in foregoing sections. This must be done with care, however. In essence, all of the concepts involved in constructing incomplete block designs become relevant. Considerations from the analysis of variance discussed in Section 9.4.2 are also relevant.

To begin with, one must have sufficient units (columns from our designs) to provide degrees of freedom for the direct effects after allowing for unit effects, period effects, and carryover effects. Generally, this means at least four units or columns. In addition, the units or columns must be selected so that all treatment differences of the form $\tau_j - \tau_{j'}$ for $j \neq j'$ are estimable. The key requirement here is that the design be connected (Section 8.2). Following this, one needs to check for variances of $\widehat{\tau}_j - \widehat{\tau}_{j'}$ for $j \neq j'$. For this, one looks at the system of reduced normal equations

$$\boldsymbol{X}_\tau^t(\boldsymbol{I} - \boldsymbol{X}_0(\boldsymbol{X}_0^t\boldsymbol{X}_0)^+\boldsymbol{X}_0^t)\boldsymbol{X}_\tau\widehat{\boldsymbol{\tau}} = \boldsymbol{X}_\tau^t(\boldsymbol{I} - \boldsymbol{X}_0(\boldsymbol{X}_0^t\boldsymbol{X}_0)^+\boldsymbol{X}_0^t)\boldsymbol{y}, \tag{9.2}$$

where $\boldsymbol{X}_0 = \boldsymbol{1}_k|\boldsymbol{X}_p|\boldsymbol{X}_u|\boldsymbol{X}_\gamma$ (see Sections 8.1.3 and 9.4.1) and $\boldsymbol{H}^+$ is a generalized inverse of $\boldsymbol{H}$ defined by $\boldsymbol{H} = \boldsymbol{H}\boldsymbol{H}^+\boldsymbol{H}$. Equation (9.2) is analogous to the equation given in Section 8.1.3. If $\boldsymbol{X}_\tau^t(\boldsymbol{I}-\boldsymbol{X}_0(\boldsymbol{X}_0^t\boldsymbol{X}_0)^+\boldsymbol{X}_0^t)\boldsymbol{X}_\tau$ has rank $(t-1)$, the design is connected (i.e., all treatment differences are estimable). The variances of estimated differences are obtainable from $(\boldsymbol{X}_\tau^t(\boldsymbol{I} - \boldsymbol{X}_0(\boldsymbol{X}_0^t\boldsymbol{X}_0)^+\boldsymbol{X}_0^t)\boldsymbol{X}_\tau)^+$.

Finding a suitable design with r units involves selecting subsets of r columns and evaluating the resulting designs. At this point there is little guidance for selecting columns. Typically, in practice the number of choices of columns is limited and evaluation of all possible designs is not excessive.

9.3.8 Plans Based on Mutually Orthogonal Latin Squares

An alternative class of plans is based on orthogonal sets of Latin squares. Consider an example of a set of three mutually orthogonal 4×4 Latin squares:

1	2	3	4	1	2	3	4	1	2	3	4
2	1	4	3	3	4	1	2	4	3	2	1
3	4	1	2	4	3	2	1	2	1	4	3
4	3	2	1	2	1	4	3	3	4	1	2

Notice that in this set of 12 columns each treatment follows every other treatment exactly three times; that is, the plan is balanced for carryover effects. In fact, the design is balanced for *all orders* of carryover effects. In a plan that is balanced for all orders of carryover effects, each treatment follows every other treatment the same number of times, after a lag of one period, or two periods, or three periods, or whatever number of periods you choose.

Mutually orthogonal Latin squares can be used to construct designs balanced for carryover effects for 3, 4, 5, 7, 8, or 9 treatments. Any $t \times t$ Latin square where t is a prime number or a power of a prime has $t - 1$ mutually orthogonal Latin squares (see Appendix 6B). It is important to choose mutually orthogonal Latin squares such that the treatments appear in the same order across the top row of the $t - 1$ squares.

This sort of plan requires a large number of experimental units, but not many time periods. These plans are useful in sample survey problems where there are many subjects to be interviewed and there is concern that the order of items in the instrument will be important. Seeing one item may predispose the interviewee toward certain types of responses on a subsequent item. Another logical place for these plans is in educational research, where the columns are classrooms and the treatments are teaching methods or styles of presenting material. One can possibly get many classrooms, but probably will not be able to try more than two or three different techniques on a class in a semester or year.

It is also possible to construct designs that are balanced for carryover effects that are not required to have a full set of treatments on every subject. Eliminate the last row and note that every treatment follows every other treatment exactly two times. Eliminate the last two rows and note that every treatment follows every other treatment exactly once. In fact, it doesn't matter which rows you delete; the design is still balanced for carryover effects. For example, using the second and fourth rows of the three mutually orthogonal Latin squares for $t = 3$ treatments gives the following two-period design that is balanced for first-order carryover effects:

2 1 4 3 3 4 1 2 4 3 2 1
4 3 2 1 2 1 4 3 3 4 1 2.

Before using this design the columns are randomly ordered and randomly assigned to subjects, and the labels 1, 2, 3, 4 are randomly assigned to treatments. More information on designs for fewer than a full set of treatments in every block can be found in Chapter 8. We note in passing that this class of designs has appeal when the investigator is uncertain about the budget. The study can be terminated any time after two or more periods have been completed and yield properly analyzable and interpretable data.

9.3.9 Plans with $n = t$ and $p = 2t$

There exists a class of designs for t treatments using only t experimental units for $2t$ periods of time in which every treatment follows itself once and every other treatment an equal number of times. These plans are advantageous in situations where experimental units are relatively scarce and it is desirable to estimate direct and residual effects with roughly equal precision. The plans are constructed by starting with a plan for $2t$ treatments. Since $2t$ is an even number, it is always possible to construct a $2t \times 2t$ square balanced for carryover effects. The design for t treatments is obtained by deleting the last t columns, then reducing the remaining treatment numbers modulo t. The result is an array for t treatments in t columns and $2t$ rows.

We demonstrate the procedure for the case of five treatments. First set up a Latin square for 10 treatments balanced for first-order carryover effects, as shown in Table 9.3. Delete the last five columns. Reduce the remaining values modulo 5. The resulting design is shown in Table 9.4. In this design each treatment follows itself once and every other treatment twice. The statistical analysis of data obtained via these designs using least squares and model (9.1) can be obtained using any general-purpose regression program. However, a detailed discussion of the analysis of this class of designs and relative efficiencies relative to other designs can be found in Sharma (1981).

Table 9.3 10 × 10 Latin Square Balanced for Carryover

Period										
1	1	2	3	4	5	6	7	8	9	0
2	0	1	2	3	4	5	6	7	8	9
3	2	3	4	5	6	7	8	9	0	1
4	9	0	1	2	3	4	5	6	7	8
5	3	4	5	6	7	8	9	0	1	2
6	8	9	0	1	2	3	4	5	6	7
7	4	5	6	7	8	9	0	1	2	3
8	7	8	9	0	1	2	3	4	5	6
9	5	6	7	8	9	0	1	2	3	4
10	6	7	8	9	0	1	2	3	4	5

Table 9.4 Design for Five Treatments in 10 Periods

Period					
1	1	2	3	4	0
2	0	1	2	3	4
3	2	3	4	0	1
4	4	0	1	2	3
5	3	4	0	1	2
6	3	4	0	1	2
7	4	0	1	2	3
8	2	3	4	0	1
9	0	1	2	3	4
10	1	2	3	4	0

9.4 STATISTICAL ANALYSIS OF REPEATED TREATMENTS DESIGN DATA

9.4.1 Normal Equations

Since the unbalance in the RTDs leads to some complications in the statistical analysis of data, we spell out the normal equations in detail. It is convenient to write the model (9.1) with responses sorted by units within periods as

$$\boldsymbol{y} = (\mathbf{1}_p \otimes \mathbf{1}_n)\mu + (\boldsymbol{I}_p \otimes \mathbf{1}_n)\boldsymbol{\pi} + (\mathbf{1}_p \otimes \boldsymbol{I}_n)\boldsymbol{u} + \boldsymbol{X}_\tau \boldsymbol{\tau} + \boldsymbol{X}_\gamma \boldsymbol{\gamma} + \boldsymbol{\epsilon}, \tag{9.3}$$

where

$$\begin{aligned}
\boldsymbol{y} &= \text{column of responses, } (y_{11}, y_{12}, \ldots, y_{pn})^t \\
\mathbf{1}_k &= \text{column of } k \text{ ones} \\
\boldsymbol{I}_k &= k \times k \text{ identity matrix} \\
\boldsymbol{\pi} &= \text{column } (\pi_1, \ldots, \pi_p)^t \\
\boldsymbol{u} &= \text{column } (u_1, \ldots, u_n)^t \\
\boldsymbol{X}_\tau &= pn \times t \text{ design matrix for the direct effects} \\
\boldsymbol{\tau} &= \text{column } (\tau_1, \ldots, \tau_t)^t \\
\boldsymbol{X}_\gamma &= pn \times t \text{ design matrix for the carryover effects} \\
\boldsymbol{\gamma} &= \text{column } (\gamma_1, \ldots, \gamma_t)^t \\
\boldsymbol{\epsilon} &= \text{column } (\epsilon_{11}, \epsilon_{12}, \ldots, \epsilon_{pn})^t.
\end{aligned}$$

It is often convenient to write $\boldsymbol{X}_p$ in place of $\boldsymbol{I}_p \otimes \mathbf{1}_n$ and $\boldsymbol{X}_u$ in place of $\mathbf{1}_p \otimes \boldsymbol{I}_n$. If $\boldsymbol{X} = \mathbf{1}_{np}|\boldsymbol{X}_p|\boldsymbol{X}_u|\boldsymbol{X}_\tau|\boldsymbol{X}_\gamma$, the $\boldsymbol{X}^t\boldsymbol{X}$ matrix in the normal equations can be written directly.

Recall from Chapter 3 that the normal equations are singular and that one way to obtain solutions is to impose restrictions on the equations. Up to this point the equations are completely general, except for p and n being multiples of t. Convenient parameter restrictions are $\mathbf{1}_p^t\widehat{\boldsymbol{\pi}} = 0$, $\mathbf{1}_n^t\widehat{\boldsymbol{u}} = 0$, $\mathbf{1}_t^t\widehat{\boldsymbol{\tau}} = 0$, and $\mathbf{1}_t^t\widehat{\boldsymbol{\gamma}} = 0$. The $\boldsymbol{X}^t\boldsymbol{X}$ matrix, substituting zeros for elements corresponding to parameters subject to sum to zero constraints, is

$$\begin{bmatrix} np & \mathbf{0} & \mathbf{0} & \mathbf{0} & \mathbf{0} \\ \mathbf{0} & n\boldsymbol{I}_p & \mathbf{0} & (\boldsymbol{I}_p \otimes \mathbf{1}_n^t)\boldsymbol{X}_\tau & (\boldsymbol{I}_p \otimes \mathbf{1}_n^t)\boldsymbol{X}_\gamma \\ \mathbf{0} & \mathbf{0} & p\boldsymbol{I}_n & (\mathbf{1}_p^t \otimes \boldsymbol{I}_n)\boldsymbol{X}_\tau & (\mathbf{1}_p^t \otimes \boldsymbol{I}_n)\boldsymbol{X}_\gamma \\ \mathbf{0} & \boldsymbol{X}_\tau^t(\boldsymbol{I}_p \otimes \mathbf{1}_n) & \boldsymbol{X}_\tau^t(\mathbf{1}_p \otimes \boldsymbol{I}_n) & \boldsymbol{X}_\tau^t\boldsymbol{X}_\tau & \boldsymbol{X}_\tau^t\boldsymbol{X}_\gamma \\ \mathbf{0} & \boldsymbol{X}_\gamma^t(\boldsymbol{I}_p \otimes \mathbf{1}_n) & \boldsymbol{X}_\gamma^t(\mathbf{1}_p \otimes \boldsymbol{I}_n) & \boldsymbol{X}_\gamma^t\boldsymbol{X}_\tau & \boldsymbol{X}_\gamma^t\boldsymbol{X}_\gamma \end{bmatrix}.$$

Also, the elements of the $\boldsymbol{X}^t\boldsymbol{y}$ column are $(\mathbf{1}_p^t \otimes \mathbf{1}_n^t)\boldsymbol{y} = \sum_i \sum_j y_{ij}$, $(\boldsymbol{I}_p \otimes \mathbf{1}_n^t)\boldsymbol{y} = (y_{1\cdot}, \ldots, y_{p\cdot})^t$, and $(\mathbf{1}_p^t \otimes \boldsymbol{I}_n)\boldsymbol{y} = (y_{\cdot 1}, \ldots, y_{\cdot n})^t$.

From this point on we must look at specific designs. For example, the minimal balanced RTD with $t = p = n$, which is just a special Latin square with carryover effects, the normal equations can be reduced to

$$\begin{bmatrix} t^2 & \mathbf{0} & \mathbf{0} & \mathbf{0} & \mathbf{0} \\ \mathbf{0} & t\boldsymbol{I}_t & \mathbf{0} & \mathbf{0} & \mathbf{0} \\ \mathbf{0} & \mathbf{0} & t\boldsymbol{I}_t & \boldsymbol{A} & \boldsymbol{B} \\ \mathbf{0} & \mathbf{0} & \boldsymbol{A}^t & t\boldsymbol{I}_t & \boldsymbol{C} \\ \mathbf{0} & \mathbf{0} & \boldsymbol{B}^t & \boldsymbol{C}^t & (t-1)\boldsymbol{I}_t \end{bmatrix} \begin{bmatrix} \widehat{\mu} \\ \widehat{\boldsymbol{\pi}} \\ \widehat{\boldsymbol{u}} \\ \widehat{\boldsymbol{\tau}} \\ \widehat{\boldsymbol{\gamma}} \end{bmatrix} = \begin{bmatrix} \text{Total} \\ \boldsymbol{P} \\ \boldsymbol{U} \\ \boldsymbol{T} \\ \boldsymbol{C} \end{bmatrix}, \tag{9.4}$$

where P is the column of period totals, U is the column of unit totals, T is the column of direct treatment totals, and C is the sum of the responses by carryover effect. The five $t \times t$ matrices A, B, and C depend on the actual randomization of the design. The simple way to obtain this system of equations is to look at the expected values of the totals on the right-hand side of the equation. For example, the expected value of the first period total will include $t\mu$, $t\pi_1$, the sum of the unit effects, the sum of the direct effects, but no carryover effects. The second period total will be similar, except for the second-period effect and the sum of all carryover effects. Notice that at this point we have more equations than unknowns. However, the system is consistent, and can be solved.

9.4.2 Analysis of Variance

The analysis of variance for each of the RTDs discussed so far is obtained by least squares, based on the linear model given in equation (9.1). The lack of balance at various points in the designs causes some complications in the solution of the normal equations and means that some care must be exercised since not all sums of squares are orthogonal. However, the analysis of variance can be obtained using general purpose linear models software such as SAS® `Proc GLM`. The general analysis of variance table for the RTDs is given in Table 9.5.

Table 9.5 Analysis of Variance for a Repeated Treatment Design with First-Order Carryover Effects

Source	Symbolic Representation	df
Mean	$R(\mu) = CT$	1
Units	$R(u_1, \ldots, u_n \mid \pi_1, \ldots, \pi_p, \tau_1, \ldots, \tau_t, \gamma_1, \ldots, \gamma_t, \mu)$	$n-1$
Period	$R(\pi_1, \ldots, \pi_p \mid \mu)$ $= R(\pi_1, \ldots, \pi_p \mid u_1, \ldots, u_n, \tau_1, \ldots, \tau_t, \gamma_1, \ldots, \gamma_t, \mu)$	$p-1$
Direct treatment	$R(\tau_1, \ldots, \tau_t \mid u_1, \ldots, u_n, \pi_1, \ldots, \pi_p, \gamma_1, \ldots, \gamma_t, \mu)$	$t-1$
Carryover effect	$R(\gamma_1, \ldots, \gamma_t \mid u_1, \ldots, u_n, \pi_1, \ldots, \pi_t, \tau_1, \ldots, \tau_t, \mu)$	$t-1$
Error	By subtraction	$(n-1)(p-1)$ $-2(t-1)$
Total		np

Some special care is required to obtain these sums of squares from a general purpose regression program such as `proc GLM` in SAS®, because it is necessary to account for the fact that there is no carryover effect in the first period. The standard program commands such as the `class` statement in SAS® are set up to generate dummy variables for effects, including carryover effects for all periods, including the first. A solution in SAS® is to use a `data` step before the `GLM` step and manually generate appropriate columns of dummy variables. Since there are t effects, implying t carryover effects, our advice is to create $t-1$ columns of dummy variables. Let the columns be $c_1, \ldots, c_{t-1}$. For each experimental unit in each period, define

$$c_j = \begin{cases} 1 & \text{if treatment in preceding period was } j \\ -1 & \text{if treatment in preceding period was } t \\ 0 & \text{otherwise} \end{cases}$$

for $j = 1, \ldots, t-1$. These $t-1$ columns are then entered into the `model` statement to account for the carryover sum of squares. There will be $t-1$ single-degree-of-freedom sums of squares that need to be added together to yield the sum of squares for carryover effects with $t-1$ degrees of freedom.

Example 9.1: A Carburetor Icing Study Four gasoline additives were studied to compare their ability to prevent carburetor icing (Robert Hader and H. Lucas, personal communication). The purpose of a carburetor in an automobile engine is to mix gasoline into the air being drawn into the engine. Inside the carburetor there is a throttle blade, which almost closes the carburetor throat or annulus. When the accelerator is depressed, the blade is tilted, opening a larger passage for air. The engine draws in more air, which aspirates more gasoline and this causes the engine to run faster. When the accelerator is released, the blade snaps back to the horizontal position, decreasing the air flow, and consequently the engine slows down. If the annulus becomes partially blocked, still less air gets through and the engine idles roughly or stalls.

What could block the annulus? A major culprit under some conditions is an accumulation of ice crystals (or snow). As the gasoline evaporates in the carburetor, it absorbs heat from the surroundings and cools the surrounding air. If the air, hardware, and gasoline are below 50° F to begin with, a short drive can cool the environment to 32° F or less. If the relative humidity of the air is 60% or higher, this cooling will precipitate water, which turns to snow if the temperature drops below freezing. The snow drifts across the annulus, blocks the air flow, and stalls the engine. Enough driving will eventually warm the engine sufficiently to prevent the problem. However, in cold engines the problem is severe enough that oil companies (gasoline manufacturers) had to do something about it.

There are two ways to fix the problem. The first is to add an anti-freeze, say, isopropyl alcohol, to the gasoline. This leaches from the fuel into the condensed water and prevents the water from freezing. The second is to add an anti-stick, a surface-active agent to the fuel. This coats the metal surfaces with an oily coating, one or more molecules thick, that prevents snow and ice from clinging to the throttle blade and building up to damaging proportions. The purpose of the study was to compare four additives.

Four carburetors appropriately equipped were available (Table 9.6). The engineers could never be sure that all of the fuel additive from one test had been cleaned off the metal surfaces in the carburetor before a second test could be performed. The Latin square, balanced for carryover effects was set up and the data collected. The letters denote the different de-icing additives.

The estimated means (estimated by ordinary least squares) for the four fuel additives are A: 91.8, B: 81.7, C: 86.2, and D: 92.3. The standard error of a difference between two treatment means is 2.00. There are significant differences among the four fuel additives (Table 9.7, F-test, p-value $= .03$). It appears that the engineers successfully cleaned the previous treatment out of the carburetor before each new fuel additive was tested (F-test for carryover effects, p-value $=$ 0.30).

Table 9.6 Design and Engine Smoothness Responses

Carburetor I	Carburetor II	Carburetor III	Carburetor IV
A 88.0	*B* 78.0	*C* 87.5	*D* 90.5
B 76.0	*D* 94.0	*A* 95.5	*C* 78.5
C 88.0	*A* 90.0	*D* 95.5	*B* 82.5
D 92.0	*C* 90.0	*B* 87.5	*A* 94.5

Table 9.7 Analysis of Variance for Carburetor Example

Source	df	Partial SS	F-ratio	p-value
Carburetor	3	101.31	4.69	0.12
Period	3	72.00	3.33	0.17
Direct trt	3	275.85	12.77	0.03
Carryover	3	42.40	1.96	0.30
Error	3	21.60		

```
proc glm data = two ;
 class period carburetor deicer ;
 model y = period carburetor carryo_A
           carryo_B carryo_C deicer;
 contrast 'carryover' carryo_A 1,
           carryo_B 1, carryo_C 1;
 lsmeans deicer / stderr;
 estimate 'A vs B' deicer 1 -1 0 0;
run;
```

Display 9.1 SAS® code for carburetor analysis of variance.

The SAS® code for the analysis of variance is shown in Display 9.1 The data step coding the dummy variables for carryover effects (labeled carryo_A, carryo_B, and carryo_C), along with the data and output, are available on the companion website. The sum of squares for carryover effects was obtained by summing SScarryo_A, SScarryo_B, and SScarryo_C (using the `contrast` statement). All of the other sums of squares in Table 9.9 are printed out automatically as Type III SS.

Some insight is to be gained by considering how one might compute the sums of squares in Table 9.7. by comparing error sums of squares from different models. To do this, the analysis of variance given in Table 9.5 will require fitting five models.

- The first run, with the full model including units, periods, direct effects, and carryover effects, yields the error sum of squares, $R(\mu) = CT$, and the full-model sum of squares $R(u_1, \ldots, u_n, \pi_1, \ldots, \pi_p, \tau_1, \ldots, \tau_t, \gamma_1, \ldots, \gamma_t|\mu)$.
- A second run, with a model including only periods, direct effects, and carryover effects, yields $R(\pi_1, \ldots, \pi_p, \tau_1, \ldots, \tau_t, \gamma_1, \ldots, \gamma_t|\mu)$. Subtraction yields the units sum of squares.
- The third regression run requires only period effects in the model, since in all designs considered periods are orthogonal to all other effects. This analysis yields $R(\pi_1, \ldots, \pi_p|\mu)$ directly.

- The fourth regression analysis uses a model that includes units, periods, and carryover effects and yields $\mathrm{R}(u_1, \ldots, u_n, \pi_1, \ldots, \pi_p, \gamma_1, \ldots, \gamma_t|\mu)$. This provides $\mathrm{R}(\tau_1, \ldots, \tau_t|u_1, \ldots, u_n, \pi_1, \ldots, \pi_p, \gamma_1, \ldots, \gamma_t, \mu)$.
- The fifth regression analysis requires a model for units, periods, and direct effects. This yields $\mathrm{R}(u_1, \ldots, u_n, \pi_1, \ldots, \pi_p, \tau_1, \ldots, \tau_t, |\mu)$. Subtraction then yields $\mathrm{R}(\gamma_1, \ldots, \gamma_t|u_1, \ldots, u_n, \pi_1, \ldots, \pi_t, \tau_1, \ldots, \tau_t, \mu)$.

Note that this analysis assumes the minimal amount of balance. Most designs contain much more balance and consequently can be analyzed in fewer steps. For example, the RTD with $t = p = n$ has unit, periods, and direct treatment effects all mutually orthogonal to one another. Only carryover and direct effects are not orthogonal. On the other hand, the RTD with $t = n$ and $p = t + 1$ has units, periods, and carryover effects mutually orthogonal, and only direct effects and units are not orthogonal. We recognize orthogonality by noting that all levels of one factor appear with all levels of the other. An easy way to see that this is sufficient is to consider any one of the set of all possible contrasts among levels of the first factor and any one of the contrasts among levels of the second. Note also that this analysis assumes that no planned observations are missing. Loss of observations is serious in these designs and means detailed least squares analyses.

In practice, the experimenter will probably also want to examine estimates of the parameters and test subhypotheses about various parameters.

In many cases, where an investigator has set up an RTD to ensure against possible carryover effects, there is the temptation to use a preliminary test of the carryover effects as a guide for the final statistical analysis. Abeyasekera and Curnow (1984) investigate the implications of such a procedure. Their conclusion is that even though including carryover effects in the model when they are not necessary inflates the variance of other estimates somewhat, the safest policy is always to allow for the carryover effects.

9.5 CARRYOVER DESIGN FOR TWO TREATMENTS

A scenario that occurs frequently enough to warrant a special note is the case where the experimenter has just two treatments. The subjects are assigned at random into a group of s_1 units and a second of s_2 units (usually, $s_1 = s_2$, although that is not absolutely necessary). The first group receives one sequence of treatments, A first and then B, while the second group receives B first and then A. We can display this as

$$\begin{array}{lcc} \text{period 1} & \underbrace{\begin{matrix} A & A & \cdots & A \\ B & B & \cdots & B \end{matrix}}_{s_1 \text{ subjects}} & \underbrace{\begin{matrix} B & B & \cdots & B \\ A & A & \cdots & A \end{matrix}}_{s_2 \text{ subjects}} \\ \text{period 2} & & \end{array}$$

The two-period, two-treatment crossover design has a special problem not encountered with other numbers of treatments. Like the other designs in this chapter, it is balanced with respect to rows and columns; i.e., both treatments

Table 9.8 Contrast Coefficients for a Two-Period Two-Treatment Crossover Design

Effect Contrast	Subject 1		Subject 2		Subject 3		Subject 4	
	Per. 1	Per. 2	Per. 1	Per. 2	Per. 1	Per. 2	Per. 1	Per. 2
Subject	1	1	0	0	0	0	−1	−1
	0	0	1	1	0	0	−1	−1
	0	0	0	0	1	1	−1	−1
Period	1	−1	1	−1	1	−1	1	−1
Treatment	1	−1	1	−1	−1	1	−1	1
Carryover	0	1	0	1	0	−1	0	−1

appear in every column and both appear an equal number of times in each row. This makes the treatment assignments orthogonal to rows and columns. It is also balanced for first-order carryover effects. However, it is not possible to estimate subject, period, treatment, *and* carryover effects simultaneously. The effects are *confounded* with each other. This means that the contrasts for one effect are linear combinations of the contrasts for some of the other effects.

To see the confounding, examine the contrasts for subject, period, treatment, and carryover effects for a two-period design for four subjects. The contrast coefficients are shown in Table 9.8. Notice that the contrast for carryover effects can be obtained as (subject 1 + subject 2 − subject 3 − treatment)/2 contrast. It is a linear combination of the other terms in the model. The carryover effect cannot be estimated separately from the period and treatment effects; thus it is confounded with them.

This design is very efficient when there are no carryover effects. An example of this type of situation, where it is reasonable to assume that there are no carryover effects, appears in a study of the effect of food ingestion on absorption of a drug for seasonal allergic rhinitis by Stoltz et al. (1997). In this study, each of 24 healthy male subjects was given a single dose of two 40-mg capsules either after a 10-hour fast (treatment A) or after eating a breakfast of eggs, bacon, toast, hash browns, and milk (treatment B). Concentration of the drug in the blood was measured for 48 hours after taking the capsules. Six days later each subject was given the second treatment. Half of the subjects were randomly assigned to receive the treatments in the order AB, and half in the order BA. In this case, the washout period is so long compared to the length of time that the drug stays in the bloodstream that it is reasonable to assume that the rate of absorption of the drug in the second period is not affected by whether the subject fasted before taking the drug 6 days previously. It is reasonable to assume that there is no carryover effect.

If there is no carryover effect, the term $\gamma_{d(i,j-1)}$ is omitted from model (9.1) and analysis by least squares proceeds as usual. The test for direct treatment effects can also be done easily by hand when there are no carryover effects. To test direct treatment effects, compute the difference between the first- and second-period response for each subject, $z_j = y_{1j} - y_{2j}$. Let $\{z(AB)_j\}$ and $\{z(BA)_j\}$ represent the subset with treatment A first and treatment B first, respectively. They are two independent samples with

$$\mathrm{E}[z(AB)_j] = \pi_1 - \pi_2 + \tau_A - \tau_B$$

$$\mathrm{E}[z(BA)_j] = \pi_1 - \pi_2 - \tau_A + \tau_B$$

and

$$\mathrm{Var}[z(AB)_j] = \mathrm{Var}[z(BA)_j] = 2\sigma_e^2 \qquad \text{for all } j.$$

Under the $\mathrm{H}_0 : \tau_A = \tau_B$, the two samples $\{z(AB)_j\}$ and $\{z(BA)_j\}$ represent samples from populations with the same mean and variance and can be tested with the standard two-sample t-test or the nonparametric two-sample Wilcoxon test.

Exercise 9.6: Comparing F and t. Show that the t-statistic for comparing the means of $\{z(AB)_j\}$ and $\{z(BA)_j\}$ coincides with the square root of the F-statistic for testing treatment effects in the model $y_{ij} = \mu + \pi_i + u_j + \tau_{d(i,j)} + \epsilon_{ij}$. □

Possibly the best way to check for carryover effects is to take covariate measurements immediately before administration of the treatment in each period. This way, it is possible to measure whether the subject has returned to baseline directly. This may, however, still not answer questions about whether the mechanism, speed, or magnitude of response to a treatment is changed by the treatment in the preceding period.

It is possible to make a test of carryover effects in the two-period design by comparing the means of the two sequence groups. If we write a model with no sequence effects, it is possible to estimate period, direct treatment, and carryover effects. Under this model the carryover effects are tested by comparing the mean for the AB sequence with the mean for the BA sequence. The test is based on the sums for the individual subjects: $s(AB)_j = y_{1j} + y_{2j}$ for those subjects with treatment A first and $s(BA)_j = y_{1j} + y_{2j}$ for those subjects with treatment B first. The $\{s(AB)_j\}$ and the $\{s(BA)_j\}$ can be interpreted as two independent samples with

$$\mathrm{E}[s(AB)_j] = 2\mu + \pi_1 + \pi_2 + \tau_A + \tau_B + \gamma_A$$

$$\mathrm{E}[s(BA)_j] = 2\mu + \pi_1 + \pi_2 + \tau_A + \tau_B + \gamma_B$$

and

$$\mathrm{Var}[s(AB)_j] = \mathrm{Var}[s(BA)_j] = 2(\sigma_e^2 + 2\sigma_u^2) \qquad \text{for all } j.$$

Under $\mathrm{H}_0 : \gamma_A = \gamma_B$ the two samples, $\{s(AB)_j\}$ and $\{s(BA)_j\}$, have the same mean and variance and the hypothesis can be tested using the standard two sample t-test or the nonparametric two-sample Wilcoxon test. A discussion of the latter test may be found in *Nonparametric Statistical Methods* by Hollander and Wolf (1973).

If the test on carryover effect, $\text{H}_0 : \gamma_A = \gamma_B$, is used as a preliminary test for testing the hypothesis of equal treatment effects, Grizzle (1965, 1974) recommends insisting on a higher than conventional level of significance for the test on carryover effects, say probability equal to 0.1. If the null hypothesis of $\gamma_A = \gamma_B$ is rejected, a test of $\text{H}_0 : \tau_A = \tau_B$ is obtained by a two-sample test that ignores the data from the second period altogether. This procedure is the subject of considerable controversy and criticism. A good discussion of some of the difficulties can be found in Section 2.13 of *Design and Analysis of Cross-Over Trials* by Jones and Kenward (1989). The major difficulty centers around the problem that the power of the preliminary test for carryover effects is quite inadequate in realistically sized trials. The type II error, failure to reject the null hypothesis of equal carryover effects, may well lead to a biased estimator of the within-subject differences.

The recent work on bioequivalence testing Berger and Hsu (1996) has some bearing on the problem encountered here. The failure to reject $\text{H}_0 : \gamma_A = \gamma_B$ does not give strong evidence that H_0 is true, even approximately. Failure to reject H_0 could be the result of a test with low power. The equivalence testing approach would suggest that you test $\text{H}_0 : \gamma_A - \gamma_B \geq \delta$ or $\gamma_A - \gamma_B \leq -\delta$ vs. $\text{H}_A : -\delta < \gamma_A - \gamma_B < \delta$. Rejection with this test would be convincing evidence that γ_A is close to γ_B, as measured by δ. This is not entirely satisfactory either, however, because we want the carryover effects to be equal at the second stage, not just close to each other.

At least one author (Freeman (1989)) warns that this design and Grizzle's statistical analysis of it should never be used. This criticism comes despite the fact that the plan and analysis are used frequently. The criticism appears to be too harsh. In fairness, it should be pointed out that this is a very good, very efficient design in the absence of carryover effects. In this case Grizzle's procedure (and probably the nonparametric one as well) deviates only mildly from its nominal properties. There is a collection of papers in the Volume 3, number 4, 1994 issue of *Statistics in Medicine* devoted to this topic. A recent paper that discusses statistical analyses of two-period crossover trials is by Wang and Hung (1997). The bottom line, as we see it, is that the design is useful, especially in situations where the researcher really does not expect the carryover effect to be important but still wants to check for possible carryover effects.

The confounding between carryover effects and the other terms in the model can be avoided simply by adding an extra period to the two-period two-treatment design, either as a switchback design in which the first row of the design is repeated in a third period, or as an extra period design, in which the second period is repeated. Two such designs are shown in Tables 9.9 and 9.10.

Notice that in the switchback design, B always follows A (if it follows anything). In the extra period design, B follows B twice, and also follows A twice.

Table 9.9 Switchback Design

	Subject			
Period	1	2	3	4
1	A	A	B	B
2	B	B	A	A
3	A	A	B	B

Table 9.10 Extra-Period Design

	Subject			
Period	1	2	3	4
1	A	A	B	B
2	B	B	A	A
3	B	B	A	A

If an extra period is added, all effects in the model are estimable. The extra period design is superior to the switchback design because in the switchback design the direct and carryover effects are highly correlated with each other and hard to disentangle. When the columns of the $\boldsymbol{X}$ matrix are highly correlated, the variance of the estimated treatment and carryover effects are large. This is the collinearity problem in regression. The result is that the extra-period design gives more precise estimates of treatment and carryover effects than does the switchback design.

Many variations of three-period designs for two treatments have been proposed. Carriere (1994) compares the efficiency of some different three-period designs under various assumptions about the types of carryover effects present. When second-order carryover effects are present, some other designs are much more efficient for estimating treatment effects than the two designs presented here. This paper also makes the practical point that it is important to have a design that has good properties, even if some subjects drop out of the study before the third period. In clinical trials, or any trials with human subjects, this is an important consideration. If you use the extra-period design with sequences ABB and BAA, and if a large proportion of the subjects drop out after the second period, you are left with something like the two-period design and no good way to test carryover effects. On the other hand, a design that has four sequence groups: ABB, BAA, AAB, and BBA, still enables estimation of carryover effects if the third period is lost. The interested reader is urged to see Kershner and Federer (1981) for an extensive discussion of two-treatment crossover designs with two or more periods.

9.5.1 Binary Data

The designs discussed in this chapter are frequently used in medical and pharmaceutical research, where the response is binary, i.e., the treatment made the patient feel better, or it did not. Jones and Kenward (1989) discuss the statistical analysis of such data for two-period, two-treatment designs. Becker and Balagtas (1993) and Balagtas et al. (1995) discuss a general, maximum likelihood–based statistical analysis of binary data from multiperiod, multitreatment designs.

9.6 CORRELATED ERRORS

We pointed out in Section 9.2.1 that the responses on an experimental unit in different periods are not independent. Up to this point we have assumed that the lack of independence is accounted for by the location parameters in the model. We now consider the case where the errors have a correlation structure in addition to the location parameter structure. We repeat model (9.3), which is equivalent to (9.1) but in matrix notation as

$$\boldsymbol{y} = (\mathbf{1}_p \otimes \mathbf{1}_n)\mu + (\boldsymbol{I}_p \otimes \mathbf{1}_n)\boldsymbol{\pi} + (\mathbf{1}_p \otimes \boldsymbol{I}_n)\boldsymbol{u} + \boldsymbol{X}_\tau \boldsymbol{\tau} + \boldsymbol{X}_\gamma \boldsymbol{\gamma} + \boldsymbol{\epsilon}. \quad (9.5)$$

However, we now make the assumption that elements of $\boldsymbol{\epsilon}$ have mean zero and variance–covariance matrix $\sigma_e^2 \boldsymbol{V} = \sigma_e^2 \boldsymbol{V}_p \otimes \boldsymbol{I}_n$, where the within-unit correlation matrix $\boldsymbol{V}_p$ is a $p \times p$ positive definite matrix. Note that if we were to sort the responses by period within units, $\boldsymbol{V}$ would be block diagonal with $p \times p$ blocks along the diagonal. The $\boldsymbol{V}_p \otimes \boldsymbol{I}_n$ structure, i.e., independence of errors on distinct units, can be justified by the random assignment of experimental units to treatment sequences. Various more restrictive assumptions about $\boldsymbol{V}_p$ are possible. Common assumptions are:

- A fully general, unstructured covariance matrix.
- The Huynh and Feldt (1970) $\boldsymbol{H}$-structure, which says that we can write $(\boldsymbol{V}_p)_{ij} = \alpha_i + \alpha_j + \theta\delta_{ij}$, where $\theta > 0$ and $\delta_{ij} = 1$ if $i = j$ and zero otherwise.
- The equal correlation model, in which we write $\boldsymbol{V}_p = (1-\rho)\boldsymbol{I}_p + \rho \boldsymbol{J}_p$, where $\boldsymbol{J}_p$ is the $p \times p$ matrix with all values equal to 1, and ρ is the correlation between repetitions on a unit. This can be thought of as a special case of the $\boldsymbol{H}$-structure.
- The first-order autoregressive stationary model, which we write as $(\boldsymbol{V}_p)_{ij} = \rho^{|i-j|}/(1-\rho^2)$.
- The first-order stationary moving-average model, which we write as $(\boldsymbol{V}_p)_{ij} = \rho^{|i-j|} f(|i-j|)$, where $f(x) = 1$ if $x = 0, 1$ and zero otherwise.

The $\boldsymbol{V}_p = (1-\rho)\boldsymbol{I}_p + \rho \boldsymbol{J}_p$ model appears too idealistic. It seems reasonable to assume that units change slowly over time. This implies that periods closer together have a higher correlation than do periods farther apart. It is also quite possible that the error variance changes with time.

We note immediately that *ordinary least squares* estimates are unbiased, regardless of the error structure. Our concern is with efficiency, standard errors, and error rates for tests. One approach to the general problem of what to do about possible complex error structures is to see how bad things can be if one uses ordinary least squares, as we have done, when $\boldsymbol{V}_p$ is fully general. This is the approach taken in Kunert (1987) and Kunert and Utzig (1993). They show that while the error terms are generally underestimated and the statistical tests too liberal, upper bounds on the ratio of the true error terms to the ones provided in the ordinary least squares analysis are generally less than 2.0 and often considerably less. However, as Huynh and Feldt (1970) and Bellavance et al. (1996) point out, if $\boldsymbol{V}_p$ has the $\boldsymbol{H}$-structure, ordinary least squares analysis is appropriate for direct treatment effects, and yields correct standard errors. This does not hold for tests on carryover effects. However, a simulation study by Bellavance et al. (1996) indicates that the level of the test remains close to the nominal level.

A somewhat different approach by Matthews (1990) is to compare ordinary least squares and *generalized least squares*. The immediate problem here is that generalized least squares is an unattainable ideal since, in general, $\boldsymbol{V}_p$ is unknown. This leaves *estimated generalized least squares*. The problem then hinges on errors in the estimate of $\boldsymbol{V}_p$. Estimating $\boldsymbol{V}_p$ is not a trivial matter, since one must estimate error parameters from a relatively short series of numbers. Matthews examines a number of three- and four-period, two-treatment designs with the first-order stationary autoregressive model and the first-order stationary moving-average model and finds that ordinary least squares is generally within 10% of full efficiency. However, it must be remembered that the ordinary least squares analysis will underestimate the error variance when errors do not have the $\boldsymbol{V}_p = (1-\rho)\boldsymbol{I}_p + \rho\boldsymbol{J}_p$ structure. Again, Matthews's work suggests that the problem is not too serious.

If using estimated generalized least squares, we recommend using modified or restricted maximum likelihood (Patterson and Thompson 1971) to estimate the parameters of the covariance matrix. Simulation studies by Goad and Johnson (2000) and by Bellavance et al. (1996) indicate that for small and moderate numbers of subjects per sequence, estimated generalized least squares still leads to F-tests that are too liberal, i.e., that reject the null hypothesis too often when the null hypothesis is really true. Based on their simulation study of two designs, Goad and Johnson recommend a multivariate analysis that provides good control of the size of test. Simulations reported by Bellavance et al. show that ordinary least squares followed by F-tests with degrees of freedom modified using an approximation proposed by Box (1954a,b) provide better control on the type I error rate than does estimated generalized least squares.

It seems reasonable that when the number of subjects is large, good estimates of the variances and covariances will be available, and estimated generalized least squares should approach the efficiency of generalized least squares. Brown and Kempton (1994) give a number of references to examples of estimated generalized least squares analyses employing restricted maximum likelihood to estimate variances and covariances. However, in many situations, using the $\boldsymbol{V}_p = (1-\rho)\boldsymbol{I}_p + \rho\boldsymbol{J}_p$ model results in an analysis that is nearly fully efficient, and there is little practical difference between the results from the two types of analysis. At this time, we recommend using the $\boldsymbol{V}_p = (1-\rho)\boldsymbol{I}_p + \rho\boldsymbol{J}_p$ model but urge the investigator to keep in mind that the statistical tests may err somewhat on the side of being too liberal, and the standard error estimates may be slightly too small. However, this is an area of ongoing statistical research.

9.7 DESIGNS FOR MORE COMPLEX MODELS

9.7.1 Designs Balanced for First- and Second-Order Carryover Effects

Sharma (1977) gives a recipe for constructing designs balanced for both first- and second- order residual effects. These plans have parameters $p = 3t$ and $n = t^2$. The construction is relatively straightforward. Lay out three base rows and t^2 columns:

$$
\begin{array}{cccccccccccccccc}
0 & 1 & \cdots & (t-1) & 0 & 1 & \cdots & (t-2) & (t-1) & \cdots & 0 & 1 & \cdots & (t-1) \\
0 & 0 & \cdots & 0 & 1 & 1 & \cdots & 1 & 1 & \cdots & (t-1) & (t-1) & \cdots & (t-1) \\
0 & 1 & \cdots & (t-1) & 1 & 2 & \cdots & (t-1) & 0 & \cdots & (t-1) & 0 & \cdots & (t-2)
\end{array}
$$

Then construct $3(t-1)$ more rows by adding $i = 1, 2, \ldots, (t-1)$ to each element and reducing modulo t where necessary.

Consider the specific example with $t = 4$. The three base rows are

$$
\begin{array}{cccccccccccccccc}
0 & 1 & 2 & 3 & 0 & 1 & 2 & 3 & 0 & 1 & 2 & 3 & 0 & 1 & 2 & 3 \\
0 & 0 & 0 & 0 & 1 & 1 & 1 & 1 & 2 & 2 & 2 & 2 & 3 & 3 & 3 & 3 \\
0 & 1 & 2 & 3 & 1 & 2 & 3 & 0 & 2 & 3 & 0 & 1 & 3 & 0 & 1 & 2
\end{array}.
$$

The final plan with 12 periods is

$$
\begin{array}{cccccccccccccccc}
0 & 1 & 2 & 3 & 0 & 1 & 2 & 3 & 0 & 1 & 2 & 3 & 0 & 1 & 2 & 3 \\
0 & 0 & 0 & 0 & 1 & 1 & 1 & 1 & 2 & 2 & 2 & 2 & 3 & 3 & 3 & 3 \\
0 & 1 & 2 & 3 & 1 & 2 & 3 & 0 & 2 & 3 & 0 & 1 & 3 & 0 & 1 & 2 \\
1 & 2 & 3 & 0 & 1 & 2 & 3 & 0 & 1 & 2 & 3 & 0 & 1 & 2 & 3 & 0 \\
1 & 1 & 1 & 1 & 2 & 2 & 2 & 2 & 3 & 3 & 3 & 3 & 0 & 0 & 0 & 0 \\
1 & 2 & 3 & 0 & 2 & 3 & 0 & 1 & 3 & 0 & 1 & 2 & 0 & 1 & 2 & 3. \\
2 & 3 & 0 & 1 & 2 & 3 & 0 & 1 & 2 & 3 & 0 & 1 & 2 & 3 & 0 & 1 \\
2 & 2 & 2 & 2 & 3 & 3 & 3 & 3 & 0 & 0 & 0 & 0 & 1 & 1 & 1 & 1 \\
2 & 3 & 0 & 1 & 3 & 0 & 1 & 2 & 0 & 1 & 2 & 3 & 1 & 2 & 3 & 0 \\
3 & 0 & 1 & 2 & 3 & 0 & 1 & 2 & 3 & 0 & 1 & 2 & 3 & 0 & 1 & 2 \\
3 & 3 & 3 & 3 & 0 & 0 & 0 & 0 & 1 & 1 & 1 & 1 & 2 & 2 & 2 & 2 \\
3 & 0 & 1 & 2 & 0 & 1 & 2 & 3 & 1 & 2 & 3 & 0 & 2 & 3 & 0 & 1
\end{array}
$$

It is actually possible to perform considerable randomization:

- Randomize the first three rows before generating the other rows.
- Next, generate t rows from each of the three. Randomize rows within the t sets of three.
- Randomly assign experimental units to the columns.
- Randomly assign treatments to symbols $0, 1, \ldots, t-1$.

Table 9.11 ANOVA Table

Source	df	
Mean	1	
Rows	$3t - 1$	
Columns ignoring direct effects and residuals	$t - 1$	
Direct effects eliminating columns and ignoring residuals	$t - 1$	
First residuals ignoring second residuals	$t - 1$	Both calculated eliminating columns and direct effects
Second residuals eliminating first residuals	$t - 1$	
First residuals eliminating second residuals	$t - 1$	Both calculated eliminating columns and direct effects
Second residuals ignoring first residuals	$t - 1$	
Error	$(t-1)[(t+1)(3t-1)-3]$	
Total	$3t^2$	

The model for data collected is

$$y_{ij} = \mu + \pi_i + u_j + \tau_{d(i,j)} + \gamma_{d(i-1,j)} + \nu_{d(i-2,j)} + \epsilon_{ij}, \tag{9.6}$$

where

$$\begin{aligned}
\pi_i &= \text{period effect, } i = 1, \ldots, p \\
u_j &= \text{unit effect, } j = 1, \ldots, n \text{ and } u_j \sim \text{iid}(0, \sigma_u^2) \\
\tau_{d(i,j)} &= \text{(direct) treatment } d(i, j) \text{ effect in the } i\text{th period for the } j\text{th unit} \\
&\qquad d(i, j) = 1, \ldots, t \\
\gamma_{d(i-1,j)} &= \text{carryover of treatment effect from period } i - 1 \text{ to period } i \text{ on unit } j \\
\nu_{d(i-2,j)} &= \text{carryover of treatment effect from period } i - 2 \text{ to period } i \text{ on unit } j \\
\epsilon_{ij} &= \text{ residual error } \sim \text{iid}(0, \sigma_e^2).
\end{aligned}$$

The analysis of variance table is shown in Table 9.11.

9.7.2 Designs for Models with Lasting Residual Effects

Lakatos and Raghavarao (1987) discuss a class of designs where one or more of the treatments have a permanent carryover effect. This type of design and model would be appropriate in a medical setting where some of the treatments are surgical procedures that will leave permanent residual effects. A second application is in designing a questionnaire where some items may either irritate the respondent or give some additional information. Typically, one would consider this type of model if one is concerned about some treatments having lasting residual effects and wants to remove that as a source of error. Lakatos and Raghavarao give an example of a study to identify and rank lasting residual effects to provide guidance for the eventual questionnaire design.

These designs can be constructed for t treatments, t experimental units, and t periods, for t even. They do not exist for t odd. The requirements for balance are that:

- Every treatment appear in every period.
- Every unit receive every treatment.
- Every ordered pair of treatments occur equally frequently on units.

The construction is relatively simple. Assign treatment $j + (i - 1)/2$ modulo t to unit j in period i if i is odd and treatment $j - i/2$ modulo t if period i is even. We illustrate with a six-period, six-unit design,

$$\begin{array}{cccccc} 1 & 2 & 3 & 4 & 5 & 0 \\ 0 & 1 & 2 & 3 & 4 & 5 \\ 2 & 3 & 4 & 5 & 0 & 1 \\ 5 & 0 & 1 & 2 & 3 & 4 \\ 3 & 4 & 5 & 0 & 1 & 2 \\ 4 & 5 & 0 & 1 & 2 & 3 \end{array}.$$

Notice that, for example, treatment 0 follows treatment 1 in three columns and precedes it in three columns. In practice, this design would be repeated with r sets of experimental units. The detailed model for this example will then include

$$y_{11} = \mu + \pi_1 + u_1 + \tau_1 + \epsilon_{11}$$

$$\vdots$$

$$y_{16} = \mu + \pi_1 + u_6 + \tau_0 + \epsilon_{16}$$

$$y_{21} = \mu + \pi_2 + u_1 + \tau_0 + \gamma_1 + \epsilon_{21}$$

$$\vdots$$

$$y_{31} = \mu + \pi_3 + u_1 + \tau_2 + \gamma_1 + \gamma_0 + \epsilon_{31}$$

$$\vdots$$

$$y_{66} = \mu + \pi_6 + u_6 + \tau_3 + \gamma_0 + \gamma_5 + \gamma_1 + \gamma_4 + \gamma_2 + \epsilon_{66}.$$

The statistical analysis is again obtained via ordinary least squares or estimated generalized least squares discussed in a previous section.

When t is odd, $2t$ experimental units are required. For example, if $t = 5$, construct the first square with elements $i + j$ modulo 5 and the second square by subtracting each element from $t - 1 = 4$. In the resulting 10 columns, each treatment will precede every other treatment exactly five times.

REVIEW EXERCISES

9.1 It is often said that seven digits is the longest string that humans can easily store in short-term memory. Another factor of interest is the effect of background noise on short-term memory. In this exercise you will design, conduct, and analyze an experiment to compare success rates for memorizing seven- and 10-digit strings of digits with and without background noise.

This is a group exercise. The response variable is proportion of digits recalled correctly out of 10 seven-digit strings or out of seven 10-digit strings. The basic experimental protocol is this: group member 1 writes a seven- or 10-digit number down and shows it to group member 2, who can study the number for 10 seconds and then returns the number. Ten seconds after the number is returned, member 2 writes down the string and the number of digits recalled correctly (in the right position in the string) is recorded. For the background noise factor, test two different conditions: radio off or radio on during memorization period. Thus, there are a total of four treatment combinations.

Your tasks are the following:

(a) Construct a suitable design for this experiment and conduct the experiment.

(b) Estimate the mean success rates (and standard errors) for the four combinations of string length and background noise.

(c) Compute the analysis of variance table and test all hypotheses of interest.

(d) Write a few paragraphs describing the experimental layout, the model, and the statistical methods.

(e) Write up your conclusions.

(f) Evaluate the experimental design. Include discussion (and computations where appropriate) of power of tests, additional factors that might have been important, validity of assumptions, and how you would improve the experiment if you were to repeat it.

9.2 An education professor wants to study reading times for two systems (A and B) of writing for the blind. The professor is considering two types of crossover designs and a completely randomized design. The possible sequences for the two crossover designs are as follows:

Design 1: sequence 1: ABB
sequence 2: BAA
Design 2: sequence 1: AAB
sequence 2: BBA
Design 3: completely randomized

In the crossover designs, each subject is assigned to one of the sequences and receives samples to read from each method in the order specified, with an appropriate washout period between periods. In the completely randomized design (design 3), each subject receives a sample of only one writing system, and period 1 is the only period used.

(a) Is there any reason to prefer design 1 to design 2, or vice versa? Show computations to justify your answer. For your computations assume that subject effects are fixed effects. (Remember that for fixed effects models, $\text{Var}[\widehat{\boldsymbol{L\beta}}] = \sigma^2 \boldsymbol{L}(\boldsymbol{X}^t\boldsymbol{X})^+\boldsymbol{L}^t$.)

(b) Are there any circumstances under which you would recommend design 3 over designs 1 and 2?

(c) How many subjects would you need in design 3 to have the same standard error for $\widehat{\mu}_A - \widehat{\mu}_B$ as obtained by design 2?

9.3 Construct a crossover design balanced for first-order carryover effects for the following numbers of treatments, subjects, and periods per subject.

- 3 treatments, 6 subjects, 2 periods per subject
- 4 treatments, 8 subjects, 4 periods per subject
- 7 treatments, 42 subjects, 3 periods per subject

9.4 Neuroscientists and linguists are interested in understanding which parts of the brain are used for language, for interpreting visual stimuli, and for interpreting conflicting messages. Positron emission tomography (PET) scans can be used to determine the regions of activity in the brain while a subject looks at different combinations of words and colors. In one experiment each subject looks at four different combinations of words and ink colors ("blue" written with blue ink, "blue" with red ink, "red" with blue ink, and "red" with red ink) one at a time while a PET scan is done of the brain activity.

(a) Design a crossover experiment that is balanced for first-order carryover effects using the smallest possible number of subjects.

(b) Explain how to randomize the design.

CHAPTER 10

Factorial Experiments: The 2^N System

10.1 INTRODUCTION

Usually, there is a long list of factors that may affect any process under study. It immediately becomes obvious that a complete factorial with all the factors, each at several levels or forms, i.e., $n_A n_B \cdots n_N$ treatment combinations, requires a completely unreasonable number of runs. This is the first of a number of chapters that address this problem. In this chapter we reduce the problem by reducing each factor to two levels.

10.2 2^N FACTORIALS

We begin our discussion with a classical approach to factorials, which can be traced back to an underlying theory in finite geometry. We augment this theory with a class of factorial plans based on Hadamard matrices, and then eventually introduce a very flexible approach based on algebraic structures known as *orthogonal arrays*.

10.2.1 Background

The 2^N factorial system, factorial experiments with several factors, each at two levels, is important for several reasons. One of these is the sheer elegance of these experiments. They maximize the number of factors for a given size experiment. The construction of the plans is relatively straightforward and the interpretation of the results from experiments tends to be clearcut. These are the most widely used plans in industry.

Planning, Construction, and Statistical Analysis of Comparative Experiments,
by Francis G. Giesbrecht and Marcia L. Gumpertz
ISBN 0-471-21395-0

For a simple example, a pharmaceutical company may want to produce tablets that reliably dissolve within 10 minutes of ingestion. Several factors may affect the rate of dissolution of the tablets, such as shape of the tablet, temperature of storage, and length of storage. In this example, $N = 3$ factors and a 2^3 factorial would consist of all eight combinations of two tablet shapes, two temperatures, and two lengths of storage.

Much of the structure and theoretical development of the 2^N system is very dependent on notation. Several systems are commonly used, and it turns out that some things are much easier using one system of notation, while others are easier with another system. Consequently, a fair amount of time will be devoted to developing several systems. Our aim is to get the reader to the stage where shifting from one system to another becomes quite natural. The ability to do this is particularly important if the reader finds it necessary to consult papers in scientific journals.

We begin with the simplest case, two factors, A and B, each at two levels:

A at the low level	$\Rightarrow$	denoted by a_0 or a_1
B at the low level	$\Rightarrow$	denoted by b_0 or b_1
A at the high level	$\Rightarrow$	denoted by a_1 or a_2
B at the high level	$\Rightarrow$	denoted by b_1 or b_2.

We note in passing that while the terminology "high level" or "low level" is standard, it may well be that "high level" may correspond to presence of a factor and "low level" to absence, as, for example, a catalyst in a mixture of chemicals, or it may refer to a switch being on or off. Taking both treatments together in all possible combinations leads to four treatment combinations:

A low	B low	$\Rightarrow$	a_1b_1	a_0b_0	(1)	00
A high	B low	$\Rightarrow$	a_2b_1	a_1b_0	a	10
A low	B high	$\Rightarrow$	a_1b_2	a_0b_1	b	01
A high	B high	$\Rightarrow$	a_2b_2	a_1b_1	ab	11

four different systems of notation

Notice that the special symbol (1) is used to represent the control, or the treatment combination with both factors at the low level.

All four systems are used in practice, and we will use at least the last three of them. In addition, there is at least one more system that we plan to introduce eventually.

10.2.2 2 x 2 Factorial in Detail

In this section we develop a basic model for the 2^N system, beginning with the simplest case from first principles, the 2×2 factorial.

Effects as Parameters

Definition of Effect A

Consider defining the effect of treatment A. We let $\mu_{(1)}$, μ_a, μ_b, and μ_{ab} represent the true responses to the four treatment combinations and $y_{(1)}$, y_a, y_b, and y_{ab} the responses observed. The effect of shifting from the low level of A to the high level of A can be obtained in two ways or two places, in the presence of B or the absence of B. We have

$$\mu_{ab} - \mu_b \qquad \text{when } B \text{ is present or at the high level}$$

$$\mu_a - \mu_{(1)} \qquad \text{when } B \text{ is absent or at the low level}$$

Now define the average of these quantities,

$$\tfrac{1}{2}[(\mu_{ab} - \mu_b) + (\mu_a - \mu_{(1)})] = \tfrac{1}{2}(\mu_{ab} + \mu_a) - \tfrac{1}{2}(\mu_b + \mu_{(1)})$$

as the effect of A. Note that this definition is twice that of the A main effect given in Chapter 2. This is traditional in the context of 2^N factorials.

Estimating Effect A

Now consider estimating the effect of treatment A. Since $y_{(1)}$, y_a, y_b, and y_{ab} represent the responses to the four treatment combinations, we observe the effect of shifting from the low level of A to the high level of A at two places, in the presence of B or the absence of B. We compute

$$y_{ab} - y_b \qquad \text{in the presence of } B \text{ or when } B \text{ is at the high level}$$
$$y_a - y_{(1)} \qquad \text{when } B \text{ is absent or at the low level}$$

Estimate A by the average of the two,

$$\tfrac{1}{2}[(y_{ab} - y_b) + (y_a - y(1))].$$

Common practice is to write $y_{ab} - y_b$ and $y_a - y_{(1)}$ symbolically as $ab - b$ and $a - (1)$, respectively. The estimate of the A effect can then be written symbolically as

$$\tfrac{1}{2}[(ab - b) + (a - 1)]$$

or even more compactly as

$$\tfrac{1}{2}(a - 1)(b + 1).$$

Notice also that

$$\widehat{A} = \tfrac{1}{2}[(y_{ab} - y_b) + (y_a - y_{(1)})] = \tfrac{1}{2}(y_{ab} + y_a) - \tfrac{1}{2}(y_b + y_{(1)}),$$

i.e., a contrast based on the marginal means. In this form it is easy to see that the variance of $\widehat{A}$ is

$$\text{Var}[\widehat{A}] = \frac{\sigma^2}{r},$$

where r is the number of times each treatment combination is replicated.

Effect B

From symmetry, we see immediately that the effect of B occurs in two places,

$$\mu_{ab} - \mu_a \qquad \text{when } A \text{ is present or at the high level}$$

$$\mu_b - \mu_{(1)} \qquad \text{when } A \text{ is absent or at the low level}$$

Now define the average of these quantities,

$$\tfrac{1}{2}[(\mu_{ab} - \mu_a) + (\mu_b - \mu_{(1)})] = \tfrac{1}{2}(\mu_{ab} + \mu_b) - \tfrac{1}{2}(\mu_a + \mu_{(1)})$$

as the effect of B.

Then we estimate the B effect as

$$\tfrac{1}{2}[(y_{ab} - y_a) + (y_b - y_{(1)})].$$

Symbolically this can be written as

$$\tfrac{1}{2}[(ab - a) + (b - 1)]$$

or even more compactly as

$$\tfrac{1}{2}(a + 1)(b - 1).$$

Also note that

$$\text{Var}[\widehat{B}] = \frac{\sigma^2}{r},$$

where r is the number of times each treatment combination is replicated.

Interaction Effect

In the definitions of both the A and B effects, there is the question of whether the two pieces being combined are equal or not. Is $(\mu_{ab} - \mu_b)$ really equal to $(\mu_a - \mu_{(1)})$? The difference between the two, $\frac{1}{2}[(\mu_{ab} - \mu_b) - (\mu_a - \mu_{(1)})]$, is known as the AB *interaction*. Notice that this can also be written as $\frac{1}{2}[(\mu_{ab} - \mu_a) - (\mu_b - \mu_{(1)})]$.

The interaction between A and B is estimated as $\widehat{AB} = \frac{1}{2}[(y_{ab} - y_b) - (y_a - y_{(1)})]$ or as $\frac{1}{2}[(y_{ab} - y_a) - (y_b - y_{(1)})]$. This can be written symbolically and compactly as $\frac{1}{2}(a-1)(b-1)$. It is also clear that

$$\text{Var}[\widehat{AB}] = \frac{\sigma^2}{r},$$

where r is the number of times each treatment combination is replicated.

Some may wonder about the coefficient $\frac{1}{2}$ in the definition of the interaction. The only answer to this is: It is a common convention, it follows a pattern, and using these definitions the variances of estimates of all main effects and interactions are the same. We will encounter situations where not having that coefficient seems more natural.

Interpretation as Contrasts

In the development so far we have interpreted A, B, and AB as parameters with estimates $\widehat{A}$, $\widehat{B}$, and $\widehat{AB}$, respectively. Alternatively, one can think of $\widehat{A}$, $\widehat{B}$, and $\widehat{AB}$ as linear contrasts. We look at the table of contrasts, Table 10.1. This relationship can be inverted:

$$\begin{aligned}
y_{(1)} &= \text{mean} - \tfrac{1}{2}\widehat{A} - \tfrac{1}{2}\widehat{B} + \tfrac{1}{2}\widehat{AB} \\
y_a &= \text{mean} + \tfrac{1}{2}\widehat{A} - \tfrac{1}{2}\widehat{B} - \tfrac{1}{2}\widehat{AB} \\
y_b &= \text{mean} - \tfrac{1}{2}\widehat{A} + \tfrac{1}{2}\widehat{B} - \tfrac{1}{2}\widehat{AB} \\
y_{ab} &= \text{mean} + \tfrac{1}{2}\widehat{A} + \tfrac{1}{2}\widehat{B} + \tfrac{1}{2}\widehat{AB}.
\end{aligned}$$

Notice the rule on the coefficients of the parameter estimates. The sign on the main effect, $\widehat{A}$ or $\widehat{B}$, is + if the factor is at a high level and − if it is at the low level or absent. The sign on the interaction parameter is the product of the signs of the main-effect terms. All parameters have the coefficient $\frac{1}{2}$.

We can develop a more formal model by writing

$$A = \tfrac{1}{2}[(\mu_{ab} - \mu_b) + (\mu_a - \mu_{(1)})] \tag{10.1}$$

$$B = \tfrac{1}{2}[(\mu_{ab} - \mu_a) + (\mu_b - \mu(1))] \tag{10.2}$$

$$AB = \tfrac{1}{2}[(\mu_{ab} - \mu_b) - (\mu_a - \mu(1))] \tag{10.3}$$

$$\mu = \tfrac{1}{4}[\mu_{ab} + \mu_b + \mu_a + \mu_{(1)}]. \tag{10.4}$$

Table 10.1 Contrasts

	Observed Responses and Coefficients			
Contrast	y_{ab}	y_a	y_b	$y_{(1)}$
$2\widehat{A}$	+	+	−	−
$2\widehat{B}$	+	−	+	−
$2\widehat{AB}$	+	−	−	+
4×mean	+	+	+	+

These relationships can be inverted to give

$$\mu_{(1)} = \mu - \tfrac{1}{2}A - \tfrac{1}{2}B + \tfrac{1}{2}AB$$
$$\mu_a = \mu + \tfrac{1}{2}A - \tfrac{1}{2}B - \tfrac{1}{2}AB$$
$$\mu_b = \mu - \tfrac{1}{2}A + \tfrac{1}{2}B - \tfrac{1}{2}AB$$
$$\mu_{ab} = \mu + \tfrac{1}{2}A + \tfrac{1}{2}B + \tfrac{1}{2}AB.$$

We can then make correspondences with the linear model in the effects coding form given in Section 3.4,

$$y_{ijk} = \overline{\mu}_{..} + A_i + B_j + (A \times B)_{ij} + \epsilon_{ijk},$$

with $i = 1, 2$, $j = 1, 2$, and $k = 1, \ldots, r$, subject to the side conditions $\sum_i A_i = 0$, $\sum_j B_j = 0$, and $\sum_i (A \times B)_{ij} = \sum_j (A \times B)_{ij} = 0$. The correspondence between effects coding parameters and the 2^N notation is then

$$A_1 = -\tfrac{1}{2}A$$
$$A_2 = +\tfrac{1}{2}A$$
$$B_1 = -\tfrac{1}{2}B$$
$$B_2 = +\tfrac{1}{2}B$$
$$(A \times B)_{11} = +\tfrac{1}{2}AB$$
$$(A \times B)_{12} = -\tfrac{1}{2}AB$$
$$(A \times B)_{21} = -\tfrac{1}{2}AB$$
$$(A \times B)_{22} = +\tfrac{1}{2}AB.$$

Generalized Interaction

We now examine the contrasts corresponding to the definitions (10.1) to (10.3). We write

$$\widehat{A} = \tfrac{1}{2}y_{ab} + \tfrac{1}{2}y_a - \tfrac{1}{2}y_b - \tfrac{1}{2}y_{(1)}$$
$$\widehat{B} = \tfrac{1}{2}y_{ab} - \tfrac{1}{2}y_a + \tfrac{1}{2}y_b - \tfrac{1}{2}y_{(1)}$$
$$\widehat{AB} = \tfrac{1}{2}y_{ab} - \tfrac{1}{2}y_a - \tfrac{1}{2}y_b + \tfrac{1}{2}y_{(1)}.$$

Notice that the coefficients of the $\widehat{AB}$ contrast are (except for the coefficient $\frac{1}{2}$) the products of the corresponding coefficients of the coefficients for $\widehat{A}$ and $\widehat{B}$. This is an example of the general rule that the coefficients of the interaction of two contrasts are obtained as products of the corresponding coefficients of the contrasts. This is suggested by the notation of AB as the interaction of A and B.

However, we notice further that the coefficients of the $\widehat{B}$ contrast can be obtained as the products of corresponding elements of the $\widehat{A}$ and $\widehat{AB}$ contrasts.

We say that $\widehat{B}$ is the generalized interaction of $\widehat{A}$ and $\widehat{AB}$. The rule is that we take the product of the symbols, A and AB, to get A^2B and then reduce the exponents modulo 2 to obtain B. In a similar manner, A is the generalized interaction of B and AB.

Exercise 10.1: Estimator of the B Effect. Show that the estimate of the B effect can be written as the average of observations with the B factor at the high level minus the average of observations with the B factor at the low level. □

Exercise 10.2: Contrast for Estimating the AB Interaction. The estimate of the AB interaction can be expressed as the mean of two observations minus the mean of two other observations. Display this contrast. □

10.2.3 Extensions of the Notation to 2^3

The notation developed extends directly to larger experiments with more factors, all at two levels. We illustrate the eight treatment combinations obtained from three factors, each at two levels.

A low, B low, C low	$\Rightarrow$	$a_1b_1c_1$	$a_0b_0c_0$	(1)	000
A high, B low, C low	$\Rightarrow$	$a_2b_1c_1$	$a_1b_0c_0$	a	100
A low, B high, C low	$\Rightarrow$	$a_1b_2c_1$	$a_0b_1c_0$	b	010
A high, B high, C low	$\Rightarrow$	$a_2b_2c_1$	$a_1b_1c_0$	ab	110
A low, B low, C high	$\Rightarrow$	$a_1b_1c_2$	$a_0b_0c_1$	c	001
A high, B low, C high	$\Rightarrow$	$a_2b_1c_2$	$a_1b_0c_1$	ac	101
A low, B high, C high	$\Rightarrow$	$a_1b_2c_2$	$a_0b_1c_1$	bc	011
A high, B high, C high	$\Rightarrow$	$a_2b_2c_2$	$a_1b_1c_1$	abc	111

four different systems of notation

Let $y_{(1)}$, y_a, y_b, y_{ab}, y_c, y_{ac}, y_{bc}, and y_{abc} denote responses for the eight treatment combinations, respectively. The estimate for the effect of treatment A can now be obtained from four sources: when both B and C are at the low level $(y_a - y_{(1)})$, when B is at the high level and C is at the low level $(y_{ab} - y_b)$, when B is at the low level and C is at the high level $(y_{ac} - y_c)$, and when B and C are both at the high level $(y_{abc} - y_{bc})$. The four are averaged. Symbolically, we can summarize the definition of the main effect A as

$$\tfrac{1}{4}(a-1)(b+1)(c+1) = \left(\tfrac{1}{2}\right)^{(3-1)}(a-1)(b+1)(c+1).$$

Note also that this can be interpreted as the average of the four responses where a is present minus the average of the four responses where a is absent, A at the high level vs. A at the low level.

Exercise 10.3: Estimator of the A Main Effect. Expand the symbolic expression for A given above and then show that the estimate of the A main effect can be expressed as the average of the observations with the A factor at the high level minus the average of the observations with the A factor at the low level. □

Similarly, the B main effect is written symbolically as

$$\tfrac{1}{4}(a+1)(b-1)(c+1) = \left(\tfrac{1}{2}\right)^{(3-1)}(a+1)(b-1)(c+1),$$

which is the average of the four responses where b is present minus the average of the four responses where b is absent. The AB interaction, written as

$$\tfrac{1}{4}(a-1)(b-1)(c+1) = \left(\tfrac{1}{2}\right)^{(3-1)}(a-1)(b-1)(c+1),$$

is a bit different. It is the average of the four responses that have either both a and b present or neither a nor b present minus the average of the four that have one of either a or b, but not both, present. If we treat zero as an even number, we can construct a rule counting whether there is an even or odd number of letters in common.

It is also convenient to denote the responses by $y_{000}, y_{100}, \ldots, y_{111}$, respectively. We then refer to the subscripts as i_A, i_B, and i_C. With this notation the AB interaction is the mean of the responses for which $i_A + i_B$ is even minus those responses for which $i_A + i_B$ is odd.

The remaining main effect and interactions are given symbolically:

$$C \quad \text{as} \quad \left(\tfrac{1}{2}\right)^{(3-1)}(a+1)(b+1)(c-1)$$

$$AC \quad \text{as} \quad \left(\tfrac{1}{2}\right)^{(3-1)}(a-1)(b+1)(c-1)$$

$$BC \quad \text{as} \quad \left(\tfrac{1}{2}\right)^{(3-1)}(a+1)(b-1)(c-1)$$

$$ABC \quad \text{as} \quad \left(\tfrac{1}{2}\right)^{(3-1)}(a-1)(b-1)(c-1).$$

The rule of even and odd number of letters common between parameter and treatment combination quoted for the AB interaction can be applied to the three-factor interaction ABC, except that the signs must be reversed. This is the mean of those responses for which $i_A + i_B + i_C$ is odd minus the mean of those for which $i_A + i_B + i_C$ is even.

Exercise 10.4: Relationship between Symbolic Expression for ABC Interaction and Even–Odd Rule For Contrasts. Show that expanding the symbolic expression for the ABC interaction leads to the same contrast as the rule of even and odd number of letters common between parameter and treatment combination. Rewrite the expression using the $y_{000}, y_{001}, \ldots, y_{111}$ notation system. □

Just as for the 2×2 case, these definitions or relationships can be inverted.

$$\begin{aligned}
y_{(1)} &= \text{mean} - \tfrac{1}{2}\widehat{A} - \tfrac{1}{2}\widehat{B} + \tfrac{1}{2}\widehat{AB} - \tfrac{1}{2}\widehat{C} + \tfrac{1}{2}\widehat{AC} + \tfrac{1}{2}\widehat{BC} - \tfrac{1}{2}\widehat{ABC} \\
y_a &= \text{mean} + \tfrac{1}{2}\widehat{A} - \tfrac{1}{2}\widehat{B} - \tfrac{1}{2}\widehat{AB} - \tfrac{1}{2}\widehat{C} - \tfrac{1}{2}\widehat{AC} + \tfrac{1}{2}\widehat{BC} + \tfrac{1}{2}\widehat{ABC} \\
y_b &= \text{mean} - \tfrac{1}{2}\widehat{A} + \tfrac{1}{2}\widehat{B} - \tfrac{1}{2}\widehat{AB} - \tfrac{1}{2}\widehat{C} + \tfrac{1}{2}\widehat{AC} - \tfrac{1}{2}\widehat{BC} + \tfrac{1}{2}\widehat{ABC} \\
y_{ab} &= \text{mean} + \tfrac{1}{2}\widehat{A} + \tfrac{1}{2}\widehat{B} + \tfrac{1}{2}\widehat{AB} - \tfrac{1}{2}\widehat{C} - \tfrac{1}{2}\widehat{AC} - \tfrac{1}{2}\widehat{BC} - \tfrac{1}{2}\widehat{ABC} \\
y_c &= \text{mean} - \tfrac{1}{2}\widehat{A} - \tfrac{1}{2}\widehat{B} + \tfrac{1}{2}\widehat{AB} + \tfrac{1}{2}\widehat{C} - \tfrac{1}{2}\widehat{AC} - \tfrac{1}{2}\widehat{BC} + \tfrac{1}{2}\widehat{ABC} \\
y_{ac} &= \text{mean} + \tfrac{1}{2}\widehat{A} - \tfrac{1}{2}\widehat{B} - \tfrac{1}{2}\widehat{AB} + \tfrac{1}{2}\widehat{C} + \tfrac{1}{2}\widehat{AC} - \tfrac{1}{2}\widehat{BC} - \tfrac{1}{2}\widehat{ABC} \\
y_{bc} &= \text{mean} - \tfrac{1}{2}\widehat{A} + \tfrac{1}{2}\widehat{B} - \tfrac{1}{2}\widehat{AB} + \tfrac{1}{2}\widehat{C} - \tfrac{1}{2}\widehat{AC} + \tfrac{1}{2}\widehat{BC} - \tfrac{1}{2}\widehat{ABC} \\
y_{abc} &= \text{mean} + \tfrac{1}{2}\widehat{A} + \tfrac{1}{2}\widehat{B} + \tfrac{1}{2}\widehat{AB} + \tfrac{1}{2}\widehat{C} + \tfrac{1}{2}\widehat{AC} + \tfrac{1}{2}\widehat{BC} + \tfrac{1}{2}\widehat{ABC}.
\end{aligned}$$

We also have the generalized interactions. For example, A is the generalized interaction of B and AB and also the generalized interaction of BC and ABC. Similarly, AB is the generalized interaction of C and ABC.

10.3 GENERAL NOTATION FOR THE 2^N SYSTEM

In this section we develop the system or systems of notation that can be used for plans with N factors, each at two levels.

10.3.1 Treatment Combinations

At this point there are what amounts to four different systems of notation to identify treatment combinations for factorials with factors each at two levels. All will continue to be used, but we begin formally by extending the 00, 10, 01, and 11 notation system for the 2^2 case to the N-factor case by denoting a typical treatment combination by a set of indices $t = (i_A, i_B, \ldots, i_N)$, where i_X is either 0 or 1 for factor $X = A, B, \ldots, N$. A true response and an observed response are written as μ_t and y_t, respectively.

10.3.2 Effects and Interactions

The typical effect or interaction is then written as $A^{i_A} B^{i_B} \cdots N^{i_N}$ and the special case with all $\{i_X\}$ equal to zero is μ. Main effects have $\sum_{X=A}^{N} i_X = 1$, i.e., exactly one of the indices $\{i_X\}$ is equal to 1, and the remainder are equal to zero. Two-factor or second-order interactions have two nonzero indices, $\sum_{X=A}^{N} i_X = 2$, and so on. We occasionally follow the convention of referring to main effects as first-order interactions.

The Xth main effect is defined as the mean of all μ_t with $i_X = 1$ minus the mean of all responses with $i_X = 0$. Since there are exactly 2^N responses, we have the mean of 2^{N-1} minus the mean of the remaining 2^{N-1} responses. An example of a main effect in a 2^5 factorial would be $A^0B^1C^0D^0E^0$, which is more commonly written as B and is defined as

$$B = \left(\frac{1}{2}\right)^{5-1}\left[\sum_{i_A}\sum_{i_C}\sum_{i_D}\sum_{i_E}\mu_{i_A 1 i_C i_D i_E} - \sum_{i_A}\sum_{i_C}\sum_{i_D}\sum_{i_E}\mu_{i_A 0 i_C i_D i_E}\right].$$

The i_B subscript has been replaced by 1 in the first sum and by 0 in the second sum; i.e., we have the sum of all responses with $i_B = 1$ minus the sum of all responses with $i_B = 0$.

The two-factor interaction identified by $i_{X^*} = 1$ and $i_{X^{**}} = 1$ with $X^* \neq X^{**}$ and all other $\{i_X\}$ equal to zero is defined as the mean of the μ_t with $i_{X^*} + i_{X^{**}} = 0$ modulo 2 minus the remaining responses, i.e., those with $i_{X^*} + i_{X^{**}} = 1$ modulo 2. An example of a two-factor interaction in the 2^5 would be $A^0B^1C^0D^1E^0$, which would typically be written as the BD interaction. This interaction is defined as $(\frac{1}{2})^{(5-1)}$ times

$$\sum_{i_A}\sum_{i_C}\sum_{i_E}(\mu_{i_A 1 i_C 1 i_E} + \mu_{i_A 0 i_C 0 i_E}) - \sum_{i_A}\sum_{i_C}\sum_{i_E}(\mu_{i_A 0 i_C 1 i_E} + \mu_{i_A 1 i_C 0 i_E}).$$

Now both the i_B and the i_D subscripts have been replaced by specific numbers.

An example of a three-factor interaction, say ACD, would be the sum over all treatment combinations with $i_A + i_C + i_D = 1$ (modulo 2) minus the sum over treatment combinations with $i_A + i_C + i_D = 0$ modulo 2. Notice that the 0 or 1 sum is related to whether one is dealing with an interaction involving an even number of factors or an odd number of factors.

In general, the p-factor interaction, i.e., $A^{i_A^*}B^{i_B^*}\cdots N^{i_N^*}$ with $\sum_{X=A,\cdots,N} i_X^* = p$, is defined as the mean of the responses with $\sum_{X=A,\cdots,N} i_X i_X^* = \frac{1}{2}(1-(-1)^p)$ (modulo 2) minus the remaining responses. This is simply the formalization of the even–odd rule of number of letters. Estimates are defined in a similar manner with $\{y_t\}$ replacing the $\{\mu_t\}$. Notice that we can interpret the effects and interactions as parameters or as contrasts. Similarly, the corresponding linear functions of the y_t can and will be interpreted as estimates of parameters and also as linear contrasts among the data points.

The relationship between effects and interactions and the true responses can be inverted and written as

$$\mu_{i_A i_B \cdots i_N} = \mu + \tfrac{1}{2}\sum_{i_A^*, i_B^*, \cdots, i_N^* \neq 0,0,\cdots,0} (-1)^{\sum_X (1-i_X) i_X^*} A^{i_A^*}B^{i_B^*}\cdots N^{i_N^*},$$

where $i_X^* = 0$ or 1 for $X = A, B, \ldots, N$. This expression holds for all $\{i_X\}$ and is normally regarded as the linear model. There is a corresponding formula linking the $\{y_t\}$ and the parameter estimates. Notice also that now we are thinking in terms of parameters and parameter estimates, while earlier we were thinking in terms of linear contrasts.

We can use the symbolic notation

$$A = \left(\frac{1}{2}\right)^{N-1} (a-1)(b+1)(c+1)\cdots$$
$$= \frac{1}{2^{N-1}}\left[(\mu_a + \mu_{ab} + \mu_{ac} + \mu_{abc} + \cdots) - (\mu_{(1)} + \mu_b + \mu_c + \mu_{bc} + \cdots)\right]$$

as well as, for example,

$$ACE = \left(\tfrac{1}{2}\right)^{N-1} (a-1)(b+1)(c-1)(d+1)(e-1)(f+1)\cdots.$$

In each case there will be 2^{N-1} responses with a "+" coefficient and 2^{N-1} with a "−" coefficient. It follows that if we have r observations (replicates) for each treatment combination, each effect or interaction will be estimated with variance

$$\frac{\sigma^2}{2^{N-2}r}. \tag{10.5}$$

Exercise 10.5: Variance of Treatment Effect. Work out in detail the variance for the $\widehat{ACD}$ contrast in the 2^5 with r replicates. □

Generalized Interaction

The reader whose interest is largely in applying the techniques presented in this book may wish to skip this and the next subsection and simply accept the rules as demonstrated.

We now consider the expression for the generalized interaction of pairs of interactions. Let $A^{i_A^*}B^{i_B^*}\cdots N^{i_N^*}$ represent a p^*-order interaction, i.e., $\sum_X i_X^* = p^*$ and $A^{i_A^{**}}B^{i_B^{**}}\cdots N^{i_N^{**}}$ represent a different interaction, a p^{**}-order interaction. Note that it is possible that either one or both of these could be main effects. The generalized interaction is then written as $A^{i_A^+}B^{i_B^+}\cdots N^{i_N^+}$, where $i_X^+ = i_X^* + i_X^{**}$ (modulo 2), the sum of the indices, for $X = A, B, \ldots, N$. This is a $p^+ = \sum_X i_X^+$-order interaction.

To see the rationale for calling this the generalized interaction, consider the following 2×2 table in which each cell contains 2^{N-2} responses defined by the row and column headings. All subscript arithmetic is modulo 2.

	Resp. with $\sum i_X i_X^* = \frac{1}{2}(1-(-1)^{p^*})$	Resp. with $\sum i_X i_X^* = \frac{1}{2}(1+(-1)^{p^*})$
Resp. with $\sum i_X i_X^{**} = \frac{1}{2}(1-(-1)^{p^{**}})$		
Resp. with $\sum i_X i_X^{**} = \frac{1}{2}(1+(-1)^{p^{**}})$		

Table 10.2 2 x 2 Table

μ_{01000}	μ_{01001}	μ_{00000}	μ_{00001}
μ_{00100}	μ_{00101}	μ_{01100}	μ_{01101}
μ_{10010}	μ_{10011}	μ_{11010}	μ_{11011}
μ_{11110}	μ_{11111}	μ_{10110}	μ_{10111}
μ_{10000}	μ_{10001}	μ_{11000}	μ_{11001}
μ_{11100}	μ_{11101}	μ_{10100}	μ_{10101}
μ_{01010}	μ_{01011}	μ_{00010}	μ_{00011}
μ_{00110}	μ_{00111}	μ_{01110}	μ_{01111}

The $A^{i_A^*} B^{i_B^*} \cdots N^{i_N^*}$ interaction is defined as the difference between the mean of the responses in the second column and the mean of the responses in the first column and the $A^{i_A^{**}} B^{i_B^{**}} \cdots N^{i_N^{**}}$ interaction as the difference between the mean of responses in the second row and the mean of the responses in the first row. Now the interaction effect, i.e., the generalized interaction $A^{i_A^+} B^{i_B^+} \cdots N^{i_N^+}$, is defined as the difference between the mean of the responses in the cells on the main diagonal and the mean of the responses in the off-diagonal cells. Note that the responses in the upper left quadrant satisfy $\sum i_X i_X^* = \frac{1}{2}[1 - (-1)^{p^*}]$ (modulo 2) and $\sum i_X i_X^{**} = \frac{1}{2}[1 - (-1)^{p^{**}}]$ (modulo 2). One can think of this as a two-way table with row effect, column effect, and interaction.

As a specific example, we can consider the generalized interaction of ABC and AD in the 2^5. Our rule tells us that the generalized interaction is BCD. The 2×2 table is shown in Table 10.2.

It is straightforward to verify that the ABC interaction is the mean of the first column minus the mean of the second column and the AD interaction is the mean of the first row minus the second row. Finally, the BCD interaction is the mean of the responses in the cells on the main diagonal minus the mean of the responses in the off-diagonal cells. We also note that in the upper left cell the subscripts must satisfy the two conditions, $i_A + i_B + i_C = 1$ (modulo 2) and $i_A + i_D = 0$ (modulo 2). These two conditions together imply that $i_B + i_C + i_D = 1$ (modulo 2) since $2i_A$ is always 0 in modulo 2 arithmetic. Similarly, in the lower right cell the subscripts must satisfy the two conditions $i_A + i_B + i_C = 1$ (modulo 2) and $i_A + i_D = 0$ (modulo 2). These two conditions again imply that $i_B + i_C + i_D = 1$ (modulo 2). But these define exactly the responses that enter BCD with a positive sign. The remaining two cells contain the responses that enter BCD with a negative sign.

Orthogonality of Effects and Interactions

Finally, we also need to establish the orthogonality of estimates of different effects and interactions. Consider two distinct effect or interaction contrasts, say $A^{i_A^\#} B^{i_B^\#} \cdots N^{i_N^\#}$ and $A^{i_A^*} B^{i_B^*} \cdots N^{i_N^*}$ with $\sum i_X^\# = p^\#$ and $\sum i_X^* = p^*$. These are calculated as

$$\left(\tfrac{1}{2}\right)^{N-1}\left[\left[\sum\cdots\sum\right]_{+}^{\#} y_{i_A\cdots i_N} - \left[\sum\cdots\sum\right]_{-}^{\#} y_{i_A\cdots i_N}\right] \tag{10.6}$$

and

$$\left(\tfrac{1}{2}\right)^{N-1}\left[\left[\sum\cdots\sum\right]_{+}^{*} y_{i_A\cdots i_N} - \left[\sum\cdots\sum\right]_{-}^{*} y_{i_A\cdots i_N}\right], \tag{10.7}$$

respectively, where the symbol $\left[\sum\cdots\sum\right]_{+}^{\theta}$ means $\sum_{i_A}\cdots\sum_{i_N}$ subject to the constraint that $\sum_X i_X i_X^{\theta} = .5(1-(-1)^{p^{\theta}})$ (modulo 2) and $\left[\sum\cdots\sum\right]_{-}^{\theta}$ the similar sum subject to the constraint that $\sum_X i_X i_X^{\theta} = .5(1+(-1)^{p^{\theta}})$ (modulo 2). It is now a matter of verifying that the elements of $\{y_{i_A\cdots i_N}\}$ fall into four groups of 2^{N-2}. One group appears in equations (10.6) and (10.7) with a "+" sign. The second appears in both with a "−" sign. The third is in (10.6) with a "+" sign and in (10.7) with a "−" sign and the fourth group the converse. Establishing orthogonality then entails adding up products of coefficients and observing the result zero.

10.4 ANALYSIS OF VARIANCE FOR 2^N FACTORIALS

The structure of the analysis of variance comes from ordinary least squares. The actual arithmetic is not difficult and can be performed in several different ways. We will assume here that the experimenter has repeated each treatment combination r times; that is, there are r replications in one large completely randomized design. If $r \geq 2$, the first step is to compute the error sum of squares. This amounts to nothing more than a one-way analysis of variance, among- and within-treatment combinations. The remainder of the analysis of variance table can be computed in one of several ways. For the general 2^N factorial, the form of the analysis of variance table is shown in Table 10.3.

10.4.1 Calculating Sums of Squares

The sums of squares can be computed using general linear models software such as SAS® `Proc GLM`. However, we show the underlying calculations.

$$\text{SSA} = \frac{(\text{sum all obs. with } a \text{ at the high level})^2}{r2^{N-1}} + \frac{(\text{sum all obs. with } a \text{ at the low level})^2}{r2^{N-1}} - \frac{(\text{sum all obs.})^2}{r2^N}.$$

Table 10.3 Analysis of Variance Table

Source	SS	df
A		1
B		1
AB		1
C		1
AC		1
$\vdots$		$\vdots$
Error	By subtraction	$(r-1)2^N$
Total	$\sum y^2 - \text{CT}$	$r2^N - 1$

Alternatively, this sum of squares can be computed as

$$\text{SSA} = r2^{N-2}(\widehat{A})^2,$$

where $\widehat{A}$ is computed as

$$\text{ave. of all obs. with } a \text{ at the high level}$$
$$- \text{ ave. of all obs. with } a \text{ at the low level.}$$

Similarly, the sum of squares due to B can be computed as

$$\text{SSB} = \frac{(\text{sum all obs. with } b \text{ at the high level})^2}{r2^{N-1}} + \frac{(\text{sum all obs. with } b \text{ at the low level})^2}{r2^{N-1}} - \frac{(\text{sum all obs.})^2}{r2^N}$$

or as

$$\text{SSB} = r2^{N-2}(\widehat{B})^2,$$

where $\widehat{B}$ is computed as

$$\text{ave. of all obs. with } b \text{ at the high level}$$
$$- \text{ ave. of all obs. with } b \text{ at the low level.}$$

The sum of squares for the $A \times B$ interaction can be computed from the two-way $A \times B$ table using standard hand computing formulas,

$$\text{SSAB} = \frac{(\text{sum all obs. with } a \text{ and } b \text{ at the high level})^2}{r2^{N-2}} + \frac{(\text{sum all obs. with } a \text{ at the high level and } b \text{ low level})^2}{r2^{N-2}}$$

$$+\frac{(\text{sum all obs. with } a \text{ at the low level and } b \text{ at the high level})^2}{r2^{N-2}}$$

$$+\frac{(\text{sum all obs. with } a \text{ and } b \text{ at the low level})^2}{r2^{N-2}}$$

$$-\frac{(\text{sum all obs.})^2}{r2^N} - \text{SSA} - \text{SSB}$$

or as

$$\text{SSAB} = r2^{N-2}(\widehat{AB})^2,$$

where the interaction contrast $\widehat{AB}$ is

ave. of all obs. with both a and b at the high level or both at the low level

− ave. of all obs. with exactly one of a or b at the high level.

Exercise 10.6: Derive SSA. Show that SSA = $r2^{N-2}(\widehat{A})^2$. □

When the calculations are done by hand, the three-way and higher-order interaction sums of squares are usually computed most easily by computing the estimate, squaring, and multiplying by the constant $r2^{N-2}$. There is a very systematized algorithm for computing the estimates of main effects and all interactions in the 2^N system, known as the *Yates algorithm.* Those interested can find it illustrated in Snedecor and Cochran (1989) or in John (1971). In the complete statistical analysis of the 2^N there will be N main effects, $\binom{N}{2}$ two-factor interactions, $\binom{N}{3}$ three-factor interactions, $\binom{N}{4}$ four-factor interactions, and so on.

Calculating Specific Interaction Sums of Squares

There is a direct, relatively simple way to calculate the sum of squares for any specific interaction in the 2^N system. Consider computing the sum of squares due to the BDE interaction in a complete replicate of a 2^N. First compute the estimate of BDE. One way to begin is to expand the symbolic expression $(a+1)$ $(b-1)(c+1)(d-1)(e-1)(f+1)\cdots$, collect the appropriate terms, and then divide by the necessary constant. Terms that contain none of the letters b, d, or e will of necessity have a "−" sign. Also any term that contains exactly one of b, d, or e will have a "+" sign. Similarly, any term that contains exactly two of d, b, or e will have a "−" sign. Finally, any term that contains all three will have a "+" sign. To summarize, terms that have an odd number of letters common with BDE will have a "+" sign and terms with an even number of letters common with BDE will have a "−" sign. (We use the convention throughout that zero is treated as an even number.) It follows that the sum of squares due to BDE can be computed as

$$\frac{(\text{sum all obs. with an even number of letters common with } BDE)^2}{r2^{N-1}}$$
$$+\frac{(\text{sum all obs. with an odd number of letters common with } BDE)^2}{r2^{N-1}}$$
$$-\frac{(\text{sum all obs.})^2}{r2^N}.$$

A similar argument can be constructed for an interaction with an even number of letters. The difference is that in the estimate the rule on signs is reversed. However, this has no effect on the final sum of squares formula rule.

Exercise 10.7: Hand Computation of Analysis of Variance for a 2^3 Factorial. Use the formulas developed to compute the estimates $\widehat{A}$, $\widehat{B}$, $\widehat{AB}$, etc. and complete the analysis of variance table for the synthetic data in Table 10.4. To present the data in a reasonably systematic manner, the data have been listed as observations in the columns replication 1, 2, and 3. However, in the analysis you are to assume that the data come from a completely randomized design with 24 observations and no block structure at all. □

Table 10.4 Analysis of Variance Table

Factor			Replication		
A	B	C	1	2	3
Low	Low	Low	59	39	15
High	Low	Low	99	56	77
Low	High	Low	86	36	39
High	High	Low	11	85	91
Low	Low	High	95	25	82
High	Low	High	54	06	88
Low	High	High	16	46	34
High	High	High	92	87	46

CHAPTER 11

Factorial Experiments: The 3^N System

11.1 INTRODUCTION

In this chapter we develop the concepts and notation for factorial plans with all factors at three levels. The development in this chapter is more general than that given in Chapter 10, in the sense that all concepts extend directly to the p^N system, where p is prime or a power of a prime. There are no special tricks that apply only in the 3^N case as there were in the 2^N case. We again encounter the problem of several sets of notation. However, there is the additional problem of defining a natural model. In the 2^N case there was only one natural model. Also, there is the additional question of spacing of levels. Unless stated otherwise, it will be assumed that levels are equally spaced.

11.2 3 x 3 FACTORIAL

For the treatment combinations we find three different systems of notation used. All are direct extensions from the two-level case. We illustrate:

A low	B low	$\Rightarrow$	a_1b_1	a_0b_0	00
A intermediate	B low	$\Rightarrow$	a_2b_1	a_1b_0	10
A high	B low	$\Rightarrow$	a_3b_1	a_2b_0	20
A low	B intermediate	$\Rightarrow$	a_1b_2	a_0b_1	01
A intermediate	B intermediate	$\Rightarrow$	a_2b_2	a_1b_1	11
A high	B intermediate	$\Rightarrow$	a_3b_2	a_2b_1	21
A low	B high	$\Rightarrow$	a_1b_3	a_0b_2	02
A intermediate	B high	$\Rightarrow$	a_2b_3	a_1b_2	12
A high	B high	$\Rightarrow$	a_3b_3	a_2b_2	22

three different systems of notation

Planning, Construction, and Statistical Analysis of Comparative Experiments,
by Francis G. Giesbrecht and Marcia L. Gumpertz
ISBN 0-471-21395-0

As in the 2^N system, the true responses are denoted by μ's, and the observed experimental responses are denoted by y's, with appropriate subscripts from one of the three systems.

In the 2^N system there was really only one natural way of defining the components of the linear model, i.e., of decomposing the main effects and interactions, while in the 3^N system there are two very natural types of decompositions. It turns out that one of these is natural for interpreting results of an experiment but does not lead to a nice system for confounding, while the second, which does not lend itself to good interpretations, does lead to a nice system for confounding and setting up fractional replications. We are forced to make a compromise.

11.2.1 Linear and Quadratic Effects Model

Consider a 3^2 experiment, two factors, each at three levels, with each factor at equally spaced intervals. The common analysis of variance with polynomial decomposition takes the form shown in Table 11.1.

The underlying model for this analysis is developed in stages. Consider first the linear effect of factor A. We can think of estimating the effect of factor A by looking at the change in response resulting from going from a_0 to a_1 and going from a_1 to a_2. Using the symbolic notation, the linear effect of factor A is defined as the average of $a_1 - a_0$ and $a_2 - a_1$, i.e.,

$$A_L = \frac{[(a_1 - a_0) + (a_2 - a_1)]}{2} = \frac{(a_2 - a_0)}{2}.$$

Here the symbol a_2 is interpreted as the mean of all responses to treatment combinations that involved a_2, and the symbol a_0 the mean of all responses to treatment combinations that involve a_0. The result is the contrast estimating the linear effect of factor A. Using the same symbolic notation, the quadratic effect of factor A is

Table 11.1 Analysis of Variance Table, 3 x 3 Factorial

Source	df	
A	2	linear quadratic
B	2	linear quadratic
$A \times B$	4	linear × linear linear × quadratic quadratic × linear quadratic × quadratic

defined as half of the difference between $a_1 - a_0$ and $a_2 - a_1$, i.e.,

$$A_Q = \frac{[(a_2 - a_1) - (a_1 - a_0)]}{2} = \frac{(a_2 - 2a_1 + a_0)}{2}.$$

This translates to the contrast estimating the quadratic effect of factor A. There are analogous definitions for the linear and quadratic effects of factor B. These are the standard definitions of the linear and quadratic effects usually referred to as A_L, A_Q, B_L, and B_Q.

This notation extends directly to the interaction contrasts. Since the A_L contrast can be measured with B at the low level, the intermediate level, and at the high level, we can inquire about changes in A_L as B changes. In particular, using the symbolic notation, the change in A_L as B shifts from b_0 to b_1 is $(a_2b_1 - a_0b_1)/2 - (a_2b_0 - a_0b_0)/2$ and as B shifts from b_1 to b_2 is $(a_2b_2 - a_0b_2)/2 - (a_2b_1 - a_0b_1)/2$. The average of the two is $\frac{1}{2}\big((a_2b_2 - a_0b_2)/2 - (a_2b_0 - a_0b_0)/2\big)$. This is the A_LB_L contrast, which is conveniently written as $(a_2 - a_0)(b_2 - b_0)/4$. The difference between the two A_L values is A_LB_Q, which is conveniently written as $(a_2 - a_0)(b_2 - 2b_1 + b_0)/4$. Analogous reasoning gives us A_QB_L and A_QB_Q. We symbolically define

$$\begin{aligned}
A_LB_L \quad \text{as} \quad & (a_2 - a_0)(b_2 - b_0)/4 \\
& = (a_2b_2 - a_2b_0 - a_0b_2 + a_0b_0)/4 \\
A_QB_L \quad \text{as} \quad & (a_2 - 2a_1 + a_0)(b_2 - b_0)/4 \\
& = (a_2b_2 - a_2b_0 - 2a_1b_2 + 2a_1b_0 \\
& \qquad + a_0b_2 - a_0b_0)/4. \\
A_LB_Q \quad \text{as} \quad & (a_2 - a_0)(b_2 - 2b_1 + b_0)/4 \\
& = (a_2b_2 - 2a_2b_1 + a_2b_0 - a_0b_2 \\
& \qquad + 2a_0b_1 - a_0b_0)/4 \\
A_QB_Q \quad \text{as} \quad & (a_2 - 2a_1 + a_0)(b_2 - 2b_1 + b_0)/4 \\
& = (a_2b_2 - 2a_2b_1 + a_2b_0 - 2a_1b_2 + 4a_1b_1 \\
& \qquad - 2a_1b_0 + a_0b_2 - 2a_0b_1 + a_0b_0)/4.
\end{aligned}$$

The four interaction contrasts are then estimated as

$$(y_{22} - y_{20} - y_{02} + y_{00})/4$$

$$(y_{22} - y_{20} - 2y_{12} + 2y_{10} + y_{02} - y_{00})/4$$

$$(y_{22} - 2y_{21} + y_{20} - y_{02} + 2y_{01} - y_{00})/4$$

$$(y_{22} - 2y_{21} + y_{20} - 2y_{12} + 4y_{11} - 2y_{10} + y_{02} - 2y_{01} + y_{00})/4.$$

This system of linear and quadratic effects extends directly to more factors, each at three levels. This system leads to quantities that are easy to interpret. All components have a nice meaning. Unfortunately, this does not lead to a "nice" scheme for further development of experimental plans. The alternative system that will be developed leads to a "nice" confounding scheme but unfortunately leads to a system of parameters that cannot in general be interpreted easily.

***Exercise 11.1: Linear and Quadratic Contrasts for* 3 x 3 *Factorial*.** Show by working from the symbolic definitions for the linear and quadratic effects of factor B that for the responses y_{00}, y_{10}, y_{20}, y_{01}, y_{11}, y_{21}, y_{02}, y_{21}, and y_{22}, the actual estimates of these contrasts are

$$\left[(y_{02} + y_{12} + y_{22}) - (y_{00} + y_{10} + y_{20})\right]/6$$

and

$$\left[(y_{02} + y_{12} + y_{22}) - 2(y_{01} + y_{11} + y_{21}) + (y_{00} + y_{10} + y_{20})\right]/6,$$

respectively. □

11.2.2 Alternative General Definition of Main Effects in 3^2

We begin with definitions of the main effects. These are the same as the general definitions in Section 3.4. We define

$$A_0 = (\text{mean of all trt. comb. with } a_0) - (\text{mean of all trt. comb.})$$

$$A_1 = (\text{mean of all trt. comb. with } a_1) - (\text{mean of all trt. comb.})$$

$$A_2 = (\text{mean of all trt. comb. with } a_2) - (\text{mean of all trt. comb.})$$

Notice that these three contrasts are not linearly independent. They sum to zero and represent two degrees of freedom. Now recall that SSA, the two-degree-of-freedom sum of squares due to A, is the sum of squares among

treatment combinations with a_0

treatment combinations with a_1

treatment combinations with a_2.

The sum of squares for the A effect is

$$\text{SS}A = ((A_0)^2 + (A_1)^2 + (A_2)^2) \times \text{the number of observations at each level.}$$

For hand computations it is much easier to compute this sum of squares as

$$\begin{aligned}&(\text{sum of all trt. comb. with } a_0)^2/(\text{no. obs. in sum})\\&\quad+(\text{sum of all trt. comb. with } a_1)^2/(\text{no. obs. in sum})\\&\quad+(\text{sum of all trt. comb. with } a_2)^2/(\text{no. obs. in sum})\\&\quad-(\text{sum of all trt. comb.})^2/(\text{no. obs. in sum}).\end{aligned}$$

SS B follows the same pattern. The definitions of other main effects for higher-order factorials and their associated sums of squares follow directly.

11.2.3 General Definition of the Interaction Effects

To develop the definitions of interaction components that will be useful for constructing fractional factorials (Chapter 13) and the associated sums of squares, we need a slightly more general definition or form to identify treatment combinations. Write the symbol for a general treatment combination as $a_{i_A}b_{i_B}$, where i_A and i_B take on values 0, 1, or 2. For example, a_1b_0 would be obtained by setting $i_A = 1$ and $i_B = 0$ in the 3^2 system. Alternatively, we could denote that treatment combination by the symbols 10. This turns out to be quite useful and is parallel to the 2^2 system.

Now return to the 3^2 case. Define

$$\begin{aligned}AB_0 &= [\text{mean of all treat. comb. with } i_A + i_B = 0 \text{ (modulo 3)}]\\&\quad-(\text{mean of all treat. comb.}) &(11.1)\\AB_1 &= [\text{mean of all treat. comb. with } i_A + i_B = 1 \text{ (modulo 3)}]\\&\quad-(\text{mean of all treat. comb.}) &(11.2)\\AB_2 &= [\text{mean of all treat. comb. with } i_A + i_B = 2 \text{ (modulo 3)}]\\&\quad-(\text{mean of all treat. comb.}) &(11.3)\end{aligned}$$

These three contrasts are also linearly dependent. They sum to zero, and consequently, are not orthogonal. They lead to a sum of squares with two degrees of freedom. This can be computed as

$$\text{SS}AB = \left[(AB_0)^2 + (AB_1)^2 + (AB_2)^2\right] \times \text{some coefficient.}$$

We do not need to worry about the coefficient, since this sum of squares is easily computed as

$$\begin{aligned}&[\text{sum of all trt. comb. with } i_A + i_B = 0 \text{ (modulo 3)}]^2/(\text{no. obs. in sum})\\&\quad+[\text{sum of all trt. comb. with } i_A + i_B = 1 \text{ (modulo 3)}]^2/(\text{no. obs. in sum})\end{aligned}$$

$$+ \left[\text{sum of all trt. comb. with } i_A + i_B = 2 \text{ (modulo 3)}\right]^2/(\text{no. obs. in sum})$$
$$- (\text{sum of all trt. comb.})^2/(\text{no. obs. in sum}).$$

This accounts for two of the four $A \times B$ interaction degrees of freedom. It is usually referred to as SS AB. Note the distinction between SS$A \times B$, which has four degrees of freedom, and SS AB, which has two degrees of freedom. There are still two degrees of freedom that need to be accounted for. These are normally referred to as SS AB^2. To define these we define another three components as

$$\begin{aligned}
AB_0^2 &= [\text{mean of all treat. comb. with } i_A + 2i_B = 0 \text{ (modulo 3)}] \\
&\quad -(\text{mean of all treat. comb.}) \\
AB_1^2 &= [\text{mean of all treat. comb. with } i_A + 2i_B = 1 \text{ (modulo 3)}] \\
&\quad -(\text{mean of all treat. comb.}) \\
AB_2^2 &= [\text{mean of all treat. comb. with } i_A + 2i_B = 2 \text{ (modulo 3)}] \\
&\quad -(\text{mean of all treat. comb.}).
\end{aligned}$$

These three contrasts (again linearly dependent and not orthogonal) lead to a sum of squares with two degrees of freedom. This can be computed as

$$\text{SS } AB^2 = \left[(AB_0^2)^2 + (AB_1^2)^2 + (AB_2^2)^2\right] \times \text{some coefficient}$$

or more easily as

$$\begin{aligned}
&[\text{sum of all trt. comb. with } i_A + 2i_B = 0 \text{ (modulo 3)}]^2/(\text{no. obs. in sum}) \\
&\quad + [\text{sum of all trt. comb. with } i_A + 2i_B = 1 \text{ (modulo 3)}]^2/(\text{no. obs. in sum}) \\
&\quad + [\text{sum of all trt. comb. with } i_A + 2i_B = 2 \text{ (modulo 3)}]^2/(\text{no. obs. in sum}) \\
&\quad - (\text{sum of all trt. comb.})^2/(\text{no. obs. in sum}).
\end{aligned}$$

This sum of squares is usually denoted by SS AB^2. The two statistics, SS AB and SS AB^2, are independently distributed under standard conditions. However, it is not possible to attach meaning to SS AB that is distinct from SS AB^2. They both represent part of the SS $A \times B$ interaction. We will find that their great appeal is that they can be sacrificed independently in confounding schemes. Confounding main effects and interactions is a technique for constructing incomplete block designs for factorial treatments and is the subject of Chapter 13.

Exercise 11.2: Orthogonality of Contrasts in 3 × 3 Interaction. Verify that any one of the three linear contrasts AB_0, AB_1, and AB_2 is orthogonal to any one (all) of AB_0^2, AB_1^2, and AB_2^2, thus verifying that SS AB and SS AB^2 are independent sums of squares under the usual assumptions of normal independent identically distributed errors. □

11.2.4 Extensions to the 3^3 System

The definitions from the 3^2 system extend directly. For the 3^3 system we have three pairs of two-factor interactions, SS AB and SS AB^2; SS AC and SS AC^2; and SS BC and SS BC^2. The three factor interaction splits up into

SS ABC based on ABC_0, ABC_1, and ABC_2 accounting for two df
SS ABC^2 based on ABC_0^2, ABC_1^2, and ABC_2^2 accounting for two df
SS AB^2C based on AB^2C_0, AB^2C_1, and AB^2C_2 accounting for two df
SS AB^2C^2 based on $AB^2C_0^2$, $AB^2C_1^2$, and $AB^2C_2^2$ accounting for two df

for the total of eight degrees of freedom in $A \times B \times C$. We repeat that there is no logical distinction or interpretation that can be applied to SS ABC as distinct from SS ABC^2, SS AB^2C, or SS AB^2C^2.

11.2.5 Linear Model: The Inverse Relationship

Just as in the 2^N system, these relationships can be inverted to give a linear model. For the 3^3 case we have

$$\begin{aligned} y_{i_A i_B i_C} = \mu &+ A_{i_A} + B_{i_B} + AB_{i_A+i_B} + AB^2_{i_A+2i_B} + C_{i_C} \\ &+ AC_{i_A+i_C} + AC^2_{i_A+2i_C} + BC_{i_B+i_C} \\ &+ BC^2_{i_B+2i_C} + ABC_{i_A+i_B+i_C} \\ &+ AB^2C_{i_A+2i_B+i_C} + ABC^2_{i_A+i_B+2i_C} \\ &+ AB^2C^2_{i_A+2i_B+2i_C} + e_{i_A i_B i_C}, \end{aligned} \tag{11.4}$$

where all subscript arithmetic is modulo 3. With this system we are free to confound any two-degree-of-freedom main effects or interactions, just as in the 2^N system.

Exercise 11.3: Notation for Components of Interactions in 3^N Factorials. At this stage there is the question: Why not A^2B and A^2B^2? The answer is that if one tries to define a SSA^2B following the pattern established, one simply gets SS AB^2 over again. Similarly, working from SSA^2B^2 yields SS AB over again. Demonstrate the truth of these two statements by working through the details for a 3^2 factorial. The nine treatment combinations are y_{00}, y_{01}, y_{02}, y_{10}, y_{11}, y_{12},

y_{20}, y_{21}, and y_{22}. The SS AB^2 is

$$\begin{aligned}
&[\text{sum of the three obs. for trt. comb. with } i_A + 2i_B = 0 \text{ (modulo 3)}]^2/3 \\
&\quad + [\text{sum of the three obs. for trt. comb. with } i_A + 2i_B = 1 \text{ (modulo 3)}]^2/3 \\
&\quad + [\text{sum of the three obs. for trt. comb. with } i_A + 2i_B = 2 \text{ (modulo 3)}]^2/3 \\
&\quad - (\text{sum of all nine observations})^2/9
\end{aligned}$$

Following the same pattern the SSA^2B would be

$$\begin{aligned}
&[\text{sum of the three obs. for trt. comb. with } 2i_A + i_B = 0 \text{ (modulo 3)}]^2/3 \\
&\quad + [\text{sum of the three obs. for trt. comb. with } 2i_A + i_B = 1 \text{ (modulo 3)}]^2/3 \\
&\quad + [\text{sum of the three obs. for trt. comb. with } 2i_A + i_B = 2 \text{ (modulo 3)}]^2/3 \\
&\quad - (\text{sum of all nine observations})^2/9
\end{aligned}$$

Show that these two expressions really give you exactly the same thing. A similar demonstration holds for SS AB and SSA^2B^2. The fact is that SS(AB^x) and SS$(AB^x)^2$ give exactly the same sum of squares, and the convention is to insist on unit power for the first letter that appears. □

11.3 GENERAL SYSTEM OF NOTATION FOR THE 3^N SYSTEM

The reader whose interest is mainly in applying the techniques presented in this book may wish to skip this section and simply accept statements as found.

11.3.1 Definitions of Effects and Interactions

The general notation for the 3^N factorial system has many similarities with the 2^N system. Again a treatment combination is represented by an N-tuple $t = i_A i_B \cdots i_N$. However, now the $\{i_X\}$ take on the values 0, 1, or 2. Again the true and observed responses are $\{\mu_t\}$ and $\{y_t\}$, respectively. The interactions are written as $A^{i_A} B^{i_B} \cdots N^{i_N}$ with the restriction that the first nonzero superscript must be 1. We will see that this restriction on the first superscript is simply a device to ensure uniqueness. If in some calculations an N-tuple with first nonzero term equal to two is generated, all elements in the N-tuple must be multiplied by 2 and reduced modulo 3 if necessary. The special case with all superscripts equal to zero is again replaced by μ.

The 3^N system differs from the 2^N system in that each effect or interaction consists of three contrasts or interaction components. These are identified by a final subscript on the term, which takes on value 0, 1, or 2. For example, the interaction $A^{i_A} B^{i_B} \cdots N^{i_N}$ consists of three contrasts, $A^{i_A} B^{i_B} \cdots N_0^{i_N}$, $A^{i_A} B^{i_B} \cdots N_1^{i_N}$, and $A^{i_A} B^{i_B} \cdots N_2^{i_N}$. The interaction component $A^{i_A^*} B^{i_B^*} \cdots N_k^{i_N^*}$ is defined as the mean of the 3^{N-1} responses with $\sum_X i_X i_X^* = k$ (modulo 3) minus the mean of all of the responses. Notice that the three components corresponding to an N-tuple $(i_A^* i_B^* \cdots i_N^*)$ sum to zero. The corresponding estimates (sums of squares) account for two degrees of freedom.

11.3.2 Generalized Interaction

In the 3^N system every pair of interactions (including main effects) has two generalized interactions. The two interactions $A^{i_A^*} B^{i_B^*} \cdots N^{i_N^*}$ and $A^{i_A^{**}} B^{i_B^{**}} \cdots N^{i_N^{**}}$ have two generalized interactions, $A^{i_A^+} B^{i_B^+} \cdots N^{i_N^+}$ and $A^{i_A^{++}} B^{i_B^{++}} \cdots N^{i_N^{++}}$, where $i_X^+ = i_X^* + i_X^{**}$ (modulo 3) and $i_X^{++} = i_X^* + 2i_X^{**}$ (modulo 3) for $X = A, B, \ldots, N$. The rationale for calling these the generalized interactions can be seen from the 3×3 array in Table 11.2, where each cell contains the 3^{N-1} responses defined by the row and column headings.

We can think of this as a two-way table with two degrees of freedom for rows, two degrees of freedom for columns, and four degrees of freedom for interaction. For convenience we think of the rows and columns numbered 0, 1, 2. The $A^{i_A^*} B^{i_B^*} \cdots N_k^{i_N^*}$ interaction components for $k = 0, 1, 2$ are deviations of the column means from the grand mean and $A^{i_A^{**}} B^{i_B^{**}} \cdots N_k^{i_N^{**}}$ interaction components for $k = 0, 1, 2$ are deviations of the row means from the grand mean. The $A^{i_A^+} B^{i_B^+} \cdots N_0^{i_N^+}$ interaction component is the mean of the elements in cells (0,0), (1,2), and (2,1) minus the grand mean, $A^{i_A^+} B^{i_B^+} \cdots N_1^{i_N^+}$ is the mean of the elements in cells (0,1), (1,0), and (2,2) minus the grand mean and $A^{i_A^+} B^{i_B^+} \cdots N_2^{i_N^+}$ is the mean of the elements in cells (0,2), (1,1), and (2,0) minus the grand mean. The $A^{i_A^{++}} B^{i_B^{++}} \cdots N_0^{i_N^{++}}$ interaction component is the mean of the elements in cells (0,0), (1,1), and (2,2) minus the grand mean, $A^{i_A^{++}} B^{i_B^{++}} \cdots N_1^{i_N^{++}}$ is the mean of the elements in cells (0,2), (1,0), and (2,1) minus the grand mean, and $A^{i_A^{++}} B^{i_B^{++}} \cdots N_2^{i_N^{++}}$ is the mean of the elements in cells (0,1), (1,2), and (2,0) minus the grand mean. One can think of the analogous sums of squares for the two factors in a 3×3 Graeco–Latin square.

Table 11.2 3 x 3 Array Defining Nine Sets of Responses

	Responses with $\sum i_X i_X^* = 0$ (modulo 3)	Responses with $\sum i_X i_X^* = 1$ (modulo 3)	Responses with $\sum i_X i_X^* = 3$ (modulo 3)
Responses with $\sum i_X i_X^{**} = 0$ (modulo 3)			
Responses with $\sum i_X i_X^{**} = 1$ (modulo 3)			
Responses with $\sum i_X i_X^{**} = 2$ (modulo 3)			

Table 11.3 Example of 3 x 3 Array in Table 11.2

	$i_A + i_B + i_C = 0$ (modulo 3)	$i_A + i_B + i_C = 1$ (modulo 3)	$i_A + i_B + i_C = 2$ (modulo 3)
$i_B + i_C = 0$ (modulo 3)	μ_{000}, μ_{012}, μ_{021}	μ_{100}, μ_{112}, μ_{121}	μ_{200}, μ_{212}, μ_{221}
$i_B + i_C = 1$ (modulo 3)	μ_{222}, μ_{210}, μ_{201}	μ_{022}, μ_{010}, μ_{001}	μ_{122}, μ_{110}, μ_{101}
$i_B + i_C = 2$ (modulo 3)	μ_{111}, μ_{120}, μ_{102}	μ_{211}, μ_{220}, μ_{202}	μ_{011}, μ_{020}, μ_{002}

It is interesting to notice that the two generalized interactions that make up the interaction in the 3×3 table do not need to be of the same order. In fact, it is quite possible for one of them to be a main effect. For example, the generalized interactions of AB^2C^2 and BC are A and ABC.

As a specific example, we can consider the generalized interaction of ABC and BC in the 3^3. The rule tells us that the generalized interaction consists of the two pieces, AB^2C^2 and A. To look at this in the 3×3 table, assign ABC to the columns and BC to the rows in Table 11.3.

Now we notice that the elements in cell 00, 11, and 22 have $i_A = 0$, elements in 01, 12, and 20 have $i_A = 1$, and elements in cells 02, 10, and 21 have $i_A = 2$. Hence A is part of the generalized interaction of ABC and BC. Similarly, we notice that $i_A{+}2i_B{+}2i_C = 0$ (modulo 3) in cells 00, 12, and 21, $i_A{+}2i_B{+}2i_C = 1$ (modulo 3) in cells 01, 10, and 22, and finally, $i_A + 2i_B + 2i_C = 3$ (modulo 3) in cells 02, 11, and 20. This tells us that AB^2C^2 is part of the interaction of ABC and BC.

The linear model illustrated in equation (11.4) extends directly to the 3^N case, but the notation is messy.

11.3.3 Orthogonality of Interaction Component Sets

The typical interaction $A^{i_A^*} B^{i_B^*} \cdots N^{i_N^*}$ is based on the three components

$$\left(\frac{1}{3}\right)^{N-1} \left[\sum \cdots \sum\right]^*_{(0)} \mu_{i_A \cdots i_N} - \text{mean of all responses} \quad (11.5)$$

$$\left(\frac{1}{3}\right)^{N-1} \left[\sum \cdots \sum\right]^*_{(1)} \mu_{i_A \cdots i_N} - \text{mean of all responses} \quad (11.6)$$

$$\left(\frac{1}{3}\right)^{N-1} \left[\sum \cdots \sum\right]^*_{(2)} \mu_{i_A \cdots i_N} - \text{mean of all responses,} \quad (11.7)$$

where the symbol $\left[\sum \cdots \sum\right]^{*}_{(k)}$ represents $\sum_{i_A} \cdots \sum_{i_N}$ subject to the constraint that

$$\sum_{X} i_X i_X^{*} = k \text{ (modulo 3).}$$

Clearly, these are dependent. They sum to zero.

Consider a second interaction, $A^{i_A^{\#}} B^{i_B^{\#}} \cdots N^{i_N^{\#}}$, with components

$$\left(\frac{1}{3}\right)^{N-1} \left[\sum \cdots \sum\right]^{*}_{(0)} \mu_{i_A \cdots i_N} - \text{mean of all responses} \tag{11.8}$$

$$\left(\frac{1}{3}\right)^{N-1} \left[\sum \cdots \sum\right]^{*}_{(1)} \mu_{i_A \cdots i_N} - \text{mean of all responses} \tag{11.9}$$

$$\left(\frac{1}{3}\right)^{N-1} \left[\sum \cdots \sum\right]^{*}_{(2)} \mu_{i_A \cdots i_N} - \text{mean of all responses,} \tag{11.10}$$

These again sum to zero. Now by orthogonality we mean that each of the contrasts (11.5), (11.6), or (11.7) is orthogonal to each and every one of the contrasts (11.8), (11.9), or (11.10). Establishing the validity of this is simply a matter of noting that there are 3^N responses in all, and that, for example, 3^{N-2} of the responses in the left-hand term in (11.5) are also in the left-hand term in each of (11.8), (11.9), and (11.10).

CHAPTER 12

Analysis of Experiments without Designed Error Terms

12.1 INTRODUCTION

In a sense this chapter is out of place, but it turns out that there is really no logical place for it. It is an answer to a problem that we confront in the analysis of unreplicated experiments, experiments that leave no measure of pure error. At this stage the reader has encountered factorial experiments. As the number of treatments increases, the number of treatment combinations and the number of potential parameters in the model both become very large. Replication to provide a measure of experimental error can make these experiments prohibitively large. Consequently, these experiments are often set up without replication. The general assumption is that many of the potential parameters (interactions) are so small that they can be treated as zero. The contrasts that estimate these parameters will then be used as measures of error. The problem is that the exact identity of these, and conversely, the large interactions is unknown. We present several approaches.

In addition, there are cases where the investigator feels that some factors in the experiment affect the variance; i.e., the residual variance is larger for one level of a treatment than at another level. This problem is most frequently encountered in quality control work. In this chapter we present techniques to examine the effects of factors on variances, even though designs specifically constructed to counter this problem have not yet been discussed. The reason for including this topic here is that in a sense the statistical techniques used in the analysis are related. The first-time reader may wish to skip the latter parts of this chapter and return to them later as the need arises.

All of our analyses in this chapter assume as a base the classical linear model and look for evidence of deviation from that model. The first section deals with location parameters and the second with deviations from homoscedasticity.

Planning, Construction, and Statistical Analysis of Comparative Experiments,
by Francis G. Giesbrecht and Marcia L. Gumpertz
ISBN 0-471-21395-0

12.2 TECHNIQUES THAT LOOK FOR LOCATION PARAMETERS

The first place where the problem of no clean estimate of experimental error is encountered is large factorial experiments with no replication. A common practice, which often tends to be too conservative, is to pool some of the higher-order interactions to form an error term. This is done under the assumption that the corresponding interactions do not exist (are negligible) and that the estimates are simply measures of error. This tends to be conservative in that the error is overestimated if the assumption of negligible interactions is not satisfied. Fortunately, there are alternatives. We present a number with reasonable statistical properties. Good reviews of this and related material are found in Haaland and O'Connell (1995), Hamada and Balakrishnan (1998), and Janky (2000).

12.2.1 Daniel's Half-Normal Plot

We illustrate the computations involved in the construction and interpretation of the half-normal plot (Daniel 1959) with some simulated data for a 2^4 factorial. These data are also available on the companion Web site. The 16 observations given in Table 12.1 were constructed from a model with $\mu = 1000$, $B = -100$, $C = -200$, $D = -300$, $CD = 200$, and errors normally distributed with mean zero and standard deviation 100. The 15 contrasts (estimates of main effects and interactions) are shown in Table 12.2. The procedure is to plot the sorted absolute values of the estimates against the quantiles from the folded normal distribution. If one has n contrasts that are to be plotted, the n quantiles $\{q_i\}$ can be computed as solutions to

$$\frac{2n + 2i - 1}{4n} = \int_{-\infty}^{q_i} N(0, 1)\, dx$$

for $i = 1, \ldots, n$. In SAS® one can compute $q_i = \texttt{PROBIT}((2*n+2*i-1)/(4*n))$. Then draw a straight line through these points anchored at (0,0).

The actual half-normal plot is shown in Figure 12.1. In the original Daniel (1959) paper the absolute contrasts were plotted as the abscissa and the order number of the contrast as the ordinate on normal probability paper. We follow the more common practice to plot the contrasts as the ordinate and the quantiles

Table 12.1 Simulated Data for a 2^4 System

y_{0000}	1535	y_{1000}	1393
y_{0001}	924	y_{1001}	916
y_{0010}	1044	y_{1010}	938
y_{0011}	941	y_{1011}	736
y_{0100}	1332	y_{1100}	1500
y_{0101}	861	y_{1101}	796
y_{0110}	948	y_{1110}	959
y_{0111}	720	y_{1111}	796

Table 12.2 Estimates Computed from Data in Table 12.1

		D	−369.875
A	−33.875	AD	−16.625
B	−64.375	BD	−21.625
AB	81.375	ABD	−25.375
C	−271.875	CD	195.875
AC	−22.125	ACD	8.125
BC	5.375	BCD	.125
ABC	18.125	$ABCD$	66.375

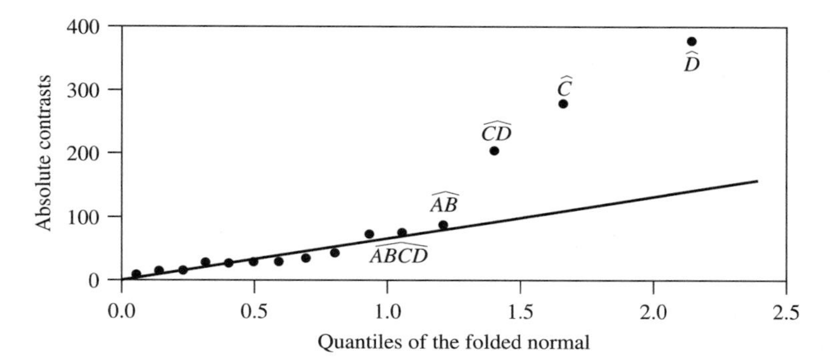

Figure 12.1 Half-normal plot of contrasts in Table 12.2.

of the folded normal as the abscissa on conventional linear×linear graph paper. Notice that all but the three contrasts, $\widehat{D}$, $\widehat{CD}$, and $\widehat{C}$, fall in a straight line. Except for the contrast $\widehat{B}$, we are looking at exactly the parameters that we know are nonzero. The value for B is exactly one standard deviation and our ability to detect it with such a small study is marginal. In this case we have failed.

The theory behind the half-normal plot is that under the null hypothesis there are no effects; the 15 contrasts are all independent normal with mean zero and variance $\sigma^2/4$ [eq. (10.5)]. Plotting these on normal probability paper or against the equally spaced quantiles on regular graph paper should give a straight line. In our case, we are dealing with contrasts, and the sign on contrasts is arbitrary. For example, in the 2^N reversing the designations of high and low for a factor changes the sign of the contrast. Changing signs on some values changes the appearance of the plot. The way around this is to plot absolute values. We still get the straight line if the absolute values are plotted on folded normal probability paper or against the quantiles of the half normal distribution on regular paper. Half normal plots are produced by design of experiments software such as SAS® `Proc ADX`. Daniel's half-normal plot tends to work best if only a relatively small number of the contrasts measure something other than zero.

Figure 12.1 shows a diagonal that is used to judge whether points fall along a line or not. We have fitted by eye. The line is anchored at the (0, 0) point and follows the points until they obviously deviate from the line. This is clearly subjective. There exist a number of objective procedures for obtaining such a line. However, even then there is still subjectivity left with respect to judging whether or not a point is on the line. The slope of this line can be used as an estimate of σ. The subjectivity of the method can be mitigated somewhat by dividing each of the absolute contrasts by the absolute contrast ranked most closely to $.683n$ (n is the number of contrasts) before plotting. The line is then drawn through points (0, 0) and (1.0, 1.0). This makes the tacit assumption that fewer than 25% of the contrasts measure something other than pure error.

The lack of objectivity is the standard criticism of the method. However, the method is fast, easy, and tends to be quite informative. The reader interested in more detail is urged to consult the Daniel paper and Zahn (1975a,b).

12.2.2 Lenth's Technique

While the pictorial nature of the half-normal plot of the contrasts has much to offer, some find the subjective aspects of the procedure unacceptable. The plotted data points will not lie on a perfectly straight line even when there are no effects and the contrasts are pure white noise. Deciding when a point is sufficiently far off a straight line to signal a real effect is clearly subjective. An alternative procedure that avoids these difficulties and has good statistical properties is given by Lenth (1989).

Following Lenth, we let $\widehat{c_1}, \widehat{c_2}, \ldots, \widehat{c_n}$ represent n estimated contrasts of interest. Assume that $\widehat{c_i}$ is approximately normally distributed with mean c_i and variance σ^2 for $i = 1, \ldots, n$. Under H_0 the $\{c_i\}$ are all zero. Now define

$$s_0 = 1.5 \times \operatorname*{median}_{i} |\widehat{c_i}| \tag{12.1}$$

$$s_1 = 1.5 \times \operatorname*{median}_{|\widehat{c_i}|<2.5s_0} |\widehat{c_i}|. \tag{12.2}$$

Lenth refers to s_1 as the *pseudostandard error* (*PSE*). Notice the similarity between expressions (12.1) and (12.2). The only difference is that the median in (12.2) is over a reduced set. The idea is to use a two-step process. One first calculates an initial robust estimate of scale, s_0, from all of the contrasts and then uses it to find an improved robust estimate of scale, s_1, based on a selected subset of contrasts.

Next define what Lenth calls the *margin of error*, *ME*, as

$$ME = t_{.975,df} \times s_1,$$

where $t_{.975,df}$ is the .975th quantile of a t-distribution with df degrees of freedom. Lenth shows that a good choice for df is $n/3$. The ME is used to test the $\widehat{c_i}$ or to set confidence intervals, i.e., $\widehat{c_i} \pm ME$.

Notice that we are doing n simultaneous tests. Since the contrasts are orthogonal, one can adjust for that by computing a *simultaneous margin of error*, *SME*, as

$$SME = t_{\gamma,df} \times s_1,$$

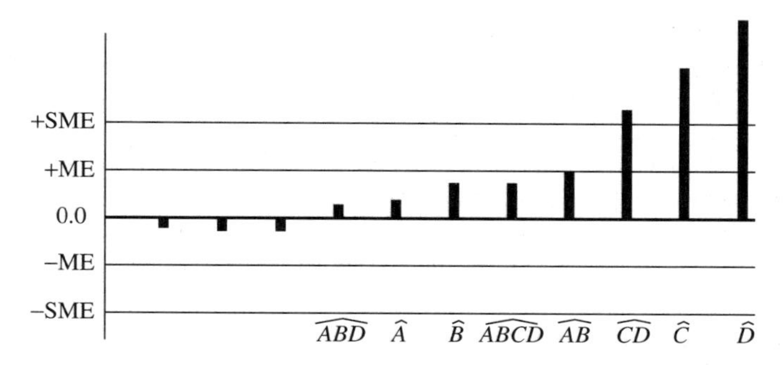

Figure 12.2 Display of Lenth's analysis, estimates, and tests for contrasts in Table 12.2.

where

$$\gamma = (1 + .95^{1/n})/2.$$

Notice that one typically has to evaluate quantiles for a t-distribution with fractional degrees of freedom. This can be done quite easily in SAS® by using the function TINV(p, df, 0).

For the 2^4 example in Table 12.1 (estimated contrasts in Table 12.2) the median absolute contrast is 25.375 and s_0 is 38.06. The median of the restricted set is between 22.125 and 21.625. We take the mean, giving us the *PSE* as 32.81. Since n was 15, df is 5 and, the ME is $2.571 \times 32.81 = 84.35$. Also, γ is $.5(1 + .95^{1/15}) = .9983$ giving a t value of 5.219 and an *SME* of 171.24. Using the *SME* identifies the same three contrasts, C, D, and CD, as active. The results are displayed in Figure 12.2.

12.2.3 Dong's Modification

Dong (1993) modified the Lenth procedure slightly. In place of the trimmed median procedure in (12.2), he used a trimmed mean procedure. Dong computed

$$s_1^2 = \sum_{|\widehat{c_i}| \le 2.5 s_0} \widehat{c_i^2}/k,$$

where

$$k = \text{number of } |\widehat{c_i}| \le 2.5 s_0$$

with s_0 computed as in (12.1). He then computed a margin of error, $ME = t_{.975,k} \times s_1$ to test the c_i or to set 95% confidence intervals. For the simultaneous margin of error Dong used $t_{\gamma,k} s_1$ with $\gamma = (1 + .98^{1/n})/2$. His simulation studies show that this modification results in error rates somewhat closer to the $\alpha = .05$ than the original Lenth suggestion. For the example in Table 12.1 we find that $s_1 = 39.39$, $ME = 85.83$, and $SME = 163.48$. These results can be displayed graphically as in Figure 12.2.

A recent simulation study by Haaland and O'Connell (1995) indicates that both methods have good power and reasonable control of error rate under rather wide conditions. Their simulations indicate that the Lenth procedure tends to perform a little better when there are relatively few nonzero effects as in early exploratory work and the Dong procedure better when there are more nonzero effects such as one would encounter in a study after most nonactive factors have been eliminated and the study is focused more on elucidating interactions. The differences are not major. Little is known about relative robustness to nonnormality.

12.2.4 Gamma Plot

Occasionally, one encounters data where there is no obvious objective scheme for constructing single-degree-of-freedom contrasts. An example would be data from a main-effects plan with all factors at three levels. If the factors are all quantitative, one can compute linear and quadratic contrasts for each factor and proceed to one of the methods we have presented. The only point of concern is that all contrasts be scaled to have common variance under the assumed hypothesis of no effects.

However, when factors are qualitative, there is no obvious system of single-degree-of-freedom contrasts. We can, however, follow Wilk and Gnanadesikan (1968) and construct a gamma plot. Assume that we have a set of n sums of squares, $\{SS_i\}$, each with k degrees of freedom. Under the null hypothesis, all of the corresponding mean squares estimate σ^2, and all sums of squares have a scaled k-degree-of-freedom χ^2 distribution. The procedure is to plot the $\{SS_i\}$ against the order statistics $\{q_i\}$ for the gamma distribution, defined as solutions to

$$\frac{2i-1}{2n} = \int_0^{q_i} \text{gamma with } k \text{ df } dx.$$

In SAS® these order statistics can easily be computed as $q_i = \texttt{GAMINV}((2i - 1)/2n, 0)$. Just as in the half-normal plot, the points should fall on a straight line anchored at (0, 0) if all sums of squares have a common scaled χ^2 distribution. Constructing the line and deciding when a point deviates from the line are subjective. However, the simplicity of the method contributes to its appeal.

12.3 ANALYSIS FOR DISPERSION EFFECTS

With the growing awareness that reduction in variance in a product is a key component in improving quality, there is a need for statistical methods for identifying factors that may be responsible for changes in amount of variation. Analysis for *dispersion effects* addresses the question of whether different factors affect the *variance* of the response. Analyses discussed up to this point study whether the *location* (i.e., *the mean*) is affected by the various factors.

Before proceeding with details we point out that looking for dispersion effects using data from screening experiments is an inherently difficult problem. In part the problem is difficult because it is almost always confounded with a lack of sufficient observations. The fact that some factors may also affect location also contributes to making the problem more difficult. In general, estimating a second-order moment, a variance, requires more data than estimating a location parameter. We also warn the reader that this is currently an area of active research and that recommendations will be sure to change. At best the results of the analyses suggested here should be treated by the investigator as indications of factors to be aware of and as subjects for future investigations. We present two suggested methods.

12.3.1 Box–Meyer Analysis

Box and Meyer (1986) address this problem for the case where the data consist of observations from a single replicate of fraction of a 2^N factorial. The base model used assumes that in place of homogeneous variance, we have

$$\mathrm{Var}[y_i] = \sigma_i^2 \tag{12.3}$$

for $i = 1, 2, \ldots, n$, where n is the number of observations. In addition, define

$$\sigma_{X+}^2 = \frac{2}{n} \sum_{X+} \sigma_i^2,$$

where the notation $\sum_{X+}$ means that the sum is over those variances for responses that enter contrast X with a positive coefficient. There is a corresponding definition for σ_{X-}^2. Box and Meyer define a pair of these for each of the $2^N - 1$ treatment contrasts.

It follows directly from the result in Exercise 12.1 that even with the nonhomogeneous error model (12.3), the error contrasts have common variance, and consequently, the half-normal plot and the Lenth–Dong techniques are valid.

Box and Meyer assume that the factors (including interactions) that affect the location part of the model constitute a small set and have all been identified. The remaining contrasts are available to estimate error. The next step in the analysis after this identification is to fit a model containing these effects and compute residuals. Next split the residuals into subsets based on the levels of the factors. Begin with factor A. Form two subsets, one with A high and the other with A low. From these subsets compute S_{A+}^2 and S_{A-}^2 as the (uncorrected) sums of squares of the residuals. Finally, compute $bm_A = \ln(S_{A+}^2/S_{A-}^2)$. Next, split on factor B and compute bm_B, and so on. Under the null hypothesis of homogeneous errors, the $bm_A, bm_B, \ldots$ can be interpreted as realizations of random variables that are approximately normally distributed. It is then a matter of examining these quantities to see if any differ sufficiently from the remainder to lead one to reject a homogeneity of errors assumption. We prefer to work with absolute values of these statistics, because of the arbitrariness introduced by the labeling of factor levels. A half-normal plot of the absolute values is a reasonable tool for examining these statistics. A criticism of this analysis that should readily be apparent to the reader is that sums of squares can be unduly inflated if the original analysis failed to remove all factors that affected the location.

To illustrate the mechanics of this dispersion effects analysis, we return to the data in Table 12.1. We recall that these data were generated with a homogeneous error model, and our analysis should show no variations in dispersion attributable to treatment factors. The first step in our analysis is to perform the Lenth analysis to check for location effects. As we recall, this analysis suggested that effects C, D, and CD were real. Consequently, the next step in the analysis is to fit a reduced model with only a mean and these three parameters, and compute residuals. The residuals are then split into two sets, those corresponding to A at the high level and those corresponding to A at the low level. Compute (uncorrected) sums of squares of the values in these sets. These are S^2_{A+} and S^2_{A-}. Finally, compute bm_A as the natural logarithm of the ratio of the larger to the smaller. These calculations are repeated, splitting the residuals on the basis of B high and B low, C high and C low, and D high and D low. The results for our data set are 1.078, .481, .012, and .197, respectively. There are no extreme values. If these are sorted and plotted against the quantiles from the folded normal, i.e., .157, .489, .887, and 1.534 (half-normal plot), there is no obvious deviation from a straight line. This is, as we expected, no reason to suspect non-homogeneity. We must bear in mind, however, that we are dealing with a very small example, and consequently, the method has little power.

Exercise 12.1: Variances of Contrasts. Show that the variance of any treatment contrast can be written as

$$n^{-1}\sum_{i=1}^{n}\sigma_i^2 = (\sigma_{X+}^2 + \sigma_{X-}^2)/2.$$

Note that this can be interpreted as σ_0^2, the mean of the $\{\sigma_i^2\}$. □

12.3.2 Bergman–Hynén Analysis

Bergman and Hynén (1997) present a slightly different analysis. They also assume that the factors (including interactions) that affect the location part of the model constitute a small set and have all been identified. They also proceed column by column through the model. They split the data into two subsets on the basis of the column for A. After the split, they fit the model containing the factors originally judged to be important and compute error sums of squares for the two subsets separately. Call the ratio of the two sums of squares F_A. Under the null hypothesis both mean squares estimate σ^2 and F_A will be a realization of a random variable from the F-distribution with appropriate degrees of freedom. One can then compute the significance level of this F statistic, i.e., a probability value. For the split on column A, call this value p_A.

This procedure is repeated with splits based on the column for B, C, and so on. The result is a set of values, $p_A, p_B, \ldots$. Notice that not all of the F statistics have the same number of degrees of freedom. If the split is on one of the columns corresponding to a factor being fitted, the degrees of freedom left for error will be larger.

Generally, it is adequate to examine the F or, equivalently, the p values for statistical significance. However, since under the null hypothesis, all of the p-values should have a uniform (0, 1) distribution, one can proceed a little further. There are m of these where m is the number of factors of fewer depending on the fraction in the design and they can be examined as a group by plotting the sorted values against $1/2m, 3/2m, \ldots, (2m-1)/2m$. Under the null hypothesis, these points should fall roughly on a straight line on the main diagonal.

For a numerical example we return to our simulated data set. We again proceed as though μ, C, D, and CD are contributors to location and must be accounted for in the model. Splitting the data into two subsets on the basis of level of A and fitting a model with μ, C, D, and CD gives F_A with four and four degrees of freedom. The ratio is 3.45. The probability $F_{4,4} \geq 3.45 = .129$ under the null hypothesis. This is p_A. We continue with a split on B and obtain $F_B = 1.92$, again with four and four degrees of freedom. We have $p_B = .272$. When we split on C, we find that C and CD drop out of the model. The result is that F_C has six and six degrees of freedom. We find that $F_C = 1.01$ and $p_C = .497$. Similarly, $F_D = 1.22$ has six and six degrees of freedom. We have $p_D = .426$. There is no reason to suspect any lack of homogeneity because they fall close to a line that goes through (0, 0).

Occasionally, we see both the Box and Meyer and Bergman and Hynén analyses extended to splits on interaction columns. In the Bergman and Hynén analysis in particular, this can be misleading. It can be shown that if the analysis indicates that two effects (or interactions) contribute to differences in dispersion, the method will usually also flag the generalized interaction of the two effects as affecting dispersion. McGrath and Lin (2001) give a modification of the Bergman and Hynén method that does not suffer from this problem.

12.3.3 Comments on Analysis of Dispersion Effects

To summarize, we reemphasize that this is an area of current research, and it is quite possible that the recommendations for analysts will evolve. It appears that one of the major problems is the potential contamination of the tests by unaccounted factors affecting location. Pan (1999) points out that the simple strategy of including or excluding borderline location effects does not significantly improve the performance of the dispersion methods.

12.3.4 Techniques Suitable for Data from Unreplicated 3^N Factorials

In quality design projects it is not unusual to use highly fractionated plans with factors at three levels. The Box and Meyer method can be extended to factors at three levels if we think of the ratios, say $bm_A = |\ln(S^2_{A+}/S^2_{A-})|$ as $bm_A =$

$|\ln(S^2_{A+}) - \ln(S^2_{A-})|$, i.e., as a contrast between two $\ln(S^2)$ values. If we are examining factors at three levels, we split the residuals into three subsets and compute $\ln(S^2_{A1})$, $\ln(S^2_{A2})$, and $\ln(S^2_{A3})$. Finally, compute $SS(A)$ from the three values. Do this for all of the factors at three levels. The full set of these sums of squares can then be examined in a gamma plot (Section 12.2.4).

REVIEW EXERCISES

12.1 In the process of recycling paper products, deinking the paper requires use of enzymes. Commercially prepared enzymes are expensive, contributing to the cost of recycling paper. Jackson et al. (1996) performed a series of experiments to determine if enzyme activity can be recovered using just a low level of the enzymes cellulase or xylanase. The results of some of these experiments are reproduced below. Four measures of enzyme activity were reported: Y1 = endoglucanase activity, Y2 = exoglucanase activity, Y3 = filter paper activity, and Y4 = xylanase activity. The effects of five factors were considered: A = enzyme charge (.2% on fiber, 2% on fiber); B = NaOH (.001 *M*, .01 *M*); C = incubation temperature (4°C, 25°C); D = addition of the surfactant Tween 80 (0, .005%); and E = incubation time (5 minutes, 15 minutes). The data are shown in Table 12.3 and are available on the companion Web site.

Table 12.3 Enzyme Activity Stability in Alkali-Retained Activity (% of Applied)

	Cellulase Preparation Activities			Xylanase
Treatment	Endoglucanase	Exoglucanase	Filter Paper	Activity
(1)	95	97	80	81
a	100	100	98	105
b	78	65	79	82
ab	97	49	56	69
c	93	61	91	76
ac	100	88	89	97
bc	93	57	61	60
abc	94	37	56	52
d	129	187	101	105
ad	99	107	90	110
bd	115	155	78	94
abd	100	60	68	69
cd	115	167	75	103
acd	103	97	73	90
bcd	113	110	78	77
abcd	95	41	69	70
e	98	90	78	89
ae	103	104	90	71
be	102	57	64	86

Table 12.3 ***(continued)***

Treatment	Cellulase Preparation Activities			Xylanase
	Endoglucanase	Exoglucanase	Filter Paper	Activity
abe	97	45	56	89
ce	84	65	58	86
ace	98	73	81	94
bce	91	55	75	52
abce	89	29	49	46
de	120	185	98	100
ade	102	108	91	112
bde	113	125	77	89
abde	97	50	62	86
cde	118	138	91	105
acde	98	80	82	104
bcde	109	96	76	54
abcde	91	32	55	47

(a) Check the xylanase activity data for variance homogeneity using residual plots from an analysis of variance fitting terms up to the two-factor interactions, and transform the data if necessary.

(b) Estimate the main effects and two-factor interactions and their standard errors for xylanase activity (in the transformed scale if so indicated).

(c) Draw a half- normal probability plot for (transformed) xylanase activity and determine which factors and interactions appear to be active.

(d) Use the Lenth–Dong method to determine which effects are active for xylanase activity at the 5% level.

(e) Summarize your findings about which combinations of these five factors would be optimal for recovery of xylanase activity.

CHAPTER 13

Confounding Effects with Blocks

13.1 INTRODUCTION

In Chapter 10 we introduced factorial experiments with all factors at two levels, i.e., the 2^N system, and in Chapter 11, all factors at three levels, i.e., the 3^N system. We will eventually introduce the p^N system with all factors at p levels. So far we have always made the assumption that the experiment was to be conducted in a completely randomized design or in randomized complete block designs with the blocks large enough to accommodate complete replicates. The object of this chapter is to introduce techniques for assigning parts of replicates to smaller blocks, with the assignment based on the factorial treatment structure. We look at ways to keep blocks small and homogeneous while retaining the desirable features and efficiencies of large factorial experiments. The emphasis is on the fundamentals of confounding. Some more subtle and complex issues that may become important in large, very expensive experiments are discussed in later chapters.

Example 13.1. Consider a hypothetical example using young rats. The scientist wants to study three factors, each at two levels, i.e., a 2^3 factorial with young male rats as experimental units. A number of litters of rats, all born roughly, but not exactly, the same day are available. Suppose that most litters have eight to 10 rats, a mix of males and females. One possibility is to select sets of eight male rats that are as alike as possible, but probably not all from the same litter, and assign the treatment combinations to the rats. An alternative would be to consider sets of four male rats, taken from those litters with four or more males. These sets should be quite uniform. Not only are they all males, they have the same parents, are the same age, but they also shared a common prenatal environment.

Planning, Construction, and Statistical Analysis of Comparative Experiments,
by Francis G. Giesbrecht and Marcia L. Gumpertz
ISBN 0-471-21395-0

The basic idea of this chapter is to assign treatment combinations to blocks in such a way that contrasts of most interest can be estimated from within-block information while sacrificing estimates of effects of lesser importance. In a factorial experiment we are often most interested in main effects and two- or three-factor interactions. Experience shows that the higher-order interactions are often much smaller in magnitude than the main effects. This is fortunate since they also tend to be much more difficult to interpret. We must, however, keep in mind, that this may not hold for any particular application. It is also worth repeating that presence or absence of interactions is not a function of the experimental plan but is a function of the scientific problem under investigation. The existence of interactions is not a shortcoming of factorial experiments.

Usually, estimating the main effects, two-factor interactions, and possibly three-factor interactions is a good starting place in a research program. The strategy, then, is to assign treatment combinations to blocks in such a way that main effects and lower-order interactions can be estimated from the treatments combinations within a block and to sacrifice higher-order interactions to contrasts that include block effects. If a treatment contrast also involves a contrast among block effects, the two contrasts are said to be *confounded*. This confounding means that the effects cannot be separated. For those more comfortable thinking in terms of the design matrix $\boldsymbol{X}$, the columns corresponding to the confounded contrasts are identical or differ only by a multiplicative constant. We exploit this idea of confounding in designing incomplete block experiments for factorial treatments by confounding the block contrasts with high-order interactions. In this way we decrease the size of blocks, and consequently, the within-block variance, and only sacrifice information on high-order (limited-interest) interactions.

13.2 CONFOUNDING 2^3 FACTORIALS

13.2.1 Two Blocks of Four

Simple Confounding

While the idea of confounding can be applied to experiments involving factors with any number of levels, the details are most easily and elegantly illustrated for 2^N factorials. To demonstrate the ideas without excessive notation, consider a 2^3 factorial experiment, an experiment involving eight treatment combinations. Assume that while the experimenter has only eight experimental units, there is some reason to believe that the units can be put into two blocks of four experimental units each, in such a manner that the variability among units within blocks is much smaller than the variability among the eight units as a set. As a concrete example, we can consider an experiment that uses individual animals as experimental units. It is reasonable to think that it will be possible to organize them into sets of four that are more uniform within sets than sets of eight. The groups may consist of litter mates, groups of four males and four females. In another example, the experimental units could be sets of four that fit into an oven at one time. A third example could be one to test design variables in the manufacture of tires for automobiles. Now the block could be a set of four tires on one automobile.

The eight treatment combinations must be split into two groups of four. One way of doing this is to divide them on the basis of the three-factor interaction. The three-factor interaction *ABC* corresponds to the contrast obtained by expanding $(\frac{1}{2})^2(a-1)(b-1)(c-1)$. The scientist conducts the experiment by placing those treatment combinations that enter into this contrast with a "+" sign into one block of four experimental units and the remaining four into the other block. The resulting plan can be illustrated as

Block I	*abc*	*a*	*b*	*c*
Block II	*ab*	*ac*	*bc*	(1)

In practice, rather than expand $(a-1)(b-1)(c-1)$, it may be simpler to look at the even–odd relationships of letters common to the interaction selected, ABC in this case, and the symbols for the treatment combinations, (1), $a, b, ab, \ldots, abc$. Treatments with an odd number of symbols in ABC are put in one block and those with an even number are put in the other block. Note that we follow the convention of always treating zero as an even number.

When the results of this experiment are analyzed, we see the following: The contrast between sets, the block contrast, accounting for one degree of freedom in the analysis of variance is

$$\frac{y_{abc}+y_a+y_b+y_c}{4}-\frac{y_{ab}+y_{ac}+y_{bc}+y_{(1)}}{4}.$$

The A contrast, accounting for one degree of freedom in the analysis of variance, is

$$\widehat{A}=\frac{y_{abc}+y_a+y_{ab}+y_{ac}}{4}-\frac{y_b+y_c+y_{bc}+y_{(1)}}{4}.$$

Note that of the four responses y_{abc}, y_a, y_{ab}, and y_{ac}, two come from block I and the other two come from block II. Similarly, of the four responses y_b, y_c, y_{bc}, and $y_{(1)}$, two from block I and two come from block II. If there should be a difference in response between the two sets or blocks, an additive effect, i.e., one block better than the other, this effect will not appear in the A contrast. In fact, this contrast is orthogonal to the contrast between blocks. The same argument can be made for the B, C, AB, AC, and BC contrasts. However, the ABC contrast,

$$\widehat{ABC}=\frac{y_{abc}+y_a+y_b+y_c}{4}-\frac{y_{ab}+y_{ac}+y_{bc}+y_{(1)}}{4},$$

is different. This contrast is identical to the block contrast in the sense that the four observations y_{abc}, y_a, y_b, and y_c come from the first block and the remaining four from the second block. We say that the two contrasts, ABC and the block contrast, are confounded with one another. The ABC interaction and the effect due to blocks cannot be separated using the data from this experiment. The six contrasts which are not confounded with blocks are said to be *estimable*. They can be estimated free of block effects. We can summarize this as a good news–bad news situation. The good news is that since the experimental units should be more uniform within blocks of size 4, this should result in smaller variances for the six estimable contrasts, while the bad news is that all information on the ABC interaction is lost.

It is also informative to consider what happens to degrees of freedom. In one complete replication of the 2^3 there are eight observations, yielding seven degrees of freedom for seven independent contrasts. If the observations are grouped into two blocks of four, one of these degrees of freedom goes with the difference between the two blocks and six remain. These six go with six linearly independent contrasts, the six estimable functions.

The choice of ABC as the interaction to be confounded with blocks was arbitrary in the sense that any one of the seven contrasts could have been selected. The theory holds just as well if any other contrast is selected for sacrifice. In practice, an experimenter would probably select ABC as the first interaction to give up, but this need not always be true. There may be good reason to feel that one of the other interactions or possibly one of the main effects is of less interest than ABC. Occasionally, these other considerations dictate an alternative choice.

This example can be extended to the 2^N in two blocks of 2^{N-1} for $N > 3$. Also, no new concepts or difficulties come up if the blocks are replicated.

Exercise 13.1: Contrasts. For the simple confounding example with two blocks of four

(a) Display the contrast between blocks I and II.
(b) Display the $\widehat{AB}$ contrast.
(c) Show that the block contrast and the $\widehat{AB}$ are orthogonal. □

Exercise 13.2: Variance of Estimated Effects in 2^3 in Blocks of Size 4. Verify that the $\widehat{A}$ contrast in the 2^3 with ABC confounded with blocks has variance $\sigma^2/2$. How does the variance of $\widehat{A}$ in this experiment compare with the variance of $\widehat{A}$ that would have been relevant if the experiment had not been set up in two blocks of four but in one block of eight instead? Use the notation $\sigma^2_{(4)}$ and $\sigma^2_{(8)}$ to refer to the error variance within blocks of size 4 and 8, respectively. □

Exercise 13.3: Assignment of Treatments. Illustrate the treatment assignment if $\widehat{BC}$ were to be confounded with blocks. □

Analysis of Variance, Blocks Fixed

The analysis of variance for an r replicate design in blocks of four of the 2^3 with ABC confounded with blocks is illustrated in Table 13.1. The sums of squares are computed as in Section 10.4.1. Fitting these using general purpose software for linear models introduces no new problems.

Table 13.1 Analysis of Variance Table, Blocks Fixed

Source	df
Mean	1
Replications	$r - 1$
Blocks (within replications)	r
A	1
B	1
AB	1
C	1
AC	1
BC	1
Error	$6(r - 1)$
Total	$8r$

Analysis of Variance, Blocks Random

If it is reasonable to assume that blocks within replicates are random, and sets of treatment combinations are assigned at random to the blocks, it is reasonable to partition the sum of squares for blocks (within reps.) into two parts, one degree-of-freedom piece due to ABC and the remaining part with $r - 1$ degrees of freedom. These can be used to construct a test of ABC. This test will have reduced power because the "experimental unit" for the ABC interaction is the block and there is random variation among blocks. The analysis of variance would then have the terms and degrees of freedom shown in Table 13.2.

The F-ratio for testing the ABC effect would be MS ABC/MS rep. $\times$ ABC with 1 and $r - 1$ degrees of freedom. The relationship between this design and

Table 13.2 Analysis of Variance Table, Blocks Random

Source	df
Mean	1
Replications	$r - 1$
ABC	1
Replication $\times$ ABC	$r - 1$
A	1
B	1
AB	1
C	1
AC	1
BC	1
Error	$6(r - 1)$
Total	$8r$

a split-plot design should be clear. In some experiments there is no reason to separate out a replicate sum of squares, and a reasonable test for the ABC interaction can be constructed with one and $2r - 2$ degrees of freedom.

Exercise 13.4: Pie Baker. A pie baker is trying to improve the recipe for the apple pies that he bakes and sells. He has three changes or factors in mind. These are not major changes. He anticipates that all of the pies baked as a part of the investigation can still be sold and that at least four or five replicates will be needed to establish conclusively that one recipe is superior. The three factors that he is contemplating are (a) to add a small amount of allspice to the recipe (b) to increase the amount of cinnamon in the filling by 10% (c) to use a small amount of tapioca in place of cornstarch to soak up excess juice in the filling. He has one oven that can accommodate four pies at one time, and his market can absorb eight apple pies per day. After considerable thought he has decided that the three-factor interaction is probably negligible, but there is a real possibility that all of the two-factor interactions may be important. The eight treatment combinations are:

(1)	no allspice,	low cinnamon,	cornstarch
a	some allspice,	low cinnamon,	cornstarch
b	no allspice,	high cinnamon,	cornstarch
ab	some allspice,	high cinnamon,	cornstarch
c	no allspice,	low cinnamon,	tapioca
ac	some allspice,	low cinnamon,	tapioca
bc	no allspice,	high cinnamon,	tapioca
abc	some allspice,	high cinnamon,	tapioca

Since his oven can hold only four pies, he has decided to use oven loads as a blocking factor, and since the three-factor interaction is expected to be negligible, the two blocks are $\{(1), ab, ac, bc\}$ and $\{a, b, c, abc\}$. Call these blocks types I and II. Note that conditions within an oven are not perfectly uniform and the assignment of pies to positions in the oven must be at random.

The first strategy that comes to mind is to perform the entire study in one week. Each morning he bakes four pies, block type I, serves them to his customers, and gets a quality score for each pie. Each afternoon he bakes the four block type II pies and obtains scores on them. Notice that block type I is always in the morning, and block type II is always in the afternoon.

(a) Describe a suitable randomization strategy for the experiment as planned. Why is it a good idea to use a new random assignment of treatment combinations in the oven each day?
(b) Construct the appropriate analysis of variance table for the resulting data if pies were baked on Monday, Tuesday, Wednesday, and Thursday. Notice that one can think of days as replicates.
(c) List some possible reasons why one cannot interpret the ABC interaction as due to only treatment effects in this experiment, if one were to compute it.

(d) Consider an alternate experimental protocol for the apple pie baker. Still use blocks of type I and II as before. However let him write I, II on two pieces of paper and II, I on two other pieces of paper. On Monday he randomly selects one of the pieces of paper from a box and if it says I, II he bakes the block type I in the morning and block type II in the afternoon. Alternatively, if he draws a II, I piece of paper, the two sets are reversed. On Tuesday he reverses the order. On Wednesday he again draws a slip of paper to determine the order for two days. Note that at the end of the four days, each treatment combination will have been baked twice in the morning and twice in the afternoon. Construct the appropriate analysis of variance table for the data resulting from this plan.

(e) Write a short note comparing and contrasting the two approaches. Give advantages and disadvantages for each approach. □

Partial Confounding

If more than one replication is included in the experiment, it is possible to confound a different effect or interaction in each replicate. In this way, it is possible to obtain estimates of all the effects and interactions, while still using the small blocks.

Example 13.2. In a dental study, treatments consisting of a 2^3 factorial were applied to the canines of human subjects. Since each subject has only four canines, the experiment had to be set up in blocks of four. The scientist did not really want to sacrifice the three-factor interaction. Since many subjects were available, it was possible to use partial confounding. In the first two replicates (two blocks in each replicate) confound the ABC interaction with blocks. These blocks can be illustrated as

Block 1	abc	a	b	c
Block 2	ab	bc	ac	(1)

and

Block 3	abc	a	b	c
Block 4	ab	bc	ac	(1)

The third replicate is based on the AB interaction,

$$\frac{y_{abc} + y_{ab} + y_c + y_{(1)}}{4} - \frac{y_a + y_b + y_{ac} + y_{bc}}{4},$$

and so the third replicate would contain

Block 5	*abc*	*ab*	*c*	(1)
Block 6	*a*	*b*	*ac*	*bc*

The fourth replicate could be based on the AC interaction and the fifth replicate on the BC interaction. The final design would have the following fourth and fifth replicates.

Block 7	*ac*	*abc*	*b*	(1)
Block 8	*a*	*c*	*ab*	*bc*

Block 9	*bc*	*abc*	*a*	(1)
Block 10	*b*	*c*	*ab*	*ac*

13.2.2 Analysis of Variance

The analysis of variance for the partially confounded case has to be modified slightly. We consider the above example of the 2^3 in five replicates, with ABC confounded in replicates 1 and 2 and AB, AC, and BC each confounded in one replicate, i.e., replicates 3, 4, and 5. We illustrate the analysis in Table 13.3.

Since A, B, and C are never confounded, these sums of squares can be computed from the complete data set. The ABC sum of squares is computed from data in replicates 3, 4, and 5 only since those are the replicates where it is not confounded. Similarly, AB sum of squares is computed from replicates 1, 2, 4,

Table 13.3 Analysis of Variance Table for the Dental Study

Source	df
Mean	1
Replications	4
Blocks (within replications)	5
A	1
B	1
AB	1
C	1
AC	1
BC	1
ABC	1
Error	23
Total	40

and 5, AC sum of squares from replicates 1, 2, 3, and 5, and BC from data in replicates 1, 2, 3, and 4.

General-Purpose linear models software handles all of these modifications automatically. A small example with simulated data is shown on the companion Web site.

Exercise 13.5: Expected Mean Squares. Using the techniques from Section 5.4 and the model adapted from Section 10.2.3, develop the expected mean squares for SS A and SS ABC. What does this tell you about differences in power for the corresponding F-tests? □

Exercise 13.6: Pie Baker (continued)

(a) Explain how the pie baker with one oven that has room for only four pies at one time in Exercise 13.4 could take advantage of partial confounding to get an experiment that would get information on all main effects and interactions in the four days allotted.

(b) Construct the analysis of variance table suitable for the data from such an experiment. Include sources, degrees of freedom, and sums of squares. *Note*: the sum of squares for block effects is more challenging and may be omitted.

(c) Assume that it is possible to modify the pie baker's oven so that it is possible to bake eight pies at one time by adding an extra rack inside the oven. However, this raises more questions about uniformity of the conditions within the oven. Write a short note discussing the pros and cons of the various plans that are now possible. Notice that position in the oven, especially with respect to the added rack, is one of those factors frequently encountered in experimental work where the investigator must first recognize the factor and then decide whether to use it to construct blocks or to randomize over it. Correct decisions with respect to questions like this lead to good efficient experimental work. □

Exercise 13.7: Partial Confounding Analysis. Let y_{ijknm} represent an observation at level k of factor A, level n of factor B, and level m of factor C in the jth block of replicate i in Example 13.2.

(a) Give the formula to use to compute the sum of squares for A.

(b) Give the sum of squares to use to compute the AB sum of squares.

(c) Give the sum of squares to use to compute the ABC sum of squares. □

13.2.3 Four Blocks of Two

Now extend the notion of confounding to the 2^3 factorial in four blocks of size 2. In this case the eight experimental units are matched in pairs, such that members of a pair are as alike as possible. Occasionally, the experimental units come naturally in pairs, as twins in some animal experiments or two halves of a fruit or the two sides of an automobile or the two arms of a person in a medical experiment. The problem is to assign the eight treatment combinations in the 2^3 to four pairs of experimental units so that the resulting data can be analyzed and interpreted. In the example in the preceding section we used one degree of freedom for blocks (ABC confounded with blocks), leaving six degrees of freedom to estimate contrasts of interest. Now with four blocks, we lose three degrees of freedom to differences among blocks and have only four left for estimable functions.

Consider confounding the ABC and the AB interactions with blocks. Looking at these two contrasts, we notice that a useful thing happens. In each contrast each treatment is multiplied by a plus or minus. For two contrasts there are four possible pairs of coefficients that a treatment could take: $++$, $+-$, $-+$, or $--$. That is, two contrasts break the set of treatments into four different groups of coefficient pairs. In both contrasts (ABC and AB), treatments abc and c appear with a $+$ sign, and ac and bc appear with a $-$ sign. Also, the pair ab and (1) appear in AB with a $+$ sign and in ABC with a $-$ sign, and a and b appear in ABC with a $+$ sign and in AB with a $-$ sign. Hence we have our four pairs of treatment combinations. Assigning these pairs to blocks gives us

Block 1	abc	c
Block 2	a	b
Block 3	ac	bc
Block 4	ab	(1)

It is straightforward to verify that the ABC contrast is also the contrast of blocks 1 and 2 vs. blocks 3 and 4, the contrast AB is the contrast of blocks 1 and 4 vs. blocks 2 and 3, and finally, contrast C is the contrast of blocks 1 and 3 vs. blocks 2 and 4. What started as an attempt to assign treatment combinations sacrificing ABC and AB ended up also sacrificing C. There are four blocks of two and consequently, three degrees of freedom, three contrasts confounded with blocks. The generalized interaction of ABC and AB; i.e., C is also confounded with blocks. It is a simple exercise to show that contrasts A, B, AC, and BC are free of block effects.

13.3 GENERAL CONFOUNDING PATTERNS

The concept of generalized interactions becomes important whenever one constructs blocking systems with more than two blocks. Setting one replicate of a 2^N factorial into a design with two blocks involves sacrificing one interaction. Putting one replicate of a 2^N factorial into a design with four equal-sized blocks involves sacrificing three effects or interactions. Of these, two can be chosen arbitrarily, but the third is their generalized interaction.

Putting one replicate of a 2^N factorial into a design with eight equal-sized blocks involves sacrificing seven effects or interactions. Of these, three can be chosen; however, the choice is restricted in the sense that none can be the generalized interaction of the other two. The remaining four are the generalized interactions. In general, it is possible to put one replicate of a 2^N factorial into 2 blocks, into $2^2 = 4$ blocks, into $2^3 = 8$ blocks, and so on. Constructing a blocking scheme with 2^M blocks requires that $2^M - 1$ effects and interactions be confounded. Of these, M can be selected freely with the proviso that none is the generalized interaction of two others. Once the M effects or interactions have been selected, the additional $2^M - 1 - M$ generalized interactions will be sacrificed automatically. The set of interactions selected and all of their generalized interactions is closed under the multiplication rule illustrated in Section 10.3.2 to construct generalized interactions. In other words, the generalized interaction of any two effects or interactions in the set is again a member of the set. The complete set of $2^M - 1$ effects and interactions confounded with 2^M blocks can be produced from M independent generators.

13.3.1 Confounding the 2^5 in Blocks of Size 4

We illustrate with an example: Consider the problem of putting a 2^5 factorial in blocks of size 4. The first thing to notice is that $2^5 = 32$ treatment combinations in all. Putting this into blocks of four means that there will be eight blocks in a complete replicate. Confounding one contrast, either an effect or interaction, with blocks allows the experiment to be split into two blocks. Splitting each block into two equal parts leads to four blocks with three degrees of freedom for block effects, which means that we give up two more degrees of freedom. Finally, splitting each block again will entail giving up four more degrees of freedom. For the first split to blocks of 16, we were free to select any contrast we wanted. For the next split, from blocks of size 16 to blocks of eight, we could select any other contrast we want. However, the generalized interaction of the two selected is also sacrificed automatically. Finally, to split blocks down to size 4, another generator must be selected. However, with it, all generalized interactions with previous contrasts are also confounded automatically. A reasonable choice for the first contrast to be sacrificed in this example is $ABCD$. One can reasonably expect that four-factor interactions are not important. Then for the second generator one can select BDE. This implies that $ABCD \times BDE \Rightarrow AB^2CD^2E \Rightarrow ACE$ is also confounded. ACE is the generalized interaction of $ABCD$ and BDE. Now for the third generator we select ADE. This implies immediately that

$$\begin{aligned} ADE \times ABCD &\Rightarrow BCE \\ ADE \times BDE &\Rightarrow AB \\ ADE \times ACE &\Rightarrow CD \end{aligned}$$

are also confounded. To conclude, to put the 2^5 factorial into eight blocks of four, we need to confound a total of seven contrasts. These result from selecting three generators and then finding all their generalized interactions.

The set $\{ABCD, BDE, ACE, ADE, BCE, AB, CD\}$ is closed under the system of multiplication that we are using. The reader should verify that selecting AB as a generator in place of ADE would yield exactly the same set. Of course, a different set of contrasts could be obtained by selecting an entirely different set of generators. For example, one could select $ABCDE$, ABC, and BCD. The full set including all generalized interactions is $\{ABCDE, ABC, DE, BCD, AE, AD, BCE\}$.

Some sets of generators are better than others in the sense that they lead to sacrificing fewer potentially interesting effects and interactions. In general, it is felt that main effects are usually more important than two-factor interactions, two-factor interactions more important than three-factor interactions, and so on. Once a set of contrasts to be confounded has been selected, we have the properties of the plan. However, we still need to tell the experimenter exactly which treatment combinations go into each block.

13.3.2 Assigning Treatments to Blocks: Intra block Subgroup

Design of experiments software such as JMP®, SAS® `Proc ADX`, and `Proc FACTEX` can be used to construct 2^N and 3^N factorials with confounding. In this section we present an algebraic method of assigning treatments to blocks and show how to implement it in a `SAS® Data` step. When only one interaction is being confounded with blocks, the even–odd method for selecting treatment combinations is perfectly adequate for determining the treatment combinations that go into each block. However, when more interactions are being confounded, this method becomes confusing and a more systematic scheme is needed. First, we can always tell how many treatment combinations go into a block by dividing 2^N by number of blocks. The method we use to obtain the identity of the treatment combinations in specific blocks utilizes what is known as the *intra-block subgroup*. This terminology comes from modern algebra, where a *group* or *subgroup* is a set that is closed under some operation and includes a unit element. In the context of design of experiments with confounding, the intra-block subgroup is the block of elements or treatments that are used to generate the rest of the design. We illustrate with the 2^5 in eight blocks discussed above. Simple division tells us that there are four treatment combination in each block. The intrablock subgroup contains four elements. Once we have this intrablock subgroup, the rules for obtaining the full set of blocks for the experiment are simple.

There are several methods for obtaining the intrablock subgroup. The method we choose appears at first to be rather complex, but has the advantage that it is rather mechanical and is easily extended to more complex cases when the need arises. For the confounding scheme developed above with the three generators,

```
data abcde;
   do iA=0 to 1; do iB=0 to 1; do iC=0 to 1;
   do iD=0 to 1; do iE=0 to 1;
   c1=mod(iA+iB+iC+iD,2); c2=mod(iB+iD+iE,2);
   c3=mod(iA+iD+iE,2);
   if (c1=0 & c2=0 & c3=0) then output;
   end; end; end; end; end;
proc print data=abcde noobs; run;
```

Display 13.1 SAS® code to obtain the intrablock subgroup.

$ABCD$, BDE, and ADE, set up the system of equations:

$$\begin{array}{cccccll}
a & b & c & d & e & & \\
\hline
i_A & +i_B & +i_C & +i_D & & = 0 \text{ (modulo 2)} & \text{(from } ABCD) \\
 & i_B & & +i_D & +i_E & = 0 \text{ (modulo 2)} & \text{(from } BDE) \\
i_A & & & +i_D & +i_E & = 0 \text{ (modulo 2)} & \text{(from } ADE).
\end{array}$$

There are exactly four solutions to this system, i.e., four treatment combinations satisfy the conditions. These are easily obtained. In SAS® the solutions could be obtained using syntax that closely follows the notation for modulo 2 arithmetic. Suitable code is illustrated in Display 13.1.

The four solutions are

$$\begin{array}{cccccccc}
0 & 0 & 0 & 0 & 0 & \Rightarrow & (1) \\
1 & 1 & 0 & 0 & 1 & \Rightarrow & abe \\
0 & 0 & 1 & 1 & 1 & \Rightarrow & cde \\
1 & 1 & 1 & 1 & 0 & \Rightarrow & abcd.
\end{array} \qquad (13.1)$$

Note that the first solution puts the low level of all factors, i.e., the control treatment, identified as (1), into the intrablock subgroup. If you do this by hand the second and third solutions are obtained by trial and error. Since this is a homogeneous system of equations and sums of solutions to homogeneous equations are also solutions, the fourth is obtained by adding the second solution to the third solution modulo 2. Observe that we really only needed to search for two solutions.

The fact that there are exactly four solutions follows from some theorems in modern algebra or, alternatively, from the fact that we selected generators to give us blocks of size 4.

The set of four treatment combinations can be looked at in two equivalent ways, either by adding the exponents using modulo 2 arithmetic or by symbolic multiplication of the treatment combinations. Each corresponds to using one of the systems of notation for the 2^N system. The solutions in the 0–1 notation and adding solutions modulo 2 gives a closed system. In the alternate, lowercase letter notation for treatment combinations, the corresponding manipulation is a symbolic multiplication similar to that used to compute generalized interactions of effects and interactions. This gives the equivalent closed system. In either case, it is clear that the set of treatment combinations can be treated as a group. Also, the control, indicated by a set of 0's or by the symbol (1), serves as the unit element. Usually, it is simpler to multiply treatment combinations identified by sets of lowercase letters (often referred to as *words*).

Exercise 13.8: Set of Elements Is Closed under Multiplication. Verify that the set $\{(1), abe, cde, abcd\}$ is closed under our system of multiplication by showing that the product of any two elements in the set is again in the set. □

13.3.3 Finding the Remaining Blocks

The intrablock subgroup specifies the treatment combinations in the first block and the remaining seven blocks are constructed from this block.

Block:	1	2	3	4	5	6	7	8
	(1)	a	b	c	d	e	ac	ad
	abe	be	ae	$abce$	$abde$	ab	bce	bde
	cde	$acde$	$bcde$	de	ce	cd	ade	ace
	$abcd$	bcd	acd	abd	abc	$abcde$	bd	bc

The treatment combinations in block two are obtained from those in block one by multiplying by a. The treatment a was selected simply because it was not in the intrablock subgroup. Block three was generated by multiplying by b, where b was selected because it was not in the intrablock subgroup and had not appeared in the first block generated. The remaining blocks are generated in a similar manner. In each case an initial treatment combination, which is not in the intrablock subgroup and has not appeared in any of the blocks generated up to that time, must be selected.

We leave it to the reader to demonstrate that selecting different treatment combinations for multipliers at the various stages will generate the same final set of blocks, but in a different order. The block compositions are unique.

Exercise 13.9: Blocks from Different Starting Subgroups. Show that you can begin with the intrablock subgroup $\{(1), abe, cde, abcd\}$ and the sequence of multipliers ab, ac, ad, ae, be, ce, and de and obtain the same set of blocks as above, but in a different order. □

13.3.4 Randomization

The final step in the design task is the random assignment of the treatment combinations to the experimental units. First, the treatment combinations are assigned at random to the experimental units within blocks on a block-by-block basis. In some cases it may also be a good idea to assign the treatments or factors randomly to the symbols, i.e., A, B, ..., used to generate the confounding scheme.

However, in other cases where the experimenter "knows" that two factors interact and that interaction is to be studied, or alternatively, that a certain interaction does not exist or is not of interest and a good candidate for confounding, it is not possible to assign treatments or factors to symbols randomly. It is good practice, however, to assign sets to blocks at random.

13.3.5 Analysis of Variance

The basic outline of the analysis of variance for a 2^N factorial in a randomized block design with 2^M blocks takes the form shown in Table 13.4. In this design, $2^M - 1$ effects or interactions are completely confounded with blocks. The sums of squares for effects and interactions for nonconfounded effects are found in exactly the same way, as if no blocking had been used. In practice this would be done with linear models software such as SAS® `proc GLM`. Effects and interactions are estimated as

$$\widehat{\text{Effect}} = \frac{1}{2^{N-1}}(a \pm 1)(b \pm 1)\ldots,$$

where lowercase letters $a, b, \ldots$ indicate which treatment responses to sum together. For example, in a 2^3 factorial, the contrast for the A effect is

$$\begin{aligned}\widehat{A} &= \frac{1}{4}(a-1)(b+1)(c+1)\\ &= \frac{1}{4}[abc + ab + ac + a - b - c - bc - (1)]\\ &= \frac{1}{4}(y_{111} + y_{110} + y_{101} + y_{100} - y_{010} - y_{001} - y_{011} - y_{000}).\end{aligned}$$

The sums of squares for nonconfounded effects and interactions can be found by squaring the estimated effect and multiplying by 2^{N-2},

$$\begin{aligned}\text{SS effect} &= \sigma^2 \frac{(\widehat{\text{Effect}})^2}{\text{Var(Effect)}}\\ &= \sigma^2 \frac{(\widehat{\text{Effect}})^2}{\sigma^2/2^{N-2}}\\ &= 2^{N-2}(\widehat{\text{Effect}})^2.\end{aligned}$$

Table 13.4 Analysis of Variance Table, 2^N Factorial in 2^M Blocks

Source	df
Mean	1
Blocks	$2^M - 1$
Unconfounded effects	$2^N - 2^M$
Total	2^N

The sum of squares for blocks is most easily found by subtraction from the corrected total sum of squares.

Note that this type of experiment provides no proper estimate of error unless treatments are replicated. However, it is not uncommon to assume that there are no true four-, five- factor or higher-order interactions. The sums of squares due to these terms are then pooled to form an error term. Occasionally, the three-factor interactions are also deemed unimportant and pooled into the error term. If one does not wish to make assumptions about higher-order interactions being negligible and providing a measure of error without further testing, the methods outlined in Chapter 12, developed by Daniel (1959), Lenth (1989), and Dong (1993) can be used. These methods identify factors and interactions whose magnitudes are much larger than would be expected if they really had no effect and, consequently, become candidates for further study.

Exercise 13.10: Confounding a 2^5 Factorial. Earlier in this chapter it was suggested that the four-factor interaction contrast $ABCD$ was a reasonable choice to use in developing a confounding scheme for a 2^5 in blocks of size 4. Try developing a confounding scheme for (a) four blocks of eight and (b) eight blocks of four by starting first with the five-factor interaction $ABCDE$ and then the four-factor interaction $ABCD$ and thereby establish why the original recommendation was a good one. □

13.4 DOUBLE CONFOUNDING

The common thread in the confounding schemes used so far is that there is only one basis for stratifying or grouping experimental units, i.e., a randomized block design. Double confounding provides a strategy to use when experimental units are stratified in two directions. The Latin square designs discussed previously come to mind. The idea is that one can select one set of interactions to confound with the row stratification of experimental units and then select another set of interactions to confound with the column stratification. These designs belong to the broad class of row–column designs.

As an example of the technique, consider the 2^5 factorial with 32 treatment combinations considered earlier in experimental material that can be thought of as four rows and eight columns. Consider first the eight columns. We think of this as eight blocks of four. Recall that, this requires confounding seven interactions. Three (generator interactions) to be selected and their four generalized interactions are confounded automatically. The set of generators selected was $ABCD$, BDE, and ADE. The intrablock subgroup was obtained as solutions to the system of equations

a	b	c	d	e		
i_A	$+i_B$	$+i_C$	$+i_D$		$= 0$ (modulo 2)	(from $ABCD$)
	i_B		$+i_D$	$+i_E$	$= 0$ (modulo 2)	(from BDE)
i_A			$+i_D$	$+i_E$	$= 0$ (modulo 2)	(from ADE).

The four solutions are

$$\begin{array}{cccccccc} 0 & 0 & 0 & 0 & 0 & \Rightarrow & (1) \\ 1 & 1 & 0 & 0 & 1 & \Rightarrow & abe \\ 0 & 0 & 1 & 1 & 1 & \Rightarrow & cde \\ 1 & 1 & 1 & 1 & 0 & \Rightarrow & abcd. \end{array}$$

The intrablock subgroup specifies the treatment combinations in the first block. The contents of the remaining seven blocks are constructed from this first block.

Block:	1	2	3	4	5	6	7	8
	(1)	*a*	*b*	*c*	*d*	*e*	*ac*	*ad*
	abe	*be*	*ae*	*abce*	*abde*	*ab*	*bce*	*bde*
	cde	*acde*	*bcde*	*de*	*ce*	*cd*	*ade*	*ace*
	abcd	*bcd*	*acd*	*abd*	*abc*	*abcde*	*bd*	*bc*

Note that at this point the order of treatment combinations in each block is random. Double confounding puts restrictions on this randomization.

Now it is necessary to construct a second confounding scheme for the four rows. There are 32 treatment combinations to be placed in four rows of eight. A set of generators to produce this second confounding scheme is required. Any two interactions that have not been used previously can be chosen. Select ABE and BCD. The generalized interaction is $ACDE$.

To obtain the assignment of treatment combinations to positions, it is again necessary to find an intrablock subgroup. Obtain this as the solutions to the system

$$\begin{array}{cccccll} a & b & c & d & e & & \\ \hline i_A & +i_B & & & +i_E & = 0 \text{ (modulo 2)} & \text{(from } ABE\text{)} \\ & i_B & +i_C & +i_D & & = 0 \text{ (modulo 2)} & \text{(from } BCD\text{).} \end{array}$$

The solutions are

$$\left.\begin{array}{rccccccl} 1. & 0 & 0 & 0 & 0 & 0 & (1) \\ 2. & 1 & 0 & 0 & 0 & 1 & ae \\ 3. & 0 & 0 & 1 & 1 & 0 & cd \\ 2+3 \Rightarrow 4. & 1 & 0 & 1 & 1 & 1 & acde \\ 5. & 0 & 1 & 1 & 0 & 1 & bce \\ 2+5 & 1 & 1 & 1 & 0 & 0 & abc \\ 3+5 & 0 & 1 & 0 & 1 & 1 & bde \\ 4+5 & 1 & 1 & 0 & 1 & 0 & abd \end{array}\right\} \text{“the intrablock subgroup”}$$

The treatment combinations assigned to row 1 are (1), *ae*, *cd*, *acde*, *bce*, *abc*, *bde*, and *abd*. Notice that each one of these was in a different column in the previous assignment.

The treatment combinations assigned to row 2 are a, e, acd, cde, $adce$, bc, $abde$, and bd. These are obtained by multiplying the intrablock subgroup by a. Again, each one of these was in a different column in the preceding table.

The treatment combinations assigned to row 3 are b, abe, bcd, $abcde$, ce, ac, de, and ad. These are obtained by multiplying the intrablock subgroup by b.

The treatment combinations assigned to row 4 are c, ace, d, ade, be, ad, $bcde$, and $abcd$. These are obtained by multiplying the intrablock subgroup by c.

The final design is

(1)	*acde*	*ae*	*abd*	*abc*	*cd*	*bce*	*bde*
cde	*a*	*acd*	*abce*	*abde*	*e*	*bd*	*bc*
abe	*bcd*	*b*	*de*	*ce*	*abcde*	*ac*	*ad*
abcd	*be*	*bcde*	*c*	*d*	*ab*	*ade*	*ace*

The rows represent blocks in one direction and columns represent blocks in the other direction.

One can also obtain the assignment in the table above by doing a symbolic multiplication of the two intrablock subgroups. The analysis of variance for one replicate of this plan is shown in Table 13.5.

Designs of this type are occasionally referred to as *quasi Latin square designs*. The source of the name is obviously from the fact that they permit control on the variability in two directions. They are really examples of a large class of row–column designs.

While designs of this type have considerable appeal, in the sense that they make good use of resources and at the same time allow the removal of two extraneous sources of variation (row and column effects), the randomizations have an unfortunate tendency to turn up patterns that are or at least appear to be too systematic. Experimenters have a tendency simply to discard these unfortunate random assignments when they appear. Unfortunately, under the randomization paradigm, this practice introduces biases into the statistical analysis of the resulting data. The obvious thing to do is to use a randomization scheme that is restricted or constrained not to allow these unfortunate randomizations. If the experimenter does not discard seemingly systematic patterns but always accepts the randomization produced, the statistical analysis and inference go smoothly. This entire topic is beyond the level of this book and is discussed under the heading of constrained randomization in Kotz, Johnson, and Read (1988).

13.5 3^N SYSTEM

13.5.1 Model

In the 3^N system, confounding schemes are limited to block size and number of blocks equal to some power of three, i.e., three, nine, 27, and so on. This can be compared to the 2^N system, where block size and number of blocks were limited

Table 13.5 Analysis of Variance Table for the Quasi Latin Square

Source	df		
Mean	1		
Columns	7	Pool SS for $ABCD$, BDE, ACE, ADE, BCE, AB, and CD	
Rows	3	Pool SS for ABE, BCD, and $ACDE$	
A	1	main effects	
B	1		
C	1		
D	1		
E	1		
AC	1	other two-factor interactions are confounded	
AD	1		
AE	1		
BC	1		
BD	1		
BE	1		
CE	1		
DE	1		
ABC	1	other three-factor interactions are confounded	usually pooled to form error
ABD	1		
ACD	1		
CDE	1		
$ABCE$	1	other four-factor interactions are confounded	
$ABDE$	1		
$BCDE$	1		
$ABCDE$	1		
Total	32		

to powers of 2. In Section 11.2 we developed two different models for use in the 3^N system. It turns out that the model based on linear and quadratic effects, although very useful for interpreting results from an experiment, is not at all suitable for constructing blocking schemes. For this, one must use the projective geometry-based approach outlined in Sections 11.2.2 and 11.2.3. The definitions of main effects given there correspond with the general definitions of main effects in Chapter 3, while definitions of the interactions differ somewhat.

Instead of the notation $a_0b_0, a_0b_1, \ldots, a_2b_2$ to denote treatment combinations for, say, the 3^2 case, we use the more general form $a_{i_A}b_{i_B}$ to represent treatment combinations, with i_A and i_B taking on the values 0, 1, or 2. This has the obvious extension to the 3^N case.

We recall that the main effect of, for example, factor A was based on splitting the responses into three groups, those with $i_A = 0$, those with $i_A = 1$, and those with $i_A = 2$. Similarly for the two-factor interaction, the split is defined by $i_A + i_B = 0$ modulo 3, $i_A + i_B = 1$ modulo 3, and $i_A + i_B = 2$ modulo 3 and for the AB^2 interaction by $i_A + 2i_B = 0$ modulo 3, $i_A + 2i_B = 1$ modulo 3, and

$i_A + 2i_B = 2$ modulo 3. This nature of splitting extends directly to higher-order three-level factorials. Splittings of this type are used to develop a confounding scheme.

13.5.2 Confounding 3^2 in Three Blocks of Three

Structure

Consider a simple example of a 3^2 system in three blocks of three. If we wish to keep main effects unconfounded, we can confound either AB or AB^2. Both of these have two degrees of freedom. Since neither has a clean interpretation, the choice is arbitrary. Assume that we select AB^2 as the interaction to be confounded. The blocks are constructed by assigning treatment combinations to blocks so that the three contrasts (11.1), (11.2), and (11.3) are confounded with blocks. A systematic way to obtain the intrablock subgroup is to find the solutions to the equation

$$\frac{a \qquad b}{i_A \quad +2i_B} = 0 \text{ (modulo 3).}$$

There are exactly three solutions. These are

$$\left.\begin{array}{llc} 0 & 0 & a_0b_0 \\ 1 & 1 & a_1b_1 \\ 2 & 2 & a_2b_2 \end{array}\right\} \text{block 1}$$

This is called the *intrablock subgroup*. The set for the second block is obtained by adding 1 to the symbol for level of A and for the third by adding 2. In all cases numbers are reduced modulo 3, if necessary.

$$\text{block 2} \quad \left.\begin{array}{c} a_1b_0 \\ a_2b_1 \\ a_0b_2 \end{array}\right\} \quad \text{add 1 (modulo 3) to the } i_A \text{ col. in the intrablock subgroup}$$

$$\text{block 3} \quad \left.\begin{array}{c} a_2b_0 \\ a_0b_1 \\ a_1b_2 \end{array}\right\} \quad \text{add 2 (modulo 3) to the } i_A \text{ col. in the intrablock subgroup.}$$

If a second replicate, i.e., three more blocks, are available, it is suggested that AB be confounded with these. This would be an example of partial confounding.

Analysis of Variance

For purposes of illustration we assume partial confounding of AB and AB^2 with replicates r of each type ($2r$ in all). The analysis of variance takes the form shown in Table 13.6. The sum of squares attributed to "mean" is often called the *correction term*. The sum of squares among replicates and blocks (replicates) follow the usual pattern and often are not separated but called simply *blocks*.

Table 13.6 Analysis of Variance Table for the 3 x 3 Factorial in Blocks of Size Three

Source	df	SS
Mean	1	
Replicates	$2r - 1$	
Blocks(replicates)	$4r$	
A	2	
B	2	
AB	2	
AB^2	2	
Error	$12r - 8$	By subtraction
Total	$18r$	$\sum y^2 - \text{CT}$

The A sum of squares is computed as

$$\sum_{i=0,1,2} (\text{sum of all obs. with } a_i)^2 / 6r - (\text{sum of all obs.})^2 / 18r.$$

The B sum of squares is computed using an analogous formula. The next two are slightly different. The AB sum of squares is

$$\sum_{k=0,1,2} \left[\text{sum of all obs. with } i_A + i_B = k \text{ (modulo 3)}\right]^2 / 3r$$

$$- \left(\text{sum of all obs.}\right)^2 / 9r,$$

where the sums are restricted to those replicates where AB is not confounded with blocks. Similarly, the AB^2 sum of squares is computed from observations in replicates where AB^2 is not confounded with blocks.

13.5.3 More Complex 3^4 in Nine Blocks of Nine

Now consider a more complex example, a 3^4 in nine blocks of nine. Nine blocks implies eight degrees of freedom among blocks. We can confound ABC with blocks, but that accounts for only two degrees of freedom. This is not enough. Something else has to be confounded with blocks. Select another interaction with two degrees of freedom, say ABD^2. From previous experience with the 2^N system, we know that generalized interactions will be involved. First, ABC and ABD^2 each involve two degrees of freedom. Therefore, their generalized interaction will involve $2 \times 2 = 4$ degrees of freedom. The two for ABC and two for ABD^2 and the four for their generalized interaction account for the necessary eight degrees of freedom for contrasts among the nine blocks. The remaining question is: Which interaction or interactions are compromised by confounding ABC and ABD^2 with blocks?

It turns out that ABC and ABD^2 have two generalized interactions. First, we take $ABC \times ABD^2 = A^2B^2CD^2$. Before proceeding to find the other generalized interaction, we must first take care of a matter of notation. In order to have unique names for all effects and interactions, it is necessary to require that the first letter to appear in the name have exponent 1. If a name with first exponent equal to 2 is generated, the rule is to square the entire expression (multiply all exponents by 2) and reduce modulo 3. The justification for this is by example. First, our rule says to replace $A^2B^2CD^2$ by ABC^2D. It is a simple exercise to verify that the three linear contrasts defining the ABC^2D interaction compare exactly to the three contrasts one would associate with $A^2B^2CD^2$.

Finally, we need to locate the remaining two degrees of freedom. There are various ways to find this interaction. A simple rule is to take $ABC \times ABD^2 \times ABD^2 = A^3B^3CD^4 \equiv CD$ after reducing modulo 3. This interaction accounts for the remaining two degrees of freedom. The set $\{ABC, ABD^2, ABC^2D, CD\}$ is closed under multiplication, using reduction of exponent modulo 3, and the rule to force unit power on the first symbol. Any two can be interpreted as the generalized interactions of the other two.

The experimenter needs to know which treatment combinations to assign to the nine blocks. We obtain the intrablock subgroup by finding the solutions for the system

a	b	c	d		
i_A	$+i_B$	$+i_C$		$= 0$ (modulo 3)	from ABC.
i_A	$+i_B$		$+2i_D$	$= 0$ (modulo 3)	from ABD^2.

As in the 2^N system, there is some arbitrariness in the choice of system of equations. Any two of the four interactions in the set can be used. It is straightforward to demonstrate that any choice leads to the same set of solutions, i.e., the same set of treatment combinations.

Solutions are

0	0	0	0	$a_0b_0c_0d_0$	This set of treatments constitutes the intrablock subgroup
1	2	0	0	$a_1b_2c_0d_0$	
2	1	0	0	$a_2b_1c_0d_0$	
1	0	2	1	$a_1b_0c_2d_1$	
2	0	1	2	$a_2b_0c_1d_2$	
2	2	2	1	$a_2b_2c_2d_1$	
1	1	1	2	$a_1b_1c_1d_2$	
0	1	2	1	$a_0b_1c_2d_1$	
0	2	1	2	$a_0b_2c_1d_2$	

The contents of block two are obtained by adding something (modulo 3) not found in this set, say 1 0 0 0 to each of the solutions in turn. The remaining blocks are found in a similar manner, by adding something that has not yet appeared. The fact that ABC, ABD^2, ABC^2D, and CD are actually all confounded with blocks can be demonstrated by displaying the contrasts associated with ABC, ABD^2, ABC^2D, and CD using the methods discussed in Section 13.5.1.

Exercise 13.11: Subgroup for 3^N Factorial. Consider the set of four interactions ABC, ABD^2, ABC^2D, and CD. Notice that it is possible to select six different pairs of interactions from this set. Show that every one of these six pairs represents a basis for the whole set in the sense that the two and their generalized interactions make up the whole set of four. □

13.6 DETAILED NUMERICAL EXAMPLE: 3^3 FACTORIAL IN BLOCKS OF NINE

In this section we present a detailed numerical example of a 3^3 factorial in blocks of size 9 with two replications, beginning with the construction of the plan and continuing with the statistical analysis. First of all, 3^3 implies an experiment with 27 treatment combinations. The block size of 9, in turn, implies three blocks in each replicate, with each block containing nine, treatment combinations. We will assume that the experimenter has used partial confounding, confounding AB^2C with blocks in replicate 1 and AB^2C^2 with blocks in replicate 2. The data and the SAS® code for generating the plan are available on the companion Web site. SAS® `proc ADX`, `proc Factex`, JMP®, and other design of experiments software also make constructing this type of plan easy.

13.6.1 Generating the Plan and Simulated Data

For replicate 1, obtain the intrablock subgroup (block 1) as solutions to the equation $i_A + 2i_B + i_C = 0$ (modulo 3). For replicate 2, obtain the intrablock subgroup as solutions to the equation $i_A + 2i_B + 2i_C = 0$ (modulo 3).

The three blocks for replicates 1 and 2 with randomly generated yields are given in Table 13.7. In replicate 1, AB^2C is confounded with blocks, and in replicate 2 AB^2C^2 is confounded with blocks.

13.6.2 Analysis of Variance

The model and the analysis follow the same pattern as for a randomized complete block design. The analysis can be conveniently done using general-purpose software for fitting linear models. The hand computations for effects that are not confounded with blocks are exactly the same as in the RCBD. For effects that are confounded with blocks, the computations have the same form but are based only on replicates in which they are not confounded. The details are given in Section 13.6.3. The SAS® `proc glm` code for the analysis of variance is shown in Display 13.2. This code and its output are available on the companion Web site.

Table 13.7 Simulated Data for 3^3 Factorial in Blocks of Nine

Block 1		Block 2		Block 3	
Trt.	Yield	Trt.	Yield	Trt.	Yield
		Replicate 1			
000	34	100	39	010	54
110	42	210	75	120	78
220	51	020	73	200	40
011	56	111	65	021	53
121	72	221	91	101	45
201	82	001	55	211	85
022	45	122	92	002	60
102	58	202	62	112	69
212	69	012	56	222	70
		Replicate 2			
000	50	100	56	001	48
110	52	210	49	111	81
220	61	020	98	221	79
101	83	201	86	102	95
202	101	002	68	200	56
211	65	011	75	212	103
021	89	121	74	022	65
012	90	112	109	010	64
122	122	222	92	120	84

```
proc glm ;
  classes block a b c ;
  model y = block a b c a*b a*c b*c a*b*c ;
  run ;
```

Display 13.2 SAS® code to analyze data, partial confounding in the 3^3.

Note that all main effects and interactions are estimable because different components of the interaction were confounded with blocks in the replicates. In practice, it is usually easier to interpret the results if one resorts to linear and quadratic effects for main effects and interactions. Note also that the analysis assumes that the six blocks are assigned at random. If the three blocks are randomized separately for each replicate, the model (and analysis of variance) should contain a replicate effect (1 df) and a term for blocks within replicates (4 df).

13.6.3 Hand Computations of the Analysis of Variance

It is instructive to go through parts of the hand computation of the analysis of variance to see which parts of the design provide information on the different effects. The two-factor interactions are direct and present no problems. The three-factor interaction which is partially confounded with blocks is more involved.

Consider first SS ABC, the two-degree-of-freedom component due to ABC from the eight-degree-of-freedom interaction sum of squares SS $A \times B \times C$. There is full information. It is computed from data in both replicates. The three components are based on

$$i_A + i_B + i_C = 0, 1, \text{ or } 2 \text{ (modulo 3)}.$$

		Treatment combination	Total
$i_A + i_B + i_C =$	0	000, 012, 021, 102, 111, 120, 201, 210, 222	1287
	1	001, 010, 022, 100, 112, 121, 202, 211, 220	1175
	2	002, 011, 020, 101, 110, 122, 200, 212, 221	1304

$$\text{SS } ABC = \frac{1287^2 + 1175^2 + 1304^2}{18} - \text{CT} = 545.81.$$

Similarly, the two-degree-of-freedom sum of squares due to ABC^2 is computed from both replicates. This component are based on

$$i_A + i_B + 2i_C = 0, 1, \text{ or } 2 \text{ (modulo 3)}.$$

		Treatment combination	Total
$i_A + i_B + 2i_C =$	0	000, 011, 022, 101, 112, 120, 202, 210, 221	1250
	1	100, 111, 122, 201, 212, 220, 002, 010, 021	1295
	2	001, 012, 020, 102, 110, 121, 200, 211, 222	1221

$$\text{SS } ABC^2 = \frac{1250^2 + 1295^2 + 1221^2}{18} - \text{CT} = 154.48.$$

The sum of squares due to the component AB^2C^2 is available only from replicate 1. The treatments in each block are based on

$$i_A + 2i_B + 2i_C = 0, 1, \text{ or } 2 \text{ (modulo 3)}.$$

		Treatment combination	Total
$i_A + 2i_B + 2i_C =$	0	000, 110, 220, 101, 202, 211, 021, 012, 122	520
	1	100, 210, 020, 201, 002, 011, 121, 112, 222	596
	2	001, 111, 221, 102, 200, 212, 022, 010, 120	555

$$\text{SS } AB^2C^2 = \frac{520^2 + 596^2 + 555^2}{9} - \frac{(520 + 596 + 555)^2}{27} = 321.56.$$

The sum of squares due to the component AB^2C is available only from replicate 2. The block assignments are based on

$$i_A + 2i_B + i_C = 0, 1, \text{ or } 2 \text{ (modulo 3)}.$$

	Treatment combination	Total
	0 000, 110, 220, 011, 121, 201, 022, 102, 212	661
$i_A + 2i_B + i_C =$	1 100, 210, 020, 111, 221, 001, 122, 202, 012	724
	2 010, 120, 200, 021, 101, 211, 002, 112, 222	710

$$\text{SS } AB^2C = \frac{661^2 + 724^2 + 710^2}{9} - \frac{(661 + 724 + 710)^2}{27} = 243.19.$$

The total sum of squares is equal to $34^2 + 42^2 + \cdots + 84^2 - \text{CT} = 20,446.37$.

The four sums of squares for components of the $A \times B \times C$ interaction sum up to SS$A \times B \times C$ with eight degrees of freedom.

Typically, this analysis would be broken down further into linear and quadratic effects and their interactions in order to interpret the results. For example, to break up the $A \times C$ interaction, we work from the table of $A \times C$ totals. To illustrate the computations, we construct the following table, where the first two rows specify treatment levels, row 3 shows the yield totals, and the remaining lines, coefficients for orthogonal contrasts.

A:	0	0	0	1	1	1	2	2	2
C:	0	1	2	0	1	2	0	1	2
Yield:	373	376	384	351	420	545	332	488	497
A_L:	−1	−1	−1	0	0	0	1	1	1
A_Q:	1	1	1	−2	−2	−2	1	1	1
C_L:	−1	0	1	−1	0	1	−1	0	1
C_Q:	1	−2	1	1	−2	1	1	−2	1
$A_L \times C_L$:	1	0	−1	0	0	0	−1	0	1
$A_L \times C_Q$:	−1	2	−1	0	0	0	1	−2	1
$A_Q \times C_L$:	−1	0	1	2	0	−2	−1	0	1
$A_Q \times C_Q$:	1	−2	1	−2	4	−2	1	−2	1

Sum of squares for

$$A_L = \frac{(-373 - 376 - 384 + 332 + 488 + 497)^2}{6 \times 6}$$
$$= 940.44.$$

Sum of squares for

$$A_Q = \frac{(373 + 376 + 384 - 2 \times 351 - 2 \times 420 - 2 \times 545 + 332 + 488 + 497)^2}{6 \times 18}$$
$$= 306.70.$$

Note that SS A_L + SS A_Q = SS A. Sum of squares for $A_Q \times C_Q$ is computed as

$$\frac{(373 - 2 \times 376 + 384 - 2 \times 351 + 4 \times 420 - 2 \times 545 + 332 - 2 \times 488 + 497)^2}{6 \times 36} = 298.69.$$

13.6.4 Estimating Linear and Quadratic Contrasts from Partially Confounded Effects

The example in Section 13.6.3 illustrated the use of linear and quadratic contrasts in the interpretation of data from a 3^N factorial. A slight further complication arises if one has used partial confounding and it turns out that the interactions that were partially confounded, i.e., estimable from only part of the experiment, are important. We illustrate the calculations with a small example of partial confounding of a 3^2 factorial in blocks of size 3: In replicate 1 confound AB, and in replicate 2 confound AB^2.

Construct blocks in replicate 1 by working with equation $i_A + i_B = 0$ (modulo 3).

Block 1	Block 2	Block 3
0 0	1 0	2 0
1 2	2 2	0 2
2 1	0 1	1 1

Construct blocks in replicate 2 by working with equation $i_A + 2i_B = 0$ (modulo 3).

Block 1	Block 2	Block 3
0 0	1 0	2 0
1 1	2 1	0 1
2 2	0 2	1 2

This leads to the analysis of variance shown in Table 13.8.

Finding A_L, A_Q, B_L, and B_Q presents no difficulty. However, finding $A_L B_L$, $A_L B_Q$, $A_Q B_L$, and $A_Q B_Q$ is not that obvious. The key fact to keep in mind

Table 13.8 Analysis of Variance Table for 3^2 with Partial Confounding

Source	df	
Blocks	5	
SS A	2	(linear + quadratic)
SS B	2	(linear + quadratic)
SS AB (from rep. 2)	2	
SS AB^2 (from rep. 1)	2	
Error	4	
Total	17	

is that AB is estimable only in replicate two and AB^2 is estimable only in replicate 1. Also these are both orthogonal to A and B. We must "synthesize" the appropriate observations from uncontaminated information:

$$\begin{aligned}
\widehat{y}_{a_0b_0} &= \widehat{\mu} + \widehat{A}_0 + \widehat{B}_0 + \widehat{AB}_0 + \widehat{AB}{}^2_0 \\
\widehat{y}_{a_0b_1} &= \widehat{\mu} + \widehat{A}_0 + \widehat{B}_1 + \widehat{AB}_1 + \widehat{AB}{}^2_2 \\
\widehat{y}_{a_0b_2} &= \widehat{\mu} + \widehat{A}_0 + \widehat{B}_2 + \widehat{AB}_2 + \widehat{AB}{}^2_1 \\
&\vdots \\
\widehat{y}_{a_2b_2} &= \widehat{\mu} + \widehat{A}_2 + \widehat{B}_2 + \widehat{AB}_1 + \widehat{AB}{}^2_0.
\end{aligned}$$

Actually, in this case we can ignore $\widehat{\mu}$ because we consider only contrasts and $\widehat{A}_i$ and $\widehat{B}_j$ since interactions are orthogonal to main effects. The contrasts A_LB_L, A_LB_Q, A_QB_L, and A_QB_Q can now be estimated as in the example of Section 13.6.3.

13.7 CONFOUNDING SCHEMES FOR THE 3^4 SYSTEM

13.7.1 Nine Blocks

The general confounding schemes are available just as in the 2^N system. As an extension of the preceding example, consider a 3^4 in blocks of nine. Since $3^4 = 81$, this implies that there will be nine blocks needed for one complete replicate. Select ABD and ACD^2 as candidates for confounding. Confounding two interactions, each with two degrees of freedom, implies that their generalized interactions—four—degrees of freedom will also be confounded. This means that two more interactions, each with two degrees of freedom, will be confounded. These are AB^2C^2 and BC^2D^2. The first is easy to find.

$$ABD \times ACD^2 \Rightarrow A^2BCD^3 \Rightarrow A^2BC \Rightarrow A^4B^2C^2 \Rightarrow AB^2C^2$$

The second may be more difficult. Consider first

$$ABD \times AB^2C^2 \Rightarrow A^2B^3C^2D \Rightarrow A^2C^2D \Rightarrow A^4C^4D^2 \Rightarrow ACD^2.$$

This is a failure, since it did not produce a new term. An old term was regenerated. Now try

$$ABD \times ABD \times ACD^2 \Rightarrow A^3B^2CD^4 \Rightarrow B^2CD^4 \Rightarrow B^2CD \Rightarrow BC^2D^2.$$

This time we got a new term. The key point, $3^4 = 81$ treatments combinations in blocks of size 9, implies nine blocks. This, in turn, implies eight degrees of freedom for blocks, or equivalently, four interactions, each with two degrees of freedom confounded with blocks. This set of four interactions forms a closed set under the sort of multiplication being performed.

The final step is to obtain the intrablock subgroup in order to establish the actual treatment assignments. We need solutions to the system

$$\begin{array}{cccccl} a & b & c & d & & \\ \hline i_A & +i_B & & +i_D & = 0 \text{ (modulo 3)} \\ i_A & & +i_C & +2i_D & = 0 \text{ (modulo 3)} \end{array}$$

13.7.2 Twenty-Seven Blocks

Confounding 3^4 in 27 blocks of three is possible. Clearly, 13 interactions will need to be confounded. Three of these can be chosen directly and the remaining 10 generated. Begin with the set generated in the preceding example and add one more interaction. Add AB^2C^2D:

$$\begin{array}{llll} AB^2C^2D \times ABD & \Rightarrow A^2B^3C^2D^2 & \Rightarrow ACD & \\ AB^2C^2D \times ABD \times ABD & \Rightarrow A^3B^4C^2D^3 & \Rightarrow BC^2 & \\ AB^2C^2D \times ACD^2 & \Rightarrow A^2B^2C^3D^3 & \Rightarrow AB & \\ AB^2C^2D \times ACD^2 \times ACD^2 & \Rightarrow A^3B^2C^4D^5 & \Rightarrow B^2CD^2 & \Rightarrow BC^2D \\ AB^2C^2D \times AB^2C^2 & \Rightarrow A^2B^4C^4D & \Rightarrow AB^2C^2D^2 & \\ AB^2C^2D \times AB^2C^2 \times AB^2C^2 & \Rightarrow A^3B^6C^6D & \Rightarrow D & \\ AB^2C^2D \times BC^2D^2 & \Rightarrow AB^3C^4D^3 & \Rightarrow AC & \\ AB^2C^2D \times BC^2D^2 \times BC^2D^2 & \Rightarrow AB^4C^6D^5 & \Rightarrow ABD^2. & \end{array}$$

Three generators have been used: ABD, ACD^2, and AB^2C^2D. The others have all been generated as generalized interactions. Note that in the confounding scheme, two two-factor interactions and one main effect have been sacrificed. It is possible to do better than this.

REVIEW EXERCISES

13.1 (a) Devise a suitable confounding scheme for a 2^6 factorial in four blocks of size 16, being sure to keep all main effects and two-factor interactions clear of confounding.

(b) Is it possible to construct a design in eight blocks of 8 keeping all main effects and two-factor interactions clear of confounding? If so, provide such a design.

13.2 Suppose that for some reason only 16 treatments can be run at a time in the deinking experiment of review exercise 12.1, so that you want to run the experiment in blocks of size 16. Suppose also that it is important to estimate all main effects and two-factor interactions and to estimate the interaction between enzyme charge, incubation temperature, and the addition of the surfactant Tween. Design a suitable experiment.

13.3 A plant pathologist is studying the effects of ozone on how plant tissue decays over time. Soybean plants have been exposed to one of three levels of ozone (labeled A, E, or C). The plant stems will be put into soil to decay for different lengths of time (3 months, 6 months, or 9 months). There will be one bag of stems for each of the nine treatment combinations. Each block of soil is large enough to hold only three bags of stems.

(a) For each of the three experimental plans shown below, give a name or description of the design and state what effect(s) are confounded with blocks.

Block	Plan 1			Plan 2			Plan 3		
1	E3	C6	A9	E3	A6	C9	E9	C9	A9
2	A6	E9	C3	A3	C6	E9	A3	C3	E3
3	C9	E6	A3	A9	C3	E6	C6	A6	E6

(b) Which of the three plans do you recommend? Give your reasons.

CHAPTER 14

Fractional Factorial Experiments

14.1 INTRODUCTION

The great advantage of factorial experiments is that they allow the experimenter to study a number of factors simultaneously and, in particular, to see how the factors interact with one another. Until factors are tested together in various combinations, there is no way to know whether the effect produced by one factor depends on the level of another. Consequently, there is a risk involved in leaving out any factor that could possibly have an effect. In fact, one cannot really leave out a factor. By "ignoring" a factor, the experimenter really selects one level of the factor for the entire study. This then raises the question of whether this is really the appropriate level of the factor. The negative side of this reasoning is that taking all possible combinations of even a modest number of factors, even if at only two levels each, leads to an unreasonably large experiment. The strategy to follow is to look at only a fraction of the full set of all possible treatment combinations. The trick is to take not a haphazard fraction of the total set of possible treatment combinations, but to take the fraction in such a manner that the resulting data can easily be analyzed and interpreted. Fractional factorial plans have their place when the experimenter has a large number of factors to be studied, but the resources do not permit even one complete replicate.

Fractional factorial experiments are especially useful when the goal is to screen a large number of factors to determine which might be most important. Often, at the early stage of a research program, or in exploratory research, the experimenter is faced with having to make choices among many potential factors but has limited resources. It is at this stage of the research work that the application of principles of fractional factorial experiments can be of most benefit. This is contrary to a view that is commonly expressed. It is not unusual to hear an investigator claim that not enough is known about a process, that the process is too variable to benefit from a designed experiment, or that there are too many ethical

Planning, Construction, and Statistical Analysis of Comparative Experiments,
by Francis G. Giesbrecht and Marcia L. Gumpertz
ISBN 0-471-21395-0

constraints. Nothing can be further from the truth. Designing an experiment is *carefully planning a strategy to maximize the amount of information gained at the lowest cost while respecting all side constraints, such as cost or limitations on material, time limits, time trends, ethical considerations, and so on.* Factorial arrangements of treatments are a large part of this.

One common strategy, which we do not recommend, is to make a control run and then test a sequence of factors one at a time. If a response differs appreciably from the control, that factor is judged to be important. In part, the belief in this strategy comes from our experience with small experiments in our own education. This strategy is entirely appropriate in the early education setting, where the instructor who guides the work has a sound understanding of the principles at work and which are to be demonstrated. The instructor knows exactly what will, or at least should happen. Careful thought, however, reveals several shortcomings of this strategy in real-world investigations. To illustrate the reasoning and advantages of the techniques proposed by the principles of design of experiments, we consider a small hypothetical example.

A researcher is studying the process of roasting peanuts to obtain an improved flavor. To simplify our example, we ignore the nontrivial problem of measuring quality. A characteristic of peanut plants is that not all nuts mature at the same time. Consequently, all lots of peanuts contain a mixture of kernels at various stages of maturity, and this has some affect on the roasting process. It is not possible to sort the kernels sufficiently to get rid of this variability. There is, however, a standard time and temperature roasting process that yields an acceptable product. We now assume that our researcher wants to test whether slightly increased roasting temperature or slightly longer roasting time will improve the flavor. Since there is variability in the material, the researcher roasts four samples of kernels at the standard time–temperature combination (the controls) and four samples for the proposed longer time. The results for the test will be either better or poorer than the control. Assume better. Next, the experimenter roasts four samples for the new roast time, but at the proposed higher roast temperature. There are now 12 quality responses, which we denote $y_1, y_2, \ldots, y_{12}$. The effect of time change on quality is evaluated as

$$\frac{y_5 + y_6 + y_7 + y_8}{4} - \frac{y_1 + y_2 + y_3 + y_4}{4}$$

and the effect of temperature change on quality by

$$\frac{y_9 + y_{10} + y_{11} + y_{12}}{4} - \frac{y_1 + y_2 + y_3 + y_4}{4},$$

or possibly by

$$\frac{y_9 + y_{10} + y_{11} + y_{12}}{4} - \frac{y_5 + y_6 + y_7 + y_8}{4}.$$

Before looking at this experiment in detail, we present the alternative, a 2×2 factorial that involves only eight lots of peanuts. Two lots are roasted at the control time–temperature, two at the control temperature but the longer time, two at the higher temperature but for the control time, and two at the higher temperature and for the longer time period. The eight actual roastings are done in random order. Note that four lots are roasted at each temperature and four lots are subject to each of the roasting times. The estimate of the effect of roasting temperature involves the difference between means of two sets of four, as does the estimate of the effect of roasting time. In addition, there is the estimate of interaction effect, the mean of the control and high–high vs. the mean of the low–high and the high–low combinations. If this interaction contrast should turn out to be large, the experimenter is warned to construct the 2×2 table of means, roasting time by roasting temperature, to examine what is going on. Examination of the table may suggest looking at lower, or even possibly higher temperatures, at the current roast times.

We now look critically at the results of the one-factor-at-a-time experiment and emphasize four points. First, we note that when factors are studied one at a time, there is no option for randomization. If the environmental conditions are changing throughout the experiment, or the technician is learning during the experiment, the results in the one-factor-at-a-time experiment will be distorted. Second, we note that estimates of the effect of both time and temperature change are estimated with variance,

$$(4\sigma^2 + 4\sigma^2)/16 = \sigma^2/2$$

in both experimental protocols. However, the two pieces of information are not independent in the one-factor-at-a-time experiment. The two estimates have covariance

$$\pm 4\sigma^2/16 = \pm\sigma^2/4$$

and consequently, a correlation of $\pm\frac{1}{2}$ with sign depending on which estimate of effect of temperature is used. There is a lack of independence of the two pieces of information. In the factorial experiment, the three pieces, two main effects and interaction, are uncorrelated, making them easier to interpret.

Next, there is the question of interaction. The one-factor-at-a-time study fails to take into account the very real possibility of an interaction between time and temperature. In essence, the example assumed that the experimenter tested the higher temperature at the longer roast time with the implicit assumption that the two effects were additive. It may well be that at the longer roast times, the old control roast temperature gave better results, and at the old control roast time, the higher roast temperature gave better results. The experiment as described does not permit the investigator to "untangle" the phenomena and understand the process. The current study gives little guidance for further investigation.

Finally, there is the question of use of resources. The one-factor-at-a-time experiment required 12 samples of kernels to get the same variances of estimates of effects of time and temperature as the factorial experiment, which required only eight samples. And this assumes no interaction. While the extra samples may not be serious in a peanut roasting study, they do represent a 50% increase in resources. In other types of experiments involving animal or human subjects, a 50% increase in the number of subjects may raise serious questions of ethics when dealing with experiments that cause pain or even require the sacrifice of animals.

14.2 ORGANIZATION OF THIS CHAPTER

The organization of this chapter rests on three considerations, all based on the realization that it is impossible for most researchers to carry in mind the full range of possible factorial plans that can be utilized in any particular problem. One approach to dealing with this is to try equip the reader (potential experimenter) with the tools required to derive or generate plans as needed. In practice, this tends to fail because of time pressures and the inevitable forgetting of details.

The second approach is to equip the researcher with algorithms and computer programs that simplify the process of generating plans. The problem here is to develop algorithms that are sufficiently broad to handle the complications that inevitably arise in practice and yet are sufficiently simple to handle. The third approach is to provide the researcher with tables of plans. This also has some obvious shortcomings.

We present a compromise of all three. We present the minimal basic theory that the experimenter will need to understand the use of fractional factorials. We provide an algorithm for finding plans that satisfy specific, well-formulated requirements. Finally, for the broader range of requirements, we provide a series of tables of plans. Enough theory is presented to provide the tools to use the algorithm and the tables efficiently.

The discussion in this chapter begins with three distinct approaches to constructing experimental plans that employ fractional replication. All three have advantages and disadvantages. We begin with Section 14.3, which spells out the concepts of half replicate, quarter replicate, and so on, and in the process develops a method of finding fractions by examining the defining contrasts. This method is probably the most direct, but tends to fail for large or complex experiments. In Section 14.4 we develop the concepts of resolution and aberration. We also introduce a sequence of $1/16, 1/32, \ldots, 1/256$ fractions presented in Appendixes 14A and 14B. In Section 14.5 we present a method for constructing fractions by superimposing factors on a given structure. Not only is this a very useful general technique, especially when there are constraints on the selection of interactions to use for defining contrasts, it also introduces methodology used in some of the more advanced constructions involving orthogonal arrays. In Section 14.7 we present the Franklin–Bailey algorithm, which will find a suitable set of defining contrasts, if one exists. This method is most convenient if the other two fail. It is also the method used extensively in Chapter 20.

We then present a series of sections that illustrate a large number of plans that cannot be obtained by straightforward application of the three standard methods. These include three-fourths replicate plans and a series of compromise plans in which the concept of resolution is compromised, and two-factor interactions involving subsets of factors are retained. In the final two major sections in the chapter we discuss problems introduced by blocking and then extend the notions to the 3^N case.

14.3 FRACTIONAL REPLICATION IN THE 2^N SYSTEM

14.3.1 One-Half Replicate

The idea of fractional factorials is simple: to select a subset of the 2^N treatment combinations in such a manner that the results of the experiment yield maximal information. The techniques used occasionally employ some clever mathematics.

To illustrate the principles involved without too much complexity, consider a small study, an experiment with three factors each at two levels. A complete replicate involves eight treatment combinations. Now assume that the experimenter can afford only four experimental units. The question is: Which four treatment combinations should be looked at? One approach is to select the four treatment combinations that enter into ABC with a "+" sign. This is similar to what we would do if we wanted to confound the ABC interaction with blocks, except that in a fractional factorial we use only one of the blocks, i.e., half of the treatment combinations. We identify the treatment combinations by expanding $(a-1)(b-1)(c-1)$ into $abc+a+b+c-ab-ac-bc-(1)$. We pick those with a "+" sign.

The experimenter decides to look at only y_{abc}, y_a, y_b, and y_c. Now, estimate A using $(y_{abc}+y_a-y_b-y_c)/2$. This looks good, but now the question is: What does this really estimate?

To answer this question, write out y in terms of the model:

$$
\begin{aligned}
y_a &= \mu + \tfrac{1}{2}A - \tfrac{1}{2}B - \tfrac{1}{2}AB - \tfrac{1}{2}C - \tfrac{1}{2}AC + \tfrac{1}{2}BC + \tfrac{1}{2}ABC + \epsilon_a \\
y_b &= \mu - \tfrac{1}{2}A + \tfrac{1}{2}B - \tfrac{1}{2}AB - \tfrac{1}{2}C + \tfrac{1}{2}AC - \tfrac{1}{2}BC + \tfrac{1}{2}ABC + \epsilon_b \\
y_c &= \mu - \tfrac{1}{2}A - \tfrac{1}{2}B + \tfrac{1}{2}AB + \tfrac{1}{2}C - \tfrac{1}{2}AC - \tfrac{1}{2}BC + \tfrac{1}{2}ABC + \epsilon_c \\
y_{abc} &= \mu + \tfrac{1}{2}A + \tfrac{1}{2}B + \tfrac{1}{2}AB + \tfrac{1}{2}C + \tfrac{1}{2}AC + \tfrac{1}{2}BC + \tfrac{1}{2}ABC + \epsilon_{abc}.
\end{aligned}
$$

The expectation of the contrast for the A effect is

$$
\begin{array}{r|l}
E[y_{abc}] & +(\mu + \tfrac{1}{2}A + \tfrac{1}{2}B + \tfrac{1}{2}AB + \tfrac{1}{2}C + \tfrac{1}{2}AC + \tfrac{1}{2}BC + \tfrac{1}{2}ABC) \\
+E[y_a] & +(\mu + \tfrac{1}{2}A - \tfrac{1}{2}B - \tfrac{1}{2}AB - \tfrac{1}{2}C - \tfrac{1}{2}AC + \tfrac{1}{2}BC + \tfrac{1}{2}ABC) \\
-E[y_b] & -(\mu - \tfrac{1}{2}A + \tfrac{1}{2}B - \tfrac{1}{2}AB - \tfrac{1}{2}C + \tfrac{1}{2}AC - \tfrac{1}{2}BC + \tfrac{1}{2}ABC) \\
-E[y_c] & -(\mu - \tfrac{1}{2}A - \tfrac{1}{2}B + \tfrac{1}{2}AB + \tfrac{1}{2}C - \tfrac{1}{2}AC - \tfrac{1}{2}BC + \tfrac{1}{2}ABC) \\
\hline
 & \quad +2A \qquad\qquad\qquad\qquad\qquad +2BC
\end{array}
$$

Thus, $(y_{abc}+y_a-y_b-y_c)/2$ really estimates $A+BC$, not just A.

If we had tried to estimate BC, we would have obtained exactly the same contrast. Recall that to estimate BC in a full factorial, we expand $(a + 1)(b - 1)(c - 1)$ into $abc - ab - ac - b - c + bc + a + (1)$. Of the eight treatment combinations, only y_{abc}, y_a, y_b, and y_c are observed in our fractional factorial. Hence, the experimenter is again left with $(y_{abc} + y_a - y_b - y_c)$. The conclusion is that the contrast $(y_{abc} + y_a - y_b - y_c)$ has two names. A and BC are *aliases* for the same contrast. We can go through an analogous argument to show that B and AC, and C, and AB are aliases. Hence, in a fractional factorial treatment contrasts are confounded with *each other*, in contrast to the situation in Chapter 13, where effects were confounded with block contributions.

We picked the *defining contrast* ABC to select the fraction that was to be used. Standard convention is to identify the fraction as $\mathcal{I} = ABC$. Pairs of effects that are aliased can then be found using the special type of multiplication developed in Chapters 10 and 13 for 2^N factorials.

For A multiply both sides of the expression $\mathcal{I} = ABC$ by A:

$$A = A^2BC \rightarrow BC,$$

implying that A and BC are aliases, i.e., two names for the same contrast. Similarly, we compute

$$B = AB^2C \rightarrow AC$$

implying that B and AC are aliases, i.e., two names for the same contrast. Also, C and AB are aliases for the same contrast.

Finally, there is still a question about the actual set of treatment combinations to be selected. We expanded $(a - 1)(b - 1)(c - 1)$, the symbolic expression for ABC, and selected those with a positive sign. We could just as easily have selected those with a negative sign, i.e., $y_{(1)}$, y_{ab}, y_{ac}, and y_{bc}. This fraction is identified as $\mathcal{I} = -ABC$. It is straightforward to check that the estimate of A, $(y_{ab} + y_{ac} - y_{bc} - y_{(1)})/2$, is actually an estimate of the difference $A - BC$, or we could say, of A and BC with a negative sign. Similarly, it can be shown that the estimate of B is really the estimate of the difference $B - AC$, and the estimate we call C is the estimate of the difference $C - AB$.

There is no real reason to favor one fraction over the other. The convention is to use a $+$ or $-$ sign to indicate the fraction being selected for definiteness. We selected $\mathcal{I} = +ABC$, implying that A and BC, B and AC, and C and AB were aliased. If we had selected the other half, we would have $\mathcal{I} = -ABC$, implying that A and $-BC$, B and $-AC$, and C and $-AB$ were aliases.

In this relatively simple case there really was not much choice available for the defining contrast. In more complex cases, one typically tries to alias main effects and two-factor interactions with higher-order interactions, which in turn can safely be assumed to be negligible.

14.3.2 One-Fourth Replicate

For a more complex example, consider selecting eight observations from a 2^5 system, commonly written as $2^{(5-2)}$, or, in other words, a one-fourth replicate of the 2^5. Each effect or interaction defines a division of the set of all treatment combinations into two halves. A half-replicate is defined by selecting one defining contrast. For example, choose $\mathcal{I} = -ABC$. To reduce down to a one-fourth replicate we must select another defining contrast and then take the treatment combinations in the intersection of two selected halves. We can choose $-CDE$. This means that we choose those treatment combinations that enter into both the ABC contrast and the CDE contrast with a "$-$." As in the construction of blocks, the generalized interaction must be considered. The generalized interaction of ABC and CDE is $ABDE$. In fact, it turns out that the one-fourth replicate that we have selected comes from the treatment combinations that enter into $ABDE$ with a "+." Symbolically, we write $\mathcal{I} = -ABC = -CDE = +ABDE$. Note that we are free to think of any one of ABC, CDE, or $ABDE$ as the generalized interaction of the other two. We were free to select any sign on two of the defining contrasts, but the sign on the generalized interaction is fixed as the product of the other two. To find the aliases we again use the multiplication scheme

$$A \times (\mathcal{I} = -ABC = -CDE = ABDE)$$

$$A = -A^2BC = -ACDE = A^2BDE$$

$$A = -BC = -ACDE = BDE.$$

Similarly,

$$B = -AC = -BCDE = ADE,$$

and so on. Every effect or interaction, i.e., every contrast computed from the data, has three aliases.

Finally, one must obtain the actual fraction, the actual set of treatment combinations that are to be used by the scientist. A formal procedure is to obtain solutions to the homogeneous system of equations

$$\begin{array}{ccccc} a & b & c & d & e \\ \hline i_A & +i_B & +i_C & & \\ & & +i_C & +i_D & +i_E \end{array} \begin{array}{l} \\ = 0 \text{ (modulo 2)} \quad \text{(from } ABC) \\ = 0 \text{ (modulo 2)} \quad \text{(from } CDE). \end{array} \qquad (14.1)$$

Adding the extra equation corresponding to $ABDE$ contributes nothing. We note in passing that while there are $2^5 = 32$ treatment combinations, each equation can be thought of as a restriction that reduces the number of solutions by a factor of 2. In this case there are eight solutions.

The solutions to the system are

$$\begin{array}{cccccl} 0 & 0 & 0 & 0 & 0 & (1) \\ 1 & 1 & 0 & 0 & 0 & ab \\ 0 & 0 & 0 & 1 & 1 & de \\ 1 & 1 & 0 & 1 & 1 & abde \\ 1 & 0 & 1 & 0 & 1 & ace \\ 0 & 1 & 1 & 0 & 1 & bce \\ 1 & 0 & 1 & 1 & 0 & acd \\ 0 & 1 & 1 & 1 & 0 & bcd. \end{array}$$

At this stage we must still check to see which fraction the design actually contains. As far as the experiment is concerned, we can use any one of the four fractions. However, if we want to communicate to someone else the exact nature of the fraction selected, we need to be specific. Look at the fraction listed, and note that, for example, *ab* enters into ABC with a $-$ sign. This is in the fraction $\mathcal{I} = -ABC$. The same treatment, *ab*, also enters into CDE with a $-$ sign, so the fraction $\mathcal{I} = -CDE$ is the second design generator. The generalized interaction is $\mathcal{I} = +ABDE$ by multiplication, and the treatments in this quarter fraction do satisfy the requirement of entering into $ABDE$ with a positive sign. We have the $\mathcal{I} = -ABC = -CDE = ABDE$ fraction. Display 13.1 illustrates using SAS® to generate the treatment combinations.

Occasionally, one is presented with a fraction and needs to determine the exact identity. For example, one may know that ABC is one of the defining contrasts, but it is necessary to establish whether it is the + or the $-$ fraction. A reasonable technique is to consider the expansion of $(a-1)(b-1)(c-1)(d+1)(\dots)$. If any term in the set in question would appear in this expansion with a "+," one knows immediately that it is part of the $\mathcal{I} = +ABC$ fraction.

14.3.3 Identifying Aliases

The rule used to identify aliases says that, for example, in a half replicate of the 2^5 system defined by $\mathcal{I} = -ABCDE$, the AB interaction is aliased with $-CDE$. Specifically, the rule is to multiply symbols,

$$AB(\mathcal{I} = -CDE)$$

$$AB = -CDE,$$

showing that AB is aliased with $-CDE$. The reason for this rule becomes clear when one considers that the $\mathcal{I} = -ABCDE$ fraction consists of solutions to

$$i_A + i_B + i_C + i_D + i_E = 0 \text{ (modulo 2)}$$

and notices that any solutions will also satisfy both equations in the pair

$$i_A + i_B = 0 \text{ (modulo 2)}$$

$$i_C + i_D + i_E = 0 \text{ (modulo 2)}$$

or in the pair

$$i_A + i_B = 1 \text{ (modulo 2)}$$

$$i_C + i_D + i_E = 1 \text{ (modulo 2)}.$$

This means that the contrast named AB corresponding to $i_A + i_B = 0$ or 1 (modulo 2) and the contrast named CDE corresponding to $i_C + i_D + i_E = 0$ or 1 (modulo 2) are identical, except for the sign.

14.3.4 One-Eighth Fraction

Consider eight treatment combinations from a 2^6 factorial, i.e., a $2^{(6-3)}$ fraction from a factorial with six factors each at two levels. Trial-and-error searching reveals the defining contrasts

$$\mathcal{I} = ACE = BDE = ABCD = BCF = ABEF = CDEF = ADF$$

where

$$\begin{aligned}
ABCD &= \text{generalized interaction of } ACE \text{ and } BDE \\
ABEF &= \text{generalized interaction of } BCF \text{ and } ACE \\
CDEF &= \text{generalized interaction of } BCF \text{ and } BDE \\
ADF &= \text{generalized interaction of } BCF \text{ and } ABCD.
\end{aligned}$$

Since there will be only eight observations, it follows that only seven independent contrasts can be estimated. The aliases of the six main effects are

$$\begin{aligned}
A &= CE = ABDE = BCD = ABCF = BEF = ACDEF = DF \\
B &= ABCE = DE = ACD = CF = AEF = BCDEF = ABDF \\
C &= AE = BCDE = ABD = BF = ABCEF = DEF = ACDF \\
D &= ACDE = BE = ABC = BCDF = ABDEF = CEF = AF \\
E &= AC = BD = ABCDE = BCEF = ABF = CDF = ADEF \\
F &= ACEF = BDEF = ABCDF = BC = ABE = CDE = AD.
\end{aligned}$$

The fact that all interactions in the set of defining contrasts have three or more letters dictates that no main effects are aliases for other main effects. Six contrasts have been identified. There is one more contrast that can be estimated, one more contrast that is orthogonal to all the above main-effect contrasts. It is

$$AB = BCE = ADE = CD = ACF = EF = ABCDEF = BDF.$$

Statistical Analysis

Even if one assumes no two-factor interactions, there is only one degree of freedom for error, so no good F-test of hypothesis is possible. The only good options for statistical analysis are methods such as Daniel's (see Section 12.2.1) normal probability plot of the estimates or the Lenth–Dong (see Section 12.2.2) method for identifying active factors. In an exploratory experiment, one may simply pick out the effects that are larger than the others and decide that further study is necessary. An important feature of this plan is that main effects are clear of each other, in the sense that they are not aliased with one another and they are uncorrelated. This is often referred to as a *main-effects clear plan*. In exploratory work, one would typically follow an experiment of this type with more detailed work.

14.3.5 One-Sixteenth Fraction

It is possible to construct a $2^{(7-4)}$ fraction, i.e., an experiment to examine seven factors each at two levels using only eight treatment combinations, and yet have all main effects clear of one another. This is a one-sixteenth fractional factorial. A suitable set of defining contrasts is given by

$$\begin{aligned}\mathcal{I} &= ABC = ADE = BCDE = BDF = ACDF = ABEF \\ &= CEF = ABCDEFG = DEFG = BCFG = AFG \\ &= ACEG = BEG = CDG = ABDG. \end{aligned} \tag{14.2}$$

No two main effects are aliases for the same contrast, since all of the generalized interactions or defining contrasts contains at least three letters.

***Exercise 14.1:* 2^{7-4}**

(a) Set up the system of four equations that must be solved to obtain the symbols for a set of eight treatment combinations that constitute a fraction defined by the defining contrasts listed in expression (14.2). Notice that there is still some ambiguity about the exact fraction, since we have not specified signs on the various interactions in the list of defining contrasts. Notice that each treatment appears at the high level in the set selected exactly four times.

(b) Express the estimates of A and B as linear functions of the eight observations.

(c) Obtain expressions for Var$[\widehat{A}]$, Var$[\widehat{B}]$, and Cov$[\widehat{A}, \widehat{B}]$. Notice that the covariance is zero; the two are orthogonal contrasts. □

14.3.6 Control and Seven One-at-a-Time Trials

In this section we digress to reconsider some aspects of a problem that is encountered frequently in any statistical consulting practice, and was alluded to in an earlier section. It typically appears at the early stage of a research program, or in exploratory research, where the experimenter is faced with having to make choices among many potential factors but has limited resources. A much too common strategy is to make a control run and then test a sequence of factors one at a time. If the response differs appreciably from the control, that factor is judged to be important. The alternative proposed is to perform a fraction of a factorial plan.

To be specific, consider a case where the scientist has seven factors to test. The control and the seven individual tests require a total of eight runs. In our notation the responses would be denoted by $y_{(1)}$, y_a, y_b, y_c, y_d, y_e, y_f, and y_g. The estimate of the effect due to A is computed as

$$y_a - y_{(1)}.$$

This contrast has variance $2\sigma^2$. Similarly, the estimate of the effect due to B is

$$y_b - y_{(1)}.$$

This contrast also has variance $2\sigma^2$. The other estimates are similar.

Now compare with using the fractional factorial plan developed from defining contrast (14.2). This plan also requires eight runs and allows for seven factors. Exactly four of the treatment combinations involve a and four do not, four involve b and four do not, and so on. Now estimate A:

$$\frac{\text{sum of four trt. comb. with } a - \text{ sum of four trt. comb. without } a}{4}$$

This contrast has variance

$$(4\sigma^2 + 4\sigma^2)/16 = \sigma^2/2.$$

The corresponding estimate of B is

$$\frac{\text{sum of four trt. comb. with } b - \text{ sum of four trt. comb. without } b}{4}$$

This contrast also has variance $\sigma^2/2$. In addition, this contrast is orthogonal to the estimate of the A effect. In fact, all of the contrasts of the main effects $A, B, \ldots, G$ are mutually orthogonal. Also, all have variance $\sigma^2/2$.

When we compare these contrasts with the "simpleminded" approach, which is to use the eight observations, $y_{(1)}$, y_a, y_b, y_c, y_d, y_e, y_f, and y_g, we see that both strategies give estimates of the effects of the seven factors. In the case of the fractional factorial, the estimates all have variance $\sigma^2/2$, while in the other scheme, they have variance $2\sigma^2$, a fourfold increase in information, in return for no increase in amount of laboratory work.

The plan based on the factorial scheme is much more efficient at the price of a slightly more complex experiment. As far as interactions are concerned, assumptions are required in both cases. Consider the two cases if two-factor interactions should be present. In the factorial-based scheme, it can easily be shown that the estimate of A really estimates the sum of A and $BC + DE + FG$. Alternatively, in the "simple" scheme the estimate of A really estimates A minus the sum $AB + AC + AD + AE + AF + AG$. It is not clear where one has the advantage here. However, remember that one has four times the variance of the other.

An additional factor that is difficult to quantify, but favors the fractional factorial, is that all contrasts in the factorial are orthogonal (uncorrelated), while the contrasts in the other scheme are all correlated.

Exercise 14.2: One-at-a-Time Experiment

(a) Write out the models (μ, main effects, and two-factor interactions) for the set of eight observations at the treatment combinations selected via the defining contrasts in expression (14.2). Use these to find the expected value of the $\widehat{A}$ contrast (main effect and two-factor interactions only).

(b) Write out the models (μ, main effects, and two-factor interactions) for the set $y_{(1)}$, y_a, y_b, y_c, y_d, y_e, y_f, and y_g. Use these to find the expected value of the $\widehat{A}$ contrast (main effect and two-factor interactions only). Compare with the results in part (a). □

14.4 RESOLUTION

There is a need for a system to classify fractional factorial plans according to the types of aliasing relationships. The concept of resolution of a plan was developed to fill this need.

We define a *resolution* III plan as a fractional factorial plan in which no main effect is aliased with another main effect, and at least one main effect is aliased with at least one two-factor interaction. In the same vein, we define a *resolution* IV plan as a fractional factorial plan in which no main effect is aliased with a two-factor interaction, and at least one two-factor interaction is aliased with at least one other two-factor interaction, and a *resolution* V plan as one in which no two-factor interaction is aliased with another two-factor interaction, and at least one two-factor interaction and one three-factor interaction are aliased. In the context of the discussions in this chapter, this means that at least one word in the defining contrast contains three, four, or five letters, and no word contains fewer than three, four, or five letters, respectively.

Although some statisticians will consider this definition of resolution to be inadequate or possibly incomplete, it is not misleading and conveys the essence of the concept. A brief discussion of the difficulties with the definition can be found on page 88 of *Factorial Designs* by Raktoe et al. (1981) and under the entry for resolution in the *Encyclopedia of Statistical Sciences* (Kotz et al. 1988).

14.4.1 Projection to Smaller Factorials

An interesting implication of this definition of resolution is that if it is possible to ignore all factors in the experiment except for a (any) subset of $R-1$ factors in a resolution R fraction, a complete replicate of a factorial with $R-1$ factors remains. More formally, when a resolution R fraction is projected onto any subset of $R-1$ factors, it is a full factorial design. As a simple illustration, we consider the quarter replicate of the 2^5, defined by

$$\mathcal{I} = ABCD = CDE = ABE.$$

The treatment combinations selected are $\{(1), ab, cd, abcd, ace, bce, ade, bde\}$. This is a resolution III plan. It can be projected onto two full replicates of a 2^2 for any subset of two factors. Consider first A and B. If we remove (or disregard) factors C, D, and E, we are left with $\{(1), ab, (1), ab, a, b, a, b\}$.

Similarly, focusing on A and E and removing B, C, and D from the plan gives $\{(1), a, (1), a, ae, e, ae, e\}$, which we recognize as two complete replicates of a 2^2 system.

An alternative and more general way of seeing the projection property of a resolution R plan is to consider the expression for the defining contrast directly. This expression will contain at least one word with R letters and no word with fewer letters. Assume N factors in all. Now assume that only $R-1$ factors are real. Select a subset of $R-1$ letters. Two cases are possible. Either all are in one of the words of length R or more, or some number less than $R-1$ are in all of the words. Assume that all happen to be in one word. Looking at this situation tells one that the $(R-1)$-factor interaction is aliased with the effect or interaction defined by the remaining letters. However it was assumed that factors corresponding to these additional letters had no effect. Hence the $(R-1)$-factor interaction is estimable, as claimed. A similar argument holds for the alternate case as well. This projection property implies that within reason, one should be willing to use higher fractional factorials and include factors that may appear doubtful, without increasing the size of the experiment.

This is actually one of the features of the property of hidden replication in factorial experiments. If a resolution III, 2^{5-2} fraction and the statistical analysis of the data reveals that only two of the factors are active, i.e., have any important affect, the investigator is left with two replicates of a full factorial with the important factors.

Exercise 14.3: Projection of a Resolution III Plan onto Two Factors. Consider the main-effect plan generated in Exercise 14.1.

(a) Verify that if all factors but A and B are assumed to be nonactive, you have two complete replicates of the 2^2 factorial with factors A and B.

(b) Verify that if all factors but C and D are assumed to be nonactive, you have two complete replicates of the 2^2 factorial with factors C and D.

(c) Convince yourself that the same statement holds for any pair of factors selected. □

14.4.2 Plans with Minimum Aberration

It turns out that in the search for plans with good all-around properties, the concept of resolution needs to be refined somewhat. Two basic ideas underlying the concept of resolution are that main effects are more important than two-factor interactions, two-factor interactions are more important than three-factor interactions, and so on, and that all main effects are equally important, all two-factor interactions are equally important, all three-factor interactions are equally important, and so on. Also, the concept generally implies that once the investigator is forced to give up an rth-order interaction, all other rth-order interactions are equally expendable. We illustrate the inadequacy of this with the example of three 2^{7-2} plans defined by

$$\mathcal{I} = ABCF = BCDG = ADFG$$
$$\mathcal{I} = ABCF = ADEG = BCDEFG$$
$$\mathcal{I} = ABCDF = ABCEG = DEFG.$$

All are resolution IV. However, they carry with them distinctly different aliasing implications. Under the assumption that three-factor interactions (or higher) are negligible, the plans lead to the following aliasing relationships:

First plan	$AB = CF$, $AC = BF$, $AF = BC = DG$, $BD = CG$, $BG = CD$, $AD = FG$, $AF = DG$, $AG = DF$
Second plan	$AB = CF$, $AC = BF$, $AF = BC$, $AD = EG$, $AE = DG$, $AG = DE$
Third plan	$DE = FG$, $DF = EG$, $DG = EF$

In all cases the two-factor interactions not mentioned are estimable, provided that one assumes that all higher-order interactions are negligible. The third plan provides information on many more two-factor interactions than either of the other two. We should have suspected this by noting that this plan has only one four-letter word (besides $\mathcal{I}$) in the set of defining contrasts, while the other two have three and two four-letter words, respectively. If one wants to classify or organize fractional factorial plans, there is an obvious need to extend the notion of resolution to take into account the numbers of words of various lengths in addition to the simple criterion of length of shortest word.

Fries and Hunter (1980) formally define a plan as having *minimum aberration* if the plan minimizes the number of words in the set of defining contrasts that are of minimum length. More formally, if D_1 and D_2 are two 2^{N-m} fractional factorial plans with w_{1u} and w_{2u} words of length u ($1 \leq u \leq N$), and r is the smallest value of u for which $w_{1u} \neq w_{2u}$, then D_1 has less aberration than D_2 if $w_{1r} < w_{2r}$. If there exists no plan with less aberration than D_1, then D_1 has minimum aberration. This is a natural extension of the concept of resolution. However, this concept is defined in terms of the lengths of the words in the set of defining contrasts, and consequently, really applies only to the regular p^N (with p a prime or a power of a prime), or equivalently, the geometric system of plans. Up to this point in this chapter we have considered only the geometric 2^N system.

Catalog of Minimum Aberration Fractional Factorial Plans

It is not difficult to construct minimum aberration plans by hand using the catalogs described in this section. Design of experiments software such as SAS® Proc Factex also makes it easy to construct these designs. To present the catalog of minimum aberration 2^{N-k} plans compactly, we need to adjust our notation somewhat. In place of the symbols A, B, C, ... we use F_1, F_2, F_3, ... to represent the factors. With this notation, the symbols $F_1F_2F_6$ and $F_2F_5F_{14}$ represent two different three-factor interactions. We look at separate cases:

- We note immediately that the minimum aberration 2^{N-1} plan is obtained using $\mathcal{I} = F_1F_2 \cdots F_N$.
- Robillard (1968), in a thesis at the University of North Carolina at Chapel Hill, established the following rules for finding the minimum aberration 2^{N-2} plan. He considered the three cases $r = 0$, 1, or 2, where $N - 2 = 3m + r$. For $r = 0$, define

$$\mathcal{I} = \left(\prod_{i=1}^{2m} F_i\right) F_{N-1} = \left(\prod_{i=m+1}^{3m} F_i\right) F_N.$$

For $r = 1$, define

$$\mathcal{I} = \left(\prod_{i=1}^{2m+1} F_i\right) F_{N-1} = \left(\prod_{i=m+1}^{3m+1} F_i\right) F_N. \tag{14.3}$$

For $r = 2$, define

$$\mathcal{I} = \left(\prod_{i=1}^{2m+1} F_i\right) F_{N-1} = \left(\prod_{i=m+1}^{3m+2} F_i\right) F_N.$$

The resolution of these plans is $[2N/3]$, where $\lfloor x \rfloor$ is the largest integer less than or equal to x. These are all 1/4 replicate plans.
- Chen and Wu (1991) provide plans for the 2^{N-3} 1/8 replicate plans. They write $N = 7m + r, \quad 0 \le r \le 6$. Then for $j = 1, \ldots, 7$ they write

$$B_j = \begin{cases} \left(\prod_{i=jm-m+1}^{jm} F_i\right) F_{7m+j} & \text{where } j \le r \\ \prod_{i=jm-m+1}^{jm} F_i & \text{otherwise.} \end{cases} \tag{14.4}$$

These split the factors into seven approximately equal groups. The defining contrasts are then written as

$$\begin{aligned} \mathcal{I} &= B_1B_2B_3B_4 = B_1B_2B_6B_7 = B_3B_4B_6B_7 \\ &= B_1B_3B_5B_7 = B_2B_4B_5B_7 = B_2B_3B_5B_6 = B_1B_4B_5B_6. \end{aligned}$$

The resolution of this plan can be shown to be $[4N/7]$ when $r \ne 2$ and $([2N/3] - 1)$ otherwise.

- Chen and Wu (1991) also provide plans for the 2^{N-4} cases, the one-sixteenth replicate plans. They write $N = 15m + r$, $1 \le r < 15$ and then for $j = 1, \ldots, 15$,

$$B_j = \begin{cases} \left(\prod_{i=jm-m+1}^{jm} F_i\right) F_{15m+j} & \text{where } j \le r \\ \prod_{i=jm-m+1}^{jm} F_i & \text{otherwise.} \end{cases}$$

These split the factors into 15 approximately equal groups. When $r \neq 5$, the base set of defining contrasts can then be written as

$$\begin{aligned} \mathcal{I} &= B_1 B_6 B_7 B_8 B_9 B_{12} B_{14} B_{15} = B_2 B_5 B_7 B_8 B_9 B_{11} B_{13} B_{15} \\ &= B_3 B_5 B_6 B_8 B_{10} B_{11} B_{14} B_{15} = B_4 B_5 B_6 B_7 B_{10} B_{12} B_{13} B_{15}. \end{aligned}$$

The full set of 15 defining contrasts involves all of the generalized interactions of this base set. When $r = 5$, B_5 and B_{15} are interchanged. These plans have resolution $[8N/15]$ when $r \neq 2, 3, 4, 6$, or 10 and $([4N/7]-1)$ otherwise.

- Chen (1992) extends the foregoing to 2^{N-5} plans that have minimum aberration for all values of N. However, the results are left in a form that is not readily accessible.
- Fries and Hunter (1980) conclude their paper with a general algorithm for finding minimum aberration 2^{N-k} plans, but state that as plans become large, this algorithm becomes inefficient. Franklin (1984) gives a description of another algorithm to construct minimum aberration plans. Rather than give this algorithm, we present a summary of the catalog of minimum aberration plans he has constructed for $1/32, \ldots, 1/512$ fractions in Appendix 14A.

Exercise 14.4: Minimum Aberration 2^{10-2} Plan. Consider $N = 10$. This implies that $m = 2, r = 2$. Verify that the plan is defined by $\mathcal{I} = F_1F_2F_3F_4F_5F_9 = F_3F_4F_5F_6F_7F_8F_{10} = F_1F_2F_6F_7F_8F_9F_{10}$. □

Exercise 14.5: Minimum Aberration 2^{11-2} and 2^{12-2} Plans. Display the defining contrasts for the minimum aberration 2^{11-2} and 2^{12-2} plans. □

Exercise 14.6: Minimum Aberration 2^{9-4} Plan

(a) Using the material given, complete the defining contrast for the minimum aberration plan for nine factors in $2^5 = 32$ runs.
(b) What is the resolution of the plan?
(c) Display the equations that must be solved to obtain the 32 treatment combinations. □

14.5 CONSTRUCTING FRACTIONAL REPLICATES BY SUPERIMPOSING FACTORS

The fractional factorial plans given in Section 14.4 are adequate for cases where limits on resources rule out the use of a complete replicate, but not enough is known about the problem to permit the investigator to be specific about interactions that can and cannot be safely ignored. The investigator must rely on general guidelines to select a plan of appropriate resolution. An alternative construction procedure, which is useful for 2^{N-k} fractions with a large number of factors and allows better control on what is sacrificed, is described in this section. The method is based on setting up the design matrix (model) for a full factorial with $N-k$ factors and then relabeling the columns. This is particularly simple if the objective is to construct a resolution III plan.

14.5.1 Resolution III Plans

Resolution III plans have a very important place in the early stages of any experimental program. They permit the exploration of many factors with the expenditure of the least effort. The emphasis in resolution III plans is on main effects under the assumption of no two-factor or higher interactions. Investigators are often turned off by the severity of the assumption of no two-factor interactions. In the geometric 2^N system we are presenting here, the only real alternative is to enlarge the study, i.e., include more observations. We will find in Chapter 16, however, that there are some alternatives with somewhat less severe assumptions. We will also find in Section 14.6 that resolution III plans can be converted to resolution IV quite easily. We recall that resolution IV plans have main effects free (not aliased with) of two-factor interactions.

Consider, now, constructing a main-effect plan with seven factors in eight experimental units a 2^{7-4} fractional factorial. Begin by noting that the number of experimental units, eight, is exactly 2^3. Now write out the model in full detail for the eight treatment combinations in a 2^3 factorial.

$$
\begin{aligned}
\mathrm{E}[y_{(1)}] &= \mu - \tfrac{1}{2}A - \tfrac{1}{2}B + \tfrac{1}{2}AB - \tfrac{1}{2}C + \tfrac{1}{2}AC + \tfrac{1}{2}BC - \tfrac{1}{2}ABC \\
\mathrm{E}[y_{a}] &= \mu + \tfrac{1}{2}A - \tfrac{1}{2}B - \tfrac{1}{2}AB - \tfrac{1}{2}C - \tfrac{1}{2}AC + \tfrac{1}{2}BC + \tfrac{1}{2}ABC \\
\mathrm{E}[y_{b}] &= \mu - \tfrac{1}{2}A + \tfrac{1}{2}B - \tfrac{1}{2}AB - \tfrac{1}{2}C + \tfrac{1}{2}AC - \tfrac{1}{2}BC + \tfrac{1}{2}ABC \\
\mathrm{E}[y_{ab}] &= \mu + \tfrac{1}{2}A + \tfrac{1}{2}B + \tfrac{1}{2}AB - \tfrac{1}{2}C - \tfrac{1}{2}AC - \tfrac{1}{2}BC - \tfrac{1}{2}ABC \\
\mathrm{E}[y_{c}] &= \mu - \tfrac{1}{2}A - \tfrac{1}{2}B + \tfrac{1}{2}AB + \tfrac{1}{2}C - \tfrac{1}{2}AC - \tfrac{1}{2}BC + \tfrac{1}{2}ABC \\
\mathrm{E}[y_{ac}] &= \mu + \tfrac{1}{2}A - \tfrac{1}{2}B - \tfrac{1}{2}AB + \tfrac{1}{2}C + \tfrac{1}{2}AC - \tfrac{1}{2}BC - \tfrac{1}{2}ABC \\
\mathrm{E}[y_{bc}] &= \mu - \tfrac{1}{2}A + \tfrac{1}{2}B - \tfrac{1}{2}AB + \tfrac{1}{2}C - \tfrac{1}{2}AC + \tfrac{1}{2}BC - \tfrac{1}{2}ABC \\
\mathrm{E}[y_{abc}] &= \mu + \tfrac{1}{2}A + \tfrac{1}{2}B + \tfrac{1}{2}AB + \tfrac{1}{2}C + \tfrac{1}{2}AC + \tfrac{1}{2}BC + \tfrac{1}{2}ABC
\end{aligned}
$$

Now consider the seven columns of the design matrix, using effects coding $(+1, -1)$ and omitting the columns of ones (for the intercept). The following array shows the + and − symbols obtained from the columns:

a	b	ab	c	ac	bc	abc
−	−	+	−	+	+	−
+	−	−	−	−	+	+
−	+	−	−	+	−	+
+	+	+	−	−	−	−
−	−	+	+	−	−	+
+	−	−	+	+	−	−
−	+	−	+	−	+	−
+	+	+	+	+	+	+
A	B	C	D	E	F	G

(14.5)

We treat the rows as the eight experimental units and associate the seven factors $A, B, \ldots, G$ in the planned experiment with the seven columns. The factors in the 2^{7-4} fractional factorial are labeled with capital letters at the bottom of each column. A "+" means that the factor is at the high level and a "−" means that the factor is at the low level. Note that each factor is at the high level four times and at the low level four times. It is easy to show that contrasts for all seven factors are orthogonal—hence, a main-effect plan. A main-effect plan for six or fewer factors in eight experimental units can be obtained by dropping any column. Also note that the + and − symbols in column 3 are the products of symbols in columns 1 and 2. This tells us that the AB interaction is aliased with C, or equivalently, the AC interaction with B, or A with BC—hence, a resolution III plan. Similarly, the symbols in column 5 are the product of the symbols in columns 1 and 4, and the symbols in column 4 are the product of symbols in columns 1 and 5. The easy way to see or remember this is to think of the columns as the a column, the b column, the ab column, and so on (from the original 2^3 factorial) and then think of ab as the interaction of a and b, c as the interaction of a and ac, and so on. The factors in the 2^3 full factorial are designated with lower case letters at the top of each column. A neat thing about this is that it is immediately clear that the factor assigned to column 3 is aliased with the two-factor interaction of the factors assigned to columns 1 and 2. Similarly, the factor assigned to column 1 is aliased with the two-factor interaction between the factors assigned to columns 6 and 7, as well as with the two-factor interaction between the factors assigned to columns 4 and 5. These insights become important when one is setting up a fractional factorial and is trying to make use of prior information. For example, one may know from experience or basic science that certain subsets of factors do not interact, while one suspects that certain other pairs of factors (for example, A and C) may interact. The plan is easily modified to accommodate this by leaving the appropriate interaction column unused.

The reader must be careful not to confuse the two different sets of labels we have used for the columns. We have labels $A, B, \ldots, G$ for the factors to be used in the experiment and the labels $a, b, ab, c, \ldots$ derived from the base plan. The latter set of labels is used only to identify aliasing relationships quickly. In fact, we can use the association between the two sets of labels to find the defining contrast for the fraction developed. We know that we have a 2^{7-4} fraction. This tells us that there are four independent interactions to form a basis for the set of defining contrast. The fact that A was assigned to a, B to b, and C to ab tells us that ABC is one word in the defining contrast. Similarly, we find that ADE, BDF, and $ABDG$ are words in the defining contrast. These four form a basis for the set of defining contrasts. We conclude that the fraction that we have constructed must be

$$\mathcal{I} = ABC = ADE = BCDE = BDF = ACDF = ABEF = CEF = ABDG$$
$$= CDG = BEG = ACEG = AFG = BCFG = DEFG = ABCDEFG.$$

From this we can get the full set of aliasing relationships. Notice that in this scheme, we do not need to construct the columns with + and − symbols until ready to perform the experiment.

14.5.2 Resolution IV Plans

The big feature of the resolution IV plans is that main effects are not aliased with two-factor interactions. The price for fewer assumptions is a larger number of runs.

This superimposing of factors can also be used to produce resolution IV plans. The trick is to assign new treatments only to columns that are labeled with an odd number of letters in the original 2^k system (the lowercase letters above). In this scheme, only columns a, b, c, abc, d, and so on are used and columns ab, ac, bc, and so on are not used. The defining contrasts of the plan actually constructed are obtained as before.

This method of superimposing factors can also be used to construct resolution V plans but becomes rather cumbersome. Note that resolution V plans are required if it is important to estimate two-factor interactions. We also note, in passing, that in our example the array of eight lines constructed from the models for the eight treatment combinations in a 2^3 contains a lot of "stuff" that we do not need when constructing the fractional factorial. All we really need is the 8×7 array of + and − symbols. Any device or trick that would give us such an array with the acceptable distribution of symbols is sufficient.

14.5.3 Additional Examples of the Superimposing Factors Method

As we have already seen, it is common practice to specify fractions by means of defining contrasts. Up to this point, our method of constructing fractional replicates from such a set of defining contrasts has been to set up the system of homogeneous equations and then search for solutions. An alternative and often simpler technique for finding the treatment combinations that satisfy a particular aliasing structure is to superimpose the factors onto an existing array. Rather than try to develop a general method or algorithm, we present the method by way of a series of examples.

Consider first the simple 2^{4-1} plan with the defining contrast $\mathcal{I} = ABCD$. This plan is a half replicate. It has four factors applied in eight treatment combinations. Since there are eight treatment combinations, begin with the 8×7 array obtained from the 2^3. The column labels in Yates' or standard order are a, b, ab, c, ac, bc, and abc. Now assign treatment A to column a, treatment B to column b, treatment C to column c, and treatment D to column abc. Notice that this last assignment is obtained from $\mathcal{I} = ABCD$, which implies that $D = ABC$. Also notice that as before, one does not really need to write the full 8×7 array. Writing column headings is sufficient. One can then easily generate the columns required.

A second example is the Robillard 2^{N-2} plan in expression (14.3). Consider $N = 6$, implying that $m = 1$ and $r = 1$ in the notation of the example. The basis set of defining contrasts for this 2^{6-2} plan consists of $F_1F_2F_3F_5$ and $F_2F_3F_4F_6$. The fraction constructed is defined by

$$\mathcal{I} = F_1F_2F_3F_5 = F_2F_3F_4F_6 = F_1F_4F_5F_6.$$

There are 16 treatment combinations. The treatment combinations are assigned to the columns of a 2^4 factorial. The column labels are a, b, ab, c, $\ldots$, $abcd$. First, assign F_1 to column a, F_2 to column b, F_3 to column c, and F_4 to column d. Then, assign F_5 to column abc. This fifth assignment comes from the word $F_1F_2F_3F_5$ in the defining contrast, which implies that F_5 is aliased with $F_1F_2F_3$, i.e., $F_5 = F_1F_2F_3$. Similarly, F_6 is assigned to column bcd, which comes from $F_2F_3F_4F_6$ or equivalently, $F_6 = F_2F_3F_4$.

For a slightly more complex example, consider the 2^{N-5} example for 11 factors by Franklin (1984) in Appendix 14A. The base set of contrasts given in place of the full set of defining contrasts consists of $\{F_1F_7F_8F_9F_{10}F_{11}$, $F_2F_6F_8F_9F_{10}F_{11}$, $F_3F_6F_7F_9F_{10}$, $F_4F_6F_7F_8F_{10}$, $F_5F_6F_7F_8F_9F_{10}\}$. This experimental plan involves $2^6 = 64$ treatment combinations. The 63 columns obtained from the 2^6 can be identified in standard order as a, b, ab, c, ac, bc, abc, $d, \ldots, abcdef$. The plan involves 11 factors. Assign the six factors, F_6, F_7, F_8, F_9, F_{10}, F_{11} to the six single-letter or main-effect columns. Notice that F_1 is in only in the first contrast, F_2 in only in the second contrast, $\ldots$, F_5 in only in the fifth, and these are not in the set to be assigned to single-letter columns. Assume the assignment $F_6 \to a$, $F_7 \to b, \ldots, F_{11} \to f$. To complete the plan, assign F_1 to column $bcdef$, F_2 to $acdef$, F_3 to $abde$, F_4 to $abce$, and F_5 to $abcde$. Again, one does not really need to write the full set of 63 columns. The columns corresponding to a–f are easy to write. The five interaction columns corresponding to $bcdef, \ldots, abcde$ for factors $F_1, \ldots, F_5$ can then be generated as combinations of appropriate columns.

An example such as the 2^{N-3} plan from equation (14.4) requires a bit more thought but can also be handled easily. Consider $N = 8$. This implies, in the notation of the plan, that $m = 1$ and $r = 1$. This is a 2^{8-3} plan. We have

$$\begin{aligned} B_1 &= F_1F_8 \\ B_2 &= F_2 \\ B_3 &= F_3 \\ B_4 &= F_4 \\ B_5 &= F_5 \\ B_6 &= F_6 \\ B_7 &= F_7. \end{aligned}$$

The defining contrasts are

$$\begin{aligned} \mathcal{I} &= F_1F_2F_3F_4F_8 = F_1F_2F_6F_7F_8 = F_3F_4F_6F_7 = F_1F_3F_5F_7F_8 \\ &= F_2F_4F_5F_7 = F_2F_3F_5F_6 = F_1F_4F_5F_6F_8. \end{aligned}$$

A basis for this set is $\{F_1F_2F_3F_4F_8, F_1F_2F_6F_7F_8, F_2F_4F_5F_7\}$. In this set F_3 is unique to the first, F_6 to the second, and F_5 to the third. Now, assign factors F_1, F_2, F_4, F_7, and F_8 to the single-letter columns of the 2^5. Assume that these are a, b, c, d, and e. Then, finally, assign F_3, F_5, and F_6 to the appropriate interaction columns, i.e., to $abce$, $abde$, and bcd, respectively.

Given a set of defining contrasts, it is always possible to select a basis set of contrasts such that each will contain a factor not found in any of the other contrasts in the basis set. From that point on, the construction is mechanical. Once the basis set has been constructed, it is simple to find the full set of defining contrasts and the aliasing relationships.

14.6 FOLDOVER TECHNIQUE

A third technique, developed in a seminal paper by Box and Wilson (1951) is to add the *foldover* to an existing plan. Although this technique is normally referred to as adding the foldover (the original Box–Wilson term), we will eventually see (Section 19.3) that it would be better to use the term *complete foldover*. If a 2^{N-p} plan is given using the (0–1) notation, the foldover or complete foldover is obtained by replacing all 0's by 1's and 1's by 0's. Equivalently, we can think of switching high levels of all factors to low levels and low levels to high levels. Adding the complete foldover to a resolution III plan converts it to at least a resolution IV plan. A common practice is to follow a resolution III experiment with the complete foldover to break up alias relationships between main effects and two-factor interactions. A particularly nice feature is that the combination of original plan and the complete foldover form two orthogonal blocks. If extraneous conditions have changed slightly, their effect gets eliminated with the block effect. We must warn the reader that use of the term *complete foldover* (foldover) is somewhat confusing. Some use the term *foldover* to refer to the additional points only, while others use the term for the original set plus the additional. One must rely on the context.

We also point out that if a fraction in the geometric 2^N system is identified by a defining contrast consisting of a series of words, each with appropriate sign, the complete foldover fraction is defined by the same set of words, but with signs reversed on all words with an odd number of letters. For example, the foldover of the fraction defined by $\mathcal{I} = -ABC = -BCDE = ADE$ is defined by $\mathcal{I} = ABC = -BCDE = -ADE$. The foldover technique is explored more fully in Chapter 19.

Exercise 14.7: Proof of Foldover. The object of this exercise is to demonstrate that adding the foldover to a resolution III plan creates a resolution IV plan. Begin with the 8×7 array in (14.5) and add the eight rows of the foldover on the bottom, creating a 16×7 array. Now, construct the interaction column for any two columns in the plan. Verify that if you consider only the eight rows of the original plan, the interaction is perfectly correlated with one of the other factors in the experiment, and if you consider all 16 rows, these two are orthogonal. Convince yourself that this also holds for any other pair of factors in the plan. The 16×7 array is the base for a resolution IV plan. □

14.7 FRANKLIN–BAILEY ALGORITHM FOR CONSTRUCTING FRACTIONS

14.7.1 Algorithm

The methods we have used in the preceding sections are useful once we have specified a set of contrasts or *design generators* for determining which factors are aliased with each other and for finding the treatment set that satisfies the particular aliasing structure. They are not helpful, however, for telling us how to choose the design generators that will give a particular aliasing structure. When actually faced with the task of designing an experiment in which considerable information exists about effects and interactions that may be important, it is more natural to think about what main effects and interactions we are interested in estimating and testing than to think about which effects to use as defining contrasts. For example, assume that we want the smallest possible fraction of the 2^5 which allows us to estimate all main effects as well as the AB and BE interactions, while assuming that all other interactions are negligible. Greenfield (1976) presented a discussion of this problem, and Franklin and Bailey (1977) and Franklin (1985) presented a systematic algorithm that always generated the smallest possible fraction that would allow estimation of these effects.

We present the algorithm in terms of a small specific example. Assume that the study involves five factors, A, B, C, D, and E. In addition to the five main effects, the investigator wishes to estimate interactions AB and BE and is willing to assume that all other interactions are negligible. The experiment will require at least $2^3 = 8$ treatment combinations, since seven contrasts are to be estimated. There is no guarantee, at this point, that this will be sufficient. Following Franklin and Bailey, we define the *requirements set*,

$$\{A, B, C, D, E, AB, BE\}.$$

In words, this is simply the set of effects and interactions that are required, because the investigator wants to estimate them.

The experimenter's requirements imply that certain interactions cannot be used as defining contrasts. In fact, none of the elements in the requirements set nor the generalized interaction of any pair can be used in the set of defining contrasts. This defines the *ineligible set* as

$$\{\mathcal{I}, A, B, AB, C, AC, BC, ABC, D, AD, BD, ABD,$$
$$CD, E, AE, BE, ABE, CE, BCE, DE, BDE\}$$

and the *eligible set* as $\mathcal{I}$ plus the complement of the ineligible set. In this discussion, the symbol $\mathcal{I}$ will be interpreted at various points as the mean and included as an effect or interaction. It is included in both the eligible and the ineligible sets.

We now attempt to find a set of 2^3 treatment combinations that satisfies the experimenter's needs. First, generate a set of $2^3 = 8$ basic effects $\{\mathcal{I}, A, B, AB, C, AC, BC, ABC\}$. The set of basic effects should be chosen to include only effects and interactions from the ineligible set, if possible. In the example this accounts for three of the five factors in the experiment. Then, construct a table with rows labeled with the basic effects and the columns with the two additional factors (Table 14.1). Next, fill in the table by examining, in turn, all of the generalized interactions of the row label (basic effect) and the column label (added factor). At this stage we are looking for interactions that can be used as generators for defining contrasts. If the generalized interaction is in the ineligible set, insert a "$\star$" in the cell. Otherwise, put the interaction in the cell. The interaction is eligible.

In our case, the resulting table has the form shown in Table 14.2.

Recall that we are attempting to construct a 2^{5-2} fraction and that there will be three interactions, two forming a basis and their generalized interaction, which together with $\mathcal{I}$, constitute the group of defining contrasts. The possible sets of defining contrasts that lead to suitable fractions are obtained by selecting interactions, one from each of the columns labeled "Added Factors." Any pair, such that their generalized interaction is not in the ineligible set, provides a suitable basis. In this case, the possible choices are as shown in Table 14.3.

The conclusion is that we have found two suitable plans, two plans for a 2^{5-2} that preserve all five main effects and the two two-factor interactions, AB and

Table 14.1 Treatment Combination Setup

Basic Effects	Added Factors	
	D	E
$\mathcal{I}$		
A		
B		
AB		
C		
AC		
BC		
ABC		

Table 14.2 Final Table Form

Basic Effects	Added Factors	
	D	E
$\mathcal{I}$	⋆	⋆
A	⋆	⋆
B	⋆	⋆
AB	⋆	⋆
C	⋆	⋆
AC	ACD	ACE
BC	BCD	⋆
ABC	$ABCD$	$ABCE$

Table 14.3 Possible Choices

From Col. D	From Col. E	Gen. Int.	
ACD	ACE	DE	Not suitable
ACD	$ABCE$	BDE	Not suitable
BCD	ACE	$ABDE$	Suitable
BCD	$ABCE$	ADE	Suitable
$ABCD$	ACE	BDE	Not suitable
$ABCD$	$ABCE$	DE	Not suitable

BE. The two defining relationships are

$$\mathcal{I} = BCD = ACE = ABDE$$

$$\mathcal{I} = BCD = ABCE = ADE.$$

If the choice of basic effects can be restricted to only ineligible effects and interactions, this algorithm will produce all possible sets of defining contrasts.

The proof of the algorithm consists of showing that it is really nothing more than a systematic way of looking at all possible choices for defining contrasts.

A strong feature of the Franklin–Bailey algorithm is not only that it finds a plan, if one exists, but it finds the full set of suitable plans, if several exist. The investigator is then able to select the best of those possible.

Exercise 14.8: Franklin–Bailey to Find a 2^{5-2}. Show that it is not possible to find a 2^{5-2} fraction that preserves all five main effects and the AB and CD interactions. □

14.7.2 Programming the Franklin–Bailey Algorithm

Although we have illustrated the Franklin–Bailey algorithm for the 2^N system, the algorithm is quite general in the sense that it extends to factorials with many levels. However, it is particularly well suited for programming on a computer in the 2^N system. The trick is to use the conventional $\mathcal{I}$, A, B, AB, ... notation in Yates' order but replace the words with numbers 0, 1, 2, ..., $2^N - 1$ written in binary. The generalized interaction between any two can then be obtained as the "exclusive or" function of the two numbers. We illustrate for the 2^6 in the following array:

$$
\begin{array}{ccccc}
\mathcal{I} & \Rightarrow & 0 & \Rightarrow & 000000 \\
A & \Rightarrow & 1 & \Rightarrow & 000001 \\
B & \Rightarrow & 2 & \Rightarrow & 000010 \\
AB & \Rightarrow & 3 & \Rightarrow & 000011 \\
C & \Rightarrow & 4 & \Rightarrow & 000100 \\
\vdots & \Rightarrow & \vdots & \Rightarrow & \vdots \\
ABCDEF & \Rightarrow & 2^6 - 1 = 63 & \Rightarrow & 111111.
\end{array}
$$

The third column presents each effect and interaction as an integer in binary. The generalized interactions of, say, AB and A is obtained using the "exclusive or" function of 000011 and 000001, i.e., 000010.

The set of 2^r basic effects in the Franklin–Bailey algorithm are represented by the integers from 0 to $2^r - 1$, and the ineligible set, the appropriate subset of these integers. The table of eligible generators for the defining contrasts is then obtained by a series of loops, first to generate the interactions, and then a set of loops to check which interactions form an eligible set. At each step, generalized interactions are obtained by using the "exclusive or" function. The algorithm is fast, simple to program, and the only real limitation to the implementation is the word size in the computer.

14.8 IRREGULAR FRACTIONS OF THE 2^N SYSTEM

One of the problems with fractions of the 2^N system is that one is often rather limited in experiment size. For example, with six factors, one has available eight of 64 or 16 of 64 or 32 of 64 and nothing in between. An experimenter may well be in a position where he needs a plan for 48 units in the 2^6. In this section we look at a class of techniques that can be used in cases such as this. We begin with a general example and then specialize to a number of small specific plans to clarify the details. Two key references in this material are John (1961, 1962).

14.8.1 Three-Fourths Replicate Fraction

We begin with the 2^N factorial. A quarter replicate is defined by the relationship

$$\mathcal{I} = +P = +Q = +PQ,$$

where P, Q, and PQ represent effects or interactions in the 2^N system. Clearly, each of these terms represents the generalized interaction of the other two. It follows that if we examine the fraction defined, we will find that each effect has three aliases; i.e., the complete set of effects and interactions is divided into what we can call alias sets of size 4. For example, the effect (or interaction) R is a member of the alias set $\{R, RP, RQ, RPQ\}$. Notice that unless specially mentioned, we will use the term *effect* to denote either a main effect or an interaction. Similarly, unless qualified, the term *interaction* can represent either a main effect or an interaction among two or more factors.

There are three remaining quarter replicates, defined by

$$\mathcal{I} = +P = -Q = -PQ \quad \text{(i)}$$

$$\mathcal{I} = -P = +Q = -PQ \quad \text{(ii)}$$

$$\mathcal{I} = -P = -Q = +PQ. \quad \text{(iii)}$$

Now, consider the three-fourths replicate formed by combining the fractions (i), (ii), and (iii). We look at pieces of the plan, in turn. First, consider the half replicate provided by (i) and (ii). We see that this half replicate is also defined by

$$\mathcal{I} = -PQ.$$

Now, examine what happens to the members of the alias set that included R. Looking at the four effects, in turn, we have the alias relationships $R = -RPQ$ and $RP = -RQ$.

Next, look at the half provided by (i) and (iii). This half replicate is also defined by

$$\mathcal{I} = -Q.$$

The alias relationships now are $R = -RQ$ and $RP = -RPQ$.

Finally, the last half replicate defined by (ii) and (iii) or

$$\mathcal{I} = -P$$

provides the alias relationships $R = -RP$ and $RQ = -RPQ$.

We now notice that in the three-fourths replicate, any three of the effects in the alias set that included R become estimable if one is able to assume a priori that the remaining effect is negligible. For example, if RQ is assumed negligible, R, RP, and RPQ all become estimable. Each is estimable from the portion of the experiment where it is aliased with RQ. It is estimated as the

$$\frac{\text{sum of } 2^{N-2} \text{ responses}}{2^{N-2}} - \frac{\text{sum of } 2^{N-2} \text{ different responses}}{2^{N-2}}.$$

Under the usual assumptions of uncorrelated homogeneous errors, the variance of this estimate is $\sigma^2/2^{N-3}$. We can also look at covariances between estimates. Clearly, this is

$$\text{Cov}\left(\left[\frac{\text{sum of } 2^{N-2} \text{ resp.}}{2^{N-2}} - \frac{\text{sum of } 2^{N-2} \text{ resp.}}{2^{N-2}}\right],\right.$$
$$\left.\left[\frac{\text{sum of } 2^{N-2} \text{ resp.}}{2^{N-2}} - \frac{\text{sum of } 2^{N-2} \text{ resp.}}{2^{N-2}}\right]\right).$$

Now, notice that we are dealing with four sets of responses. Either the first and third, or the first and fourth, have 2^{N-3} elements in common (because they come from the same quarter replicate) and correspondingly, either the second and fourth, or second and third, have 2^{N-3} elements in common. The covariance will be $+\sigma^2/2^{N-2}$ in the first instance and $-\sigma^2/2^{N-2}$ in the second.

The summarizing rule is simple. If the defining relationship for quarter fraction being dropped contains an effect T (either P, Q or PQ) with a positive sign, estimates of R and RT will be positively correlated, and if T has a negative sign, estimates of R and RT will be negatively correlated. If two of the four, say RPQ and RP, are both assumed negligible, R and PQ both become estimable in the (i) plus (ii) and the (ii) plus (iii) half replicates and remain aliased in the (i) plus (iii) half.

Two other cases need to be considered. First, consider two effects, U_1 and U_2, which are not in the set $\{R, RP, RQ, RPQ\}$ and are in different alias sets. Contrasts associated with $\{U_1, PU_1, QU_1, PQU_1\}$ are orthogonal to contrasts associated with $\{U_2, PU_2, QU_3, PQU_4\}$ in all quarters and also to contrasts associated with $\{R, RP, RQ, RPQ\}$. They are fully orthogonal to one another. Finally, the estimates of the set P, Q, and PQ are special. In the three half replicates examined, we find the alias relationships $P = -Q$, $P = -PQ$, and $Q = -PQ$, respectively. One of the three must be assumed negligible a priori for the other two to become estimable. Each becomes estimable from a half replicate. Clearly, the estimates are correlated. The statistical analysis appears to be complex. However, it is easily obtained using a general least squares computer program, such as `Proc GLM` in SAS®.

14.8.2 Three-Fourths Replicate of the 2^3 Case

The three-fourths replicate of the 2^3 consists of six of the eight treatment combinations. Consider leaving out the fourth (two treatment combinations), defined by the relationship

$$\mathcal{I} = +AB = +BC = +AC.$$

This fourth consists of the pair of treatment combinations $\{(1), abc\}$. We construct the plan by combining the three-fourths replicates, (i) defined by $\mathcal{I} = -AB = +BC = -AC$, consisting of $\{a, bc\}$; (ii) defined by $\mathcal{I} = -AB = -BC = +AC$, consisting of $\{b, ac\}$; and (iii) defined by $\mathcal{I} = +AB = -BC = -AC$, consisting of $\{c, ab\}$. The half replicate from combining (i) and (ii) corresponds to $\mathcal{I} = -AB$ consists of $\{a, b, bc, ac\}$, the half replicate from combining (i) and (iii) corresponds to $\mathcal{I} = -AC$ consists of $\{a, c, bc, ab\}$, and the half from combining (ii) and (iii) corresponds to $\mathcal{I} = -BC$ consists of $\{b, c, ac, ab\}$. We assume that the experimenter has decided a priori that ABC is negligible.

Now look at contrasts in detail. Note that in the fraction left out, A is in an alias set with B, ABC, and C. Consequently, in the estimation, they behave as a set. We will construct contrasts and evaluate them using the model developed in Section 10.2.3.

In the $\mathcal{I} = -AB$, i.e., (i) plus (ii) half replicate, we can compute three mutually orthogonal contrasts:

$$\mathrm{I}_1 : (y_a + y_{ac})/2 - (y_b + y_{bc})/2,$$

which has expected value $A - B$,

$$\mathrm{I}_2 : (y_{ac} + y_{bc})/2 - (y_a + y_b)/2,$$

which has expected value C, and

$$\mathrm{I}_3 : (y_a + y_{bc})/2 - (y_b + y_{ac})/2,$$

which has expected value $BC - AC$.

Similarly, in the $\mathcal{I} = -AC$, i.e., (i) plus (iii) half replicate, we can compute three mutually orthogonal contrasts:

$$\mathrm{II}_1 : (y_a + y_{ab})/2 - (y_c + y_{bc})/2,$$

which has expected value $A - C$,

$$\mathrm{II}_2 : (y_{ab} + y_{bc})/2 - (y_a + y_c)/2,$$

which has expected value B, and

$$\mathrm{II}_3 : (y_a + y_{bc})/2 - (y_c + y_{ab})/2,$$

which has expected value $BC - AB$.

Finally, in the $\mathcal{I} = -BC$, i.e., (ii) plus (iii) half replicate, we can compute another set of three mutually orthogonal contrasts:

$$\text{III}_1 : (y_b + y_{ab})/2 - (y_c + y_{ac})/2,$$

which has expected value $B - C$,

$$\text{III}_2 : (y_{ab} + y_{ac})/2 - (y_b + y_c)/2,$$

which has expected value A, and

$$\text{III}_3 : (y_b + y_{ac})/2 - (y_c + y_{ab})/2,$$

which has expected value $AC - AB$.

At this point, the claims for estimability are clear. Linearly independent (not orthogonal) contrasts I_2, II_2, and III_2 establish that estimates of A, B, and C are available. The AB, AC, and BC interactions appear only in expected values of I_3, II_3, and III_3. Consequently, their estimates are available from only these contrasts. However, $\text{I}_3 - \text{II}_3 = \text{III}_3$, i.e., there are only two linearly independent contrasts. If we assume that any one of AB, AC, or BC is negligible, the remaining two become estimable. This establishes the five parameters (contrasts) which are estimable from the six observations. These are exactly the conclusions established in the general case in Section 14.8.1. Finding *best linear unbiased estimates* requires a bit more work.

A simple way of obtaining the best linear unbiased estimates is to rewrite the model developed in Section 10.2.3, replacing the $\frac{1}{2}$'s with 1's. Then any least squares regression program will yield proper estimates of the mean, A, B, C, and any two of AB, AC, or BC directly. The change in coefficients will cause the parameters to be reduced by a factor of .5. We illustrate the computations using SAS® in Displays 14.1 and 14.2. The values for y_a, y_b, y_c, y_{ab}, y_{ac}, and y_{bc} have been generated using $\mu = 10$, $A = 3$, $B = 2$, $C = 2$, $AB = 1$, and normal(0,.5) errors.

The AC estimate in the analysis represents one degree of freedom for error. The experiment is obviously too small to yield good tests of significance and really has a place only if observations are costly and error variance is small. The half normal plot (not shown) of the estimates with a line going through the points and anchored at (0, 0) indicates that A, B, and C are real and there is a question about $A * B$.

14.8.3 Three-Fourths Replicate of the 2^4 Case

We next consider the 12-run, three-fourths replicate of the 2^4 obtained by leaving out the quarter replicate defined by $\mathcal{I} = +AC = +ABD = +BCD$. We assume that all three- and four-factor interactions are negligible. Our plan consists of the union of quarter (i), defined by

$$\mathcal{I} = -AC = +ABD = -BCD.$$

```
data one ;
 input a b c y @@ ;
datalines ;
 1 -1 -1  8.6081   -1  1 -1  5.8093
-1 -1  1  9.0531    1  1 -1 14.6488
 1 -1  1 13.0362   -1  1  1  9.5838
ods output parameterestimates = est ;
proc glm data = one ;
 model y = a b c a*b a*c ;
* Delete intercept in preparation of half  ;
*                    normal plot.          ;
data two ; set est ;
 if parameter = 'Intercept' then delete ;
 estimate = abs(estimate) ;
proc sort data = two ; by estimate ;
* Add quantiles to data set ;
data three ; set two ;
 q = probit((5+_n_-.5)/(2*5)) ;
* Note data set has 5 observations.  ;
proc print ;
proc gplot data = three ; plot  estimate*q ; run ;
```

Display 14.1 Using SAS® to analyze data from three-fourths replicate.

Parameter	Estimate	q
a*c	0.16340000	0.12566
a*b	1.37750000	0.38532
b	1.64285000	0.67449
c	2.05065000	1.03643
a	3.20565000	1.64485

Display 14.2 Output from code in Display (14.1).

consisting of *cd*, *a*, *bc*, and *abd*, quarter (ii) defined by

$$\mathcal{I} = -AC = -ABD = +BCD$$

consisting of *c*, *ab*, *ad*, *bcd* and quarter (iii) defined by

$$\mathcal{I} = +AC = -ABD = -BCD$$

and consisting of (1), *acd*, *bd*, and *abc*.

The alias sets in the quarters are (the mean, AC, ABD, and BCD), (A, C, BD, and $ABCD$), (B, ABC, AD, and CD), and (AB, BC, D, and ACD). The combination of (i) plus (ii) gives the half replicate $\mathcal{I} = -AC$. The combination of (i) plus (iii) gives the half replicate $\mathcal{I} = -BCD$, and the combination (ii) plus (iii) gives the half replicate $\mathcal{I} = -ABD$.

Now look at the alias set (A, C, BD, and $ABCD$). The A effect can be estimated from the (i) plus (iii) half, C from the (ii) plus (iii) half, and BD from the (i) plus (ii) half. Each estimate has a variance $\sigma^2/2$. The pairwise covariances are $\sigma^2/4$. Recall that $ABCD$ is negligible and is not estimated. Similarly, for the alias set (B, ABC, AD, and CD), we see that B is estimated from the (i) plus (ii) half, AD from the (i) plus (iii) half, and CD from the (ii) plus (iii) half. These estimates each have variance $\sigma^2/2$ and pairwise covariances $\sigma^2/4$. They are uncorrelated with $\widehat{A}$, $\widehat{C}$, and $\widehat{BD}$. The one special case is AC. It can be estimated from both the (i) plus (iii) half and the (ii) plus (iii) half. Again, the user is advised to use a general-purpose least squares program to perform the statistical analysis.

Exercise 14.9: Contrasts in the Three-Fourths Replicate of 2^3

(a) Show that contrasts I_1, I_2, and I_3 have variance σ^2 and are uncorrelated.
(b) Show that contrasts I_1, I_2, and I_3 have covariances either $+.5\sigma^2$ or $-.5\sigma^2$ with contrasts II_1, II_2, and II_3.
(c) Show that it is not possible to construct contrast III_2 as a linear combination of contrasts I_2 and II_2. This establishes linear independence.
(d) Verify that $\text{III}_3 = \text{I}_3 - \text{II}_3$. □

14.8.4 Adding Additional Factors

The three-fourths replicate of the 2^3 can be converted to a resolution III plan. In the base plan, there are three factors, and it is possible to estimate main effects A, B, and C and two interactions, say AB and AC, in six runs. Let i_A, i_B and i_C represent the levels of the three factors, i.e., 0's and 1's. Additionally, let $i_D = i_A + i_B$ (modulo 2) (from AB) and $i_E = i_A + i_C$ (modulo 2) (from AC) represent levels of two additional factors. The result is a main-effect or resolution III plan for five factors in six runs. Recall, however, that these are not orthogonal and the plan provides no estimate of error. This plan can be of value in specific problems where for some reason the investigator is restricted to exactly six experimental units. The statistical analysis of the data from this plan follows the analysis spelled out in Section 14.8.2 or methods in Section 12.2. The usefulness of this plan is limited to cases where error variance is very small and observations are costly.

Similar additions to the three-fourths replicate of the 2^4 can yield a main-effect plan for 10 factors in 12 runs. This plan appears to have little to recommend it, since there exists an 11-factor Plackett–Burman plan in 12 runs (see Chapter 16) that has better efficiency. However, it is conceivable that an experimenter may need a plan that provides up to seven main effects and estimates of BC, BD, and CD in 12 runs. This is available by associating new factors E with AC, F with AD, and G with AB. Similar fractions can be generated for larger factorials.

Exercise 14.10: Five Factors in Six Runs. Construct a three-fourths replicate of the 2^3 plan by deleting the fourth defined by $\mathcal{I} = +AB = +AC = +BC$. Show by displaying the contrasts that the three estimators $\widehat{A}$, $\widehat{B}$, and $\widehat{C}$ are correlated. Now add two extra treatments, D and E, on two columns generated by $i_A + i_B$ (modulo 2) and $i_A + i_C$ (modulo 2). Show that while the contrasts $\widehat{D}$ and $\widehat{E}$ are correlated, both are orthogonal to $\widehat{A}$, $\widehat{B}$, and $\widehat{C}$. Try to think of a situation where one would consider such a plan. □

Exercise 14.11: Three-Fourths Replicate of a 2^4 Factorial. Construct a three-fourths replicate of the 2^4 plan by deleting the fourth defined by $\mathcal{I} = +AB = +AC = +BC$. Show that if one assumes that AB, ABC, BCD, and $ABCD$ are negligible, one has a 12-run plan that provides estimates of all the remaining effects and interactions. Show further that if one associates a new factor E with ABD and F with ACD, one has a main-effect plan with all main effects clear of two-factor interactions, i.e., a resolution IV plan for six factors in 12 runs. □

14.9 FRACTIONAL FACTORIALS WITH COMPROMISED RESOLUTION

The original concept of resolution traces back to an old classic paper by Box and Wilson (1951). Although this concept has proved to be of major value in the development and use of fractional factorial plans, over the years a number of shortcomings have become evident. In a sense, the procedures we have outlined for finding fractions provide a complete answer to the problem of constructing fractional factorial plans. However, in many experimental settings, the classification by resolution is too broad for the experimenter, yet prior knowledge may not be sufficiently specific to apply the Franklin–Bailey algorithm. Addelman (1962) presents a useful way of thinking about factors in an experiment and gives a very convenient catalog of fractional plans (Table 14.4).

Table 14.4 Addelman's Three Classes of Plans

Class	First Subset	Second Subset
One	Interact among themselves only.	Do not interact with anyone.
Two	Interact among themselves only.	Interact among themselves only.
Three	Interact among themselves and with all other factors.	Do not interact among themselves but do interact with factors in other sets.

14.9.1 Addelman Compromise Plans

One of the major strengths of the construction of fractional factorials by superimposing factors (Section 14.5) on an existing factorial structure is that it is easy to separate factors into groups of different importance. It is a messy method but provides a good "hands-on" approach to plan construction. It allows experimenters to "protect" interactions that they feel are important and sacrifice others that a priori knowledge leads them to suspect are of limited importance.

Addelman (1962) used the superimposing construction to develop a very useful catalog of what he called *compromise plans*. His plans were compromise plans in the sense that they relaxed the assumption that all R-order interactions are of equal importance. For example, his plans allowed the investigator to estimate some two-factor interactions without having to spend the resources required to go to a resolution V plan. In all of the plans, he assumed that the experimenter wanted uncorrelated estimates of all N main effects in the experiment. Addelman split the set of factors into two sets, $\mathcal{X}$ containing $\nu < N$ and $\mathcal{Y}$ the remaining $N - \nu$ factors. His three sets of plans are summarized in Table 14.4.

Addelman's First Set of Compromise Plans

In his first class of compromise plans, it was assumed that all main effects and all two-factor interactions between pairs of factors in $\mathcal{X}$ were important and that all interactions between factors in $\mathcal{X}$ and $\mathcal{Y}$, as well as interactions among factors in $\mathcal{Y}$, could be ignored as negligible. This set of plans is given in Table 14.5. The entry in the first column in the table specifies the number of runs. The second specifies the number of factors in $\mathcal{X}$, the third column the total number of factors in the study, and the fourth column the actual factor assignments. The semicolon separates the two sets.

As an illustration of a plan in this class, we look at a 16-run plan based on the 2^4 factorial. The 2^4 factorial gives us a base 16×15 matrix. For convenience, we label the columns as a, b, ab, $c, \dots, abcd$. If there were no interactions, there would be room to assign up to 15 factors to orthogonal columns. Assume that the experimental problem involves four factors which may possibly interact. As a precaution, all six two-factor interactions must be estimable. One solution is to assign the four, A, B, C, and D to columns a, b, c, and d, respectively. The six two-factor interactions are all estimable. However, there are still five unused columns. It is possible to assign noninteracting factors E, F, G, H, and I to columns abc, abd, acd bcd, and $abcd$, respectively, for a total of nine factors in the experiment. In the statistical analysis of the resulting data the experimenter can use the columns identified by ab, ac, ad, bc, bd, and cd to estimate interactions AB, AC, AD, BC, BD, and CD, respectively.

The actual treatment assignments for the plan are illustrated in Table 14.6. For convenience, the column labels and the labels for the treatment factors assigned are given.

If one uses a model that includes the AB, AC, BC, AD, BD, and CD interactions, this is a saturated plan. It provides estimates of nine main effects and

Table 14.5 Addelman's Class One Compromise Plans

No. of Runs	No. of Interacting Factors (ν)	Total No. of Factors (N)	Column Assignment
8	2	6	a, b; c, ac, bc, abc
8	3	4	a, b, c; abc
16	2	14	a, b; c, d, all interactions excluding ab
16	3	12	a, b, c; d, all interactions excluding ab, ac, and bc
16	4	9	a, b, c, d; abc, abd, acd, bcd, $abcd$
16	5	5	a, b, c, d, $abcd$
32	2	30	a, b; c, d, e, all interactions excluding ab
32	3	28	a, b, c; d, e, all interactions excluding ab, ac, and bc
32	4	25	a, b, c, d; e, ae, be, ce, de, all three-, four-, and five-factor interactions
32	5	21	a, b, c, d, e; all three-, four- and five-factor interactions
32	6	16	a, b, c, d, e, $abcde$; all three-factor interactions
64	2	62	a, b; c, d, e, f, and all interactions except ab
64	3	60	a, b, c; d, e, f, all interactions excluding ab, ac, and bc
64	4	57	a, b, c, d; e, f, all interactions excluding ab, ac, ad, bc, bd, and cd
64	5	53	a, b, c, d, e; f, af, bf, cf, df, ef, all three-, four-, five-, and six-factor interactions
64	6	48	a, b, c, d, e, f; all three-, four-, five-, and six-factor interactions
64	7	42	a, b, c, d, e, f, $abcdef$; all three- and four-factor interactions
64	8	35	a, b, c, d, e, f, $abcd$, $abef$; ace, acf, ade, adf, bce, bcf, bde, bdf, cde, cdf, cef, def, $acdef$, $bcdef$, $abcdef$, and all remaining four-factor interactions, excluding $cdef$

Source: Abstracted from Addelman (1962); reprinted with permission.

six two-factor interactions and no estimate of error. Statistical analysis involves using one of the graphical techniques illustrated in Chapter 12.

Exercise 14.12: Defining Contrasts in an Addelman Plan. The plan in Table 14.6 is a 2^{9-5} fraction. Show that a basis for the set of defining contrasts consists of $ABCE$, $ABDE$, $ACDG$, $BCDG$, and $ABCDI$. From this one can develop the full set of alias relationships. □

Table 14.6 Addelman Compromise Plan with Nine Factors in 16 Runs

Factor:	A	B		C			E	D			F		G	H	I
	a	b	ab	c	ac	bc	abc	d	ad	bd	abd	cd	acd	bcd	$abcd$
	−1	−1	+1	−1	+1	+1	−1	−1	+1	+1	−1	+1	−1	−1	+1
	+1	−1	−1	−1	−1	+1	+1	−1	−1	+1	+1	+1	+1	−1	−1
	−1	+1	−1	−1	+1	−1	+1	−1	+1	−1	+1	+1	−1	+1	−1
	+1	+1	+1	−1	−1	−1	−1	−1	−1	−1	−1	+1	+1	+1	+1
	−1	−1	+1	+1	−1	−1	+1	−1	+1	+1	−1	−1	+1	+1	−1
	+1	−1	−1	+1	+1	−1	−1	−1	−1	+1	+1	−1	−1	+1	+1
	−1	+1	−1	+1	−1	+1	−1	−1	+1	−1	+1	−1	+1	−1	+1
	+1	+1	+1	+1	+1	+1	+1	−1	−1	−1	−1	−1	−1	−1	−1
	−1	−1	+1	−1	+1	+1	−1	+1	−1	−1	+1	−1	+1	+1	−1
	+1	−1	−1	−1	−1	+1	+1	+1	+1	−1	−1	−1	−1	+1	+1
	−1	+1	−1	−1	+1	−1	+1	+1	−1	+1	−1	−1	+1	−1	+1
	+1	+1	+1	−1	−1	−1	−1	+1	+1	+1	+1	−1	−1	−1	−1
	−1	−1	+1	+1	−1	−1	+1	+1	−1	−1	+1	+1	−1	−1	+1
	+1	−1	−1	+1	+1	−1	−1	+1	+1	−1	−1	+1	+1	−1	−1
	−1	+1	−1	+1	−1	+1	−1	+1	−1	+1	−1	+1	−1	+1	−1
	+1	+1	+1	+1	+1	+1	+1	+1	+1	+1	+1	+1	+1	+1	+1

Addelman's Second Set of Compromise Plans

In the second class of compromise plans, Table 14.7, it is assumed that all two-factor interactions between members of $\mathcal{X}$ and all two-factor interactions between members of $\mathcal{Y}$ are required. Interactions between factors from different sets are assumed negligible. These plans are again constructed by assigning treatments to columns in a base matrix. The trick now is to select the ν columns from the base matrix in such a manner that the interaction of any two or three of the selected columns is not among those chosen. The reason for not selecting columns that are two-factor interaction columns is immediately obvious, since those columns (contrasts) are to be used to estimate interactions. The restriction on the three column interactions is a bit more subtle, but also follows from the restriction that two-factor interactions are to remain estimable, i.e., not be aliased with other two-factor interactions. The remaining $N - \nu$ treatments in $\mathcal{Y}$ are then assigned to the remaining columns with the provision that no treatment is assigned to a column that represents either a two- or three-factor interaction of the columns in the first set, or a two- or three-factor interaction of the columns in the second set. The reason for the three-factor interactions getting involved can be seen from the following hypothetical example. Say that the investigator wants to look at four effects and their six possible two-factor interactions. If A, B, and C are assigned to the columns a, b, and c and D to abc, the AD interaction ($a \times abc \Rightarrow bc$) will be aliased with the BC interaction. This violates the condition that the two-factor interactions are to be estimable.

As an illustration of the construction in the second set of compromise plans, we can consider a 32-run experiment. Begin with the 32×31 base matrix constructed

Table 14.7 Addelman's Class 2 Compromise Plans

No. of Runs	No. of Factors in First Set (ν)	Total No. of Factors (N)	Column Assignment
32	2	7	a, b; c, d, e, acd, bce
32	3	7	a, b, c; d, e, abc, ade
64	2	10	a, b; c, d, e, f, acd, bce, aef, $bcdf$
64	3	10	a, b, c; d, e, f, abc, ade, bdf, $cdef$
64	4	10	a, b, c, d; e, f, abc, $bcde$, $acdf$, $bcef$
64	5	10	a, b, c, d, e; f, abc, ade, $bdef$, $acef$
128	2	13	a, b; c, d, e, f, g, acd, bce, $abcde$, $cdef$, $adeg$, $abcfg$
128	3	13	a, b, c; d, e, f, g, abc, $abde$, $acfg$, $bcdf$, $defg$, $bdeg$
128	4	13	a, b, c, d; e, f, g, abc, cef, deg, $adfg$, $bceg$, $bdefg$
128	5	13	a, b, c, d, e; f, g, abc, ade, $acef$, $bcdg$, $abfg$, $defg$
128	6	13	a, b, c, d, e, f; g, abc, ade, bdf, $aefg$, $bcdg$, $abdeg$

Source: Abstracted from Addelman (1962); reprinted with permission.

from the 2^5 factorial. It is possible to put seven factors into this plan, three in $\mathcal{X}$ and four in $\mathcal{Y}$. In the first set, assign A to a, B to b, and C to c. This commits columns ab, ac, and bc. In the second set assign D to abc, E to d, F to e, and G to ade. The resulting plan is shown in Table 14.8. The full model involves the mean and 16 effect and interaction parameters, leaving 15 degrees of freedom for error.

Exercise 14.13: Limit in Addelman Plan. Verify, by listing all the column headings, i.e., $a, b, ab, \ldots, abcde$ and assigning factors to the columns, that you cannot insert a fifth factor into the second group in the plan in Table 14.8 and still maintain all two-factor interactions among the effects in the first group and all two-factor interactions among the effects in the second group. □

Exercise 14.14: Estimate in Addelman Plan. Show how the experimenter will estimate the AB interaction in the plan in Table 14.8. □

Addelman's Third Set of Compromise Plans

In the third class of compromise plans, in Table 14.9, it is again assumed that all main effects are required and that the factors can be split into two sets, the first set $\mathcal{X}$ consisting of $\nu < N$ factors and $\mathcal{Y}$, the remaining $N - \nu$ factors. The requirement now is that all two-factor interactions involving at least one member of set $\mathcal{X}$ are required. Remaining interactions are assumed negligible. Notice that $\mathcal{X}$ empty corresponds to the resolution III plans. Also $\mathcal{Y}$ empty implies a resolution IV plan.

Table 14.8 Seven Factor Addelman Plan in 32 Runs

Factor:	A	B	C	D	E	F	G	A	B	C	D	E	F	G
	0	0	0	0	0	0	0	1	0	0	1	0	0	1
	0	0	0	0	0	1	1	1	0	0	1	0	1	0
	0	0	0	0	1	0	1	1	0	0	1	1	0	0
	0	0	0	0	1	1	0	1	0	0	1	1	1	1
	0	0	1	1	0	0	0	1	0	1	0	0	0	1
	0	0	1	1	0	1	1	1	0	1	0	0	1	0
	0	0	1	1	1	0	1	1	0	1	0	1	0	0
	0	0	1	1	1	1	0	1	0	1	0	1	1	1
	0	1	0	1	0	0	0	1	1	0	0	0	0	1
	0	1	0	1	0	1	1	1	1	0	0	0	1	0
	0	1	0	1	1	0	1	1	1	0	0	1	0	0
	0	1	0	1	1	1	0	1	1	0	0	1	1	1
	0	1	1	0	0	0	0	1	1	1	1	0	0	1
	0	1	1	0	0	1	1	1	1	1	1	0	1	0
	0	1	1	0	1	0	1	1	1	1	1	1	0	0
	0	1	1	0	1	1	0	1	1	1	1	1	1	1

To illustrate the construction of a member of this class, we again assign treatments to columns in the 8×7 base matrix obtained from the 2^3 factorial. Assume we have two factors in the first set. We can assign them to columns a and b. Since their interaction is wanted, this uses up column ab. Additionally, we can put C on column c. This commits column ac and bc since we want the AC and BC interactions. Only column abc is left and we cannot use it because there are no more columns for interactions.

14.10 A CAUTION ABOUT MINIMUM ABERRATION

We saw in the Addelman work that there are cases where the experimenter has good reason not to accept the two principles that led to the work on minimum aberration plans. However, Chen et al. (1993) show, by example, that even under the two principles: (1) that main effects are more important than two-factor interactions, which in turn are more important than three-factor interactions, and so on, and (2) that all q-level interactions are equally important, minimum aberration plans are not always the most desirable. Their example has some merit. It is a case where the minimum aberration plan does not minimize the number of two-factor interactions sacrificed.

It can be shown that the 2^{9-4} fraction D_1 given by defining contrasts,

$$\begin{aligned}
\mathcal{I} &= ABCF = ACDG = ACHI = BDFG = BGHI = DGHI \\
&= ABDEH = ABEGI = AEFGH = ADEFI = BCDEI = BCEGH \\
&= CDEFG = CDFGI = ABCDFGHI,
\end{aligned}$$

Table 14.9 Addelman's Class 3 Compromise Plans

No. of Runs	No. of Factors in First Set (ν)	Total No. of Factors (N)	Column Assignment
8	0	7	a, b, c, ab, ac, bc, abc
8	1	4	$a; b, c, bc$
8	2	3	$a, b; c$
8	3	3	$a, b, c;$
16	0	15	a, b, c, d, and all interactions
16	1	8	$a; b, c, d, bc, bd, cd, bcd$
16	2	5	$a, b; c, d, cd$
16	3	5	$a, b, c; d, abcd$
16	4	5	$a, b, c, d; abcd$
16	5	5	$a, b, c, d, abcd;$
64	0	31	a, b, c, d, e, and all interactions
64	1	16	$a; b, c, d, e$, and all interactions not containing a
64	2	9	$a, b; c, d, e, cd, ce, de, cde$
64	3	9	$a, b, c; d, e, de, abcd, abce, abcde$
64	4	7	$a, b, c, d; e, abcd, abcde$
64	5	6	$a, b, c, d, e; abcde$
64	6	6	$a, b, c, d, e, abcde;$
128	0	63	a, b, c, d, e, f, and all interactions
128	1	32	$a; b, c, d, e, f$, and all interactions not containing a
128	2	17	$a, b; c, d, e, f$, and all interactions not containing a or b
128	3	17	$a, b, c; d, e, f, de, df, ef, def, abcd, abce, abcf, abcde, abcdf, abcef, abcdef$
128	4	11	$a, b, c, d; e, f, ef, abcd, abcde, abcdf, abcdef$
128	5	10	$a, b, c, d, e; f, abcf, adef, bcde, bcdef$
128	6	9	$a, b, c, d, e, f; abcd, abef, cdef$

Source: Abstracted from Addelman (1962); reprinted with permission.

has minimum aberration. Under the assumption that all three-factor and higher interactions are negligible, all main effects and the eight two-factor interactions $\{AE, BE, CE, DE, EF, EG, EH$, and $EI\}$ are estimable. Notice that all factors except E appear in at least one of the four-factor interactions in the set of defining contrasts.

Now, consider a second 2^{9-4} fraction, D_2, given by the defining contrasts

$$\begin{aligned}\mathcal{I} &= ABCF = ABGH = ACDG = ADFH = BCDH = BDFG \\ &= CFGH = ABDEI = ACEHI = AEFGI = BCEGI = BEFHI \\ &= DEGHI = CDEFI = ABCDEFGHI.\end{aligned}$$

This set of defining contrasts includes seven four-factor interactions, while D_1's set of defining contrasts included only six four-factor interactions. In fact, the four-factor interactions in the set of defining contrasts for D_2 contain neither E nor I. It follows that all main effects and the 15 two-factor interactions $\{AE, BE, CE, DE, EF, EG, EH, EI, AI, BI, CI, DI, FI, GI$, and $HI\}$ are all estimable. This example suggests going back to direct enumeration and careful thinking to find optimal fractions.

The conclusion from this example is that one cannot rely just on the existing criteria for good plans. As Chen et al. (1993) claim, there is a need for a catalog of good plans. The Addelman plans were a step in that direction.

Use of the term *optimal* in experimental designs is really somewhat questionable. A plan can be termed optimal if it exceeds all other plans in some characteristic. The deeper question is whether this characteristic is really most important for the problem at hand. Our view is that any optimality claim should always be questioned from the viewpoint of whether it is really optimal for the research question. Careful examination of the problem and the plan are always required.

14.11 DIRECT ENUMERATION APPROACH TO CONSTRUCTING PLANS

The example above suggests that the only approach to finding optimal plans is direct enumeration. However, this is not without difficulties. The procedure to construct a 2^{N-k} plan is first to write down a full factorial plan, a basic plan with $N-k$ factors, each at two levels. This gives a matrix with 2^{N-k} rows or equivalently, 2^{N-k} runs. Then associate the remaining k factors, the indicator factors with columns involving interactions among the first $(N-k)$ factors. Each such assignment results in a word or defining contrast equal to the symbol $\mathcal{I}$. Taking products of the k defining contrasts, two at a time, three at a time, and so on, gives the complete set of defining contrasts. For example, in the 2^{5-2} plan, we begin with a 2^3 plan. Then one must assign the two additional actors, D and E, to two of the interaction columns in the initial 2^3 factorial. We can assign D to AB giving $\mathcal{I} = ABD$, and E to BC giving $\mathcal{I} = BCE$. The full set of defining contrasts consists of ABD, BCE, and their product $ABD \times BCE = ACDE$. Unfortunately, this simple enumeration technique gets out of hand very quickly. Consider a 2^{15-10} fraction. This has 32 runs. Begin with the basic 2^5 plan. There are $\binom{31-5}{15-5} = 5{,}311{,}735$ possible ways to assign the 10 remaining indicator factors.

The minimum aberration and the moments criteria were attempts to get around the problem of looking at so many possibilities. The Chen et al. (1993) example shows that both of these criteria are flawed. There is need for a rapid screening technique that keeps one from having to look at so many possibilities. One proposal was to look at complete *word-length patterns*. By *word-length pattern* is meant the list of numbers, the number of main effects, the number of two-factor interactions, the number of three-factor interactions, and so on in the defining contrast. The word-length patterns for D_1 and D_2 given previously are (0, 0, 0, 6, 8, 0, 0, 1, 0) and (0, 0, 0, 7, 7, 0, 0, 0, 1). These two plans are not equivalent, since one cannot be obtained from the other by interchanging letters. But just identical word-length pattern is also not sufficient. Chen and Lin (1991) give the following two nonequivalent 2^{12-3} plans:

$$\mathcal{I} = ABCFGJ = ABCHIK = FGHIJK = ADEFIL$$
$$= BCDEGIJL = BCDEFHKL = ADEGHJKL$$
$$\mathcal{I} = ACDFIJ = ACEGHK = BEHIJL = BDFGKL$$
$$= DEFGHIJK = ABCDEFHL = ABCGKIJKL.$$

A simple way to see that these two are not equivalent is to look at the eight-letter words. There are three of these in each plan. In the first plan, the letter A appears in only one eight-letter word, while the letter L appears in all three. In the second plan, each of the 12 letters appears in exactly two of the three eight-letter words. Clearly, the plans are not equivalent.

The conclusion from these examples is that if one wants to be sure not to miss good plans, one must enumerate, but one must be clever about the enumeration so that things do not get too big. Chen et al. (1993) have performed a complete search of 16- and 32-run plans. They also examined the 64-run plans but eventually retained only the resolution IV or better plans. They give a catalog of nonisomorphic plans. There is also an old, very extensive catalog of 1/2, 1/4, 1/16, 1/32, 1/64, 1/128, and 1/256 fractions for the 2^N system initially published by the National Bureau of Standards (1957) and republished by McLean and Anderson (1984).

14.12 BLOCKING IN SMALL 2^{N-k} PLANS

Adding blocking to the 2^{N-k} plans is relatively simple. However, there are minor complications that must be considered. The number of blocks must be a power of 2, i.e., 2^b blocks. It follows that after assigning the N treatments to the $2^{N-k} - 1$ columns, one must assign b blocking factors to remaining columns. The key point to consider at this stage is that the blocking factors must be treated as though they do not interact with other treatments, but all interact among one another, and all interactions must be accounted for. For example, if $b = 3$, i.e., three blocking factors, there are eight blocks, seven degrees of freedom, and the three-factor (blocking) interaction cannot fall on a column that is being used for something else. On the other hand, the b blocking factors do not interact with the N treatment factors in the experiment (the usual assumptions of no block $\times$ treatment interaction). Bisgaard (1994) presents a small catalog of 2^{N-k} plans that incorporate blocking.

In the discussion Bisgaard makes an interesting observation about the value of blocking in industrial experimentation. It is not uncommon for authors of statistics books for engineers and industrial applications simply to ignore blocking, or sometimes even make the statement that blocking is only of value in agricultural work. The point by Bisgaard is that in the early stages of an experimental study, at the screening stage, it is not at all unusual to find that technicians do not have very good control of the process, and that blocking may be very necessary. Once technicians become familiar with the process, the need for blocking may not be nearly as crucial.

We illustrate the mechanics of setting up a blocking scheme for a fractional factorial with the following four examples of two blocks of four and two examples of four blocks of two. In a sense, blocking just introduces new factors into the system. To distinguish clearly from treatment factors, we use the symbol $\mathcal{B}$ with subscript, if necessary, to denote blocking factors. We also introduce the notation 2_R^{N-k} to represent a 2^{N-k} fraction with resolution R.

1. Full replicate of the 2^3 in two blocks of four:

$$\mathcal{I} = ABC\mathcal{B}.$$

   All three main effects and their assorted two-factor interactions are available.

2. The half replicate 2_{III}^{4-1} in two blocks of four:

$$\mathcal{I} = ABCD = AB\mathcal{B} = CD\mathcal{B}.$$

   The two two-factor interactions AB and CD are aliased with the block difference.

3. The quarter replicate 2_{III}^{5-2} in two blocks of four:

$$\begin{aligned}\mathcal{I} &= ABD = ACE = BCDE \\ &= BC\mathcal{B} = ACD\mathcal{B} = ABE\mathcal{B} = DE\mathcal{B}.\end{aligned}$$

   Four interactions are aliased with the block difference.

4. The eighth replicate 2_{III}^{6-3} in two blocks of four:

$$\begin{aligned}\mathcal{I} &= ABD = ACE = BCF = \text{four generalized interactions} \\ &= ABC\mathcal{B} = \text{seven generalized interactions}.\end{aligned}$$

5. Full replicate of the 2^3 in four blocks of two:

$$\mathcal{I} = AB\mathcal{B}_1 = AC\mathcal{B}_2 = BC\mathcal{B}_1\mathcal{B}_2.$$

   Three two-factor interactions are aliased with block contrasts.

Table 14.10 Blocking by Superimposing of Factors

A	B	$\mathcal{B}_1$	C	$\mathcal{B}_2$		
a	b	ab	c	ac	bc	abc
−	−	+	−	+	+	−
+	−	−	−	−	+	+
−	+	−	−	+	−	+
+	+	+	−	−	−	−
−	−	+	+	−	−	+
+	−	−	+	+	−	−
−	+	−	+	−	+	−
+	+	+	+	+	+	+

6. The half replicate 2_{III}^{4-1} in four blocks of two:

$$\begin{aligned}\mathcal{I} &= ABCD = AB\mathcal{B}_1 = AC\mathcal{B}_2\\ &= CD\mathcal{B}_1 = BC\mathcal{B}_1\mathcal{B}_2 = BD\mathcal{B}_2 = AD\mathcal{B}_1\mathcal{B}_2.\end{aligned}$$

Bisgaard (1994) gives a table of all possible combinations of factors in 16-run two-level fractional factorial plans. The simple way to obtain the actual treatment combinations and block assignments is to use the assignment technique outlined in Section 14.5.3. The $\mathcal{B}$'s are assigned to columns just like regular treatment factors. Then the final block assignment is based on the combinations of levels in the $\mathcal{B}$ columns.

Example 14.1. We illustrate the details of the example in case 5. There are three treatment factors and two blocking factors, eight runs in four blocks of two. Considering only the treatment factors, there is a full replicate. However, formally, we construct this as a 2^{5-2}. We use the superimposing of factors technique shown in Table 14.10. The blocks are defined by columns 3 and 5. The first block is obtained from lines 2 and 7 in the table. The four blocks are $\{a, bc\}$, $\{b, ac\}$, $\{ab, c\}$, and $\{(1), abc\}$.

14.12.1 Blocking and Resolution

In one sense it appears that blocking can be thought of as simply introducing extra factors into the defining contrasts. This is true up to a point. If we do this, problems appear when we look at word length (the defining contrasts) to determine resolution. Recall that resolution is not defined in terms of word length but in terms of main effects being aliased with two-factor interactions, two-factor interaction aliased with other two-factor interactions, and so on. We illustrate the problem with an example. Consider the 16-run 2^{6-4} in four blocks of four. First, select the fraction by choosing

$$\mathcal{I} = ABCE = BCDF = ADEF.$$

Then add the two blocking generators

$$\mathcal{I} = ACD\mathcal{B}_1 = ABD\mathcal{B}_2 = BC\mathcal{B}_1\mathcal{B}_2.$$

Multiplying these generators together gives the full set of defining contrasts:

$$\begin{aligned}\mathcal{I} &= ABCE = BCDF = ADEF = ACD\mathcal{B}_1 = BDE\mathcal{B}_1 = ABF\mathcal{B}_1 \\ &= CEF\mathcal{B}_1 = ABD\mathcal{B}_2 = CDE\mathcal{B}_2 = ACF\mathcal{B}_2 = BEF\mathcal{B}_2 \\ &= BC\mathcal{B}_1\mathcal{B}_2 = AE\mathcal{B}_1\mathcal{B}_2 = DF\mathcal{B}_1\mathcal{B}_2 = ABCDEF\mathcal{B}_1\mathcal{B}_2.\end{aligned}$$

If we count letters mechanically, this appears to be a resolution IV plan. However, note that the $\mathcal{B}_1\mathcal{B}_2$ interaction is really a contrast among blocks. In fact, it should really be thought of as a main effect. It follows that $BC\mathcal{B}_1\mathcal{B}_2$, $AE\mathcal{B}_1\mathcal{B}_2$, and $DF\mathcal{B}_1\mathcal{B}_2$ should be thought of as three-factor interactions. If we are counting word lengths, these three are really words of length 3.

14.12.2 New Definition of Word Length

This question of resolution becomes an important factor when we develop a system for classifying plans. It also has an impact on the concept of aberration. Sitter et al. (1997) get around the problem by modifying the definition of word length.

Definition: If we let c_i be the count of the number of treatment factors (letters) present in the ith word and b_i be the count of the number of blocking factors (letters) present in the ith word, then, they say, let the length of the ith word be defined as

$$c_i + 1.5I_{[b_i \geq 1]},$$

where $I_{[\cdot]}$ is the indicator function, taking on the value 1 or 0, depending on whether $[\cdot]$ is true or not true.

It follows that with this definition, we can have plans with resolution III, III.5, IV, IV.5, and so on. To illustrate, in the previous example, we have $ABCE$ of length 4, $ACD\mathcal{B}_1$ of length 4.5, and $BC\mathcal{B}_1\mathcal{B}_2$ of length 3.5. As a consequence of this, they also redefine the word-length pattern as $(\ell_3,\ \ell_{3.5},\ \ell_4,\ \ell_{4.5},\ \ell_5, \ldots)$.

Sitter et al. (1997) use the notation $2^{(N+b)-(b+k)}$ to denote a 2^{N-k} fractional factorial in 2^b blocks. For example, a one-eighth replicate of an eight-factor factorial in four blocks would be indicated as a $2^{(8+2)-(2+3)}$ plan. To illustrate the concept of minimum aberration with this new definition of word length, consider the following two $2^{(4+1)-(1+1)}$ plans. Both consist of a half replicate of a 2^4 in two blocks:

$$D_1 : \mathcal{I} = ABCD = AB\mathcal{B} = CD\mathcal{B}$$

$$D_2 : \mathcal{I} = ABC\mathcal{B} = ABD = CD\mathcal{B}.$$

Table 14.11 Table of Minimum Aberration Eight-Run Plans

Plan	Defining Contrasts	Word-Length Pattern[a]	Resolution
$2^{(4+1)-(1+1)}$	$ABCD, AB\mathcal{B}$	0 2 1	III.5
$2^{(4+2)-(2+1)}$	$ABCD, AB\mathcal{B}_1, AC\mathcal{B}_2$	0 6 1	III.5
$2^{(5+1)-(1+2)}$	$ABD, ACE, BC\mathcal{B}$	2 2 1 2	III
$2^{(6+1)-(1+3)}$	$ABD, ACE, BCF, ABC\mathcal{B}$	4 3 3 4 0 0 0 0 0 1	III

[a] Word-length pattern $(\ell_3, \ell_{3.5}, \ell_4, \ell_{4.5}, \ell_5, \ell_{5.5}, \ldots)$.
Source: Based on a table of plans in Sitter et al. (1997); adapted with permission.

Under the Sitter et al. definition, plan D_2 is resolution III and plan D_1 is resolution III.5.

Using this definition of word length, they construct a large catalog consisting of all minimum aberration eight- and 16-run fractional factorial plans and a large selection of minimum aberration 32-, 64-, and 128-run fractional factorial plans.

All of the eight-run plans are constructed from a base plan consisting of the seven columns from the full 2^3 factorial. Table 14.11 gives a basis for the full set of defining contrasts. The actual treatment assignments to the columns of the underlying 2^3 structure are easily obtained using the superimposing technique outlined in Section 14.5.3. For the first plan in the table, A is assigned to column a, B to column b, D to column c, C to column abc, and column $\mathcal{B}$ to column ab. The first block consists of $\{(1), ab, cd, abcd\}$ and the second block of $\{bd, bc, ad, ac\}$. The entries under "word-length pattern" are the word-length patterns for the full set of defining contrasts beginning with words of length 3.

Tables for 16-, 32-, 64-, and 128-run plans are given in Appendix 14B.

14.13 FRACTIONAL REPLICATION IN THE 3^N SYSTEM

14.13.1 One-Third Replicate of the 3^3 Case

As we saw in confounding three-level factors with blocks in Section 13.5, we can assign treatments to incomplete blocks with block sizes that are powers of 3. Similarly, it is possible to construct fractional factorials for 3^N factorials that use $1/3, 1/9, 1/27, \ldots$ fractions of the total number of possible treatment combinations. Recall from Section 11.2.3 that two-factor interactions in a 3^N factorial consist of two pieces; e.g. $A \times B$ can be decomposed into AB and the AB^2 components, each representing two contrasts. Three-factor interactions can be decomposed into four pieces, e.g., ABC, ABC^2, AB^2C, and AB^2C^2, each accounting for two degrees of freedom. We can base the selection of a one-third replicate on any one of these pairs of contrasts. For example, choosing $\mathcal{I} = ABC^2$ splits the 27 treatment combinations into three sets of nine. The three fractions can be obtained as solutions to

$$\begin{array}{cccccc} a & b & c & & \\ \hline i_A & +i_B & +2i_C & = & \alpha \text{ (modulo 3),} \end{array}$$

where α is one of 0, 1, or 2. For example, if we select $\alpha = 1$ we obtain the set 100, 010, 220, 201, 021, 111, 002, 122, and 212. Notice that we do not have the convenient $+/-$ notation to identify the fraction.

We must warn the reader that while this fractional plan using nine of the total of 27 treatment combinations has proved to be quite popular in some circles, it is not really a good experimental plan. The problem is that the main effects are all aliased with two-factor interactions. To see this, it is necessary to recall that there are two generalized interactions for each pair of effects: for example,

$$A \times ABC^2 \Rightarrow A^2BC^2 \Rightarrow AB^2C$$

$$A \times ABC^2 \times ABC^2 \Rightarrow A^3B^2C^4 \Rightarrow BC^2.$$

The main effect A is aliased with part of the $B \times C$ interaction. With this experimental design, the contrast between the low and high levels of A would be estimated from

$$y_{010} + y_{021} + y_{002} - y_{220} - y_{201} - y_{212}.$$

This is the same as the contrast for the BC^2 component comparing treatments with $i_B + 2i_C = 1$ to those with $i_B + 2i_C = 2$ (modulo 3):

$$y_{010} + y_{021} + y_{002} - y_{220} - y_{201} - y_{212}.$$

In like manner, B is aliased with part of the $A \times C$ interaction. Finally,

$$C \times ABC^2 \Rightarrow ABC^3 \Rightarrow AB$$

$$C \times ABC^2 \times ABC^2 \Rightarrow A^2B^2C^5 \Rightarrow ABC.$$

The main effect C is aliased with part of the $A \times B$ interaction.

This design has only nine runs, and it is not possible safely to interpret the results of the experiment unless we are willing to assume that there are no two-factor interactions. It is important to be aware that if you have three factors or fewer, the main effects in a $\frac{1}{3}$ fraction will always be aliased with two-factor interactions, as well as with the three-factor interaction of the design generator.

14.13.2 One-Third Replicate of the 3^4 Case

When there are four factors, each at three levels, the obvious choice to use to split the treatment combinations into three sets is a four-factor interaction. Consider

$\mathcal{I} = ABCD$. A suitable fraction is obtained via solutions to

$$\begin{array}{cccccc} a & b & c & d & & \\ \hline i_A & +i_B & +i_C & +i_D & = & \alpha \text{ (modulo 3)}, \end{array}$$

where α is one of 0 or 1 or 2. The fraction consists of 27 treatment combinations.

The aliases of A are $AB^2C^2D^2$ and BCD, of B are AB^2CD and ACD, of C are ABC^2D and ABD, and of D are $ABCD^2$ and ABC, all free of two-factor interactions. Two-factor interactions are aliased with parts of other two-factor interactions. For example, part of $A \times B$ is aliased with part of $C \times D$, part of $A \times C$ is aliased with part of $B \times D$, and part of $A \times D$ with part of $B \times C$. This accounts for all of the degrees of freedom available in the plan. Note also that if 27 experimental units is too large to handle in one group, the experimenter is free to use one of the two-factor interaction components to split the units into three subsets of nine, three blocks of nine.

This is not a very good plan. The experimenter is faced with a relatively large number of experimental runs, and yet is left with a plan that can only screen main effects. Also, the second-order interactions must be negligible if the experiment is to provide a valid source of error. This is a continuing problem with fractional replication in the 3^N system.

14.13.3 Fractions by Superimposing Factors

The technique illustrated in Section 14.5 for the 2^N system can also be applied in the 3^N system. We illustrate with a 3^{13} main-effect plan in 27 runs. We begin with the model for the complete 3^3 factorial using 27 runs. The columns are identified as $a, b, ab, ab^2, c, ac, ac^2, bc, bc^2, abc, abc^2, ab^2c$, and ab^2c^2. Then assign the factors $F_1, \ldots, F_{13}$ to these columns. If one has fewer factors, then by judiciously selecting columns, one can keep certain potential interactions free. However, this must be done with full realization that the interaction of any pair of columns utilizes two columns.

As an illustration, consider again the one-third replicate of the 3^4. Assign the four main effects $A, \ldots, D$ to the columns from the 3^3 labeled a, b, c, and abc, respectively. Inspection shows that

- $A \times B$ interaction falls on ab and ab^2.
- $A \times C$ on ac and ac^2.
- $B \times C$ on bc and bc^2.
- $A \times D$ on bc and ab^2c^2.
- $B \times D$ on ac and ab^2c.
- $C \times D$ on ab and abc^2.

If the investigator is willing to assume that D does not interact with any other factors, this plan provides estimates of $A \times B$, $A \times C$, and $B \times C$ interactions. It may well be advantageous to consider linear $\times$ linear, linear $\times$ quadratic, and so on, contrasts if the factors are continuous. In fact, it is even possible to estimate a linear $\times$ linear $\times$ linear component of the $A \times B \times C$ interaction, although it is not orthogonal to $\widehat{D}$.

14.13.4 Example of a 3^{5-2} Plan

As an example of the technique of superimposing factors in the 3^N system, we refer the reader to a study to improve the cutting operation by a metal turning tool given by Taguchi (1987a). The study involved five factors, each at three levels. The factors are:

A: feed rate (mm/rev) $a_0 = .30$, $a_1 = .42$, $a_3 = .53$

B: depth of cut (mm) $b_0 = 2$, $b_1 = 3$, $b_2 = 4$

C: cutting speed (m/min) $c_0 = 65$, $c_1 = 90$, $c_2 = 115$

D: material being cut $d_0 = SS34$, $d_1 = SS41$, $d_2 = SS50$

E: type of cutting tool e_0: JIS turning tool, e_1: new SWC turning tool, e_2: old SWC turning tool.

The response observed was a variable called *cutting power*.

The plan is obtained by superimposing factors on the columns from a 3^3. The factor assignment was based on the assumption that factors A, B, and C could possibly interact. The assignment was A to a, B to b, C to c, D to abc, and E to ab^2c^2. This leaves columns ab and ab^2 free for possible $A \times B$ interaction, columns ac and ac^2 free for possible $A \times C$ interaction, and columns bc and bc^2 free for possible $B \times C$ interaction. If we allow for two columns for each of the three two-factor interactions, only two columns or four degrees of freedom are left for error. Rather than perform a conventional analysis of variance, we recommend the techniques discussed in Chapter 12.

14.13.5 Minimum Aberration Plans

The concept of minimum aberration also extends to the 3^N system. However, it turns out that there is much more flexibility in choice of interactions for use in defining fractions. Franklin (1984) presents the following set of interactions that serve as a basis to define a wide range of minimum aberration fractions:

$$\begin{aligned} \mathcal{I} &= F_1F_7^2F_8F_9F_{11}F_{12} = F_2F_7F_8F_{10}^2F_{11}F_{12}^2 = F_3F_7F_9^2F_{10}F_{11}F_{12} \\ &= F_4F_8^2F_9F_{10}F_{11}F_{12}^2 = F_5F_7F_8F_9F_{10}F_{11}^2 = F_6F_7F_8^2F_9F_{10}^2F_{12}. \end{aligned}$$

To obtain a 3^{12-6} fraction, use the full set of six generators. Other fractions are obtained by deleting generators (to increase the fraction) or deleting factors, beginning with F_{12}, to decrease the size of experiment. Examination of the potential generators given reveals that two sets of plans suggested by the operation above should clearly not be used. These are 3^{5-1}, 3^{6-1}, and 3^{7-1} and the set 3^{5-4}, 3^{6-5}, and 3^{7-6}. To illustrate, consider the 3^{5-1} fraction. The rule suggests using

$$\mathcal{I} = F_1 F_7^2 F_8 F_9$$

as the defining interaction for a study that involves factors F_1, F_7, F_8, F_9, and F_{10}. Obviously, one can do better. Franklin claims that all plans have been checked and all except the six identified are good plans.

14.13.6 Foldover

The complete foldover in the 3^N system is somewhat more complex than in the 2^N. The foldover set is generated by interchanging symbols in accordance with the operations of a symmetric group (Ryser 1963) on three symbols, i.e., {e, (012), (021), (01), (02), (12)}. These operations are applied as shown in the following example. Treatment combination t leads to the six treatment combinations {t, $(012)t$, $(021)t$, $(01)t$, $(02)t$, $(12)t$}, where $(012)t$ represents the treatment combination obtained from t by changing 0 to 1, 1 to 2, and 2 to 0. The set obtained by applying the complete foldover to any specific treatment combination often includes duplicates. We illustrate the technique with the main-effect 3^{4-2} plan. The base plan consists of {0000, 0112, 0221, 1011, 1120, 1202, 2022, 2101, and 2210}. In this plan, all main effects are aliased with two-factor interactions. The complete foldover of this set consists of two copies of the treatment combinations {0000, 0112, 0221, 1011, 1120, 1202, 2022, 2101, 2210, 1111, 1220, 1002, 2122, 2201, 2010, 0100, 0212, 0021, 2222, 2001, 2110, 0200, 0012, 0121, 1211, 1020, and 1102}. It is a simple exercise to show that in this plan, all main effects are free of two-factor interactions.

14.13.7 Catalog of Plans

Conner and Zellin (1959) contains a catalog of 1/3, 1/9, 1/27, 1/81, and 1/243 fractions of the 3^N factorials, where N ranges from four to 10 factors. This monograph has been republished in McLean and Anderson (1984).

REVIEW EXERCISES

14.1 Use the method of superimposing factors to construct a main-effect plan for three factors each at two levels in four runs.

√ **14.2** Construct a one-sixteenth replicate of a 2^8 factorial. Show that it is possible to do this in a manner that keeps all main effects free of two-factor interactions. However, two-factor interactions will be aliased with other two-factor interactions. This is a resolution IV plan. Show that if it should happen that only factors A, B, and C were active and all other factors had no effect, either singly or in interactions with other factors, the scientist would in effect have two complete replicates of a 2^3 factorial with factors A, B, and C.

14.3 A consumer group wants to assess the factors affecting the firmness of a mattress. They include two levels of each of the following factors: A, number of coils; B, design of coils; C, wire gauge; D, padding material; and E, thickness of padding. They can only test eight mattresses. One member of the group proposes testing the following set of eight mattresses: (1), ad, b, ce, abd, bce, $acde$, $abcde$.

(a) What are the defining contrasts and generalized interactions?

(b) What effects are aliased with the A main effect?

(c) Write out the sum of squares for the $A \times B$ interaction.

(d) Write out the sources and degrees of freedom for the analysis of variance table.

(e) Propose a better design for this experiment. Give its defining contrasts and write out the treatments that will be included.

14.4 Monte Carlo simulations are computer experiments to study the statistical properties of different estimators and test statistics. Experimental design principles can be used in designing simulation studies. Suppose that we want to evaluate the width and coverage of confidence intervals of estimated treatment effects when the assumptions of normality, independence, and equal variance are violated. The factors under study are: A, distribution of Y (normal, lognormal); B, sample size ($n = n_1 + n_2 = 10$ or $n = 20$); C, equality of sample sizes ($n_1 = n_2$, $n_1 \neq n_2$); D, variance homogeneity ($\sigma_1 = \sigma_2$, $\sigma_1 \neq \sigma_2$); and E, correlated observations (yes, no). Design an experiment to study these five factors using sixteen treatment combinations.

(a) Give the design generators and explain why they were chosen. Show the treatments in the design and the aliases.

(b) Try designing an experiment with 12 runs. Can you come up with a satisfactory plan?

14.5 Experiment. The solubility of pills is one property of a drug product that is important to control in the manufacture of pharmaceuticals. You are to conduct an experiment to study the time it takes to dissolve a tablet or caplet of aspirin or Tylenol. There are five factors in this study: A, drug type (aspirin, Tylenol); B, shape (tablet, caplet); C, solvent (water, vinegar); D, stirring (no, yes); and E, water temperature (cold, warm). Your task is to use two tablets and caplets of each type (for a total of eight pills), and you are to design and conduct an experiment to test the effects of these factors on the time it takes to dissolve a pill. The objectives of the exercise are to gain experience and understanding of the process of experimentation, to try your hand at designing and analyzing a real experiment, to evaluate the suitability of the design, and to gain experience writing up methods and results.

14.6 Construct a resolution IV, one-eighth replicate of a 2^7 factorial, i.e., a 2_{IV}^{7-3} plan. Notice that there must be three generators, each consisting of words with at least four letters each. In this case all of the words in the expression for the defining contrast will contain exactly four letters.

√**14.7** Consider constructing an experimental plan for a 2^{7-2} fractional factorial.

(a) Construct an argument to show that it is not possible to find a resolution V plan. You can accomplish this by examining possible subsets of five of seven letters for words in the defining contrast.

(b) Now consider three alternative resolution IV plans:

$$(1)\ \mathcal{I} = ABCF = BCDG = ADFG$$

$$(2)\ \mathcal{I} = ABCF = ADEG = BCDEFG$$

$$(3)\ \mathcal{I} = ABCDF = ABCEG = DEFG.$$

Assume that all three-factor and higher-order interactions are negligible. Note that in all cases the main effects are not aliased with two-factor interactions, a property of the resolution IV plans. Now examine the alias relationships among two-factor interactions. There are $\binom{7}{2} = 21$ two-factor interactions in all. How many two-factor interactions are aliased with at least one other two-factor interaction in plan (1), in plan (2), and in plan (3). Which is the minimum aberration plan?

14.8 Fill in the degrees of freedom for the sequential (SAS® Type I) sums of squares in the following analysis of variance table for a 3^{3-1} fractional factorial with design generator $\mathcal{I} = ABC$. How many runs are there in this design? Also fill in the aliases for each effect.

Source	Type I df	Alias with Effect Already in the Model
A		
B		
C		
AB		
AB^2		
AC		
AC^2		
BC		

BC^2		
ABC		
ABC^2		
AB^2C		
AB^2C^2		

14.9 Consider a half fraction of a factorial with seven factors, each at two levels, arranged in blocks of size 8 that was constructed using the design generators $\mathcal{I} = ABCG$, block $= ABD$, block $= ACE$, and block $= BCF$.

(a) Draw a diagram of the treatments and blocks.

(b) What effects are aliased with main effects?

(c) What effects are confounded with blocks?

APPENDIX 14A: MINIMUM ABERRATION FRACTIONS WITHOUT BLOCKING*

14A.1 Plans for 2^{N-5}

The general forms of the defining contrasts for the 1/32 fractions for $N = 10$ and 11 are

$$\begin{aligned}\mathcal{I} &= F_1F_7F_8F_9F_{10}F_{11} = F_2F_6F_8F_9F_{10}F_{11} = F_3F_6F_7F_9F_{10} \\ &= F_4F_6F_7F_8F_{10} = F_5F_6F_7F_8F_9F_{10}.\end{aligned}$$

If $N = 10$, the F_{11} factor is dropped. The defining contrasts for $N = 12, 13, 14, 15$, and 16 are

$$\begin{aligned}\mathcal{I} &= F_1F_6F_8F_9F_{10}F_{13}F_{14}F_{16} = F_2F_6F_9F_{10}F_{11}F_{12}F_{14}F_{15} \\ &= F_3F_6F_7F_{10}F_{11}F_{13}F_{15}F_{16} = F_4F_6F_7F_8F_{11}F_{12}F_{14}F_{16} \\ &= F_5F_6F_7F_8F_9F_{12}F_{13}F_{15}.\end{aligned}$$

If $N < 16$, factors $> N$ are dropped.

*Based on plans in Franklin (1984); adapted with permission.

14A.2 Plans for 2^{N-6}

The defining contrasts for $N = 11$ are

$$\begin{aligned}\mathcal{I} &= F_1F_8F_9F_{10} = F_2F_9F_{10}F_{11} = F_3F_7F_{10}F_{11} = F_4F_7F_8F_{11}\\ &= F_5F_7F_8F_9 = F_6F_7F_8F_9F_{10}F_{11}.\end{aligned}$$

The defining contrasts for $N = 12$ are

$$\begin{aligned}\mathcal{I} &= F_1F_8F_9F_{10}F_{12} = F_2F_7F_9F_{10}F_{11} = F_3F_8F_{10}F_{11}F_{12}\\ &= F_4F_7F_9F_{11}F_{12} = F_5F_7F_8F_{10}F_{12} = F_6F_7F_8F_9F_{11}.\end{aligned}$$

The defining contrasts for $N = 13$ are

$$\begin{aligned}\mathcal{I} &= F_1F_9F_{10}F_{11}F_{12} = F_2F_{10}F_{11}F_{13} = F_3F_8F_9F_{10}F_{13}\\ &= F_4F_7F_8F_9F_{11}FF_{12}F_{13} = F_5F_7F_{11}F_{12}F_{13} = F_6F_7F_8F_{10}F_{11}.\end{aligned}$$

The defining contrasts for $N = 14$ are

$$\begin{aligned}\mathcal{I} &= F_1F_7F_8F_9F_{10}F_{11} = F_2F_7F_8F_9F_{12} = F_3F_7F_8F_{10}F_{12}F_{13}\\ &= F_4F_7F_8F_{11}F_{12}F_{13}F_{14} = F_5F_8F_9F_{10}F_{11}F_{12}F_{13}F_{14} = F_6F_7F_{10}F_{12}F_{14}.\end{aligned}$$

The defining contrasts for $N = 15$ are

$$\begin{aligned}\mathcal{I} &= F_1F_7F_8F_9F_{10}F_{11} = F_2F_7F_8F_9F_{12}F_{13} = F_3F_7F_8F_9F_{14}F_{15}\\ &= F_4F_7F_8F_{10}F_{12}F_{14} = F_5F_7F_9F_{11}F_{13}F_{15} = F_6F_7F_{10}F_{11}F_{12}F_{13}F_{14}F_{15}.\end{aligned}$$

14A.3 Plans for 2^{N-7}

The defining contrasts for $N = 13$ are

$$\begin{aligned}\mathcal{I} &= F_1F_{11}F_{12}F_{13} = F_2F_{10}F_{11}F_{13} = F_3F_8F_{10}F_{11} = F_4F_8F_9F_{10}F_{13}\\ &= F_5F_8F_9F_{11}F_{12}F_{13} = F_6F_8F_{11}F_{12} = F_7F_9F_{10}F_{11}F_{12}.\end{aligned}$$

The defining contrasts for $N = 14$ and 15 are

$$\begin{aligned}\mathcal{I} &= F_1F_8F_9F_{10}F_{15} = F_2F_8F_9F_{12}F_{14}F_{15} = F_3F_8F_{10}F_{12}F_{13}F_{15}\\ &= F_4F_9F_{10}F_{12}F_{13}F_{14}F_{15} = F_5F_8F_9F_{11}F_{13}F_{15} = F_6F_8F_{10}F_{11}F_{13}F_{14}\\ &= F_7F_9F_{10}F_{11}F_{14}.\end{aligned}$$

The contrasts for $N = 14$ are obtained by dropping F_{15}. The defining contrasts for $N = 16$ are

$$\begin{aligned}\mathcal{I} &= F_1F_8F_9F_{10}F_{12}F_{16} = F_2F_9F_{10}F_{11}F_{13}F_{16} = F_3F_{10}F_{11}F_{12}F_{14}F_{16}\\ &= F_4F_{11}F_{12}F_{13}F_{15}F_{16} = F_5F_8F_{12}F_{13}F_{14}F_{16} = F_6F_9F_{13}F_{14}F_{15}F_{16}\\ &= F_7F_8F_{10}F_{14}F_{15}F_{16}.\end{aligned}$$

14A.4 Plans for 2^{N-8}

The defining contrasts for $N = 14$ are

$$\begin{aligned}\mathcal{I} &= F_1F_9F_{10}F_{11}F_{13} = F_2F_9F_{10}F_{11}F_{14} = F_3F_9F_{10}F_{11}F_{12}F_{13}F_{14}\\ &= F_4F_9F_{12}F_{13} = F_5F_{10}F_{12}F_{14} = F_6F_{10}F_{13}F_{14} = F_7F_{11}F_{12}F_{14}\\ &= F_8F_{11}F_{13}F_{14}.\end{aligned}$$

The defining contrasts for $N = 15$ are

$$\begin{aligned}\mathcal{I} &= F_1F_9F_{10}F_{11}F_{13}F_{14} = F_2F_9F_{10}F_{12}F_{13}F_{15} = F_3F_9F_{11}F_{12}F_{14}F_{15}\\ &= F_4F_{13}F_{14}F_{15} = F_5F_{10}F_{11}F_{12} = F_6F_{11}F_{12}F_{13}F_{14}\\ &= F_7F_{10}F_{12}F_{14}F_{15} = F_8F_9F_{10}F_{11}F_{12}F_{13}.\end{aligned}$$

The defining contrasts for $N = 16$ and 17 are

$$\begin{aligned}\mathcal{I} &= F_1F_9F_{10}F_{11}F_{13}F_{17} = F_2F_{10}F_{11}F_{12}F_{14}F_{17} = F_3F_{11}F_{12}F_{13}F_{15}F_{17}\\ &= F_4F_{12}F_{13}F_{14}F_{16}F_{17} = F_5F_9F_{13}F_{14}F_{15}F_{17} = F_6F_{10}F_{14}F_{15}F_{16}F_{17}\\ &= F_7F_9F_{11}F_{15}F_{16}F_{17} = F_8F_9F_{10}F_{12}F_{16}F_{17}.\end{aligned}$$

The set of contrasts for $N = 16$ is obtained by dropping F_{17}.

14A.5 Plans for 2^{N-9}

The defining contrasts for $N = 15$ are

$$\begin{aligned}\mathcal{I} &= F_1F_{10}F_{11}F_{12}F_{13}F_{14}F_{15} = F_2F_{10}F_{11}F_{12}F_{13} = F_3F_{10}F_{11}F_{12}F_{14}\\ &= F_4F_{10}F_{13}F_{15} = F_5F_{10}F_{14}F_{15} = F_6F_{11}F_{13}F_{15}\\ &= F_7F_{11}F_{14}F_{15} = F_8F_{12}F_{13}F_{15} = F_9F_{12}F_{14}F_{15}.\end{aligned}$$

The defining contrasts for $N = 16$, 17, and 18 are

$$\mathcal{I} = F_1F_{10}F_{11}F_{12}F_{14}F_{18} = F_2F_{11}F_{12}F_{13}F_{15}F_{18} = F_3F_{12}F_{13}F_{14}F_{16}F_{18}$$
$$= F_4F_{13}F_{14}F_{15}F_{17}F_{18} = F_5F_{10}F_{14}F_{15}F_{16}F_{18} = F_6F_{11}F_{15}F_{16}F_{17}F_{18}$$
$$= F_7F_{10}F_{12}F_{16}F_{17}F_{18} = F_8F_{10}F_{11}F_{13}F_{17}F_{18}$$
$$= F_9F_{10}F_{11}F_{12}F_{13}F_{14}F_{15}F_{16}F_{17}F_{18}.$$

The defining contrasts for $N < 18$ are obtained by dropping factors $> N$.

APPENDIX 14B: MINIMUM ABERRATION FRACTIONS WITH BLOCKING*

Minimum Aberration 16-Run Plans with Blocking

Plan	Defining Contrasts	WLP[a]	Resolution
$2^{(5+1)-(1+1)}$	$ABCE$, $ABD\mathcal{B}$	0 0 1 2	IV
$2^{(5+2)-(2+1)}$	$ABCE$, $ACD\mathcal{B}_1$, $BCD\mathcal{B}_2$	0 2 1 4	III.5
$2^{(5+3)-(3+1)}$	$ABCE$, $AD\mathcal{B}_1$, $BD\mathcal{B}_2$, $CD\mathcal{B}_3$	0 10 1 0 0 4	III.5
$2^{(6+1)-(1+2)}$	$ABCE$, $ABDF$, $ACD\mathcal{B}$	0 0 3 4	IV
$2^{(6+2)-(2+2)}$	$ABCE$, $ABDF$, $ACD\mathcal{B}_1$, $BCD\mathcal{B}_2$	0 3 3 8	III.5
$2^{(6+3)-(3+2)}$	$ABCE$, $ABDF$, $AC\mathcal{B}_1$, $AD\mathcal{B}_2$, $ABCD\mathcal{B}_3$	0 15 3 0	III.5
$2^{(7+1)-(1+3)}$	$ABCE$, $ABDF$, $ACDG$, $BCD\mathcal{B}$	0 0 7 7 0 0	IV
$2^{(7+2)-(2+3)}$	$ABCE$, $ABDF$, $ACDG$, $AB\mathcal{B}_1$, $AC\mathcal{B}_2$	0 9 7 0 0 12	III.5
$2^{(7+3)-(3+3)}$	$ABCE$, $ABDF$, $ACDG$, $AB\mathcal{B}_1$, $AC\mathcal{B}_2$, $AD\mathcal{B}_3$	0 21 7 0 0 28	III.5
$2^{(8+1)-(1+4)}$	$ABCE$, $ABDF$, $ACDG$, $BCDH$, $AB\mathcal{B}$	0 4 14 0 0	III.5
$2^{(8+2)-(2+4)}$	$ABCE$, $ABDF$, $ACDG$, $BCDH$, $AB\mathcal{B}_1$, $AC\mathcal{B}_2$	0 12 14 0 0 24	III.5
$2^{(8+3)-(3+4)}$	$ABCE$, $ABDF$, $ACDG$, $BCDH$, $AB\mathcal{B}_1$, $AC\mathcal{B}_2$, $AD\mathcal{B}_3$	0 28 14 0 0 56	III.5
$2^{(9+1)-(1+5)}$	ABE, ACF, ADG, $BCDH$, $ABCDI$, $BC\mathcal{B}$	4 4 14 4 8 8	III
$2^{(9+2)-(2+5)}$	ABE, ACF, BDG, CDH, $ABCDI$, $BC\mathcal{B}_1$, $ABD\mathcal{B}_2$	4 12 14 12 8 24	III
$2^{(10+1)-(1+6)}$	ABE, ACF, BCG, ADH, $BCDI$, $ABCDJ$, $BD\mathcal{B}$	8 4 18 8 16 12	III
$2^{(10+2)-(2+6)}$	ABE, ACF, BCG, ADH, $BCDI$, $ABCDJ$, $BD\mathcal{B}_1$, $ACD\mathcal{B}_2$	8 13 18 24 16 32 8 48	III
$2^{(11+1)-(1+7)}$	ABE, ACF, BCG, ADH, BDI, $ACDJ$, $BCDK$, $CD\mathcal{B}$	12 4 26 13 28 20 24 24	III

(*continued*)

*Based on tables of plans in Sitter et al. (1997); adapted with permission.

Minimum Aberration 16-Run Plans with Blocking (*continued*)

Plan	Defining Contrasts	WLP[a]	Resolution
$2^{(11+2)-(2+7)}$	ABE, ACF, BCG, ADH, BDI, $ACDJ$, $BCDK$, $ABC\mathcal{B}_1$, $ABD\mathcal{B}_2$	12 15 26 36 28 48 24	III
$2^{(12+1)-(1+8)}$	ABE, ACF, BCG, ADH, BDI, $ACDJ$, $BCDK$, $ABCDL$, $ABC\mathcal{B}$	16 6 39 16 48 24 48	III
$2^{(12+2)-(2+8)}$	ABE, ACF, BCG, ADH, BDI, $ACDJ$, $BCDK$, $ABCDL$, $ABC\mathcal{B}_1$, $ABD\mathcal{B}_2$	16 18 39 48 48 72 48	III
$2^{(13+1)-(1+9)}$	ABE, ACF, BCG, ADH, BDI, $ABDJ$, $ACDK$, $BCDL$, $ABC\mathcal{B}$	22 6 55 22 72 40 96 72	III
$2^{(14+1)-(1+10)}$	ABE, ACF, BCG, ADH, BDI, $ABDJ$, CDK, $ACDL$, $BCDM$, $ABC\mathcal{B}$	28 7 77 28 112 56 168 112	III

[a] Word-length pattern ($\ell_3, \ell_{3.5}, \ell_4, \ell_{4.5}, \ell_5, \ell_{5.5}, \ldots$). Some have been truncated.

Minimum Aberration 32-Run Plans with Blocking

Plan	Defining Contrasts	WLP[a]	Resolution
$2^{(6+1)-(1+1)}$	$ABCDEF$, $ABC\mathcal{B}$	0 0 0 2 0 0 1	IV.5
$2^{(6+2)-(2+1)}$	$ABCF$, $ABD\mathcal{B}_1$, $BCDE\mathcal{B}_2$	0 0 1 4 0 2	IV
$2^{(6+3)-(3+1)}$	$ABCDEF$, $ACE\mathcal{B}_1$, $BCE\mathcal{B}_2$, $ADE\mathcal{B}_3$	0 3 0 8 0 3 1	III.5
$2^{(6+4)-(4+1)}$	$ABCDEF$, $AB\mathcal{B}_1$, $AC\mathcal{B}_2$, $AD\mathcal{B}_3$, $AE\mathcal{B}_4$	0 15 0 0 0 15 1	III.5
$2^{(7+1)-(1+2)}$	$ABCF$, $ABDEG$, $ACD\mathcal{B}$	0 0 1 2 2 2	IV
$2^{(7+2)-(2+2)}$	$ABCF$, $ABDG$, $ABE\mathcal{B}_1$, $BCDE\mathcal{B}_2$	0 0 3 7 0 4	IV
$2^{(7+3)-(3+2)}$	$ABCF$, $ABDEG$, $BCD\mathcal{B}_1$, $BCE\mathcal{B}_2$, $ACDE\mathcal{B}_3$	0 5 1 12 2	III.5
$2^{(7+4)-(4+2)}$	$ABCF$, $ADEG$, $BD\mathcal{B}_1$, $CD\mathcal{B}_2$, $ABCE\mathcal{B}_3$, $BCDE\mathcal{B}_4$	0 21 2 0 0 33	III.5
$2^{(8+1)-(1+3)}$	$ABCF$, $ABDG$, $ACDEH$, $ABE\mathcal{B}$	0 0 3 3 4 4	IV
$2^{(8+2)-(2+3)}$	$ABCF$, $ABDG$, $ACDEH$, $ABE\mathcal{B}_1$, $BCDE\mathcal{B}_2$	0 1 3 10 4 8	III.5
$2^{(8+3)-(3+3)}$	$ABCF$, $ABDG$, $ACEH$, $ACD\mathcal{B}_1$, $CDE\mathcal{B}_2$, $ABCDE\mathcal{B}_3$	0 7 5 18 0 10	III.5
$2^{(8+4)-(4+3)}$	$ABCF$, $ABDG$, $ACEH$, $DE\mathcal{B}_1$, $ABDE\mathcal{B}_2$, $ACDE\mathcal{B}_3$, $BCDE\mathcal{B}_4$	0 28 5 0 0 65 2	III.5
$2^{(9+1)-(1+4)}$	$ABCF$, $ABDG$, $ABEH$, $ACDEI$, $BCDE\mathcal{B}$	0 0 6 4 8 8	IV
$2^{(9+2)-(2+4)}$	$ABCF$, $ABDG$, $ACEH$, $ADEI$, $BDE\mathcal{B}_1$, $CDE\mathcal{B}_2$	0 2 9 14 0 9 6 12	III.5
$2^{(9+3)-(3+4)}$	$ABCF$, $ABDG$, $ACEH$, $ADEI$, $ACD\mathcal{B}_1$, $BCD\mathcal{B}_2$, $ABCDE\mathcal{B}_3$	0 9 9 27 0 18 6 27	III.5

Minimum Aberration 32-Run Plans with Blocking (*continued*)

Plan	Defining Contrasts	WLP[a]	Resolution
$2^{(9+4)-(4+4)}$	$ABCF$, $ABDG$, $ACEH$, $ADEI$, $BC\mathcal{B}_1$, $BD\mathcal{B}_2$, $ACDE\mathcal{B}_3$, $BCDE\mathcal{B}_4$	0 36 9 0 0 117 6	III.5
$2^{(10+1)-(1+5)}$	$ABCF$, $ABDG$, $ACEH$, $ADEI$, $ABCDEJ$, $ACD\mathcal{B}$	0 0 15 10 0 0 15 12	IV.5
$2^{(10+2)-(2+5)}$	$ABCF$, $ABDG$, $ABEH$, $ACDEI$, $BCDEJ$, $ACD\mathcal{B}_1$, $BCD\mathcal{B}_2$	0 3 15 20 0 13 15 24	III.5
$2^{(10+3)-(3+5)}$	$ABCF$, $ABDG$, $ACDH$, $ABEI$, $ACEJ$, $ADE\mathcal{B}_1$, $BDE\mathcal{B}_2$, $CDE\mathcal{B}_3$	0 12 16 36 0 30 12 60	III.5
$2^{(10+4)-(4+5)}$	$ABCF$, $ABDG$, $ACEH$, $ADEI$, $ABCDEJ$, $AB\mathcal{B}_1$, $AC\mathcal{B}_2$, $AE\mathcal{B}_3$, $ABDE\mathcal{B}_4$	45 15 0 0 195 15	III
$2^{(11+1)-(1+6)}$	$ABCF$, $ABDG$, $ACDH$, $ABEI$, $ACEJ$, $ADEK$, $BCD\mathcal{B}$	0 0 25 13 0 0 27 25	IV
$2^{(11+2)-(2+6)}$	$ABCF$, $ABDG$, $ACDH$, $ABEI$, $ACEJ$, $ADEK$, $BCD\mathcal{B}_1$, $BDE\mathcal{B}_2$	0 4 25 26 0 19 27 50	III.5
$2^{(11+3)-(3+6)}$	$ABCF$, $ABDG$, $ACDH$, $BCDI$, $ABEJ$, $ACEK$, $ADE\mathcal{B}_1$, $BDE\mathcal{B}_2$, $CDE\mathcal{B}_3$	0 15 26 48 0 48 24 112	III.5
$2^{(11+4)-(4+6)}$	$ABCF$, $ABDG$, $ACDH$, $ABEI$, $ACEJ$, $ADEK$, $ABCE\mathcal{B}_1$, $ABDE\mathcal{B}_2$, $ACDE\mathcal{B}_3$, $BCDE\mathcal{B}_4$	55 25 0 0 305 27	III
$2^{(12+1)-(1+7)}$	$ABCF$, $ABDG$, $ACDH$, $BCDI$, $ABEJ$, $ACEK$, $ADEL$, $BCE\mathcal{B}$	0 0 38 17 0 0 52 44	IV
$2^{(12+2)-(2+7)}$	$ABCF$, $ABDG$, $ACDH$, $BCDI$, $ABEJ$, $ACEK$, $BCE\mathcal{B}_1$, $BDE\mathcal{B}_2$	0 5 38 34 0 28 52 88 0 62	III.5
$2^{(12+3)-(3+7)}$	$ABCF$, $ABDG$, $ACDH$, $BCDI$, $ABEJ$, $ACEK$, $BCEL$, $ADE\mathcal{B}_1$, $BDE\mathcal{B}_2$, $CDE\mathcal{B}_3$	0 18 39 64 0 72 48 192	III.5
$2^{(12+4)-(4+7)}$	$ABCF$, $ABDG$, $ACDH$, $BCDI$, $ABEJ$, $ACEK$, $ADEL$, $ABCD\mathcal{B}_1$, $AE\mathcal{B}_2$, $BE\mathcal{B}_3$, $CE\mathcal{B}_4$	66 38 0 0 457 52	III
$2^{(13+1)-(1+8)}$	$ABCF$, $ABDG$, $ACDH$, $BCDI$, $ABEJ$, $ACEK$, $ACEL$, $ADEM$, $BDE\mathcal{B}$	0 0 55 22 0 0 96 72	IV
$2^{(13+2)-(2+8)}$	$ABCF$, $ABDG$, $ACDH$, $BCDI$, $ABEJ$, $ACEK$, $ACEL$, $ADEM$, $CDE\mathcal{B}_1$, $ABCDE\mathcal{B}_2$	0 6 55 44 0 40 96 144	III.5
$2^{(13+3)-(3+8)}$	$ABCF$, $ABDG$, $ACDH$, $BCDI$, $ABEJ$, $ACEK$, $ACEL$, $ADEM$, $AD\mathcal{B}_1$, $AE\mathcal{B}_2$, $ABDE\mathcal{B}_3$	0 36 55 0 0 310 96	III.5
$2^{(14+1)-(1+9)}$	$ABCF$, $ABDG$, $ACDH$, $BCDI$, $ABEJ$, $ACEK$, $ACEL$, $ADEM$, $BDEN$, $CDE\mathcal{B}$	0 0 77 28 0 0 168 112	IV

(*continued*)

Minimum Aberration 32-Run Plans with Blocking (*continued*)

Plan	Defining Contrasts	WLP[a]	Resolution
$2^{(14+2)-(2+9)}$	$ABCF$, $ABDG$, $ACDH$, $BCDI$, $ABEJ$, $ACEK$, $ACEL$, $ADEM$, $BDEN$, $AB\mathcal{B}_1$, $CDE\mathcal{B}_2$	0 7 77 56 0 56 168 224	III.5
$2^{(14+3)-(3+9)}$	$ABCF$, $ABDG$, $ACDH$, $BCDI$, $ABEJ$, $ACEK$, $ACEL$, $ADEM$, $BDEN$, $AD\mathcal{B}_1$, $CD\mathcal{B}_2$, $AE\mathcal{B}_3$	0 42 77 0 0 434 168	III.5
$2^{(15+1)-(1+10)}$	$ABCF$, $ABDG$, $ACDH$, $BCDI$, $ABEJ$, $ACEK$, $ACEL$, $ADEM$, $BDEN$, $CDEO$, $ABCDE\mathcal{B}$	0 0 105 35 0 0 280 168	IV
$2^{(15+2)-(2+10)}$	$ABCF$, $ABDG$, $ACDH$, $BCDI$, $ABEJ$, $ACEK$, $ACEL$, $ADEM$, $BDEN$, $CDEO$, $AB\mathcal{B}_1$, $AC\mathcal{B}_2$	0 21 105 0 0 252 280	III.5
$2^{(15+3)-(3+10)}$	$ABCF$, $ABDG$, $ACDH$, $BCDI$, $ABEJ$, $ACEK$, $ACEL$, $ADEM$, $BDEN$, $CDEO$, $AB\mathcal{B}_1$, $AC\mathcal{B}_2$, $AD\mathcal{B}_3$	49 105 0 0 588 280	III

[a]Word-length pattern ($\ell_3, \ell_{3.5}, \ell_4, \ell_{4.5}, \ell_5, \ell_{5.5}, \ldots$). Some have been truncated.

Minimum Aberration 64-Run Plans with Blocking

Plan	Defining Contrasts	WLP[a]	Resolution
$2^{(7+1)-(1+1)}$	$ABCDEG$, $ABCF\mathcal{B}$	0 0 0 0 0 2 1	V.5
$2^{(7+2)-(2+1)}$	$ABCDG$, $ABE\mathcal{B}_1$, $BCEF\mathcal{B}_2$	0 0 0 2 1 3 0 1	IV.5
$2^{(7+3)-(3+1)}$	$ABCDEFG$, $ABC\mathcal{B}_1$, $ADE\mathcal{B}_2$, $BDF\mathcal{B}_3$	0 0 0 7 0 7	IV.5
$2^{(7+4)-(4+1)}$	$ABCDEG$, $ACDF\mathcal{B}_1$, $BCDF\mathcal{B}_2$, $CEF\mathcal{B}_3$, $DEF\mathcal{B}_4$	0 5 0 12 0 7 1 4	III.5
$2^{(7+5)-(5+1)}$	$ABCDEG$, $ABCF\mathcal{B}_1$, $ABDF\mathcal{B}_2$, $ACDF\mathcal{B}_3$, $ABEF\mathcal{B}_4$, $BCEF\mathcal{B}_5$	0 21 0 0 0 35 1	III.5
$2^{(8+1)-(1+2)}$	$ABCDG$, $ABEFH$, $ACE\mathcal{B}$	0 0 0 1 2 2 1 1	IV.5
$2^{(8+2)-(2+2)}$	$ABCDG$, $ABEFH$, $ACE\mathcal{B}_1$, $BDF\mathcal{B}_2$	0 0 0 4 2 5 1 2	IV.5
$2^{(8+3)-(3+2)}$	$ABCG$, $ABDEFH$, $ACF\mathcal{B}_1$, $BCDF\mathcal{B}_2$, $BCEF\mathcal{B}_3$	0 1 1 10 0 10 2 4	III.5
$2^{(8+4)-(4+2)}$	$ABCDG$, $ABEFH$, $BCDE\mathcal{B}_1$, $BCDF\mathcal{B}_2$, $ACEF\mathcal{B}_3$, $ADEF\mathcal{B}_4$	0 7 0 18 2 15 1 10	III.5
$2^{(8+5)-(5+2)}$	$ABCG$, $ABDEFH$, $ACDE\mathcal{B}_1$, $BCDE\mathcal{B}_2$, $ACDF\mathcal{B}_3$, $BCDF\mathcal{B}_4$, $ACEF\mathcal{B}_5$	0 28 1 0 0 69 2	III.5
$2^{(9+1)-(1+3)}$	$ABCG$, $ABDEH$, $ACDFI$, $ABEF\mathcal{B}$	0 0 1 1 4 4 2 2	IV
$2^{(9+2)-(2+3)}$	$ABCG$, $ABDEH$, $ACDFI$, $AEF\mathcal{B}_1$, $ABCDEF\mathcal{B}_2$	0 0 1 6 4 8 2 5	IV

Minimum Aberration 64-Run Plans with Blocking (*continued*)

Plan	Defining Contrasts	WLP[a]	Resolution
$2^{(9+3)-(3+3)}$	$ABCG$, $ABDEH$, $ACDFI$, $ABF\mathcal{B}_1$, $ACEF\mathcal{B}_2$, $BCDEF\mathcal{B}_3$	0 2 1 14 4 17 2 8	III.5
$2^{(9+4)-(4+3)}$	$ABCG$, $ABDEH$, $ACDFI$, $BCD\mathcal{B}_1$, $BCF\mathcal{B}_2$, $CEF\mathcal{B}_3$, $ADEF\mathcal{B}_4$	0 9 1 27 4 26 2 23	III.5
$2^{(9+5)-(5+3)}$	$ABCG$, $ABDH$, $ACDEFI$, $AE\mathcal{B}_1$, $BCDE\mathcal{B}_2$, $CF\mathcal{B}_3$, $BCDF\mathcal{B}_4$, $ABEF\mathcal{B}_5$	0 36 3 0 0 123 4	III.5

[a]Word-length pattern ($\ell_3, \ell_{3.5}, \ell_4, \ell_{4.5}, \ell_5, \ell_{5.5}, \ldots$). Some have been truncated.

Minimum Aberration 128-Run Plans with Blocking

Plan	Defining Contrasts	WLP[a]	Resolution
$2^{(8+1)-(1+1)}$	$ABCDEH$, $ABCFG\mathcal{B}$	0 0 0 0 0 0 2 1	VI
$2^{(8+2)-(2+1)}$	$ABCDEH$, $ABCE\mathcal{B}_1$, $CDFG\mathcal{B}_2$	0 0 0 0 0 5 1	V.5
$2^{(8+3)-(3+1)}$	$ABCDEFGH$, $ABCD\mathcal{B}_1$, $ABEF\mathcal{B}_2$, $ACEG\mathcal{B}_3$	0 0 0 0 0 14	V.5
$2^{(8+4)-(4+1)}$	$ABCDEFH$, $BDF\mathcal{B}_1$, $CDFG\mathcal{B}_2$, $BEFG\mathcal{B}_3$, $ACEFG\mathcal{B}_4$	0 1 0 10 0 11 0 4	III.5
$2^{(8+5)-(5+1)}$	$ABCDEFH$, $ADG\mathcal{B}_1$, $ABCEG\mathcal{B}_2$, $ABCFG\mathcal{B}_3$, $BDEFG\mathcal{B}_4$, $CDEFG\mathcal{B}_5$	0 7 0 18 0 15 0 12	III.5
$2^{(9+1)-(1+2)}$	$ABCDEH$, $ABCFGI$, $ABDF\mathcal{B}$	0 0 0 0 0 3 3	V.5
$2^{(9+2)-(2+2)}$	$ABCDEH$, $ABCFGI$, $ABDF\mathcal{B}_1$, $ACEG\mathcal{B}_2$	0 0 0 0 0 9 3	V.5
$2^{(9+3)-(3+2)}$	$ABCDH$, $AEFGI$, $ABE\mathcal{B}_1$, $BCEF\mathcal{B}_2$, $BDEG\mathcal{B}_3$	0 0 0 4 2 14 0 6	IV.5
$2^{(9+4)-(4+2)}$	$ABCDEH$, $ABCFGI$, $ABDF\mathcal{B}_1$, $CEG\mathcal{B}_2$, $ADEG\mathcal{B}_3$, $AEFG\mathcal{B}_4$	0 2 0 14 0 18 3 12	III.5
$2^{(9+5)-(5+2)}$	$ABCDEH$, $ABCFGI$, $ABEF\mathcal{B}_1$, $ACDEF\mathcal{B}_2$, $CDG\mathcal{B}_3$, $ADEG\mathcal{B}_4$, $BCEFG\mathcal{B}_5$	0 9 0 27 0 27 3 27	III.5

[a]Word-length pattern ($\ell_3, \ell_{3.5}, \ell_4, \ell_{4.5}, \ell_5, \ell_{5.5}, \ldots$). Some have been truncated.

CHAPTER 15

Response Surface Designs

15.1 INTRODUCTION

Response surface exploration techniques are appropriate when the independent controllable variables are quantitative and interest is in a response variable that is somehow related to the independent variables. The true relationship is unknown. Often, the goal is either to find sets of conditions that maximize yield, minimize the proportion of defects, or some combination of the two. Response surface designs are also appropriate when interest centers more on examining the relationship between inputs and response, often with the objective of learning the ranges of the input variables that yield acceptable results. Occasionally, the objective is to find ranges of the input variables under which the process is relatively insensitive to changes. It may be possible to find conditions under which it is much cheaper or easier to manufacture a particular item with adequate quality. Readers who require more detailed treatment of topics covered in this chapter are urged to consult Box and Draper (1987), Atkinson and Donev (1992), Myers and Montgomery (1995), Khuri and Cornell (1996), and Cornell (2002).

In the first part of this chapter the emphasis will be on designs for fitting response surfaces in cases where input variables can be changed independently. Next we discuss designs for the special situation in which the input variables are ingredients in a mixture that are combined in different proportions to form a product. The special feature here is that the proportions of the inputs must sum to 1. These designs are called *mixture designs*. There is a large class of problems that involve combinations of input variables, some of which can be varied independently and some of which are constrained as proportions. These problems are beyond the scope of this book, although the methods given here provide a basis for methods of attack. The next parts of this chapter focus on techniques for finding optimal conditions. The underlying problem is to find combinations of the input variables that maximize (or equivalently, minimize)

Planning, Construction, and Statistical Analysis of Comparative Experiments,
by Francis G. Giesbrecht and Marcia L. Gumpertz
ISBN 0-471-21395-0

the response. Finally, we conclude with a discussion of some of the problems likely to be encountered in practice. With some exception we limit ourselves to two- or three-input-variables cases. The concepts and techniques all extend to more dimensions, but the presentation becomes more difficult because we lose the ability to construct nice pictures.

The focus of the designs in this chapter is on the shape of the surface rather than on the values of the parameters. The designs are intended for situations where the experimenter wants to determine the response to various input factors, to understand the relationships of the factors to one another, to find factor levels that provide optimum responses, or to find factor regions that produce the best combinations of several different responses. This translates into more interest in estimating y, i.e., $\widehat{y}$, and controlling $\text{Var}[\widehat{y}]$ than in obtaining parameter estimates with small variances. The factorial-based designs we looked at in other chapters concentrated more on variances of estimates of parameters and contrasts. In many cases the statistical analyses of the data resulting from experiments discussed in this chapter are relatively simple. A very common tool is to construct contour plots of the fitted surface. Conclusions are often obvious after looking at the contours.

15.2 BASIC FORMULATION

The basic premise underlying the approach to designing experiments to be pursued in this chapter is that there is a continuous region in the factor space in which it is possible to conduct experiments. We refer to this as the *operability region*. It is reasonable to assume that some true response function exists which relates the response (say yield or fraction defective) to the levels of the input factors. If the form of the response function were known, there would be no need for experimental work. The problem is that, in practice, the function is not only unknown, but also probably very complex. Generally, the best we can hope for is that we can approximate the response function, or at least learn something about its shape, such as possibly the location of the minimum or a maximum, or a ridge, in some region of interest. The region of interest is typically a modest subregion of the operability region. When an experiment is conducted, we obtain observations at various combinations of levels of the input or independent variables, where $x_{i1}, x_{i2}, \ldots, x_{ip}$ denote the levels of the first, second, $\ldots$, pth independent variable in the ith trial in a set of n trials. The responses observed in the experiment are then analyzed, with the ith trial modeled as

$$y_i = f(x_{i1}, x_{i2}, \ldots, x_{ip}) + \epsilon_i,$$

where f is some conveniently chosen function.

The simplest function that comes to mind to approximate a response surface is the first-order polynomial

$$y_i = \beta_0 + \sum_j \beta_j x_{ij} + \epsilon_i \qquad \text{for } i = 1, 2, \ldots, n.$$

Next in complexity, we have the quadratic or second-order model,

$$y_i = \beta_0 + \sum_j \beta_j (x_{ij} - \overline{x}_{\cdot j}) + \sum_{j \le k} \gamma_{jk} (x_{ij} - \overline{x}_{\cdot j})(x_{ik} - \overline{x}_{\cdot k}) + \epsilon_i \qquad (15.1)$$

for $i = 1, 2, \ldots, n$. There is a direct extension to higher-order polynomials. Those comfortable with concepts from calculus can think of polynomial models as Taylor's series approximations to more complex functions. Many surfaces can be adequately approximated by a second-order polynomial, provided that interest is focused on a sufficiently small region of the surface. Many experimental strategies take advantage of this. Unfortunately, second-order polynomial models and, in general, all polynomial models can become very inadequate for describing large surface areas. The risks are particularly severe if an experimenter tries to extrapolate beyond the range of experimental data. This risk must always be kept in mind. In practice, investigators are urged to use functions derived from basic understanding of the complete process whenever possible. Polynomial models can be thought of as basic "know nothing" models, useful to get started.

In the first part of this chapter we begin with techniques that are particularly useful for describing the response surface in the region of an optimum. The simplicity of second-order polynomial models gives them particular appeal. A full second-order polynomial in p variables has $1+2p+p(p-1)/2$ parameters. It follows that at least three levels of each factor and $1+2p+p(p-1)/2$ distinct design points are required to fit the full model. One possible class of designs is obtained from fractions in the 3^N system with $N = p$ factors. A problem with these plans is that as p becomes large, the number of treatment combinations and design points become excessive. For example, for $p = 3$ factors there are 10 parameters. A one-third replicate of the 3^3 has only nine treatment combinations and therefore is inadequate. A full replicate can be used but requires $3^3 = 27$ runs. We examine some alternatives in a relatively general setting. In addition, in Chapter 20 we examine special plans for fitting second-order polynomial models using more reasonable numbers of design points.

Exercise 15.1: Parameters in a Polynomial Model. List the full set of parameters in a second-order polynomial model with three input variables. What is the minimal number of data points required to fit this model and also provide six degrees of freedom for an estimate of experimental error? □

15.3 SOME BASIC DESIGNS

15.3.1 Experiments on a Cube or Hypercube

A large class of simple designs, commonly known as *central composite designs*, can be built up from 2^N factorials. Their major advantages are simplicity and the fact that they lend themselves to experimentation in stages, i.e., blocking. For N factors the central composite design consists of a 2^N factorial or fractional factorial augmented by $2N$ axial points and some m center points. It

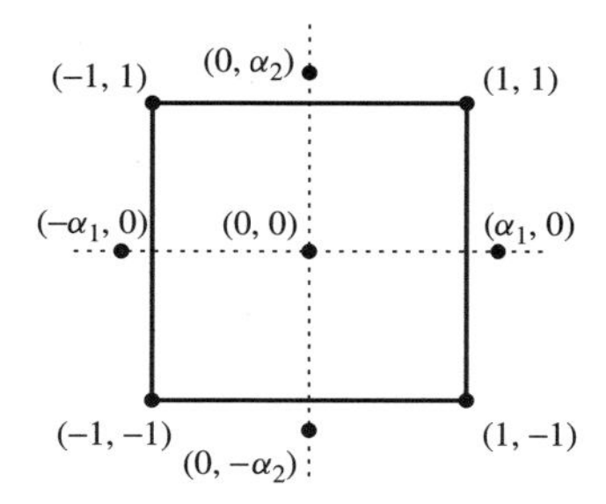

Figure 15.1 Central composite design for two factors with all points labeled.

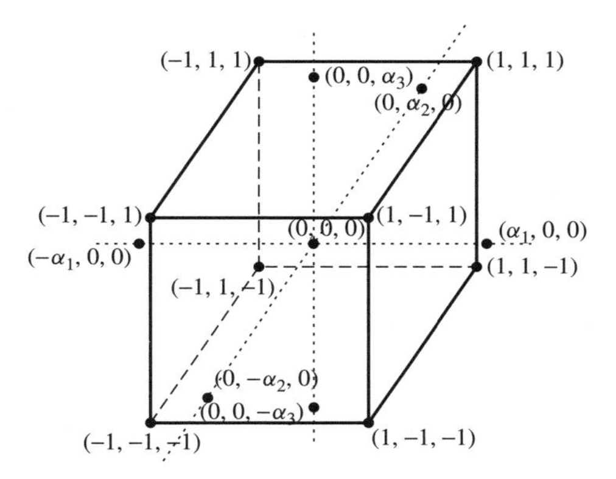

Figure 15.2 Central composite design for three factors with all points labeled.

is common practice to code the factor levels so that the high and low values for the factorial points are ± 1. The actual points are then $(\pm 1, \pm 1, \ldots, \pm 1)$. The center point is $(0, 0, \ldots, 0)$. The axial points are coded as $(\pm\alpha_1, 0, \ldots, 0)$, $(0, \pm\alpha_2, 0, \ldots, 0), \ldots, (0, \ldots, 0, \pm\alpha_N)$, where the $\{\alpha_i\}$ are constants to be determined. Most authors insist on a common α value for all axis. We see no reason for that restriction. We illustrate such plans with $N = 2$ and $N = 3$ factors in Figures 15.1 and 15.2.

At this point this design allows for considerable choice. First, there is the choice of number of center points. Then, the locations of the axial points must be chosen; i.e., the distances $\{\alpha_i\}$ must be chosen. A popular choice is to set all α equal to 1, giving *face-centered designs*.

Example 15.1: Planning Exercise. We illustrate the details of setting up an experiment with two factors using the plan illustrated in Figure 15.1. We emphasize that all the computations in this example are done before the experiment is conducted, before any data are collected. We consider a plan to study yield as a function of temperature of a reaction vessel and amount of enzyme added. We assume that temperatures in the range 75 to 80°C and enzyme additions between 5 and 9 oz per batch are of interest. For the purpose of this example, we let $(0, 0)$ correspond to the 77°C and 7 oz levels. At this point we reserve judgment on number of center points. Let the low- and high-temperature levels correspond to 75 and 79°C and the low- and high-enzyme levels to 5 and 9 oz for the corner points. Then there are the axial points. We begin by considering $\alpha_1 = \alpha_2 = 1$. This leads to points at 75°C and 7 oz of enzyme, 79°C and 7 oz of enzyme, 77°C and 5 oz of enzyme, and 77°C and 9 oz of enzyme. Note that with only one center point, this is exactly a 3^2. However, this is just a coincidence for us at this time.

Assume a first-order linear model

$$\text{yield} = \beta_0 + \beta_T \times {}^\circ\text{C} + \beta_E \times \ \text{oz.} + \epsilon$$

or in general matrix notation

$$\boldsymbol{y} = \boldsymbol{X}\boldsymbol{\beta} + \boldsymbol{\epsilon},$$

where $\boldsymbol{X}$ has the form

$$\begin{bmatrix} 1 & 77 & 7 \\ 1 & 75 & 5 \\ 1 & 75 & 9 \\ 1 & 79 & 5 \\ 1 & 79 & 9 \\ 1 & 75 & 7 \\ 1 & 79 & 7 \\ 1 & 77 & 5 \\ 1 & 77 & 9 \end{bmatrix}.$$

From multiple regression theory we have $\widehat{\boldsymbol{\beta}} = (\boldsymbol{X}^t\boldsymbol{X})^{-1}\boldsymbol{X}^t\boldsymbol{y}$, $\text{Var}[\widehat{\boldsymbol{\beta}}] = (\boldsymbol{X}^t\boldsymbol{X})^{-1}\sigma_\epsilon^2$, and $\widehat{\boldsymbol{y}} = \boldsymbol{X}\widehat{\boldsymbol{\beta}}$. Also,

$$\text{Var}[\widehat{\boldsymbol{y}}] = \boldsymbol{X}(\boldsymbol{X}^t\boldsymbol{X})^{-1}\boldsymbol{X}^t\sigma_\epsilon^2.$$

Now let $\boldsymbol{X}_p$ represent a matrix of points where we want to predict the surface. An example would be

$$\begin{bmatrix} 1 & 76 & 5 \\ 1 & 76 & 6 \\ 1 & 76 & 7 \\ 1 & 78 & 5 \\ 1 & 78 & 6 \\ 1 & 78 & 7 \end{bmatrix}$$

for six extra points. It follows that $\widehat{\boldsymbol{y}}_p = \boldsymbol{X}_p\widehat{\boldsymbol{\beta}}$ and $\text{Var}[\widehat{\boldsymbol{y}}_p] = \boldsymbol{X}_p(\boldsymbol{X}^t\boldsymbol{X})^{-1}\boldsymbol{X}_p^t\sigma_\epsilon^2$.

```
data points ;
 int = 1 ; input temp enz @ ;
datalines ;
 77 7 75 5 75 9 79 5 79 9 75 7 79 7 77 5 77 9
data pred ; int = 1 ;
 do temp = 74 to 80 by .25 ;
 do enz = 4 to 10 by .25 ;
 output ;  end ;  end ;
proc iml ;
 use points ; read all into X ;
 use pred ; read all into XP ;
 D = XP*inv(X'*X)*XP' ;  V = vecdiag(D) ;
 INF = j(nrow(V),1,1)/V ;  XP = XP||V||INF ;
 NAMES = { "INT" "TEMP" "ENZ" "VAR" "INF" } ;
 create surf from XP (| colname = NAMES |) ;
 append from XP ;
proc gcontour data = surf ;
 plot temp*enz=inf /
 levels =   2  3  4  5  6  7  8  9 10 11 12
 llevels = 35 36 34 33  2  3  4  5  6  1  1; run ;
```

Display 15.1 SAS® code to display information contours.

We are really interested in the variances of the predicted y's, i.e., the elements on the main diagonal of Var[$\widehat{y}_p$]. We are still at the planning stage and do not have data and an estimate for σ_ϵ^2. For our purposes here, we can ignore σ_ϵ^2.

We pick a design, compute predicted variances on a grid in the region of interest and plot contours of the variance or equivalently, contours of *information*, the reciprocal of the variance. We give the SAS® code in Display 15.1 and on the companion website to illustrate the calculations we have just described for computing the information contours. We use nine data points.

The task now is to examine the information contour plot in Figure 15.3 and decide if the plan is adequate. Remember, high values for information mean that the experiment provides more information at that point. Also, the contours represent quality of information provided by the experiment, not the yield response, which is the real interest for the experiment. In general, doubling the number of observations doubles the information. For our hypothetical example, the real question is whether the experiment concentrates the information in the region that is really important.

The plan we have selected to examine concentrates the information rather tightly about the 77°C, 7 oz point. Adding extra center points will exacerbate this concentration. This phenomenon is illustrated in Figure 15.4, which uses the same contour levels as Figure 15.3. In Figure 15.4a we show the information contours for the same plan as used in Figure 15.3 with four extra center points added. Three extra contour levels in the center emphasize the increased concentration of information. The right panel was obtained by moving the axial points to 74°C and 7 oz of enzyme, 80°C and 7 oz of enzyme, 77°C and 4 oz of enzyme, and 77°C and 10 oz of enzyme and duplicating each. The broader spread of

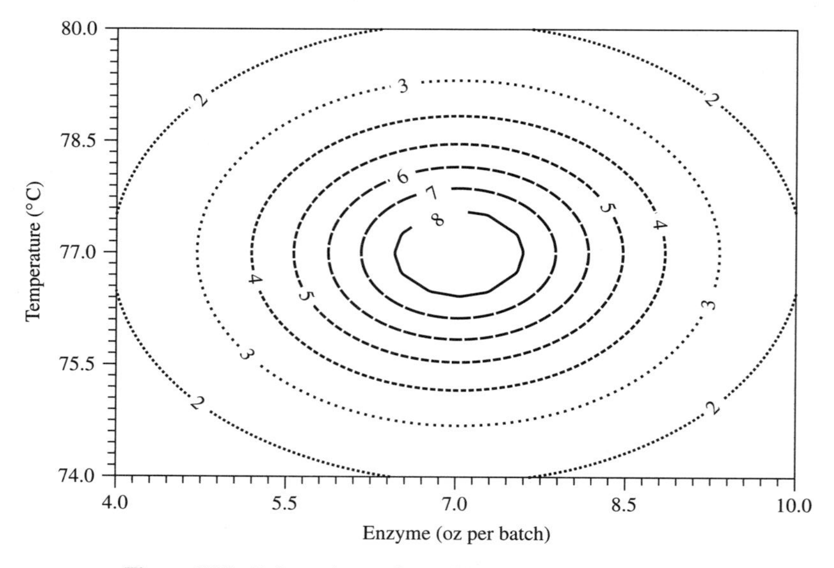

Figure 15.3 Information surface with equally spaced contours.

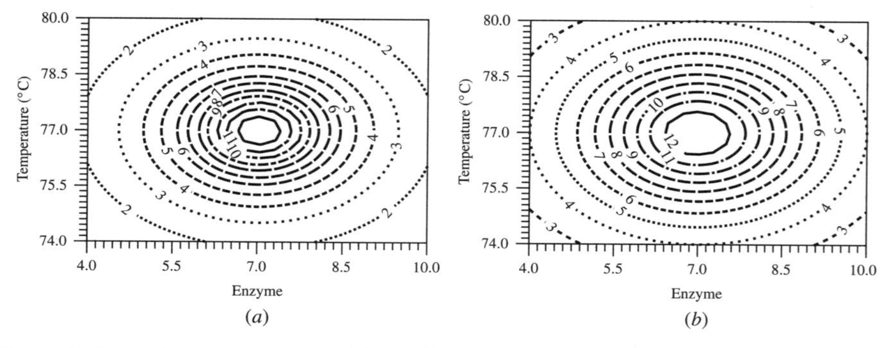

Figure 15.4 (*a*) Information surface with equally spaced contours and with four extra center points; (*b*) axial points moved out and duplicated. Each design has 13 points.

information is marked. Note that both panels are based on an experiment with 13 points.

Our discussion and information contour plots are based on fitting a first-order model. In general, all the comments made so far hold for more complex models, such as the second-order model. The information surface becomes somewhat more complex. In Figure 15.5 we illustrate contours for the same designs as used in Figure 15.4 but with a second-order model. Notice that there is a depression at the center in Figure 15.5*b*, indicating less information—hence, the common recommendation that one include three to five center points in a response surface design.

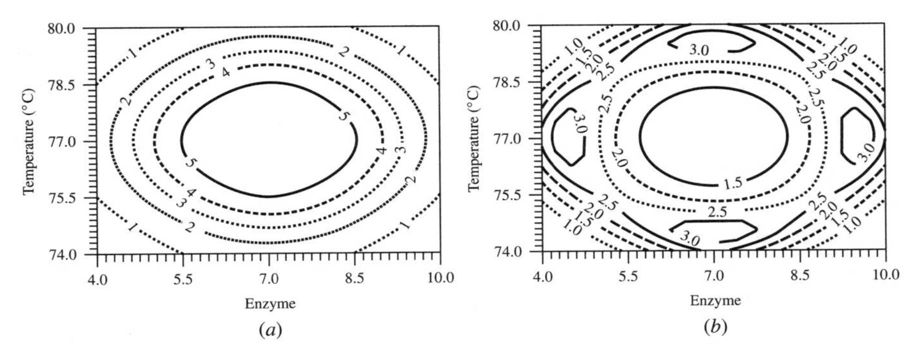

Figure 15.5 (*a*) Information surface based on a second-order model with equally spaced contours and with four extra center points; (*b*) axial points moved out and duplicated.

Inferences to be drawn

One of the first aspects of central composite designs that becomes clear is the importance of the placement of the design with respect to the operability region. By operability region we mean the region in the factor space in which experiments can be performed. In general, exploration of the entire operability region is neither feasible nor sensible. A choice must be made. Frequently, the investigator is interested in maximizing response or at least studying the surface in the neighborhood of the maximum. (Note that we are ignoring the minimization problem since it is conceptually equivalent to the maximization problem.) If the center of the design selected happens to be on or very near the point of maximal response, the central composite design with many center points is a very good choice. The weakness of central composite designs is that they are not as good if the investigator errs in the choice of location. We return to this problem in Section 15.10.

In addition, there are questions concerning placement of the axial and center points in the central composite design. From Figure 15.5 we can conclude that if analysis is to be based on a second-order model, additional replication of the axial points may be more valuable than replication of the center point unless the investigator is very sure about placement of the entire design. However, a more conservative approach may well be to consider duplicating axial points and moving them out from the face of the cube defined by the factorial. If a second-order model is contemplated, three to five replications of the center point in addition to duplicates of the axial points looks reasonable.

15.3.2 Central Composite Designs

A general form of the central composite design was given above. It is common practice to insist that all the $\{\alpha_i\}$ are equal. Generally, it is a good idea to select values for $\{\alpha_i\}$ larger than 1. However, there is no overriding reason to select all equal. In fact, a restriction such as all $\{\alpha_i\}$ equal seems rather artificial since the distances in the directions are scale dependent.

When using central composite designs in higher-dimension problems, i.e., more factors, it is possible to use fractions of the basic 2^N cube. However, restrictions apply. If the analysis is to be with a second-order model, two-factor interactions must be estimable [to provide estimates of the $\{\gamma_{jk}\}$ terms in model (15.1)]. This implies that in general a resolution V plan is required. Special cases where knowledge of the factors permits the investigator to assume a priori that some $\{\gamma_{jk}\}$ are zero permit smaller fractions.

In general, central composite designs are to be recommended in situations where the investigator has sound reasons to study the local surface in a specific region. Examples include cases where the optimum is known from experience and the investigator seeks to examine the surface to see how robust the process is to deviations in the input variables. Similarly, there are cases where standard operating conditions are prescribed for some reason or other and the investigation is to observe the effect of small changes in conditions. Central composite designs are also used occasionally to check the sensitivity of fitted models to changes in parameter values.

Example 15.2: Application of a Two-Factor Central Composite Design.* The object of this study was to examine the relationship between one of the basic taste modifiers, sodium chloride (NaCl), and the commonly used flavor enhancer, monosodium glutamate (MSG), the sodium salt of a naturally occurring amino acid, glutamic acid. There are legal restrictions on the amount of NaCl that may be added to foods, but usually, flavor is the limiting factor. It is generally recognized that there is an optimal concentration of MSG in food, beyond which palatability decreases. Establishing the proper proportions of these ingredients in food products to obtain maximum palatability is often tedious and expensive. The objective of the study being quoted here was to examine the joint effects of MSG and NaCl on the hedonic score of chicken broth, both with and without the addition of extra spices. The basic plan for the experiment was as shown in Table 15.1. Note that the relationships between coded and actual levels are:

$$\text{coded level} = (\text{actual level} - .25)/.18 \text{ for MSG}$$

$$\text{coded level} = (\text{actual level} - .80)/.28 \text{ for NaCl.}$$

Sensory evaluations were conducted by 12 panelists. Although the paper does not say so, the statistical analysis given indicates that each panelist scored each broth two times. Mean values for the 18 broths are as given in Table 15.2. The paper reports fitted models

$$\text{nonspiced broth score} = -5.89 + 12.18 \times x_1 + 26.96 \times x_2 - 14.87 \times x_1^2 - 15.71 \times x_2^2 - 2.89 \times x_1 x_2$$

*From Chi and Chen (1992), with permission from *Journal of Food Processing*, Vol. 16 (1992).

Table 15.1 Plan for Experiment

Treatment Number	Coded Levels		Actual Levels (%)	
	MSG	NaCl	MSG	NaCl
1	+1.414	0.0	0.50	0.80
2	+1.0	+1.0	0.43	1.08
3	0.0	+1.414	0.25	1.20
4	−1.0	+1.0	0.07	1.08
5	−1.414	0.0	0.00	0.80
6	−1.0	−1.0	0.07	0.52
7	0.0	−1.414	0.25	0.40
8	+1.0	−1.0	0.43	0.52
9	0.0	0.0	0.25	0.80

Table 15.2 Values Recorded for Each Broth

Treatment Number	Nonspiced Broth	Spiced Broth	Treatment Number	Nonspiced Broth	Spiced Broth
1	6.88	7.75	6	4.58	3.83
2	6.08	6.88	7	4.25	3.50
3	5.00	5.88	8	5.67	5.50
4	5.58	6.04	9	7.17	7.42
5	5.54	5.21			

and

$$\begin{aligned}\text{spiced broth score} = {} & -8.87 + 15.20 \times x_1 + 31.56 \times x_2 - 15.26 \times x_1^2 \\ & - 17.15 \times x_2^2 - 4.13 \times x_1 x_2.\end{aligned}$$

Contours for both functions with solid lines for the nonspiced broth and dashed lines for spiced broth are shown in Figure 15.6. The striking feature is the correspondence between the two. The extra spice has caused the optimum to shift to slightly higher NaCl and MSG levels. The main difference, which is not obvious from the contour plots but is clear from the data, is that the spiced broth lead to higher taste scores.

15.4 ROTATABILITY

In the response surface literature we often encounter references to a design being rotatable. A design is said to be rotatable if $\text{Var}[\widehat{y}]/\sigma_\epsilon^2$ is constant for all points equidistant from the design center. This means that the information contours form perfect circles in two dimensions, spheres in three dimensions, and so on, about

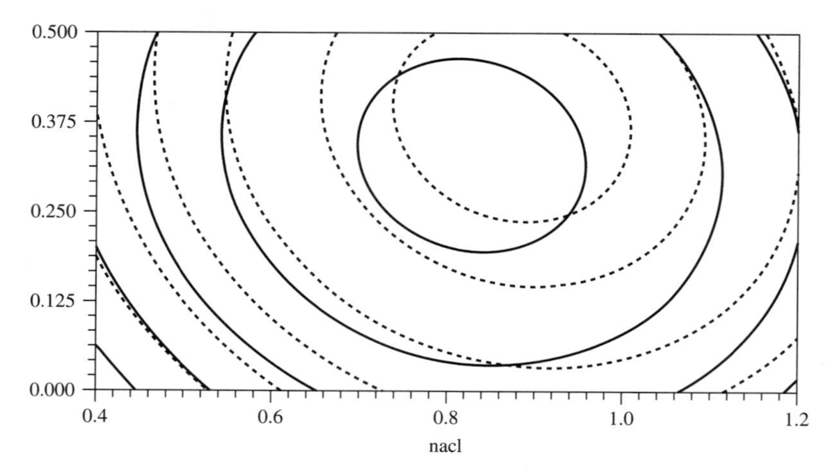

Figure 15.6 Sensory evaluation of effects of salt and MSG in broth. Dashed-line contours are with spice and solid contours without extra spices.

the center point of the design. The central composite designs introduced earlier can be made rotatable by selecting

$$\alpha = (n_f n_t / n_a)^{1/4},$$

where $n_t = 2^N$ is the total number of factorial points, n_f the number of points accounted for by the factorial part of the plan, and n_a the number of points accounted for by the axial part of the plan. The objective of the concept of rotatability is to impose some sort of regularity on the information surface. The difficulty with the concept of rotatability is that it is scale dependent. For example, if one factor in a design is concentration and another is temperature, requiring rotatability requires one somehow to relate distance in the concentration direction with distance in the temperature direction. Also, a design that is set up to be rotatable if temperature is recorded in degrees Celsius ceases to be rotatable when temperature is reexpressed in degrees Fahrenheit.

Although the concept of rotatability is a good starting point in the development of a response surface design, given the ease of exploring information surfaces, there is little excuse for not conducting such an examination before conducting an experiment.

15.5 STATISTICAL ANALYSES OF DATA FROM RESPONSE SURFACE DESIGNS

Generally the analysis begins by fitting a model by least squares and obtaining representations of the response surface and the matching information surface (inverse of the variance of points predicted). It must be pointed out that a variety of response surfaces can emerge. First, the fitted surface may or may not have a stationary point. If there is a stationary point, it may be a minimum, maximum, or a saddle point. The stationary point can be found by differentiating the estimated response function with respect to the regressor variables, setting the derivatives equal to zero, and solving for x. For second order response functions the solution has a simple form in matrix notation. We write, for example,

$$\widehat{y} = \widehat{\beta}_0 + \widehat{\beta}_1 x_1 + \widehat{\beta}_2 x_2 + \widehat{\beta}_{11} x_1^2 + \widehat{\beta}_{22} x_2^2 + \widehat{\beta}_{12} x_1 x_2$$

as

$$\widehat{y} = \widehat{\beta}_0 + \mathbf{x}^t \widehat{\mathbf{b}} + \mathbf{x}^t \widehat{\mathbf{B}} \mathbf{x},$$

where $\mathbf{x}^t = [x_1 x_2]$, $\widehat{\mathbf{b}} = \begin{bmatrix} \widehat{\beta}_1 \\ \widehat{\beta}_2 \end{bmatrix}$, and $\widehat{\mathbf{B}} = \begin{bmatrix} \widehat{\beta}_{11} & .5\widehat{\beta}_{12} \\ .5\widehat{\beta}_{12} & \widehat{\beta}_{22} \end{bmatrix}$. Then taking derivatives with respect to $\mathbf{x}$ yields the stationary point

$$\mathbf{x}_s = -\frac{1}{2} \widehat{\mathbf{B}}^{-1} \widehat{\mathbf{b}},$$

and the estimated response at that point is

$$\widehat{y}_s = \widehat{\beta}_0 + \frac{1}{2} \mathbf{x}_s^t \widehat{\mathbf{b}}.$$

The second partial derivatives of the response function, which are contained in the matrix $2\widehat{\mathbf{B}}$, determine whether the stationary point locates a maximum, minimum, or saddle point. The surface has a minimum if $\widehat{\mathbf{B}}$ is positive definite and a maximum if $\widehat{\mathbf{B}}$ is negative definite. Positive definiteness means that $\mathbf{u}^t \widehat{\mathbf{B}} \mathbf{u} > 0$ for any vector $\mathbf{u} \neq \mathbf{0}$, hence if $\widehat{\mathbf{B}}$ is positive definite the rate of change increases as $\mathbf{x}$ moves away from the stationary point and the stationary point is a minimum. Much design of experiments software has the capability to plot and overlay response surfaces, which is very helpful for visualizing low dimensional surfaces. Canonical analyses of the estimated parameters, i.e., study of the eigenvalues and eigenvectors of $\widehat{\mathbf{B}}$ are also useful for determining the nature of the surface. It is also possible to encounter a ridge in the response surface, with the result that there is no stationary point in the experimental region. More detailed discussions of the statistical analysis of data from response surface designs can be found in Myers and Montgomery (1995) and Khuri and Cornell (1996).

15.6 BLOCKING IN RESPONSE SURFACE DESIGNS

In all experimental design work whenever there is much variability among experimental units and it is possible to categorize units, one should try to use blocking to remove excess variability from among experimental units. In addition, if at

Table 15.3 Blocking Schemes for Central Composite Designs

Number of Factors:	2	3	4	5	5	6	6	7	7
Factorial blocks									
Nature of factorial	2^2	2^3	2^4	2^5	2^{5-1}	2^6	2^{6-1}	2^7	2^{7-1}
No. points in factorial	4	8	16	32	16	64	32	128	64
No. blocks in factorial	1	2	2	4	1	8	2	16	8
No. added center points	3	2	2	2	6	1	4	1	1
No. points per block	7	6	10	10	22	9	20	9	9
Axial block									
No. axial points	4	6	8	10	10	12	12	14	14
Additional center points	3	2	2	4	1	6	2	11	4
No. points per block	7	8	10	14	11	18	14	25	18
Total points in design	14	20	30	54	33	90	54	169	80
α for orthogonality	$\sqrt{2}$	$\sqrt{8/3}$	$\sqrt{4}$	$\sqrt{28/5}$	$\sqrt{4}$	$\sqrt{8}$	$\sqrt{28/5}$	$\sqrt{100/9}$	$\sqrt{8}$
α for rotatability	$\sqrt[4]{4}$	$\sqrt[4]{8}$	$\sqrt[4]{16}$	$\sqrt[4]{32}$	$\sqrt[4]{16}$	$\sqrt[4]{64}$	$\sqrt[4]{32}$	$\sqrt[4]{128}$	$\sqrt[4]{64}$

Source: Box and Hunter (1957), with permission of the Institute of Mathematical Statistics.

all possible, the blocks should all be the same size. This also holds for central composite designs. Unfortunately, the nature of the models used and the restrictions on placement of points in the central composite designs led to difficulties in constructing orthogonal blocking schemes. However, Box and Hunter (1957) have constructed a table of orthogonal blocking schemes for central composite designs. In some cases it was necessary to relax the condition of equal block sizes. Their table is reproduced in Table 15.3.

To illustrate the construction, consider the five-factor plan using the full replicate of the factorial plan. This plan calls for four blocks for the factorial part and one block for the axial points. We must select a blocking scheme. Blocking on ABC, CDE, and the generalized interaction $ABDE$ is satisfactory. It leaves all main effects and two-factor interactions estimable. This gives eight treatment combinations per block. Add two additional center points to each block. The result is four blocks of 10. There is the additional block containing the 10 axial points plus another four center points. The final selection is α for the exact location of the axial points. We must choose between $\sqrt{28/5}$ for orthogonality and $\sqrt[4]{32}$ to rotatability. In the case of the design for the 2^{5-1} fraction we encounter the happy circumstance that selecting $\alpha = 2$ gives both block orthogonality and rotatability. In the case of the design based on the 2^{6-1} fraction one must again choose between $\alpha = \sqrt{28/5}$ for orthogonality and $\alpha = \sqrt[4]{32}$ for rotatability. There are four blocks of 10 and one block of 14. We also note that in the case of the seven-factor design utilizing the half replicate, the table says that eight blocks of eight plus one points are possible. The fraction is based on $\mathcal{I} = ABCDEFG$. A suitable blocking scheme can be based on $ABCD$, $CDEF$, and $ACFG$ and their generalized interactions.

A modest lack of orthogonality is not especially serious. Also, the investigator is well advised to think carefully about whether rotatability is really a good thing in the design. The concept of rotatability is not important enough to warrant placing a point at an unreasonable place.

Before leaving this section we must also point out that if an investigator encounters a situation where blocking is necessary and the schemes given in Table 15.3 are unsatisfactory, it is always possible to fit the treatment combinations into some sort of incomplete block design. This may require adding more center points or duplicating some of the other points.

Exercise 15.2: Unsatisfactory Blocking in a Central Composite Design. Explain why it would not be satisfactory to use a blocking scheme based on ABC, DEF, and ADG and all of the generalized interactions for the central composite design for seven factors based on the half replicate defined by $\mathcal{I} = ABCDEFG$. □

15.7 MIXTURE DESIGNS

There is a class of experimental problems in which the levels of quantitative factors are constrained to sum to a constant. For example, a food scientist tries to develop a fruit juice mixture that has good taste and acceptable cost. The fruit juice may consist of, say, three or four different juices. The key restriction that makes the problem different is that the fractions must sum to 1. A significant increase in the amount of one constituent implies a corresponding decrease in one or a combination of the other components. In general, one thinks about k components in the mixture. The constraint that the proportions sum to a constant has the consequence that although the experiment has k factors, the experimental space actually has only $k - 1$ dimensions.

Consider the example of three factors. The design space in a mixture problem is the triangle bounded by $(0, 0, 1)$, $(1, 0, 0)$, and $(0, 1, 0)$. This is actually a two-dimensional subspace of the three-dimensional space. All treatment combinations must lie within this triangle. In practice, this space can be represented in two dimensions by an equilateral triangle. However, it turns out to be convenient to use a notation system that reflects the three axes and the restriction. In general, this kind of coordinate system is called a *simplex* coordinate system. A simplex is a regularly sided figure with k vertices in k dimensions which can be interpreted as a figure with k vertices in a $(k - 1)$-dimensional subspace.

Exercise 15.3: Design Space for a Mixture with Three Components. Sketch the design space; i.e., the possible treatment combinations, for a mixture with three factors or components. □

A common strategy is to select uniformly spaced points on a k-dimensional simplex, a *simplex lattice*. These are known as *simplex-lattice* $\{k, m\}$ *designs* and are suitable to fit up to mth-order polynomials to data from a k-component mixture. A *simplex lattice* is a set of uniformly spaced points on a simplex. In general, a $\{k, m\}$ design has $n = \binom{k+m-1}{m}$ design points. For the three-component mixture the $\{3, 2\}$ simplex-lattice design suitable for estimating a second-order polynomial has the combinations (1, 0, 0), (0, 1, 0), (0, 0, 1), (.5, .5, 0), (.5, 0, .5), (0, .5, .5). Note that the constraint that the component fractions must sum to 1 makes it impossible to estimate all $1 + 2k + k(k-1)/2$ parameters of a standard second-order polynomial separately. This type of three-component design allows one to fit a second-order model of the form

$$y = \beta_1 x_1 + \beta_2 x_2 + \beta_3 x_3 + \beta_{12} x_1 x_2 + \beta_{13} x_1 x_3 + \beta_{23} x_2 x_3 + \epsilon. \tag{15.2}$$

If the experimenter needs or wants a measure of true error, some of the points must be duplicated, because this design has just enough points for a quadratic function to fit perfectly. If a cubic polynomial model for three components is required, the minimal design is the $\{3, 3\}$ simplex-lattice design. It consists of the design points (1, 0, 0), (0, 1, 0), (0, 0, 1), (.33, .67, 0), (.33, 0, .67), (0 , .33, .67), (.67, .33, 0), (.67, 0, .33), (0, .67, .33), and (.33, .33, .33). This design allows a model of the form

$$\begin{aligned} y = {} & \beta_1 x_1 + \beta_2 x_2 + \beta_3 x_3 + \beta_{12} x_1 x_2 + \beta_{13} x_1 x_3 + \beta_{23} x_2 x_3 + \beta_{123} x_1 x_2 x_3 \\ & + \gamma_{12} x_1 x_2 (x_1 - x_2) + \gamma_{13} x_1 x_3 (x_1 - x_3) + \gamma_{23} x_2 x_3 (x_2 - x_3) + \epsilon. \end{aligned}$$

Again, this gives a perfect fit, and repetitions at some of the points are needed to provide a measure of experimental error. As in the second-order case, the constraint that the component mixtures sum to 1 prohibits fitting a full third-order polynomial.

***Exercise 15.4: General Quadratic Model for the Simplex-Lattice* {3, 2} *Design*.** Show that the model shown in expression 15.2 is the fully general form of a quadratic on the simplex. □

These designs and models extend directly to more components and higher-order polynomials. Readers that require a thorough treatment of mixture designs are urged to consult Cornell (2002).

15.8 OPTIMALITY CRITERIA AND PARAMETRIC MODELING

Historically, some classes of designs were derived simply on the basis of making the normal equations easy to solve. With the advent of the computer and cheap arithmetic, this criterion has lost most of its appeal. Subsequent criteria for selecting designs or design points that have been suggested tend to focus on the statistical properties of the estimates of the parameters or on the properties of the estimated response function. These criteria are based on concepts such as *D-optimality* (optimal in the sense of minimizing the determinant of the variance–covariance matrix of the parameter estimates), *A-optimality* (minimize the average of the variances of the parameter estimates), *E-optimality* (minimize the largest latent root or eigenvalue of the variance–covariance matrix of the parameter estimates), and *I-optimality* (minimizing the average, or integrated, prediction variance in the response surface region). The D-optimality criterion seems to focus on the overall Var$[\widehat{y}]$ function, while the A-optimality criterion focuses on statistical properties of the individual regression coefficients, minimizing the average variance of the regression coefficients. E-optimality minimizes the largest variance of a linear combination of parameter estimates. I-optimality minimizes the average variance of the predicted values in the entire region. `Proc OPTEX` in SAS® is a very useful tool for selecting design points that satisfy prescribed optimality criteria. A computer program called *Gosset* has been developed by Hardin and Sloane (1993) which constructs D-, A-, E-, and I-optimal designs and has the flexibility to handle constraints on the factors and very irregular experimental regions.

A common feature of all the criteria mentioned is that they rest on fitting a parametric model. Often, polynomial models are used. A more recent development has been nonparametric modeling. A very flexible design approach is to select points that in some sense cover the operability region.

15.9 RESPONSE SURFACES IN IRREGULAR REGIONS

The designs we have examined up to this point are suitable when there are no effective constraints on the operability region or when the constraints are very regular, as in the mixture problems on a simplex. In practice, it is more common to find severe constraints that limit the investigation to irregular regions. These problems typically involve many input variables, i.e., many dimensions. We will illustrate an approach that has considerable flexibility. To retain the ability to illustrate concepts in pictures, we limit our discussion to two dimensions, i.e., two input variables and a modest region. The necessary extensions to more input variables and more complex regions are direct. We consider a hypothetical example involving two input variables, each ranging from zero to five with the constraint that the sum of the two may not exceed seven. A chemical reaction that is not safe under some conditions would be a good example. We also assume a nonparametric model.

One experimental strategy is to conduct experiments at points defined by a grid placed on the feasible region. We assume, however, that the investigator has limited resources and is limited to fewer than the number dictated by a regular grid. The question then becomes one of placing a limited number of points in an irregular region so that they provide maximal information. Obviously, placing all in a group in the center or off in a corner is not good.

```
data source ;
 do x1 = 0 to 5 by .2 ;
  do x2 = 0 to 5 by .2 ;
    if x1 + x2 < 7 then output ;
       end ; end ;
proc optex data = source seed = 377 ;
 model x1 x2 ;
 generate n = 17 criterion = s ;
 examine design ;
 output out = design ;  run ;
```

Display 15.2 Example using `Proc OPTEX` in SAS® to pick points in a region.

Scattering the points as much as possible has much intuitive appeal. In some sense, one would like to make the sum of the interpoint distances as large as possible. One criterion is to minimize the harmonic mean of the minimum distance from each design point to its nearest neighbor. A nice feature of this criterion is that there exists computer software to accomplish this selection. We illustrate this with procedures available in SAS®. In our example we assume arbitrarily that the investigator is limited to 17 points.

The design process begins by defining a regular grid of points over the feasible region. Think of these as potential points for experimentation. Then use the computer algorithm to select a subset of 17 points that satisfies the maximum distance criterion. Display 15.2 illustrates the SAS® code. The resulting plan for the experiment is shown in Figure 15.7.

Notice that there is a random element in the search. Repeating the search with a different "`seed`" for the search algorithm will result in a different set of points. However, all sets will be roughly equivalent.

For purposes of our example, we continue and show the information contours for the design generated in Figure 15.7. Notice that the information is spread relatively evenly across the feasible region. In practice, one does not need to construct this information surface, although it is recommended if experimental work is costly. The information contours in Figure 15.7 were constructed by generating 1000 data sets under the 17-point design, fitting a nonparametric surface using splines, and computing the empirical variance of the fitted values at each point on the surface. The information plotted is the inverse of this empirical variance. The code we used to generate the surface is available on the companion Web site.

Statistical analysis of the resulting data can be accomplished by fitting a polynomial function by multiple regression or a more complex function, possibly a segmented polynomial, using nonlinear least squares or a nonparametric function using splines to the surface. In Display 15.3 we illustrate SAS® code to fit and draw a nonparametric surface on the complete rectangular region. The procedure automatically extrapolates to the region that has no data. Predicted values in the nonexperimental region will be suspect. The nonparametric interpolation option is not available if there are more than two input variables, i.e., more than two dimensions.

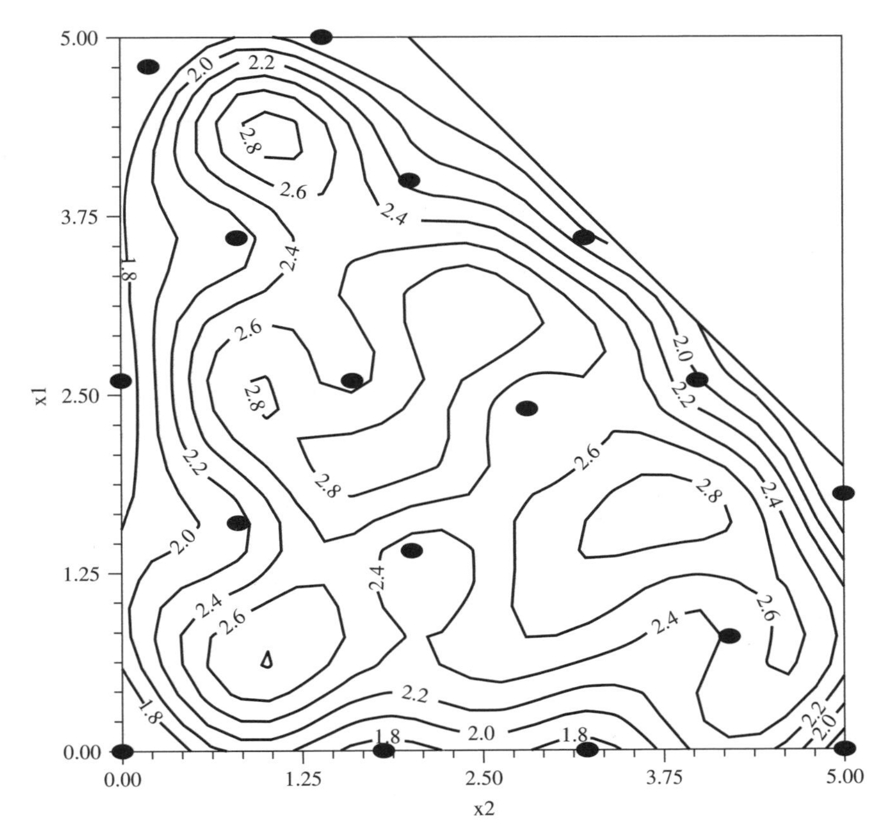

Figure 15.7 Experimental plan for 17 design points on an irregular region and the information surface contours.

```
proc g3grid data = YOURDATA  out = smooth ;
  grid x1*x2=y / spline smooth = .05
    axis1 = 0 to 5 by .2
    axis2 = 0 to 5 by .2 ;
proc gcontour data = smooth ;
  plot x1*x2=y ;   run ;
```

Display 15.3 Example of SAS® code to fit and draw a nonparametric response surface.

To conclude this section, we need to point out that occasionally an investigator has special interest in a particular subregion of the experimental space. This can be accommodated easily. One can add extra points to a plan, but a better strategy is to take advantage of a feature in `Proc OPTEX` that allows the user to require that a specific subset of points appears in the plan and then let the algorithm place the remainder to satisfy the maximization criterion.

15.10 SEARCHING THE OPERABILITY REGION FOR AN OPTIMUM

The methods of Sections 15.1 to 15.6 assume that the researcher has already narrowed down the ranges of the input variables to a fairly small region near the optimum. How does one find the *region* of the optimum in the first place? In three dimensions, two independent variables and one response variable, we have a simple analogy. Imagine a blind person trying to get to the top of a hill. With some difficulty it is possible to explore the region in the immediate neighborhood and pick a direction to move. If the surface is very irregular, in the sense that there are local relative maxima, i.e., small hilltops that are separated from the true hilltop by small valleys, the blind person cannot possibly know which direction to move to get to the top of the highest hill. Being blind, it is impossible to see whether the part of the hill being climbed leads to a relative maximum or if other hills are higher. In the experimental problem, one cannot see the entire surface without exploring, i.e., evaluating the function at all possible points. In higher dimensions, the same problems persist. The major difference is the loss of the benefit of nice pictures or physical models. We will assume that the surface is sufficiently regular, with one global optimum and no local optima that can trap the search process.

15.10.1 Searching One Variable at a Time

The simplest approach to the experimental problem, and not a very good one, is to search for an optimum first in the x_1 direction, then in the x_2 direction, then x_3, and so on, searching in all N directions in turn. Once all directions have been searched, cycle through all directions again. This is repeated until one is satisfied that the optimum has been attained. To illustrate the experimental application of this procedure and, incidentally, exhibit the difficulty with the method, we simulate an example with a nonlinear function that depends on two variables, $0 \leq x_1 \leq 1$ and $0 \leq x_2 \leq 1$. The function is

$$y = (5x_1 - 3.125x_1^2) \exp\left[-50\left(x_2 - \frac{4 + 5x_1 + 25x_1^2 - 25x_1^3}{15}\right)^2\right]$$

We assumed that the hypothetical experimenter was aware of the bounds at 0 and 1 for both variables, but was not aware of the actual form of the function. The hypothetical experimenter started by holding x_2 constant and varying x_1. Three points are required to determine some sort of optimum on a line. The three points selected were (.01, .1), (.1, .1), and (.2, .1). The responses (y values) at these three points were .0177, .0465, and .0144, respectively. A quadratic interpolation formula indicated an optimum at (.115, .1). The next step for the experimenter was to hold x_1 constant and search in the x_2 direction, starting from (.115, .1). The three trial points were (.115, .2), (.115, .3), and (.115, .4), giving y values equal to .2458, .5179, and .4013, respectively. The strategy is to begin from the optimum point and take a series of steps until one has a sequence of three points, with the middle one having the largest y value, indicating that the maximum on that line has been bracketed. In this case the quadratic interpolation indicated an optimum at (.115, .303).

Since there are only two independent variables, this completes one cycle of the search. The next step is to search again in the x_1 direction while holding x_2 constant. The three points checked are (.1, .303), (.2, .303), and (.3, .303). The responses were .4654, .6166, and .2939, respectively. Interpolation gave the maximum at (.183, .303). The next search, in the x_2 direction, involved points (.183, .25), (.183, .35), and (.183, .45), with responses .3791, .7887, and .6037, respectively. Interpolation indicated the optimum at (.183, .352). The fifth search, in the x_1 direction, utilized points (.15, .352), (.25, .352), and (.35, .352) (responses .6793, .7894, and .3560, respectively) and indicated an optimum at (.229, .352). The sixth search, in the x_2 direction, utilized points (.229, .35), (.229, .4), and (.229, .45) (responses .8176, .9758, and .9071, respectively) and indicated an optimum at (.229, .401). The hypothetical experimenter continued this through 11 linear searches. The trend established from the first 11 searches is one of continuing increases but at a decreasing rate. The result of these searches is shown in Figure 15.8. Superimposed on these searches are the contours of the function as obtained from SAS®'s `PROC GCONTOUR`.

The reason for the decreasing rate of progress is immediately obvious. If the response surface being explored has a relatively narrow ridge not parallel to one of the axes, the procedure will gravitate to that ridge and then send the experimenter back and forth across the ridge. Exactly the same thing happens in the corresponding mathematical optimization problem. It can be proved that while the one-factor-at-a-time strategy will make progress, it will find the worst possible path toward the optimum. Many experimenters, who obviously cannot see the contours in practice, have encountered the slow progress and wrongly assumed that they were at the optimum long before they were even close. The one-factor-at-a-time strategy works well only if there are no ridges in the surface. This implies no interactions among the factors, the independent variables.

15.10.2 Method of Steepest Ascent

The next strategy the hypothetical experimenter tried was the method of steepest ascent. If we return to the blind hiker on the hillside analogy, the hiker is now not restricted to moving only in the north–south or east–west directions but is allowed to explore a bit before selecting a direction. Once a direction has been selected, the hiker proceeds as long as progress is being made. The hiker then explores, finds a new direction, and proceeds. Since the scientist does not know the mathematical form of the function, calculating partial derivatives of the response function is not possible. The experimenter must conduct small factorial experiments to determine directions. From a 2^N factorial a first-order polynomial

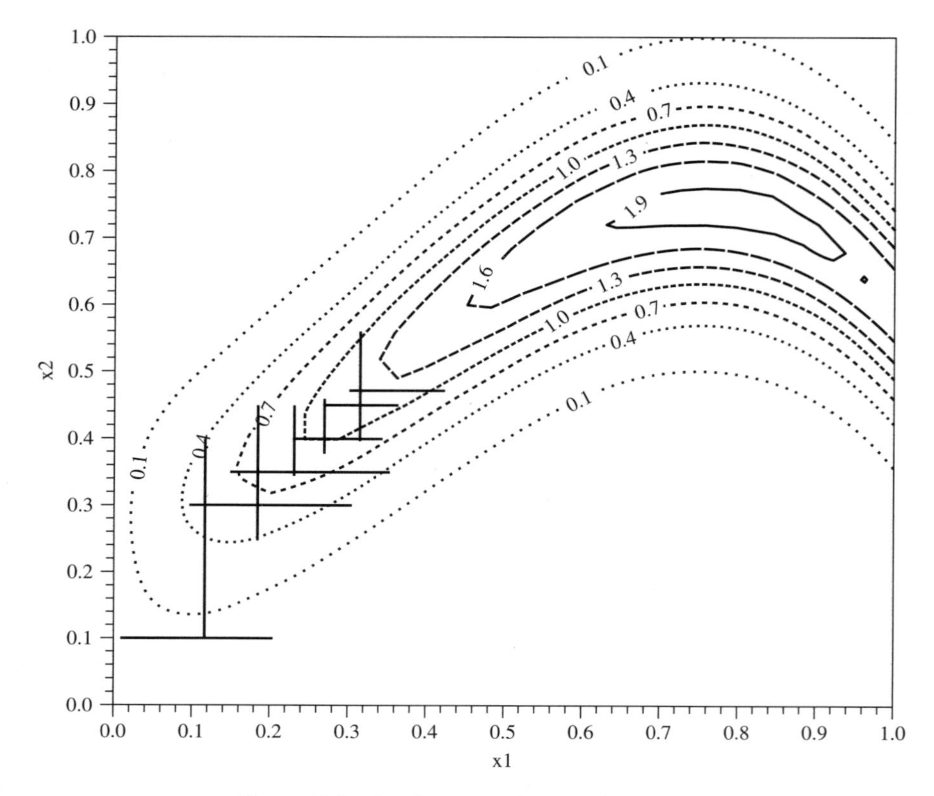

Figure 15.8 One-factor-at-a-time search strategy.

model can be fitted and the path of steepest ascent determined. The model

$$y_i = \beta_0 + \beta_1 x_{i1} + \beta_2 x_{i2} + \epsilon_i$$

is a planar surface. The gradient,

$$\frac{\partial y_i}{\partial \boldsymbol{x}_i} = \begin{bmatrix} \dfrac{\partial y_i}{\partial x_{i1}} \\ \dfrac{\partial y_i}{\partial x_{i2}} \end{bmatrix} = \begin{bmatrix} \beta_1 \\ \beta_2 \end{bmatrix},$$

gives the direction of steepest ascent.

The first experiment, a 2^2 factorial utilizing the points (.1, .1), (.1, .2), (.2, .1), and (.2, .2), gave y values .0465, .2420, .0144, and .1532, respectively. Next, the experimenter fits the planar model

$$y = \beta_0 + \beta_1 x_1 + \beta_2 x_2 + \epsilon.$$

The ordinary least squares estimates are

$$\widehat{\beta_1} = \frac{.0144 + .1532}{2} - \frac{.0465 + .2420}{2} = -.06045$$

and

$$\widehat{\beta_2} = \frac{.2420 + .1532}{2} - \frac{.0465 + .0144}{2} = .16715.$$

Note that the estimate for β_1 should be divided by the difference between levels in the x_1 direction and the estimate for β_2 by the difference between levels in the x_2 direction. Since the two scale factors are equal in our example, they can be ignored. The best point available at this time is (.1, .2) with y value .2420. The strategy is to move in the direction defined by $\widehat{\beta_1}$ and $\widehat{\beta_2}$. The points along this line,

$$(.1, .2) + \theta(-.060, .167),$$

are obtained by selecting values for θ. Selecting $\theta = 1$ gives (.040, .367) with y value .1365 . Since this relatively large step resulted in a net decrease in y, the thing to do is to select a smaller θ. Selecting $\theta = .5$ gives (.070, .284) with y value .3316. Quadratic interpolation suggests an optimum at (.075, .269). This completes the first step in the steepest ascent search procedure.

The next step begins again with a 2^2 factorial experiment. Note that if the search were in higher dimension, one would probably use a fractional factorial, to conserve resources. Select the points for the experiment about the optimum obtained in the preceding step. The experimenter selects (.05, .25), (.05, .30), (.10, .25), and (.10, .30). The corresponding y values are .2259, .2402, .3795, and .4635. Now compute the directions.

$$\widehat{\beta_1} = \frac{.3795 + .4635}{2} - \frac{.2259 + .2402}{2} = .18845$$

$$\widehat{\beta_2} = \frac{.2402 + .4635}{2} - \frac{.2259 + .3795}{2} = .04915.$$

Note that one could again argue that both of these should be divided by .05 since the step size in both x_1 and x_2 is .05. However, since both are equal, this can be ignored. Now beginning from the best point available at this time, the point (.10, .30), search in the direction defined by the $\widehat{\beta}$'s, points defined by choices of θ in

$$(.10, .30) + \theta(.1885, .049).$$

Selecting $\theta = 2$ gives the point (.477, .0398) and y value .1302. Since the response has fallen, the experimenter knows that θ is too large. $\theta = 1$ gives the point (.2885, .349) and y value .6277. Quadratic interpolation now indicates an optimum at (.252, .340). This completes step 2.

The next step again begins with a 2^2 factorial about the most recent optimum. The selected points are (.22, .31), (.22, .36), (.27, .31), and (.27, .36). The

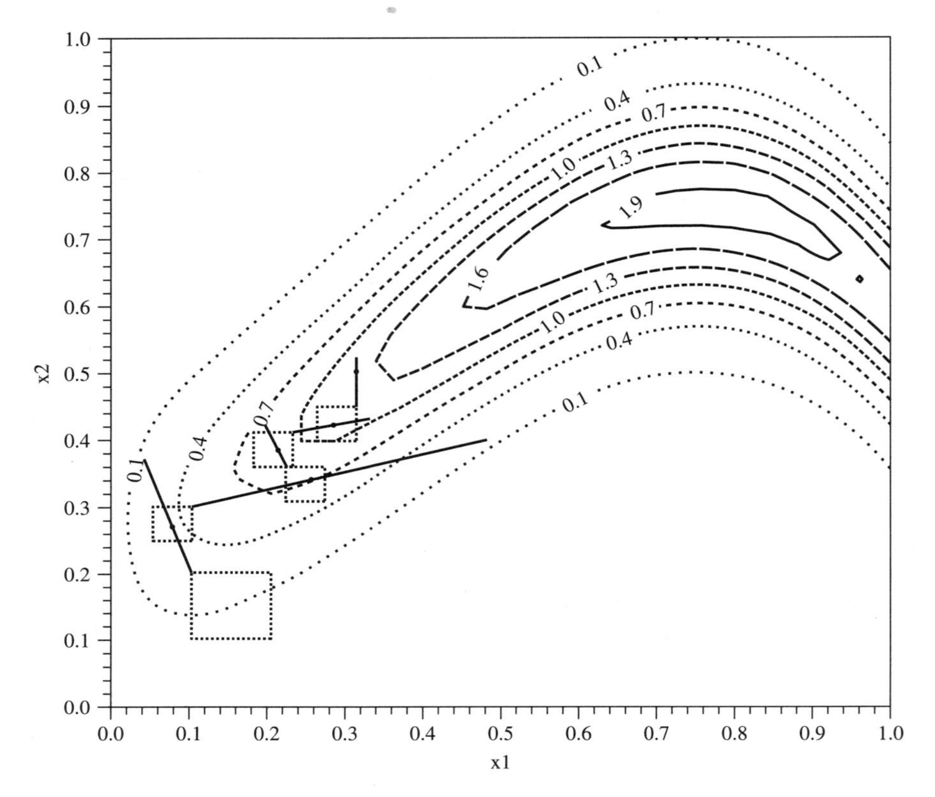

Figure 15.9 Steepest ascent strategy.

corresponding y values are .6161, .8653, .4489, and .7795. The resulting $\widehat{\beta}$ values specifying the search direction are $-.253$ and .58. Points for θ equal to .05 and .1 gave responses equal to .9006 and .8027, respectively. The optimum indicated in that direction is at (.210, .384).

This procedure was carried through two more steps and the results plotted in Figure 15.9. Again, the contours for the function are overlaid on the same graph. This method suffers from the same criticism as the previous method: that the procedure again gravitates to the poorest possible route and then sends the experimenter back and forth across the ridge with only limited progress. In practice, the steepest ascent method is little better than the one-factor-at-a-time method. Both are inefficient and tend to frustrate the experimenter. Unfortunately, we still see the steepest ascent method recommended.

15.10.3 Conjugate Gradient Method

The difficulty with the steepest ascent strategy is that there is no opportunity to learn from experience. There are several methods for getting around this problem.

We describe one simple method in the class of methods commonly referred to as *conjugate gradient methods*. *Conjugate* means "join together." The method outlined in this section uses the sum of the two most recent gradients and in this way steers a course between the extremes of either. Thus, if there is a tendency to tack back and forth across a ridge in the steepest ascent method, the conjugate gradient method will make a straighter path to the maximum. It works spectacularly well in the example we have been following.

The algorithm is:

1. Let P_0 be the initial point. This point is on some contour. Determine the gradient, i.e., the direction of steepest ascent, at this point. Search for the optimum on that line.
2. Let P_2 be this point. Determine the gradient, the direction of steepest ascent at this point. Search along that direction to find an optimum.
3. Let P_3 be the optimum on this line. Now rather than determining the gradient, extend the line connecting P_0 and P_3 as the new search direction.
4. Let P_4 be the optimum on this line. If we have only two independent variables, loop back to step 1 with P_4 in place of P_0. Otherwise, determine the gradient and search for the optimum.
5. Let this point be P_5. Search for the optimum along the line joining P_2 and P_5.
6. Let this point be P_6.
7. Continue until obtaining P_{2N} (N variables); always search in the steepest ascent direction from the even-numbered points and connect P_{2j-2} and P_{2j+1} to give a search direction to find P_{2j+2} .

Our hypothetical experimenter illustrates this method. Begin with the same steepest ascent step as in Section 15.10.2. The P_0 point is the best available before the first search. Conduct the search along the direction of steepest ascent to find P_2. Next conduct the 2^2 factorial about P_2. Determine the search direction. Our recipe tells us to search along the line determined by the previous optimum, i.e., P_2, and the search direction rather than the best point. The search was along the line

$$(.075, .269) + \theta(.1885, .049).$$

The points evaluated were $\theta = 1 \Rightarrow (.2635,\ .318)$ and $\theta = 2 \Rightarrow (.452,\ .367)$. The corresponding y values are .5246 and .0979. The interpolation gave (.230, .309) as the optimum on that line. This is P_3. The recipe now calls for a search along the line joining P_0 and P_3. The direction is easily obtained by taking differences between x_1 values and between x_2 values for the two points. The extension to more dimensions is immediate. The points for the search along the line are obtained for θ values in

$$(.230, .309) + \theta(.130, .109).$$

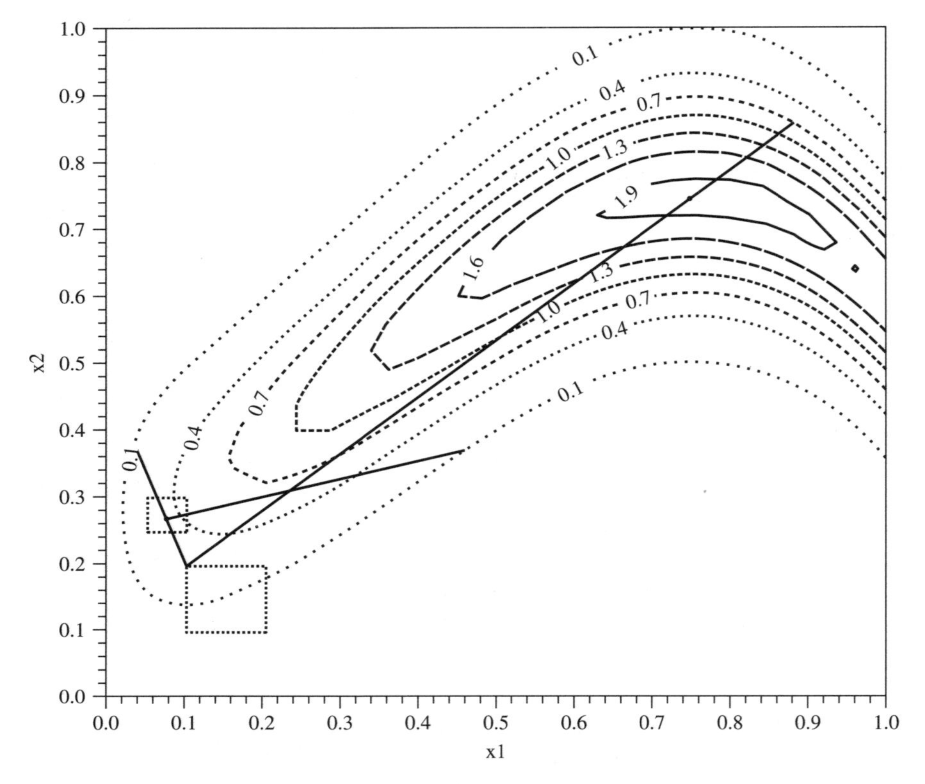

Figure 15.10 Conjugate gradient search.

The experimenter now tried $\theta = 1$, corresponding to (.360, .418) and $y =$ .7878; $\theta = 2$, corresponding to (.490, .527) and $y = .9580$; $\theta = 4$, corresponding to (.750, .745) and $y = 1.9886$, and $\theta = 5$, corresponding to (.880, .854) and $y = .7523$. $\theta = 5$ is clearly too far. The results of this search are plotted in Figure 15.10. One obviously cannot always count on quite this sort of performance. However, the method is a major improvement over the steepest ascent method. Also, the implementation is not overly complex. There are several variations that are about as good.

15.10.4 Simplex Search Method

A feature common to the last two of the optimum seeking methods discussed is that the response must be measured as a real number. The response values are used in further calculations to determine directions. The last method we discuss is different. We first discuss the method for the general case where there are N independent variables. Think of this as a problem in N dimensions even though there is another dimension, the response or dependent variable. In this

N-dimensional space we define a simplex defined by $N + 1$ points. (A minimum of $N + 1$ points are needed to define a figure in N-dimensional space.) It is convenient, although not necessary, to think of this simplex in N-space as the multidimensional analog of the equilateral triangle (three points) in two-space.

The experimental procedure begins by getting y values corresponding to each of the $N + 1$ points. Compare these y's. Find the poorest and eliminate the point corresponding to it. Geometrically, we reflect the simplex across the side opposite the poorest point. In two-space this is simply flipping the triangle away from the poorest point. To generate the coordinates for the new point, we use the formula

$$\text{new point} = 2 \times \text{average [all points but point}(p)] - \text{point}(p), \tag{15.3}$$

where p is the point in the current simplex that is to be eliminated. The calculation is done for each coordinate in turn.

As an example, consider the triangle in two-space with corners at (1, 1), (3, 1), and (2, 3). Now reflect the triangle to get rid of the (2, 3) point. The first coordinate is $2\times(1+3)/2-2 = 2$ and the second is $2\times(1+1)/2-3 = -1$. The reflected triangle has corners at (1, 1), (3, 1), and (2, $-$ 1).

The question now is: How does one get started?

1. Pick a reasonable starting value, a point that you feel is near the optimum. Let this point have coordinates $(x_1^0, x_2^0, \ldots, x_N^0)$. Let this be point (0). The assumption is that we are working with N factors.
2. Select reasonable step sizes. Reasonable amounts to selecting changes one would reasonably be able to make in each of the coordinates. This depends on the units of measurement used. Let these values be $d_1, d_2, \ldots, d_N$.
3. Set point (1):

$$(x_1^1, x_2^1, \ldots, x_N^1) = (x_1^0 + .9d_1, x_2^0 + .2d_2, x_3^0 + .2d_3, \ldots, x_N^0 + .2d_N),$$

point (2):

$$(x_1^2, x_2^2, \ldots, x_N^2) = (x_1^0 + .2d_1, x_2^0 + .9d_2, x_3^0 + .2d_3, \ldots, x_N^0 + .2d_N),$$

and so on, to point (N):

$$(x_1^N, x_2^N, \ldots, x_N^N) = (x_1^0 + .2d_1, x_2^0 + .2d_2, x_3^0 + .2d_3, \ldots, x_N^0 + .9d_N).$$

The two constants introduced, .2 and .9, respectively, come very close to defining a simplex with all sides equilateral triangles in N-dimensional space. In practice, this is good enough, since there is also the question of what is meant by *symmetrical figure* or even *equilateral triangle* when one of the axes is temperature and another is pressure. Note that the choice of $\{d_i\}$ must be made with the constants .2 and .9 in mind. There may be limits to the amount that one can change one of the factors.

Once the N starting values have been defined, proceed with the algorithm, discarding the poorest point at each step. Usually, in practice one does not need to worry very much about a stopping rule. Typically, one rapidly gets into an area where performance is much better than it was before and one is ready to quit and report successful improvement or one makes no progress at all, indicating that the initial start was very close to the optimal to begin with.

We next consider the geometry of the problem to deal with several special situations that need to be taken care of. If at any step the algorithm removes point p (poorest of the $N+1$ y values) and replaces it with p^* only to find that this is now the poorest y value (and scheduled for removal on the next step), you know that the simplex is straddling a narrow ridge or, alternatively, the step sizes are just too big. The recommendation is: first, keep the better of p and p^* and move the other point half the distance toward the center of the simplex. This should get an improved point. Then continue with the main algorithm. Note that even if we had started with a figure that had all edges of equal length, one or more of these shrinking steps will destroy the "nice" shape. We really didn't need the nice shape at all. The formula to shrink the simplex by bringing a point p half the distance to the center is

$$\text{new point} = .5 \times \text{point}(p) + .5 \times \text{average [all points but point}(p)]. \tag{15.4}$$

The calculations are again done for each coordinate or each of the x values in turn. It may be necessary to shrink more than halfway to the center. You do not go to the center because that means the loss of a search direction.

The geometry also suggests that if a particular point has not been removed after $N + 1$ steps, it is a good idea to rerun that point and possibly replace it with the new value. This is to guard against accidentally getting stuck at one point simply because, by chance, it was a very large observation. If by chance that point really was at an optimum, repeating that point should inform the investigator of that happy circumstance. Accidental low points get eliminated by the algorithm.

In general, one must remember that if the step sizes are too large, one will step over the optimum, and if the step sizes are too small, progress will be slow.

To illustrate the mechanics we return to the function used previously. The simplex is in two dimensions, i.e., a triangle. The three points defining the initial triangle are

$$(.500, .250), (.400, .075) \text{ and } (.600, .075).$$

The corresponding y values are .0008, .000012, and .000000. The third is clearly the poorest y value. Consequently, the triangle is to be reflected across the side opposite the third point. Simple arithmetic gives the new point (.300, .250) to replace (.600, .075). The new triangle is defined by the three points

$$(.500, .250), (.400, .075), \text{ and } (.300, .250).$$

The three y values are .0008, .000012, and .1045. Only one function evaluation (experimental run) was required. The (.400, .075) point now gives the poorest y value. This point is next replaced by (.400, .425). The resulting triangle is defined by the three points

$$(.500, .250), (.400, .425), \text{ and } (.300, .250).$$

The y responses of the current points are .0008, .6030, and .1045. Now the first is replaced by (.200, .425). The triangle is now defined by the three points

$$(.200, .425), (.400, .425), \text{ and } (.300, .250).$$

The responses are .8130, .6030, and .1045, respectively. The third is to be replaced. The new set is

$$(.200, .425), (.400, .425), \text{ and } (.300, .600).$$

The responses are .8130, .6030, and .5349, respectively. The third is to be replaced. But this indicates trouble. The new y value is still the smallest and reflection will simply return us to the old triangle. Of the two points (.300, .250) and (.300, .600), the second is clearly better. The recommended strategy is to move from the better of the two toward the center. The new point is (.300, .5125). [Note that $.5125 = .5 \times .600 + .5 \times$ (average of .425 and .600).] The new (shrunken) triangle is

$$(.200, .425), (.400, .425), \text{ and } (.300, .5125).$$

The corresponding y values are .8130, .6030, and 1.1213, respectively. The next step in the regular algorithm is to replace (.400, .425) by (.100, .5125). The corresponding y value is .0667. This is much poorer than the (.400, .425) point. The recommended strategy is to move from (.400, .425) toward the center of the triangle. The new point is (.325, .446875) with corresponding y value 1.1598. The current triangle is

$$(.200, .425), (.325, .446875), \text{ and } (.300, .5125),$$

with corresponding y values .8130, 1.1598, and 1.1213. The first point is the candidate for removal. This algorithm was continued through 11 more cycles, and the results plotted in Figure 15.11 together with the contours of the function. Notice how the sequence of triangles marches toward the ridge, then shrinks in size, reorients itself, and marches along the ridge.

The experimenter will notice that the triangles (simplexes in general) become very small as the optimum is approached. In its full generality the algorithm also has a feature that allows one to explore outside the simplex and expand under certain conditions. Our feeling is that a better strategy is to work with the shrinking simplex until the experimenter thinks it is too small and then begin

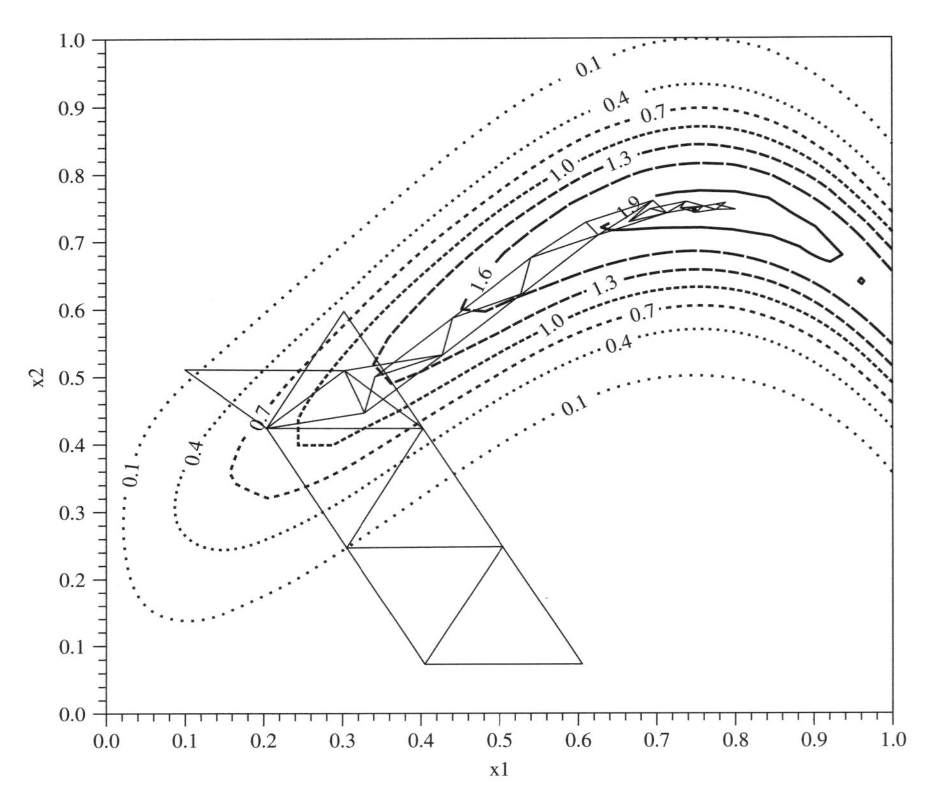

Figure 15.11 Simplex search algorithm.

with a new larger simplex defined on the basis of the accumulated knowledge. If the optimum has been located, the new simplex will immediately shrink in size because all reflections will be away from the optimum.

15.11 EXAMINATION OF AN EXPERIMENTAL PROBLEM

In this section we consider a hypothetical research program to produce a drink with maximal flavor by mixing four distinct juices. We can consider several distinct approaches. We attempt to evaluate some of these, much in the manner that a food scientist would. Our aim is to discuss considerations in the context of a simple problem rather than to present a specific recipe for an experimental strategy.

If the investigator has reasonably good a priori information about where the optimum lies and the study region can be limited to a small subregion, the mixture features of the problem can be ignored and a central composite design used. There are at least two potential problems with this. First, the a priori guess for the optimal location may be all wrong, leading to an experiment in a region that is really of little interest. Second, there is the problem of measuring the flavor response. Using an "objective" numerical flavor score from a taste panel carries with it some problems that we need not discuss here. However, the number of points required is too large to present samples of all points, all mixtures to a panel at once, and have them pick the best. Since taste panels are generally limited to about six or fewer samples at a time, this approach will require an incomplete block design to present samples to the taste panel and a numerical taste score.

Now assume that the investigator has limited faith in a priori predictions of an optimum. A larger region must be examined and the fact that the proportions of the four juices sum to 1 cannot be ignored. A possible strategy is to work from a lattice of points in the four-dimensional mixture. The lattice structures and the associated models described in Section 15.7 can be used as guides to select a reasonable and feasible number of mixtures to examine. Clearly, one would restrict the region of study if it were known, for example, that more than, say, 20% of juice x introduces an objectionable flavor. Problems again come to mind. If the flavor response surface is at all complex, a rather dense pattern of points is required to give a reasonable chance of locating a reasonable approximation of the optimum. The large number of points takes us back to the problems concerning measurement discussed earlier.

A third possible strategy is to treat this as an optimization problem. If we have an objective quantitative measure of flavor, or even if it is only possible to rank the flavors, a good choice is a simplex search strategy (Section 15.10.4). The investigator constructs a simplex about this point. The panel reports back the least desirable and most desirable points. The next step is either to flip the simplex or construct a new simplex about the most desirable point. Either strategy will lead to an improvement without requiring a numerical score. Both are fully nonparametric. As with all sequential searches, there is no guarantee that this one will not get caught on a local optimum.

REVIEW EXERCISE

15.1 The FDA is concerned about harmful bacteria such as *E. coli* and *Salmonella* contaminating food products such as apple juice and pickles. In response to this concern, a consortium of pickle producers wants to determine whether the naturally occurring acid in pickles kills harmful bacteria and how long it takes to kill a sufficient number of bacteria to ensure that the product is safe.

The consortium proposes to do an experiment in which *E. coli* is introduced into specially packed jars of dill pickles. Storage temperature and pH are potentially important factors in the length of time required to kill bacteria. The pickles can be produced at a range of pH levels from 3.0 to 4.7

and can be stored from 1 to 14 days. The objective of the experiment is to determine the pH and length of storage time necessary to kill at least 40,000 bacteria at 10°C and at 25°C. Assume that each jar can be opened just once to test the number of bacteria. Discuss possible experimental approaches to choosing the pH levels and the times to open jars.

CHAPTER 16

Plackett–Burman Hadamard Plans

16.1 INTRODUCTION

The two-level main-effect factorial plans that we have examined up to this point are limited to plans with 2^N experimental units. This class of plans was extended greatly in a classic paper by Plackett and Burman (1946), in which they introduced plans based on Hadamard matrices. Plackett–Burman plans are main-effect plans for $4m-1$ factors, each at two levels in $4m$ runs. This represents a very large increase in the range of possible size experiments. The actual Plackett and Burman paper includes catalogs for factorial plans with three, five, and seven levels, in addition to the plans for factors at two levels. However, common practice is to ignore everything but the two-level plans when referring to Plackett–Burman plans. There is a much larger set of plans based on Hadamard matrices than given in the catalog presented by Plackett and Burman. We follow a convention of using the term *Plackett–Burman plans* for plans given in the original Plackett and Burman paper and the term *Hadamard plans* for the larger class of plans based on Hadamard matrices.

16.2 HADAMARD MATRIX

A *Hadamard matrix* $\boldsymbol{H}$ of order n is an $n \times n$ matrix with the property that

$$\boldsymbol{H}\boldsymbol{H}^t = nI$$

or equivalently, that

$$\boldsymbol{H}^t\boldsymbol{H} = nI,$$

with all elements either $+1$ or -1. It is clear from the definition that one can change the signs of all elements in any row or any column of a Hadamard matrix

Planning, Construction, and Statistical Analysis of Comparative Experiments,
by Francis G. Giesbrecht and Marcia L. Gumpertz
ISBN 0-471-21395-0

without destroying the basic property. Common practice is to organize the matrix so that the first column has all entries equal to +1. Two Hadamard matrices are said to be *equivalent* if one can be obtained from the other by permuting rows and/or columns or by changing all signs in some rows and/or columns. Two examples of Hadamard matrices are

$$\begin{bmatrix} +1 & +1 \\ +1 & -1 \end{bmatrix} \quad \text{and} \quad \begin{bmatrix} +1 & +1 & +1 & +1 \\ +1 & +1 & -1 & -1 \\ +1 & -1 & +1 & -1 \\ +1 & -1 & -1 & +1 \end{bmatrix}.$$

Except for the trivial $n = 1$, and the special case with $n = 2$, it can be shown that n must be a multiple of 4 for a Hadamard matrix to exist.

Plackett and Burman (1946) presented a catalog of Hadamard matrices for $n = 4, 8, 12, 16, \ldots, 100$, except for $n = 92$. The missing case was found and added later. It has never been proved that $n = 4m$ is sufficient to guarantee the existence of a Hadamard matrix, but there are no known counterexamples. For our purposes, we can assume that at least one matrix exists for any n that is a multiple of four.

There is no single technique for constructing Hadamard matrices. However, a very powerful technique is to use the fact that if $\boldsymbol{H}$ is a Hadamard matrix of order n, then $\begin{bmatrix} \boldsymbol{H} & \boldsymbol{H} \\ \boldsymbol{H} & -\boldsymbol{H} \end{bmatrix}$ is a Hadamard matrix of order $2n$, or more generally that if $\boldsymbol{H}$ and $\boldsymbol{G}$ are $n \times n$ and $m \times m$ Hadamard matrices, then $\boldsymbol{H} \otimes \boldsymbol{G}$, where $\otimes$ is the Kronecker product, is an $nm \times nm$ Hadamard matrix.

16.3 PLACKETT–BURMAN PLANS

16.3.1 Four-Run Plackett–Burman Plan

The simplest Plackett–Burman plan is constructed from

$$\begin{bmatrix} +1 & +1 & +1 & +1 \\ +1 & +1 & -1 & -1 \\ +1 & -1 & +1 & -1 \\ +1 & -1 & -1 & +1 \end{bmatrix}$$

by deleting the first column and then assigning three treatments to the remaining three columns. The resulting rows represent treatment combinations. The result is the four-run plan with treatment combinations abc, a, b, and c. This is a main-effect plan for three factors. We recognize this plan as the half replicate of the 2^3.

Exercise 16.1: Hadamard Matrices

(a) Show that if $\boldsymbol{H}$ is an $n \times n$ Hadamard matrix, then $\begin{bmatrix} \boldsymbol{H} & \boldsymbol{H} \\ \boldsymbol{H} & -\boldsymbol{H} \end{bmatrix}$ is a $2n \times 2n$ Hadamard matrix.

(b) Show by rearranging columns that

$$\begin{bmatrix} +1 & +1 & +1 & +1 \\ +1 & +1 & -1 & -1 \\ +1 & -1 & +1 & -1 \\ +1 & -1 & -1 & +1 \end{bmatrix}$$

can be written in the form $\begin{bmatrix} \boldsymbol{H} & \boldsymbol{H} \\ \boldsymbol{H} & -\boldsymbol{H} \end{bmatrix}$, where $\boldsymbol{H}$ is a 2×2 Hadamard matrix. This construction can be thought of as the Kronecker product,

$$\begin{bmatrix} 1 & 1 \\ 1 & -1 \end{bmatrix} \otimes \begin{bmatrix} 1 & 1 \\ 1 & -1 \end{bmatrix}.$$

(c) Show that if $\boldsymbol{H}$ and $\boldsymbol{G}$ are $n \times n$ and $m \times m$ Hadamard matrices, then $\boldsymbol{H} \otimes \boldsymbol{G}$, where $\otimes$ is the Kronecker product, is an $nm \times nm$ Hadamard matrix. □

16.3.2 Eight-Run Plackett–Burman Plan

For this plan we need the 8×8 Hadamard matrix. We can construct this matrix by using $\begin{bmatrix} \boldsymbol{H} & \boldsymbol{H} \\ \boldsymbol{H} & -\boldsymbol{H} \end{bmatrix}$, where $\boldsymbol{H}$ is

$$\begin{bmatrix} +1 & +1 & +1 & +1 \\ +1 & +1 & -1 & -1 \\ +1 & -1 & +1 & -1 \\ +1 & -1 & -1 & +1 \end{bmatrix}.$$

However, there is a simpler method. Consider the following 8×7 array:

$$\begin{array}{ccccccc} +1 & +1 & +1 & -1 & +1 & -1 & -1 \\ -1 & +1 & +1 & +1 & -1 & +1 & -1 \\ -1 & -1 & +1 & +1 & +1 & -1 & +1 \\ +1 & -1 & -1 & +1 & +1 & +1 & -1 \\ -1 & +1 & -1 & -1 & +1 & +1 & +1 \\ +1 & -1 & +1 & -1 & -1 & +1 & +1 \\ +1 & +1 & -1 & +1 & -1 & -1 & +1 \\ -1 & -1 & -1 & -1 & -1 & -1 & -1 \end{array}. \tag{16.1}$$

Notice that the rows 2, 3, ..., 7 are each obtained from the row above by shifting to the right one position and moving the symbol from the last position to the front. Repeating the process one more time would produce a duplicate of the first row. Finally, the last row consists of seven -1's. If we were to add an extra column of $+1$'s, we would have a Hadamard matrix of order 8. Now, if we identify the rows of array (16.1) with observations and the columns with factors, $+1$ denotes the high level and -1 the low level, we have a suitable plan for seven factors with eight observations and all main effects estimable and uncorrelated. Alternatively, one column can be used to define blocks to give a randomized block design with six factors in two blocks.

Since eight is a power of 2, this is a plan that could have been developed using the factorial confounding scheme developed previously. In fact, if we had done that, the aliasing relationships between main effects and two-factor interactions would also be clear.

***Exercise 16.2:* 8 × 8 *Hadamard*.** Show that appending a column of eight $+1$'s to the 8×7 array shown in (16.1) gives an 8×8 Hadamard matrix. □

***Exercise 16.3: Alternative construction of* 8 × 8 *Hadamard*.** Show that the 8×8 Hadamard matrix constructed by using $\begin{bmatrix} \boldsymbol{H} & \boldsymbol{H} \\ \boldsymbol{H} & -\boldsymbol{H} \end{bmatrix}$, where $\boldsymbol{H}$ is

$$\begin{bmatrix} +1 & +1 & +1 & +1 \\ +1 & +1 & -1 & -1 \\ +1 & -1 & +1 & -1 \\ +1 & -1 & -1 & +1 \end{bmatrix}$$

is equivalent to the Hadamard matrix obtained from array (16.1) by adding the column of "+" symbols. By equivalent, we mean that one matrix can be obtained from the other by rearranging rows and columns and, if necessary, changing *all* signs in any rows or columns. □

The statistical analysis of the data from an experiment using this plan is rather simple. The effect of any treatment is estimated by computing the mean of the four observations with that treatment at the high level minus the mean of the four observations with that treatment at the low level. Computing these estimates for all treatments completes the analysis since no estimate of other interactions or experimental error is possible. One must use an external source of error, or resort to either a half-normal plot (Daniel 1959) or the Dong (1993) or Lenth (1989) procedure discussed in Chapter 12. If fewer than seven factors are to be studied, the extra columns then provide an estimate of experimental error.

16.3.3 12-Run Plackett–Burman Plan

The 12×11 array

$$
\begin{array}{ccccccccccc}
+ & + & - & + & + & + & - & - & - & + & - \\
- & + & + & - & + & + & + & - & - & - & + \\
+ & - & + & + & - & + & + & + & - & - & - \\
- & + & - & + & + & - & + & + & + & - & - \\
- & - & + & - & + & + & - & + & + & + & - \\
- & - & - & + & - & + & + & - & + & + & + \\
+ & - & - & - & + & - & + & + & - & + & + \\
+ & + & - & - & - & + & - & + & + & - & + \\
+ & + & + & - & - & - & + & - & + & + & - \\
- & + & + & + & - & - & - & + & - & + & + \\
+ & - & + & + & + & - & - & - & + & - & + \\
- & - & - & - & - & - & - & - & - & - & -
\end{array}
$$

provides a plan for up to 11 factors, each at two levels using 12 observations, or 10 factors in two blocks. Since 12 is not a power of 2, this plan cannot be obtained using the confounding schemes developed in Chapter 13. This array is easily converted to the $n = 12$ Hadamard matrix by replacing "+" and "−" with +1 and −1, respectively, and adding the extra column. In general, there is this simple correspondence between a Hadamard matrix and the Hadamard plan.

16.3.4 Small Catalog of Plackett–Burman Plans

From the patterns of + and − symbols, it is clear that a reasonable catalog of two-factor Plackett–Burman plans need contain only the one generator row for each value of n, provided that such a row can be found. It appears that it is not always possible. A catalog for the Plackett–Burman plans for $n \leq 64$ follows.

$n = 8$ $\quad + + + - + - -.$

$n = 12$ $\quad + + - + + + - - - + -.$

$n = 16$ $\quad + + + + - + - + + - - + - - -.$

$n = 20$ $\quad + + - - + + + + - + - + - - - - + + -.$

$n = 24$ $\quad + + + + + - + - + + - - + + - - + - + - - - -.$

$n = 28$ For $n = 28$ there is no single generator row. Instead, the first nine rows consist of three 9×9 blocks. The remaining 18 rows are obtained by cycling these three blocks, that is, moving blocks in positions 1 and 2 to the right and putting the block in position 3 into the space vacated by the block in position 1.

+ − + + + + − − − : − + − − − + − − + : + + − + − + + − +
+ + − + + + − − − : − − + + − − + − − : − + + + + − + + −
− + + + + + − − − : + − − − + − − + − : + − + − + + − + +
− − − + − + + + + : − − + − + − − − + : + − + + + − + − +
− − − + + − + + + : + − − − − + + − − : + + − − + + + + −
− − − − + + + + + : − + − + − − − + − : − + + + − + − + +
+ + + − − − + − + : − − + − − + − + − : + − + + − + + + −
+ + + − − − + + − : + − − + − − − − + : + + − + + − − + +
+ + + − − − − + + : − + − − + − + − − : − + + − + + + − +.

$n = 32$ − − − − + − + − + + + − + + − − − + + + + + − − + + −
+ − − +

$n = 36$ − + − + + + − − − + + + + + − + + + − − + − − − − + −
+ − + + − − + −

$n = 40$ The plan for $n = 40$ can be developed from the $n = 20$ plan by utilizing the fact that $\begin{bmatrix} +1 & +1 \\ +1 & -1 \end{bmatrix} \otimes \boldsymbol{H}$ is a Hadamard matrix of order 40 if $\boldsymbol{H}$ is a Hadamard matrix of order 20.

$n = 44$ + + − − + − + − − + + + − + + + + + − − − + − + + + −
− − − − + − − − + + − + − + + −

$n = 48$ + + + + + − + + + + − − + − + − + + + − − + − − + + −
+ + − − − + − + − + + − − − − + − − − −

$n = 52$ The first row of this plan is

+ + − + − + − + − + − + − + − + − + − + − + − + − + − + − + −
+ − + − + − + − + − + − + − + − + − + −

The next 10 rows consist of the following column and five 10×10 blocks:

\+ : − + − − − − − − − − : + + − − − − + + + + :

− : + + − + − + − + − + : + − − + − + + − + − :

\+ : − − − + − − − − − − : + + + + − − − − + + :

− : − + + + − + − + − + : + − + − − + − + + − :

\+ : − − − − − + − − − − : + + + + + + − − − − :

− : − + − + + + − + − + : + − + − + − − + − + :

\+ : − − − − − − − + − − : − − + + + + + + − − :

− : − + − + − + + + − + : − + + − + − + − − + :

\+ : − − − − − − − − − + : − − − − + + + + + + :

− : − + − + − + − + + + : − + − + + − + − + − :

\+ + + + − − + + − − : + + − − + + − − + + : + + + + + + − − − −

\+ − + − − + + − − + : + − − + + − − + + − : + − + − + − − + − +

− − + + + + − − + + : + + + + − − + + − − : − − + + + + + + − −

− + + − + − − + + − : + − + − − + + − − + : − + + − + − + − − +

\+ + − − + + + + − − : − − + + + + − − + + : − − − − + + + + + +

\+ − − + + − + − − + : − + + − + − − + + − : − + − + + − + − + −

− − + + − − + + + + : + + − − + + + + − − : + + − − − − + + + +

− + + − − + + − + − : + − − + + − + − − + : + − − + − + + − + −

\+ + − − + + − − + + : − − + + − − + + + + : + + + + − − − − + +

\+ − − + + − − + + − : − + + − − + + − + − : + − + − − + − + + −

The next 40 rows are obtained by repeating the first column and cycling through the blocks.

$n = 56$ This plan is developed just as $n = 40$ but using $n = 28$ in place of $n = 20$.

$n = 60$ + + − + + + − + − + − − + − − + + + − + + + + − − + +

\+ + + − − − − − + + − − − − + − − − + + − + + − + − +

− − − + −

$n = 64$ This plan is developed just as $n = 56$ but using $n = 32$ in place of $n = 28$.

16.4 HADAMARD PLANS AND INTERACTIONS

It is easy to verify that Hadamard plans are main-effect plans, i.e., they provide uncorrelated estimates of main effects and no clear estimates of interaction and consequently, are resolution III plans. However, there are still questions about two-factor interactions. What if they exist? How do they affect the estimates? The situation is somewhat complex, and for this reason many have recommended against the use of these plans, especially in exploratory situations. We will show that this recommendation is wrong, and that the complex interaction confounding patterns found in many Hadamard plans can be exploited to great advantage. First, Hedayat and Wallis (1978) state that all 2×2, all 4×4, all 8×8, and all 12×12 Hadamard matrices are equivalent. There are five equivalence classes of 16×16 Hadamard matrices and three equivalence classes of 20×20 Hadamard matrices. Beyond $n = 20$, it is probably safe to assume that there exist nonequivalent $n \times n$ Hadamard matrices, provided, of course, that n is a multiple of four. The point is that, at least for $n \leq 20$, it is possible to use a computer to examine all possible cases. We will look at some of these.

The reason for looking at interactions in Hadamard plans is based on the commonly observed phenomenon of effect heredity (Hamada and Wu 1992), that two-factor interactions are very rare when neither of the main effects is real. If the research is exploratory, and it is reasonable that only few factors are important (effect sparsity; Box and Meyer 1985), the complex partial aliasing patterns of the Hadamard plans permit discovering and disentangling main effects and two-factor interactions with minimal experimental effort. We present more on this later.

The four- and eight-run Plackett–Burman plans are part of the 2^N system examined in Chapter 10. In the four-run 2^{3-1} plan, the A, B, and C effects are aliased with the BC, AC, and AB interactions, respectively. The eight-run plan can be obtained by taking the 2^{7-4} fraction defined by

$$\mathcal{I} = ABC = ADE = AFG = BDF.$$

Each main effect is perfectly aliased with at least three two-factor interactions. For example, A is aliased with BC, DE, and FG. One of the five 16-run plans can be obtained as a 2^{15-11} fraction. In this plan each main effect is aliased with some two-factor interactions. In the remainder of this section we discuss plans that are not part of the 2^N system. In Chapter 23 we find a discussion of some nonorthogonal 10- and 14-run plans, clearly not part of the Hadamard system, but which have some similar properties.

16.4.1 12-Run Plackett–Burman Plan

In the 12-run Plackett–Burman, the interactions are not perfectly correlated with any main effects. We illustrate this in Table 16.1 with the factor labels at the top of the columns and an additional column identifying the AB interaction at the right. The symbols for the interaction column are obtained using the rules for

Table 16.1 Illustration of Interaction in a 12-Run Plackett–Burman Plan

Runs	A	B	C	D	E	F	G	H	I	J	K	AB
1	+	+	−	+	+	+	−	−	−	+	−	+
2	−	+	+	−	+	+	+	−	−	−	+	−
3	+	−	+	+	−	+	+	+	−	−	−	−
4	−	+	−	+	+	−	+	+	+	−	−	−
5	−	−	+	−	+	+	−	+	+	+	−	+
6	−	−	−	+	−	+	+	−	+	+	+	+
7	+	−	−	−	+	−	+	+	−	+	+	−
8	+	+	−	−	−	+	−	+	+	−	+	+
9	+	+	+	−	−	−	+	−	+	+	−	+
10	−	+	+	+	−	−	−	+	−	+	+	−
11	+	−	+	+	+	−	−	−	+	−	+	−
12	−	−	−	−	−	−	−	−	−	−	−	+

interactions developed in Chapter 10. Notice that this column is not the duplicate of any of the 11 main-effect columns (even if we reverse the + and − symbols in some columns). However, it is orthogonal only to the A and B columns. We know that the estimate of the A effect, for example, is obtained as the mean of the runs with A at the + level minus the mean of the runs with A at the − level. These are easily identified by looking at the columns of + and − symbols. It follows that the estimates of A and B will be free of the AB interaction, and estimates of the remaining main effects will not be free of the AB interaction. In fact, each of the $C, \ldots, K$ contrasts has either a $+\frac{1}{3}$ or $-\frac{1}{3}$ correlation with the AB contrast. For example, C has a $-\frac{1}{3}$ correlation with AB and J has a $+\frac{1}{3}$ correlation. Similar relations hold for all other two-factor interactions and for all factors. In general, the estimate of any main effect, say X, will be free of any two-factor interaction that involves X and another factor, and will contain part of every other two-factor interaction that happens to exist. Also, the correlation coefficients will always be either $+\frac{1}{3}$ or $-\frac{1}{3}$. In a sense, the two-factor interactions get spread across the set of effects estimated. This is in direct contrast to the situation in the case of the main-effect plans in the 2^N system, where each two-factor interaction loads completely on one specific main effect and no others.

16.4.2 16-Run Hadamard Plans

We encounter an additional problem with the 16-run plan. There are five different, i.e., not equivalent, Hadamard matrices. The plan given in the Plackett–Burman catalog corresponds to one of these matrices. The corresponding plan given in the catalog is actually part of the 2^N system. It is the 2^{15-11} fraction defined by

$$\mathcal{I} = ABC = ADE = AHI = ALM = BDF = BHJ$$
$$= BLN = FHN = DGEF = HIJK = LMNO.$$

When this basis set of defining contrasts is expanded, the full set contains the following 35 three-factor interactions:

$$ABC,\ ADE,\ AFG,\ AHI,\ AJK,\ ALM,\ ANO,\ BDF,\ BEG,$$
$$BHJ,\ BIK,\ BLN,\ BMO,\ CDG,\ CEF,\ CHK,\ CIJ,\ CLO,$$
$$CMN,\ DHL,\ DIM,\ DJN,\ DKO,\ EHM,\ EIL,\ EJO,\ EKN,$$
$$FHN,\ FIO,\ FJL,\ FLM,\ GHO,\ GIN,\ GJM,\ \text{and}\ GKL.$$

From this set it is simple to establish the alias relationships among the main effects and the two-factor interactions. Representatives of the four remaining equivalence classes of 16×16 Hadamard matrices are given in Hall (1961). The plans corresponding to these matrices are not part of the 2^N system. Consequently, it is not possible to find sets of defining contrasts that yield these plans. It turns out that for these plans some of the interactions are aliased with individual main effects, while some are partially aliased with two main effects, as is the case in, for example, the 11 factor in a 12-run Plackett–Burman plan. If one drops the first column of the following 16-run Hadamard plan, no main effect will be aliased with any two-factor interaction in the remaining plan. However, some main effects will have correlations of $\pm 1/2$ with some two-factor interactions.

$$\begin{array}{ccccccccccccccc}
- & - & - & - & - & - & - & - & - & - & - & - & - & - & - \\
- & - & - & - & - & - & - & + & + & + & + & + & + & + & + \\
- & - & - & + & + & + & + & - & - & - & - & + & + & + & + \\
- & - & - & + & + & + & + & + & + & + & + & - & - & - & - \\
- & + & + & - & - & + & + & - & - & + & + & - & - & + & + \\
- & + & + & - & - & + & + & + & + & - & - & + & + & - & - \\
- & + & + & + & + & - & - & - & - & + & + & + & + & - & - \\
- & + & + & + & + & - & - & + & + & - & - & - & - & + & + \\
+ & - & + & - & + & - & + & - & + & - & + & - & + & - & + \\
+ & - & + & - & + & + & - & - & + & + & - & + & - & + & - \\
+ & - & + & + & - & - & + & + & - & + & - & - & + & + & - \\
+ & - & + & + & - & + & - & + & - & - & + & + & - & - & + \\
+ & + & - & - & + & - & + & + & - & + & - & + & - & - & + \\
+ & + & - & - & + & + & - & + & - & - & + & - & + & + & - \\
+ & + & - & + & - & - & + & - & + & - & + & + & - & + & - \\
+ & + & - & + & - & + & - & - & + & + & - & - & + & - & +
\end{array} \qquad (16.2)$$

Both of these 16-run plans are resolution III plans. However, the plan in array (16.2) provides a much better possibility for detecting the presence of two-factor interactions if more than two factors are active. This is particularly true if there is room to avoid using the first column.

A detailed examination reveals that the first column main effect is perfectly aliased with two-factor interactions of factors on columns 2 and 3, columns 4 and 5, ..., columns 14 and 15. Also, each of the two-factor interactions involving the factor on column 1 is perfectly aliased with one of the main effects. Main effects for factors assigned to columns 2, ..., 15 have $\pm\frac{1}{2}$ correlations with 24 two-factor interactions. In summary, if one can assume that the factor assigned to the first column does not interact with any of the other factors, there is a good chance that one should be able to disentangle main effects and some two-factor interactions.

16.4.3 20-Run Plackett–Burman Plan

The 20-run cyclic Plackett–Burman plan given in the catalog has a very disperse alias pattern. The main effect for factor X is not aliased with any two-factor interaction involving X, but is partially aliased with all other two-factor interactions. The correlations are mostly $\pm.2$, although there are a few $-.6$ correlations. Lin and Draper (1993) give the details.

16.5 STATISTICAL ANALYSES

First, we recall that since a Hadamard plan is saturated, there is no true estimate of error unless some columns are intentionally left free. Consequently, an experimenter should be hesitant to use these plans unless a good a priori sense of the error variance is available, or the experiment is exploratory and there is good reason to believe that few (but unidentified) factors are important or active. In this case the argument seems quite clear. With those Hadamard plans that are of the 2^N system, individual two-factor interactions are completely aliased with individual main-effect degrees of freedom. If one of these interactions is large, its effect will appear as a main effect. If the experimenter keeps this possibility in mind when interpreting the results of an experiment, it is usually possible to get things sorted out properly.

Consider a hypothetical example of a Plackett–Burman main-effect plan, eight runs with seven factors A, B, C, D, E, F, and G assigned to the columns

$$\begin{bmatrix} -1 & -1 & +1 & -1 & +1 & +1 & -1 \\ +1 & -1 & -1 & -1 & -1 & +1 & +1 \\ -1 & +1 & -1 & -1 & +1 & -1 & +1 \\ +1 & +1 & +1 & -1 & -1 & -1 & -1 \\ -1 & -1 & +1 & +1 & -1 & -1 & +1 \\ +1 & -1 & -1 & +1 & +1 & -1 & -1 \\ -1 & +1 & -1 & +1 & -1 & +1 & -1 \\ +1 & +1 & +1 & +1 & +1 & +1 & +1 \end{bmatrix}.$$

Now we know that the AB interaction is perfectly aliased with the C factor. If the experiment were to show a large C effect in addition to large A or B effects, the experimenter would be well advised to consider the possibility of an AB interaction. On the other hand, a large C effect and no A or B effects could probably be quite safely accepted as only a C main effect. The logic of this is based on what Hamada and Wu (1992) refer to as *effect heredity*. If the experimenter has serious doubts, there is usually the option of adding a few extra points. In the example above, adding the four extra points $(-1 - 1 - 1)$, $(+1 - 1 + 1)$, $(-1 + 1 + 1)$, and $(+1 + 1 - 1)$ makes both AB and C estimable. Note two points. First, it is now assumed that factors D through G are not important but must be assigned for these extra runs. Since they are not important, they can be set arbitrarily to either ± 1 or an intermediate level. Second, the full set of 12 observations has four treatment combinations with four replicates each and four with only one replicate. A general least squares program will give appropriate estimates. In fact, if the error variance is small and observations expensive, one could even try a statistical analysis with only two of the four additional points.

In the case of the 11-factor, 12-run Hadamard plan, matters are different. Now, a possible two-factor interaction will be spread across a number of single-degree-of-freedom main-effect contrasts. Hamada and Wu (1992) have proposed a method of analysis that takes advantage of this phenomenon to look for possible interactions. The basic idea is that if an interaction is important, it will appear as part of a number of main effects. The first step in their analysis is the conventional check on main effects to form a list of possibly significant main effects. Then using the effect heredity principle, they define an expanded model that includes this list and possible two-factor interactions (must have at least one component appearing in the preceding list). At this stage, they also include the option of the experimenter adding interactions that he suspects may be real. The third step consists of a forward stepwise regression analysis applied to the data and this expanded model. On the basis of this analysis, they revise the model constructed in step 2 and continue to alternate the two steps until the model stops changing. They claim that typically, few cycles are required.

The key feature in the standard Plackett–Burman or Hadamard plans is that they are fully saturated if all the columns are used, i.e., provide no clean estimate of experimental error. In analysis of variance terminology, "There are no degrees of freedom for error." This is not a problem if the experimenter is sufficiently familiar with the experimental material and has good prior knowledge of the experimental error from other similar experiments. A possible strategy is to use a larger plan with more runs and consequently, obtain an estimate of error. An alternative is to resort to the techniques discussed in Chapter 12. Such graphical methods work well when most of the factors in the experiment have relatively little effect, and a small number have major effects. This phenomenon is typically referred to as *effect sparsity*. These plans have their greatest appeal in experimental settings where the experimenter sees a large number of potential factors but is also quite sure that most of them have little or no effect, and his problem is to locate those few that have major effect, while using limited resources.

Example 16.1. This example illustrates both the strengths and weaknesses of the Hadamard plans. Hamada and Wu (1992) illustrate their methods in the reanalysis of data from an experiment reported by Hunter et al. (1982) The experiment used a 12-run Hadamard plan, even though it involved only seven

Table 16.2 Experimental Factors and Levels

	Factor	Level +	Level −
A	Initial structure	β treatment	As received
B	Bead size	Large	Small
C	Pressure treatment	HIP	None
D	Heat treatment	Solution treatment/age	Anneal
E	Cooling rate	Rapid	Slow
F	Polish	Mechanical	Chemical
G	Final treatment	Peen	None

Table 16.3 Experimental Plan and Data

	Plan											Logged
Run	A	B	C	D	E	F	G	ϵ_1	ϵ_2	ϵ_3	ϵ_4	Data
1	+	+	−	+	+	+	−	−	−	+	−	6.058
2	+	−	+	+	+	−	−	−	+	−	+	4.733
3	−	+	+	+	−	−	−	+	−	+	+	4.625
4	+	+	+	−	−	−	+	−	+	+	−	5.899
5	+	+	−	−	−	+	−	+	+	−	+	7.000
6	+	−	−	−	+	−	+	+	−	+	+	5.752
7	−	−	−	+	−	+	+	−	+	+	+	5.682
8	−	−	+	−	+	+	−	+	+	+	−	6.607
9	−	+	−	+	+	−	+	+	+	−	−	5.818
10	+	−	+	+	−	+	+	+	−	−	−	5.917
11	−	+	+	−	+	+	+	−	−	−	+	5.863
12	−	−	−	−	−	−	−	−	−	−	−	4.809

factors. The object was to study the effects of seven factors on fatigue life of weld-repaired castings. The factors and levels in the experiment are shown in Table 16.2. The plan of the experiment and the data are given in Table 16.3. Notice that the four unused columns are assigned to error. The estimates for the seven treatment contrasts are .326, .294, −.246, −.516, .150, .915, and .183, respectively. The estimates of the four error contrasts are .446, .452, .080, and −.242. The fact that there are four error contrasts permits an analysis of variance and statistical tests of significance. The F-test, testing the null hypothesis for factor F, is highly significant and the test for factor D yields a probability of .2. The half-normal plot of the 11 contrasts is shown in Figure 16.1.

Without closer inspection, the F-tests and the half-normal plot seem to tell us that only polish has an effect. The statistical analysis with this model accounts for only 45% of the total variance. Including factor D increased this to 59%. Based on the principle of effect heredity, Hamada and Wu (1992) investigate a model including factor F and all of the two-factor interactions involving F.

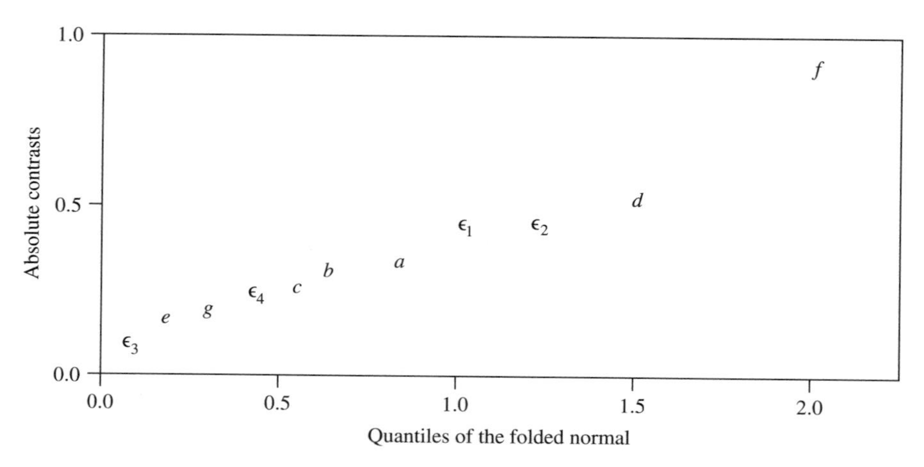

Figure 16.1 Half-normal plot.

Adding FG interaction to the model with F increases R^2 to .89. Estimates of F and FG under this model are .458 and $-.459$. With D, F, and FG in the model, R^2 becomes .92. Going back to the analysis of aliasing relationships, we find that under the model containing only seven main effects and the FG interaction, $\widehat{D}$ estimates $D + \frac{1}{3}FG$; $\widehat{\epsilon_1}, \widehat{\epsilon_2}$ and $\widehat{\epsilon_3}$ each estimate $-\frac{1}{3}FG$; and $\widehat{\epsilon_4}$ estimates $+\frac{1}{3}FG$. The estimate of D may have been reduced by the large negative FG interaction effect. This would seem to corroborate the original experimenters' (Hunter et al. 1982) feelings when they included D in their analysis, despite their questionable statistical evidence. At this point it is necessary to perform a few more experimental runs to establish definitely that there is an FG interaction and the possibility of a D effect. Notice that it would not have been possible to untangle the possible FG interaction if it had been aliased with a main effect in an eight- or 16-run fraction in the 2^N system.

16.6 WEIGHING DESIGNS

The problem of finding suitable combinations of several objects to be weighed in a scale to determine their individual weights with maximum possible precision is known as the *weighing problem*, and the suitable combinations as the *weighing designs*. Weighing designs have applicability in any problem of measurement, where the measure of a combination is expressible as a sum of the separate measures. There are really two types of problems that arise in practice: the spring balance (one-pan balance) problem and the chemical balance (two-pan balance) problem.

Formally, the chemical balance problem can be stated as follows: The results of n weighing operations to determine the individual weights of k objects fit into the general linear model $y = X\beta + \epsilon$, where y is the $n \times 1$ column vector of observations (the observed weights), $X = (x_{ij})$, $i = 1, 2, \ldots, n$ and $j = 1, 2, \ldots, k$ is an $n \times k$ matrix, with $x_{ij} = +1$ or -1, depending on whether the jth object is placed in the left or right pan for the ith weighing and ϵ is the vector of random errors. Consistent with the above, the elements of y have positive sign if balancing weights are placed in the right pan, and negative sign if the balancing weights are placed in the left pan. It is understood that each item must be in one pan, either the left or the right, for each weighing. Main-effect plans of resolution III are ideally suited to the weights problem since there is no chance of any interactions. Also, the problem is one of estimation rather than testing hypotheses. The Hadamard plans can be shown to be the optimal plans for this problem. They are the plans that minimize the variances of the estimates of β, the weights of the individual objects. Individual runs of the Hadamard plan correspond to weighings; a "+" symbol means that the object goes into the left pan; a "−" symbol means that the object goes into the right pan and the y, the response, is the sum of the weights that need to be added to provide a balance, with a positive sign, if placed in the right pan, and a negative sign if placed in the left pan.

The solution for the spring balance problem is somewhat more complicated, but it also solved by the Hadamard plans. Formally, the problem can be stated as follows. Let n weighing operations be used to determine the individual weights of k objects. Use the data to fit a linear model of the form $y = X\beta + \epsilon$, where y is the $n \times 1$ column vector of observations, $X = (x_{ij})$, $i = 1, 2, \ldots, n$ and $j = 1, 2, \ldots, k$ is a $n \times k$ matrix with $x_{ij} = +1$, if the jth object is in the pan for the ith weighing and 0 otherwise, and ϵ is the vector of random errors. The problem now is to select the subsets of objects that must be in the pan for the individual weighings.

Although the complete solution is beyond the scope of this book, we illustrate the nature of a good design. Consider a case of seven objects and seven weighings. One could weigh each object separately. However, we will weigh them in groups of three. The groups selected to weigh are dfg, aeg, abf, bcg, acd, bde, and cef. The corresponding X matrix is

$$\begin{bmatrix} 0 & 0 & 0 & 1 & 0 & 1 & 1 \\ 1 & 0 & 0 & 0 & 1 & 0 & 1 \\ 1 & 1 & 0 & 0 & 0 & 1 & 0 \\ 0 & 1 & 1 & 0 & 0 & 0 & 1 \\ 1 & 0 & 1 & 1 & 0 & 0 & 0 \\ 0 & 1 & 0 & 1 & 1 & 0 & 0 \\ 0 & 0 & 1 & 0 & 1 & 1 & 0 \end{bmatrix}.$$

Notice that this matrix is obtained from the first seven rows of matrix (16.1). Direct calculations (inverse of the X^tX matrix) show that estimates of the weights of the individual objects have variance $.444\sigma_\epsilon^2$ and covariances $-.055\sigma_\epsilon^2$. This is to be compared to separate weighing of the seven individual objects, which leads to independent estimates, but with variance σ_ϵ^2. The variance of the difference between weights of two objects is $.9964\sigma_\epsilon^2$ when the weighing design is used and $2\sigma_\epsilon^2$ when objects are weighed individually. Notice that in the pattern of grouping items, all pairs of objects appear together in the pan exactly once, so this is a balanced incomplete block design. Federer et al. (1976) describe an example of the application of such a weighing design to research in agronomy.

Weighing designs are used frequently in optics in spectroscopy and imaging. The problem there can be described as follows. Suppose that an experimenter wants to determine the spectrum of a beam of light. When the error distribution is independent of the strength of the incident signal, more accurate estimates may be obtained by combining different frequency components in groups rather than measuring the intensity of each frequency component separately. An optical separator (such as a prism) separates the different frequency components of the incident light and focuses them into different locations (slots) of a mask. Each slot can be open or closed. If a slot is open, the light is transmitted; if it is closed, the light is absorbed. The total intensity of the frequency components not blocked is then measured. If there are n frequency components, at least n different measurements must be taken. The decisions of which frequencies to combine is based on Hadamard matrices (weighing designs). The initial vectors that are given in the catalog are used to design the masks that control the light sources. Harwit and Sloan (1979) present a number of examples of the application to problems in optics.

16.6.1 Plans That Look Like Hadamard Plans

There is a class of weighing designs that on the surface may appear to be similar to the Hadamard plans but are not at all related. These are designs for weighing $4m + 2$ factors in $4m + 2$ weighings. These exist only for select values of m and would probably not be recommended except in cases where there is a high premium on minimizing the number of weighings. The plans achieve almost, but not quite, the full efficiency of the Hadamard-based designs. The plans are constructed from the a and b vectors given in Table 16.4. The design matrix (Yang 1966, 1968) is constructed by first constructing circulant matrices $\boldsymbol{A}$ and $\boldsymbol{B}$ from a and b, respectively. The design matrix is then obtained by assembling $\begin{bmatrix} \boldsymbol{A} & \boldsymbol{B} \\ -\boldsymbol{B}^t & \boldsymbol{A}^t \end{bmatrix}$. For the $n = 10$ case, for example, the $\boldsymbol{A}$ and $\boldsymbol{B}$ are both equal to

$$\begin{bmatrix} - & + & + & + & + \\ + & - & + & + & + \\ + & + & - & + & + \\ + & + & + & - & + \\ + & + & + & + & - \end{bmatrix}.$$

Table 16.4 Vectors to Define Weighing Designs

No.	id	Vector
6	*a*	+ + +
	b	− + +
10	*a*	− + + + +
	b	− + + + +
14	*a*	− + + + + + +
	b	− − + − + + +
18	*a*	− − + + + + + + +
	b	− + − + + − + + +
	a	− + − + + + + + +
	b	− − + + − + + + +
26	*a*	− − + − + + + + + − + + +
	b	− − + − + + + + + − + + +
	a	− − + + − + − + + + + + +
	b	− − + + − + − + + + + + +
30	*a*	− − + + − + − + + + + + + + +
	b	− − + − − + + + − + − + + + +

Items to be weighed are assigned to the columns "+" in the left pan and "−" in the right pan, and rows represent individual weighings. The results of weighings are the sum of the balancing weights if they are in the right pan and negative if the balancing weights are placed in the left pan.

In the extreme case when limits on number of units are severe, these matrices can be used to construct main-effect plans. One can assign factors to columns 2 to n. The first, $n/2 - 1$, will have almost full efficiency. The last, $n/2$, will have somewhat lower efficiency. A possible application could be a case where the experimental material presents sets of exactly 6, 10, 14, or 18 units and there are potential noninteracting factors.

16.7 PROJECTION PROPERTIES OF HADAMARD PLANS

Hadamard plans are resolution III, saturated, main-effect plans. The big advantage of these plans in an exploratory situation is that in a sense they maximize the number of factors that one can examine in a fixed number of runs. The down side is that the alias structure inherent in these plans may cause main effects to become blurred if two-factor (or higher) interactions should occur. If the investigator really believes that all factors will be important, the only realistic advice is to perform a larger experiment. However, in the standard screening situation the problem is that there are some *active* factors dispersed among a much larger number of *inert* factors. The problem is to identify the active factors. This is where the Hadamard plans excel. We look at some special cases.

16.7.1 Eight-Run Plan

This plan is too small to permit much flexibility. Standard projection considerations (Section 14.4.1) tell us that if any two factors are active and the remaining five are inert, the plan becomes two complete replicates of the 2^2. To the investigator, this means that if exactly two factors appear active, the interpretation is clear. If three are active, the possibility that the third is the two-factor interaction of the other two must be considered. Further experimentation may be required. Similarly, if more than three are active, the possibility of two-factor interactions must be considered.

If there are doubts, the obvious first step is to add the eight runs in the complete foldover (Section 14.6) to separate main effects from two-factor interactions. This eliminates the ambiguity with respect to main effects. Notice that the foldover can be interpreted as the second block for a randomized block design. Uncertainty with respect to two-factor interactions that remain aliased demands additional experimental work.

16.7.2 12-Run Plan

In a screening situation, this plan has much to offer. In fact, Box and Tyssedal (1996) claim that this plan is better for screening than the 16-run Hadamard plan. While the simple projection argument based on word length cannot be used, it can be shown that any Hadamard plan "projects" to a 2^2 factorial with every point replicated equally. Lin and Draper (1992) discuss the projection properties of Hadamard plans in considerable detail. They show that if any two factors in the 12-run experiment are active and the remaining nine are inert, the reduced plan has three replicates of a full 2^2 factorial.

If any three factors are active and eight inert, the reduced plan consists of a complete 2^3 plus a half replicate defined by the three-factor interaction. We illustrate by selecting columns 1, 2, and 4 from the array in Section 16.4.1. We label the factors as A, B, and D. If all other factors are inert, we have a plan with treatment combinations $\{abd, b, ad, bd, (1), d, a, ab, ab, bd, ad, (1)\}$. We recognize the full 2^3 plus the half replicate defined by $\mathcal{I} = -ABD$. We could have selected any three columns and observed a similar set of treatment combinations. This is an appealing plan that provides estimates of all three main effects and all four interactions, as well as an estimate of residual error with four degrees of freedom. Since the plan is unbalanced, one must use a general-purpose regression program to perform the statistical analysis. Unfortunately, since the two-factor interactions are each spread across nine columns, it is sometimes difficult to recognize that a simple model, say one with three main effects and two two-factor interactions, describes the data adequately. When in doubt, the advice is to try to fit the competing models and compare. Guidance from expert knowledge of the subject matter must not be underrated in this examination. Box and Meyer (1985, 1986) discuss formal methods for making such comparisons. If resources permit, an obvious strategy is to add more runs. In Chapter 19 we look at strategies for selecting additional runs to clear up ambiguities.

If a set of four factors is active, matters are slightly more complex. The projected plan now is always one run short of having a resolution IV, 2^{4-1} plan. We illustrate by selecting the first four columns from the array in Section 16.4.1. We observe $\{abd, bc, acd, bd, c, d, a, ab, abc, bcd, acd, (1)\}$. Selecting runs 1, 3, 5, 6, 7, 9, 10, and 11 from this set gives us seven of the eight runs required for the fraction $\mathcal{I} = -ABCD$, with treatment combination acd duplicated and its foldover treatment b missing. Adding the extra, thirteenth run gives the 2^{4-1} resolution IV plan, with four extra runs. A standard least squares analysis gives unbiased estimates of main effects. When adding the thirteenth point, two minor difficulties arise. First, there is the question of levels to select for the factors in the discarded columns. One can argue that since the experiment has shown that these factors are not important, i.e., are not active, the choice does not matter. Second, there is the question of which of the duplicate points to use in the statistical analysis. This is an open question, but our view is that data are expensive and one should rely on the least squares analysis to take care of the unbalance. If a full 2^4 factorial is wanted, four more points in addition to the one added previously are needed. An alternative is to analyze as a $3/4$ replicate of the 2^4.

When five factors are retained, the situation becomes considerably more messy. The situation now depends on the actual five columns selected. There are $\binom{11}{5} = 462$ possibilities, but all correspond to one or the other of the two cases in Table16.5. Note that it may be necessary to interchange $+$ and $-$ in all entries in one or more columns to obtain an exact match. The two treatment combinations listed as runs 13 and 14 must be added to complete the 2^{5-2} resolution III plans. In the case of plan A, one also has the choice of adding the bottom six runs, listed as runs 13 to 18, to the original runs 1 to 12 to obtain a 2^{5-1} resolution V plan. In the case of plan B, one must add the sequence identified as runs 13 to 22 to runs 2 to 12 to obtain a 2^{5-1} resolution V plan.

Adding the Complete Foldover

A very appealing alternative option to use when the results of an experiment are confusing, and resources permit, is to add the complete 12-run foldover as another block. Now the projection to a plan with three factors becomes four replicates of a 2^3. Projection to a plan with four factors becomes a complete 2^4 plus the half replicate defined by $\mathcal{I} = ABCD$. The five-factor projections all give essentially equivalent plans. Diamond (1995) uses an argument beyond the level of this book to show that the five-factor projections are of resolution V. This tells us that we can fit a model with the five main effects and their 10 two-factor interactions to the 24 observations. This can be done for any subset of five factors.

16.7.3 16-Run Plan

For the 16-run Hadamard, there are five distinct choices. We give only two. The first, the plan in the Plackett–Burman catalog, is a resolution III plan in the 2^N

Table 16.5 Projections of 12-Run Hadamard Plan into Five Factors

Run	Plan A					Run	Plan B				
1	−	−	−	−	−	1	−	−	−	−	−
2	−	−	−	−	−	2	+	+	+	+	+
3	−	−	+	+	+	3	−	−	+	+	+
4	−	+	−	+	+	4	−	+	+	−	+
5	−	+	+	−	+	5	+	−	−	+	+
6	−	+	+	+	−	6	+	+	−	+	−
7	+	−	−	+	+	7	+	+	+	−	−
8	+	−	+	−	+	8	+	−	+	−	−
9	+	−	+	+	−	9	+	−	−	−	+
10	+	+	−	−	+	10	−	+	−	+	−
11	+	+	−	+	−	11	−	+	−	−	+
12	+	+	+	−	−	12	−	−	+	+	−
13	−	+	−	−	+	13	−	−	−	+	−
14	+	−	−	+	−	14	−	+	+	−	−
13	+	−	−	−	−	13	−	−	+	+	+
14	−	+	−	−	−	14	−	+	+	−	+
15	−	−	+	−	−	15	+	−	−	+	+
16	−	−	−	+	−	16	+	+	−	+	−
17	−	−	−	−	+	17	+	+	+	−	−
18	+	+	+	+	+	18	+	−	−	−	−
						18	−	+	−	−	−
						20	−	−	+	−	−
						21	−	−	−	+	−
						22	−	−	−	−	+

Source: Reprinted with permission from Lin and Draper (1992).

system. It projects to four replicates of the 2^2 if only two factors are active. When three or more factors are active, the projection properties are poor. The alternative, Hall's plan shown in array (16.2), has reasonable projective properties when only three factors are active, provided that the first column is not used. Main effects and all three two-factor interactions are estimable, although correlations run as high as $\pm.5$. If four or more factors are active, the chances of being able to obtain a reasonable analysis are poor.

From general considerations, adding the complete foldover to these plans converts both to resolution IV plans. These project into 2^3 plans when three factors are active.

16.7.4 Larger Hadamard Plans

For larger Hadamard plans, the question of projections onto subsets of columns becomes much more complex because those plans are not unique. In essence, one must investigate the nature of the projection and the points that may need to be added as the need arises. Two good references for further guidance are Lin and Draper (1992) and Box and Tyssedal (1996).

16.8 VERY LARGE FACTORIAL EXPERIMENTS

The techniques that we have looked at in the standard 2^N system are adequate when an experimenter is faced with a modest number of factors, say up to 30. However, there are cases when the number of factors to be considered can range up to 200 or even more. Experimenters will concentrate on factors at just two levels each in such cases. Also in a preliminary experiment at least, they will concentrate on main effects and take their chances with possible interactions. The prominent example that comes to mind here is the design of electronic chips. In the design, the engineer must make many (hundreds) of decisions about how to connect parts. These decisions are typically of a yes–no type, and many will affect the performance of the chip. Standard practice is to simulate the performance of the newly designed chip on a computer before fabrication begins. In this simulation stage, many of the choices made by the designer can be tested.

The fact that there are so many possible combinations to be tested immediately rules out any thought of anything but a resolution III factorial. Because arbitrarily large Hadamard plans exist, they provide the obvious choice for these evaluations. The fact that Hadamard plans provide no measure of experimental error is not relevant. We will examine some ways of constructing large Hadamard matrices. Two short segments of computer code that generate the appropriate plans are available on the companion Web site.

There is no single method for constructing a Hadamard matrix of a particular size. There are a number of techniques or tricks for constructing infinite sets of Hadamard matrices. We present two closely related techniques that yield matrices for almost all the cases in the range $44 \leq n \leq 252$ which can easily be extended to larger n. The two techniques are based on work by Paley (1933) and spelled out in two theorems by Hedayat and Wallis (1978). The first theorem states that a Hadamard matrix of order n, with n a multiple of 4, exists if $n - 1$ is a prime number or a power of a prime. The second states that a Hadamard matrix of order n exists if $(n/2) + 2$ is a multiple of 4 and $(n/2) - 1$ is a prime number or a power of a prime. Both theorems are constructive in the sense that they illustrate the actual construction of the matrix. In both cases, the proof and the construction depend on being able to locate a primitive element in the finite field GF(n). Although not difficult, the proofs are not given here, since they depend on some concepts from Galois field theory. Also, for our purposes, we assume that we deal only with primes, not with powers of primes. This causes us to lose a few cases.

Two program segments, which we will refer to as I and II, are given on the companion Web site. For most cases either one code or the other will work. In a few cases, both codes will work. Hadamard matrices for the remaining cases in the table exist, but other techniques are required to generate them.

CHAPTER 17

General p^N and Nonstandard Factorials

17.1 INTRODUCTION

This chapter is rather special, and those readers with a main interest in using fractional factorial plans may want to omit it on first reading. The purpose of the chapter is twofold. First, we introduce a general system of notation that can be used for symmetric factorials with p levels, where p is either a prime number or the power of a prime number. We actually do very little with the case where p is other than just a prime number, except for one section on the 4^N system. Then we extend some of the methodology to the nonprime 6^N case. The overall purpose is to illustrate general methodology, and in addition, to respond to questions that students commonly ask: why do you ignore factorials with more than three levels? in your course? Why do you do so little with the mixed factorials? Although we do not go into much detail, the careful reader will notice the close connection between the material in this chapter and the Latin squares.

17.2 ORGANIZATION OF THIS CHAPTER

We begin by establishing the notation for the p^N system with p prime. To provide a more familiar example, we continually relate this back to the 3^N system. This is followed by a short section on the case where p is the power of a prime number. We then consider the 4^N, first using Galois fields and then not using Galois fields. The non-Galois field treatment is to illustrate, in detail, the problems with p not prime and some of the peculiarities that can be exploited in some experimental situations. This is followed by a section dealing with the 6^N case. There is no Galois field with six elements. The purpose here is twofold. First, we illustrate

Planning, Construction, and Statistical Analysis of Comparative Experiments,
by Francis G. Giesbrecht and Marcia L. Gumpertz
ISBN 0-471-21395-0

that some of the nice things from the p^N with p prime case simply do not carry over to the nonprime case. Then we illustrate some strange things that happen. Some of these turn out to be quite useful in particular experimental situations. We conclude the chapter with examples from the asymmetrical $2^N 3^M$ system. Other asymmetrical fractions will appear in Chapters 23 and 24, in which we discuss orthogonal arrays, related structures, and factorial plans that can be obtained from them.

17.3 p^N SYSTEM WITH p PRIME

17.3.1 Definitions

The p^N treatment combinations are represented by N-tuples, $i_A i_B \cdots i_N$, where the $\{i_X\}$ are integers ranging from 0 to $p - 1$ and represent the level of the factors. True responses are represented by $\mu_{i_A i_B \cdots i_N}$ and observed responses by $y_{i_A i_B \cdots i_N}$. When convenient, we write these simply as μ_t and y_t, where t is one of the N-tuples, $i_A i_B \cdots i_N$.

The typical effect or interaction term in the model in the p^N system is represented by the symbol $A^{i_A} B^{i_B} \ldots N^{i_N}$, where the $\{i_X\}$ are integers, each ranging from 0 to $p-1$, subject to the restriction that the first nonzero exponent must be 1. The special case with all $\{i_X\}$ zero is replaced by μ. This is the general mean in the model. If we exclude the special μ, there are exactly $(p^N - 1)/(p - 1)$ effects and interactions. Each accounts for $p - 1$ degrees of freedom.

Corresponding to each $i_A i_B \cdots i_N \neq 00 \cdots 0$ there is a set of linear contrasts of the form

$$\left(\frac{1}{p}\right)^{N-1} \left[\sum \cdots \sum\right]_{(0)} \mu_{i_A^* i_B^* \cdots i_N^*} - \text{mean of all responses}$$

$$\vdots \tag{17.1}$$

$$\left(\frac{1}{p}\right)^{N-1} \left[\sum \cdots \sum\right]_{(p-1)} \mu_{i_A^* i_B^* \cdots i_N^*} - \text{mean of all responses,}$$

where the symbol $\left[\sum \cdots \sum\right]_{(k)} \mu_{i_A^* i_B^* \cdots i_N^*}$ represents the sum $\sum_{i_1^*} \cdots \sum_{i_N^*}$ subject to the constraint that $\sum_X i_X^* i_X = k$ (modulo p). These contrasts can also be interpreted as parameters with labels $A^{i_A} B^{i_B} \ldots N^{i_N}{}_{(0)}, \ldots, A^{i_A} B^{i_B} \ldots N^{i_N}{}_{(p-1)}$ and referred to collectively as the set $A^{i_A} B^{i_B} \ldots N^{i_N}$. This notation is fully consistent with the notation used in Section 11.3.

The estimates of the $\{A^{i_A} B^{i_B} \ldots N^{i_N}{}_{(j)}\}$ are computed with the same (17.1) formulas as the contrasts or parameters, except with $\{\mu_t\}$ replaced by $\{y_t\}$. If all observations have independent errors with common variance (homogeneity of variance assumption), the contrasts all have a common variance.

The $A^{i_A} B^{i_B} \cdots N^{i_N}$ sum of squares is computed as

$$\sum_{k=0}^{p-1} \left(\left[\sum \cdots \sum \right]_{(k)} y_{i_A^* i_B^* \cdots i_N^*} \right)^2 \Big/ p^{(N-1)} - (\text{sum of all responses})^2 / p^N.$$

Notice that the inner sums are over all $\{y_t\}$ such that $\sum_X i_X^* i_X = k$ (modulo p) and the outer sum is over $k = 0, \ldots, p - 1$.

While the p contrasts corresponding to $A^{i_A} B^{i_B} \cdots N^{i_N}$ are linearly dependent (they sum to zero), individual contrasts in this set are orthogonal to contrasts in the set $A^{i_A^+} B^{i_B^+} \cdots N^{i_N^+}$, if $i_A i_B \cdots i_N \neq i_A^+ i_B^+ \cdots i_N^+$. This orthogonality implies that contrasts in different sets are uncorrelated under the usual independence and homogeneity of variance of observations assumptions. In addition, if the errors are normally distributed, estimators of contrasts among members of $\{A^{i_A} B^{i_B} \cdots N^{i_N}{}_{(j)}\}$ and $\{A^{i_A^+} B^{i_B^+} \cdots N^{i_N^+}{}_{(j)}\}$ are distributed independently.

Alternatively, we can think of an effect or interaction term $A^{i_A} B^{i_B} \cdots N^{i_N}$ as splitting the responses (either true or observed) into p equal-sized groups, i.e., the treatment combinations $i_A^* i_B^* \cdots i_N^*$ that satisfy $\sum_{X=1}^N i_X i_X^* = 0$ (modulo p), $\sum_{X=1}^N i_X i_X^* = 1$ (modulo p), $\cdots$, and $\sum_{X=1}^N i_X i_X^* = p - 1$ (modulo p). Differences among the p groups account for $p - 1$ degrees of freedom.

17.3.2 Generalized Interactions

A key fact that we will use repeatedly is that one of the properties of the set of numbers (modulo p) where p is a prime number is that for any number $0 < x \leq p - 1$ there exists a unique number $0 < z \leq p - 1$ such that $xz = 1$ (modulo p). This means that there is unique multiplicative inverse. Alternatively, it means that division is defined for any nonzero element.

We will follow standard convention and abbreviate the notation somewhat by using the vector $(i_A i_B \cdots i_N)$ to denote the effect or interaction $A^{i_A} B^{i_B} \cdots N^{i_N}$. Each of these vectors account's for $(p - 1)$ degrees of freedom. We also let $\mathcal{A}$ represent the full set of $(p^N - 1)/(p - 1)$ effect and interaction vectors. At this point we still insist on the usual restriction that in all the vectors $(i_A i_B \cdots i_N)$ in $\mathcal{A}$, the first nonzero i_X symbol is 1. Recall that this simply guarantees uniqueness. The generalized interaction between two distinct effects or interactions in $\mathcal{A}$, say $(i_A i_B \cdots i_N)$ and $(i_A^+ i_B^+ \cdots i_N^+)$, must account for $(p - 1) \times (p - 1)$ degrees of freedom. This interaction consists of $p - 1$ parts, effects or interactions, each with $p - 1$ degrees of freedom.

The $p - 1$ generalized interactions can be written as $(i_A i_B \cdots i_N) + k(i_A^+ i_B^+ \cdots i_N^+)$ for $k = 1, 2, \ldots, p - 1$. In all cases the arithmetic is done term by term, and for the resulting interaction the entire symbol is "fixed" to satisfy the rule that the first nonzero element is equal to 1, if necessary. This may involve doubling, tripling, and so on, and reducing (modulo p), where necessary, the elements of the interaction labels developed.

The implication of the existence of the multiplicative inverse is that should one ever generate a vector symbol for an interaction with the first nonzero element $i_X > 1$, there is always a nonzero number q such that $qi_X = 1$ (modulo p), and the "fix" to get uniqueness in our interaction names is simply to multiply all elements by q and reduce (modulo p), if necessary.

Exercise 17.1: Prime Numbers—Inverses. Seven is a prime number. The set of elements in the field (see Section 17.3.5) order seven is $\{0, 1, 2, 3, 4, 5, 6\}$. Show that if you perform arithmetic (multiplication) on these elements (modulo 7), each value other than 0 has a unique inverse; i.e., for each element there exists another such that the product is 1. □

Exercise 17.2: Prime Numbers—Prime Roots. Show that if you let $x = 3$, work (modulo 7), and compute x^2, x^3, ..., x^6, you generate the set of nonzero elements encountered in Exercise 17.1. Three is a prime root of the Galois field (see Section 17.3.5) with seven elements. Show that two is not a prime root. Can you find another prime root of this field? □

17.3.3 Fractions and Blocking

For blocking and selecting fractions we utilize the fact that the interactions split the set of treatment combinations into subsets. If a p^{-k} fraction or a blocking scheme in p^k blocks is required, then k interactions, $(i_A^{(1)} i_B^{(1)} \cdots i_N^{(1)})$, $(i_A^{(2)} i_B^{(2)} \cdots i_N^{(2)})$, ..., $(i_A^{(k)} i_B^{(k)} \cdots i_N^{(k)})$ must be selected, subject to the rule that none can be the generalized interaction of any of the remainder. These and all their possible generalized interactions form a set of $(p^k - 1)/(p - 1)$ interactions. These are the interactions lost to blocks in a blocking scheme and are used to determine the aliasing relationships in the case of fractional replication. The intrablock subgroup is obtained by finding the p^{N-k} solutions to

$$
\begin{array}{ccccccccl}
1 & & 2 & & \cdots & & N & & \\
\hline
i_A^{(1)} i_A & + & i_B^{(1)} i_B & + & \cdots & + & i_N^{(1)} i_N & = 0 \text{ (modulo } p) \\
i_A^{(2)} i_A & + & i_B^{(2)} i_B & + & \cdots & + & i_N^{(2)} i_N & = 0 \text{ (modulo } p) \\
 & & \vdots & & & & & \\
i_A^{(k)} i_A & + & i_B^{(k)} i_B & + & \cdots & + & i_N^{(k)} i_N & = 0 \text{ (modulo } p).
\end{array}
$$

The p^{N-k} solutions to this system form a group, that is, the set of solutions is closed in the sense that if any two solutions are added, term by term and reduced (modulo p), the result is also in the set. This subgroup and its cosets then define the assignment of treatment combinations to the blocks. The intrablock subgroup or any of its cosets can also be used as a p^{-k} fraction. If this intrablock subgroup or any of its cosets is used as a fraction, the full list of aliases of, for example, the first A main effect or $(10 \cdots 0)$ is obtained by adding $(10 \cdots 0)$, $(20 \cdots 0)$,

$\ldots$, $((p-1)\cdots 0)$ to the symbols for selected interactions and all their possible generalized interactions. Note that we have ignored the question of signs on the list of aliases.

17.3.4 Notation Demonstrated on a 3^4 Example

We now illustrate the notation in a 3^4 to be performed in nine blocks of nine with the condition that all two-factor interactions are to be preserved or as a one-ninth fraction. Confounding the four three-factor interactions AB^2C, AC^2D, ABD^2, and BCD leads to a suitable plan. This set of defining contrasts can be written more compactly as (1210), (1021), (1102), and (0111). Notice that we find the generalized interaction of (1210) and (1021) as $(1210)+(1021) = (2231) \to (1102)$ and $(1210)+2(1021) = (3252) \to (0222) \to (0111)$. It is simple to demonstrate that any pair in the set of four has the remaining pair as generalized interactions. The four interactions formed a closed set, closed in the sense that the generalized interactions of any two in the set consists of the remaining two in the set. Any two could be shown to represent a basis for the complete set.

The intrablock subgroup consists of the nine solutions to the system of equations

$$\begin{array}{ccccccccl} a & & b & & c & & d & & \\ \hline i_A & + & 2i_B & + & i_C & & & = 0 \text{ (modulo 3)} \\ i_A & & & + & 2i_C & + & i_D & = 0 \text{ (modulo 3)} \end{array} \quad (17.2)$$

Notice that only two equations are used in the system. We also note that the intrablock subgroup obtained as the set of solutions to this system is closed in the sense that adding any two members of the subgroup element by element and reducing (modulo 3) produces another member. Since the set includes the 0000 element, which acts as a unit element under addition, the set is actually a group.

The intrablock subgroup can be used as a one-ninth fraction. The aliases of the A main effect are found as

$$A \times (\mathcal{I} = AB^2C = AC^2D = ABD^2 = BCD)$$
$$A^2 \times (\mathcal{I} = AB^2C = AC^2D = ABD^2 = BCD).$$

The full set of aliases is ABC^2, ACD^2, AB^2D, $ABCD$, BC^2, CD^2, BD^2, and $AB^2C^2D^2$. Aliases of B, C, and D are found in a similar manner. We see that the large number of aliases creates problems with fractions of factorials with more than two levels.

Alternatively, one can simply think of one of the interactions, say (1201), as splitting the complete set of 3^4 treatment combinations into three sets and the two interactions (1201) and (1021) together splitting them into nine equal-sized sets.

17.3.5 Extension to p a Power of a Prime

A *field* is a set of elements that satisfies commutativity, associativity, and distributivity for addition and multiplication and contains an identity and an inverse operation. A *Galois field* is a field with a finite number of elements (often called *marks*). Galois fields exist for any prime number and any power of a prime. To avoid going into details of mathematics, we give the bare essentials. We simply say that we have a field with n unique elements. Common notation is GF(p^m) for the field with p^m elements, where p is a prime number. We denote the elements by $\alpha_0, \alpha_1, \ldots, \alpha_{n-1}$. Actually, $\alpha_0 = 0$ and usually $\alpha_{n-1} = 1$. For convenience, we let $\alpha_1 = \alpha$ be a prime root of the field. All of the manipulations illustrated in Sections 17.3.1, 17.3.2, and 17.3.3 can be performed, using the elements $\alpha_0, \alpha_1, \ldots, \alpha_{n-1}$ in place of the integers $0, 1, \ldots, p-1$. The rule that the first nonzero exponent must be 1 leads to some potential confusion. Recall that the rule was arbitrary and served only to provide uniqueness for names and sums of squares. To keep the uniqueness and make the notation look similar, we impose the condition that the first non-α_0 exponent must be α_1. The difficult step in constructing prime power fields is establishing the appropriate addition and multiplication rules. For our examples we provide the necessary addition and multiplication tables, so that $\alpha_i = \alpha^i$ for $i = 1, \ldots, n-1$. *Introduction to the Theory of Groups of Finite Order* by Carmichael (1956) provides a detailed discussion of the construction and properties of Galois fields.

17.4 4^N SYSTEM

We now proceed to look at the 4^N system. In a formal sense we can say that the 4^N system is treated just as the 3^N system. However, some modifications are required since 4 is the power of a prime rather than a prime number. There are two distinct possibilities. The first is elegant and probably most common in textbooks. It requires the use of the Galois field [GF(4)] arithmetic. The disadvantage of the Galois field approach is that it is not sufficiently flexible to deal with the asymmetric situations that we eventually plan to examine. The alternative approach is to modify some of the rules developed up to this point.

17.4.1 4^N Factorials Using Galois Fields

For completeness we illustrate briefly the nature of the computations in the 4^N system using elements from GF(4). Without going into detail, the basic idea is to perform the arithmetic using a system based on four elements, with addition and multiplication defined so that the mathematical requirements for a field are satisfied. In particular, we let the four elements be denoted by α_0, α_1, α_2, and α_3 and define addition and multiplication in a special way by means of Tables 17.1 and 17.2. These tables were constructed by letting $\alpha_0 = 0$, $\alpha_1 = x$, $\alpha_2 = 1 + x$, and $\alpha_3 = 1$ and then reducing all expressions generated [modulo $(1 + x + x^2)$] and (modulo 2).

Table 17.1 Addition Table for GF(2^2)

Addition	α_0	α_1	α_2	α_3
α_0	α_0	α_1	α_2	α_3
α_1	α_1	α_0	α_3	α_2
α_2	α_2	α_3	α_0	α_1
α_3	α_3	α_2	α_1	α_0

Table 17.2 Multiplication Table for GF(2^2)

Multiplication	α_0	α_1	α_2	α_3
α_0	α_0	α_0	α_0	α_0
α_1	α_0	α_2	α_3	α_1
α_2	α_0	α_3	α_1	α_2
α_3	α_0	α_1	α_2	α_3

Details for the 4^2 Case

In the four-level factorial, the individual treatment levels are represented by the values $\alpha_0, \alpha_1, \alpha_2$, and α_3. The 16 treatment combinations in the 4^2 factorial are represented by $\alpha_0\alpha_0, \alpha_0\alpha_1, \ldots, \alpha_3\alpha_3$. The two vectors $(\alpha_1\ \alpha_0)$ and $(\alpha_0\ \alpha_1)$ represent main effects, and the three vectors $(\alpha_1\ \alpha_1)$, $(\alpha_1\ \alpha_2)$, and $(\alpha_1\ \alpha_3)$ represent interactions, i.e., parts of the $A \times B$ interaction. Each of these vectors corresponds to a set of three orthogonal contrasts and represents a set of three degrees of freedom. It is straightforward to verify that contrasts associated with different vectors are orthogonal, but we must use the addition and multiplication tables provided to set up the contrasts. Notice that α_0 corresponds to the symbol "0" and α_3 to the symbol "1" in the 3^N system. However, we see in our labeling of the interaction components that the first non-α_0 symbol is α_1 in each of the vectors. Also notice that α_3 acts as the unit element in multiplication and α_1 and α_2 are the multiplicative inverses of each other, i.e., $\alpha_1 \times \alpha_3 = \alpha_1$, and $\alpha_1 \times \alpha_2 = \alpha_3$.

To illustrate use of the addition and multiplication tables, notice that the parts of the generalized interaction of the two main effects $(\alpha_1\ \alpha_0)$ and $(\alpha_0\ \alpha_1)$ are obtained as

$$\begin{aligned}(\alpha_1\ \alpha_0) + \alpha_1(\alpha_0\ \alpha_1) &= (\alpha_1\ \alpha_0) + (\alpha_0\ \alpha_2)\\ &= (\alpha_1\ \alpha_2),\end{aligned}$$

which corresponds to

$$\begin{aligned}A^{\alpha_1}B^{\alpha_0} \times (A^{\alpha_0}B^{\alpha_1})^{\alpha_1} &= A^{\alpha_1}B^{\alpha_0} \times A^{\alpha_0}B^{\alpha_2}\\ &= A^{\alpha_1}B^{\alpha_2},\\ (\alpha_1\ \alpha_0) + \alpha_2(\alpha_0\ \alpha_1) &= (\alpha_1\ \alpha_0) + (\alpha_0\ \alpha_3)\\ &= (\alpha_1\ \alpha_3),\end{aligned}$$

which corresponds to

$$\begin{aligned} A^{\alpha_1} B^{\alpha_0} \times (A^{\alpha_0} B^{\alpha_1})^{\alpha_2} &= A^{\alpha_1} B^{\alpha_0} \times A^{\alpha_0} B^{\alpha_3} \\ &= A^{\alpha_1} B^{\alpha_3}, \end{aligned}$$

and

$$\begin{aligned} (\alpha_1\ \alpha_0) + \alpha_3(\alpha_0\ \alpha_1) &= (\alpha_1\ \alpha_0) + (\alpha_0\ \alpha_1) \\ &= (\alpha_1\ \alpha_1), \end{aligned}$$

which corresponds to

$$\begin{aligned} A^{\alpha_1} B^{\alpha_0} \times (A^{\alpha_0} B^{\alpha_1})^{\alpha_3} &= A^{\alpha_1} B^{\alpha_0} \times A^{\alpha_0} B^{\alpha_1} \\ &= A^{\alpha_1} B^{\alpha_1}. \end{aligned}$$

Notice in particular that one cannot simply take $A^{\alpha_1} B^{\alpha_0} \times A^{\alpha_0} B^{\alpha_1} \times A^{\alpha_0} B^{\alpha_1}$ and $A^{\alpha_1} B^{\alpha_0} \times A^{\alpha_0} B^{\alpha_1} \times A^{\alpha_0} B^{\alpha_1} \times A^{\alpha_0} B^{\alpha_1}$ in the manner developed for the 3^N and, in fact the p^N system, whenever p was prime.

To actually exhibit the contrasts associated with a specific vector, say $(i_A\ i_B)$, we partition the treatment combinations into four groups on the basis of whether the expression

$$i_A i_A^* + i_B i_B^*$$

is equal to α_0, α_1, α_2, or α_3 where the arithmetic is performed as indicated in the addition and multiplication tables. We must find solutions for i_A^* and i_B^*. The contrasts associated with $(i_A\ i_B)$ are then simply the contrasts among the four groups.

For example, the contrasts associated with $(\alpha_1\ \alpha_3)$ are the contrasts among the four sets of observations $(y_{\alpha_0\alpha_0}, y_{\alpha_1\alpha_2}, y_{\alpha_2\alpha_3}, y_{\alpha_3\alpha_1})$, $(y_{\alpha_0\alpha_1}, y_{\alpha_1\alpha_3}, y_{\alpha_2\alpha_2}, y_{\alpha_3\alpha_0})$, $(y_{\alpha_0\alpha_2}, y_{\alpha_1\alpha_0}, y_{\alpha_2\alpha_1}, y_{\alpha_3\alpha_3})$, and $(y_{\alpha_0\alpha_3}, y_{\alpha_1\alpha_1}, y_{\alpha_2\alpha_0}, y_{\alpha_3\alpha_2})$.

Exercise 17.3: Main-Effect Contrasts in 4 × 4 Factorial. Find the four groups of treatment combinations that are specified by $(\alpha_1\ \alpha_0)$ and show that the contrasts among the corresponding responses are main effect, in fact A contrasts. □

Exercise 17.4: Interaction Contrasts in 4 × 4. Verify that the contrasts among the four groups of responses defined by $(\alpha_1\ \alpha_3)$ in the preceding discussion are interaction contrasts, i.e., verify that they are contrasts in the 4×4 table of responses and are orthogonal to the A and B contrasts. □

If an investigator wishes to put this experiment into four blocks of four and keep both main effects clear, the options are to construct a blocking scheme based on $(\alpha_1\ \alpha_1)$, $(\alpha_1\ \alpha_2)$, or $(\alpha_1\ \alpha_3)$.

4^3 *Case*

The concepts illustrated extend directly to more factors. For example, in the 4^3, the main effects are associated with $(\alpha_1\ \alpha_0\ \alpha_0)$, $(\alpha_0\ \alpha_1\ \alpha_0)$, and $(\alpha_0\ \alpha_0\ \alpha_1)$. The $A \times B$ interaction, the $A \times C$ interaction, and the $B \times C$ interaction have parts $\big((\alpha_1\ \alpha_1\ \alpha_0), (\alpha_1\ \alpha_2\ \alpha_0)$, and $(\alpha_1\ \alpha_3\ \alpha_0)\big)$, $\big((\alpha_1\ \alpha_0\ \alpha_1), (\alpha_1\ \alpha_0\ \alpha_2)$, and $(\alpha_1\ \alpha_0\ \alpha_3)\big)$, and $\big((\alpha_0\ \alpha_1\ \alpha_1), (\alpha_0\ \alpha_1\ \alpha_2)$, and $(\alpha_0\ \alpha_1\ \alpha_3)\big)$, respectively. The $A \times B \times C$ interaction has nine pieces, $(\alpha_1\ \alpha_1\ \alpha_1)$, $(\alpha_1\ \alpha_1\ \alpha_2)$, $(\alpha_1\ \alpha_1\ \alpha_3)$, $(\alpha_1\ \alpha_2\ \alpha_1)$, $(\alpha_1\ \alpha_2\ \alpha_2)$, $(\alpha_1\ \alpha_2\ \alpha_3)$, $(\alpha_1\ \alpha_3\ \alpha_1)$, $(\alpha_1\ \alpha_3\ \alpha_2)$, and $(\alpha_1\ \alpha_3\ \alpha_3)$. It is possible to run the 4^3 factorial in 16 blocks of four by confounding two of the interaction components, but there are problems. For example, if $(\alpha_1\ \alpha_1\ \alpha_1)$ and $(\alpha_1\ \alpha_1\ \alpha_2)$ are confounded with blocks, the calculations

$$
\begin{aligned}
(\alpha_1\ \alpha_1\ \alpha_1) + \alpha_1(\alpha_1\ \alpha_1\ \alpha_2) &= (\alpha_1\ \alpha_1\ \alpha_1) + (\alpha_2\ \alpha_2\ \alpha_3)\\
&= (\alpha_3\ \alpha_3\ \alpha_2)\\
&\rightarrow (\alpha_1\ \alpha_1\ \alpha_3)\\
(\alpha_1\ \alpha_1\ \alpha_1) + \alpha_2(\alpha_1\ \alpha_1\ \alpha_2) &= (\alpha_1\ \alpha_1\ \alpha_1) + (\alpha_3\ \alpha_3\ \alpha_1)\\
&= (\alpha_2\ \alpha_2\ \alpha_0)\\
&\rightarrow (\alpha_1\ \alpha_1\ \alpha_0)\\
(\alpha_1\ \alpha_1\ \alpha_1) + \alpha_3(\alpha_1\ \alpha_1\ \alpha_2) &= (\alpha_1\ \alpha_1\ \alpha_1) + (\alpha_1\ \alpha_1\ \alpha_2)\\
&= (\alpha_0\ \alpha_0\ \alpha_3)\\
&\rightarrow (\alpha_0\ \alpha_0\ \alpha_1)
\end{aligned}
$$

show that the generalized interactions confounded with blocks are $(\alpha_0\ \alpha_0\ \alpha_1)$, which is the C main effect, $(\alpha_1\ \alpha_1\ \alpha_0)$, which is part of the $A \times B$ interaction, and the $(\alpha_1\ \alpha_1\ \alpha_3)$ component from the $A \times B \times C$ interaction.

If confounding a main effect is not acceptable, an alternative is to confound $(\alpha_1\ \alpha_1\ \alpha_1)$ and $(\alpha_1\ \alpha_2\ \alpha_3)$. This choice confounds a part of each of the three two-factor interactions. The intrablock subgroup for this case is obtained by solving the analog to equation (17.2) by finding the four solutions to the system

$$
\begin{aligned}
\alpha_1 i_A^* + \alpha_1 i_B^* + \alpha_1 i_C^* &= \alpha_0\\
\alpha_1 i_A^* + \alpha_2 i_B^* + \alpha_3 i_C^* &= \alpha_0
\end{aligned}
$$

using the arithmetic defined in Tables 17.1 and 17.2. The intrablock subgroup consists of $(\alpha_0\ \alpha_0\ \alpha_0)$, $(\alpha_1\ \alpha_2\ \alpha_3)$, $(\alpha_2\ \alpha_3\ \alpha_1)$, and $(\alpha_3\ \alpha_1\ \alpha_2)$. Notice that we are dealing with a homogeneous system of equations and that the first solution is obvious, the second and third must be found, and the fourth is the sum of the second and third. As usual, the remaining blocks are obtained by adding, componentwise, to the treatment combinations in the intrablock subgroup, any

```
proc iml ;
* Note, all subscripts are 1-4 rather than 0-3. ;
* a = addition table
 m = multiplication table;
 a = {1 2 3 4 , 2 1 4 3 , 3 4 1 2 , 4 3 2 1 };
 m = {1 1 1 1 , 1 3 4 2 , 1 4 2 3 , 1 2 3 4 };
 block = 0 ;
 do ista = 1 to 4 ;  do istb = 1 to 4 ;
   block = block + 1 ;  print ista istb block ;
   do ia = 1 to 4 ;  do ib = 1 to 4 ;
     do ic = 1 to 4 ;
      if (a[a[m[2,ia],m[2,ib]],m[2,ic]]=ista) then
      if (a[a[m[2,ia],m[3,ib]],m[4,ic]]=istb) then
         print ia ib ic ;
     end; end; end; end; end;
```

Display 17.1 Generate 16 blocks of four for 4^3 factorial.

treatment combination that has not yet appeared. The addition must be according to Table 17.1.

To see the confounding pattern, we compute

$$(\alpha_1\alpha_1\alpha_1) + \alpha_1(\alpha_1\alpha_2\alpha_3) \rightarrow (\alpha_1\alpha_3\alpha_0),$$

$$(\alpha_1\alpha_1\alpha_1) + \alpha_2(\alpha_1\alpha_2\alpha_3) \rightarrow (\alpha_1\alpha_0\alpha_2),$$

$$(\alpha_1\alpha_1\alpha_1) + \alpha_3(\alpha_1\alpha_2\alpha_3) \rightarrow (\alpha_0\alpha_1\alpha_2).$$

Parts of each of $A \times B$, $A \times C$, and $B \times C$ are confounded with blocks. All main-effect degrees of freedom are clear.

It can be shown that one cannot do any better than these confounding choices for 16 blocks of four in the 4^3. Other choices sacrifice either a main effect or parts of all of the two-factor interactions. Also, recall that in the 3^N system we were concerned that individual parts of, say, the two-factor interaction could not be interpreted separately. This same problem appears here. There is no clean interpretation for the separate interaction pieces.

The segment of SAS® code in Display 17.1 generates the 16 blocks of four for this confounding scheme. It is also reproduced on the companion Web site.

***Exercise 17.5: Generalized Interactions in the* 4^3.** Verify that the generalized interaction of $(\alpha_1\ \alpha_1\ \alpha_1)$ and $(\alpha_1\ \alpha_2\ \alpha_3)$ consists of $(\alpha_0\ \alpha_1\ \alpha_3)$, $(\alpha_1\ \alpha_3\ \alpha_0)$, and $(\alpha_1\ \alpha_0\ \alpha_2)$. □

17.4.2 4^N System Not Using Galois Fields

We now proceed to look at the same case but without the introduction of the GF(4) arithmetic. Instead of using the special arithmetic and elements from the four-element GF(4), we simply perform standard arithmetic and reduce (modulo 4). This is in anticipation of the important cases coming where the option of using finite field arithmetic is not available because the requisite finite fields do not exist. However, as mentioned previously, some modifications are necessary. First, it turns out that we need to replace our set of effects and interactions, denoted by $\mathcal{A}$, by a new set which we denote by $\mathcal{A}^*$. In particular, we need to relax the restriction that the first non-a_0 (nonzero) element in each vector be equal to a_1 (1). This leads to some uniqueness problems, which will need to be addressed when they arise. The set $\mathcal{A}^*$ associated with the 4^2 experiment, for example, now consists of 15 vectors. Recall that the vector of coefficients (00) is part of neither the set $\mathcal{A}$ nor $\mathcal{A}^*$ and will be ignored whenever it is generated. We examine the vectors in $\mathcal{A}^*$ in some detail. It turns out that in contrast to the p^N with p prime, the sets of contrasts associated with different individual vectors need not necessarily be disjoint and also that the sets do not need to be homogeneous. By *not being homogeneous*, we mean that, for example, some sets will consist of some interaction contrasts and some main-effect contrasts. *Not disjoint* means that two sets will have common contrasts.

4^2 *Factorial*

We examine the 15 vectors in $\mathcal{A}^*$ in some detail. Consider first the vector (10). This vector will partition the treatment combinations into four sets, and orthogonal contrasts among the sets of responses are the main-effect contrasts for factor A. The sets satisfy $i_A + 0 \times i_B = 0$ (modulo 4), $i_A + 0 \times i_B = 1$ (modulo 4), $i_A + 0 \times i_B = 2$ (modulo 4), and $i_A + 0 \times i_B = 3$ (modulo 4). In words, the first set consists of all treatment combinations with A at the first level, ..., and the fourth set consists of all treatment combinations with A at the fourth level. Differences among these sets account for three main-effect degrees of freedom. We also look at (30). The sets satisfy $3 \times i_A + 0 \times i_B = 0$ (modulo 4), ..., $3 \times i_A + 0 \times i_B = 3$ (modulo 4). It is a simple exercise to show that using (30) only reorders the sets of treatment combinations defined by (10) and still leads to the same set of main-effect contrasts and the same degrees of freedom. However, (20) splits the treatment combinations into only two sets, i.e., accounts for only one A main-effect degree of freedom. This is part of the (10) or (30) set. This is what was meant in the preceding paragraph by the comment on being disjoint. The key point to notice is that 3 is not a divisor of 4, but 2 is a divisor of 4. This sort of thing does not happen when dealing with prime numbers. A similar series of comments applies to vectors (01), (02), and (03).

Next consider interactions. Here matters become even more complex. First, (11) and (33) define the same grouping of treatment combinations and consequently are associated with the same set of contrasts, the same set of three $A \times B$ interaction degrees of freedom. The vector (22) splits the treatment combinations into two sets, i.e., accounts for one interaction degree of freedom. It is easy to show that this degree of freedom is one of the three associated with (11) and (33).

The vector (12) in $\mathcal{A}^*$ also splits the 16 treatment combinations into four groups and hence accounts for three degrees of freedom. However, this set is not homogeneous. Notice that multiplying by 2 and reducing (modulo 4) collapses the split into two sets of eight, and the contrast between the responses in these sets is in fact an A main-effect contrast. It follows that the three degrees of freedom associated with (12) consist of two $A \times B$ interaction degrees of freedom and one A main-effect degree of freedom. Multiplying (12) by 3 and reducing (modulo 4) gives us (32) and simply reorders sets. The two vectors (12) and (32) are associated with the same set of degrees of freedom. It is straightforward to show that the contrasts (degrees of freedom) associated with (12) and (32) are orthogonal to the contrasts (degrees of freedom) associated with (11) and (33). It is often convenient to simply talk about orthogonal sets of degrees of freedom when we are really referring to orthogonal sets of contrasts.

Next, look at the contrasts associated with (13) and (31). Since $3 \times (13) \to (31)$, these just provide alternative arrangements of sets of responses and define the same set of contrasts and three degrees of freedom. However, since multiplication by 2 and reducing (modulo 4) in both cases leads to (22), which defines one contrast, it follows that one degree of freedom from the set defined by either (13) or (31) is common with the set defined by (11) and (33).

To summarize, examining all vectors in $\mathcal{A}^*$ located three degrees of freedom for A, three degrees of freedom for B, and $3 + 2 + 2 + (3 - 1) = 9$ degrees of freedom for interaction. One can define a blocking system for four blocks of four, keeping all main effects clear by using the sets defined by either (11) or (33). Also, one can use (22) to set up a system for two blocks of eight that keeps main-effect contrasts clear of block effects.

Finally, there is the question of generalized interactions. Consider the generalized interaction of (11) and (13). We know that (11) is equivalent to (33) and contains (22) as a subset. Similarly, (13) is equivalent to (31) and also contains (22) as a subset. In a sense, the two sets (11) and (13) are not completely clear of one another, and one should expect their generalized interaction to have fewer than nine degrees of freedom. A systematic approach is to make up the table, assigning all variations on (11) to columns and the variations on (13) to rows:

	(11)	(22)	(33)
(13)	·	·	·
(22)	·	·	·
(31)	·	·	·

and proceed to fill in the symbols for the generalized interactions. Using the approach that worked in the 2^N and 3^N cases, we add element by element [and reduce (modulo 4) if necessary]. We combine (13) with (11) and obtain (20), combine (13) with (22), and obtain (31), and so on. This produces the table

	(11)	(22)	(33)
(13)	(20)	(31)	(02)
(22)	(33)	(00)	(11)
(31)	(02)	(13)	(20)

Table 17.3 Contrasts Associated with (11) and (13) and the Interaction

Trt. Comb.	(11) Contrasts	(13) Contrasts	Interaction Contrasts
00	+ + +	+ + +	+ + + + + + + + +
01	+ − −	− − +	− + + − + + + − −
02	− + −	− + −	+ − + − + − + − +
03	− − +	+ − −	− − + + + − + + −
10	+ − −	+ − −	+ − − − + + − + +
11	− + −	+ + +	− + − − + − − + −
12	− − +	− − +	+ + − + + − − − +
13	+ + +	− + −	− − − + + + − − −
20	− + −	− + −	+ − + − + − + − +
21	− − +	+ − −	− − + + + − + + −
22	+ + +	+ + +	+ + + + + + + + +
23	+ − −	− − +	− + + − + + + − −
30	− − +	− − +	+ + − + + − − − +
31	+ + +	− + −	− − − + + + − − −
32	+ − −	+ − −	+ − − − + + − + +
33	− + −	+ + +	− + − − + − − + −

Notice that (20) and (02) both appear twice in the table. Each accounts for one main-effect degree of freedom. The (00) is ignored. The remaining four interactions, in the table, (31), (13), (33) and (11) are repeats of entries in row and column headings and contribute nothing new. It follows that the generalized interaction of (11) and (13) has two degrees of freedom, one coming from A and the other from B.

To demonstrate the correctness of this assertion, we examine the partitions among treatment combinations induced by using both $i_A + i_B$ and $i_A + 3i_B$ where both expressions are reduced (modulo 4) when necessary. Both of these lead to three linear orthogonal contrasts. Such contrasts are given in columns 2 to 4 and 5 to 7, respectively, in Table 17.3. They are arbitrary, in the sense that any other contrasts could be used in their place. Columns 8 to 16 contain the interaction contrasts obtained by multiplying coefficients from appropriate 2 to 7 columns. Inspection of the last nine columns, the interaction columns of Table 17.3, reveals a B main-effect contrast in the first and last interaction columns and an A main-effect contrast in the third and seventh interaction columns and four columns duplicating contrasts found in (11) and (13). The nine interaction columns account for only two degrees of freedom beyond those accounted for by (11) and (13), and these are both main effects.

The actual block assignments based on (11) and (13) can be determined by finding the solutions to the system

$$
\begin{array}{ccccl}
a & & b & & \\
\hline
i_A & + & i_B & = 0 \text{ (modulo 4)} \\
i_A & + & 3i_B & = 0 \text{ (modulo 4).}
\end{array}
$$

The intrablock subgroup is $(00,\ 22)$. The remaining seven blocks are obtained in the usual manner by symbolically adding treatment combinations that have not appeared previously.

We can also determine the block assignment from Table 17.3 by looking at the $+/-$ patterns in the columns corresponding to (11) and (13). Treatment combinations that have similar patterns go to common blocks. There are eight different $+/-$ patterns. The first block consists of 00 and 22. These have $+$ values in all six columns, corresponding to the (11) and (13) contrasts. Another block consists of 13 and 31. These both have "$+\ +\ +\ -\ +\ -$" in the six columns corresponding to (11) and (13) contrasts. The remaining blocks are (10, 32), (01, 23), (11, 33), (02, 20), (03, 21), and (12, 30).

At first glance this plan seems to have little merit, since one degree of freedom from A and one from B are confounded with blocks. However, at this point, the assignment of actual levels of the factors to the symbols 0, 1, 2, 3 in the experiment is arbitrary. If the factors are both quantitative and we assign equally spaced levels to the symbols, $0 \Leftrightarrow$ lowest level, $1 \Leftrightarrow$ second level, $2 \Leftrightarrow$ next level, and $3 \Leftrightarrow$ highest level for both factors, we find that full information on the quadratic response components of both main-effect factors is preserved, and only a small amount of the information on the linear contrast is sacrificed. If, on the other hand, the experimenter assigns the equally spaced levels to the symbols, $0 \Leftrightarrow$ lowest level, $1 \Leftrightarrow$ second level, $3 \Leftrightarrow$ next level, and $2 \Leftrightarrow$ highest level for both factors, full information is retained on the linear components of both main effects and the quadratic effect is lost. The quadratic effect is fully confounded with blocks. Both of these plans, not available if one uses the special Galois field arithmetic, are useful in some situations. Plans of this type are discussed in more detail in Chapter 20.

We must, however, point out that since we are now using $\mathcal{A}^*$, which relaxes the side condition that the first nonzero element in the vectors defining effects and interactions be 1 and allows (22) as a valid member, we find that block number $2i_A + 2i_B$ (modulo 4) defines a blocking scheme with two blocks of eight and with main effects clear of block effects. This is a plan that is not available to us when we use GF(4) arithmetic, unless blocks are combined.

Exercise 17.6: Partitioning a 4 × 4 Factorial. Use the vector (10) to partition the 16 treatment combinations in the 4^2 factorial into four sets of four by finding the solutions to

$$\begin{array}{cc} a & b \\ \hline i_A & \end{array} = 0 \text{ (modulo 4)}$$

$$\begin{array}{cc} a & b \\ \hline i_A & \end{array} = 1 \text{ (modulo 4)}$$

$$\begin{array}{cc} a & b \\ \hline i_A & \end{array} = 2 \text{ (modulo 4)}$$

$$\begin{array}{cc} a & b \\ \hline i_A & \end{array} = 3 \text{ (modulo 4)}. \qquad \square$$

***Exercise 17.7: Partitioning the* 4 x 4 *Another Way*.** Repeat Exercise 17.6 with vector (20). Notice now that you can find eight solutions to

$$\frac{a \quad b}{2i_A \quad\quad} = 0 \text{ (modulo 4)}$$

$$\frac{a \quad b}{2i_A \quad\quad} = 2 \text{ (modulo 4)}$$

and no solutions to

$$\frac{a \quad b}{2i_A \quad\quad} = 1 \text{ (modulo 4)}$$

$$\frac{a \quad b}{2i_A \quad\quad} = 3 \text{ (modulo 4)}$$

□

***Exercise 17.8: Arithmetic for* 4 x 4.** To illustrate the problem with non-prime numbers, construct a plan for a 4^2 experiment in four blocks of four by confounding the AB interaction. The simple way to proceed is to generate the block assignment for each of the 16 treatment combinations denoted by the 16 possible values of $i_A i_B$ by using $i_A + 2i_B$ (modulo 4) to generate block assignments. This is confounding the interaction corresponding to (12) in the notation developed above. Examination of the block assignment reveals that the intra-block subgroup and one of the other blocks contain only treatment combinations that involve $i_A = 0$ and $i_A = 2$, while the remaining two blocks contain only treatment combinations that involve $i_A = 1$ and $i_A = 3$. One of the main-effect A contrasts is confounded with blocks. In the notation developed in this chapter, there is a contrast that is common to the sets corresponding to (12) and (10).

Repeat the exercise, but use $i_A + i_B$ (modulo 4) and then $i_A + 3i_B$ (modulo 4) to define the blocking. Now it can easily be shown that both main effects are free of block effects. We know from our study of the Latin square designs that it is possible to construct a set of three mutually orthogonal 4×4 Latin squares. It follows that it must be possible to find three distinct blocking schemes that do not confound any part of the main effects. Since four is a power of a prime, the problem we have encountered can easily be fixed by using elements from the four-element finite field known as GF(4). The object of this exercise is to illustrate that the methodology being developed can get into trouble if one uses conventional arithmetic and is not dealing with factors with levels that are prime numbers. Further examples will illustrate cases where one encounters problems and does not have the option of resorting to a modified arithmetic. □

***Exercise 17.9:* 4 x 4 *in SAS*.** Use SAS® to provide an easy demonstration of the points illustrated in Exercise 17.8. Use the following code to generate a data set with 16 observations:

```
data one ;
 do f1 = 0 to 3 ;  do f2 = 0 to 3 ;
  block = mod((f1+2*f2),4) ;
  y = normal(0) ;  output ;
  end ;  end ;
```

Next, use the following code to perform the analysis. Note that the "Y" variable is generated by simulation as a trick to get `proc GLM` to run.

```
proc glm data=one ;   class f1 f2 block ;
 model y = block f1 f2 f1*f2 /e1 ;
proc glm data=one ;   class f1 f2 block ;
 model y = f1 f2 f1*f2 block /e1 ;  run ;
```

□

4^3 *Factorial*

Example 17.1. The extension to more factors is direct. In the 4^3 case we have 64 treatment combinations and $\mathcal{A}^*$ with 63 vectors. Of these vectors, 27 are associated with parts of the $A \times B \times C$ interaction. Some are associated with three degrees of freedom, others with a single degree of freedom. Generalized interactions are computed as in the 4^2 case. For example, both (111) and (133) have three degrees of freedom. However, they are not disjoint. To observe this, we split the 64 treatment combinations into four sets of 16 by using $i_A + i_B + i_C = k$ (modulo 4) for $k = 0, 1, 2, 3$ and again into four sets using $i_A + 3i_B + 3i_C = k$ (modulo 4) for $k = 0, 1, 2, 3$. Then observe that one of the contrasts between groups in both cases is exactly the contrast between the two groups obtained by using $2i_A + 2i_B + 2i_C = k$ (modulo 4) for $k = 0, 2$.

To compute the generalized interaction, we first note that $2 \times (111) \rightarrow (222)$, $3 \times (111) \rightarrow (333)$, $2 \times (133) \rightarrow (222)$, and $3 \times (133) \rightarrow (311)$, and then construct the table

	(133)	(222)	(311)
(111)	(200)	(333)	(022)
(222)	(311)	(000)	(133)
(333)	(022)	(111)	(200)

Entries in the table are obtained by adding the row and column headings, term by term, and always reducing (modulo 4). Now examine the entries in this table to see the generalized interactions. First, note that (000) contributes nothing. Then the (200) entry (which appears twice) tells us that there is one degree of freedom from A. The (022) entry (which also appears twice) tells us that there is one degree of freedom from $B \times C$. The (333), (311), (133), and (111) entries are already in the row and column headers and therefore add nothing. The total split based on (111) and (133) accounts for $(3+3-1)+1+1 = 7$ degrees of freedom. The split defines eight blocks of eight.

To obtain the actual blocking assignments based on (111) and (133), we need to find the intrablock subgroup by finding the eight solutions to

$$\begin{array}{ccccc} a & & b & & c \\ \hline i_A & + & i_B & + & i_C \\ i_A & + & 3i_B & + & 3i_C \end{array} \begin{array}{l} \\ = 0 \text{ (modulo 4)} \\ = 0 \text{ (modulo 4).} \end{array}$$

The intrablock subgroup is

000	211
220	031
202	013
022	233.

The remaining seven blocks are generated in the usual manner.

Five degrees of freedom from $A \times B \times C$ and one degree of freedom from A and one from the $B \times C$ interaction are confounded with blocks. Again, if A is continuous and levels are properly assigned to symbols, one can have a plan that allows one to estimate both linear and quadratic A effects (the former with somewhat reduced efficiency), or one can have a plan that provides full efficiency on the linear contrast and no estimate of the quadratic effect.

Exercise 17.10: Confounding a 4^3 Factorial. Find the intrablock subgroup for the plan based on confounding (111) and (133). □

Example 17.2. Consider a blocking scheme based on (111) and (123). These correspond to two mutually orthogonal sets of degrees of freedom. This will generate 16 blocks of four. To find the generalized interaction, we examine the following table:

	(123)	(202)	(321)
(111)	(230)	(313)	(032)
(222)	(301)	(020)	(103)
(333)	(012)	(131)	(210)

The row headings consist of (111) and the two alternative forms $2 \times (111)$ and $3 \times (111)$. The column headings include (123) and the two alternative forms obtained as $2 \times (123)$ (modulo 4) and $3 \times (123)$ (modulo 4). Notice that one contrast from the $A \times C$ interaction is confounded via the column headings. On examining the body of the table, we immediately find that part of the main effect B, part of the $A \times B$, part of the $A \times C$, part of the $B \times C$, and part of the $A \times B \times C$ interactions form the generalized interaction of (111) and (123).

We look at the pieces in detail:

(a) (020) accounts for one degree of freedom from B

(b) (230) and (210) account for only two degrees of freedom from $A \times B$ since the (020) degree of freedom is a subset. Note that $3 \times (230) \rightarrow (210)$.

(c) (032) and (012) account for only two degrees of freedom from $B \times C$ since the (020) degree of freedom is a subset. Note that $3 \times (032) \rightarrow (012)$.

(d) (301) and (103) account for only two additional degrees of freedom from $A \times C$, since (202) in the columns is a subset.

(e) (313) and (131) account for three degrees of freedom from $A \times B \times C$. However, one degree of freedom associated with (222) in the row headings is a subset.

The row and column headings account for six degrees of freedom and the generalized interaction for nine more, for a total of 15. It follows that the confounding scheme will yield 16 blocks of four. The intrablock subgroup is obtained by solving

$$\begin{array}{ccccccl} a & & b & & c & & \\ \hline i_A & + & 3i_B & + & 3i_C & & = 0 \text{ (modulo 4)} \\ i_A & + & i_B & + & i_C & & = 0 \text{ (modulo 4).} \end{array}$$

This plan can be compared with the two plans available for 4^3 in 16 blocks of four when GF(4) arithmetic is used, where only two-factor interaction degrees of freedom were lost. In this plan we lose the two-factor interactions and one degree of freedom from B.

Example 17.3. As an alternative plan, we can consider confounding both (111) and (113) with blocks. This gives the table

	(113)	(222)	(331)
(111)	(220)	(333)	(002)
(222)	(331)	(000)	(113)
(333)	(002)	(111)	(220)

We notice immediately that the contrasts corresponding to (111) and (113) are not fully orthogonal. They have one common contrast, one common degree of freedom. In the body of the table we find (002) and (220), both of which account for single degrees of freedom. All of the remaining vectors are duplicates of vectors in column or row headings. We account for a total of seven degrees of freedom among block differences. This tells us that we have a plan for 4^3 in eight blocks of eight. This plan confounds part of the C main effect and part of the $A \times B$ interaction as well as contrasts from the three-factor interaction with blocks. This plan could have some appeal, especially if the C factor is continuous and one can select treatment levels.

An alternative plan that could have appeal in some cases would be three replicates, 24 blocks of eight with partial confounding. A reasonable scheme would confound (111) and (113) in the first replicate, (111) and (131) in the second, and (111) and (311) in the third. This would make all main effects and two-factor interactions estimable.

17.5 4^N SYSTEM USING PSEUDOFACTORS AT TWO LEVELS

We have developed the 4^N system using GF(4) arithmetic and using standard (modulo 4) arithmetic. An alternative method that is messy but has many practical applications is again to use the fact that four is 2 squared, and base the development on pseudofactors in the 2^N system.

The idea of using pseudofactors in the 2^N system is simple. Consider two factors, each at two levels. Together they provide four combinations. Now make the association

two factors each at two levels		one factor at four levels
0 0		0
0 1	$\Rightarrow$	1
1 0		2
1 1		3.

In fact, we can make any association between levels of the pseudofactors and the four-level factor we choose, provided that we are consistent throughout the entire plan. The two main-effect degrees of freedom, one from each pseudofactor and one degree of freedom from the their interaction combine to form the three main-effect degrees of freedom for the four-level factor.

17.5.1 4^2 Factorial

We illustrate the technique with two factors, each at four levels. For convenience, we label the two-level factors $A1$, $A2$, $B1$, and $B2$. This is the standard 2^4. Now combine the two factors $A1$ and $A2$ into the four-level factor A and $B1$ and $B2$ into the four-level factor B using the rules illustrated in the preceding paragraph.

Formal structures, such as blocking are set up in the 2^N system. The key point to remember is that $A1A2$, which looks like a two-factor interaction, is really a main effect, and $A1A2B1$ and $A1A2B1B2$, which look like three- and four-factor interactions, are really parts of the two-factor AB interaction. The method is somewhat messy but is extremely powerful since it rests on the very flexible 2^N system. For example, it is possible to conduct a 4^2 study in 12 blocks of four by partial confounding. Confound $A1B1$, $A2B2$, and $A1A2B1B2$ in one replicate, $A1B2$, $A1B1B2$, and $A1A2B1$ in a second replicate, and $A2B1$, $A1B1B2$, and

$A1A2B2$ in a third replicate. Two-thirds information is available on all two-factor degrees of freedom. Another possibility is in blocks of eight.

17.5.2 4^3 Factorial

For a 4^3 plan we begin with six two-level factors, $A1$, $A2$, $B1$, $B2$, $C1$, and $C2$. We combine these to form the three factors, A, B, and C. For any pair of two-level factors being combined, we must be consistent in the association throughout the entire plan. However, different associations can be used for different factors. For example, for $A1$ and $A2$ we can use the association $0\ 0 \Rightarrow 0$, $0\ 1 \Rightarrow 1$, $1\ 0 \Rightarrow 2$, and $1\ 1 \Rightarrow 3$, while for $B1$ and $B2$ we use $0\ 0 \Rightarrow 0$, $1\ 0 \Rightarrow 1$, $0\ 1 \Rightarrow 2$, and $1\ 1 \Rightarrow 3$. The only requirement is consistency throughout the plan.

As an example of the flexibility provided by the method, we can consider placing the 64 treatment combinations into four blocks of 16 by blocking on part of the ABC interaction (pseudofactor interactions $A1B1C1$, $A2B2C2$, and the generalized interaction $A1A2B1B2C1C2$). This uses three of the 27 degrees of freedom for ABC. An alternative blocking configuration that may be of interest at times is eight blocks of eight, obtained by confounding $A1B1C2$, $A2B1C1$, $A1A2C1C2$, $A1B2C1C2$, $B1B2C1$, $A1A2B1B2C2$, and $A2B2$. This sacrifices four degrees of freedom from the three-factor interaction and one from each of the two-factor interactions. This can be followed by a second replicate that confounds $A1B1C1$, $A2B2C2$, $A1A2B1B2C1C2$, $A2B1C1C2$, $A1A2C2$, $B1B2C1$, and $A1B2$. At this stage partial information is available on all parts of all the interactions. This is a plan not available in the GF(4)-based system.

In general, it is not possible to attribute special meaning to individual degrees of freedom in the interactions of four-level factors. There is no reason to select, say, $A1A2B1B2C1C2$ as a candidate for sacrifice over $A1B2C2$. If the four levels are continuous and equally spaced, however, we will eventually see that special interpretations are possible. We examine this in Chapter 20.

An additional feature in the use of pseudofactors that should not be overlooked is the ease of combining in one plan factors with four levels with factors at two levels. The only problem (minor inconvenience) is that when looking at the pseudofactor structure, one needs care to recognize contrasts or degrees of freedom that become part of main effects, become parts of two-factor interactions, and so on.

17.5.3 One-Fourth Fraction of the 4^3 Factorial: Comparison of the Three Methods

Finding suitable fractions to preserve main effects is very straightforward in either the GF(4), the simple (modulo 4) approach or the pseudofactor approach. Consider constructing a one-fourth replicate of the 4^3. In the GF(4) approach, one would select, say, $(a_1a_2a_3)$ as the defining contrast. The major consideration here is that we do not want α_0 in any of the three positions. Remember that α_0 acts like (is) a zero. The desired treatment combinations in the intrablock subgroup are the 16 solutions $i_A i_B i_C$ to the equation

$$a_1 i_A + a_2 i_B + a_3 i_C = \alpha_0,$$

where all the arithmetic operations are performed using Tables 17.1 and 17.2. The aliasing relationships are computed as in the prime-level case. For example, the aliases of A, when using $(\alpha_1\ \alpha_2\ \alpha_3)$ as the defining contrast, are computed as follows:

$$\begin{aligned}
A^{\alpha_1} B^{\alpha_0} C^{\alpha_0} \times (A^{\alpha_1} B^{\alpha_2} C^{\alpha_3})^{\alpha_1} &= A^{\alpha_1} B^{\alpha_0} C^{\alpha_0} \times A^{\alpha_2} B^{\alpha_3} C^{\alpha_1} \\
&= A^{\alpha_3} B^{\alpha_3} C^{\alpha_1} \\
&\rightarrow A^{\alpha_1} B^{\alpha_1} C^{\alpha_2} \\
A^{\alpha_1} B^{\alpha_0} C^{\alpha_0} \times (A^{\alpha_1} B^{\alpha_2} C^{\alpha_3})^{\alpha_2} &= A^{\alpha_1} B^{\alpha_0} C^{\alpha_0} \times A^{\alpha_3} B^{\alpha_1} C^{\alpha_2} \\
&= A^{\alpha_2} B^{\alpha_1} C^{\alpha_2} \\
&\rightarrow A^{\alpha_1} B^{\alpha_3} C^{\alpha_1} \\
A^{\alpha_1} B^{\alpha_0} C^{\alpha_0} \times (A^{\alpha_1} B^{\alpha_2} C^{\alpha_3})^{\alpha_3} &= A^{\alpha_1} B^{\alpha_0} C^{\alpha_0} \times A^{\alpha_1} B^{\alpha_2} C^{\alpha_3} \\
&= A^{\alpha_0} B^{\alpha_2} C^{\alpha_3} \\
&\rightarrow A^{\alpha_0} B^{\alpha_1} C^{\alpha_2}.
\end{aligned}$$

We conclude that A is aliased with part of the $B \times C$ interaction and parts of the $A \times B \times C$ interaction. Similarly, it can be shown that B is aliased with part of the $A \times C$ interaction and parts of the $A \times B \times C$ interaction, and C is aliased with part of the $A \times B$ interaction and parts of the $A \times B \times C$ interaction.

Now using the (modulo 4) arithmetic, consider the fraction defined by (113). Again, the actual treatment combinations are found as the 16 solutions to

$$i_A + i_B + 3i_C = 0 \text{ (modulo 4).}$$

The aliases of A are found by examining the table

	(113)	(222)	(331)
(100)	(213)	(322)	(031)
(200)	(313)	(022)	(131)
(300)	(013)	(122)	(231)

A is aliased with part of the $B \times C$ interaction and parts of the $A \times B \times C$ interaction. Similarly, B is aliased with part of the $A \times C$ interaction and parts of the $A \times B \times C$ interaction, and C is aliased with part of the $A \times B$ interaction and parts of the $A \times B \times C$ interaction.

Notice that we must exercise some care in selection of the defining contrast. For example, selecting (123) to define the fraction leads to a plan with a contrast from A being aliased with a contrast from C.

In the pseudofactor approach, care must be exercised to avoid aliasing main-effects contrasts with other main-effects contrasts. For example, $A1A2$ and $B1B2$ both look like two-factor interactions but really are main effects. A 4^{3-1} fraction can be obtained via the defining contrast

$$\mathcal{I} = A1B1B2C1 = A2B1C1C2 = A1A2B2C2.$$

In this fraction, the $B1B2$ degree of freedom from B is aliased with one degree of freedom from AC, $C1C2$ from C with $A2B1$ from AB, and $A1A2$ from A with $B2C2$. In addition all three are aliased with parts of ABC.

17.6 6^N FACTORIAL SYSTEM

17.6.1 Blocking

In the 6^N case we have no choice, we must work with the (modulo 6) arithmetic. There is no finite field with six elements. We consider first the 6×6 factorial. If we now attempt to construct a blocking scheme with six blocks of six, confounding the two-factor interaction, we find that we must confound either (11), (15), (51), or (55). All other choices confound at least one main-effect degree of freedom with blocks. Notice that $3 \times (12)$ (modulo 6) gives (30), and this identifies a main-effect contrast and that $2 \times (13)$ (modulo 6) gives (20), which identifies two main-effect degrees of freedom. The vectors (22) and (44) generate three blocks, and the vector (33) generates only two blocks. Note that (15) and (51) lead to the same blocks, as do (11) and (55). Also, it turns out that the sets of contrasts associated with (11) and (15) are not fully mutually orthogonal. This should also come as no surprise, since we know that there are 6×6 Latin squares and no 6×6 Greco–Latin squares.

There are cases where an experimenter may require a blocking scheme for a 6^3 factorial in six blocks of 36. Now it is a matter of constructing possible plans and then examining for unwanted confounding. It turns out that one can construct suitable confounding schemes that preserve main effects and two-factor interactions by using one of (111), (115), or (155) to construct blocks. All others result in some loss of main effect or two-factor interaction information. To see this, examine the variations of (111). We have $2 \times (111) \rightarrow (222)$, $3 \times (111) \rightarrow (333)$, $4 \times (111) \rightarrow (444)$, and $5 \times (111) \rightarrow (555)$. All of these have three nonzero values and consequently, all represent parts of the three-factor interaction. On the other hand, if one of the values in the initial vector is "2," "3," or "4", a vector containing a "0" will be generated, indicating something less than a three-factor interaction. For example, from (123), we get $2 \times (123) \rightarrow (240)$, $3 \times (123) \rightarrow (303)$, $4 \times (123) \rightarrow (420)$, and $5 \times (123) \rightarrow (543)$ indicating that

(123) is associated with a set of contrasts that includes contrasts from $A \times B$ and from $A \times C$ as well as from $A \times B \times C$.

The intrablock subgroup for the plan based on confounding (111) is obtained by finding solutions to

$$\begin{array}{ccccc} a & & b & & c \\ \hline i_A & + & i_B & + & i_C \end{array} = 0 \text{ (modulo 6).}$$

The intrablock subgroup for the plan based on (123) is obtained by finding solutions to

$$\begin{array}{ccccc} a & & b & & c \\ \hline i_A & + & 2i_B & + & 3i_C \end{array} = 0 \text{ (modulo 6).}$$

Exercise 17.11: Confounding a 6^3 Factorial. Construct the intrablock subgroup for the six blocks of 36 that sacrifices the (115) interaction. □

It is possible to construct plans with smaller blocks. To illustrate, consider a blocking scheme based on confounding (111) and (115). As a first step, we examine the generalized interaction. We construct the array

	(115)	(224)	(333)	(442)	(551)
(111)	(220)	(335)	(444)	(553)	(002)
(222)	(331)	(440)	(555)	(004)	(113)
(333)	(442)	(551)	(000)	(115)	(224)
(444)	(553)	(002)	(111)	(220)	(335)
(555)	(004)	(113)	(222)	(331)	(440)

First, notice that (333) is common to both row headings and column headings. This implies that these two interactions together account for only nine degrees of freedom. We examine the generalized interaction by looking at elements within the table. Begin with (335). This has alternative forms (004), (333), (002), and (331). The conclusion here is that since (333) already appeared in the row headings, this vector accounts for four new degrees of freedom. Next examine (553). This has alternative forms (440), (333), (220), and (113) and again accounts for four degrees of freedom. At this point we have accounted for all elements in the table and a total of 5 (for rows) + 4 (for columns) + 4 for (335) + 4 for (553) = 17 degrees of freedom. Our confounding scheme gives us a plan with 18 blocks of 12 with full information on A, B, $A \times C$, and $B \times C$. Unfortunately, there is considerable loss of information on C. If, however, C is a continuous factor and the experimenter assigns levels appropriately, a more acceptable plan can be obtained. If we denote the successive levels of factor C by $i_C(0), i_C(1), \ldots, i_C(5)$ and assign treatment levels according to Table 17.4, full information will be

Table 17.4 Treatment Level Assignment

Symbol	Level Assigned
$i_C(0)$	x units
$i_C(1)$	$x+1$ units
$i_C(2)$	$x+2$ units
$i_C(3)$	$x+5$ units
$i_C(4)$	$x+4$ units
$i_C(5)$	$x+3$ units

available on the linear and cubic contrasts of C. However, information on the quadratic contrast will be severely compromised.

We obtain the intrablock subgroup for the plan by finding the solutions to the system of equations

$$\begin{array}{ccccc} a & & b & & c \\ \hline i_A & + & i_B & + & i_C & = 0 \text{ (modulo 6)} \\ i_A & + & i_B & + & 5i_C & = 0 \text{ (modulo 6).} \end{array}$$

The 12 solutions to this system are (000, 150, 240, 330, 420, 510, 303, 453, 543, 033, 123, and 213).

Another example is to confound (151) and (115) with blocks. Now construct the array

	(115)	(224)	(333)	(442)	(551)
(151)	(200)	(315)	(424)	(533)	(042)
(242)	(351)	(400)	(515)	(024)	(133)
(333)	(442)	(551)	(000)	(115)	(224)
(424)	(533)	(042)	(151)	(200)	(315)
(515)	(024)	(133)	(242)	(351)	(400)

In this example, there again is one degree of freedom common to the two interactions being confounded. When we examine the body of the table, we find (315) with alternative forms (024), (333), (042), and (351). Of these (333) is already accounted for, leaving four degrees of freedom. The next vector not accounted for is (533). This has forms (400), (333), (200), and (133), accounting for another four degrees of freedom. This accounts for all of the vectors in the generalized interaction. In total, there are 5 (for rows) + 4 (for columns) + 4 [for (315)] + 4 [for (533)] = 17 degrees of freedom. In this plan B, C, $A \times B$, and $A \times C$ are clear. These two plans are equivalent, in the sense that both keep two two-factor interactions clear. The intrablock subgroup is (000, 111, 222, 333, 444, 555, 330, 341, 352, 303, 314, and 324).

17.6.2 One-Sixth Fraction of the 6^3 Factorial

Finding suitable fractions of the total set of 216 treatment combinations in the 6^3 follows exactly the same pattern as in the 4^3 case. Again, some care must be exercised in selecting the appropriate vector to define the fraction. A possible choice is (115). The rule is that no value in the selected vector be divisible by one of the factors of 6. To find the aliases of A, we need to find the generalized interaction of A and ABC^5. This means that we need to examine the table

	(115)	(224)	(333)	(442)	(551)
(100)	(215)	(324)	(433)	(542)	(051)
(200)	(315)	(424)	(533)	(042)	(151)
(300)	(415)	(524)	(033)	(142)	(251)
(400)	(515)	(024)	(133)	(242)	(351)
(500)	(015)	(124)	(233)	(342)	(451)

From this table we see that A is aliased with the (015) interaction, i.e., part of the $B \times C$ interaction. It is also aliased with parts of the $A \times B \times C$ interaction. Similarly, it can be shown that B is aliased with part of the $A \times C$ interaction, and parts of the $A \times B \times C$ interaction and C is aliased with part of the $A \times B$ interaction and parts of the $A \times B \times C$ interaction.

17.7 ASYMMETRICAL FACTORIALS

To introduce the necessary modifications in our techniques to handle the asymmetric case, we first digress to the more general case. We then return to a sequence of special cases that have some merit in practice.

In general, we can think of an $s_1^{n_1} \times s_2^{n_2} \times \cdots \times s_k^{n_k}$ factorial experiment with n_i factors, each with s_i levels for $i = 1, \ldots, k$ and a total of $N = \sum n_i$ factors. The factors are denoted by F_{ij}. There are factors $F_{i1}, F_{i2}, \ldots, F_{in_i}$ each with s_i levels for $i = 1, 2, \ldots, k$. A treatment combination is represented by the N-tuple $f_{11} \cdots f_{1n_1} f_{21} \cdots f_{2n_2} \cdots f_{kn_k}$, where $0 \leq f_{ij} < s_i$ for $j = 1, \ldots, k$. In general, the interactions are denoted by either $F_{11}^{a_{11}} \cdots F_{1n_1}^{a_{1n_1}} F_{21}^{a_{21}} \cdots F_{kn_k}^{a_{kn_k}}$ or more compactly, by the vector $(a_{11} \cdots a_{1n_1} \cdots a_{kn_k})$ with the restriction that not all values may be zero simultaneously. We recall from previous sections in this chapter that there is a question of uniqueness of names if we do not insist on the first nonzero value being the unit value. In fact, one should really think of the symbol $(a_{11} \;\; \cdots a_{1n_1} \; a_{11} \;\; \cdots \;\; a_{kn_k})$ as a "splitter," in the sense that it can be used to split the responses into equal-sized subsets. The differences among these subsets then account for sets of degrees of freedom. These sets need not be homogeneous in the sense that they may represent contrasts from different interactions or even from main effects. Also, two different splitters need not be disjoint, in the sense that a contrast found among the sets from one splitter can also appear among the contrasts found among the sets from another splitter.

In the symmetrical case, with s levels for each factor, the set of contrasts defined by an n-tuple $(a_1\ a_2\ \cdots\ a_n)$ was determined by computing $(aa_1\ aa_2\ \cdots\ aa_n)$ and reducing modulo s whenever necessary for $a = 1, 2, \ldots, s-1$. In the asymmetrical case, the corresponding operation is to compute

$$(aa_{11}\gamma s_1^{-1}\ \cdots\ aa_{1n_1}\gamma s_1^{-1}\ aa_{21}\gamma s_2^{-1}\ \cdots\ aa_{kn_k}\gamma s_k^{-1}), \tag{17.3}$$

first reducing (modulo γ) and then dividing by the appropriate γs_i^{-1} for $a = 2, \ldots, \gamma - 1$, where γ is the least common multiple of $s_1, s_2, \ldots, s_k$.

Computing the generalized interactions between two interactions or splitters involves setting up the two-way table with columns represented by the distinct variations of one interaction and rows the distinct forms of the other, and then filling the table by adding term by term and reducing [modulo (the appropriate s_i)].

Example 17.4. To illustrate the general notation established, we look at a $2 \times 3 \times 3 \times 5 \times 5$ factorial. In our notation this has $s_1 = 2$, $s_2 = 3$, and $s_3 = 5$, $n_1 = 1$, $n_2 = 2$, and $n_3 = 2$. Also, $\gamma = 2 \times 3 \times 5 = 30$. Now try using the splitter (1 1 2 2 1) to define blocks. Clearly, this is part of the five-factor interaction. However, some work is needed to establish the nature of all the degrees of freedom confounded with blocks. Following equation (17.3), we write

$$(a \times 1 \times 30/2\quad a \times 1 \times 30/3\quad a \times 2 \times 30/3\quad a \times 2 \times 30/5\quad a \times 1 \times 30/5)$$

for $a = 2, 3, \ldots, 29$. For $a = 2$ we get

$$(2 \times 1 \times 30/2\quad 2 \times 1 \times 30/3\quad 2 \times 2 \times 30/3\quad 2 \times 2 \times 30/5\quad 2 \times 1 \times 30/5),$$

which evaluates to

$$(30\quad 20\quad 40\quad 24\quad 12).$$

Now all elements must first be reduced (modulo 30) and then multiplied by the appropriate s_i/γ factor. The final result is (0 2 1 4 2). Notice the s_i/γ factors are 15, 10, 10, 6, and 6, respectively. Remaining terms include

$$(3 \times 1 \times 30/2\quad 3 \times 1 \times 30/3\quad 3 \times 2 \times 30/3\quad 3 \times 2 \times 30/5\quad 3 \times 1 \times 30/5)$$

$$\rightarrow\quad (10013)$$

$$(4 \times 1 \times 30/2\quad 4 \times 1 \times 30/3\quad 4 \times 2 \times 30/3\quad 4 \times 2 \times 30/5\quad 4 \times 1 \times 30/5)$$

$$\rightarrow\quad (01234)$$

```
data blocks ;
 do ia = 0 to 1 ; do ib = 0 to 2 ;
 do ic = 0 to 2 ; do id = 0 to 4 ;
 do ie = 0 to 4 ;
  block = 15*0*ia+10*1*ib+10*2*ic+6*3*id+6*1*ie;
  block = mod(block,30) ;
  output ;
 end ; end ; end ; end ; end ;
proc sort ; by block ;
proc print ; run;
```

Display 17.2 Construct 15 blocks of 30 for a $2 \times 3 \times 3 \times 5 \times 5$ factorial.

$$(5 \times 1 \times 30/2 \quad 5 \times 1 \times 30/3 \quad 5 \times 2 \times 30/3 \quad 5 \times 2 \times 30/5 \quad 5 \times 1 \times 30/5)$$
$$\rightarrow \quad (12100)$$
$$(6 \times 1 \times 30/2 \quad 6 \times 1 \times 30/3 \quad 6 \times 2 \times 30/3 \quad 6 \times 2 \times 30/5 \quad 6 \times 1 \times 30/5)$$
$$\rightarrow \quad (00021)$$
$$(7 \times 1 \times 30/2 \quad 7 \times 1 \times 30/3 \quad 7 \times 2 \times 30/3 \quad 7 \times 2 \times 30/5 \quad 7 \times 1 \times 30/5)$$
$$\rightarrow \quad (11242)$$
$$\vdots$$
$$(15 \times 1 \times 30/2 \quad 15 \times 1 \times 30/3 \quad 15 \times 2 \times 30/3 \quad 15 \times 2 \times 30/5 \quad 15 \times 1 \times 30/5)$$
$$\rightarrow \quad (10000)$$
$$\vdots$$
$$(29 \times 1 \times 30/2 \quad 29 \times 1 \times 30/3 \quad 29 \times 2 \times 30/3 \quad 29 \times 2 \times 30/5 \quad 29 \times 1 \times 30/5)$$
$$\rightarrow \quad (12134).$$

It turns out that the A main effect and parts of the $B \times C$ and $D \times E$ two-factor interactions are confounded with blocks. Inspection reveals that it is not possible to confound a five-factor interaction without sacrificing the A main effect. The best one can do is to begin with, say, (0 1 2 3 1). This leads to 15 blocks of 30. Display 17.2 illustrates finding the block assignment in SAS®.

Comments

Since $\prod_{j \neq h}^{k} s_j < \gamma$ and is also a factor of γ, it follows that any interaction that involves exactly one factor from one of the subsets that have exactly s_h levels will involve one contrast from the corresponding main effect. In the example we saw that (1 1 2 2 1) involved the A main effect. A similar argument shows that if a selected interaction involves exactly two factors from a subset, all of which have the same number of levels, part of the corresponding two-factor interaction will be compromised. The conclusion is that the asymmetrical factorial experiments generally are not nice. They tend to be large and blocking is impossible, unless it can be assumed a priori that certain interactions contrasts are negligible.

17.8 $2^N \times 3^M$ PLANS

In practice, there are frequent demands for plans from this subfamily from the general class of asymmetrical factorials. Consequently, they have been the object of many investigations, and a number of very useful plans have been discovered. We will look at a number of these. We begin with a number of special cases and then proceed to a number of general approaches that have been developed. Three basic methods that we mention are typically referred to as *collapsing levels, replacing factors*, and *conjoining fractions*. Cases where the confounding or fractional replication is restricted to just two- or just three-level factors present no new problems and are not discussed in this chapter. They are really part of the material on either the 2^N or 3^N system.

17.8.1 Some Special Cases

2 x 2 x 3 *Plan*

Yates (1935) gives the $2 \times 2 \times 3$ plan with three replicates and blocks of six given in Table 17.5. The statistical analysis of data from this plan can be analyzed most easily be using a general-purpose least squares program such as `proc GLM` in SAS®. This plan provides full information sums of squares which can be computed using standard formulas for all main effects and AC and BC interactions. The AB and ABC interactions are partially confounded with blocks.

2 x 3 x 3 *Plan*

Kempthorne (1952) presents the plan for the $2 \times 3 \times 3$ factorial in six blocks of six shown in Table 17.6. This plan retains all the main effects and AB and AC information, all the BC^2 information, part of the $A \times BC$ information, and all of the $A \times BC^2$ information. It follows that a least squares analysis will recover

Table 17.5 Yates' 2 x 2 x 3 Plan

Replicate I		Replicate II		Replicate III	
Block 1	Block 2	Block 1	Block 2	Block 1	Block 2
000	100	100	000	100	000
110	010	010	110	010	110
011	001	001	101	101	001
101	111	111	011	011	111
012	002	102	012	002	102
102	112	012	112	112	012

Table 17.6 Kempthorne's 2 x 3 x 3 Plan

Replicate I			Replicate II		
Block 1	Block 2	Block 3	Block 1	Block 2	Block 3
010	020	000	020	000	010
001	011	012	011	012	001
022	002	021	002	021	022
120	100	110	110	120	100
111	112	101	101	111	112
102	121	122	122	102	121

partial information on the $B \times C$ and $A \times B \times C$ interactions and full information on all main effects and other interactions.

2 x 2 x 3 x 3 ***Plan***

This plan presents a number of possibilities. One can construct three blocks of 12 by confounding CD or CD^2. This is really part of the 3^M system. One can construct two blocks of 18 by confounding AB. This is part of the 2^N system. Then one can construct six blocks of six by confounding AB, CD, and CD^2 or AB, CD^2, and $ABCD^2$ with blocks.

One can generate confounding schemes using (1111), (1112), or (1122) as splitters. Consider (1111) in detail. First, we need to find the alternative forms. Notice that the least common multiple of the levels is six. We examine

$$2 \times (1 \times 3, 1 \times 3, 1 \times 2, 1 \times 2) \rightarrow (0, 0, 2, 2)$$

after reducing (modulo 6) and then dividing each term by its appropriate coefficient. Similarly,

$$3 \times (1 \times 3, 1 \times 3, 1 \times 2, 1 \times 2) \rightarrow (1, 1, 0, 0)$$

$$4 \times (1 \times 3, 1 \times 3, 1 \times 2, 1 \times 2) \rightarrow (0, 0, 1, 1)$$

$$5 \times (1 \times 3, 1 \times 3, 1 \times 2, 1 \times 2) \rightarrow (1, 1, 2, 2).$$

The conclusion is that using (1111) as a splitter to assign treatment combinations to six blocks of six results in confounding the $A \times B$ interaction and two degrees of freedom from each of $C \times D$ and $A \times B \times C \times D$ interaction. The $A \times C$, $A \times D$, $B \times C$, and $B \times D$ interactions are preserved. This is a plan that may well be useful in some situations where the experimenter is sure from the nature of the experimental factors that there are no $A \times B$ or $C \times D$ interactions of interest. The actual block assignment is obtained by computing $(3 \times 1 \times i_A + 3 \times 1 \times i_B + 2 \times 1 \times i_C + 2 \times 1 \times i_D)$ (modulo 6). The resulting plan is shown in Table 17.7. This plan provides full information on all main effects and two-factor interactions, except the $A \times B$ and $C \times D$. Only two degrees of freedom

Table 17.7 2 x 2 x 3 x 3 Plan

Block I	Block II	Block III	Block IV	Block V	Block VI
0000	0102	0001	0100	0002	0101
0012	0120	0010	0112	0020	0110
0021	0111	0022	0121	0011	0122
1100	1002	1101	1000	1102	1001
1112	1020	1110	1012	1120	1010
1121	1011	1122	1021	1111	1022

are available for $C \times D$. A least squares analysis fitting only main effects and two-factor interactions leaves 14 degrees of freedom for error.

It is also possible to construct this plan by conjoining fractions. The AB and CD interactions are used to split the treatment combinations into subsets, and these are then joined to form the plan. Construct block I from the six solutions to $i_A + i_B = 0$ (modulo 2) and $i_C + i_D = 0$ (modulo 3), block III from the solutions to $i_A + i_B = 0$ (modulo 2) and $i_C + i_D = 1$ (modulo 3), ..., and block II from $i_A + i_B = 1$ (modulo 2) and $i_C + i_D = 2$ (modulo 3).

Trying to use the contents of the principal block, the block containing (0000) as a one-sixth replicate results in A being aliased with B, and C being aliased with D.

Exercise 17.12: Confounding a $2^2 \times 3^2$ Factorial. Show that using (1122) as a splitter in place of (1111) results in exactly the same confounding pattern. Also show that using (1112) as a splitter in place of (1111) results in a different treatment assignment pattern but a similar analysis of variance. □

17.8.2 Collapsing Levels

Collapsing levels provides a very flexible method of constructing $2^N 3^M$ factorial plans. We illustrate with the $2^2 3^2$ factorial. Begin with the full set of 81 treatment combinations in the 3^4. Since the desire is to have factors A and B at two levels, the technique is to collapse the first two factors using a pattern such as that shown in Table 17.8. or any one of its variations. The only real requirement is that the same replacement pattern be used throughout for a given factor. The result is a $2^2 3^2$ plan with 81 runs. There are duplicate runs. However, the simple analysis of variance still holds, because of the proportional frequencies.

Table 17.8 Collapsing Technique

Old Level		New Level
0	→	0
1	→	1
2	→	0

Putting two-factor interactions into the model retains the estimability of effects and interactions.

There are immediate modifications. One can block on one of the components of the four-factor interaction to obtain three blocks of 27. Alternatively, one can block on $ABCD$ and AB^2CD^2 (and the generalized interactions AC and BD) to obtain nine blocks of nine. One loses the ability to estimate the $A \times C$ and $B \times D$ interactions but retains all other two-factor interactions. One can reduce the size of the study by dropping duplicate treatment combinations, but then orthogonality suffers. Main effects and the four two-factor interactions are still estimable.

As a second illustration of the possibilities, construct a main-effect plan with 27 runs by superimposing up to 13 factors on a 3^3. Then produce other main-effect plans by collapsing any number of factors to two levels. Since frequencies will be proportional, orthogonality will still hold. These are all resolution III plans. If it is necessary to protect against two-factor interactions, matters become messy. It is reasonable to begin with a resolution IV plan. A simple way to obtain an eight-factor, three-level resolution IV plan in 81 runs is to use `proc FACTEX` in SAS®:

```
proc factex ;
 factors i1-i8 /nlev = 3 ;
 size design = minimum ;
 model resolution = 4 ;
 run ;
```

Alternatively, one can superimpose factors on A, B, C, D, ABC, ABD, ACD, and BCD on a 3^4 with 81 runs. Fractions of 2^N3^M with $N + M \leq 8$ can be obtained by collapsing any of the factors to two levels.

To investigate the properties of the collapsed designs, it is useful to think in terms of contrasts. A factor with three levels allows us to define two orthogonal contrasts. We denote these by α and β. Notice that this is strictly formal and does not require equal spacing of the levels or even that the factor be continuous. We let α be the contrast of the first level vs. the mean of the second and third and β the difference between the second and third. We do this for all factors in the 3^M. The contrasts are A_α, A_β, B_α, B_β, and so on. Each contrast carries with it a single degree of freedom. All contrasts are mutually orthogonal. Note also that any other estimable contrasts will be functions of these contrasts. This extends directly into the interactions. The four degrees of freedom in the $A \times B$ interaction are carried by the $A_\alpha \times B_\alpha$, $A_\alpha \times B_\beta$, $A_\beta \times B_\alpha$, and $A_\beta \times B_\beta$ contrasts. The coefficients in these interaction contrasts are obtained as products of the coefficients of the main-effect contrasts. This extends directly to all orders of interactions.

Now if two main effects or interactions are not aliased, i.e., are estimable orthogonally, all of the contrasts in the two sets are orthogonally estimable. As a specific example, if the $A \times B$ and the $C \times D$ interactions are not aliased and can be estimated, the two sets of four contrasts, i.e., $A_\alpha \times B_\alpha, \ldots, A_\beta \times B_\beta$ and

$C_\alpha \times D_\alpha, \ldots, C_\beta \times D_\beta$ are all estimable and orthogonal. It follows that in a resolution IV plan, where main effects are free of two-factor interactions, all A_α, A_β, $B_\alpha, \ldots$ contrasts will be orthogonal to the two-factor interaction contrasts.

Now consider collapsing by replacing the second and third levels by one level for a number of factors. This leaves the α-contrasts unscathed but shifts the affected β contrasts to error contrasts. If, say, A and C had both been collapsed, then A_β, C_β, $A_\alpha \times C_\beta$, $A_\beta \times C_\alpha$, and $A_\beta \times C_\beta$ all become error contrasts. The A_α, C_α, and $A_\alpha \times C_\alpha$ contrasts are untouched.

The conclusion from this is that if we begin with a resolution IV plan which keeps main effects free of two-factor interactions, then after collapsing we still have main effects free of two-factor interactions. We will have a big increase in number of degrees of freedom for error.

17.8.3 Replacing Factors

An alternative but closely related technique is to replace factors in a 2^{N-k} fraction to create some four-level factors and then collapse. For example, one can replace two factors, say A and B, using a pattern such as that shown in Table 17.9, or a variation. There is the obvious restriction that if columns corresponding to A and B are used in the original fraction, the AB interaction must also be estimable. If no interactions exist, the proportional frequencies guarantee a simple and orthogonal statistical analysis. For interactions one can again use the argument from Section 17.8.2.

17.8.4 Conjoining Fractions

Conner and Young (1961) used this technique to combine fractions from a 2^{N-k} with fractions from the 3^{M-p} to produce a series of 39 plans with N factors at two levels and M factors at three levels ranging from $N + M = 5$ to $N + M = 10$. [This monograph has been reproduced in McLean and Anderson (1984).] We illustrate the construction for a fraction of the $2^3 3^2$ plan. Let S_0 and S_1 denote the fractions of the 2^3 defined by solutions to $i_A + i_B + i_C = 0$ (modulo 2) and $i_A + i_B + i_C = 1$ (modulo 2), respectively. Similarly, let S_0', S_1', and S_2' denote the fractions $i_D + i_E = 0$ (modulo 3), $i_D + i_E = 1$ (modulo 3), and $i_D + i_E = 2$ (modulo 3), respectively. The actual treatment combinations for the complete

Table 17.9 Replacing Two Factors Each at Two Levels With One Factor at Three Levels

Old Level A	Old Level B		New Level
0	0	$\rightarrow$	0
1	0	$\rightarrow$	1
0	1	$\rightarrow$	3
1	1	$\rightarrow$	3

factorial can then be represented symbolically by the pieces S_0S_0', S_0S_1', S_0S_2', S_1S_0', S_1S_1', and S_1S_2'. Each piece consists of 12 treatment combinations. Conner and Young give S_0S_0', S_1S_1', and S_1S_2' as a suitable fraction. Their plan consists of the 36 treatment combinations shown in Table 17.10

Table 17.10 The 36 Treatment Combinations in a Fraction of a $2^3 3^2$ Plan

0 0 0 0 0	0 0 1 0 1	0 0 1 0 2
0 1 1 0 0	0 1 0 0 1	0 1 0 0 2
1 0 1 0 0	1 0 0 0 1	1 0 0 0 2
1 1 0 0 0	1 1 1 0 1	1 1 1 0 2
0 0 0 1 2	0 0 1 1 0	0 0 1 2 0
0 1 1 1 2	0 1 0 1 0	0 1 0 2 0
1 0 1 1 2	1 0 0 1 0	1 0 0 2 0
1 1 0 1 2	1 1 1 1 0	1 1 1 2 0
0 0 0 2 1	0 0 1 2 2	0 0 1 1 1
0 1 1 2 1	0 1 0 2 2	0 1 0 1 1
1 0 1 2 1	1 0 0 2 2	1 0 0 1 1
1 1 0 2 1	1 1 1 2 2	1 1 1 1 1 .

CHAPTER 18

Plans for Which Run Order Is Important

18.1 INTRODUCTION

Standard recommendation is always to randomize the order in which the trials are conducted, or equivalently in the case of field trials, the order in which trials are arranged. The object of this chapter is to examine some situations where there is good reason to deviate from this recommendation. First, there are cases where it is known that there will be a time or position trend that can seriously compromise the results. We examine strategies that can be used to protect against these trends. Time trends could be due to learning, progressive change in temperature, wear of equipment, change in concentration of some chemical reagent, change in altitude, and so on. Joiner and Campbell (1976) give a number of examples of laboratory experiments in which linear time trends were either observed or expected. When considering the systematic order of trials to counteract one of these trends, the experimenter must bear in mind the concomitant risks due to lack of randomization. In effect, the investigator must be willing to place greater faith in an assumed model. Also, in a nonrandomized study there are additional difficulties in arguing cause and effect.

Another situation where an experimenter should consider dispensing with randomization is where it is difficult, time consuming, or excessively expensive to change factor levels, or if it takes a long time for the experimental system to return to steady state after a change in factor level. In such a case the advantages of a plan that minimizes the number of level changes may well outweigh the disadvantages.

On the other hand, there are situations where observations (runs) must be made sequentially, yet there is reason to believe that the random errors have a strong correlation structure. It may be that the responses are affected by a trend

Planning, Construction, and Statistical Analysis of Comparative Experiments,
by Francis G. Giesbrecht and Marcia L. Gumpertz
ISBN 0-471-21395-0

that can be modeled as a process with autocorrelated errors. In these cases it may be advantageous to order the runs so that the number of level changes in the factors is large.

Additionally, there is a class of experiments in which there is either no random error at all, or at least the errors are sufficiently small to allow them to be neglected safely. In these cases, there may be no real need for randomization and the investigator is free to choose an order for runs that maximizes convenience and minimizes cost. Generally, there is also no need for any statistical analysis other than just fitting a model. Examples here include simulation experiments on a computer, studies to optimize performance of a computer program, or a piece of equipment and projects to synthesize chemical compounds with specific properties. In the latter case, there may well be some small random errors, but usually these can be ignored.

18.1.1 An example

Bromwell et al. (1990) describe a project to optimize the production of self-adhesive label stock material. This label stock, which is coated and later laminated, is used to make a special type of pharmaceutical labels for small-diameter vials. The label stock is shipped to customers who use automated equipment to convert the stock to labels and eventually, transfer the labels onto vials. Quality of this product relates to the smoothness of the customer's process of converting and dispensing the labels. The people involved in the study listed seven different factors in the production line that could affect the product. These were identified as bottom pressure, silicone formulation, top pressure, backing paper supplier, amount of silicone, temperature, and application rate. A fractional factorial was used to reduce the number of treatment combinations that were to be examined. Since the test stocks had to be produced on the regular production line, and changing factors tended to cause disruption, it was important to organize things to make as few changes in the production line over time as possible. Also, some factors caused less disruption than others. There was a premium on selecting an efficient order for treatment combinations.

18.2 PRELIMINARY CONCEPTS

We begin by considering the situation where it is assumed that the trend (spatial or temporal) in the experimental material can be modeled by a low-order polynomial. The object will be to assign the treatment combination in such an order that the contrasts of major interest, be they main effects or interactions, are orthogonal to the trend. In practice, we are most often concerned with linear first order or at most quadratic (second-order) trends. It is also clear that in an experiment with 2^N treatment combinations, it is not possible to have all $2^N - 1$ contrasts orthogonal to a trend. Compromises will be required. The experimenter must select the contrasts to protect against trend and the contrasts to sacrifice.

To illustrate a simple example, consider a full replicate of the 2^3 in Yates' standard order, i.e., (1), a, b, ab, c, ac, bc, and abc equally spaced in time or space. The A contrast $(-1, +1, -1, +1, -1, +1, -1, +1)$ is clearly not resistant to a linear trend, since the inner product

$$(-1, +1, -1, +1, -1, +1, -1, +1)\,(1, 2, 3, 4, 5, 6, 7, 8)^t \neq 0.$$

However, the AB contrast is resistant to a linear trend, since

$$(+1, -1, -1, +1, +1, -1, -1, +1)\,(1, 2, 3, 4, 5, 6, 7, 8)^t = 0.$$

The AC and BC contrasts are also resistant to a linear trend. The ABC contrast is resistant to both a linear trend, i.e.,

$$(-1, +1, +1, -1, +1, -1, -1, +1)\,(1, 2, 3, 4, 5, 6, 7, 8)^t = 0,$$

and a quadratic trend, i.e.,

$$(-1, +1, +1, -1, +1, -1, -1, +1)\,(1, 4, 9, 16, 25, 36, 49, 64)^t = 0.$$

In general, a contrast with coefficients $(\alpha_1, \alpha_2, \ldots, \alpha_m)$ is orthogonal to a kth-order trend, if the responses are uncorrelated, have common variance, and

$$\sum_{i=1}^{n} \alpha_i i^k = 0.$$

A contrast is said to be *m-trend-free* if it is orthogonal to all polynomial time trends of degree m or less.

Our example illustrates that in the 2^3 case, for example, it is possible to have at least one contrast that is 2-trend-free and three contrasts that are 1-trend-free. The three main-effect contrasts were affected by a linear trend. Since experimenters are typically more interested in main effects and possibly two-factor interactions, the strategy will be to rearrange the order of the treatments to make as many as possible of these trend-resistant. In the example it is possible to reorder the treatment combinations to make, for example, the A contrast 2-trend free. We examine several methods for constructing orders that have desired contrasts trend-free.

Exercise 18.1: Trend in the 2^3 Case (First Example). Verify that it is possible for a run order in a complete factorial to exist for which no main effect or interaction is resistant to a linear trend by showing that none of the main effects or interactions in the complete 2^3 factorial with run sequence ((1), a, b, c, bc, ab, ac, abc) are resistant to a linear trend. □

18.3 TREND-RESISTANT PLANS

It is well known (Cheng and Jacroux 1988) that if the treatment combinations in a full 2^N factorial are in Yates' standard order, all t-factor interactions will be resistant to all trends of order up to $t - 1$. However, main effects are not resistant to any trends. This fact and the superimposing of factors technique can be utilized to construct run orders that have desirable trend resistance properties. This can be applied to fractional factorials as well as full replicates. We illustrate with two specific examples and then discuss some principles involved. We first construct a 2^4 plan with main effects resistant to linear and quadratic trends by utilizing the fact that the four three-factor interactions ABC, ABD, ACD, and BCD are resistant to linear and quadratic trends if the treatment combinations are in Yates order. Notice that the four interactions selected are mutually independent in the sense that it is not possible to express any one of them as a generalized interaction of the others. They serve as a basis set. We begin our construction with Table 18.1 where the rows are the runs for a 2^4 in standard Yates order and the columns are labeled at the top as effects and interactions. It is simple to verify that with rows as runs in the order given, the two-factor interaction contrasts are resistant to a linear trend, the three-factor interactions are resistant to linear and quadratic trends, and the four-factor interaction is resistant to linear, quadratic, and cubic trends.

We now superimpose a new set of factors. Assign factor A to the column labeled ABC, factor B to the column labeled ABD, factor C to the column labeled ACD, and factor D to the column labeled BCD. These labels are shown at the bottom of the columns in Table 18.1 Keeping rows in the order given, as runs, we now have the sequence of runs (1), abc, abd, cd, acd, bd, bc, a, bcd, ad, ac,

Table 18.1 Effects and Interactions in a 2^4 in Standard Yates Order

A	B	AB	C	AC	BC	ABC	D	AD	BD	ABD	CD	ACD	BCD	$ABCD$
−	−	+	−	+	+	−	−	+	+	−	+	−	−	+
+	−	−	−	−	+	+	−	−	+	+	+	+	−	−
−	+	−	−	+	−	+	−	+	−	+	+	−	+	−
+	+	+	−	−	−	−	−	−	−	−	+	+	+	+
−	−	+	+	−	−	+	−	+	+	−	−	+	+	−
+	−	−	+	+	−	−	−	−	+	+	−	−	+	+
−	+	−	+	−	+	−	−	+	−	+	−	+	−	+
+	+	+	+	+	+	+	−	−	−	−	−	−	−	−
−	−	+	−	+	+	−	+	−	−	+	−	+	+	−
+	−	−	−	−	+	+	+	+	−	−	−	−	+	+
−	+	−	−	+	−	+	+	−	+	−	−	+	−	+
+	+	+	−	−	−	−	+	+	+	+	−	−	−	−
−	−	+	+	−	−	+	+	−	−	+	+	−	−	+
+	−	−	+	+	−	−	+	+	−	−	+	+	−	−
−	+	−	+	−	+	−	+	−	+	−	+	−	+	−
+	+	+	+	+	+	+	+	+	+	+	+	+	+	+
						A				B		C	D	

b, ab, c, d, and $abcd$. This is a plan with all four main effects resistant to linear and quadratic trends. We can also see that our superimposing scheme has placed the AB interaction on what was labeled as CD, AC on BD, AD on AD, BC on BC, BD on AC, and CD on AB. This tells us that all six two-factor interactions are resistant to a linear trend. The four-factor interaction is still resistant to linear, quadratic, and cubic trends, and the three-factor interactions are not resistant to any trends. If it is reasonable to assume that a model with only a linear trend and two-factor interactions is adequate, then in an analysis of variance, one can fit a linear trend, four main effects, six two-factor interactions, and have four degrees of freedom for error.

For the second example, we begin with the same table, and as before, assign factor A to the column labeled ABC, factor B to the column labeled ABD, factor C to the column labeled ACD, and factor D to the column labeled BCD. Then, in addition, assign factor E to the column labeled $ABCD$. The resulting plan, a 2^{5-1} resolution IV fraction, has the sequence of runs e, abc, abd, cde, acd, bde, bce, a, bcd, ade, ace, b, abe, c, d, and $abcde$. This plan has main effects A, B, C, and D resistant to linear and quadratic trends and main effect E resistant to cubic trends as well.

This relabeling technique is quite general but is subject to a number of restrictions. First, if we work with a full 2^N factorial, we need a basis set of N independent contrasts in order to find columns for all factors. The members of the basis set must be mutually independent, with independence defined in the sense that none is the generalized interaction of others. It is simple to verify that in our first example, one could replace the BCD contrast with the $ABCD$ contrast and still proceed. However, now a number of the two-factor interactions in the new run order will not be resistant to a linear trend. If $N = 6$, all five-factor interactions are mutually independent and can be used to construct a full replicate with all main effects resistant to linear, quadratic, cubic and, quartic trends, and all two-factor interactions resistant to a linear trend. They provide a good choice for a basis set.

For $N = 5$, the four-factor interactions are not mutually independent, and another strategy is needed. A suitable choice that leads to main effects that are resistant to at least linear and quadratic effects and all two-factor interactions resistant to linear trends is to select the five-factor interaction and four different three-factor interactions.

Exercise 18.2: Trend in the 2^3 Case (Second Example). Use the scheme outlined via Table 18.1 to construct an eight-run plan with main effects resistant to a linear trend. Begin by listing eight runs in Yates order. Then write the coefficients for the three contrasts, AB, AC, and ABC. Finally, write the sequence of eight treatment combinations such that main effects are linear trend resistant. □

Exercise 18.3: Linear and Quadratic Resistant

(a) Verify that the A contrast for the sequence of runs for the 2^4 just produced is linear and quadratic trend resistant.

(b) Verify that the AB contrast for the sequence of runs for the 2^4 just produced is linear trend resistant. □

18.4 GENERATING RUN ORDERS FOR FRACTIONS

It is clear from the discussion in Section 18.3 that there are ways to generate specific run orders, i.e., specific sequences of treatment combinations for full factorials. We now develop two additional methods, the sequence foldover and the reverse sequence foldover that can be used to generate run orders with specific properties for fractions as well as full factorials. We note that in the statistical journals the term *foldover* is generally used in place of our term *sequence foldover*. We do this to reduce confusion with the term *foldover* used other places in this book, i.e., in Chapters 14 and 16 as well as in other books. In general, a student will need to rely on context to identify the proper meaning of the term *foldover* whenever it is used.

18.4.1 Sequence Foldover Method

We begin our discussion with the full 2^N factorial, although the method works well for fractions as well. We need a symbolic method for combining treatment combinations in our manipulations to generate sequences of treatment combinations. We use the conventional symbolic multiplication of two treatment combinations with the convention that a letter is suppressed if it appears with exponent 2 in the product. For example, $ab \cdot abc \Rightarrow c$, $a \cdot b \Rightarrow ab$, and $a \cdot (1) \Rightarrow a$. This is fully analogous to the computation of generalized interaction effects in the 2^N system. Treatment combinations in a set are said to be *independent* if it is not possible to generate any member of the set by multiplying two or more of the others in the set.

Now a fractional factorial plan of 2^k runs can be generated from the control (1) and a set of k suitably chosen independent treatment combinations x_1, x_2, ..., x_k by means of the following scheme. First write (1). Next write x_1, which can be thought of as $(1) \cdot x_1$. Then write x_2 or equivalently, $(1) \cdot x_2$, followed by $x_1 \cdot x_2$. Next write x_3 or $(1) \cdot x_3$, followed by $x_1 \cdot x_3$, $x_2 \cdot x_3$ and $x_1 \cdot x_2 \cdot x_3$. This is continued. One can write the sequence as

$$(1), x_1, x_2, x_1 \cdot x_2, x_3, x_1 \cdot x_3, x_2 \cdot x_3, x_1 \cdot x_2 \cdot x_3, x_4, \text{etc.}$$

In general, once a sequence of 2^s treatment combinations has been constructed using s generators, 2^s additional combinations can be generated using the $(s+1)$st generator.

For a second illustration, consider generating the full 2^3 from the three generators ab, abc, and ac. The procedure given leads to the sequence (1), ab, abc, c, ac, bc, b, a. This is called the *sequence foldover method*. The nature of the sequence produced depends on the set (and order) of the generators.

Note in passing that the sequence foldover method applied to the generators a, b, c in order, produces the Yates standard order: (1), a, b, ab, c, ac, bc, and abc. This extends directly to N factors.

The key result [derived by Coster and Cheng (1988)] is that a necessary and sufficient condition for a run order to be m-trend free for main effect X, where X is one of $\{A, B, \ldots, N\}$, is that x appear in at least $m + 1$ of the generators. For example, consider the 2^3 generated from the generator sequence ac, abc, and ab. There is an a in all three generators. The A contrast is quadratic-trend-free in the final sequence. The B and C contrasts are both linear-trend-free.

Exercise 18.4: Sequence Foldover and Verifying Trends. Verify that the A contrast in the sequence generated from ac, abc, and ac by the sequence foldover method is resistant to both linear and quadratic trends. □

Exercise 18.5: Good Generating Sequence. Use the generating sequence abd, abc, acd, and bcd to produce a sequence in which all four main-effect contrasts are free of quadratic trends. Verify by examining the details that the A contrast is resistant to both linear and quadratic trends. □

Exercise 18.6: Failure of the Sequence Foldover Method with Nonindependent Bases. Demonstrate that attempting to use the set of generators $abcd$, $abce$, $abde$, $acde$, and $bcde$ to generate a 32-run sequence does not work. The sequence of runs generated contains duplicates, and some treatment combinations will be missing. □

Exercise 18.7: Nonindependent Set of Interactions. Demonstrate that the five contrasts $ABCD$, $ABCE$, $ABDE$, $ACDE$, and $BCDE$ are not independent i.e., any one of the five can be expressed as the generalized interaction of the other four. □

The problem of finding m-trend-free orders for a subset of N factors is now reduced to finding a suitable set of generators. Once we have the generators, the sequence foldover method gives us the sequence. Note that the order in which the generators are used has an effect on the order of the treatment combinations in the sequence produced but does not influence the trend resistance. Consequently, there is still some room for randomization. Also, it makes it possible to construct run orders that are required to satisfy other constraints in addition to trend resistance.

Cheng (1990) also shows that a two-factor interaction, say AB, is m-trend-free if and only if there are at least $m + 1$ generators that contain a and not b or b and not a.

Statistical analysis of the data resulting from a restricted order plan is by least squares. The model used must allow for the assumed trend in order to provide a suitable estimate of error. This model would include a term for linear trend (time or space) in addition to the usual main effects and interactions.

Example 18.1. To illustrate use of the sequence foldover method to construct a fractional factorial, consider 2^{8-3}. This plan consists of 32 runs. A suitable fraction defined by $\mathcal{I} = ABCD = EFGH = CDEF$ is obtained by finding solutions to

a	b	c	d	e	f	g	h	
i_A	$+i_B$	$+i_C$	$+i_D$					$= 0$ (modulo 2)
				i_E	$+i_F$	$+i_G$	$+i_H$	$= 0$ (modulo 2)
		i_C	$+i_D$	$+i_E$	$+i_F$			$= 0$ (modulo 2)

It is easy to show that $abgh$, $cdef$, $abcd$, $aceg$, and $abefgh$ form a set of five independent solutions to the system of equations. Using these five generators, the sequence foldover method gives us a sequence of 32 runs that has A free of linear, quadratic, and cubic trends; B, C, E, and G free of linear and quadratic trends; and D, F, and H free of linear trends. An alternative set of generators available is $abgh$, $abcdefgh$, $abcd$, $adeh$, and $cdefgh$. This set produces a sequence of runs where A, D, and H are free of linear, quadratic and cubic trends, B, C, E, and G are free of linear and quadratic trends and F is free of a linear trend. A very bad selection would be ab, cd, ef, gh, and $aceg$. The resulting order would have only four factors, A, C, E, and G, resistant to a linear trend.

18.4.2 Reverse Sequence Foldover Method

The reverse sequence foldover is a variation of the sequence foldover method. Begin with the set (1), x_1, x_2, ..., x_k. This time generate the sequence as (1), x_1, $x_1 \cdot x_2$, x_2, $x_2 \cdot x_3$, $x_1 \cdot x_2 \cdot x_3$, $x_1 \cdot x_3$, $(1) \cdot x_3$, and so on. The sequence produced can be written as

$$(1), x_1, x_1 \cdot x_2, x_2, x_2 \cdot x_3, x_1 \cdot x_2 \cdot x_3, x_1 \cdot x_3, x_3, x_3 \cdot x_4, \text{etc.}$$

We illustrate the reverse sequence foldover procedure by using generators ab, c, and a to produce the sequence (1), ab, abc, c, ac, bc, b, and a.

A full 2^N factorial is generated by either the sequence foldover or reverse sequence foldover method, given that the members of the initial set of N generators are independent. N generators plus (1) form a basis for the set of 2^N treatment combinations.

In passing, we note that applying the reverse sequence foldover method to the set of generators x_1, x_2, ..., x_k gives the same sequence of treatment combinations as applying the sequence foldover method to y_1, y_2, ..., y_k, where $y_1 = x_1$ and $y_i = y_{i-1} \cdot x_i$ for all $i > 1$. This is equivalent to defining $y_j = \prod_{i=1}^{j} x_i$ for $j = 1, \ldots, k$.

18.5 EXTREME NUMBER OF LEVEL CHANGES

Occasionally, one encounters a situation where there is a premium on either minimizing or maximizing the number of changes in factors as one proceeds through the sequence of runs. The eight runs in the 2^3 in Yates order, (1), a, b, ab, c, ac, bc, and abc, has a total of 11 changes in factors. As an alternative, the sequence (1), a, ab, b, bc, abc, ac, and c obtained via the reverse sequence foldover has only seven changes in factors. Reasons for wanting to minimize the number of changes could be that the cost of making a change in the factors is excessive, or it may take a long time for the process to stabilize after a change.

Alternatively, it turns out that run orders with a maximum number of level changes are highly efficient when there is a strong positive serial correlation among errors affecting successive responses. An example where this would be appealing is an industrial process where the regular quality control charting has indicated that the process is out of control. The problem is to find the factor (factors) responsible. A fraction of the 2^N seems like a good starting point. However, some control on the sequence of runs is necessary, since all runs are on the same process, and extraneous uncontrollable factors will likely cause a strong serial correlation in the observations. Before proceeding, we must, however, point out that in a regular saturated resolution III plan, such as Hadamard or Plackett–Burman plans, the number of level changes from one run to the next is constant for all run orders and the question of optimal run order is moot.

18.5.1 Minimizing Number of Level Changes

We first consider the case where there is a premium on reducing the number of level changes in factors as the study proceeds. The problem arises naturally from economic considerations when it is expensive, time-consuming, or difficult to change factor levels. A potential application could be in a marketing study where factors involve changes of placement of items in a store, and too many changes at one time, especially in some key products, can lead to too much confusion. In another application, making changes in levels of factors involving disassembling and reassembling complex pieces of apparatus can lead to excessive expense. However, we repeat that these investigations lose the protection offered by randomization. We also point out that the constraints that drive investigators to using these plans may still leave the independent, identically distributed errors assumption under an assumed normal error model undamaged and leave ordinary least squares, calculation of differences of means, and simple contrasts intact. We present a sequence of plans to illustrate the general approach.

It is probably safe to assume that the question of randomization helping to establish cause-and-effect relationships is more crucial in scientific research work, where the object is to establish basic truths, than in studying an industrial process or a marketing study. In the latter two cases, the typical investigator has a vested interest in arriving at the correct conclusion and will more likely be conscious of possible hidden biases. If hidden biases exist and are not recognized, there is usually fairly quick pain.

Eight-Run Plans

We use the reverse sequence foldover process applied to the three generators a, b, and c to create the sequence

$$(1),\ a,\ ab,\ b,\ bc,\ abc,\ ac,\ \text{and}\ c. \tag{18.1}$$

Now, corresponding to elements a to c, define (using the Kronecker product) the seven vectors

$$x_a = \begin{bmatrix}-1\\+1\end{bmatrix} \otimes \begin{bmatrix}+1\\+1\end{bmatrix} \otimes \begin{bmatrix}+1\\+1\end{bmatrix}$$

$$x_b = \begin{bmatrix}+1\\+1\end{bmatrix} \otimes \begin{bmatrix}-1\\+1\end{bmatrix} \otimes \begin{bmatrix}+1\\+1\end{bmatrix}$$

$$x_c = \begin{bmatrix}+1\\+1\end{bmatrix} \otimes \begin{bmatrix}+1\\+1\end{bmatrix} \otimes \begin{bmatrix}-1\\+1\end{bmatrix}$$

$$\vdots$$

$$x_{abc} = \begin{bmatrix}-1\\+1\end{bmatrix} \otimes \begin{bmatrix}-1\\+1\end{bmatrix} \otimes \begin{bmatrix}-1\\+1\end{bmatrix}.$$

Assemble these vectors as columns into the following base array:

a	ab	b	bc	abc	ac	c
−1	+1	−1	+1	−1	+1	−1
−1	+1	−1	−1	+1	−1	+1
−1	−1	+1	−1	+1	+1	−1
−1	−1	+1	+1	−1	−1	+1
+1	−1	−1	+1	+1	−1	−1
+1	−1	−1	−1	−1	+1	+1
+1	+1	+1	−1	−1	−1	−1
+1	+1	+1	+1	+1	+1	+1

(18.2)

Notice that this array is nothing more than the arrays that we have used to construct eight-run fractional factorials based on the full 2^3 plan, with columns in a special order. In fact, the reverse sequence foldover has done nothing more than place the columns in a specific order. The key thing to notice is that if one interprets columns as factors and rows as runs or treatment combinations in order, the first column dictates one level change in the sequence of runs; the second, two level changes; ... ; the seventh column, seven changes.

The strategy for an investigator who plans eight runs and wishes to minimize level changes for factors is now clear. A one-factor study would clearly use the first column. A two-factor study would use the first and second columns. A three-factor study would use the first, second, and fourth columns. Since the interaction of columns 1 and 2 falls on column 3, the interaction of columns 1 and 4 falls on column 5, and the interaction of columns 1 and 4 falls on column 6, all main effects are estimable, free of each other and two-factor interactions, i.e., a resolution III plan. A 2^{4-1} resolution IV plan that minimizes level changes uses columns 1, 2, 4, and 7.

The statistical analysis of the data from these plans is straightforward. The investigator must, however, always keep in mind that the lack of randomization may compromise the interpretation. This is a risk with all plans that use a deterministic scheme to control run order.

Exercise 18.8: 2^{4-1} Restricted Run-Order Plan. Exhibit the actual plan obtained by selecting columns 1, 2, 4, and 7 from the base array in (1.2). □

Exercise 18.9: Contrast of 2^{4-1} Restricted Run-Order Plan. Show that the resolution IV plan obtained by selecting columns 1, 2, 4, and 7 in the base array in (1.2) has the defining contrast $\mathcal{I} = ABCD$. □

We must note, however, that some of the columns in the plans suggested, e.g., columns 1 and 7, are sensitive to a possible linear trend. The investigator may be well advised to allow for more level changes and get some protection against a possible linear trend. Any column headed by a single letter is not resistant to linear trends. Any column that is headed by two letters or more is resistant to linear trends. A compromise three-factor plan that minimizes level changes while protecting main effects against a linear trend is obtained by assigning factors to columns 2, 4, and 5. The sequence of runs is ab, ac, c, b, bc, (1), a, and abc. We also note that if changes in one factor are more costly than changes in another factor, some column assignments will be more attractive to the investigator.

16-Run Plans

To construct the base array for the 16-run plans, apply the reverse-sequence foldover process to the four generators a, b, c, and d to create the sequence of treatment combinations

$$(1),\ a,\ ab,\ b,\ bc,\ abc,\ ac,\ c,\ cd,\ acd,\ abcd,\ bcd,\ bd,\ abd,\ ad,\ \text{and}\ d. \tag{18.3}$$

Define the vectors

$$x_a = \begin{bmatrix} -1 \\ +1 \end{bmatrix} \otimes \begin{bmatrix} +1 \\ +1 \end{bmatrix} \otimes \begin{bmatrix} +1 \\ +1 \end{bmatrix} \otimes \begin{bmatrix} +1 \\ +1 \end{bmatrix}$$

$$x_b = \begin{bmatrix} +1 \\ +1 \end{bmatrix} \otimes \begin{bmatrix} -1 \\ +1 \end{bmatrix} \otimes \begin{bmatrix} +1 \\ +1 \end{bmatrix} \otimes \begin{bmatrix} +1 \\ +1 \end{bmatrix}$$

$$\vdots$$

$$x_{abcd} = \begin{bmatrix} -1 \\ +1 \end{bmatrix} \otimes \begin{bmatrix} -1 \\ +1 \end{bmatrix} \otimes \begin{bmatrix} -1 \\ +1 \end{bmatrix} \otimes \begin{bmatrix} -1 \\ +1 \end{bmatrix}$$

and construct the base array as in (1.2). It is simple to verify that the first column has one level change, column 2 has two, ... , and column 15 has 15 level changes. Resolution III plans are now easily obtained by assigning treatments to columns, beginning with column 1. If one is concerned about resistance to possible linear trends, one must skip columns headed by single letters. The cost is an increase in number of level changes.

Selecting resolution IV plans is somewhat more complex since once some columns are selected, one must avoid columns that represent generalized interactions involving two columns of those already selected. The 2^{5-1} resolution IV plan obtained by selecting columns 2, 3, 4, 5, and 8 has 22 level changes in all. This is the least possible for a resolution IV plan. However, this plan has the second main effect at risk to a possible linear trend. An alternative resolution IV plan that does not suffer this risk is obtained by selecting columns 3, 6, 7, 12, and 13. This plan has 28 level changes and all main effects resistant to a linear trend. It is possible to construct a resolution V plan by selecting columns 2, 4, 7, 8, and 9. However, this plan involves 30 level changes.

A 2^{6-2} resolution IV plan with a minimal number of level changes is constructed by selecting columns 2, 3, 4, 5, 8, and 9. The 2^{7-3} resolution IV plan with minimal number of level changes is obtained by selecting columns 2, 3, 4, 5, 8, 9, and 14. Adding column 15 yields the 2^{8-4} resolution IV plan. Inspection of these three plans reveals that all have at least one main effect at risk of a possible linear trend. An investigator may be well advised to look for a compromise plan that protects all main effects against a possible linear trend at the cost of some extra level changes. Also, since the number of level changes per column (factor) increases as one proceeds from left to right in the base array, there may be some opportunities for decreasing the cost of experimentation by appropriate choice of factor-to-column assignment.

32-Run Plans

To construct these plans, one must first construct the 32×31 base array. This construction is based on the initial set of five generators, a, b, c, d, and e. Resolution III plans are again straightforward. The 2^{9-4} resolution IV plan with minimal number of level changes is obtained by selecting columns 4, 5, 6, 7, 8, 9, 10, 11, and 16 from the base array. The 2^{10-5} resolution IV plan with minimal number of level changes is obtained by selecting columns 4, 5, 6, 7, 8, 9, 10, 11, 16, and 17 from the base array.

A resolution V plan that keeps all main effects protected against linear and quadratic trends and then minimizes the number of level changes is given by selecting columns 5, 9, 10, 11, 13, and 17. This plan may, however, be somewhat suspect, in that it leaves two-factor interactions AC, BC, CD, and CE vulnerable to potential linear trends.

General Method of Construction

Resolution III plans are easily constructed. Cheng et al. (1998) give a general set of rules for constructing resolution IV plans with a minimal number of level changes. The rules are somewhat complex and involve a number of steps. Assume that the study is to involve N factors. The first step is to determine the number of generators needed to construct the sequence that defines the base array. Since at least $2N$ runs are required for a resolution IV plan with N factors, it follows that we need at least k generators, where k is determined from

$$2^{k-1} < 2N \leq 2^k.$$

For example, if the study involves 10 factors, then 20 must lie between 2^{k-1} and 2^k; i.e., k must be 5. The procedure is to select k generators, use the reverse sequence foldover to generate the sequence of 2^k terms, and then the $2^k \times 2^k - 1$ base array. Now there are two cases. First, if $2^k/4 < N \leq 2^k/4 + 2^k/8$, then select 2^{k-2} columns $2^{k-3}, 2^{k-3}+1, \ldots, 2^{k-3}+2^{k-2}$ and the remaining $N-2^{k-2}$ columns beginning with columns $2^{k-1}, \ldots$. If, on the other hand, $2^k/4+2^k/8 < N \leq 2^k/2$, then select the 2^{k-2} columns 2^{k-3}, $2^{k-3}+1, \ldots, 2^{k-3}+2^{k-2}-1$, then the 2^{k-3} columns $2^{k-1}, \ldots, 2^{k-1}+2^{k-3}-1$ and the remaining columns beginning with $2^k - 2^{k-3}$, $2^k - 2^{k-3} + 1, \ldots$. This rule ignores any restriction to avoid possible linear trends.

18.5.2 Maximizing Number of Level Changes

The problem of maximizing the number of level changes arises when the runs must be in sequence and the nature of the experimental material leads one to suspect that the errors have a strong serial correlation. One would suspect such serial correlations in cases where the basic experimental material consists of something like mineral ore or industrial processes that run continuously and are subject to small random changes. Complex Ev-Op programs come to mind. One would also expect serial correlations in plant responses when plants are grown in a long row in a field. One is really confronted with two problems. There is the question of order of runs, and additionally a question of appropriate statistical analysis. Ordinary least squares may not be the best. Discussions of this topic can be found in Steinberg (1988), Cheng (1990), Cheng and Steinberg (1991), Saunders and Eccleston (1992), and Saunders et al. (1995). Fortunately, any of a number of error structures with positive serial correlations, and either generalized least squares or ordinary least squares, lead to the same recommendation for ordering the runs: many level changes. The actual construction is straightforward.

One begins with the set of generators, applies the reverse sequence foldover, and then constructs the base array. Plans with a maximal number of level changes are then obtained by selecting columns, beginning from the right-hand side. If a resolution III plan is desired, all columns are eligible, although the investigator should seriously consider omitting columns that are subject to linear trends. If resolution IV plans are required, one must be somewhat more careful to avoid selecting columns that represent generalized interactions of columns selected previously. Table 1.2 presents a collection of 8-, 16-, and 32-run plans that either maximize the number of level changes or at least give a large number of changes while protecting against a possible linear trend.

Exercise 18.10: Run Order. Display the series of eight runs in appropriate order for the resolution III plan for five factors with the maximum number of level changes. □

Exercise 18.11: Resolution IV Plans with Maximal Level Changes. In Table 1.2, the claim is made that the plan with five factors in 16 runs identified as $\mathcal{I} = ABCD$ has all main effects resistant to a possible linear trend, while the plan identified as $\mathcal{I} = BCDE$ has more level changes. Verify that these statements are true. □

Table 18.2 Some Useful Plans with Many Level Changes

No. of Runs	No. of Factors	Res.	Defining Contrasts	Base Array	Columns Selected[a]
8	3		Complete	8×7	5–7
8[b]	3		Complete	8×7	4–6
8	4	IV	$\mathcal{I} = ABCD$	8×7	4–7
8	5	III	$\mathcal{I} = BCDE$	8×7	3–7
			$= ACD = ABE$		
16	4		Complete	16×15	11, 13–15
16[b]	4		Complete	16×15	11–14
16	4	IV	$\mathcal{I} = ABCD$ repeated	16×15	12–15
16	5	IV	$\mathcal{I} = BCDE$	16×15	11–15
16[b]	5	IV	$\mathcal{I} = ABCD$	16×15	10–14
16	5	V	$\mathcal{I} = ABCDE$	16×15	10–15
			$= CDEF = ABEF$		
16[b]	6	IV	$\mathcal{I} = ABEF$	16×15	9–14
16			$= BCDE = ACDF$		
	7	IV	$\mathcal{I} = ABEF$	16×15	9–15
			$= BCDE = ACDF$		
			$= ABDG = DEFG$		
			$= ACEG = BCFG$		
16[b]	7	IV	$\mathcal{I} = ABCD$	16×15	8–14
			$= ABEF = CDEF$		
			$= ACEG = BDEG$		
			$= BCFG = ADFG$		
16	8	IV	$\mathcal{I} = ABCD$	16×15	8–15
			$= ABEF = ACEG$		
			$= ACGH$ and gen. int.		
32	5		Complete	32×31	23, 27, 29–31
32[b]	5		Complete	32×31	23, 25–27, and 29
32	5	IV	$\mathcal{I} = BCDE$ repeated	32×31	27–31
32	6	IV	$\mathcal{I} = CDEF$	32×31	23, 27–31
32	6	VI	$\mathcal{I} = ABCDEF$	32×31	16, 23, 27, 29–31
32[b]	6	VI	$\mathcal{I} = ABCDEF$	32×31	13, 16, 23, 27, 29 and 30
32	6	IV	$\mathcal{I} = ABEF$	32×31	25–30
			$= BCDE = ACDF$ repeated		

(*continued*)

Table 18.2 (*continued*)

No. of Runs	No. of Factors	Res.	Defining Contrasts	Base Array	Columns Selected[a]
32[b]	6	IV	$\mathcal{I} = ABCD$ $= ABEF = CDEF$ repeated	32×31	26–31
32	7	IV	$\mathcal{I} = ABEF$ $= BCDE = ACDF$	32×31	23, 26–31
32	7	IV	$\mathcal{I} = ABEF$ $= BCFG = ACDF$ and gen. int., repeated	32×31	25–31
32[b]	7	IV	$\mathcal{I} = ABCD$ $= CDEF = ADFG$ and gen. int., repeated	32×31	24–30
32	8	IV	$\mathcal{I} = ABCD$ $= ABEF = BDEF$ and gen. int.	32×31	23, 25–31

[a]Column numbers also correspond to numbers of level changes in the column. The base array is the reverse sequence foldover.
[b]These plans have all main effects resistant to possible linear trends. The cost is somewhat fewer level changes, compared to the plan that maximizes the number of level changes that immediately precedes it.
Source: Adapted from Cheng et al. (1998) with permission from *Annals of Statistics*. Copyright 1998.

18.5.3 Compromise Plans

A criticism of much of the work on minimizing the number of level changes in factorial experiments is that it is generally assumed that all level changes are equally costly. Wang and Jan (1995) address this problem.

They present a strategy for assigning factors in two-level factorial experiments that allow the experimenter to reduce the number of level changes for selected factors. Their paper includes a number of 16- and 32-run plans that satisfy various criteria. Their approach to the design problem is also to set up 16×15 (or 32×31) array of ± 1's based on a 2^4 (or 2^5) complete with all interactions and then reassigning factors to the various columns. If a trend-resistant plan is needed, factors are assigned only to interaction columns. Also, some columns have fewer level changes than others. These are used for factors for which level changes are costly. Alternatively, one can follow a strategy based on assigning factors to columns of base arrays, analogous to array (1.2). Similar concerns appear when maximizing the number of level changes. Changing some factors may be so difficult that the investigator is forced to accept some inefficiencies due to an error structure.

18.6 TREND-FREE PLANS FROM HADAMARD MATRICES

An alternate procedure for constructing trend-free plans based on computing Kronecker or direct products of Hadamard matrices has been explored by Wang

(1990). Consider the Kronecker product

$$\begin{bmatrix} +1 & +1 \\ +1 & -1 \end{bmatrix} \otimes \begin{bmatrix} +1 & +1 \\ +1 & -1 \end{bmatrix} = \begin{bmatrix} +1 & +1 & +1 & +1 \\ +1 & -1 & +1 & -1 \\ +1 & +1 & -1 & -1 \\ +1 & -1 & -1 & +1 \end{bmatrix}.$$

The estimate of the treatment assigned to column 4 of the product matrix is orthogonal to a linear trend, i.e., $(+1, -1, -1, +1)(1, 2, 3, 4)^t = 0$. Column 4 was generated from column 2 in both initial matrices. Notice that columns 2 and 3 are not trend resistant, and of course, column 1 is not used for a factor. We can proceed further:

$$\begin{bmatrix} +1 & +1 & +1 & +1 \\ +1 & -1 & +1 & -1 \\ +1 & +1 & -1 & -1 \\ +1 & -1 & -1 & +1 \end{bmatrix} \otimes \begin{bmatrix} +1 & +1 \\ +1 & -1 \end{bmatrix}$$

$$= \begin{bmatrix} +1 & +1 & +1 & +1 & +1 & +1 & +1 & +1 \\ +1 & -1 & +1 & -1 & +1 & -1 & +1 & -1 \\ +1 & +1 & -1 & -1 & +1 & +1 & -1 & -1 \\ +1 & -1 & -1 & +1 & +1 & -1 & -1 & +1 \\ +1 & +1 & +1 & +1 & -1 & -1 & -1 & -1 \\ +1 & -1 & +1 & -1 & -1 & +1 & -1 & +1 \\ +1 & +1 & -1 & -1 & -1 & -1 & +1 & +1 \\ +1 & -1 & -1 & +1 & -1 & +1 & +1 & -1 \end{bmatrix}.$$

Estimates of factors assigned to columns 4, 6, 7, and 8 are all resistant to a linear trend, i.e.,

$$(+1, -1, -1, +1, +1, -1, -1, +1)(1, 2, 3, 4, 5, 6, 7, 8)^t = 0$$

$$(+1, -1, +1, -1, -1, +1, -1, +1)(1, 2, 3, 4, 5, 6, 7, 8)^t = 0$$

$$(+1, +1, -1, -1, -1, -1, +1, +1)(1, 2, 3, 4, 5, 6, 7, 8)^t = 0$$

$$(+1, -1, -1, +1, -1, +1, +1, -1)(1, 2, 3, 4, 5, 6, 7, 8)^t = 0.$$

In addition, note that column 8 is also resistant to a quadratic trend. In particular,

$$(+1, -1, -1, +1, -1, +1, +1, -1)(1, 4, 9, 16, 25, 36, 49, 64)^t = 0.$$

Notice that column 4 contains two stacked copies of column 4 in the first matrix, columns 6, 7, and 8 are generated from columns 2, 3 and 4 in the first matrix and column 2 in the second, and column 8 is generated from column 4 in the first matrix and column 2 in the second.

We can build up any desired level of trend resistance by repeating the process of multiplying by the $\begin{bmatrix} +1 & +1 \\ +1 & -1 \end{bmatrix}$ matrix. For example, we can construct a 16-run plan where columns 4, 6, 7, 9, 10, 11, 12, and 13 are linear trend resistant, columns 8, 12, 14, and 15 are linear and quadratic trend resistant, and column 16 is linear, quadratic, and cubic trend resistant.

We now consider the 12×12 Hadamard matrix

$$\begin{bmatrix}
+1 & +1 & +1 & -1 & +1 & +1 & +1 & -1 & -1 & -1 & +1 & -1 \\
+1 & -1 & +1 & +1 & -1 & +1 & +1 & +1 & -1 & -1 & -1 & +1 \\
+1 & +1 & -1 & +1 & +1 & -1 & +1 & +1 & +1 & -1 & -1 & -1 \\
+1 & -1 & +1 & -1 & +1 & +1 & -1 & +1 & +1 & +1 & -1 & -1 \\
+1 & -1 & -1 & +1 & -1 & +1 & +1 & -1 & +1 & +1 & +1 & -1 \\
+1 & -1 & -1 & -1 & +1 & -1 & +1 & +1 & -1 & +1 & +1 & +1 \\
+1 & +1 & -1 & -1 & -1 & +1 & -1 & +1 & +1 & -1 & +1 & +1 \\
+1 & +1 & +1 & -1 & -1 & -1 & +1 & -1 & +1 & +1 & -1 & +1 \\
+1 & +1 & +1 & +1 & -1 & -1 & -1 & +1 & -1 & +1 & +1 & -1 \\
+1 & -1 & +1 & +1 & +1 & -1 & -1 & -1 & +1 & -1 & +1 & +1 \\
+1 & +1 & -1 & +1 & +1 & +1 & -1 & -1 & -1 & +1 & -1 & +1 \\
+1 & -1 & -1 & -1 & -1 & -1 & -1 & -1 & -1 & -1 & -1 & -1
\end{bmatrix}.$$

The second column in this matrix is resistant to both linear and quadratic trends, and the remaining columns are not trend resistant. Wang (1990) gives the following matrix:

$$\begin{bmatrix}
+1 & +1 & +1 & +1 & +1 & +1 & +1 & +1 & +1 & +1 & +1 & +1 \\
+1 & +1 & +1 & +1 & +1 & +1 & -1 & -1 & -1 & -1 & -1 & -1 \\
+1 & +1 & +1 & -1 & -1 & -1 & +1 & +1 & +1 & -1 & -1 & -1 \\
+1 & +1 & -1 & +1 & -1 & -1 & +1 & -1 & -1 & +1 & +1 & -1 \\
+1 & +1 & -1 & -1 & +1 & -1 & -1 & +1 & -1 & +1 & +1 & -1 \\
+1 & +1 & -1 & -1 & -1 & +1 & -1 & -1 & +1 & -1 & +1 & +1 \\
+1 & -1 & +1 & -1 & -1 & +1 & +1 & -1 & -1 & +1 & -1 & +1 \\
+1 & -1 & +1 & -1 & +1 & -1 & -1 & -1 & +1 & +1 & +1 & -1 \\
+1 & -1 & +1 & +1 & -1 & -1 & -1 & +1 & -1 & -1 & +1 & +1 \\
+1 & -1 & -1 & -1 & +1 & +1 & +1 & +1 & -1 & -1 & +1 & -1 \\
+1 & -1 & -1 & +1 & -1 & +1 & -1 & +1 & +1 & +1 & -1 & -1 \\
+1 & -1 & -1 & +1 & +1 & -1 & +1 & -1 & +1 & -1 & -1 & +1
\end{bmatrix}.$$

which has columns 4 and 8 linear trend–free. Recall that all 12×12 Hadamard matrices are equivalent in the sense that any one can be obtained from any other by interchanging rows, interchanging columns, and/or changing signs in complete columns. We see that to an extent, the trend-resistant properties of a matrix can be modified by interchanging rows. Depending on the choice for $\boldsymbol{H}_{12}$, the multiplication $\boldsymbol{H}_2 \otimes \boldsymbol{H}_{12}$ gives either a matrix that has one column cubic trend-free and 11 linear trend-free columns, or a matrix with two columns that are quadratic trend free and 10 columns that are free of a linear trend.

Exercise 18.12: 24-Run Trend-Resistant Plan. It is possible to construct a run order for a factorial with 12 factors in 24 runs, such that all main effects are resistant to a linear trend. Give such a sequence of runs. Investigate the status of the two-factor interactions in your plan. □

Manipulation of the sort illustrated above yield many special types of plans. Consider, for example, an experiment of seven factors in 16 runs, where trends may be important. One would assign the most important factor to the column generated by $[(+1\ -1)\otimes(+1\ -1)\otimes(+1\ -1)\otimes(+1\ -1)]^t$, which is cubic trend free, the next four factors to the columns $[(+1\ +1)\otimes(+1\ -1)\otimes(+1\ -1)\otimes(+1\ -1)]^t, \ldots, [(+1\ -1)\otimes(+1\ -1)\otimes(+1\ -1)\otimes(+1\ +1)]^t$ and the remaining two factors to any columns constructed by using two $(+1\ +1)$ and two $(+1\ -1)$ terms.

A slightly more elaborate example is the experiment with seven factors in 32 runs. One must assign the seven factors to the columns:

$$[(+1\ -1)\otimes(+1\ -1)\otimes(+1\ -1)\otimes(+1\ -1)\otimes(+1\ -1)]^t$$
$$[(+1\ +1)\otimes(+1\ +1)\otimes(+1\ -1)\otimes(+1\ -1)\otimes(+1\ -1)]^t$$
$$[(+1\ -1)\otimes(+1\ +1)\otimes(+1\ +1)\otimes(+1\ -1)\otimes(+1\ -1)]^t$$
$$[(+1\ -1)\otimes(+1\ -1)\otimes(+1\ +1)\otimes(+1\ +1)\otimes(+1\ -1)]^t$$
$$[(+1\ -1)\otimes(+1\ -1)\otimes(+1\ -1)\otimes(+1\ +1)\otimes(+1\ +1)]^t$$
$$[(+1\ +1)\otimes(+1\ -1)\otimes(+1\ -1)\otimes(+1\ -1)\otimes(+1\ +1)]^t$$
$$[(+1\ -1)\otimes(+1\ +1)\otimes(+1\ -1)\otimes(+1\ +1)\otimes(+1\ -1)]^t\ .$$

It can be shown that when the two-factor interactions involving the factor assigned to the seventh column above are negligible, the remaining two-factor interactions become estimable. In fact, the plan is resolution IV. When only the first six factors are used, the plan is 2^{6-1} resolution V. Either of these experiments has main effects resistant to linear and quadratic trends and two-factor interactions resistant to linear trends.

***Exercise 18.13: Linear Trend-Free Plan—Seven Factors in* 16 *Runs*.** Exhibit the 16 runs in the plan obtained by assigning factors to the columns:

$$[(+1\ -1)\otimes(+1\ -1)\otimes(+1\ -1)\otimes(+1\ -1)]^t$$
$$[(+1\ +1)\otimes(+1\ -1)\otimes(+1\ -1)\otimes(+1\ -1)]^t$$
$$[(+1\ -1)\otimes(+1\ +1)\otimes(+1\ -1)\otimes(+1\ -1)]^t$$
$$[(+1\ -1)\otimes(+1\ -1)\otimes(+1\ +1)\otimes(+1\ -1)]^t$$
$$[(+1\ -1)\otimes(+1\ -1)\otimes(+1\ -1)\otimes(+1\ +1)]^t$$
$$[(+1\ +1)\otimes(+1\ +1)\otimes(+1\ -1)\otimes(+1\ -1)]^t$$
$$[(+1\ -1)\otimes(+1\ -1)\otimes(+1\ +1)\otimes(+1\ +1)]^t\ .$$

Verify that all main effects are free of a possible linear trend. □

18.7 EXTENSIONS TO MORE THAN TWO LEVELS

As a first step in the extension of the concepts developed to trend-free plans for factorials with more than two levels, we need to refine the notion of trend-free. We follow the definition given by Bailey et al. (1992). Recall that in the two-level factorial, an effect was said to be t-trend resistant if the least squares estimate of the effect was free of t-order trends but not free of $(t+1)$-order trends. In the extension to three-level factorials we require that least squares estimates of both linear and quadratic treatment effects be trend-free. In the case of an interaction the least squares estimates of the linear $\times$ linear, linear $\times$ quadratic, quadratic $\times$ linear, and quadratic $\times$ quadratic must all be resistant to t-order trends for the interaction to be declared t-order trend resistant. The extension to higher-order interactions and factorials with more levels is immediate. Bailey et al. prove that if N factors in a factorial are in standard Yates order, any k-factor interaction is $(k-1)$-trend-free. One of the possibilities is illustrated with the 3^3 factorial in standard order, with equally spaced levels, 0, 1, and 2 in Table 1.3.

It is a simple exercise to verify that none of the three main effects, labeled as A, B, and C, are resistant to a linear trend with the run order given. However, the two-factor interactions are resistant to a linear trend. To demonstrate this for the $A \times B$ interaction, for example, one must look at the $A_L \times B_L$, $A_L \times B_Q$, $A_Q \times B_L$, and $A_Q \times B_Q$ contrasts separately. The three-factor interaction is resistant to both linear and quadratic trends. Since an experimenter is usually much more interested in main effects than interactions, this run order is probably of limited utility.

Table 18.3 Three-Level Resistant to Trends Plans

A	B	C	A^*	B^*	C^*	D^*	A^+	B^+	C^+	D^+	E^+	F^+
0	0	0	0	0	0	0	0	0	0	0	0	0
0	0	1	1	2	1	1	0	0	1	2	1	2
0	0	2	2	1	2	2	0	0	2	1	2	1
0	1	0	1	1	2	2	1	2	0	0	1	1
0	1	1	2	0	0	0	1	2	1	2	2	0
0	1	2	0	2	1	1	1	2	2	1	0	2
0	2	0	2	2	1	1	2	1	0	0	2	2
0	2	1	0	1	2	2	2	1	1	2	0	1
0	2	2	1	0	0	0	2	1	2	1	1	0
1	0	0	1	1	1	2	1	1	1	1	0	0
1	0	1	2	0	2	0	1	1	2	0	1	2
1	0	2	0	2	0	1	1	1	0	2	2	1
1	1	0	2	2	0	1	2	0	1	1	1	1
1	1	1	0	1	1	2	2	0	2	0	2	0
1	1	2	1	0	2	0	2	0	0	2	0	2
1	2	0	0	0	2	0	0	2	1	1	2	2
1	2	1	1	2	0	1	0	2	2	0	0	1
1	2	2	2	1	1	2	0	2	0	2	1	0

Table 18.3 ***(continued)***

A	B	C	A^*	B^*	C^*	D^*	A^+	B^+	C^+	D^+	E^+	F^+
2	0	0	2	2	2	1	2	2	2	1	0	0
2	0	1	0	1	0	2	2	2	0	0	1	2
2	0	2	1	0	1	0	2	2	1	2	2	1
2	1	0	0	0	1	0	0	1	2	2	1	1
2	1	1	1	2	2	1	0	1	0	1	2	0
2	1	2	2	1	0	2	0	1	1	0	0	2
2	2	0	1	1	0	2	1	0	2	2	2	2
2	2	1	2	0	1	0	1	0	0	1	0	1
2	2	2	0	2	2	1	1	0	1	0	1	0

However, there are other possibilities. We can work from the three-factor interaction which we know is resistant to linear and quadratic trends. We define four new columns, $i_{A^*} = i_A + i_B + i_C$ (modulo 3), $i_{B^*} = i_A + i_B + 2i_C$ (modulo 3), $i_{C^*} = i_A + 2i_B + i_C$ (modulo 3), and $i_{D^*} = i_A + 2i_B + 2i_C$ (modulo 3). These columns present a plan for a 3^{4-1} fraction with all four main effects resistant to linear and quadratic trends. These columns are identified as A^*, B^*, C^*, and D^* in Table 18.3. Alternatively, one could define six columns, $i_{A^+} = i_A + i_B$ (modulo 3), $i_{B^+} = i_A + 2i_B$ (modulo 3), $i_{C^+} = i_A + i_C$ (modulo 3), $i_{D^+} = i_A + 2i_C$ (modulo 3), $i_{E^+} = i_B + i_C$ (modulo 3), and $i_{F^+} = i_B + i_C$ (modulo 3) and assign six factors to these columns. The result is a plan for a 3^{6-3} with all six main effects resistant to a linear trend. Table 18.3 shows these as columns $A^+, \ldots, F^+$. Remember, however, that both of these plans are fractions of a full replicate and have alias problems.

18.8 SMALL ONE-AT-A-TIME PLANS

In this section we consider a collection of plans suitable for a laboratory setting in which the investigator can observe the results of his experiments or individual trials quickly and can use this information as input to the decision-making process before the next trial. The investigator can react to the observations as they come, or at least there are no forces that compel him to make a series of runs before studying the results of the first run. Examples that come to mind are simulation studies on a computer or tests in a chemistry laboratory.

A second assumption inherent in the material to be considered is that the random error component is so small relative to the anticipated responses to treatments that it can be ignored. This again is consistent with a simulation study on a computer where there is no random error. However, it will be assumed that finding the proper model is important. We pay special attention to the case where there may be large two-factor interactions which the experimenter wishes to uncover. The methods extend directly to cases where the experimenter is concerned about higher-order interactions as well. A key reference for this material is Daniel (1973).

An additional constraint to observe will be to make as few changes as possible in the experimental protocol when going from one run to the next. This can be of major advantage in a laboratory or industrial setting where shifting from one level of a factor to another level may require major reconstruction of equipment. Again the price for this is no randomization. However, if random error really is small, this should not be serious. A potential problem to be considered is that there may be an overall drift in the system over time. Later observations may be favored because of experience by the technician or discriminated against by wear in the equipment. We will consider plans that guard against this type of trend.

18.8.1 Small Plans that Include Interactions

2^2 Experiment

For the first example consider the 2^2 experiment where the strategy is

Run 1 (1)
Run 2 a implies that we can compute $a - (1)$
estimate of $A - AB$.
Run 3 ab $\Rightarrow$ $ab - a$ $\Rightarrow$ estimate of $B + AB$.
Run 4 b $\Rightarrow$ A, B, and AB become estimable.

Note that only one factor was changed at each stage. At both run 2 and run 3 we observe something that can be interpreted, provided that we are willing to assume no interaction. Then once the four corners of the quadrilateral were observed, the interaction became estimable.

2^3 Experiment

The 2^3 case is more interesting. We begin by ignoring the possibility of a three-factor interaction.

Run 1 (1)

Run 2 a $\Rightarrow$ $a-(1)$ $\Rightarrow$ estimate of $A - AB - AC$.

Run 3 ab $\Rightarrow$ $ab - a$ $\Rightarrow$ estimate of $B + AB - BC$.

Run 4 abc $\Rightarrow$ $abc - ab$ $\Rightarrow$ estimate of $C + AC + BC$.

Things are nicely interpretable at this stage if there are no interactions at all. Otherwise, continue.

Run 5 bc $\Rightarrow$ $abc - bc$ $\Rightarrow$ estimate of $A + AB + AC$
$\Rightarrow$ one can estimate A and $AB + AC$.
Run 6 c $\Rightarrow$ $bc - c$ $\Rightarrow$ estimate of $B + BC - AB$
$\Rightarrow$ one can estimate B, $BC - AB$,
$BC + AC$, and C.

At this stage all main effects are clear of two-factor interactions. However, two-factor interactions are still not clear of one another.

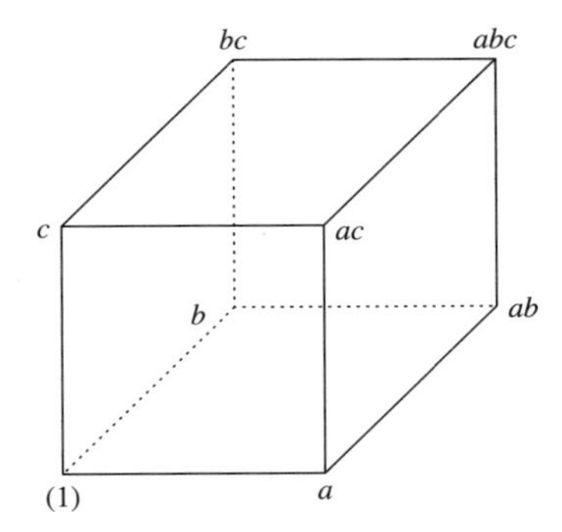

Figure 18.1 Three-factor factorial as a geometric figure.

Run 7 ac $\Rightarrow$ AB, AC, and BC are all clear of main effects and one another.

At this stage we have three faces on the cube (Figure 18.1) and consequently, can estimate all three two-factor interactions.

Run 8 b

This last run may not be needed. It is special in that two factors had to be changed. It does, however, make the three-factor interaction estimable.

2^4 Factorial

We now consider the 2^4 factorial.

Run 1 (1)
Run 2 a can now estimate $A - AB - AC - AD$.
Run 3 ab can now estimate $B + AB - BC - BD$.
Run 4 abc can now estimate $C + AC + BC - CD$.
Run 5 $abcd$ can now estimate $D + AD + BD + CD$.

At this point we are in good shape if there are no two-factor interactions.

Run 6 bcd can now estimate A and $AB + AC + AD$.

Alternative options at this stage would have been acd or abd. These would have given $B + AB + BC + BD$ to free up $B + AB$ and $BC - BD$ or $C + AC + BC + CD$ to free up $C + AC + BC$ and CD, respectively.

Run 7 cd can now estimate B and $AB - BC - BD$.
Run 8 d can now estimate C and $AC + BC - CD$.

At this point we can also take $d-(1)$ to estimate $D - AD - BD - CD$, which frees up D and $AD + BD + CD$. All main effects are estimated at this point. All changes have involved only one factor at each step.

Table 18.4 Sequence of 10 Runs in a Fraction of the 2^5

Run	Specification	Estimable Quantities
1	(1)	—
2	a	$A -$ interactions with A
3	ab	$B \pm$ interactions with B
4	abc	$C \pm$ interactions with C
5	$abcd$	$D \pm$ interactions with D
6	$abcde$	$E \pm$ interactions with E
7	$bcde$	$A,\ AB + AC + AD + AE$
8	cde	$B,\ AB - BC - BD - BE$
9	de	$C,\ AC + BC - CD - CE$
10	e	$D,\ AD + BD + CD - DE$

Table 18.5 Six Additional Runs in a Fraction of the 2^5

Run	Specification	Estimable Quantities
11	ae	AE (Completes one face on a five-dimensional figure.)
12	ade	AD (Use runs 9, 10, 11, and 12.)
13	$acde$	AC and AB (Use runs 8, 9, 12, and 13.)
14	acd	BE
15	ac	$BC,\ BD$
16	ace	$CD,\ CE,\ DE$

2^5 Factorial

Now consider a plan for a five-factor experiment, shown in Table 18.4. All main effects are estimable at this point. If we continue, the next six runs are shown in Table 18.5. All two-factor interactions estimable at this point.

18.8.2 Protected 2^3 Plan

Consider a realistic example adapted from a recent experiment. A machine has been designed to sort sweet potatoes. Three different factors were to be examined in a 2^3 factorial. Changing any one of the factors involved rather extensive disassembly and reassembly of the equipment. This presented a major advantage to changing factors one at a time. Now introduce an extra wrinkle. One lot of potatoes was to be used in all trials. The reason for this is the irregular shapes of sweet potatoes. It is impossible to assemble two lots of sweet potatoes that contain exactly the same mixture of shapes. The potatoes were run through the machine and a score assigned on the basis of how well the potatoes were sorted. The potatoes were then mixed again, to be run through the machine again. The advantage was that the same potatoes were used in all trials. The problem was that over time the potatoes would become bruised and change. In terms of the design, this meant that there would be a trend in the responses. This would have to be removed.

This required some extra observations. It was assumed that the trend was linear.
Strategy:

Run 1	(1)	
Run 2	a	Compute $[a-(1)]$ to estimate $A - AB - AC + L$.
Run 3	ab	Compute $(ab - a)$ to estimate $B + AB - BC + L$.
Run 4	abc	Compute $(abc - ab)$ to estimate $C + AC + BC + L$.
Run 5	bc	Compute $(abc - bc)$ to estimate $A + AB + AC - L$ also at this time estimate A and $AB + AC - L$.
Run 6	c	Compute $(bc - c)$ to estimate $B - AB + BC - L$ also at this time estimate B and $AB - BC + L$. Compute $[c-(1)]$ to estimate $C - AC - BC + 5L$ and an estimate of $C + 6L$.
Run 7	(1)	Use this and run 1 to estimate $6L$.

At this point all main effects can be estimated, free of two-factor interactions and free of the linear trend. This same sort of thing can be extended to protect against quadratic trends as well.

18.9 COMMENTS

Some more recent ideas related to the material in this chapter (but presented in a much more abstract setting) can be found in Srivastava and Hveberg (1992). Also, some related ideas can be found in the literature on "*search designs*." A recent reference is Gupta (1991). Both *search* and *probing designs* are similar in the sense that the experimenter can identify a set of parameters that are unknown but assumed to be important, and a second set of parameters that includes a subset of k_0 parameters which may be important. For example, it may be reasonable to assume that the mean, main effects, and block effects are real and that possibly k_0 of the two-factor interactions are important. The difficulty is that the exact identity of the members of this subset in the second set is not known. In fact, often k_0 is unknown, although one may be willing to assume that no more than, say, three of the two-factor interactions are important. Search and probing designs differ in that probing designs allow one to add more observations sequentially after one has started to examine the data, while the search design assumes that one must select the treatment combinations to be examined in advance as, for example, in an agricultural field experiment. The assumption of a *tree structure* among the elements in a set of parameters means that if a three-factor interaction exists in the set, at least one of the three two-factor interactions obtained by dropping any one of the factors is also in the set. Similarly, if a two-factor interaction is in the set, at least one of the main effects is also in the set.

CHAPTER 19

Sequences of Fractions of Factorials

19.1 INTRODUCTION

A basic fact in all research work is that the investigator does not know all of the factors that will eventually prove to be relevant. There is the famous comment by J. W. Tukey that "Every experiment is a confounded experiment" (Daniel 1962). We would like to modify this slightly to: "Every experiment is a fractional factorial experiment with some factors confounded." It is important to remember the following quote from Daniel.

> If the result (of an experiment) is to be used for any other purpose than as a record of what happened that time, then all the conditions under which it was taken should be recorded because they are relevant, or indicated to be irrelevant. If this is not done, then it may be forgotten that although the result measures the response at the desired levels of the factors controlled, it is perturbed by influences of the levels of other factors that may not be repeated in the future. It is this perturbation that is called $(\cdots)$ confounding.
>
> Confounding may come into the simplest possible experimental situation, and has little to do with experimental error. The result of the measurement may be judged by the experimenter to apply to another situation, say to another level of some factor we will call A. This judgment may be a good one, either because the variation of A produces no change under any condition, or because A does not have any effect so long as the other experimental conditions, B, C, $\ldots$, remain at their existing levels. On the other hand, the judgment may be mistaken; a change in A may indeed change the result by an important amount. In that case we say that the results of the conditions of the original experiment are confounded with the *effect* of factor A.

The problem in all research is that the identity and appropriate range of all of the relevant factors, whether controlled or not, is unknown. After all, if we had complete knowledge, there would be little or no need for most research. As work

Planning, Construction, and Statistical Analysis of Comparative Experiments,
by Francis G. Giesbrecht and Marcia L. Gumpertz
ISBN 0-471-21395-0

proceeds on a research project, more becomes known, the range of some factors needs to be extended, factors can be dropped, new factors need to be added, and so on.

In this chapter we explore a number of techniques that allow an investigator to introduce new factors and new levels of factors into a study in a systematic manner as the investigation proceeds. Dropping factors that prove to be irrelevant is always easy. In fact, one of the great strengths of the factorial experiments is that nonactive factors as well as active factors provide replication and increase the sensitivity of the experiment for other factors. Often, including additional factors has modest impact on cost. It is adding new factors, after part of an experiment has been run, without unnecessary duplication of effort that presents a challenge. Additionally, there is always the ever-present limitation on resources. Often, an investigator finds that at the beginning of a study only a fraction of the resources required for a proper study are available. Machinery is required to set up a study that is doable with existing resources, but the experiment can be extended efficiently as more resources become available.

We pursue two distinct, but related aspects of sequences of fractional factorial plans. In one, we assume that experimental results become available to the investigator relatively quickly and that the information can be incorporated into future work. This situation is often handled in practice by a series of small disconnected studies. Our objective is to illustrate techniques that lead to a large coherent study without undue waste of resources. In the other scenario, we assume that the investigator has a comprehensive plan in mind, but due to restrictions in resources, must be prepared to execute part of the plan and anticipate having to make preliminary analyses and interpretations of the results before the full study is completed. The basic idea is that interpretation under more realistic assumptions will become possible as more data become available.

Basic to our procedure are a number of recipes related to the foldover technique developed initially by Box and Wilson (1951). At one point in our discussion of sequences of fractional factorial plans we digress to a discussion of the foldover and related methodologies. A benefit of this will be a compact, yet simple system of notation. We will routinely make use of the fact that a regular fraction of a factorial can be specified in two ways. First, one can list the exact set of treatment combinations. Alternatively, one can specify the fraction by means of a defining contrast, really a collection of effects and interactions. This collection of effects and interactions is written as a series of words (each with appropriate sign). We find it convenient to use both representations, and more important, to shift back and forth between the two. The collection of treatment combinations is convenient (needed by the investigator) in the laboratory, while the defining contrast notation is useful to spell out the alias relationships.

An underlying assumption in all of our examples in this chapter will be that the experimenter has very good control of the experimental material. First, error variances are assumed to be small. Second, there is no danger of missing values, or if for some unexplained or accidental reason a trial fails, or the result is an obvious outlier, the trial can be repeated immediately. Third, it will usually be assumed that the experimental procedures are well in hand and that there is sufficient control on conditions that there is no need to introduce blocking. We also assume that the results of individual trials become available to the experimenter relatively quickly. The methods developed become especially attractive when trials are expensive or cause suffering and place a premium on minimizing the number of treatment combinations examined.

19.2 MOTIVATING EXAMPLE

We begin our discussion of sequences of fractional factorial plans with a simple example taken from Daniel (1962). The object of this example at this point is to suggest an approach and some techniques without introducing too many details. We follow this example with a section that explains notation and then a series of examples.

An investigator has observed four responses, which we denote by $y_{(1)}$, y_a, y_b, and y_c, i.e., a control and three factors, one at a time. The obvious measures of the three effects are $(y_a - y_{(1)})$, $(y_b - y_{(1)})$, and $(y_c - y_{(1)})$. We assume that the error variance is relatively small, so that these estimates are reasonably precise. As Daniel points out, at this point the major problem with this plan is not the inefficiency of the plan, but the fact that the estimates may be severely biased if there should happen to be some interactions. For example, it is simple to show that

$$E[y_a - y_{(1)}] = A - AB - AC + ABC,$$

where we are now using A, AB, AC, and ABC to represent effects and interactions, i.e., parameters. It is clearly possible that the difference could be due to the interaction of A and B or A and C. The question is: How does one correct this with minimal effort?

A first step is to add the *abc* run. An assumption here is that conditions are sufficiently stable that one can disregard any need for blocking. The investigator can now compute $(y_{abc} - y_{(1)})$ and $(y_a + y_b + y_c - 3y_{(1)})$ and more important, the difference $(y_a + y_b + y_c - y_{abc} - 2y_{(1)})$, which is an unbiased estimate of

$$-AB - AC - BC + 2ABC.$$

This provides a rough check on assumptions. If $(y_a + y_b + y_c - y_{abc} - 2y_{(1)})$ is "small," the investigator may be safe in judging that all four components are negligible. However, if this is not the case, there is need for more experimental work. Notice also that if we set $y_{(1)}$ aside, we have the main-effect plan defined by $\mathcal{I} = +ABC$ and can compute the three orthogonal contrasts

$$\begin{aligned} &y_a - y_b - y_c + y_{abc} \\ -&y_a + y_b - y_c + y_{abc} \\ -&y_a - y_b + y_c + y_{abc}, \end{aligned}$$

which provide unbiased estimates of $A + BC$, $B + AC$, and $C + AB$, respectively. The final step would be to complete the factorial by adding treatment combinations ab, ac, and bc.

19.3 FOLDOVER TECHNIQUES EXAMINED

In this section we digress to develop the formal notation that will help us keep track of what happens when the foldover and variations thereof are applied. Unfortunately, the notation becomes somewhat complicated. However, we demonstrate that the actual manipulations required in practice are very simple. Also, relatively simple rules provide guidance in the application.

We begin with the complete foldover encountered in Section 14.6. Let $\mathcal{S}$ represent a given set of treatment combinations. The *complete foldover* is the set $\mathcal{S}'$ where all treatment combinations have been replaced by the complement, i.e., high and low levels of all factors interchanged. Next, we define the *single foldover of* $\mathcal{S}$ *on factor* A as the new set of treatment combinations obtained by replacing the low levels of A with high levels of A, and high levels with low levels for all treatment combinations in $\mathcal{S}$ and leaving all other treatment levels as they were. In the same vein, we define a *double foldover on factors A and B* in which high and low levels of both factors are changed, and so on. It follows logically that the complete foldover of $\mathcal{S}$ is obtained by replacing high levels by low levels and low levels by high levels for all factors.

Example 19.1: Formal Examination of the Foldover. To illustrate the nature of the single foldover, we look at a specific example. We consider the fraction defined by

$$\mathcal{I} = -ABC = -ADE = +BCDE. \tag{19.1}$$

We recall that this fraction can also be defined as the set of solutions to the set of equations

$$\begin{array}{ccccccl} a & b & c & d & e & & \\ \hline i_A & +i_B & +i_C & & & = 0 \text{ (modulo 2)} & \\ i_A & & & +i_D & +i_E & = 0 \text{ (modulo 2)} & \\ & i_B & +i_C & +i_D & +i_E & = 0 \text{ (modulo 2).} & \end{array} \tag{19.2}$$

Equations are from $-ABC$, $-ADE$, and $+BCDE$, respectively. It is important to note that equations represent restrictions on solutions. Adding independent equations means adding more restrictions and reducing the number of solutions. We also realize that in the case presented, two equations are sufficient. The third, representing the generalized interaction, is redundant. However, for our purposes at this time it is informative to keep it in the system, although it does not reduce the number of solutions.

Now consider the single foldover on factor A. Recall that the single foldover on A has low and high levels of A interchanged. If we look at the equations, we notice that this is equivalent to finding the eight solutions to the system

$$\begin{array}{ccccccl} a & b & c & d & e & & \\ \hline i_A & +i_B & +i_C & & & = 1 \text{ (modulo 2)} & \\ i_A & & & +i_D & +i_E & = 1 \text{ (modulo 2)} & \\ & i_B & +i_C & +i_D & +i_E & = 0 \text{ (modulo 2)}. & \end{array} \tag{19.3}$$

An equivalent specification for this set of solutions is given by the defining contrast

$$\mathcal{I} = +ABC = +ADE = +BCDE.$$

From equation systems (19.2) and (19.3) we see that combining the two sets of solutions is equivalent to finding solutions to the single equation

$$\begin{array}{cccccl} a & b & c & d & e & \\ \hline & i_B & +i_C & +i_D & +i_E & = 0 \text{ (modulo 2)} \qquad \text{(from } BCDE\text{)}. \end{array}$$

We have fewer equations, fewer restrictions, and consequently, more solutions. Any treatment combination that was eliminated by the first or second equation in system (19.2) is accepted by the first and second equations in system (19.3). The simple result that we are after is to notice that combining the two fractions defined by

$$\mathcal{I} = -ABC = -ADE = +BCDE$$

and the single foldover on A defined by

$$\mathcal{I} = +ABC = +ADE = +BCDE$$

yields the fraction defined by

$$\mathcal{I} = BCDE.$$

Exercise 19.1: Foldover Sets

(a) Display the set of eight treatment combinations represented by solutions to (19.2).
(b) Find the eight solutions to the system (19.3).
(c) Show how the two sets of eight generated above differ only in levels of factor A.
(d) Show that all 16 treatment combinations found will satisfy the equation defined by $\mathcal{I} = +BCDE$. □

19.3.1 Formal Abbreviated Notation

We now develop a formal, abbreviated notation which some may wish to skip on first reading. This notation appears a bit cumbersome, but eventually provides a very compact shorthand to illustrate what is actually happening as various foldover sets of treatment combinations are added to an experimental plan.

The defining contrast that specifies a regular fraction $\mathcal{S}$ can be written formally as

$$\mathcal{I} = \mathcal{W}_{\mathcal{S}}(A) + \mathcal{W}_{\mathcal{S}}(\overline{A}), \tag{19.4}$$

where $\mathcal{W}_{\mathcal{S}}(A)$ represents the collection of words in the defining contrast (together with the algebraic signs) that contain the letter "A"; and $\mathcal{W}_{\mathcal{S}}(\overline{A})$, the words that do not that contain the letter "A." Notice that $\mathcal{W}_{\mathcal{S}}(A)$ and $\mathcal{W}_{\mathcal{S}}(\overline{A})$ in expression (19.4) are both sets and common notation would be to use the union symbol. We have chosen not to do that because as we shall see shortly, we have occasion to take unions of sets after changing signs on all words in a set. It follows that the foldover on A can be conveniently represented in our notation by the expression

$$\mathcal{I} = -\mathcal{W}_{\mathcal{S}}(A) + \mathcal{W}_{\mathcal{S}}(\overline{A})$$

and the combination of $\mathcal{S}$ and the foldover on A by

$$\mathcal{I} = \mathcal{W}_{\mathcal{S}}(\overline{A}).$$

Notice that $-\mathcal{W}$ consists of the same words as $\mathcal{W}$, but with all associated signs reversed.

For a concrete example, $\mathcal{W}_{\mathcal{S}}(A)$ in expression (19.1) consists of the two words $-ABC$ and $-ADE$, $-\mathcal{W}_{\mathcal{S}}(A)$ consists of ABC and ADE, and $\mathcal{W}_{\mathcal{S}}(\overline{A})$ constitutes one word, $+BCDE$. This notation has the extension to

$$\mathcal{I} = \mathcal{W}_{\mathcal{S}}(\overline{AB}) + \mathcal{W}_{\mathcal{S}}(A\overline{B}) + \mathcal{W}_{\mathcal{S}}(\overline{A}B) + \mathcal{W}_{\mathcal{S}}(AB),$$

where $\mathcal{W}_{\mathcal{S}}(\overline{AB})$ consists of words containing neither A nor B, $\mathcal{W}_{\mathcal{S}}(A\overline{B})$ and $\mathcal{W}_{\mathcal{S}}(\overline{A}B)$ consist of words containing A and not B and B and not A, respectively, and $\mathcal{W}_{\mathcal{S}}(AB)$ consists of the words containing both letters A and B. The double foldover on A and B can then be written as

$$\mathcal{I} = \mathcal{W}_{\mathcal{S}}(\overline{AB}) - \mathcal{W}_{\mathcal{S}}(A\overline{B}) - \mathcal{W}_{\mathcal{S}}(\overline{A}B) + \mathcal{W}(AB).$$

For the complete foldover we have to write the original definition as

$$\mathcal{I} = \mathcal{W}_{\mathcal{S}}(\text{odd number of letters}) + \mathcal{W}_{\mathcal{S}}(\text{even number of letters})$$

and the foldover set as

$$\mathcal{I} = -\mathcal{W}_{\mathcal{S}}(\text{odd number of letters}) + \mathcal{W}_{\mathcal{S}}(\text{even number of letters}).$$

19.3.2 Rules

At this point we can summarize the manipulations by two simple rules. We begin with a fraction, say $\mathcal{S}$, specified by a defining contrast consisting of $2^k - 1$ words (k words form a basis) with specific signs attached.

Rule 1. A single foldover on, say, factor A consists of the set of treatment combinations specified by the defining contrast consisting of the same set of $2^k - 1$ words, but with signs on all words containing A reversed. For a double foldover, triple foldover, and so on, the sign reversal rule is applied sequentially.

Rule 2. If two fractions, both defined by the same set of $2^k - 1$ words, but with differing complements of signs are combined into one fraction, the resulting fraction corresponds to the defining contrast specified by the subset of words that had common signs in the two fractions.

At this point we need to mention that the complete foldover involves a foldover on each factor. Applying the sign reversals in order for all factors results in an odd number of sign reversals shown above for words with an odd number of letters and an even number for words with an even number of letters. It follows that adding the complete foldover to a fraction eliminates all words with an odd number of letters from the defining contrast. Consequently, a resolution III plan will be converted to at least a resolution IV plan and a resolution V plan to at least a resolution VI plan. Also, for example, a resolution IV plan may remain a resolution IV plan.

19.3.3 Semifold

The semifold (John 2000) is related to the foldover. In any regular fraction, say $\mathcal{S}$, that permits estimation of all main effects, every factor will be at the low level for half of the treatment combinations and at the high level for half. Consider such a fraction in which the factor A is involved in at least part of the defining contrast. The half of $\mathcal{S}$ with A at the low level can be identified by adding the equation (restriction) $i_A = 0$ (modulo 2) to the system that is solved to obtain the collection of treatment combinations in $\mathcal{S}$, or alternatively, adding the word $-A$ and all of the generalized interactions to the defining contrast. Now think of generating the simple foldover on A of this subset. In practice we simply take the subset from $\mathcal{S}$ that has A at the low level, and replace the low level with the high level in all treatment combinations. The resulting set is known as the *semifold* of $\mathcal{S}$ on A. Adding the semifold on A to $\mathcal{S}$ gives an irregular fraction (see Section 14.8). However, this irregular fraction has the property that all aliases involving A are broken.

Extension of Formal Notation to the Semifold

It is possible to extend our formal notation to the semifold in an intuitively informative way. Initially, corresponding to $\mathcal{S}$, we have the defining contrast

$$\begin{aligned} \mathcal{I} &= \mathcal{W}_{\mathcal{S}} \\ &= \mathcal{W}_{\mathcal{S}}(\overline{A}) + \mathcal{W}_{\mathcal{S}}(A). \end{aligned} \tag{19.5}$$

Next, we split the set $\mathcal{S}$ into two parts, the half with a and the half without. We denote these as $\mathcal{S}(a)$ and $\mathcal{S}(\overline{a})$. Corresponding to these we have the defining contrasts

$$\begin{aligned} \mathcal{I} &= \mathcal{W}_{\mathcal{S}(a)} \\ &= \mathcal{W}_{\mathcal{S}(a)}(\overline{A}) + \mathcal{W}_{\mathcal{S}(a)}(A) \end{aligned} \tag{19.6}$$

and

$$\begin{aligned} \mathcal{I} &= \mathcal{W}_{\mathcal{S}(\overline{a})} \\ &= \mathcal{W}_{\mathcal{S}(\overline{a})}(\overline{A}) + \mathcal{W}_{\mathcal{S}(\overline{a})}(A), \end{aligned} \tag{19.7}$$

respectively. Note that $\mathcal{W}_{\mathcal{S}(a)}(A)$ includes $+A$ and $\mathcal{W}_{\mathcal{S}(\overline{a})}(A)$ includes $-A$. Also, $\mathcal{W}_{\mathcal{S}}(\overline{A})$ is a subset of both $\mathcal{W}_{\mathcal{S}(a)}(\overline{A})$ and $\mathcal{W}_{\mathcal{S}(\overline{a})}(\overline{A})$, and similarly, $\mathcal{W}_{\mathcal{S}}(A)$ is a subset of both $\mathcal{W}_{\mathcal{S}(a)}(\overline{A})$ and $\mathcal{W}_{\mathcal{S}(\overline{a})}(A)$.

We now rewrite (19.7) for the half of $\mathcal{S}$ with A at the low level as

$$\mathcal{I} = \mathcal{W}_{\mathcal{S}(\overline{a})}(\overline{A}) + \mathcal{W}_{\mathcal{S}(\overline{a})}(A^+) - A \tag{19.8}$$

and the half of $\mathcal{S}$ with A at the high level as

$$\mathcal{I} = \mathcal{W}_{\mathcal{S}(a)}(\overline{A}) + \mathcal{W}_{\mathcal{S}(a)}(A^+) + A. \tag{19.9}$$

The symbol $\mathcal{W}.(A^+)$ is defined as the collection of words that contain A and at least one other letter. The foldover of the half defined by (19.8) can then be written as

$$\mathcal{I} = \mathcal{W}_{\mathcal{S}(\overline{a})}(\overline{A}) - \mathcal{W}_{\mathcal{S}(\overline{a})}(A^+) + A. \tag{19.10}$$

Combining any two of the three sets of treatment combinations defined by (19.8), (19.9), and (19.10) gives a regular fraction, and combining all three gives an irregular fraction. The regular fractions can then be examined for estimability conditions.

Illustration of the Semifold

To illustrate the semifold and its property we consider the 2^{3-1} fraction defined by $\mathcal{I} = +ABC$. This example is too small to be of any real practical value, but is simple to manipulate and illustrates the technique. This fraction has A aliased with BC. The four treatment combinations are $\{a, b, c, abc\}$. Now add the semifold on A. This consists of ab and ac. Add this to the original set of four. We now have a three-fourths replicate. Consider the subset $\{b, c, ab, ac\}$, defined by $\mathcal{I} = BC$. In this subset, A is orthogonal to BC (which happens to be aliased with the mean). The estimate of A main effect is free of BC. The original alias chain involving A is broken. Note, however, that in the full least squares analysis that uses all of the data points and recovers full information, estimates will be correlated. This correlation is a price one pays in the irregular fractions.

To connect this example with our formal notation, $\mathcal{W}_{\mathcal{S}}(A)$ is $+ABC$, $\mathcal{W}_{\mathcal{S}}(\overline{A})$ is the null set [see (19.5)]. Also, $\mathcal{S}(a) = \{a, abc\}$ and $\mathcal{S}(\overline{a}) = \{b, c\}$. In addition, in expression (19.7), $\mathcal{W}_{\mathcal{S}(\overline{a})}(A)$ consists of $+ABC$ and $-A$, and $\mathcal{W}_{\mathcal{S}(\overline{a})}(\overline{A}) = -BC$. Also, $\mathcal{W}_{\mathcal{S}(a)}(A)$ consists of $+ABC$ and $+A$, and $\mathcal{W}_{\mathcal{S}(a)}(\overline{A}) = +BC$. Finally, $\mathcal{W}_{\mathcal{S}(a)}(A^+)$ in (19.9) is $+ABC$ and $\mathcal{W}_{\mathcal{S}(\overline{a})}(A^+)$ in (19.10) is $+ABC$.

Exercise 19.2: Foldover Exercise. Assume that an investigator has completed the 2^{4-2} fraction defined by $\mathcal{I} = +ABC = +BCD = +AD$. Show, by examining the A contrast, that adding the foldover on A frees the estimate A free of all two-factor interactions. □

19.4 AUGMENTING A 2^{4-1} FRACTION

In this example, we begin with a four-factor, eight-run experiment based on the defining contrast $\mathcal{I} = +ABCD$. The treatment combinations are (1), ab, ac, ad, bc, cd, bd, and $abcd$. We assume that on the basis of an examination of the results from an experiment, say an analysis of variance or a graphical analysis (see Chapter 12), that just estimating main effects is not sufficient. There is a need to examine the two-factor interactions.

There are two possibilities. The first is to complete the other half replicate, another eight runs. This presents no difficulties and one can incorporate a blocking factor, (the $ABCD$ interaction) if necessary. The alternative that we will look at is to add a semifold on one of the factors. This requires the addition of only four new experimental runs. If experimental runs are expensive, this becomes appealing.

Assume that on the basis of the eight runs and basic understanding of the subject matter that factor A is selected as most interesting. The semifold on A consisting of the four treatment combinations that had A at the low level are repeated with A at the high level. The four treatment combinations with a at the low level were (1), bc, bd, and cd. The four new treatment combinations are a, abc, abd, and acd. Formally, the subset of four treatment combinations with A at the low level is defined by the contrast $\mathcal{I} = +ABCD = -A = -BCD$ and the new set with level of factor A switched to the high level by $\mathcal{I} = -ABCD = +A = -BCD$. The set of treatment combinations obtained by combining these two sets of four is defined by the contrast $\mathcal{I} = -BCD$, indicating that now AB, AC, and AD become estimable, provided that three-factor interactions are negligible. Combined with the original half replicate that gave estimates of $AB+CD$, $AC+BD$, and $AD+BC$, we see that all two-factor interactions are estimable. The difference between selecting A vs. selecting one of the other factors for the semifold is not major. The correlation structure of the estimates favors selecting A slightly. However, $\widehat{A}$ is correlated with a possible block effect, while $\widehat{B}$, $\widehat{C}$, and $\widehat{D}$ are not.

A least squares analysis via, for example, `proc GLM` in SAS® provides estimates of all four main effects and all six two-factor interactions. In fact, it is even possible to incorporate a blocking factor (see the discussion of blocking in Chapter 5), one block of eight and a second block of four, if it is suspected that conditions have changed from the first set of runs to the second. Note, however, that no degrees of freedom are available to estimate experimental error. A reasonable step would be to ignore the fact that estimates of contrasts are correlated and use one of the methods outlined in Chapter 12.

It is also interesting to examine what happens if a third block of four treatment combinations, b, c, d, and bcd, is added to give a complete replicate in one block of eight and two blocks of four. The estimate of the A contrast has variance $.5\sigma_\epsilon^2$ and the other main effect and two-factor interaction contrasts have variance $.25\sigma_\epsilon^2$. The estimate of the A contrast is correlated with the estimates of block effects. The reason for this is somewhat complicated. An intuitive answer for why the A contrast is more poorly estimated can be obtained by noting that the first eight treatment combinations correspond to sets defined by $\mathcal{I} = +ABCD = -A = -BCD$ and $\mathcal{I} = +ABCD = +A = +BCD$, and the remaining two sets of four are defined by $\mathcal{I} = -ABCD = +A = -BCD$ and $\mathcal{I} = -ABCD = -A = +BCD$, respectively. Each of the four sets of four, by itself, provides some information on contrasts B, C, and D (although aliased with two-factor interactions) and no information on the A contrast. The bottom line is that if the experimenter contemplates adding the eight treatment combinations, it should *not* be done in two steps of four.

Example 19.2: Detailed Statistical Analysis. We begin by rewriting the linear model for the data by replacing the half coefficients on the parameters with 1's, as was done in Section 14.8.2 to make things fit nicely into a general least squares regression program. We will use SAS® code to illustrate. The original eight runs are coded as shown in Table 19.1. The four runs in the foldover are coded as shown in Table 19.2.

Table 19.1 SAS® Coding for $I = +ABCD$ Example

Block	a	b	c	d
1	−1	−1	−1	−1
1	1	1	−1	−1
1	1	−1	1	−1
1	1	−1	−1	1
1	−1	1	1	−1
1	−1	1	−1	1
1	−1	−1	1	1
1	1	1	1	1

Table 19.2 Coded Foldover Values

Block	*a*	*b*	*c*	*d*
2	1	−1	−1	−1
2	1	1	1	−1
2	1	1	−1	1
2	1	−1	1	1

Table 19.3 Coded Additional Set

Block	*a*	*b*	*c*	*d*
3	−1	1	−1	−1
3	−1	−1	1	−1
3	−1	−1	−1	1
3	−1	1	1	1

The final four runs are coded as shown in Table 19.3.

The analysis of variance for the initial set of eight runs is obtained via

```
proc glm ;
model y = a b c d ; run ;
```

This analysis gives least squares estimates of the four main effect parameters and tests of significance using an error term with three degrees of freedom. This test is probably inadequate, and a better option would be to fit a saturated model as in

```
proc glm ;
model y = a b c d a*b a*c a*d ; run ;
```

and then proceed with one of the methods outlined in Chapter 12. The values produced by this code as estimates of AB, AC, and AD really have no meaning, unless one assumes no CD, BD, and BC interactions. They do provide linearly independent contrasts that can be used in, say, a half-normal plot.

When the second block, the semifold, is added, the SAS® code is

```
proc glm ;  class block ;
model y = block a b c d a*b a*c
         a*d b*c b*d c*d/i solution ; run ;
```

Several points are to be noted about this analysis. First, orthogonality of estimates is gone. Next, we are fitting a saturated model, implying no degrees of freedom for error. The "`i`" option on the "`model`" statement requests the generalized inverse of the $\boldsymbol{X}^t\boldsymbol{X}$ matrix. Elements from this matrix are used to scale the contrasts for a possible half-normal plot. Details for constructing a half-normal plot or some related analyses are given in Chapter 12.

19.5 SEQUENCE STARTING FROM A SEVEN-FACTOR MAIN EFFECT PLAN

We assume that a study has begun with a 2^{7-4} main-effect plan. The defining contrast for this fraction (obtained by superimposing seven factors on a 2^3) is

$$\begin{aligned} \mathcal{I} \quad & = +ABC = +ADE = +BCDE \\ & = +BDF = +ACDF = +ABEF = +CEF \\ = +ABDG \quad & = +CDG = +BEG = +ACEG \\ & = +AFG = +BCFG = +DEFG = +ABCDEFG. \end{aligned} \tag{19.11}$$

Assume that the results look interesting and the complete foldover is added. Also assume that there is no need for blocking and that an additional factor H is introduced. One can formally think of this as the original eight treatment combinations constituting the fraction defined by expression (19.11) with the addition of $-H$ and the 15 generalized interactions involving H. The complete foldover can be thought of as the fraction defined by the string of 31 words (interactions) with the sign reversed on all words consisting of an odd number of letters. Adding the two fractions gives a 2^{8-4} fraction defined by

$$\begin{aligned} \mathcal{I} \quad & = +ACDF = +ABEF = +BCDE \\ & = +ABDG = +BCFG = +DEFG = +ACEG \\ = -BDFH \quad & = -ABCH = -ADEH = -CEFH \\ & = -AFGH = -CDGH = -BEGH = -ABCDEFGH. \end{aligned} \tag{19.12}$$

This, a resolution III plan plus the complete foldover is a resolution IV plan. Main effects are all free of two-factor interactions, but two-factor interactions are aliased with one another.

Now assume that factor A appears interesting. There is a need to examine the interactions involving A. An attractive option is to add eight treatment combinations, the semifold on A. Recall that this set consists of the eight treatment combinations in the set defined by expression (19.12) that had A at the low level, but now have it set to the high level. Let $\mathcal{S}(\overline{a})$ refer to the subset with A at the low level and $\mathcal{S}(a)$ the subset with A at the high level.

Formally, the subset with A at the low level can be thought of as defined by the collection of words in (19.12) with the addition of $-A$ and the 15 generalized interactions. Note the signs on words. The set of eight in the semifold on A is defined by the same 31 words, but with the sign on all words that contain the letter "A" reversed. Let $\mathcal{SSF}(a)$ represent this set of eight treatment combinations.

Now think of combining $\mathcal{S}(\overline{a})$ and $\mathcal{SSF}(a)$. Examining the collections of words defining these two subsets reveals that all words involving the letter "A" in the first set appear in the second set, but with sign reversed. The collection of words defining the set of 16 treatment combinations contains words with three or more letters and no words with the letter "A." Consequently, all two-factor interactions involving A are estimable, provided that we assume no three-factor or higher-order interactions. Subsets $\mathcal{S}(\overline{a})$ and $\mathcal{S}(a)$ together let us estimate all main effects free of two-factor interactions. Note that if all three subsets are combined, we have an irregular, $3/32$ fraction. Full information, least squares estimates, say those obtained via `proc GLM` in SAS®, are correlated.

19.5.1 Blocking

We commented in our development that we were assuming no need for blocking. We now illustrate that it is possible to accommodate blocking, but at the cost of not allowing the introduction of the new factor H. First, the two subsets, $\mathcal{S}(\overline{a})$ and $\mathcal{S}(a)$, form two blocks. Formally, the defining contrast with blocking factor $\mathcal{B}$ can be written as

$$\mathcal{I} = +ABC = +ADE = +BDF = +ABDG = +\mathcal{B}$$

and all generalized interactions. In practice, the experimenter simply labels the eight original observations as block 1 and the eight that constitute the complete foldover to convert to a resolution IV plan as block 2. The eight observations generated via the semifold on A are labeled as block 3. A formal notation to describe what happens becomes excessively cumbersome. However, recall that the last set of eight treatment combinations was generated by taking the eight treatment combinations with A at the low level and converting to the foldover on A. At this stage, A serves as a blocking factor. Note, however, that A remains estimable, free of $\mathcal{B}$ in blocks 1 and 2. Adding extra points cannot make something nonestimable. The SAS code to perform these manipulations is shown in Display 19.1 and on the companion Web site.

The analysis of variance is illustrated in Display 19.2. Differences among Type I, II, and III sums of squares reflect the nonorthogonality in the irregular $3/32$ fraction. We use the Type II sums of squares in this case. This measures the addition to the regression sum of squares after all other effects, except interactions that involve the factor being tested.

19.6 AUGMENTING A ONE-EIGHTH FRACTION OF 4^3

To illustrate another application of the foldover techniques, we return to an example discussed earlier. In Section 17.5.2 we noted that that it is possible to put a 4^3 in eight blocks of eight by confounding $A1B1C2$, $A2B1C1$, $A1A2C1C2$, $A1B2C1C2$, $B1B2C1$, $A1A2B1B2C2$, and $A2B2$. One can use one of these eight blocks: say, the one defined by a positive sign on each of these words as a one-eighth fraction. Assume that this fraction has been run or at least is being contemplated. This fraction, consisting of eight treatment combinations,

```
data base ;
block = 1 ; do a = 0 , 1 ;
do b = 0 , 1 ; do c = 0 , 1 ; do d = 0 , 1 ;
do e = 0 , 1 ; do f = 0 , 1 ; do g = 0 , 1 ;
      if mod(a+b+c,2) = 0 then
      if mod(a+d+e,2) = 0 then
      if mod(b+d+f,2) = 0 then
      if mod(a+b+d+g,2) = 0 then
         output ;
 end ; end ; end ; end ; end ; end ; end ;
data foldover ; set base ;
block = 2 ; a = mod(a+1,2) ;
b = mod(b+1,2) ; c = mod(c+1,2) ; d = mod(d+1,2) ;
e = mod(e+1,2) ; f = mod(f+1,2) ; g = mod(g+1,2) ;
data semifold ; set base foldover ;
block = 3 ;
if a = 0 then do ;  a = 1 ; output ; end ;
*   Generate some noise data ;
data comb ; set base foldover semifold ;
y = normal(33557) ;
proc glm ;
class block a b c d e f g ;
model y = block a b c d e f g a*b a*c
    a*d a*e a*f a*g ; run ;
```

Display 19.1 Constructing a 3/32 fraction in three blocks.

```
                                Sum of
Source                DF       Squares    Mean Square

Model                 15        12.78        0.85 Error
12.55            1.56 Corrected Total   23        25.34
```

Source	DF	Type I SS	Type II SS	Type III SS
block	2	2.708	2.252	2.252
a	1	0.003	0.003	0.003
b	1	0.792	0.792	1.341
c	1	1.019	1.019	0.986
d	1	0.077	0.077	0.053
e	1	3.083	3.083	1.785
f	1	0.049	0.049	0.213
g	1	0.139	0.139	0.028
a*b	1	0.914	0.914	0.914
a*c	1	0.015	0.015	0.015
a*d	1	2.188	2.188	2.188
a*e	1	0.918	0.918	0.918
a*f	1	0.576	0.576	0.576
a*g	1	0.299	0.299	0.299

Display 19.2 Analysis of data from Display 19.1.

has parts of all three main effects aliased with other main effects. For example, $A1A2$ is aliased with $C1C2$. These are both main effect contrasts. Similarly, $B1B2$ is aliased with $C1$ and $A2$ is aliased with $B2$. The important question now is: Which eight additional treatment combinations will be most beneficial to clear up aliases? A good choice is the double foldover, on $A1$ and $B2$. The original plus the foldover gives the fraction defined by

$$\mathcal{I} = +A2B1C1 = +A1B2C1C2 = +A1A2B1B2C2.$$

This gives a fraction with all main effects free of each other.

The original set of eight treatment combinations is shown in Table 19.4. The double foldover on $A1$ and $B2$ is shown in Table 19.5. The analysis of variance of data from such an experiment takes the form shown in Table 19.6.

The sums of squares for A, B, and C can each be decomposed into linear, quadratic, and cubic terms. In many cases it will be reasonable to assume a priori that there are no cubic effects and that these degrees of freedom (sums of squares) can be added to error.

Table 19.4 Treatment Combinations for Fraction of 4^3

Pseudofactors						Four-Level Factors		
$A1$	$A2$	$B1$	$B2$	$C1$	$C2$	A	B	C
0	0	0	0	1	1	0	0	3
1	1	1	1	1	1	3	3	3
0	0	1	0	0	0	0	2	0
1	1	0	1	0	0	3	1	0
1	0	1	0	0	1	2	2	1
0	1	0	1	0	1	1	1	1
1	0	0	0	1	0	2	0	2
0	1	1	1	1	0	1	3	2

Table 19.5 Double Foldover of Fraction Displayed in Table 19.4

Pseudofactors						Four-Level Factors		
$A1$	$A2$	$B1$	$B2$	$C1$	$C2$	A	B	C
1	0	0	1	1	1	2	1	3
0	1	1	0	1	1	1	2	3
1	0	1	1	0	0	2	3	0
0	1	0	0	0	0	1	0	0
0	0	1	1	0	1	0	3	1
1	1	0	0	0	1	3	0	1
0	0	0	1	1	0	0	1	2
1	1	1	0	1	0	3	2	2

Table 19.6 Analysis of Variance Table

Source	df	SSq
Block	1	
A	3	
B	3	
C	3	
Error	6	By subtraction
Total	15	$\sum y^2 - \text{CT}$

Exercise 19.3: Displaying the Semifold

(a) Display the defining contrast (set of words) that define $\mathcal{S}(\overline{a})$ in Section 19.5.
(b) Display the defining contrast (set of words) that define $\mathcal{SSF}(a)$ in Section 19.5.
(c) Verify the statement that all words that contain the letter "A" appearing in one set also appear in the other set, but with sign reversed. Also verify that all words, in either set, consist of at least three letters.
(d) Display the defining contrast (set of words) that define the combination of $\mathcal{S}(\overline{a})$ and $\mathcal{SSF}(a)$ in Section 19.5. Notice that this set of words is smaller, corresponding to a larger collection of treatment combinations.
(e) Obtain the set treatment combinations defined by the set of defining contrasts above. □

Exercise 19.4: 4^{3-1} Fraction

(a) Verify that all main effect contrasts for the three factors in the plan set up in Section 19.6 are free of a block effect.
(b) Verify that the three main effects are orthogonal. □

19.7 ADDING LEVELS OF A FACTOR

Sometimes it happens that one of the factors in an initial experiment appears to be very important but that the levels of this factor were set too high or too low. Consider a case with four two-level factors where factor A represents the levels of some continuous additive factor. The initial experiment was the half replicate (eight runs) defined by

$$\mathcal{I} = +ABCD. \tag{19.13}$$

Assume that amounts of factor A, say 20 and 26, which were coded as -1 and $+1$ in the experiment should have been higher. One obvious option is to rerun the complete experiment. We propose an alternative that utilizes all of the existing data and consider three alternative scenarios. We warn the reader that this section relies heavily on concepts developed more fully in Chapter 20 and may be skipped on first reading.

Table 19.7 Level Coding for New Factor A

A	E		A^*	Coded Level
a_0	e_0	$\Rightarrow$	a_0^*	-3
a_1	e_0	$\Rightarrow$	a_1^*	-1
a_0	e_1	$\Rightarrow$	a_2^*	1
a_1	e_1	$\Rightarrow$	a_3^*	3

19.7.1 Extending the Range of Levels

The investigator decides to add levels 32 and 38 for factor A. We show how to do this in a manner that also allows the estimation of quadratic response effects. The half replicate defined by (19.13) can also be interpreted as a quarter replicate defined by

$$\mathcal{I} = +ABCD = -E = -ABCDE, \tag{19.14}$$

where we will think of A and E as pseudofactors for a new four-level factor as in Chapter 17. Add the foldover on E defined by

$$\mathcal{I} = +ABCD = +E = +ABCDE. \tag{19.15}$$

The key now is to define the new factor A^* using the assignment shown in Table 19.7. Notice that with this assignment the first part of the experiment involved the first level of pseudofactor E only. Also, the AE pseudofactor interaction corresponds to the quadratic effect of the new A^* factor the contrast between the two extreme levels of A^* and the two intermediate levels. The E pseudofactor corresponds to the greater part of the linear effect of the new A^* factor.

Now, if it is assumed that there is no need for blocking and if there are no BC, BD, CD, or higher-order interactions, the experiment provides independent, unbiased estimates of pseudofactors A and E, factors B, C, and D, as well as AE, AB, BE, ABE, AC, CE, ACE, AD, DE, and ADE. A and E together give estimates of the linear and cubic effect of A^*, and AE the quadratic effect of A^*. These relationships are developed in detail in Chapter 20. The ABE, ACE, and ADE contrasts, which should possibly be written as AEB, AEC, and AED, estimate the A-quadratic by B, C, and D factors, respectively. The A-linear and A-cubic interactions with factors B, C, and D are available from AB and BE, AC and CE, and AD and DE, respectively. An ordinary least squares regression program such as `proc GLM` in SAS® can be used to obtain the complete analysis. If the investigator is not willing to assume that BC, BD, and CD are negligible but is willing to assume the cubic effect of A negligible, the appropriate analysis can still be obtained using ordinary least squares. The trick is to have BC, BD, and CD appear in the model before $A_L \times B$, $A_L \times C$, and $A_L \times D$.

```
data base ;
 do a = -1 , 1 ;  do b = -1 , 1 ;  do c = -1 , 1 ;
    do d = -1 , 1 ;  do e = -1 , 1 ;
    if a*b*c*d > 0 then
    if e < 0 then output ;
      end ; end ; end ; end ; end ;
data extend ;
 do a = -1 , 1 ;  do b = -1 , 1 ;  do c = -1 , 1 ;
    do d = -1 , 1 ;  do e = -1 , 1 ;
    if a*b*c*d > 0 then
    if e > 0 then output ;
      end ; end ; end ; end ; end ;
data comb ; set base extend ;
 if a<0 then if e<0 then astar = -3 ;
 if a>0 then if e<0 then astar = -1 ;
 if a<0 then if e>0 then astar =  1 ;
 if a>0 then if e>0 then astar =  3 ;
proc glm data = comb;
 model y = astar  a*e   b astar*b a*b*e
        c astar*c a*c*e d astar*d a*d*e  ;  run ;
```

Display 19.3 Code to illustrate extending the range of factor A.

Example 19.3. To illustrate the augmenting procedure, we construct a numerical example. The data are generated from a model based on an A factor that ranges from 20 to 40 and factors B, C, and D that have levels 5 and 10, 1 and 9, and 0 and 10, respectively, all coded as -1 and $+1$. Response to factor A has both a linear and a quadratic component, as well as an interaction with B. Errors were standard normal, mean zero, and unit variance. Specifically, the model used was

$$y = 100 - .5(a - 30) - (a - 35)^2/b + 2c + d + \epsilon. \tag{19.16}$$

The SAS code to generate the initial eight treatment combinations and the additional set of eight, followed by the code for the statistical analysis of the combined data set, are shown in Display 19.3. The analysis of variance produced and the parameter estimates are shown in Display 19.4. This analysis was followed by a final run to produce fitted values with the model reduced by deleting `astar*c`, `a*e*c`, `astar*d`, and `a*e*d`. The data values and the fitted values are shown in Display 19.5.

19.7.2 Inserting Levels

As an alternative case, assume that rather than extending the range of A, the investigator decides that two new intermediate levels, 22 and 24, are required. The procedure again is to generate the foldover on E. However, this time define the new set of levels using Table 19.8. Now notice that the pseudofactor E corresponds to the quadratic effect of the new A^* factor. Pseudofactors A and AE each contribute to the linear effect of A^*. The statistical analysis again proceeds under the assumption of no blocking effect and no BC, BD, CD, or

Source	DF	Type I&III SS	F Value	Pr > F	Estimate
Intercept					88.62
astar	1	1097.612	691.04	<.0001	3.70
a*e	1	464.364	292.36	<.0001	−5.38
b	1	298.905	188.19	0.0002	4.32
astar*b	1	269.233	169.50	0.0002	−1.83
a*e*b	1	55.353	34.85	0.0041	1.85
c	1	264.724	166.67	0.0002	4.06
astar*c	1	2.152	1.36	0.3091	−0.16
a*e*c	1	0.001	0.00	0.9803	−0.00
d	1	457.934	288.31	<.0001	5.34
astar*d	1	0.117	0.07	0.7994	0.03
a*e*d	1	0.148	0.09	0.7747	−0.09
Error	4	6.353			

Display 19.4 Analysis of data generated from Display 19.3.

Obs	a	b	c	d	e	y	astar	yhat
1	−1	−1	−1	−1	−1	51.02	−3	51.01
2	−1	−1	1	1	−1	69.94	−3	69.85
3	−1	1	−1	1	−1	84.63	−3	85.09
4	−1	1	1	−1	−1	83.60	−3	82.52
5	1	−1	−1	1	−1	87.24	−1	87.29
6	1	−1	1	−1	−1	84.48	−1	84.72
7	1	1	−1	−1	−1	83.32	−1	85.18
8	1	1	1	1	−1	104.01	−1	104.01
9	−1	−1	−1	−1	1	88.53	1	87.66
10	−1	−1	1	1	1	105.93	1	106.50
11	−1	1	−1	1	1	100.63	1	99.62
12	−1	1	1	−1	1	97.91	1	97.05
13	1	−1	−1	1	1	95.49	3	94.95
14	1	−1	1	−1	1	91.74	3	92.38
15	1	1	−1	−1	1	85.54	3	85.60
16	1	1	1	1	1	103.88	3	104.44

Display 19.5 Fitted values in the analysis in Display 19.4.

higher-order interaction. However, the estimates are somewhat different from the extending range example. Now the A-linear and A-cubic effects are part of pseudofactors A and AE. The one-degree-of-freedom contrast can be estimated directly. The A-quadratic effect is in pseudofactor E. Notice that this differs from the extending case. Estimates of B, C, and D are as before. They are not affected by the introduction of pseudofactors. In addition, we obtain A^*-linear $\times$ B from AB and ABE, A^*-linear $\times$ C from AC and ACE, and A^*-linear $\times$ D from AD and ADE. The statistical analysis again proceeds under the assumptions of no blocking effects, and no BC, BD, CD or higher-order interactions.

Table 19.8 Level Coding for New Factor A

A	E		A^*	Coded Level	A^* Level
a_0	e_0	$\Rightarrow$	a_0^*	−3	20
a_1	e_0	$\Rightarrow$	a_3^*	3	26
a_0	e_1	$\Rightarrow$	a_1^*	−1	22
a_1	e_1	$\Rightarrow$	a_2^*	1	24.

Source	DF	Type I&III SS	F Value	Pr > F	Estimate
Intercept					81.42
astar	1	661.175	2457.99	<.0001	2.87
e	1	2.414	8.98	0.0401	0.38
b	1	954.123	3547.05	<.0001	7.72
astar*b	1	141.771	527.05	<.0001	−1.33
e*b	1	0.301	1.12	0.3495	−0.13
c	1	264.724	984.14	<.0001	4.06
astar*c	1	0.000	0.00	0.9907	0.00
e*c	1	2.675	9.95	0.0344	−0.40
d	1	457.934	1702.42	<.0001	5.34
astar*d	1	1.945	7.23	0.0547	0.15
e*d	1	0.090	0.34	0.5933	−0.07
Error	4	1.075			

Display 19.6 Statistical analysis of experiment with inserted levels.

Example 19.4. To illustrate the insertion procedure, we return to the numerical example used to demonstrate augmenting. The data are again generated via the model given in expression (19.16). All levels remain the same, except that a^* in the foldover now takes on values 22 and 24. The original eight observations remain as before. The analysis of variance produced and the parameter estimates are shown in Displays 19.6 and 19.7. We notice that this analysis did not pick up the quadratic response to the A factor. If we look at the model given in (19.16), we notice that the curvature does not really become pronounced until after about $a = 30$. This is an example where it is necessary to extend the range of a factor to detect a nonlinear effect.

19.7.3 Extending and Inserting Levels

The final case that we could consider is to insert a new level of A between the two existing levels, i.e., at 23, and extending the range by adding a new level, 29. The level association table now becomes that shown in Table 19.9. We see immediately that the quadratic contrast, the two extreme levels versus the two intermediate levels, is the pseudofactor AE contrast. The A_Q^* contrast is the AE contrast. The analysis of variance follows the pattern of the case extending the range. We leave the remainder of this case as an exercise.

Obs	a	b	c	d	e	y	astar	yhat
1	−1	−1	−1	−1	−1	51.02	−3	51.25
2	−1	−1	1	1	−1	69.94	−3	70.90
3	−1	1	−1	1	−1	84.63	−3	85.38
4	−1	1	1	−1	−1	83.60	−3	83.64
5	1	−1	−1	1	−1	87.24	3	87.19
6	1	−1	1	−1	−1	84.48	3	85.44
7	1	1	−1	−1	−1	83.32	3	83.94
8	1	1	1	1	−1	104.01	3	103.60
9	−1	−1	−1	−1	1	61.53	−1	60.48
10	−1	−1	1	1	1	78.93	−1	78.50
11	−1	1	−1	1	1	89.63	−1	89.29
12	−1	1	1	−1	1	86.91	−1	85.90
13	1	−1	−1	1	1	80.09	1	79.59
14	1	−1	1	−1	1	76.34	1	76.21
15	1	1	−1	−1	1	81.34	1	81.67
16	1	1	1	1	1	99.68	1	99.69

Display 19.7 Fitted values for experiment with inserted levels.

Table 19.9 Level Coding for New Factor *A*

A	E		A^*	Coded Level	A^* Level
a_0	e_0	$\Rightarrow$	a_0^*	−3	20
a_1	e_0	$\Rightarrow$	a_2^*	1	26
a_0	e_1	$\Rightarrow$	a_1^*	−1	23
a_1	e_1	$\Rightarrow$	a_3^*	3	29.

Exercise 19.5: Modifying the Range of a Variable. Use the model given in equation (19.16) to generate data for the foldover on A to introduce levels 23 and 29 for factor A into an original half replicate as defined by defining contrast (19.13) that had A at levels 20 and 26. Since the A levels in the foldover are coded as -1 and $+1$, you must use the translation $a \Leftrightarrow 3a + 23$ into model (19.16) in the original set of eight treatment combinations and the translation $a \Leftrightarrow 3a + 26$ in the foldover set. After the 16 data values have been generated, the two sets are combined and the appropriate a^* values must be generated. From then on it is simply a matter of setting up the correct analysis of variance. Obtain the analysis and interpret the results. Reconcile the results of the least squares analysis and the known model which can be evaluated without error. □

19.8 DOUBLE SEMIFOLD

There exists the possibility of a double or even a triple semifold to help an investigator recover from an unfortunate design situation. We consider the double semifold first. Assume that an investigator has completed the resolution IV, half replicate defined by $\mathcal{I} = +ABCD$. This fraction consists of the treatment combinations (1) , ab, ac, ad, bc, bd, cd, and $abcd$. The investigator then realizes that another factor, E, is required, and adds the eight new treatment combinations e, abe, ace, ade, bce, bde, cde, and $abcde$. This is still only a resolution IV plan and has two-factor interactions aliased with other two-factor interactions. Now the question is to convert this to a resolution V plan with minimal effort. It will be assumed throughout that three-factor and higher-order interactions are negligible.

The technique is simple. First select factor E and one other factor, say B. Then generate the double semifold on B and E. Select the four treatment combinations that have both B and E at the low level; switch both to the high level. These four constitute the double semifold. The result is a set of 20 runs with all two-factor interactions estimable. Note that if the investigator had realized that E was required initially, a half replicate defined by $\mathcal{I} - ABCDE$ would have been the choice.

We now explain the details. The 16 treatment combinations can be broken into four sets of four, defined by

$$\begin{aligned} \mathcal{I} &= +ABCD = -B = -ACD \\ &= -E = -ABCDE = +BE = +ACDE \end{aligned} \tag{19.17}$$

$$\begin{aligned} \mathcal{I} &= +ABCD = +B = +ACD \\ &= -E = -ABCDE = -BE = -ACDE \end{aligned} \tag{19.18}$$

$$\begin{aligned} \mathcal{I} &= +ABCD = -B = -ACD \\ &= +E = +ABCDE = -BE = -ACDE \end{aligned} \tag{19.19}$$

$$\begin{aligned} \mathcal{I} &= +ABCD = +B = +ACD \\ &= +E = +ABCDE = +BE = +ACDE. \end{aligned} \tag{19.20}$$

The four sets are (i) (1), ac, ad, and cd; (ii) ab, bc, bd, and $abcd$; (iii) e, ace, ade, and cde; and (iv) abe, bce, bde, and $abcde$. The double semifold obtained from (i) consists of the four treatment combinations (v) be, $abce$, $abde$, and $bcde$. This set is defined by

$$\begin{aligned} \mathcal{I} &= -ABCD = +B = -ACD \\ &= +E = -ABCDE = +BE = -ACDE. \end{aligned} \tag{19.21}$$

The problem started with alias relationships $AB = CD$, $AC = BD$, and $AD = BC$. Main effects and all two-factor interactions involving E were estimable in the set of 16 observations. Now combining (iii) and (v), i.e., (19.19) and (19.21), gives the fraction defined by

$$\mathcal{I} = -ACD = +E = -ACDE,$$

which provides estimates of AB, BC, and BD. Similarly, combining (iv) and (v), i.e., (19.20) and (19.21), gives the fraction defined by

$$\mathcal{I} = +B = +E = +BE,$$

which provides estimates of CD, AC, and AD. All two-factor interactions are estimable. We have recovered a resolution V plan.

John (2000) extends this technique to the triple semifold, to recover from the case where an investigator has started with a 2^{4-1} fraction defined by $\mathcal{I} = ABCD$ and has added the 24 treatment combinations by running combinations with E and F. The technique is to add the triple semifold on one of A, B, C, or D and E and F. The result is a 36-run resolution V plan. However, again, one would only do this if one is trying to recover from a bad start.

19.9 PLANNED SEQUENCES

We now proceed to look at the case where an experimenter is forced to consider a sequence of fractions because of budgetary constraints. Recall that in previous sections we always assumed that the reason for considering sequences of fractions was that the investigator could look at early results and adjust strategy.

We have seen that in many areas of research it is advisable to group treatment combinations in a large factorial experiment into subsets called *blocks* in order to provide more homogeneous environments. In many areas of research it is common practice (often necessary) to run the entire experiment (all blocks) simultaneously. In a laboratory and/or many industrial settings this frequently is not necessary; the investigator has the option of performing part of the study, looking at the results, and then terminating the study if the results are conclusive or proceeding if more information is required. An alternative scenario is one where the investigator is unsure of resources available. The study is initiated with limited resources. If additional resources become available, the plan is in place to proceed efficiently. A factorial plan with treatment combinations arranged in blocks in an appropriate manner can be very advantageous in these situations. A particularly appealing strategy is to organize blocks so that main effects (which usually are most important) become estimable early in the process and that later blocks of experimental runs either elucidate two-factor and possibly higher-order interactions, or reduce the severity of required assumptions. For example, at an early stage it may be necessary to assume that there are no two-factor interactions. As more data become available, more of these interactions become estimable, and it becomes possible to entertain more realistic models. If the blocks (groups of treatment combinations) are organized properly, estimates of interest will be orthogonal to differences between blocks. Two major early pieces of work in this area are Addelman (1969) and Daniel (1962). We present a number of examples. The reader that requires more detail is urged to examine the original papers.

19.9.1 Three Factors in Sets of Two

We use the 2^3 with observations in sets of two to illustrate the technique. We assume throughout this example that the three-factor interaction is negligible. Notice that there are four possible grouping (blocking) schemes possible. These can be identified with

$$\mathcal{I} = +AB = +AC = +BC$$

$$\mathcal{I} = +A = +ABC = +BC$$

$$\mathcal{I} = +B = +ABC = +AC$$

$$\mathcal{I} = +C = +ABC = +AB.$$

These lead to the four possible sequences of treatment combinations:

$$\begin{array}{lllll} 1: & \{(1), abc\}, & \{b, ac\}, & \{a, bc\}, & \{c, ab\} \\ 2: & \{a, abc\}, & \{b, c\}, & \{(1), bc\}, & \{ac, ab\} \\ 3: & \{b, abc\}, & \{a, c\}, & \{(1), ac\}, & \{bc, ab\} \\ 4: & \{c, abc\}, & \{c, b\}, & \{(1), ab\}, & \{bc, ac\}. \end{array}$$

It is clear that taking only one pair of observations gives little information without assuming no two-factor interactions. Taking the two blocks $\{(1), abc\}$ and $\{b, ac\}$ from the first sequence produces a one-half replicate $\mathcal{I} = +AC$ and allows estimates of B, $A + C$, and $AB + BC$, i.e., one clear main effect. Notice that $\widehat{AB + BC}$ includes the block effect. The investigator could as easily have chosen to make one of the other main effects estimable by selecting $\{(1), abc\}$ and one of the other blocks in the sequence.

Using the first three blocks in the first sequence gives three-fourths replicate (see Section 14.8) which can be viewed as the one-half replicate $\mathcal{I} = +AC$, the one-half replicate $\mathcal{I} = +BC$, and the one-half replicate $\mathcal{I} = AB$, each with two extra treatment combinations left over. Now all main effects are estimable. However, the estimates are correlated.

A second alternative is to use the second sequence of sets. The first pair yields an estimate of $C + AB + AC$, which by itself is of limited value. The first two pairs produce the one-half replicate $\mathcal{I} = +ABC$ and the estimates of $A + BC$, $B + AC$, and $C + AB$. Of these, the first is really a between-blocks contrast and of limited value unless it is possible to assume no block effects. Using the first three pairs helps very little. Estimates of main effects are correlated and aliased with two-factor interactions. Four sets, a complete replicate, still leaves A confounded with blocks.

A third possibility that one can consider is to take the first pair from each of the sequences. The first two pairs, $\{(1), abc\}$ and $\{a, abc\}$, provide an estimate of $(A - 1/2AB - 1/2AC)$. Interpretation again requires the assumption that there are no two-factor interactions. This problem continues even as more pairs are added. One really cannot do very well here.

19.9.2 Five Factors

The 2^5 factorial in which it is possible to test eight experimental units in one block presents some much more appealing alternatives. We illustrate with the assumption that three-factor and higher-order interactions are negligible. It is also assumed that several sets will be possible. Begin considerations with the one-fourth replicate of the 2^5 factorial, defined by

$$\mathcal{I} = ABC = ADE = BCDE.$$

The actual treatment combinations making up the four possible one-fourth replicates are obtained by finding solutions to

$$\begin{aligned}
&i_A + i_B + i_C &&= 0 \quad \text{(modulo 2)}\\
&i_A + i_D + i_E &&= 0 \quad \text{(modulo 2)}\\[4pt]
&i_A + i_B + i_C &&= 0 \quad \text{(modulo 2)}\\
&i_A + i_D + i_E &&= 1 \quad \text{(modulo 2)}\\[4pt]
&i_A + i_B + i_C &&= 1 \quad \text{(modulo 2)}\\
&i_A + i_D + i_E &&= 0 \quad \text{(modulo 2)}\\[4pt]
&i_A + i_B + i_C &&= 1 \quad \text{(modulo 2)}\\
&i_A + i_D + i_E &&= 1 \quad \text{(modulo 2)}.
\end{aligned}$$

Recall that the first set is also often referred to as the intrablock subgroup. We will also find it convenient to use the notation

$$\begin{aligned}
&\text{(i)} \quad \mathcal{I} = -ABC = -ADE = +BCDE\\
&\text{(ii)} \quad \mathcal{I} = -ABC = +ADE = -BCDE\\
&\text{(iii)} \quad \mathcal{I} = +ABC = -ADE = -BCDE\\
&\text{(iv)} \quad \mathcal{I} = +ABC = +ADE = +BCDE,
\end{aligned}$$

respectively, for these four fractions or blocks.

Now assume that the data from block (i) are available. With the given assumption that all three-factor and higher interactions are negligible, the aliases of the main effects,

$$\begin{aligned}
A &= -BC = -DE\\
B &= -AC\\
C &= -AB\\
D &= -AE\\
E &= -AD,
\end{aligned}$$

imply that one can estimate $A - BC - DE$, $B - AC$, $C - AB$, $D - AE$, and $E - AD$. Additionally, it is possible to estimate $BD - CE$ and $BE - CD$.

Clearly, these estimates become informative only if one can make the further assumptions that at least some two-factor interactions are negligible. Adding block (iv) gives a half-replicate with defining contrast $\mathcal{I} = BCDE$. Estimates of the five main effects are all free of two-factor interactions. In fact, AB, AC, AD, and AE are also free of all other two-factor interactions. However, $BC + DE$, $BD + CE$, and $BE + CD$ are still aliased. All estimating contrasts are mutually orthogonal and free of possible block effects. Notice that the result of combining the fraction defined by (i) and (iv) can be observed symbolically by "averaging" the defining contrasts.

Adding block (ii) gives a three-fourths replicate (see Section 14.8) which can be viewed as the half replicate defined by $\mathcal{I} = BCDE$ plus some extra observations, the half replicate defined by $\mathcal{I} = -ABC$ plus some extra observations, or as the half replicate defined by $\mathcal{I} = ADE$ plus some extra observations. The six two-factor interactions, BC, BD, BE, CD, CE, and DE can be estimated from one or other of the half replicates. Now all main effects and two-factor interactions are estimable. However, the estimates are correlated. The simple way to obtain the most efficient estimates and standard errors is to use a general-purpose regression program. Choosing block (iii) in place of block (ii) is equivalent. Finally, if resources permit, adding the fourth block yields fully orthogonal estimates of all main effects and two-factor interactions, all free of possible block effects.

Notice that an alternative strategy would have been to use block (ii) after block (i). This would have provided uncorrelated estimates of $A - BC$, $B - AC$, $C - AB$, D, E, AD, AE, BD, BE, CD, CE, and DE. It is difficult to imagine a situation where this would be preferable to the set available from blocks (i) and (iv), although it does provide unencumbered estimates of D and E and the two-factor interaction involving these two. This option is available. Using blocks (i) and (iii) is equivalent. It would have freed up B and C and all two-factor interactions involving them. Repeating a block provides a clean estimate of experimental error but does not provide new information about parameters that were not previously estimable and consequently is usually difficult to justify.

19.9.3 Some General Comments

An important point that must be stressed is that, in general, any two proper $1/2^m$ fractions together yield a proper $1/2^{m-1}$ fraction. However, it is possible to pick four $1/2^m$ fractions which together do not yield a $1/2^{m-2}$ fraction. This also implies that not all sets of eight $1/2^m$ fractions together result in a $1/2^{m-3}$ fraction.

It is also possible to use sequences of blocks defined by several different sets of defining contrasts. We are aware of no general theory to provide guidance for the investigator at this time. These sequences can be set up to repeat certain treatment combinations and provide clean estimates of experimental error. However, in general, they lead to correlated estimates of parameters. Our advice is to use a general-purpose regression program such as `proc GLM` in SAS® to evaluate exactly what is estimable at each stage before embarking on such an experimental program.

19.9.4 2^7 Factorial in a Sequence of Blocks of Eight

At this point the reader has no doubt noticed that the techniques and concepts are somewhat of a cross between fractional factorials and blocking. One begins with the concepts of blocking to assign treatment combinations to sets. Then one uses one or typically a sequence of the sets. This is a fractional factorial. Then finally, if all sets are used, one is back to a complete replicate in blocks. We illustrate the general procedures involved with two final examples. Consider first the 2^7 factorial. This can be put into 16 blocks of eight. To do this, four interactions and all of their generalized interactions must be confounded with blocks. A suitable choice for confounding is the set $ABCDEFG$, $DEFG$, $BCFG$, and $ACEG$. This is the same plan as the starting plan in Section 19.5. We write out the full set of confounded interactions in the following systematic array:

$$\begin{array}{ccccc} & & ABCDEFG & DEFG & ABC \\ & BCFG & ADE & BCDE & AFG \\ ACEG & & BDF & ACDF & BEG \\ & ABEF & CDG & ABDG & CEF. \end{array} \qquad (19.22)$$

We think of the contents of the intrablock subgroup as representing a 2^{7-4} replicate. This set can be obtained as the solutions to

$$\begin{aligned} i_A + i_B + i_C + i_D + i_E + i_F + i_G &= 0 \text{ (modulo 2)} \\ i_D + i_E + i_F + i_G &= 0 \text{ (modulo 2)} \\ i_B + i_C \qquad\qquad + i_F + i_G &= 0 \text{ (modulo 2)} \\ i_A \qquad + i_C \qquad + i_E \qquad + i_G &= 0 \text{ (modulo 2)}. \end{aligned}$$

Other blocks can be easily obtained by finding treatment combinations not in this set and then symbolic multiplication.

Now shift to the sequential experimentation setting. We now interpret the intrablock subgroup as a one-sixteenth replicate, or as the 2^{7-4} fraction. This is the fraction defined by

$$\begin{array}{llllll} \mathcal{I} & & = -ABCDEFG & = +DEFG & = -ABC \\ & = +BCFG = & -ADE & = +BCDE & = -AFG \\ = +ACEG & \qquad = & -BDF & = +ACDF & = -BEG \\ & = +ABEF = & -CDG & = +ABDG & = -CEF. \end{array}$$

Notice in particular the signs on the interactions and that the sign on the generalized interaction of two interactions is the product of the signs. It is clear that the seven main effects are estimable and that there are aliases. For example, A is aliased with BC, DE, and FG, given that three-factor and higher interactions are negligible. The question now is to select the next block or set of eight treatment combinations. Clearly, a good strategy would be to select so as to clear up as much of the aliasing as possible. A good choice is the fraction defined by

$$\begin{aligned}
\mathcal{I} \qquad\qquad\qquad &= +ABCDEFG = +DEFG = +ABC \\
= +BCFG &= +ADE = +BCDE = +AFG \\
= +ACEG \qquad &= +BDF = +ACDF = +BEG \\
= +ABEF &= +CDG = +ABDG = +CEF.
\end{aligned}$$

The trick to finding a good fraction is to find a defining contrast that lists as many of the three-letter interactions with changed sign as possible. Finding these treatment combinations requires that one find one treatment combination that satisfies the definition and then symbolic multiplication by elements in the intrablock subgroup. These two fractions together yield a 2^{7-3} fraction defined by

$$\begin{aligned}
\mathcal{I} \qquad\qquad &= +DEFG \\
= +BCFG &= +BCDE \\
= +ACEG \qquad &= +ACDF \\
= +ABEF &= +ABDG.
\end{aligned} \tag{19.23}$$

If the study is terminated at this point, all main effects are estimable free of two-factor interactions. However, two-factor interactions are aliased with one another. The between-block contrast consists of $ABCDEFG$ and the aliases ADE, BDF, CDG, ABC, AFG, BEG, and CEF. One must realize, however, that even if all 16 blocks in this sequence are used, the 15 interactions in (19.22) will all be confounded with blocks.

Reasonable choices for the third and fourth blocks would be the fractions defined by

$$\begin{aligned}
\mathcal{I} \qquad\qquad\qquad &= -ABCDEFG = -DEFG = +ABC \\
= +BCFG &= -ADE = -BCDE = +AFG \\
= +ACEG \qquad &= -BDF = -ACDF = +BEG \\
= +ABEF &= -CDG = -ABDG = +CEF
\end{aligned}$$

and

$$\begin{aligned}
\mathcal{I} \qquad\qquad\qquad &= +ABCDEFG = -DEFG = -ABC \\
= +BCFG &= +ADE = -BCDE = -AFG \\
= +ACEG \qquad &= +BDF = -ACDF = -BEG \\
= +ABEF &= +CDG = -ABDG = -CEF.
\end{aligned}$$

The result is the 2^{7-2} fraction defined by

$$\begin{aligned}
\mathcal{I} \qquad\qquad &= +BCFG \\
= +ACEG & \\
&= +ABEF.
\end{aligned}$$

Taking only one of these last two blocks gives an irregular fraction (Section 14.8) with correlated estimates. The fourth block does not add to the collection of estimable effects and interactions, but makes the best linear unbiased estimates uncorrelated.

Now assume that after the second stage of the study, when the two 2^{7-4} fractions are available, the investigator suspects (on the basis of examination of the data) that factor A is really important and that a good move would be to examine all the two-factor interactions that involve A. To use the term from John (2000), it is time to break the alias chains that involve A. Note that all factors in the study are balanced, i.e., appear at the high and low levels equally frequently. Exactly eight of the treatment combinations have A at the low level. These eight are part of the two 2^{7-4} fractions which together make up the 2^{7-3} fraction displayed by (19.23). Selecting only the half with A at the low level introduces $-A$ (and all generalized interactions) into the set of defining contrasts. This 2^{7-4} fraction is defined by

$$\begin{array}{llllll}
\mathcal{I} & & = +DEFG = & -A & = -ADEFG \\
 & = +BCFG & = +BCDE = & -ABCFG & = -ABCDE \\
= +ACEG & & = +ACDF = & -CEG & = -CDF \\
 & = +ABEF & = +ABDG = & -BEF & = -BDG.
\end{array}$$

The set of treatment combinations is easy to locate since they all have A at the low level. Now add the 2^{7-4} fraction obtained by reversing the level of A in this fraction. This semifold is the 2^{7-4} fraction defined by

$$\begin{array}{llllll}
\mathcal{I} & & = +DEFG = & +A & = +ADEFG \\
 & = +BCFG & = +BCDE = & +ABCFG & = +ABCDE \\
= -ACEG & & = -ACDF = & -CEG & = -CDF \\
 & = -ABEF & = -ABDG = & -BEF & = -BDG.
\end{array}$$

Combining the two 2^{7-4} fractions gives a 2^{7-3} fraction with all interactions free of A, implying that there are no aliases involving interactions of A and any other factor. All two-factor interactions involving A are estimable.

At this stage the investigator has tested 24 treatment combinations. These constitute three 2^{7-4} fractions. Using a least squares program to compute best linear unbiased estimates yields correlated estimates. However, all main effects are free of two-factor interactions and all two-factor interactions involving A are free of other two-factor interactions.

Exercise 19.6: Sequence of Runs. An investigator has made four runs, which can be coded as (1), ef, $abcd$, and $abcdef$. Notice that there are six factors. The control and all six factors at the high level have been examined in addition to two runs with a subset of factors at the high level. It turns out that this is exactly a one-sixteenth fraction of the 2^6 defined by

$$\begin{array}{lllll} \mathcal{I} & & = +ABCDEF & = +CDEF = & +AB \\ & = +BDEF = & +AC & = +BC & = +ADEF \\ = +BCEF & & = +AD & = +BD & = +ACEF \\ & = +CD & = +ABEF & = +EF & = +ABCD. \end{array}$$

Note that at this stage main effects are aliased with each other. Examine whether selecting the next set of four runs based on $\mathcal{I} = +ABCDEF = -CDEF = +BDEF = -BCEF$ is a good idea or not. Are there better choices? Now assume that the set suggested has been selected. The set of eight runs can now be interpreted as a one-eighth fraction of a 2^6. Give the aliasing relationships. Suggest another set of four runs to be added to these eight. What is estimable now? Notice that with these four extra runs, the investigator has a three-sixteenth replicate of a 2^6. □

19.10 SEQUENTIAL FRACTIONS

We now discuss various routes that can be taken after a preliminary resolution III plan, a 2^{7-4} fraction has been executed. If the investigator realizes at this point that a potentially important factor has been neglected. The defining contrast consisting of 15 words can be written as

$$\begin{array}{lllll} \mathcal{I} & & = +ABC & = +ADE = & +BCDE \\ & = +AFG & = +BCFG & = +DEFG = & +ABCDEFG \\ = +CEG & & = +ABEG & = +ACDG = & +BDG \\ & = +ACEF = & +BEF & = +CDF = & +ABDF. \end{array} \tag{19.24}$$

Since a factor was neglected, the plan really was a 2^{8-5} fraction with defining contrast

$$\mathcal{I} = +ABC = +ADE = +AFG = +CED = -H$$

and all generalized interactions. Another set of eight runs is required to study the effect of H. The foldover on H, represented by

$$\mathcal{I} = +ABC = +ADE = +AFG = +CED = +H,$$

and all generalized interactions is a possible candidate set that was discussed in Section 19.5. Two questions are "Can one do better?" and "Would there be an advantage to including the foldover on another factor or factors?" We need to examine defining contrasts. Adding the single foldover on H leaves all the three-letter words in (19.24) in the defining contrast. However, the complete foldover on all factors, including H, leads to a resolution IV, 2^{8-4} plan with defining contrast

$$\begin{aligned}
\mathcal{I} &= +BCDE = +BCFG = +DEFG \\
&= +ABEG = +ACDG = +ACEF = +ABDF \\
&= -ABCH = -ADEH = -AFGH = -ABCDEFGH \\
&= -CEGH = -BDGH = -BEFH = -CDFH.
\end{aligned}$$

At this stage, with 16 observations, the investigator is probably in a position to discard some factors as not important and to investigate some two-factor interactions. There are many options at this point. If, say, A looks as though its interactions are important, adding the semifold on A is a good option. This requires another eight experimental runs. In practice this simply means selecting the subset of eight treatment combinations with A at the low level and rerunning them with A at the high level. These eight treatment combinations with A at the low level can be identified as the fraction defined by

$$\mathcal{I} = +BCDE = +BCFG = +ABEG = -ABCH = -A$$

and all generalized interactions. The eight treatment combinations in the semifold on A can then be identified as the fraction defined by

$$\mathcal{I} = +BCDE = +BCFG = -ABEG = +ABCH = +A$$

and all generalized interactions. To implement this we just reverse the sign on all words containing A, just as in the foldover. The net result is that when these two sets are combined, all words containing the letter "A" will drop out of the defining contrast.

The 24 runs, the 16 plus the eight generated by the semifold, constitute a three-sixteenth fraction of a 2^8 factorial. As outlined in Section 14.8, one looks at the three overlapping one-eighth fractions, each with 16 treatment combinations, to see what is estimable. The three one-sixteenth fractions are identified as

$$\mathcal{I} = +BCDE = +BCFG = +ABEG = -ABCH = -A \quad (19.25)$$

$$\mathcal{I} = +BCDE = +BCFG = +ABEG = -ABCH = +A \quad (19.26)$$

$$\mathcal{I} = +BCDE = +BCFG = -ABEG = +ABCH = +A \quad (19.27)$$

and all generalized interactions in each case. When fully expanded, with all generalized interactions, all contain seven four-letter words containing "A" showing that two-factor interactions involving factor A have many aliases. All three are resolution III fractions. Combining (19.25) and (19.27) eliminates all words containing the letter "A." Consequently, combining the sets defined by (19.25) and (19.27) is the one-eighth that frees all two-factor interactions involving A from their aliases. All two-factor interactions involving factor A are estimable. In practice, the best estimates, i.e., minimum variance estimates, are obtained via a least squares analysis. Since all the data combined constitute an irregular fraction, estimates are correlated. A second semifold on, say, B frees up those interactions.

Formally one can write the defining contrasts given in (19.25), (19.26), and (19.27) as

$$\mathcal{I} = \mathcal{W}(\overline{A}) = \mathcal{W}(A) = -A$$

$$\mathcal{I} = \mathcal{W}(\overline{A}) = \mathcal{W}(A) = +A$$

$$\mathcal{I} = \mathcal{W}(\overline{A}) = -\mathcal{W}(A) = +A,$$

where $\mathcal{W}(\overline{A})$ represents all of the words that do not contain the letter "A," $\mathcal{W}(A)$ all of the words that do contain the letter "A," and $-\mathcal{W}(A)$ the same set with signs reversed. These derivations should not deter the reader interested in immediate application. The semifold can be easily adopted as a tool for generating sets of treatment combinations that break aliases.

If resources are limited and the error variance is modest, it is possible to use a double semifold to free aliases involving two factors. To obtain the double semifold on, say, A and B, select all treatment combinations with both A and B at the low level and replace low levels with high levels. Adding the double semifold to the original set of treatment combinations makes all two-factor interactions involving A or B estimable.

Formally we, can think of the defining contrast for the subset of treatment combinations with both A and B at the low level as

$$\mathcal{I} = \mathcal{W}(1) = +\mathcal{W}(A) = +\mathcal{W}(B) = +\mathcal{W}(AB) = -A = -B = +AB,$$

where $\mathcal{W}(1)$, $\mathcal{W}(A)$, $\mathcal{W}(B)$, and $\mathcal{W}(AB)$ represent words containing neither "A" nor "B," "A" and not "B," "B" and not "A," and both "A" and "B," respectively. The double semifold is then given by

$$\mathcal{I} = \mathcal{W}(1) = -\mathcal{W}(A) = -\mathcal{W}(B) = +\mathcal{W}(AB) = +A = +B = +AB.$$

For one-final scenario assume that, after looking at the results of the eight runs, it is found that the levels for factor A are not what they should be. Assume that it appears that looking for a quadratic response on A is in order. We now convert the two-level factor A into a four-level factor. This leads to the methodology in Chapter 20.

CHAPTER 20

Factorial Experiments with Quantitative Factors: Blocking and Fractionation

20.1 INTRODUCTION

A common requirement in research work is to estimate a smooth and continuous model that spans a region of interest and can be used to provide information about rates of change of a response variable and/or the location of a maximal (minimal) response. In addition, there are the ever-present questions about interactions with other quantitative as well as qualitative factors. To fit a quadratic surface, the simplest possible curved surface, it is necessary to have at least three levels of each factor. Full factorials with three or four levels per factor may require too many runs to fit into homogeneous blocks, and often too many runs to complete even one complete replicate. For instance, a full factorial with four factors, each at three levels, would require 81 runs, and one replicate of a 4^4 would require 256 runs.

The object of this chapter is to provide a set of tools for constructing plans and designs that allow efficient estimation of response surfaces using small blocks and/or fractional replication. In Sections 20.3 and 20.4 we focus on how to construct fractional factorials or incomplete blocking schemes for four-level factors that will allow estimation of linear, quadratic, and interaction effects. In Section 20.5 we discuss Box–Behnken plans for factors with three levels that lend themselves well to incomplete blocking schemes.

20.2 FACTORS AT THREE LEVELS

The first design possibility that comes to mind that would allow fitting a second-order polynomial is the 3^N factorial system. A parameterization system based

Planning, Construction, and Statistical Analysis of Comparative Experiments,
by Francis G. Giesbrecht and Marcia L. Gumpertz
ISBN 0-471-21395-0

on contrasts to estimate linear and quadratic effects is available in Chapter 11. However, there it was pointed out that it is not possible to develop a coherent system for blocking and fractional replication based on these contrasts. The alternative of a blocking and fractional replication scheme based on two-degree-of-freedom interaction components, developed in Section 11.2.3 for the purpose of constructing 3^{N-k} fractional factorials, is not really satisfactory either. The contrasts estimating linear and quadratic effects do not "line up" properly with the components being confounded or even more importantly, the components being aliased. The linear and quadratic contrasts can be estimated satisfactorily only from those interactions that have all parts estimable. In practice, this restricts the experiment to very large blocks and/or fractions.

In addition, incorporating factors with only two levels into the system is problematic. In practice, many qualitative factors come in two distinct forms and do not merge well with the 3^N system, as noted in Section 17.8. Another approach to the question of designing experiments for three-level factors, Box–Behnken designs, is discussed in Section 20.5.

20.3 FACTORS AT FOUR LEVELS BASED ON THE 2^N SYSTEM

Four-level factorials are useful for estimating response surfaces and lend themselves to fractionation or to (incomplete) blocking schemes that allow estimation of linear and quadratic effects without confounding important effects with blocks. Generally, a quantitative factor at four levels allows one to define and estimate contrasts that measure linear, quadratic, and cubic terms in the relationship between levels of a factor and the response. We focus on the linear and quadratic terms.

We recall from Section 17.5 that a theory for the 4^N factorial system can be developed from the 2^N system by means of suitably defined pseudofactors. We will utilize the fact that these pseudofactors and their interactions can be related to linear and quadratic contrasts and their interactions in the four-level system to construct fractions that will preserve the ability to estimate linear and quadratic effects. We deviate somewhat from the previous development in that emphasis will always be on the pseudofactors to generate the blocking and fractional replication schemes. We begin by defining a pseudofactor scheme.

20.3.1 Pseudofactor 2^{2N} System

To establish some generality and retain flexibility to extend easily to larger experiments, we begin with the two-level factorial system with factors $A1$, $A2$, $B1$, $\ldots$, $N1$, $N2$. The pairs, $A1$ and $A2$, $B1$ and $B2$, $\ldots$, $N1$ and $N2$ are then combined to form the four-level factors A, B, $\ldots$, N. To illustrate the technique, we begin with a 2^2 factorial with pseudofactors $A1$ and $A2$. The standard representation for treatment combinations is 00, 01, 10, and 11, which correspond to the levels a_0, a_1, a_2, and a_3 of the four-level factor.

Table 20.1 Pseudofactor Contrasts

	a_0	a_1	a_2	a_3
$A1$	-1	-1	$+1$	$+1$
$A2$	-1	$+1$	-1	$+1$
$A1A2$	$+1$	-1	-1	$+1$

The pseudofactor main effects for $A1$ and $A2$ would be estimated (we ignore the divisor) by $(y_{a_2}+y_{a_3})-(y_{a_0}+y_{a_1})$ and $(y_{a_1}+y_{a_3})-(y_{a_0}+y_{a_2})$, respectively, and the $A1A2$ interaction by $(y_{a_0}+y_{a_3})-(y_{a_1}+y_{a_2})$. In terms of the levels of the four-level factor, A, the three pseudofactor contrasts are as shown in Table 20.1. These contrasts are mutually orthogonal.

Now if the levels a_0, a_1, a_2, and a_3 are equally spaced, the contrasts estimating linear, quadratic, and cubic responses are proportional to

$$A_L : -3y_{a_0} - y_{a_1} + y_{a_2} + 3y_{a_3}$$

$$A_Q : +y_{a_0} - y_{a_1} - y_{a_2} + y_{a_3}$$

$$A_C : +y_{a_0} - 3y_{a_1} + 3y_{a_2} - y_{a_3}.$$

There are three key facts to notice. First, the A_Q contrast on the four-level factor is exactly the $A1A2$ interaction contrast in the pseudofactors. Second, the A_L contrast in the four-level factor is proportional to the sum of $2A1 + A2$ of the pseudofactors. Finally, the A_C contrast is proportional to the sum $A1 - 2A2$.

The importance of these observations is to realize that constructing a confounding scheme in the pseudofactor system that keeps $A1A2$ estimable also keeps A_Q estimable. Also, if the true response is a linear and quadratic function of the treatment factor (no cubic or higher-order effect), both pseudofactor main effects provide unbiased estimates of A_L in the four-level system. The two are not equally efficient. The two uncorrelated estimates, $[(y_{a_2}+y_{a_3})-(y_{a_0}+y_{a_1})]/4$ and $[(y_{a_1}+y_{a_3})-(y_{a_0}+y_{a_2})]/2$, obtained from $A1$ and $A2$ are both unbiased for β_A the linear regression coefficient of $\boldsymbol{y}$ on input and have variances $\sigma_\epsilon^2/4$ and σ_ϵ^2, respectively. The information content is in a 4:1 ratio. The combined, unbiased estimate $(2A1 + A2)/10$ has variance $\sigma_\epsilon^2/5$. In an N-factor system, these definitions extend directly to the main effect contrasts for each factor.

20.3.2 Interactions

To illustrate the relationships and the notation, we consider the case with factors A and B, each at four levels. We consider a response that can be modeled as a quadratic function of the two input factors; i.e., the model (20.1) contains parameters μ, β_A, γ_A, β_B, γ_B, and β_{AB}. The correspondence between pseudofactors and the four-level factors is shown in Table 20.2.

The main effects are as in Section 20.3.1. In terms of the pseudofactors, the BLUE of β_A, β_B, γ_A, and γ_B are multiples of $2A1 + A2$, $2B1 + B2$, $A1A2$,

Table 20.2 Correspondence between Pseudofactors at Two Levels and Factors at Four Levels

$a1a2b1b2$			$a1a2b1b2$		
0000	$\Rightarrow$	a_0b_0	0100	$\Rightarrow$	a_2b_0
0001	$\Rightarrow$	a_0b_1	0101	$\Rightarrow$	a_2b_1
0010	$\Rightarrow$	a_0b_2	0110	$\Rightarrow$	a_2b_2
0011	$\Rightarrow$	a_0b_3	0111	$\Rightarrow$	a_2b_3
1000	$\Rightarrow$	a_1b_0	1100	$\Rightarrow$	a_3b_0
1001	$\Rightarrow$	a_1b_1	1101	$\Rightarrow$	a_3b_1
1010	$\Rightarrow$	a_1b_2	1110	$\Rightarrow$	a_3b_2
1011	$\Rightarrow$	a_1b_3	1111	$\Rightarrow$	a_3b_3

and $B1B2$, respectively. The BLUE of the additional linear $\times$ linear interaction parameter, denoted by β_{AB}, can be written symbolically as a multiple of $(2A1+A2)\circledast(2B1+B2)$, which expands into the four parts $4A1B1+2A1B2+2A2B1+A2B2$. We look at $A1B1$ in some detail. In terms of the pseudofactors this contrast consists of

$$y_{0000} + y_{0100} + y_{0001} + y_{0101} + y_{1010} + y_{1110} + y_{1011} + y_{1111}$$
$$-y_{0010} - y_{0110} - y_{0011} - y_{0111} - y_{1000} - y_{1100} - y_{1001} - y_{1101}.$$

In terms of the four-level factors, this contrast is

$$y_{a_0b_0} + y_{a_1b_0} + y_{a_0b_1} + y_{a_1b_1} + y_{a_2b_2} + y_{a_3b_2} + y_{a_2b_3} + y_{a_3b_3}$$
$$-y_{a_0b_2} - y_{a_1b_2} - y_{a_0b_3} - y_{a_1b_3} - y_{a_2b_0} - y_{a_3b_0} - y_{a_2b_1} - y_{a_3b_1}.$$

If the four A treatment levels are equally spaced as x_{A0}, $x_{A0}+\delta x_A$, $x_{A0}+2\delta x_A$, and $x_{A0}+3\delta x_A$, and the B treatment levels as x_{B0}, $x_{B0}+\delta x_B$, $x_{B0}+2\delta x_B$, and $x_{B0}+3\delta x_B$, i.e., δx_A and δx_B represent the incremental level changes in x_A and x_B, respectively, and we write the model as

$$y(x_A x_B) = \mu + \beta_A(x_A - \overline{x}_A) + \gamma_A(x_A - \overline{x}_A)^2 + \beta_B(x_B - \overline{x}_B)$$
$$+ \gamma_B(x_B - \overline{x}_B)^2 + \gamma_{AB}(x_A - \overline{x}_A)(x_B - \overline{x}_B) + \epsilon, \quad (20.1)$$

then the expected value of the $A1B1$ contrast follows as

$$\text{E}[A1B1] = 16\gamma_{AB}\delta x_A \delta x_B.$$

For convenience we simply use x_A and x_B for δx_A and δx_B and rely on the context to make the meaning clear.

Continuing in this manner, evaluating the other three contrasts, we can construct Table 20.3. The conclusion to be drawn from this table is that all four contrasts provide independent but unequal amounts of information on γ_{AB}. The

Table 20.3 Expected Values of Interaction Contrasts

Contrast	Expected Value	Variance
$A1B1/16x_Ax_B$	γ_{AB}	$\sigma_\epsilon^2/32x_A^2x_B^2$
$A1B2/8x_Ax_B$	γ_{AB}	$\sigma_\epsilon^2/8x_A^2x_B^2$
$A2B1/8x_Ax_B$	γ_{AB}	$\sigma_\epsilon^2/8x_A^2x_B^2$
$A2B2/4x_Ax_B$	γ_{AB}	$\sigma_\epsilon^2/2x_A^2x_B^2$

first, $A1B1$, provides four times as much information as $A1B2$ or $A2B1$ and 16 times as much as $A2B2$. The BLUE of γ_{AB} has the pieces combined with proper weights. If in the design, one of the pieces must be sacrificed, $A2B2$ is the obvious candidate, and $A1B1$ must be retained. Similarly, at the main effect level, the $A2$ and $B2$ pieces are the next to be sacrificed if it is necessary. An alternative approach is to show that four times the $A1B1$ contrast plus twice the $A1B2$ contrast plus twice the $A2B1$ contrast plus the $A2B2$ contrast is exactly the linear-by-linear contrast.

A regression analysis of a full replicate in a CRD, using model (20.1) with levels $x_A, x_B = 0, 1, 2, 3$ yields uncorrelated estimates of β_A and β_B with variance $.05\sigma_\epsilon^2$, of γ_A and γ_B with variance $.0625\sigma_\epsilon^2$, and of γ_{AB} with variance $.04\sigma_\epsilon^2$.

Exercise 20.1: Expected Values of Contrasts. Let treatment a_0 consist of heating the experimental material to x_0 degrees, a_1, heating to $x_0 + x$ degrees, a_2 to $x_0 + 2x$ and a_3 to $x_0 + 3x$ degrees. Let y be the response, say yield under the various conditions. The mean of the four treatments is $\overline{x} = x_0 + 1.5x$. A reasonable model is

$$y(x) = \mu + \beta_A(x - \overline{x}) + \gamma_A(x - \overline{x})^2 + \epsilon,$$

where ϵ has variance σ_ϵ^2. Show that under this model:

(a) $\mathrm{E}[(y_{a_2} + y_{a_3}) - (y_{a_0} + y_{a_1})] = 4x\beta_A$.
(b) $\mathrm{E}[(y_{a_1} + y_{a_3}) - (y_{a_0} + y_{a_2})] = 2x\beta_A$.
(c) $\mathrm{E}[(2A1 + A2)/10] = x\beta_A$.
(d) $\mathrm{E}[A1A2] = 4x^2\gamma_A$.
(e) $\mathrm{E}[(-A1 + 2A2)/10] = 0$. □

Exercise 20.2: Variances of Contrasts. Show that under the model used in Exercise 20.1:

(a) $\mathrm{Var}[A1] = 4\sigma_\epsilon^2$.
(b) $\mathrm{Var}[A2] = 4\sigma_\epsilon^2$.
(c) $\mathrm{Var}[2A1 + A2] = 20\sigma_\epsilon^2$.

(d) The variance of the BLUE of β_A is $\sigma_\epsilon^2/5x^2$.

(e) $\text{Var}[\widehat{\gamma}_A] = \sigma_\epsilon^2/16x^4$. □

Exercise 20.3: Expected Value of Interaction Contrast. Show that under the model used

$$\text{E}[A1B2] = 8\gamma_{AB}x_A x_B. \qquad \square$$

20.3.3 Blocking for the 4^2

For an experiment with two blocks of eight, the first choice is to define blocks using the $A1A2B1B2$ pseudofactor interaction. The analysis of variance can be obtained via the pseudofactors and combining contrasts or using general linear models software such as `proc GLM` in SAS® with A and B treated as continuous variables. The analysis of variance table is illustrated in Table 20.4. The variances of the estimates of β_A, β_B, γ_A, γ_B, and γ_{AB}, using model (20.1) are as in the case of the CRD, since the blocking is fully orthogonal to the trend parameters.

Four blocks of four requires that a part of the $A_L B_L$ interaction $A2B2$ be sacrificed. A suitable scheme uses $A1A2B1$, $A1B1B2$, and $A2B2$ to define blocks. The analysis of variance takes the form illustrated in Table 20.5. Since this blocking is not orthogonal to the interaction component, the variance of $\widehat{\gamma}_{AB}$ increases from $.04\sigma^2$ to $.0417\sigma^2$, a 4% loss of information.

A third possibility is two rows and two columns with four experimental units per cell, with rows based on $A1A2B1$ and columns based on $A1B1B2$. However, this plan raises the question of proper randomization and error structure. It is possible that the four cells are "whole plots," the four treatment combinations assigned to "split plots" within cells. Two rows based on $A1A2B1B2$ and four columns based on $A1A2B1$, $A1B1B2$, and $A2B2$ encounter the same problem.

An even more extreme case, but which does not have problems with the error structure, consists of a row–column design with the four rows defined by $A1A2B1$, $A1B1B2$, and $A2B2$, and four columns defined by $A2$, $A1B2$, and

Table 20.4 Analysis of Variance 4^2, Two Blocks of Eight

Source	df
Blocks	1
A_L	1
A_Q	1
B_L	1
B_Q	1
$A_L B_L$	1
Error	9
Total	15

Table 20.5 Analysis of Variance 4^2, Four Blocks of Four

Source	df
Blocks	3
A_L	1
A_Q	1
B_L	1
B_Q	1
$A_L B_L$	1
Error	7
Total	15

$A1A2B2$. This design sacrifices part of the A_L ($A2$) and part of the $A_L B_L$ ($A1B2$ and $A2B2$). The variances of $\widehat{\beta}_A$, $\widehat{\beta}_B$, $\widehat{\gamma}_A$, $\widehat{\gamma}_B$, and $\widehat{\gamma}_{AB}$ are .0625, .0625, .05, .0625, and .05, respectively. We notice that 80% of the information on β_A and γ_{AB}, and all of the information on γ_A, β_B, and γ_B, has been retained. Unfortunately, only four degrees of freedom remain for error. The advantage of the design is very tight error control. There are other blocking options, but they lead to similar losses of information.

20.3.4 Blocking for the 4^3

This is a relatively large experiment and consequently there is considerable flexibility for blocking. For example, one can block on $A1A2B1B2$, $A1A2C1C2$, and $A1B1C1$ and generalized interactions for eight blocks of eight and retain full information on the full quadratic model. Blocking on $A2$ in addition to the seven above and the additional generalized interactions leads to the loss of one-fifth of the β_A information and $8/25$ of the γ_{BC} information in an experiment with 16 blocks of size 4. It is also possible to go to 32 blocks of 2, by confounding $B1$ and all generalized interactions. However, now losses become severe. The experiment loses 80% of the information on β_B, 20% on β_A and β_C, and 32% on all three linear×linear interactions. Full information is retained on quadratic main effects.

An 8×8 row–column design can be obtained by double confounding, $A1A2B1B2$, $B1B2C1C2$, and $A1B1C1$ with rows and $A2B2$, $A2C2$, and $A1A2B1B2C1C2$ with columns. There is a 4% loss of information on each of the three interaction terms and full information on all other parameters. Display 20.1 illustrates an easy way to obtain the treatment assignment.

20.3.5 Fractional Replication and Sequences of Fractions

Two Factors

Constructing fractional replicates in the pseudofactor system provides a wide range of options. However, care must be used in the application of the theory

```
data ; drop a1 a2 b1 b2 c1 c2 ;
do a1 = 0 , 1 ;  do a2 = 0 , 1 ;   a = 2*a1+a2 ;
 do b1 = 0 , 1 ;  do b2 = 0 , 1 ;  b = 2*b1+b2 ;
  do c1 = 0 , 1 ;  do c2 = 0 , 1 ;  c = 2*c1+c2 ;
   row = 4*mod(a1+a2+b1+b2,2) + 2*mod(b1+b2+c1+c2,2)
                   + mod(a1+b1+c1,2) ;
   col = 4*mod(a2+b2,2) + 2*mod(a2+c2,2)
                  +  mod(a1+a2+b1+b2+c1+c2,2) ;
   output;
   end ; end; end; end; end; end;
```

Display 20.1 SAS® code to assign treatments to a 4^4 factorial in eight rows and eight columns.

presented in Chapter 4. Problems appear with concepts such as resolution and aberration because contrasts like $A1A2$ which appear at first glance to be interactions really are main effects.

We illustrate some of the possibilities, first with a half replicate of the 4^2 and then an eighth replicate of the 4^3. Consider the two fractions defined by $\mathcal{I} = A1A2B1B2$ and $\mathcal{I} = A2B2$. The first aliases A_Q with B_Q and some of the pieces of the $A_L \times B_L$ interaction with each other. The second aliases part of A_L with part of B_L. Now compare the variances of estimates based on model (20.1) obtained from the two plans with those obtained from the full replicate where all estimates are uncorrelated and have full information (Table 20.6). In the first fraction, there is an estimate that can be either γ_A or γ_B with variance $.125\sigma_\epsilon^2$. The first fraction sacrifices information on the main effect quadratic terms and a bit of the interaction term, while the second keeps all parameters estimable but sacrifices information mainly on the quadratic terms. Note that it is easy to compute the variances of parameter estimates for any design under consideration using SAS® `Proc IML` or `Splus`.

It is informative to consider a subsequent set of four runs to augment the first fraction. We follow the methods developed in Section 19.9.4. The fraction defined by $\mathcal{I} = +A1A2B1B2$ can be thought of as the combination of the two one-fourth replicates defined by $\mathcal{I} = +A1A2 = +B1B2 = +A1A2B1B2$ and $\mathcal{I} = -A1A2 = -B1B2 = +A1A2B1B2$. We augment the design by adding the one-fourth replicate defined by $\mathcal{I} = +A1A2 = -B1B2 = -A1A2B1B2$. We now have two partially overlapping half replicates defined by $\mathcal{I} = +A1A2$ and $\mathcal{I} = -B1B2$. Fitting model (20.1) with a block effect (one block of eight and

Table 20.6 Variances of Parameter Estimates Using Four Alternate 4^2 Plans

	β_A	γ_A	β_B	γ_B	γ_{AB}
Full replicate	$.050\sigma_\epsilon^2$	$.0625\sigma_\epsilon^2$	$.050\sigma_\epsilon^2$	$.0625\sigma_\epsilon^2$	$.0400\sigma_\epsilon^2$
First fraction	$.100\sigma_\epsilon^2$	·	$.100\sigma_\epsilon^2$	·	$.0488\sigma_\epsilon^2$
Second fraction	$.104\sigma_\epsilon^2$	$.1562\sigma_\epsilon^2$	$.104\sigma_\epsilon^2$	$.1562\sigma_\epsilon^2$	$.1250\sigma_\epsilon^2$
Augmented set	$.067\sigma_\epsilon^2$	$.1006\sigma_\epsilon^2$	$.067\sigma_\epsilon^2$	$.1006\sigma_\epsilon^2$	$.0488\sigma_\epsilon^2$

one block of four) leads to estimates of the five parameters with variances given in the last line of Table 20.6.

Three Factors

Now consider the 4^3. A full replicate requires 64 runs. However, it is possible to use a one-eighth replicate consisting of eight runs to obtain estimates of linear and quadratic effects for each factor. Unfortunately, some of the estimates will be correlated. If we use the Franklin–Bailey algorithm discussed in Section 14.7 with requirements set $\{A1, A1A2, B1, B1B2, C1, C1C2\}$, we find a number of possible plans. One that seems particularly appealing is defined as

$$\begin{aligned} \mathcal{I} &= +A2B2 = +A2C2 = +B2C2 \\ &= +A1B1C1 = +A1A2B1B2C1 = +A1A2B1C1C2 = +A1B1B2C1C2. \end{aligned}$$

Inspection of this defining contrast immediately reveals that $A2$ is aliased with $B2$, $A2$ with $C2$, and $B2$ with $C2$. All of these are parts of linear contrasts, indicating that the estimates of the three linear contrasts are somewhat correlated. Inspection also reveals that $A1A2$, $B1B2$, and $C1C2$ are not aliased with any other contrast that we want to estimate. Fitting model (20.1) to the eight responses yields estimates with variances shown in the second row of entries in Table 20.7. The $A2B2$, $A2C2$, and $B2C2$ interactions in the defining contrast imply covariance between $\widehat{\beta_A}$, $\widehat{\beta_B}$, between $\widehat{\beta_A}$, $\widehat{\beta_C}$, and between $\widehat{\beta_B}$, and $\widehat{\beta_C}$. These covariances are all equal to $-.018\sigma_\epsilon^2$. The cost, in terms of increased variance of the parameters of interest, is minor. The remaining covariances are zero. Recall that the full replicate uses eight times as many runs as the fractions, so its variances are approximately eight times smaller.

A second fraction defined by

$$\begin{aligned} \mathcal{I} &= +A2B2 = +A2C2 = +B2C2 \\ &= +A1A2B1C1 = +A1B1B2C1 = +A1B1C1C2 = +A1A2B1B2C1C2 \end{aligned}$$

yields estimates with variances displayed in the third line of Table 20.7. As in the preceding fraction, the $A2B2$, $A2C2$, and $B2C2$ interactions in the defining contrast imply the same covariance as in the first fraction.

Table 20.7 Variances of Parameter Estimates Using Five Alternate 4^3 Plans

	β_A	γ_A	β_B	γ_B	β_C	γ_C
Full replicate	$.0125\sigma_\epsilon^2$	$.0156\sigma_\epsilon^2$	$.0125\sigma_\epsilon^2$	$.0156\sigma_\epsilon^2$	$.0125\sigma_\epsilon^2$	$.0156\sigma_\epsilon^2$
Fraction I	$.107\sigma_\epsilon^2$	$.125\sigma_\epsilon^2$	$.107\sigma_\epsilon^2$	$.125\sigma_\epsilon^2$	$.107\sigma_\epsilon^2$	$.125\sigma_\epsilon^2$
Fraction II	$.107\sigma_\epsilon^2$	$.125\sigma_\epsilon^2$	$.107\sigma_\epsilon^2$	$.125\sigma_\epsilon^2$	$.107\sigma_\epsilon^2$	$.125\sigma_\epsilon^2$
Fraction III	$.156\sigma_\epsilon^2$	$.125\sigma_\epsilon^2$	$.125\sigma_\epsilon^2$	$.156\sigma_\epsilon^2$	$.125\sigma_\epsilon^2$	$.164\sigma_\epsilon^2$
Fraction IV	$.125\sigma_\epsilon^2$	$.156\sigma_\epsilon^2$	$.125\sigma_\epsilon^2$	$.156\sigma_\epsilon^2$	$.125\sigma_\epsilon^2$	$.156\sigma_\epsilon^2$

A third fraction defined by

$$\begin{aligned} \mathcal{I} &= +A1B2 = +A1A2B1C1 = +A2B1B2C1 \\ &= +A1B1C2 = +B1B2C2 = +A2C1C2 = +A1A2B2C1C2 \end{aligned}$$

yields estimates with variances displayed in the fourth line of Table 20.7. This fraction is somewhat different. The three interactions, $A1B2$, $B1B2C2$, and $A2C1C2$ imply that covariances exist between $\widehat{\beta}_A$ and $\widehat{\beta}_B$, between $\widehat{\gamma}_B$ and $\widehat{\beta}_C$, and between $\widehat{\beta}_A$ and $\widehat{\gamma}_C$. In addition, since $\widehat{\beta}_A$ is involved in the estimation of both $\widehat{\beta}_B$ and $\widehat{\gamma}_C$, there is a correlation between $\widehat{\beta}_B$ and $\widehat{\gamma}_C$ when all three are computed. It turns out that $\text{Cov}[\widehat{\beta}_A, \widehat{\beta}_B] = -.0625\sigma_\epsilon^2$, $\text{Cov}[\widehat{\beta}_A, \widehat{\gamma}_C] = .0781\sigma_\epsilon^2$, $\text{Cov}[\widehat{\beta}_B, \widehat{\gamma}_B] = -.0312\sigma_\epsilon^2$, and $\text{Cov}[\widehat{\gamma}_B, \widehat{\beta}_C] = .0625\sigma_\epsilon^2$. The remaining covariances are again zero. In this case we see considerable inflation in variance of estimates due to the correlation.

A fourth fraction defined by

$$\begin{aligned} \mathcal{I} &= +A1B1B2 = +A1B1C1 = +A1A2B2C1 \\ &= +A1A2C2 = +A1B1B2C2 = +A2B1C1C2 = +B2C1C2 \end{aligned}$$

yields estimates with variances displayed in the fifth line of Table 20.7. In this fraction, the three interactions $A1B1B2$, $A1A2C2$, and $B2C1C2$ imply that $\widehat{\beta}_A$ and $\widehat{\gamma}_B$, $\widehat{\gamma}_A$ and $\widehat{\beta}_C$, and $\widehat{\beta}_B$ and $\widehat{\gamma}_C$ are correlated. The three covariances are all $.062\sigma_\epsilon^2$. All other covariances are zero.

Now consider augmenting these fractions by adding an additional one-eighth replicate. For fraction I, it seems clear that the investigator would select one of the following three fractions:

$$\begin{aligned} \mathcal{I} &= -A2B2 = -A2C2 = +B2C2 \\ &= +A1B1C1 = -A1A2B1B2C1 = -A1A2B1C1C2 = +A1B1B2C1C2 \end{aligned}$$

$$\begin{aligned} \mathcal{I} &= +A2B2 = -A2C2 = -B2C2 \\ &= +A1B1C1 = +A1A2B1B2C1 = -A1A2B1C1C2 = -A1B1B2C1C2 \end{aligned}$$

$$\begin{aligned} \mathcal{I} &= -A2B2 = +A2C2 = -B2C2 \\ &= +A1B1C1 = -A1A2B1B2C1 = +A1A2B1C1C2 = -A1B1B2C1C2. \end{aligned}$$

If the first is selected, the result is the one-fourth replicate defined by

$$\begin{aligned} \mathcal{I} &= +B2C2 \\ &= +A1B1C1 = +A1B1B2C1C2. \end{aligned} \tag{20.2}$$

This still leaves β_B and β_C correlated, since $B2$ and $C2$ are aliased. The second choice leaves β_A and β_B correlated, and the third choice leaves β_A and β_C correlated. The investigator should, if possible, look at the data from the first eight runs before selecting the next set. Other fractions can also be selected but at this point seem to have little to recommend them, but we will return to that

Table 20.8 Variances of Parameter Estimates Using Augmented Fractions

	β_A	γ_A	β_B	γ_B	β_C	γ_C
Fraction I	$.05\sigma_\epsilon^2$	$.0625\sigma_\epsilon^2$	$.0521\sigma_\epsilon^2$	$.0625\sigma_\epsilon^2$	$.0521\sigma_\epsilon^2$	$.0625\sigma_\epsilon^2$
II	$.05\sigma_\epsilon^2$	$.0625\sigma_\epsilon^2$	$.0521\sigma_\epsilon^2$	$.0625\sigma_\epsilon^2$	$.0521\sigma_\epsilon^2$	$.0625\sigma_\epsilon^2$
III	$.05\sigma_\epsilon^2$	$.0625\sigma_\epsilon^2$	$.05\sigma_\epsilon^2$	$.0781\sigma_\epsilon^2$	$.0625\sigma_\epsilon^2$	$.0625\sigma_\epsilon^2$
IV	$.05\sigma_\epsilon^2$	$.0625\sigma_\epsilon^2$	$.0625\sigma_\epsilon^2$	$.0625\sigma_\epsilon^2$	$.05\sigma_\epsilon^2$	$.0781\sigma_\epsilon^2$

point. The variances of the estimates of the linear and quadratic effects are given in Table 20.8.

When compared to the full replicate, shown in Table 20.7, we see full information (adjusted for number of observations) on β_A, γ_A, γ_B, and γ_C. The $\text{Cov}[\widehat{\beta}_B, \widehat{\beta}_C] = -.031\sigma_\epsilon^2$ has caused a minor loss of information.

Selecting the fraction

$$\begin{aligned} \mathcal{I} \quad &= -A2B2 &&= -A2C2 &&= +B2C2 \\ = +A1A2B1C1 &= -A1B1B2C1 &&= -A1B1C1C2 &&= +A1A2B1B2C1C2 \end{aligned}$$

leads to the same set of variances and covariances. The two fractions are equivalent for the model considered at this point. We extend the model later.

The obvious choice for augmenting fraction III is

$$\begin{aligned} \mathcal{I} \quad &= -A1B2 &&= +A1A2B1C1 &&= -A2B1B2C1 \\ = -A1B1C2 &= +B1B2C2 &&= -A2C1C2 &&= +A1A2B2C1C2 \end{aligned}$$

and for fraction IV is

$$\begin{aligned} \mathcal{I} \quad &= -A1B1B2 &&= +A1B1C1 &&= -A1A2B2C1 \\ = -A1A2C2 &= +A1B1B2C2 &&= -A2B1C1C2 &&= +B2C1C2. \end{aligned}$$

In fraction III, $\text{Cov}[\widehat{\gamma}_B, \widehat{\beta}_C] = .0312\sigma_\epsilon^2$, and in fraction IV, $\text{Cov}[\widehat{\beta}_B, \widehat{\gamma}_C] = .0312\sigma_\epsilon^2$. All other covariances are zero.

Looking at the defining contrasts for one-fourth replicates I and II, we see that both alias $B2$ and $C2$, parts of the linear components of B and C. This leads to the β_B, β_C correlation and consequent loss of information. In one-fourth replicates III and IV we have either $B1B2C2$ or $B2C1C2$, leading to correlations between B_Q and C_L or B_L and C_Q. The best choice is, clearly, problem dependent.

If we extend the model by adding interaction terms, terms like $(2A1 + A2) \circledast (2B1 + B2) \Rightarrow (4A1B1 + 2A1B2 + 2A2B1 + A2B2)$, $(2A1 + A2) \circledast (2C1 + C2)$, and $(2B1 + B2) \circledast (2C1 + C2)$ become involved. In terms of the regression model, this involves adding γ_{AB}, γ_{AC}, and γ_{BC}. Now $A1B1C1$ in fraction I aliases linear and interaction terms. For example, $A1$ from A_L gets aliased with $B1C1$ from the $A \times C$ interaction. With this extended model, fraction II becomes a disaster. From the defining contrast,

$$\mathcal{I} = B2C2 = A1A2B1C1 = A1A2B1B2C1C2,$$

Table 20.9 Variances of Parameter Estimates under and Extended Model Using Four Augmented Fractions

Factor	β_A	γ_A	β_B	γ_B	β_C	γ_C	γ_{AB}	γ_{AC}	γ_{BC}
	$.0125\sigma_\epsilon^2$	$.0156\sigma_\epsilon^2$	$.0125\sigma_\epsilon^2$	$.0156\sigma_\epsilon^2$	$.0125\sigma_\epsilon^2$	$.0156\sigma_\epsilon^2$	$.01\sigma_\epsilon^2$	$.01\sigma_\epsilon^2$	$.01\sigma_\epsilon^2$
I	$.25\sigma_\epsilon^2$	$.0625\sigma_\epsilon^2$	$.1886\sigma_\epsilon^2$	$.1406\sigma_\epsilon^2$	$.1886\sigma_\epsilon^2$	$.1406\sigma_\epsilon^2$	$.1509\sigma_\epsilon^2$	$.1509\sigma_\epsilon^2$	$.3125\sigma_\epsilon^2$
III	$.05\sigma_\epsilon^2$	$.2349\sigma_\epsilon^2$	$.0776\sigma_\epsilon^2$	$.0718\sigma_\epsilon^2$	$.0625\sigma_\epsilon^2$	$.0625\sigma_\epsilon^2$	$.0678\sigma_\epsilon^2$	$.0678\sigma_\epsilon^2$	$.1724\sigma_\epsilon^2$
IV	$.1228\sigma_\epsilon^2$	$.0625\sigma_\epsilon^2$	$.2625\sigma_\epsilon^2$	$.1125\sigma_\epsilon^2$	$.1124\sigma_\epsilon^2$	$.1281\sigma_\epsilon^2$	$.0839\sigma_\epsilon^2$	$.20\sigma_\epsilon^2$	$.1138\sigma_\epsilon^2$
V	$.05\sigma_\epsilon^2$	$.0776\sigma_\epsilon^2$	$.05\sigma_\epsilon^2$	$.0776\sigma_\epsilon^2$	$.05\sigma_\epsilon^2$	$.0776\sigma_\epsilon^2$	$.0603\sigma_\epsilon^2$	$.0603\sigma_\epsilon^2$	$.0603\sigma_\epsilon^2$

we see that $B1C1$ is aliased with $A1A2$, $A1B1$ with $A2C1$, $A1C2$ with $A2B1$, and $B2C2$ with the mean. Hence, $B_L \times C_L$, i.e., γ_{BC} is not estimable. We also see that A_Q, B_Q, and C_Q are messed up with $B_L \times C_L$. The two fractions that appeared to be equivalent now are seen to be very different. Fraction III, defined by

$$\mathcal{I} = A1A2B1C1 = B1B2C2 = A1A2B2C1C2,$$

has the aliasing relationships $A1A2 = B1C1$, $A1B1 = A2C1$, $A1C1 = A2B1$, $B1B2 = C2$, $B1 = B2C2$, and $B2 = B1C2$. We see that A_L and B_Q are clear and estimable with full information. All other estimates are correlated. Fraction IV, defined by

$$\mathcal{I} = A1B1C1 = A1B1B2C2 = B2C1C2,$$

has the aliasing relationships $A1 = B1C1$, $B1 = A1C1$, $C1 = A1B1$, $A1B1 = B2C2$, $B1B2 = A1C2$, $A1B2 = B1C2$, $B2 = C1C2$, $C1 = B2C2$, and $C2 = B2C1$. From this mess, we see that only the A_Q effect, estimated by the $A1A2$ contrast, is clean.

We summarize the variances for the three fractions as well as an additional fraction (V) in Table 20.9. Fraction V has defining contrast

$$\mathcal{I} = A1A2B2C1 = A1B1B2C2 = A2B1C1C2.$$

This has the nine aliasing relationships $A1A2 = B2C1$, $A1B2 = A2C1$, $A1C1 = A2B2$, $A1C2 = B1B2$, $A1B1 = B2C2$, $A1B2 = B1C2$, $A2B1 = C1C2$, $A2C1 = B1C2$, and $A2C2 = B1C1$. We see that no contrasts which are part of the linear terms are involved. Also, the three contrasts that estimate pure quadratic effects are aliased with minor parts of the contrasts that enter into the covariance terms. This one-fourth replicate has superior statistical properties, full information on the linear main effects, almost full information on the main effects quadratic terms, and some information on the interaction effects.

Summary

To summarize the use of replication and sequences of fractions, for designing experiments with quantitative factors we note the following:

- There is tremendous flexibility in choice of fractions.
- The plans considered are most appropriate in those cases where results of experimental work become available relatively quickly and the error terms are modest relative to the effects of the factors.
- The eventual aim and potential of the project as well as constraints such as money, time, or resources must be considered early. If it is known that, say, only a one-eighth replicate of a 4^3 can be performed, one should select a different fraction than if it is likely that one may be able to add a second one-eighth. Similarly, if one can go to a one-fourth replicate directly, again the optimal choice is different.
- There is always the possibility of adding a small number of points to a plan as a "repair" after looking at the data.
- Classical measures of design optimality are not really appropriate for evaluating these plans, because there is always the question of whether the criterion really targets what the investigator has in mind.

20.4 PSEUDOFACTORS AND HADAMARD PLANS

20.4.1 Plans Based on the 12-Run Hadamard

The pseudofactor technique to introduce continuous factors into the 2^N system can also be used with the Hadamard plans. We illustrate with the 12-run Hadamard discussed in Section 16.4.1. We replace the first two columns with one column suitable for a factor at four levels. The replacement rule for the four equally spaced levels that we use is shown in Table 20.10. To see the logic behind this assignment, we need to look at the two columns and the corresponding interaction column. In this set, column 1 and the interaction take the place of, say, $A1$ and $A2$ in the 2^N system. Factor $A2 \Leftrightarrow$ the interaction column will be sacrificed directly. Column 2 takes the place of $A1A2$ in the 2^N system. Columns 1 and 2 are replaced by the four-level column.

In the 12-run Hadamard plan it is possible to insert five four-level factors and keep one two-level factor. Several points should be noted. First, interaction effects can generally not be estimated. Second, since part of the linear component in each of the main effects was sacrificed, estimates will be correlated. This correlation pattern will be complex because of the complex aliasing structure

Table 20.10 Replacement Rule for Two Columns in the 12-Run Hadamard Plan

Column 1	Column 2		
-1	-1	$\Rightarrow$	a_1
-1	$+1$	$\Rightarrow$	a_0
$+1$	-1	$\Rightarrow$	a_2
$+1$	$+1$	$\Rightarrow$	a_3

involving interactions in the Hadamard plan. However, the correlations turn out to be modest. The special model with only quadratic effects has all estimates uncorrelated.

As an upper bound, consider the 12-run case, saturated with five four-level factors and one two-level factor and fit the model

$$\begin{aligned} y(a,b,c,d,e,f) &= \mu + \beta_A a + \beta_B b + \beta_C c + \beta_D d + \beta_E e \\ &+ \gamma_A a^2 + \gamma_B b^2 + \gamma_C c^2 + \gamma_D d^2 + \gamma_E e^2 + \delta_F f + \epsilon \end{aligned}$$

where a, b, c, d, and e each take the values -1.5, $-.5$, $.5$, and 1.5 and f takes values $+1$ and -1. The variances for the $\widehat{\beta}$'s then range from $.081\sigma_\epsilon^2$ to $.092\sigma_\epsilon^2$ and for the $\widehat{\gamma}$'s range from $.093\sigma_\epsilon^2$ to $.096\sigma_\epsilon^2$. Most covariances fall in the range $-.02\sigma_\epsilon^2$ to $+.02\sigma_\epsilon^2$. For comparison, we note that if four observations are taken at each of $-1.5, 0.0$, and $+1.5$, the variance of the quadratic effect is $.074\sigma_\epsilon^2$.

Adding the foldover to the Hadamard plan eliminates the correlations among estimates of quadratic effects and greatly reduces the correlations among estimates of linear effects. If the foldover is added to the 12-run Hadamard plan, all the $\widehat{\beta}$'s have variance $.034\sigma_\epsilon^2$ and all $\widehat{\gamma}$'s have variance $.042\sigma_\epsilon^2$. Covariances among the $\widehat{\beta}$'s are $\pm.0019\sigma_\epsilon^2$. Note that if eight observations are taken at each of -1.5, 0.0, and $+1.5$, the variance of the quadratic effect is $.028\sigma_\epsilon^2$. The 24-run plan appears to have considerable merit. The number of runs relative to the number of parameters also makes it possible to extend the model to include interaction effects in the model. Correlations among parameter estimates remain modest.

20.4.2 Plans Based on the 16-Run Hadamard

It is also possible to use the pseudofactor technique in the 16-run Hadamard illustrated in array (16.2). Since the first column in this plan is somewhat different, we work only with columns 2 to 15. There is room for seven four-level factors. We use the same substitution rule as for the 12-run Hadamard. The $\widehat{\beta}$'s have variance $.05625\sigma_\epsilon^2$, and all $\widehat{\gamma}$'s have variance $.0625\sigma_\epsilon^2$. The variance for the $\widehat{\gamma}$'s can be compared with a minimal variance, $.0417\sigma_\epsilon^2$. The $\widehat{\beta}$'s have pairwise correlations equal to $-.111$. The $\widehat{\gamma}$'s are all uncorrelated. Adding the foldover is not as dramatic in this plan as in the 12-run case. The variances are reduced by half and the correlations to $-.1$. It is also possible to extend the model to allow for linear-by-linear interactions. However, the correlations among estimates again tend to be large.

20.5 BOX–BEHNKEN PLANS

The Box–Behnken plans (Box and Behnken 1960) constitute a class of fractional 3^N factorial plans that are rarely used in agricultural work but have attained a measure of popularity in industrial experimentation. Notice that the most general second-degree polynomial has 1 (constant) + k (linear) + k (pure quadratic) + $\binom{k}{2}$ (cross product) = $\binom{k+2}{2}$ terms. It follows that a plan with $\binom{k+2}{2}$

observations should be sufficient to estimate all parameters on the model. Extra observations contribute to a measure of error and for that reason may be desirable. One of the objects of the Box–Behnken plans is to provide reasonable estimates of the parameters with as few observations as possible. These plans become particularly attractive in cases where the experimental error variance is sufficiently small that a few observations will give information with adequate precision.

Many of the Box–Behnken designs are resolvable in the sense that they can be set up with an orthogonal blocking system, a system in which block contrasts are uncorrelated with the estimates of the coefficients in the polynomial model.

Example 20.1. To illustrate the Box–Behnken plans we consider a case with $k = 4$ factors. Begin with a balanced incomplete block design for four treatments in six blocks of size 2. The actual blocking for the Box–Behnken plan is different from the BIB and is shown in Table 20.11.

In addition, consider the 2^2 factorial in Table 20.12. To obtain the plan desired, these two structures are combined. The two asterisks in every row of the incomplete block design are replaced by the two columns of the 2^2 factorial. Whenever an asterisk does not appear, a column of zeros is inserted. The plan is completed by adding a number of center points (points with all factors at the zero level). In this case the resulting plan for four factors with 27 runs is as shown in Table 20.13.

With the addition of the three center points, the result is a design in three blocks of nine runs each. Block contrasts are orthogonal to estimates of the parameters in the polynomial model. Note that the model for three blocks, four factors has

Table 20.11 Schematic Structure of Box–Behnken Plan for Four Factors

Treatment:		x_1	x_2	x_3	x_4
		*	*		
	1				
				*	*
		*			*
Block:	2				
			*	*	
			*		*
	3				
		*		*	

Table 20.12 Coding for 2 x 2 Factorial

Factors:	x_i	x_j
	−1	−1
Level:	1	−1
	−1	1
	1	1

Table 20.13 27-Run Box–Behnken Design for Four Factors

Factors:	x_1	x_2	x_3	x_4
Block 1	−1	−1	0	0
	1	−1	0	0
	−1	1	0	0
	1	1	0	0
	0	0	−1	−1
	0	0	1	−1
	0	0	−1	1
	0	0	1	1
	0	0	0	0
Block 2	−1	0	0	−1
	1	0	0	−1
	−1	0	0	1
	1	0	0	1
	0	−1	−1	0
	0	1	−1	0
	0	−1	1	0
	0	1	1	0
	0	0	0	0
Block 3	0	−1	0	−1
	0	1	0	−1
	0	−1	0	1
	0	1	0	1
	−1	0	−1	0
	1	0	−1	0
	−1	0	1	0
	1	0	1	0
	0	0	0	0

17 parameters. The plan with 27 observations leaves 10 degrees of freedom for error. Box–Behnken plans for three to 10 factors, along with blocking schemes, are given in Appendix 20A.

REVIEW EXERCISES

20.1 (Courtesy of Noel Pollen) A food science class plans to do an experiment to compare formulations of an orange drink. There are three ingredients that can be manipulated: (1) amount of orange flavoring (5, 15, or 25 mL); (2) amount of citric acid (5, 15, or 25 mL); and (3) amount of sucrose (200, 300, or 400 mL). The goal of the experiment is to develop an orange drink product that matches the flavor intensity of a competitor's product. Your task is to propose two possible experimental designs for this project, each using 15 treatment combinations. For each proposed design do the following:

(a) List the treatment combinations in both coded and uncoded form. Give the name of the design.

(b) Draw a three-dimensional diagram of the treatment combinations included in the design.

(c) For the 200-mL level of sucrose, draw a contour plot of the variance of the value predicted ($\widehat{y}$) as a function of the levels of orange flavoring and citric acid. Assume that a full second-order polynomial will be fitted to the data.

20.2 (From Xiaohong Zhang, personal communication) A polymer scientist is developing a fabric intended for use in clothing for strenuous exercise. The objective is to develop a fiber that would have a high water transmission rate. The first step in producing such a fiber is to heat a polymer chip in an extruder. In the extruder the polymer takes on the consistency of a gel. It is then extruded through a 3-inch die with about 90 0.4-mm holes. Air is blown through an air hole in the top of the die; the air pressure affects how fast the gel is cooled. The very fine fiber emerging from the die is in semiliquid form. The fiber is collected on a drum, which is at room temperature, and it solidifies into a web on the drum. Several factors can be adjusted during this process: the air pressure, the extruder speed or throughput, the drum speed or takeup speed, and the distance from the die to the drum. One response variable of interest is the pore size distribution of the finished web of fiber.

Propose a suitable experimental design for modeling the effects of these factors, using the following levels of the four factors in 40 experimental runs: **(a)** throughput (75%, 100%, 150%, 175%, 200%); **(b)** takeup speed (18, 22, 26, 30 in./min); **(c)** air pressure (10, 20, 30, 40 psi); and **(d)** die-to-collecting distance (4, 6, 8, 10 in.).

APPENDIX 20A: BOX–BEHNKEN PLANS*

The following array table gives designs for investigating 3, 4, 5, 6, 7, 9, 10 variables. The symbol (±, ±, . . . , ±) means that all possible combinations of "+" and "−" levels are to be run. The blocking scheme for each design is indicated.*

*Arrays are reprinted with permission from Box and Behnken (1960).

Factors				Points	Blocking
3 factors	±1	±1	0	12 points	No orthogonal blocking possible.
	±1	0	±1		
	0	±1	±1		
	0	0	0	3 controls	

Factors					Points	Blocking
4 factors	±1	±1	0	0	8 points	3 blocks of 9
	0	0	±1	±1		
	0	0	0	0	1 control	
	±1	0	0	±1	8 points	
	0	±1	±1	0		
	0	0	0	0	1 control	
	±1	0	±1	0	8 points	
	0	±1	0	±1		
	0	0	0	0	1 control	

Factors						Points	Blocking
5 factors	±1	±1	0	0	0	20 points	2 blocks of 23
	0	0	±1	±1	0		
	0	±1	0	0	±1		
	±1	0	±1	0	0		
	0	0	0	±1	±1		
	0	0	0	0	0	3 controls	
	0	±1	±1	0	0	20 points	
	±1	0	0	±1	0		
	0	0	±1	0	±1		
	±1	0	0	0	±1		
	0	±1	0	±1	0		
	0	0	0	0	0	3 controls	

Factors							Points
6 factors	±1	±1	0	±1	0	0	48 points
	0	±1	±1	0	±1	0	
	0	0	±1	±1	0	±1	
	±1	0	0	±1	±1	0	
	0	±1	0	0	±1	±1	
	±1	0	±1	0	0	±1	
	0	0	0	0	0	0	6 controls

This experiment can be put into two blocks of 27. The first block contains zero or two "–" values and three controls. All others are in the second block.

Factors								Points
7 factors	0	0	0	±1	±1	±1	0	56 points
	±1	0	0	0	0	±1	±1	
	0	±1	0	0	±1	0	±1	
	±1	±1	0	±1	0	0	0	
	0	0	±1	±1	0	0	±1	
	±1	0	±1	0	±1	0	0	
	0	±1	±1	0	0	±1	0	
	0	0	0	0	0	0	0	6 controls

This experiment can be put into two blocks of 31. The first block contains zero or two "–" values and three controls. All others are in the second block.

	±1	0	0	±1	0	0	±1	0	0	
	0	±1	0	0	±1	0	0	±1	0	24 points
	0	0	±1	0	0	±1	0	0	±1	
	0	0	0	0	0	0	0	0	0	2 controls
	±1	±1	±1	0	0	0	0	0	0	
	0	0	0	±1	±1	±1	0	0	0	24 points
	0	0	0	0	0	0	±1	±1	±1	
	0	0	0	0	0	0	0	0	0	2 controls
	±1	0	0	0	±1	0	0	0	±1	
9 factors	0	0	±1	±1	0	0	0	±1	0	24 points
	0	±1	0	0	0	±1	±1	0	0	
	0	0	0	0	0	0	0	0	0	2 controls
	±1	0	0	0	0	±1	0	±1	0	
	0	±1	0	±1	0	0	0	0	±1	24 points
	0	0	±1	0	±1	0	±1	0	0	
	0	0	0	0	0	0	0	0	0	2 controls
	±1	0	0	±1	0	0	±1	0	0	
	0	±1	0	0	±1	0	0	±1	0	24 points
	0	0	±1	0	0	±1	0	0	±1	
	0	0	0	0	0	0	0	0	0	2 controls

This is an experiment with five blocks of 26.

	0	±1	0	0	0	±1	±1	0	0	±1	
	±1	±1	0	0	±1	0	0	0	0	±1	
	0	±1	±1	0	0	0	±1	±1	0	0	
	0	±1	0	±1	0	±1	0	0	±1	0	
	±1	0	0	0	0	0	0	±1	±1	±1	
10 factors	0	0	±1	±1	±1	0	0	0	0	±1	
	±1	0	0	±1	0	0	±1	±1	0	0	
	0	0	±1	0	±1	0	±1	0	±1	0	
	±1	0	±1	0	0	±1	0	0	±1	0	
	0	0	0	±1	±1	±1	0	±1	0	0	
	0	0	0	0	0	0	0	0	0	0	10 controls

This plan with 10 factors can be put into two blocks of 85, by putting into the first block all treatment combinations corresponding to one or three of the -1's and five controls, and the remainder into the second.

CHAPTER 21

Supersaturated Plans

21.1 INTRODUCTION

In all of the situations encountered up to this point it has always been assumed that the number of experimental units or runs available to the experimenter exceeds the number of factors to be studied. However, there are situations where for some reason or other this is not true. There are cases where experimental units or experimental runs are limited by cost, time, or laboratory facilities, forcing very small experiments. Examples of this are easy to find. Also, there are other cases where the number of treatments of interest is so large that it is simply not possible to conduct an experiment that is large enough to accommodate all in one of the plans we have studied up to this point. In addition, in many cases it is also true that most of the factors are not really effective, i.e., only few are active. The problem really is one of identifying the few active factors. In the statistical literature we find these problems discussed under titles that include the word *supersaturated* and that use the words *screening experiments*. An area of application that comes to mind is testing the robustness of a product against adverse conditions it is likely to encounter in practice. Obviously, there is a limit to the number of tests and a very large number of potential adverse conditions.

Most papers discuss both the problems of data generation or design, and data analysis. There is no doubt that of these, the first is by far the most important since there is no way that a sophisticated statistical analysis can recover information lost due to a poor design. We also emphasize that the practice of just simply ignoring some factors is not a reasonable alternative. Just because the investigator decides to ignore a factor does not mean that it goes away. The factor must still be set to some level, and the complete investigation will be conditioned on that selection. An adverse condition can still hurt a product, even if it has not been tested. This, however, is not to say that experience cannot be a tremendous potential for screening large numbers of variables.

Planning, Construction, and Statistical Analysis of Comparative Experiments,
by Francis G. Giesbrecht and Marcia L. Gumpertz
ISBN 0-471-21395-0

The phenomenon of few active effects in a larger set is commonly referred to in the statistical literature as *effect sparsity*. The object of a good experimental plan is to identify the few active factors in the larger set. In this chapter we assume that the cost of missing a truly effective factor is far greater than the cost of including a noneffective factor. The investigator should expect that in most cases, an additional set of runs in a small confirmatory experiment will be required to provide the definitive answers.

21.1.1 Short Historical Note

Random Balance

One of the first proposals in the statistical literature to deal with the problem of more factors than observations was the random balance technique. The exact technique for constructing random balance plans has never been spelled out carefully. In general terms, the technique consists of selecting at random a subset of n treatment combinations from the total of N treatment combinations from either a complete or a fractional replicate of a factorial involving f factors. Typically, $n < f$ rather than just $n < N$. The actual definition of selecting the subset at random was never established. For example, some practitioners insisted on levels of individual factors being equally represented (balance), while others did not. The major criticism of the random balance strategy is that individual treatments are randomly confounded with one another. Since the number of factors exceeds the number of observations, it is not possible to have orthogonal estimates of all treatments. The randomization or haphazard selection of treatment combinations dictates the extent to which specific treatments will be mutually confounded. It is never possible to untangle the factors that have become confounded. Note that this randomization is not to be confused with the random assignment of the selected treatment combinations to experimental units. The latter randomization is always desirable. In summary, random balance designs have questionable usefulness. The interested reader should consult the special issue of *Technometrics*, No. 2, Vol. 1, published in 1959.

Group Screening

An early alternative to the random balance is the group screening method. The essential idea here is to put the factors into groups, test these group factors, and then test in another experiment the factors that were in significant groups. No further testing is done with factors in groups that do not show significance. Although this procedure has simplicity, generality, and intuitive appeal, some rather stringent assumptions must be met before the method performs reasonably well. Two major assumptions are that directions of possible effects must be known, and no interactions among factors exist.

21.2 PLANS FOR SMALL EXPERIMENTS

Recent years have seen considerable effort to develop plans that allow for more factors than runs, but yet have no factors fully confounded with any other, i.e.,

Table 21.1 Supersaturated Plans

No. of Runs	No. of Factors	Generating Vectors
6	10	(+ − − − +) (− + − + −)
8	14	(− + + − − − +) (− + + − + − −)
10	18	(+ + − + − − − + −) (− + + + − − + − −)
12	22	(− + + + − − + − − − +) (+ − − − − + + − + − +)
14	26	(− + + + − + + − − + − − −) (+ + − + − − + − + − − − +)
16	30	(+ − + + + + − − + − − − − + −) (− + − − + + − + − − − + + + −)
18	34	(− − − + − + + − + − + − + + + − −) (+ + − − + + − − − + − − + − − + +)
20	38	(− + + − + + + + − − + − − − + − − − +) (− + − + + − − − + − + + + − − − − + +)
22	42	(+ + + + − + − − + − − − − + + − − − + − +) (− + + − − + + − − + − + + + − − − − + − +)
24	46	(+ + − − + + − + − + − − − − − + + + + − + − −) (− + − + − − − − + − − + − − + + + + − − + + +)
26	50	(− + + − + + − − + − + + + + − − + + − + − − − − −) (− + − − − + + + + − + − + + − − − + + − + − + − −)
28	54	(− − + + + − + − − + − − + − + + − + − + − + − − − + +) (+ − − − + + − − − − − − − + + − + + + − + − − + + + +)
30	58	(+ + − − − + − − − − + − + + + + + + − − + + − + − − − − +) (+ − − − + − − + − + + − − − + + − + + + − + − + − − + + −)

Source: Nguyen (1996).

estimates of individual factors are not perfectly correlated. The key idea is that despite the correlations, a careful statistical analysis of the data will reveal the active factors. Lingering ambiguities that remain must then be cleared up with a small follow-up, confirmatory experiment.

Table 21.1 gives generator vectors for constructing supersaturated plans ranging from six runs plans suitable for up to 10 factors to 30-run plans that can accommodate up to 58 factors. The generator vectors are constructed to yield good plans, good in the sense that they attempt to minimize the correlation structure among the treatment factors. We illustrate the construction by developing the plan for 14 factors in an experiment with eight runs. First, the two generator vectors given in the table are written vertically and then each is expanded

Table 21.2 Eight-Run Supersaturated Plan

Run	Factors
1	+ + + + + + + + + + + + + +
2	− + − − − + + − − − + − + +
3	+ − + − − − + + − − − + − +
4	+ + − + − − − + + − − − + −
5	− + + − + − − − + + − − − +
6	− − + + − + − + − + + − − −
7	− − − + + − + − + − + + − −
8	+ − − − + + − − − + − + + −

cyclically into seven columns. Then finally a row of +'s is added to the top of the array. The resulting array is shown in Table 21.2.

The actual plan for the experiment is obtained by assigning factors to the columns and using rows to define the treatment combinations for the eight runs. Notice that each factor appears at both the high and low levels four times in the plan and that the first run has all factors at the high level. The second run has factor 1 at the low level, factor 2 at the high level, factors 3, 4, and 5 at the low level, ... and factor 14 at the high level. Also note that the investigator is free to reverse the assignment of factor levels, i.e., low to "+" and high to "−", so that the first run with all factors at the low level is the conventional control treatment. If the experiment requires fewer than 14 factors, some columns are left unused.

In a small example such as this the investigator would proceed with the statistical analysis of the data by calculating differences, mean at the high level minus mean at the low level, for the factors in the experiment. For larger plans one could utilize a stepwise regression program or a regression program that computes analyses with all possible subsets of independent variables. A small-scale simulation study with this plan showed that if one factor were active and the difference between high and low were one standard deviation, stepwise regression with significance level for entry and staying in the model set at 0.05 would select on average just over two variables as active and include the correct variable just over 35% of the time. Under the same conditions an effect equal to two standard deviations will be selected 80% of the time, and on average two variables identified as active. In other words, even with such a small study, there is a reasonable chance of getting a signal that there is an important factor and some reasonable guidance for further work. Effects as large as three standard deviations are almost certain to be identified with even this small experiment. A second simulation with the 18-run plan and 30 factors gave slightly lower success rates for locating one active factor.

The simulation study also shows that while increasing the significance level for entry and staying in the model over 0.05 does increase the probability of selecting an active factor, it is probably not worthwhile since it also increases the number of nonactive factors selected. At this point there are still some questions about the statistical analysis to recommend. The stepwise regression appears to work reasonably well. However, Abraham et al. (1999) claim that a computer program that computes regression analyses for all possible sets of up to p independent variables for a reasonably chosen value of p performs notably better. Our experience indicates that there is little difference between the performance of a stepwise procedure and an all-possible-regressions procedure.

To put this in another perspective, we note that if the investigator randomly eliminates seven of the factors (in the eight-run study), then the upper bound on the probability of finding the effective factor is 0.5, regardless of the size of the factor. In addition, there is still the question of actually identifying the factor even if it is selected. This reduces the probability significantly. A simulation study here suggests that one active factor (effect equal to one standard deviation) is detected 64% of the time by the stepwise regression program. The net is a 32% chance of finding the active factor.

It is important to notice the differences between these supersaturated plans and the random balance and group screening plans discussed earlier in the chapter. First, there is nothing random about the way factors are partially confounded with one another. There are fixed bounds (that can be calculated) on the correlations between estimates of treatment effects. Also, there is no need to make any assumptions about the direction of potential effects as in the group screening plans. Also, possible low-order interactions are not such a serious problem. In fact, it is possible to extend the statistical analysis to allow for possible interaction columns in the larger plans. Lin (1995) gives an algorithm for constructing much larger supersaturated plans that still maintain a close bound on the size of correlations between factors.

21.2.1 Some Words of Caution

We repeat the following words of caution concerning the use of supersaturated plans found in Kettaneh-Wold and Lin (1995):

(1) Using supersaturated plans entails some risk, but have a definite place when the investigator is faced with more factors than number of runs that either time or resources permit. They are far superior to other approaches such as subjective selection of factors in the absence of information.

(2) Supersaturated plans have a real place in the early stages of an experimental investigation of complicated processes involving many factors. Other experiments must then be used to follow up and establish definite conclusions.

(3) The success of a supersaturated plan depends heavily on the *effect sparsity* assumption.

We also emphasize that there are some questions about the optimal statistical analysis of data obtained via a study utilizing a supersaturated plan. Stepwise regression appears to work reasonably well. Some authors have claimed that fitting all-possible-regression models works better. A simple alternative would be to ignore all correlations among estimates, i.e., compute all estimates simply by taking the difference between the mean factor-high and mean factor-low observations and then use either the half-normal plot proposed by Daniel (1959) or the procedure suggested by Lenth (1989). Although the simplicity of this analysis lends some appeal, simulations suggest that stepwise or all-possible-subsets regression works better. At this point there is still some room for further research.

It is fair to summarize by saying that supersaturated plans present a strategy of last resort. If one is faced with many factors, limited resources, and the effect sparsity assumption appears not unreasonable, they provide a reasonable option. In fact, there appears to be no other option that has real appeal.

21.2.2 Recommendation

We note that a particularly attractive application for these plans is to test robustness of a product or a procedure. The objective here is not to identify important factors but to test many factors to verify that the response will remain within specifications regardless of the levels of these factors. We have in mind a situation where the investigator is convinced a priori that none of the factors are really active but need a final check.

21.3 SUPERSATURATED PLANS THAT INCLUDE AN ORTHOGONAL BASE

Yamada and Lin (1997) present a class of supersaturated plans that include an orthogonal base. The idea is that the collection of factors consists of two sets. It is believed that factors in the first or dominant set are much more likely to be active than factors in the second set. Consequently, the investigator wants to keep the factors in the first set orthogonal to one another so that their effects will be easier to identify. The factors in the second set are relegated to those columns in the design matrix that tend to be intercorrelated. They explicitly spell out plans for up to 35 factors in both eight and 20 runs (seven and 19 mutually orthogonal factors) and detailed descriptions of the construction of 12-, 16-, and 24-run plans (11, 15, and 23 mutually orthogonal factors) for 66, 71, and 66 factors, respectively. For the 12-run case, with 11 mutually orthogonal columns, the construction is particularly simple. One begins with the 11 orthogonal columns in the 12-run Plackett–Burman plan as the base and then constructs the remaining up to 55 columns from the cross products of all pairs of columns in the base.

21.4 MODEL-ROBUST PLANS

A common situation is the case where an investigator feels quite certain that relatively few of the possible two-factor interactions are large enough to be important. However, the identity of the important ones is unknown. We put these plans in this chapter because there are potentially many more interactions than runs. Li and Nachtsheim (2000) give an example in which officials in an automotive company decided that there was a need to reduce the leakage of a clutch slave cylinder. Four factors were identified as potentially important. These were body inner diameter, body outer diameter, seal inner diameter, and seal outer diameter. The budget allowed for only eight runs in an experiment. It was felt that some two-factor interactions in addition to main effects could be important. Although the number was believed to be small (≤ 2), there was no real basis to select those likely to be important. Hence, the model involved main effects and possibly one or two, two-factor interactions. A conventional half replicate defined by $\mathcal{I} = ABCD$ is unsatisfactory, since it aliases $A \times B$ with $C \times D$, $A \times C$ with $B \times D$, and $A \times D$ with $B \times C$. The experiment cannot identify which of, say, $A \times B$ or $C \times D$ is important. Li and Nachtsheim developed an algorithm to search for model-robust plans that get around this problem. Table 21.3 contains a number of 8, 12, and 16 run factorial plans that permit the estimation of all main effects (each at two levels) and up to g of the $\binom{m}{2}$ possible two-factor interactions. In each array in Table 21.3, the lines represent runs and the columns factor levels, with "-1" representing the low level and "$+1$" the high level.

Table 21.3 Model-Robust Plans for m Factors Where Up to g Interactions Are Important

Eight-Run Plans

$m = 4$, $g = 3$

−1	+1	+1	−1
+1	−1	−1	+1
+1	+1	+1	+1
−1	−1	−1	−1
+1	−1	+1	−1
−1	+1	−1	−1
−1	+1	−1	+1
+1	−1	+1	+1

$m = 5$, $g = 2$

−1	+1	+1	+1	−1
−1	−1	−1	−1	−1
−1	+1	−1	−1	+1
+1	+1	+1	+1	+1
+1	+1	−1	+1	−1
−1	−1	−1	+1	+1
+1	−1	+1	−1	−1
+1	−1	+1	−1	+1

$m = 6$, $g = 1$

+1	−1	−1	−1	−1	+1
+1	−1	+1	+1	+1	−1
+1	+1	−1	−1	+1	+1
−1	−1	+1	−1	+1	−1
+1	+1	+1	−1	−1	−1
−1	+1	−1	+1	−1	−1
−1	−1	−1	+1	+1	+1
−1	+1	+1	+1	−1	+1

(*continued*)

Table 21.3 ***(continued)***

12-Run Plans, m = number of factors, g = number of interactions

$m = 5,\ g = 5$

+1	+1	+1	−1	+1
−1	+1	+1	+1	−1
−1	+1	−1	+1	−1
−1	−1	+1	−1	+1
−1	+1	−1	−1	−1
−1	−1	+1	+1	−1
+1	−1	−1	−1	+1
−1	+1	−1	−1	+1
+1	−1	+1	−1	−1
+1	−1	−1	+1	−1
+1	−1	+1	+1	+1
+1	+1	−1	+1	+1

$m = 6,\ g = 5$

−1	+1	−1	−1	−1	−1
+1	−1	+1	+1	+1	+1
−1	−1	+1	+1	−1	−1
+1	+1	+1	+1	−1	−1
−1	+1	−1	+1	+1	+1
−1	+1	+1	+1	+1	−1
+1	+1	−1	−1	+1	+1
+1	−1	−1	−1	−1	+1
+1	−1	−1	+1	−1	−1
−1	−1	+1	−1	+1	+1
−1	+1	+1	−1	−1	+1
+1	−1	−1	−1	+1	−1

$m = 7,\ g = 4$

+1	−1	+1	−1	−1	+1	−1
−1	+1	+1	+1	−1	+1	+1
+1	+1	−1	−1	+1	−1	+1
+1	−1	−1	+1	−1	−1	+1
−1	+1	+1	−1	+1	+1	+1
+1	−1	−1	−1	+1	−1	−1
−1	+1	−1	−1	−1	−1	−1
+1	−1	+1	+1	−1	−1	−1
−1	−1	+1	−1	+1	−1	+1
+1	+1	−1	+1	−1	+1	−1
−1	−1	−1	+1	+1	+1	+1
−1	+1	+1	+1	+1	+1	−1

$m = 8,\ g = 3$

+1	+1	−1	−1	+1	+1	−1	+1
+1	−1	−1	+1	−1	+1	+1	−1
+1	+1	+1	+1	+1	−1	−1	−1
−1	+1	+1	−1	+1	+1	−1	−1
−1	−1	+1	+1	−1	−1	+1	−1
+1	−1	+1	−1	+1	−1	−1	+1
−1	+1	+1	+1	+1	+1	+1	−1
−1	−1	−1	+1	−1	+1	+1	+1
−1	−1	−1	−1	−1	−1	−1	+1
+1	+1	−1	+1	−1	−1	−1	+1
−1	−1	−1	−1	+1	−1	+1	−1
+1	+1	+1	−1	−1	+1	+1	+1

$m = 9,\ g = 2$

+1	+1	+1	+1	+1	+1	−1	+1	+1
+1	+1	+1	−1	−1	+1	+1	−1	−1
+1	+1	−1	−1	+1	+1	−1	−1	+1
−1	−1	+1	+1	−1	+1	−1	+1	−1
−1	+1	+1	+1	+1	−1	−1	−1	−1
+1	−1	−1	−1	+1	+1	+1	+1	−1
−1	−1	−1	−1	−1	−1	+1	−1	−1
−1	−1	−1	+1	−1	+1	−1	+1	+1
−1	−1	−1	+1	+1	−1	+1	−1	+1
+1	+1	−1	+1	−1	−1	+1	+1	−1
+1	−1	+1	−1	−1	−1	−1	−1	+1
−1	+1	+1	−1	+1	−1	+1	+1	+1

Table 21.3 (*continued*)

16-Run Plans, m = number of factors, g = number of interactions.

$m = 7,\ g = 5$						
−1	+1	+1	−1	+1	+1	−1
−1	+1	−1	+1	−1	+1	−1
−1	−1	−1	+1	+1	−1	−1
+1	+1	+1	+1	+1	+1	+1
−1	−1	+1	−1	+1	−1	+1
+1	−1	−1	+1	+1	−1	+1
+1	+1	−1	−1	−1	−1	−1
−1	+1	+1	+1	−1	−1	−1
+1	+1	−1	−1	+1	+1	+1
+1	−1	−1	−1	−1	−1	+1
+1	−1	+1	−1	+1	+1	−1
+1	+1	−1	+1	−1	−1	+1
−1	−1	+1	−1	−1	+1	+1
−1	−1	+1	+1	−1	+1	−1
+1	+1	+1	−1	−1	−1	−1
−1	−1	−1	+1	+1	+1	+1

$m = 8,\ g = 5$							
−1	+1	−1	+1	+1	+1	+1	−1
+1	+1	+1	+1	+1	−1	−1	−1
+1	+1	+1	+1	−1	+1	+1	+1
+1	−1	+1	−1	−1	−1	−1	−1
−1	−1	−1	−1	+1	+1	+1	+1
+1	−1	−1	+1	+1	−1	+1	+1
−1	+1	−1	+1	−1	+1	−1	+1
−1	−1	−1	−1	+1	−1	−1	−1
−1	−1	+1	+1	−1	−1	−1	+1
+1	−1	+1	+1	−1	+1	−1	−1
−1	+1	−1	−1	−1	−1	+1	+1
+1	+1	−1	−1	+1	+1	+1	−1
+1	+1	+1	−1	+1	−1	+1	−1
−1	−1	+1	+1	−1	−1	+1	−1
+1	−1	−1	−1	−1	+1	−1	+1
−1	+1	+1	−1	+1	+1	−1	+1

$m = 9,\ g = 5$								
−1	−1	−1	−1	+1	−1	+1	+1	+1
+1	+1	−1	−1	+1	+1	−1	+1	+1
+1	−1	−1	−1	−1	+1	−1	−1	−1
+1	+1	−1	−1	+1	+1	+1	−1	−1
−1	−1	+1	+1	+1	−1	+1	−1	−1
−1	−1	+1	+1	−1	−1	−1	−1	+1
+1	−1	−1	+1	−1	−1	−1	+1	+1
+1	−1	+1	+1	+1	+1	−1	−1	−1
−1	+1	+1	+1	−1	+1	−1	−1	−1
−1	−1	−1	+1	−1	+1	−1	+1	+1
−1	+1	−1	−1	−1	−1	+1	−1	+1
−1	+1	+1	−1	+1	−1	−1	+1	−1
−1	+1	+1	+1	+1	+1	+1	+1	+1
+1	+1	−1	+1	−1	−1	+1	−1	−1
+1	+1	+1	−1	+1	−1	+1	+1	+1
+1	−1	+1	−1	−1	+1	+1	+1	−1

$m = 10,\ g = 3$									
+1	+1	+1	−1	+1	−1	+1	−1	+1	−1
−1	+1	−1	+1	−1	+1	+1	+1	+1	−1
+1	+1	+1	+1	−1	−1	−1	+1	−1	+1
−1	−1	−1	−1	+1	+1	+1	+1	+1	+1
−1	−1	+1	−1	−1	+1	+1	−1	−1	+1
+1	+1	−1	−1	+1	+1	+1	+1	−1	+1
+1	−1	+1	−1	+1	+1	−1	−1	+1	+1
−1	+1	+1	+1	+1	−1	+1	+1	−1	−1
−1	+1	−1	−1	−1	+1	−1	−1	+1	−1
+1	−1	−1	−1	−1	−1	−1	+1	+1	+1
−1	−1	+1	−1	−1	−1	−1	−1	−1	−1
−1	−1	−1	+1	+1	+1	+1	+1	−1	−1
+1	−1	+1	+1	−1	+1	−1	+1	+1	+1
+1	+1	−1	+1	+1	−1	−1	−1	−1	−1
+1	−1	+1	+1	−1	−1	+1	−1	+1	−1
−1	+1	−1	+1	+1	−1	−1	−1	−1	+1

Source: Reprinted with permission from Li and Nachtsheim (2000).

One must notice, however, that these are very small plans. In fact, most are either saturated, or very nearly saturated, unless very few of the factors are active. Consequently, one would expect good performance only when the errors are small. If errors are larger, the only alternative is to perform larger experiments with adequate replication.

Example 21.1. To illustrate the power and the robustness of the procedure, we work through a numerical example using simulated data. The advantage of simulated data is that the "truth" is known. When the statistical analysis is done, one knows whether it worked or not.

We consider the 12-run, $m = 7$ factor, $g = 4$ interactions model. The data are constructed with A, B, C, and D effects all equal to 4 and interactions $A \times B$, $A \times C$, and $C \times D$ all equal to 2. All other effects and interactions are zero. For errors we have .5, -1.1, .9, $-.2$, $-.8$, .4, -1.4, 1.7, 1.0, -1.5, .8, and $-.3$. These errors sum to zero and the average squared value is 1.0. A stepwise regression (`proc STEPWISE` in SAS®) using default parameters and all seven main effects as potential candidates correctly identified the four, $\widehat{A} = 2.0$, $\widehat{B} = 1.3$, $\widehat{C} = 1.8$, and $\widehat{D} = 1.9$. A second stepwise run with A, B, C, and D fixed in the model and the six two-factor interactions as candidates correctly identified $\widehat{AB} = 1.6$, $\widehat{AC} = 1.1$, and $\widehat{CD} = 1.5$. The next two steps brought in $\widehat{BD} = -.4$ and $\widehat{AD} = -.2$. A reasonable conclusion is that while the magnitudes of the estimates are off somewhat, the procedure identifies the correct model.

Multiplying the error vector by 1.414 (doubling the error variance) leads to five estimates: $\widehat{A} = 2.0$, $\widehat{B} = 1.2$, $\widehat{C} = 2.1$, $\widehat{D} = 2.1$, and $\widehat{F} = -.7$. If the user is fortunate enough to identify the four active factors, the next stepwise run identifies $\widehat{AB} = 1.8$, $\widehat{AC} = .9$, $\widehat{CD} = 1.6$, and $\widehat{BD} = -.6$. Again it is likely, although not certain, that the correct model would have been identified. If F is retained as active from the first step and the additional four two-factor interactions added as potential candidates, the second stepwise run incorrectly identifies $\widehat{BD} = -2.8$ and $\widehat{DF} = 4.1$ as the active two-factor interactions.

With the error multiplied by 1.732, the first stepwise identifies only A, C, and D as clearly active and B as very marginal. If the four are identified as active and kept, the first three two-factor interactions picked up in the second stepwise run are AB, AC, and CD. However, they are not clearly separated from BD and AD. There is clear evidence that the procedure is beginning to fail. The errors are beginning to overwhelm the signal from the active effects.

CHAPTER 22

Multistage Experiments

22.1 INTRODUCTION

22.1.1 Multistage or Split-Plot Experiment

In Chapter 7 we introduced the basic ideas and structures of the split-plot experiment along with a number of modifications and generalizations. In this chapter we revisit the same array of designs but superimpose a factorial treatment structure over and beyond the simple whole-plot and split-plot treatments. This additional factorial structure opens a broad range of possibilities that can be exploited to produce designs that are efficient and satisfy constraints on the number of treatment combinations.

A major objective of this chapter is to illustrate that the split-plot concept has many applications beyond research in agriculture. It is not unusual to find split-plot experiments ignored completely in discussions of experimental designs intended for readers in disciplines other than agriculture. This very unfortunate state of affairs is due, in part at least, to the choice of terminology—hence, our emphasis on terminology.

22.1.2 Review of Terminology

As pointed out in Chapter 7, the terms *plot* and *split-plot* have their origin in the classical agricultural field experiment (Fisher 1926; Yates 1935). The experimental material available to the scientist working in, say, agronomy consisted of a collection of r blocks of land. The scientist wished to test t treatments. Each block was divided into t plots. It is easy to envisage the agronomist with a list of t treatments, typically documented on a piece of paper and the set of $n = rt$ "plots" that are to receive the treatments. There is a definite randomization step, the step where the decision is made to assign specific treatments to specific plots. Once this decision has been made, it is a simple matter to apply the treatments.

Planning, Construction, and Statistical Analysis of Comparative Experiments,
by Francis G. Giesbrecht and Marcia L. Gumpertz
ISBN 0-471-21395-0

The identity of the plots or experimental units and the act of randomization are clear. It is equally easy to envisage the second step. Now each plot is split into s pieces and a second factor at s levels randomly assigned to these split plots. The extensions to strip plots were easy to picture.

22.1.3 Alternative Terminology

The purpose of this section is to review some of the terms often encountered in research work that are equivalent in the design of experiments context. We have already emphasized that in many ways the terms *unit* and *experimental unit* are in a sense better than the term *plot*. We have the obvious extensions *split unit* and *strip unit*. Additional terms encountered are *batches, lots*, and *assemblies*. Each of these terms can be augmented with the word *split*. We will use all of these on some occasions. We also frequently use the term *stratum* in an experiment. We talk about the *whole-plot stratum* and the *split-plot stratum* as two different strata in the experiment.

22.2 FACTORIAL STRUCTURES IN SPLIT-PLOT EXPERIMENTS

22.2.1 Complete Factorials

Motivating Example

To illustrate and motivate some of the considerations that lead experimenters to use a split-unit design for a factorial experiment, we present a number of variations of a basic 2^5 factorial experiment. This material is based, in part, on an example initially discussed by Box and Jones (1992) in the context of robust product development. Our purpose is to illustrate the potential advantages and disadvantages of the split-unit designs and various modifications without raising too many subject-matter considerations before going into more detailed discussions of the various plans possible. The 2^5 is selected for illustrative purposes only. It should be clear to the reader that one could use factors with any numbers of levels.

The objective of the experiment is to develop a cake recipe. The five factors are type of flour (two types), amount of shortening (regular or extra), amount of egg powder (low or high), baking time (two times), and baking temperature (high or low). The response is some quality measure on the resulting cakes. We use this example to show how an experiment could be performed using several different protocols and how to write the linear model corresponding to each protocol. Under specific conditions, each has its advantages and disadvantages. Our view at this point is that there is no "correct" protocol for the experiment. The investigator has the responsibility of analyzing the goals and constraints of the study, deciding which are important, and selecting the most appropriate protocol.

One protocol would be to make up eight batches of cake batter (one for each combination of flour type, amount of shortening, and amount of egg) that are large enough to split into four portions and bake each in one of the four time–temperature combinations, with one cake in the oven at a time. The result is a total of 32 different cakes, each baked separately, that must be evaluated. The appeal is that it is only necessary to mix eight batches of batter. It is still necessary to run the oven 32 times or use 32 ovens. It may well be necessary to replicate the entire procedure r times. The statistical analysis of the resulting data follows the classical split-plot pattern. Information is available in two strata, the flour–shortening–egg stratum and the time–temperature stratum. Information on interactions between factors in different strata is in the time–temperature (split-plot) stratum. A feature of this protocol is that the information at the flour–shortening–egg stratum is weaker than the information at the time–temperature stratum. The amounts of information can be modified by increasing replication.

The linear model suitable for this experiment is completely analogous to the generic split-plot model given in equation (7.19). The $\{w_i\}$ terms in the whole-unit stratum in equation (7.19) are replaced by three main effects $\{f_i\}$, $\{s_j\}$, and $\{e_k\}$ and their interactions, in the split-unit stratum, the $\{s_k\}$ are replaced by two main effects $\{tm_\ell\}$ and $\{tp_m\}$, and the interaction and the $\{(w \times s)_{i_w i_s}\}$ interactions replaced by interactions between factors in different strata. In the latter case only two- and three-factor interactions are included. In practice, one could seriously consider ignoring three-factor interactions as well, although in the split-unit stratum one normally has sufficient degrees of freedom for the error term without including treatment interactions. We assume that each replicate is randomized independently so that replicates serve as blocks. The linear model, with a term for blocks, is

$$\begin{aligned} y_{hijk\ell m} &= \mu + block_h + f_i + s_j + e_k + (f \times s)_{ij} + (f \times e)_{ik} + (s \times e)_{jk} \\ &\quad + (f \times s \times e)_{ijk} + \epsilon(1)_{hijk} + tm_\ell + tp_m + (f \times tm)_{i\ell} \\ &\quad + (f \times tp)_{im} + (s \times tm)_{j\ell} + (s \times tp)_{jm} + (e \times tm)_{k\ell} + (e \times tp)_{km} \\ &\quad + (tm \times tp)_{\ell m} + (f \times s \times tm)_{ij\ell} + (f \times s \times tp)_{ijm} \\ &\quad + (f \times e \times tm)_{ik\ell} + (f \times e \times tp)_{ikm} + (f \times tm \times tp)_{i\ell m} \\ &\quad + (s \times e \times tm)_{jk\ell} + (s \times e \times tp)_{jkm} + (s \times tm \times tp)_{j\ell m} \\ &\quad + (e \times tm \times tp)_{k\ell m} + \epsilon(2)_{hijk\ell m}, \end{aligned}$$

where $\epsilon(1)_{hijk} \sim \text{iid}(0, \sigma_1^2)$, $\epsilon(2)_{hijk\ell m} \sim \text{iid}(0, \sigma_2^2)$, and $\{\epsilon(1)_{hijk}\}$ and $\{\epsilon(2)_{hijk\ell m}\}$ are all mutually independent. The analysis of variance is again a direct extension of the analysis shown in Tables 7.5 and 7.6. The details of the analysis of variance for this experiment based on this model with r blocks are given in Table 22.1.

This protocol may have some appeal to the experimenter in that only eight batches of cake batter need to be made up for each replicate. It is still necessary to bake 32 cakes separately for each replicate. The expected values of the error (1) mean square and the error (2) mean square are $\sigma_2^2 + 4r\sigma_1^2$ and σ_2^2, respectively.

Table 22.1 Analysis of Variance for 2^5 in a Split-Plot Design

Source	df	E[MS]
Blocks	$r - 1$	$\cdots$
Type of flour	1	$\sigma_2^2 + 4r\sigma_1^2 + 16r\phi_f$
Amount of shortening	1	$\sigma_2^2 + 4r\sigma_1^2 + 16r\phi_s$
Flour × shortening	1	$\sigma_2^2 + 4r\sigma_1^2 + 8r\phi_{f\times s}$
Amount of egg	1	$\sigma_2^2 + 4r\sigma_1^2 + 16r\phi_e$
Flour × egg	1	$\sigma_2^2 + 4r\sigma_1^2 + 8r\phi_{f\times e}$
Shortening × egg	1	$\sigma_2^2 + 4r\sigma_1^2 + 8r\phi_{s\times e}$
Flour × shortening × egg	1	$\sigma_2^2 + 4r\sigma_1^2 + 4r\phi_{f\times s\times e}$
Error (1)	$7(r - 1)$	$\sigma_2^2 + 4r\sigma_1^2$
Time	1	$\sigma_2^2 + 16r\phi_{tm}$
Temperature	1	$\sigma_2^2 + 16r\phi_{tp}$
Time × temperature	1	$\sigma_2^2 + 8r\phi_{tm\times tp}$
Time × type of flour	1	$\sigma_2^2 + 8r\phi_{tm\times f}$
Time × amount of shortening	1	$\sigma_2^2 + 8r\phi_{tm\times s}$
Time × amount of egg	1	$\sigma_2^2 + 8r\phi_{tm\times e}$
Temperature × type of flour	1	$\sigma_2^2 + 8r\phi_{tp\times f}$
Temperature × amount of shortening	1	$\sigma_2^2 + 8r\phi_{tp\times s}$
Temperature × amount of egg	1	$\sigma_2^2 + 8r\phi_{tp\times e}$
Rem. three-, four- and five-factor int.	15	$\sigma_2^2 + \cdots$
Error (2)	$24(r - 1)$	σ_2^2
Total	$32r - 1$	

Since σ_1^2 is a variance and hence nonnegative, we see that tests on split-unit factors will be more sensitive than tests on whole-unit factors. The two error terms correspond directly to the two errors in Tables 7.1 and 7.5.

If resources permit only one replicate, i.e., $r = 1$, the lines labeled error (1) and error (2) will not be available. In the flour–shortening–egg stratum, the three-factor interaction can be used for error but is really inadequate. The alternative is to use the graphical techniques discussed in Section 12.2.1. In the time–temperature stratum there are 15 degrees of freedom from three-, four-, and five-factor interactions that can be used for an error term.

General Discussion

From the motivating example, we see that introducing full factorials into the split plot proceeds with little difficulty. In fact, in practice, split plots are sometimes created inadvertently. An investigator plans one replicate of a complete factorial. When the experiment is being conducted it turns out that some factors are much more difficult to change than others. An example would be a food science experiment involving ultra high-temperature processing. Changing factors such as time held at the high temperature involve disassembling and reassembling equipment that becomes very hot while being used. There are good reasons to change such factors as infrequently as possible. The usual strategy is to group the difficult-to-change factors as whole-plot treatments and easy-to-change factors as split-plot factors. Similarly, it is not unusual to encounter an experiment in which treatment combinations have been fully randomized, but then at the very end of each run, two (or more) aliquots of the resulting product are tested using slightly different techniques. In this case a split-plot factor is added at very little cost. We also point out that the material in this chapter is related to the material on run order in Chapter 18.

22.2.2 Fractional Factorials

The full range of fractional factorial plans (Chapters 14 and 17), including the Plackett–Burman plans (Chapter 16) and the more general class of plans developed via orthogonal arrays (Chapter 23), can be used in split-plot designs. The statistical analyses of the data go through without difficulty. In the nested designs, such as the basic split plot, the generalized interactions between whole-plot effects and interactions and split-plot effects and interactions fall into the split-plot stratum. In the case of crossed designs such as the strip plot with rows and columns, the generalized interactions between effects and interactions in rows and columns fall into the interaction stratum. These designs provide a very powerful design tool when interest centers on interactions between factors whose main effects are already well understood or for some reason of little interest. A feature of these designs which we will utilize eventually is that they tend to provide error terms with excess degrees of freedom.

22.2.3 Application in Robust Product Studies

In robust product studies many investigators conduct experiments utilizing Taguchi's inner or control and outer or noise arrays as split-plot experiments. Properties of these inner–outer array designs and the variations thereof are examined in Chapter 24. The nature of the control factors and the noise factors often makes it very appealing to place the former in a whole-plot stratum and the latter in a split-plot stratum. Occasionally, one also sees these experiments conducted in a strip-plot design with the control factors in the row stratum and the noise factors in the column array. These experiments are very efficient in the sense that they permit the investigation of a wide range of combinations of factors with reasonable effort. Unfortunately, the design aspects are often missed in the statistical analyses of the data from these experiments.

22.3 SPLITTING ON INTERACTIONS

Fractional factorial split-plot (FFSP) *designs* are easily constructed by combining two fractional factorial plans, one used in the whole-plot stratum and the other in the split-plot stratum. These plans require no new theory. However, better plans are possible. Bingham and Sitter (1999a,b) examined the problem of constructing minimum aberration fractional factorial plans for split-plot designs. They found that restricting attention to plans in which the defining contrast for the split-plot stratum fraction involved only factors and interactions from the split-plot stratum was unnecessarily constraining.

To illustrate the problem and the difficulties of finding minimum aberration fractional factorial plans for split-plot designs, we repeat an example given in Bingham and Sitter (1999a). The problem considered is to improve the efficiency of a ball mill. Engineers had identified seven potential factors, each at two levels:

A_w: motor speed
B_w: feed mode
C_w: feed sizing
D_w: material type
A_s: gain
B_s: screen angle
C_s: screen vibration level

The manufacturing conditions dictate that the first four factors be assigned to the whole-plot stratum and the last three can be assigned to the split-plot stratum. Cost constraints limit the number of runs. A possible plan would be a half replicate based on the defining contrast

$$\mathcal{I} = A_w B_w C_w D_w.$$

This defines a plan with 64 runs.

Before proceeding we need to define some notation. Following the notation in Chapter 14, we can describe the whole-plot fraction as a 2^{4-1} fraction. More generally, we denote the whole- and split-plot fractions as $2^{k_w - p_w}$ and $2^{k_s - p_s}$, respectively, and the complete plan as $2^{(k_w + k_s) - (p_w + p_s)}$.

Next assume that cost constraints limit the number of runs to 32. A suitable $2^{(4+3)-(1+1)}$ plan is obtained by utilizing the defining contrast

$$\mathcal{I} = A_s B_s C_s$$

in the split-plot stratum. This gives the $2^{(4+3)-(1+1)}$, resolution III plan defined by

$$\mathcal{I} = A_w B_w C_w D_w = A_s B_s C_s = A_w B_w C_w D_w A_s B_s C_s.$$

The key point to notice is that the split-plot stratum has a fraction of the form 2^{3-1}.

Now to illustrate some difficulties and possibilities, we consider a restriction to 16 runs. This will require another generator and we encounter difficulties. In the whole-plot stratum the best we can do is

$$\mathcal{I} = A_w B_w C_w = B_w C_w D_w = A_w D_w$$

and combined defining contrast

$$\mathcal{I} = A_w B_w C_w = B_w C_w D_w = A_w D_w = A_s B_s C_s = A_w B_w C_w A_s B_s C_s$$
$$= B_w C_w D_w A_s B_s C_s = A_w D_w A_s B_s C_s.$$

This is a resolution II plan, which is unacceptable. Selecting two generators in the split-plot stratum does not give a good plan either.

However, there is another possibility. A much better $2^{(4+3)-(1+2)}$ plan is obtained by using the defining contrast

$$\mathcal{I} = A_w B_w C_w = D_w A_s B_s = A_w A_s C_s = A_w B_w C_w D_w A_s B_s$$
$$= B_w C_w A_s C_s = A_w D_w B_s C_s = B_w C_w D_w B_s C_s,$$

which is resolution III. Note that there is exactly one generator from the whole-plot stratum and this defines a 2^{4-1} fraction, but there are no generators that involve only factors from the split-plot stratum. The split-plot part of the plan is not a 2^{3-2} fraction. This is a plan that cannot be obtained simply by combining two fractions of factorials.

There are important points to notice about these plans.

- Allowing split-plot stratum generators to include whole-plot stratum factors opens up more design options.
- The whole-plot stratum generators must not include any split-plot stratum factors. For a contrast to lie in the whole-plot stratum, it must consist of only whole-plot factors.
- If a fractional generator contains both whole- and split-plot stratum factors, the interaction defined by the product of the whole- and split-plot factors is moved automatically to the whole-plot stratum. For example, the generator $A_w B_w C_w A_s B_s$ makes $A_s B_s$ the alias of $A_w B_w C_w$, which is a whole-plot contrast.
- If a generator contains only one split-plot stratum factor, that factor shifts automatically to the whole-plot stratum. For example, $A_w B_w C_w D_s$ makes D_s the alias of $A_w B_w C_w$. This shifts D_s to the whole-plot stratum.

An additional point that must be kept in mind is that if the two interactions $B_w C_w$ and $A_s C_s$ are aliased and it is assumed a priori that the $B_w C_w$ interaction is negligible, the $A_s C_s$ interaction is in the whole-plot stratum and its statistical significance is assessed with the whole-plot error term.

To conclude, we note that in many cases adequate plans can be obtained by constructing separate fractions in the two strata. However, there are numerous cases where plans with much higher resolution can be obtained by working with the full set of factors. Bingham and Sitter (1999b) describe an algorithm to develop two-level fractional factorial plans for split-plot designs with minimal aberration. Notice that finding suitable minimal-aberration plans is complicated by the fact that plans that are isomorphic in the sense that one can be obtained from the other by interchanging factors need not be equivalent because factors are restricted to lying in specific strata.

22.3.1 Tables of Minimal Aberration Plans

Table 22.2 contains Bingham and Sitter's (1999a) catalog of eight-run two-level fractional factorial plans for split-plot designs. The first column in the table gives $(k_w + k_s) - (p_w + p_s)$, specifying the number of whole-plot factors, number of split-plot factors, and the fractions in each stratum. The next column gives the $p_w + p_s$ generators with the generators defining the whole-plot $2^{k_w - p_w}$ fraction listed first. These contain only whole-plot factors. In general, the remaining p_s generators contain both whole- and split-plot factors and do not define a proper $2^{k_s - p_s}$ fraction for the split-plot stratum. To conserve space the generators are given as 1's and 0's, denoting presence and absence of the factor symbol. For example, the $2^{(3+2)-(1+1)}$ plan has whole-plot factors A_w, B_w, and C_w and split-plot factors A_s and B_s with the whole-plot part of the design forming a 2^{3-1} fraction. The generators are $A_w B_w C_w$ and $A_w A_s B_s$. The defining contrast consists of two three-factor interactions ($\ell_3 = 2$) and one four-factor interaction ($\ell_4 = 1$).

Exercise 22.1: Fractional Factorial in a Split Plot. Use the information in Table 22.2 to find the defining contrast for a $2^{(3+2)-(1+1)}$ fractional factorial. Also show the composition of all of the whole plots. □

Table 22.2 Catalog of Eight-Run Fractional Factorial Plans for Split-Plot Experiments

Plan	Defining Contrasts	WLP[a]
$(1+3)-(0+1)$	1 111	0 1
$(1+3)-(0+1)$	1 101	1 0
$(1+3)-(0+1)$	0 111	1 0
$(2+2)-(0+1)$	11 11	0 1
$(2+2)-(0+1)$	10 11	1 0
$(1+4)-(0+2)$	1 1010, 1 0101	2 1
$(1+4)-(0+2)$	1 1010, 0 1101	2 1
$(2+3)-(0+2)$	10 110, 01 101	2 1
$(3+2)-(1+1)$	111 00, 100 11	2 1
$(1+5)-(0+3)$	1 10100, 1 01010, 0 11001	4 3
$(2+4)-(0+3)$	10 1100, 01 1010, 11 1001	4 3
$(3+3)-(1+2)$	111 000, 100 110, 010 111	4 3
$(1+6)-(0+4)$	1 101000, 1 010100, 0 110010, 1 110001	7 7 0 0 1
$(3+4)-(1+3)$	111 0000, 100 1100, 010 1010, 110 1001	7 7 0 0 1

[a]Word-length pattern $(\ell_3, \ell_4, \ell_5, \ldots)$.

Source: Adapted with permission from Brigham and Sitter (1999a).

Appendix 22A contains Bingham and Sitter's (1999a) catalog of 16-run two-level fractional factorial plans for split-plot designs. The general format of this table is the same as earlier. However, to conserve space the generators are padded on the right with 0's to make the length a multiple of four and then expressed in hexadecimal notation (*Note:* Many calculators and Web sites contain hexadecimal to binary conversion tables or functions that make this task easy.) Consequently, after expanding the symbol given for the generator from hexadecimal to binary notation, the excess 0's must be removed and then converted to the expression for the interaction. For example, the generators for the defining contrast for the $2^{(3+4)-(1+2)}$ plan are given as $E0$, 94, and $5A$. These expand directly to 11100000, 10010100, and 01011010. Since there are only seven factors in the plan, a 0 must be removed from the rightmost end. The generators for the defining contrast are $A_wB_wC_w$, $A_wA_sC_s$, and $B_wA_sB_sD_s$. The actual treatment combinations are obtained as the 16 solutions to the system

a_w	b_w	c_w	a_s	b_s	c_s	d_s		
i_{A_w}	$+i_{B_w}$	$+i_{C_w}$					$=$	0 (modulo 2)
i_{A_w}			$+i_{A_s}$		$+i_{C_s}$		$=$	0 (modulo 2)
	$+i_{Bw}$		$+i_{A_s}$	$+i_{B_s}$		$+i_{D_s}$	$=$	0 (modulo 2).

Since the whole-plot part of the design has a 2^{3-1} fraction, the treatment combinations are sorted into four sets of four on the basis of A_w, B_w, and C_w for the first randomization. There is example SAS® code illustrating how to generate the treatment combinations and how to use `Proc GLM` to determine the aliasing pattern on the companion website.

A possible shortcoming of our table is that most plans are resolution III. Many two-factor interactions are aliased with main effects. In product quality design problems the investigator is frequently interested in interactions, especially in interactions between control and noise factors. Split-plot designs with control factors in the whole plots and noise factors in the split plots or strip-plot designs are a natural for this type of work.

Exercise 22.2: Fractional Factorial in a Split Plot. Use the information in Appendix 22A to find the defining contrast for a $2^{(3+2)-(0+1)}$ fractional factorial. Show the composition of all of the whole plots. □

Exercise 22.3: Analysis of Variance of a Fractional Factorial in a Split Plot. Give the analysis of variance table for data that would be collected via the $2^{(4+3)-(1+2)}$ plan developed in Section 22.3. Show how one would test the difference between the mean of responses with factors A_w and A_s at the low level and A_w and A_s at the high level. □

22.3.2 Larger Plans

Bingham and Sitter (1999a) state that Bingham (1998) obtained a list of nonisomorphic 32-run fractional factorial plans containing up to 10 factors. Beyond this, the investigator faces the task of finding a suitable plan. Bingham and Sitter (1999a,b) describe their algorithm for minimal aberration plans. However, the search problem is difficult because of the large number of possibilities involved. One can always obtain a plan that meets estimability constraints by using the Franklin–Bailey algorithm described in Section 14.7.

22.4 FACTORIALS IN STRIP-PLOT OR STRIP-UNIT DESIGNS

It is also possible to incorporate fractional factorial plans into row–column strip-plot designs. The major difference is that there is an inherent symmetry between rows and columns that was not present in the split-plot design. However, there is still the third stratum nested within the interaction of rows and columns. Separate factorials, fractional or complete, in either rows and/or columns introduce no new difficulties. In fact, it is common practice to use plans based on orthogonal arrays to construct the treatment combinations for the rows and columns in product quality design experiments.

22.4.1 Motivating Example Revisited

The ability to apply treatments in sequence in a factorial structure makes the strip-plot or strip-unit design especially attractive. The cake baking example discussed in Section 22.2.1 fits into a strip-unit design very nicely. The investigator makes up eight large batches of batter, one for each of the 2^3 treatment combinations, splits each into four, and places the cakes (eight per oven) into the ovens, one for each time–temperature combination. Batches of batter make up the row stratum, runs of the oven, the column stratum, and individual cakes the nested stratum. Notice that a one-replicate experiment only requires mixing eight batters and four oven adjustments.

The analysis of variance for such an experiment replicated r times is given in Table 22.3. If the experiment is limited to only one replicate, the three error lines in the analysis of variance table go away, i.e., have zero degrees of freedom. A possible option, then, is to use the three-factor flour $\times$ shortening $\times$ egg interaction as a one-degree-of-freedom measure of error in the first stratum. A more attractive alternative is to use one of the methods outlined in Chapter 12. In the second stratum there is really no suitable measure of error. The Chapter 12 methods are not very appealing either since the number of effects and interactions is so small. In the third stratum the collection of 15 remaining three-, four-, and five-factor interactions can be used as a very adequate measure of error. If the object of this experiment is to investigate baking time or baking temperature $\times$ type of flour, amount of shortening, or amount of egg interactions, this is a very appealing design.

Note also that it is always possible and often quite desirable to add further strata. For example, in the cake baking study, one could consider cutting three pieces from each of the cakes. One piece from each cake is allowed to remain uncovered on a table, the second one is covered, and the third one is placed in a

Table 22.3 Analysis of Variance Table for Cake Example in Strip-Plot Design

Source	df	E[MS]
Replicates	$r-1$	$\cdots$
Type of flour	1	$\sigma_3^2 + 4r\sigma_1^2 + 16r\phi_f$
Amount of shortening	1	$\sigma_3^2 + 4r\sigma_1^2 + 16r\phi_s$
Flour × shortening	1	$\sigma_3^2 + 4r\sigma_1^2 + 8r\phi_{f\times s}$
Amount of egg	1	$\sigma_3^2 + 4r\sigma_1^2 + 8r\phi_e$
Flourby egg	1	$\sigma_3^2 + 4r\sigma_1^2 + 8r\phi_{f\times e}$
Shortening × egg	1	$\sigma_3^2 + 4r\sigma_1^2 + 16r\phi_{s\times e}$
Flour × shortening × egg	1	$\sigma_3^2 + 4r\sigma_1^2 + 4r\phi_{f\times s\times e}$
Error (1)	$7(r-1)$	$\sigma_3^2 + 4r\sigma_1^2$
Time	1	$\sigma_3^2 + 8r\sigma_2^2 + 16r\phi_{tm}$
Temperature	1	$\sigma_3^2 + 8r\sigma_2^2 + 16r\phi_{tp}$
Time × temperature	1	$\sigma_3^2 + 8r\sigma_2^2 + 8r\phi_{tm\times tp}$
Error (2)	$3(r-1)$	$\sigma_3^2 + 8r\sigma_2^2$
Time × type of flour	1	$\sigma_3^2 + 8r\phi_{tm\times f}$
Time × amount of shortening	1	$\sigma_3^2 + 8r\phi_{tm\times s}$
Time × amount of egg	1	$\sigma_3^2 + 8r\phi_{tm\times e}$
Temperature × type of flour	1	$\sigma_3^2 + 8r\phi_{tp\times f}$
Temperature × amount of shortening	1	$\sigma_3^2 + 8r\phi_{tp\times s}$
Temperature × amount of egg	1	$\sigma_3^2 + 8r\phi_{tp\times e}$
Rem. three-, four-, and five-factor int.	15	$\sigma_3^2 + \cdots$
Error (3)	$21(r-1)$	σ_3^2
Total	$32r-1$	

refrigerator. After some minutes the pieces are all evaluated. This added stratum contains one factor, although it is easy to imagine other additional factors. In this case, the storage condition main effect is probably of limited interest, but the interactions may well be of interest. Cakes from some recipes may retain their freshness longer than others. The storage factor is a classical "noise" factor.

22.4.2 Simple Product Quality Design Example

Hill (1990) provides an example of a manufacturing application of a strip-unit design. The purpose of the study was to develop a low-cost data transmission cable that would produce the desired output at required input frequencies, while being robust to noise factors. The experimental plan consisted of seven control factors and two noise factors. (In Chapter 24 we discuss control factors and noise factors more extensively.) The control factors, factors defining the production process, were (a) conductor size, (b) insulation type, (c) jacket type, (d) insulation thickness, (e) interaxial spacing, (f) lay type, and (g) color concentrate. Each factor consisted of two types or two levels. The noise factors were cable testing condition (two conditions, 100-ft sample of straight cable at 72°F and 1000-ft sample coiled cable at 40°F) and signal testing frequency (three levels). The precise conduct of the experiment is not clear from the discussion; however, it appears that a seven-factor (eight-run) main-effect plan was used to produce eight different types of cable. The cables were then tested in a 2×3 factorial. Our assumption is that once the cable samples were manufactured in randomly selected order, a randomly selected set of testing conditions and frequencies was set up and all eight samples tested. Then another randomly selected testing procedure was set up and again all eight samples tested. Three repeat readings of signal attenuation were made on each cable. It is convenient to think of this as a row–column design. The manufacturing processes (control factors) define the levels in the row stratum and the testing procedures (the noise factors) the levels in the column stratum. This represents a very efficient layout for the study. Only eight manufacturing and six testing processes need to be set up. At the same time, the interactions between manufacturing conditions and noise factors are measured in the interaction stratum, the stratum with smallest error. The point of the study was to examine these interactions. Which of the control factor settings would most effectively remove the effect of a noise factor?

We notice that the statistical design introduces no new complications. The row stratum treatments and the column stratum treatments are standard factorial structures.

Exercise 22.4: Analysis of Variance of a Strip-Plot Factorial. The electrical cable making example does not include any replication. Show two different modifications of the plan that would provide replication. Show the analysis of variance for both of the plans that you develop. Be sure to indicate how one would test various hypotheses of interest for the plans developed. □

22.4.3 Example of a Fraction from the Cross of Complete Factorials

To illustrate a half fraction in a strip-unit design, we consider a hypothetical plant growth experiment. Assume that the scientist has 32 similar potted plants available for the study. The plants are to be grown in small controlled chambers, each having room for four plants. The daytime conditions consist of the eight combinations in a 2^3 factorial. Similarly, the night conditions represent a 2^3. Let the symbols A_D, B_D, C_D, A_N, B_N, and C_N represent the factors. A complete cross with all factors is not possible because it requires 64 plants. First reduce the number of chambers by assigning the eight day treatments to one set of eight chambers and the eight night treatments to another set of chambers. To accommodate the plants, restrict the study to a half replicate using the defining contrast $\mathcal{I} = A_D B_D C_D A_N B_N C_N$. The layout can be illustrated as shown in Table 22.4. This is a very compact design. Unfortunately, it does not allow for testing of the day (row) or night (column) effects in the analysis of variance.

Table 22.4 Plant Assignments in a Growth Chamber Study[a]

	Night Treatments							
Day Trt.	000	001	010	011	100	101	110	111
000	×			×		×	×	
001		×	×		×			×
010		×	×		×			×
011	×			×		×	×	
100		×	×		×			×
101	×			×		×	×	
110	×			×		×	×	
111		×	×		×			×

[a] An "×" represents a plant.

Table 22.5 Analysis of Variance Table for Growth Chamber Study in Table 22.4

Source	df
Day treatments main effects	3
Two-factor interactions	3
No measure of day treatment error	—
Night treatment main effect	3
Two-factor interactions	3
Three-factor day and three-factor night interaction	1
No measure of night treatment error	—
Two-factor interactions involving day and night factors	9
Interaction stratum error	9
Total	31

However, the techniques from Chapter 12 are available. The plan does, however, provide a reasonably sensitive F-test of two-factor interactions involving one day factor and one night factor. The analysis of variance under the assumption of no three-factor or higher interactions takes the form shown in Table 22.5.

We note that a key element in this design is the ability to organize experimental units into groups and then reorganize them into different groups. In this case the plants are in eight sets of four for the day treatments and different sets of four for the night treatment. We examine this strategy in more detail in Section 22.6.

Exercise 22.5: Hypothetical Growth Chamber Experiment. For the hypothetical growth chamber experiment discussed in Section 22.6, assume that the 32 plants are labeled 1 to 32. Draw a schematic diagram of the day chambers indicating the locations of the individual plants and another diagram showing the location of the same plants in the night chamber. □

22.4.4 Example of a Cross Involving Two Fractions

As a simple example of a strip-plot involving two fractions, consider a six-factor factorial. We assign a 2^{3-1} fraction to the row stratum and another 2^{3-1} fraction

Table 22.6 Hypothetical Strip-Unit Plan

	$-A_cB_cC_c$			
$-A_rB_rC_r$	(1)	a_cb_c	a_cc_c	b_cc_c
(1)	000 000	000 110	000 101	000 011
a_rb_r	110 000	110 110	110 101	110 011
a_rc_r	101 000	101 110	101 101	101 011
b_rc_r	011 000	011 110	011 101	011 011

to the column stratum. The defining contrasts are $\mathcal{I} = A_rB_rC_r$ and $\mathcal{I} = A_cB_cC_c$. This example is too small to be of much value in practice but is sufficient to illustrate principles. Details of the actual treatment combinations used are shown in Table 22.6.

The actual experiment is performed in two steps. The rows of the table form units (consisting of sets of four), i.e., the row stratum. Treatments identified in the first column are randomly applied to these sets. Next, the columns form units (again consisting of sets of four). The four treatments defined in the first row are randomly assigned to the column groups. Row treatments are applied to sets of four units and then column treatments applied to sets of four units. The actual treatment combinations applied to the 16 experimental row–column stratum units are shown in the body of the table. Notice that one randomization is applied to rows and another to columns. At the individual unit level, both randomizations apply.

When the data are available, the statistical analysis begins by looking at contrasts among row means and among column means. Estimates of A_r, B_r, and C_r are available in the row stratum and estimates of A_c, B_c, and C_c in the column stratum. The estimates from the row (column) stratum will be subject to both the row×column stratum variance and row (column) stratum variance. Because only half replicates are available in both row and column strata, A_r is aliased with B_rC_r, B_r with A_rC_r, and so on. Similarly A_c is aliased with B_cC_c, and so on.

This aliasing has an impact on the contrasts estimable in the row×column stratum. The defining contrast for the fraction is

$$\mathcal{I} = A_rB_rC_r = A_cB_cC_c = A_rB_rC_rA_cB_cC_c.$$

It follows that, for example, A_rA_c is aliased with $B_rC_rA_c$, with $A_rB_cC_c$, and with $B_rC_rB_cC_c$. If it is assumed that all three-factor and higher interactions are negligible, then A_rA_c, A_rB_c, A_rC_c, B_rA_c, B_rB_c, B_rC_c, C_rA_c, C_rB_c, and C_rC_c are estimable. The estimates will be subject only to error coming from the row ×column stratum. There are no degrees of freedom left to estimate error. At best, the experimenter can resort to examining contrasts using the half-normal plot (Chapter 12).

Exercise 22.6: Strip-Unit Plan. Construct the analysis of variance table for the hypothetical experiment set out in Table 22.6. □

Full Replicate in Four Blocks

A more appealing plan that involves a complete replicate in four blocks (and four times the amount of work in the laboratory) is shown in Table 22.7. With this plan it is possible to estimate all main effects and two-factor interactions. Any interaction that involves at least one factor from the row stratum and at least one factor from the column stratum is estimated from the row×column stratum.

If this represents too much laboratory work, an alternative to consider is a plan that only utilizes the upper half of the first panel in Table 22.7 and the lower half of the second panel. This provides an experiment with 32 observations. It follows that A_r, B_r, C_r, A_rB_r, A_rC_r, and B_rC_r are estimable from row means and A_c, B_c, C_c, A_cB_c, A_cC_c, and B_cC_c are estimable from column means. The $A_rB_rC_r$ interaction is aliased with the $A_cB_cC_c$ interaction. The estimates account for the six row-within-block, six column-within-block, and one between-block degrees of freedom. Notice that the fraction being used is defined by $\mathcal{I} = A_rB_rC_rA_cB_cC_c$. The remaining 18 degrees of freedom provide estimates of interactions in the row×column stratum that involve at least one factor from the row stratum and one factor from the column stratum. From the defining contrast we see that A_rA_c is aliased with $B_rC_rB_cC_c$. If four-factor interactions are

Table 22.7 Alternate Strip-Unit Plan with Rows and Columns within Blocks

		$-A_cB_cC_c$			
		(1)	a_cb_c	a_cc_c	b_cc_c
$-A_rB_rC_r$	(1)	000 000	000 110	000 101	000 011
	a_rb_r	110 000	110 110	110 101	110 011
	a_rc_r	101 000	101 110	101 101	101 011
	b_rc_r	011 000	011 110	011 101	011 011
$+A_rB_rC_r$	a_r	100 000	100 110	100 101	100 011
	b_r	010 000	010 110	010 101	010 011
	c_r	001 000	001 110	001 101	001 011
	$a_rb_rc_r$	111 000	111 110	111 101	111 011
		$+A_cB_cC_c$			
		a_c	b_c	c_c	$a_cb_cc_c$
$-A_rB_rC_r$	(1)	000 100	000 010	000 001	000 111
	a_rb_r	110 100	110 010	110 001	110 111
	a_rc_r	101 100	101 010	101 001	101 111
	b_rc_r	011 100	011 010	011 001	011 111
$+A_rB_rC_r$	a_r	100 100	100 010	100 001	100 111
	b_r	010 100	010 010	010 001	010 111
	c_r	001 100	001 010	001 001	001 111
	$a_rb_rc_r$	111 100	111 010	111 001	111 111

Table 22.8 Analysis of Variance for Half Replicate of Plan in Table 22.7

Source	df	
Three main effects (A_r, B_r, C_r)	3	Row stratum
Three two-factor interactions	3	
Three main effects (A_c, B_c, C_c)	3	Column stratum
Three two-factor interactions	3	
$A_rB_rC_r = A_cB_cC_c$	1	Block stratum
Nine two-factor interactions ($A_rA_c, \ldots, C_rC_c$)	9	Row ×column stratum

negligible, the nine interactions A_rA_c, A_rB_c, A_rC_c, B_rA_c, B_rB_c, B_rC_c, C_rA_c, C_rB_c, and C_rC_c, are all estimable. Also, $A_rB_rA_c$ and $C_rB_cC_c$ are aliased. There are nine contrasts estimating contrasts of this type. If all three-factor interactions are negligible, these provide an estimate of error with nine degrees of freedom. It is important to notice that the nature of the contrasts estimable in the row×column stratum (i.e., resolution) depends on the blocks of treatment combinations selected. This becomes more obvious in larger plans. The analysis of variance table for the half replicate taken from the plan illustrated in Table 22.7 (taking the upper half of the first panel and lower half of the second panel) is given in Table 22.8.

An analogous experiment can be obtained by using the lower half of the first panel and the upper half of the second panel in Table 22.7. In this case the fraction being used is defined by $\mathcal{I} = -A_rB_rC_rA_cB_cC_c$.

Exercise 22.7: Analysis of Variance in a Strip-Plot Design with Blocking. Construct the analysis of variance table for data collected from an experiment set up as outlined in Table 22.7 with four blocks. □

Exercise 22.8: Blocking in a Strip-Plot Design. Show the consequences of selecting the two blocks in either the first or the second panel. □

22.4.5 More General Structure

For a larger example that illustrates some further possibilities and problems to be aware of, we now consider a 2^9 factorial with five factors applied to the rows and four factors applied to the columns and the restriction that the experimenter can only apply the row treatments or column treatments to sets of eight units at a time. These restrictions are satisfied by blocking on $A_rB_rC_r$, $C_rD_rE_r$, and the generalized interaction $A_rB_rD_rE_r$ in the row stratum and on $A_cB_cC_cD_c$ in the column stratum. Schematically, the design can be illustrated as shown in Table 22.9.

For a half replicate, the experimenter selects treatment combinations corresponding to four of the eight blocks in the table. Two must be from the first

Table 22.9 2^9 Factorial in Strip-Unit Design with Rows and Columns Nested in Eight Blocks

	$-A_cB_cC_cD_c$	$+A_cB_cC_cD_c$
$+A_rB_rC_r$ $-C_rD_rE_r$ $-A_rB_rD_rE_r$	1	2
$-A_rB_rC_r$ $+C_rD_rE_r$ $-A_rB_rD_rE_r$	3	4
$-A_rB_rC_r$ $-C_rD_rE_r$ $+A_rB_rD_rE_r$	5	6
$+A_rB_rC_r$ $+C_rD_rE_r$ $+A_rB_rD_rE_r$	7	8

column and two from the second, and each row of blocks must be represented. However, there is a further consideration. If the experimenter selects blocks 1, 3, 6, and 8, $\mathcal{I} = A_rB_rD_rE_rA_cB_cC_cD_c$ is the defining contrast for the half replicate. If blocks 1, 4, 5, and 8 are selected, then $\mathcal{I} = C_rD_rE_rA_cB_cC_cD_c$ is the defining contrast for the half replicate. In this case the choice probably makes little difference, since the choice is between a resolution VIII plan and a resolution VII plan. With both plans the experiment yields 256 observations which are in four blocks of 64. We assume the experimenter has elected to use blocks 1, 3, 6, and 8. We also assume that row treatments are applied before column treatments.

For the first block (cell 1 in Table 22.9) the experimenter will apply treatment combinations {(10000), (01000), (10011), (01011), (11110), (00110), (11101), and (00101)} to the eight rows of eight, and then the treatment combinations {(1000), (0100), (0010), (0001), (1110), (1101), (1011), and (0111)} to the eight columns of eight. These 64 treatment combinations are obtained as solutions to the systems

$$\begin{array}{cccccl} a_r & b_r & c_r & d_r & e_r & \\ \hline i_{A_r} + & i_{B_r} + & i_{C_r} & & & = 1 \pmod 2 \\ & & i_{C_r} + & i_{D_r} + & i_{E_r} & = 0 \pmod 2 \end{array}$$

and

$$\begin{array}{ccccl} a_c & b_c & c_c & d_c & \\ \hline i_{A_c} + & i_{B_c} + & i_{C_c} + & i_{D_c} & = 1 \pmod 2, \end{array}$$

respectively.

For the second block (cell 3 in Table 22.9) the 64 treatment combinations for rows and columns are obtained as solutions to the systems

$$\begin{array}{ccccccccccc} a_r & & b_r & & c_r & & d_r & & e_r & & \\ \hline i_{A_r} & + & i_{B_r} & + & i_{C_r} & & & & & = 0 & \text{(modulo 2)} \\ & & & & i_{C_r} & + & i_{D_r} & + & i_{E_r} & = 1 & \text{(modulo 2)} \end{array}$$

and

$$\begin{array}{cccccccc} a_c & & b_c & & c_c & & d_c & \\ \hline i_{A_c} & + & i_{B_c} & + & i_{C_c} & + & i_{D_c} & = 1 \ \text{(modulo 2)}, \end{array}$$

respectively.

The statistical analysis of the row means yields estimates of five main effects, 10 two-factor interactions, eight three-factor interactions, four four-factor interactions, and one five-factor interaction. Depending on assumptions, higher-order interactions can be pooled for an estimate of the row stratum error. This error will have expected value $\sigma_\epsilon^2 + 8\sigma_r^2$. It is the appropriate error term to use when computing standard errors and confidence intervals for row contrasts.

The analysis of column means yields estimates of four main effects, six two-factor interactions, and four three-factor interactions. Again, a column stratum error term can be constructed from the three-factor interactions. This error will have expected value $\sigma_\epsilon^2 + 8\sigma_c^2$. It is the appropriate error term to use when computing standard errors and confidence intervals for the column contrasts.

Two three-factor (row treatment) interactions and one four-factor (column treatment) interaction are confounded with blocks. All two- and three-factor interactions involving at least one row factor and at least one column factor are estimable from the row×column stratum. The row×column stratum error is estimated with 120 degrees of freedom. This error will have expected value σ_ϵ^2 and is the basis for standard errors for all contrasts in the row×column contrasts. The analysis of variance table is summarized in Table 22.10.

We notice that in return for applying 32 row treatment combinations and 32 column treatment combinations, the experimenter obtains good information on a large number of two- and three-factor interactions. In terms of the cake baking example, it is possible to examine five cake batter factors and four baking factors with a reasonable level of effort. Also note that this experiment can easily be extended to accommodate another factor. In a cake baking study, for example, another factor, say a storage factor, can easily be added. One can confound this factor with, say, the five-factor interaction $A_r B_r C_r C_c D_c$. There are tests (and estimates) for this factor and two-factor interactions involving this factor in the row×column stratum. In the case of the cake experiment, these interactions may well be of major interest since some factors may either retard or enhance changes as the cake is stored.

Exercise 22.9: Treatment Combinations. Notice from the plan illustrated in Table 22.9 that the treatment combinations that appear in cell 1 satisfy the

Table 22.10 Analysis of Variance for Plan in Table 22.9

Source	df	
Blocks	3	
Five main effects	5	Row stratum
Ten two-factor interactions	10	
Error(1)	13	
Four main effects	4	Column stratum
Six two-factor interactions	6	
Error(2)	4	
Twenty two-factor interactions	20	Row ×column stratum
Seventy three-factor interactions	70	
Error(3)	120	
Total	255	

set of equations

$$
\begin{array}{ccccccccccl}
a_r & b_r & c_r & d_r & e_r & a_c & b_c & c_c & d_c & & \\
\hline
i_{A_r} & + i_{B_r} & + i_{C_r} & & & + i_{A_c} & + i_{B_c} & + i_{C_c} & + i_{D_c} & = 0 & \text{(modulo 2)} \\
& & i_{C_r} & + i_{D_r} & + i_{E_r} & + i_{A_c} & + i_{B_c} & + i_{C_c} & + i_{D_c} & = 1 & \text{(modulo 2)} \\
i_{A_r} & + i_{B_r} & & + i_{D_r} & + i_{E_r} & + i_{A_c} & + i_{B_c} & + i_{C_c} & + i_{D_c} & = 0 & \text{(modulo 2).}
\end{array}
$$

We can write this more compactly by indicating that the fraction represented satisfies $\mathcal{I} = -A_rB_rC_rA_cB_cC_cD_c = +C_rD_rE_rA_cB_cC_cD_c = +A_rB_rD_rE_rA_cB_cC_cD_c$. Verify that if the experimenter selects cells 1, 3, 6, and 8, the resulting fraction is represented by $\mathcal{I} = -A_rB_rC_rA_cB_cC_cD_c$. □

Exercise 22.10: Computing Trick Using SAS. If one has access to SAS, there is a simple trick to verify the correctness of the tests of significance given by the rules. Consider, for example, the plan in Table 22.9. Generate a data set with 256 y observations in the four blocks of 64. Make sure that each observation is identified by the proper block number, row number, column number, and proper 0–1 value for $A_r, \ldots, E_r, A_c, \ldots, D_c$. Then run the following program:

The Ar*Br*Cr, Cr*Dr*Er, Ar*Br*Dr*Er, and Ac*Bc*Cc*Dc sum of squares will turn out to be nonestimable because of the "block" variable earlier in the

```
proc glm ;
  classes block row col ar br cr dr er ac bc cc dc ;
  model y = block ar|br|cr|dr|er@2 row
      ac|bc|cc|dc@2 col ar*br*cr cr*dr*er ar*br*dr*er
    ac*bc*cc*dc ar*ac ar*bc ar*cc ar*dc br*ac br*bc
    br*cc br*dc cr*ac cr*bc cr*cc cr*dc dr*ac dr*bc
    dr*cc dr*dc er*ac er*bc er*cc er*dc /e1 ss1;
  random row col ;  run ;
```

model. The variance components for "row" and "col" will be at the appropriate places, indicating the appropriate tests of significance and the proper error term to use in calculating standard errors and setting confidence limits. □

22.4.6 Example of a 2^{11-3} in Four 8 x 8 Blocks

Miller (1997) contains a number of examples of strip-unit factorials, including the following of an eighth replicate of a 2^{11} plan. We include this example because it illustrates some key additional points. The plan accommodates six row stratum factors and five column stratum factors. As in the preceding example, we construct the design in stages. First construct a 4×4 array of blocks. Each block will eventually consist of eight rows and columns. From the half replicate defined by $\mathcal{I} = +A_rB_rC_rD_rE_rF_r$, confound $A_rB_rC_r$ (and its alias $D_rE_rF_r$), $C_rD_rE_r$ (and its alias $A_rB_rF_r$), and the generalized interaction $A_rB_rD_rE_r$ (and its alias C_rF_r) with rows of blocks. Then confound $A_cB_cC_c$, $A_cD_cE_c$, and the generalized interaction $B_cC_cD_cE_c$ with the columns of blocks. Schematically, the design can be illustrated as shown in Table 22.11. Taking all treatment combinations in all 16 cells constitutes a half replicate of the 2^{11}.

To get to the advertised 2^{11-3} fraction, four blocks (cells) must be selected. To minimize the number of row and column factors (interactions) compromised, the four cells selected must represent four rows and all four columns of blocks. Twenty-four distinct choices are available. These are not all equivalent. Selecting

Table 22.11 2^{11-1} Factorial in Strip-Unit Design: 16 Blocks Each Containing Eight Rows and Eight Columns

	$-A_cB_cC_c$ $-A_cD_cE_c$ $+B_cC_cD_cE_c$	$-A_cB_cC_c$ $+A_cD_cE_c$ $-B_cC_cD_cE_c$	$+A_cB_cC_c$ $-A_cD_cE_c$ $-B_cC_cD_cE_c$	$+A_cB_cC_c$ $+A_cD_cE_c$ $+B_cC_cD_cE_c$
$+A_rB_rC_rD_rE_rF_r$ $-A_rB_rC_r = -D_rE_rF_r$ $-C_rD_rE_r = -A_rB_rF_r$ $+C_rF_r = +A_rB_rD_rE_r$	1	2	3	4
$+A_rB_rC_rD_rE_rF_r$ $-A_rB_rC_r = -D_rE_rF_r$ $+C_rD_rE_r = +A_rB_rF_r$ $-C_rF_r = -A_rB_rD_rE_r$	5	6	7	8
$+A_rB_rC_rD_rE_rF_r$ $+A_rB_rC_r = +D_rE_rF_r$ $-C_rD_rE_r = -A_rB_rF_r$ $-C_rF_r = -A_rB_rD_rE_r$	9	10	11	12
$+A_rB_rC_rD_rE_rF_r$ $+A_rB_rC_r = +D_rE_rF_r$ $+C_rD_rE_r = +A_rB_rF_r$ $+C_rF_r = +A_rB_rD_rE_r$	13	14	15	16

cells 1, 6, 12, and 15 yields a 2^{11-3} fraction in four blocks with defining contrasts

$$\begin{aligned}\mathcal{I} = A_rB_rC_rA_cB_cC_c = C_rD_rE_rB_cC_cD_cE_c = -A_rB_rD_rE_rA_cD_cE_c \\ = A_rB_rC_rD_rE_rF_r = D_rE_rF_rA_cB_cC_c \\ = -A_rB_rF_rB_cC_cD_cE_c = -C_rF_rA_cD_cE_c.\end{aligned}$$

This is a resolution V plan. Selecting cells 1, 6, 11, and 16 gives the resolution VI plan defined by

$$\begin{aligned}\mathcal{I} = A_rB_rC_rA_cB_cC_c = C_rD_rE_rA_cD_cE_c = A_rB_rD_rE_rB_cC_cD_cE_c \\ = A_rB_rC_rD_rE_rF_r = D_rE_rF_rA_cB_cC_c \\ = A_rB_rF_rA_cD_cE_c = C_rF_rB_cC_cD_cE_c.\end{aligned}$$

The remaining 22 possible plans are all either resolution V or VI.

This experiment has 256 treatment combinations. The statistical analysis of the row means (row stratum) yields estimates the six row main effects and 14 two-factor interactions (C_rF_r is confounded with blocks) and an estimate of error with eight degrees of freedom and expected value $\sigma_\epsilon^2 + 8\sigma_r^2$. The analysis of the column stratum yields estimates of five main effects, 10 two-factor interactions, and an estimate of column stratum error with 13 degrees of freedom and expected value $\sigma_\epsilon^2 + 8\sigma_c^2$. All of the two-factor interactions that involve one of A_r, B_r, C_r, D_r, E_r, or F_r and one of A_c, B_c, C_c, D_c, or E_c are estimable in the row×column stratum. The error term with expected value σ_ϵ^2 is estimated with 166 degrees of freedom.

Exercise 22.11: Treatment Combinations. Obtain the treatment combinations that are to be applied for Table 22.11 to the rows of the block constructed from cell 1 by finding the solutions to the system of equations

$$\begin{array}{cccccccccccl} a_r & & b_r & & c_r & & d_r & & e_r & & f_r & \\ \hline i_{A_r} & + & i_{B_r} & + & i_{C_r} & + & i_{D_r} & + & i_{E_r} & + & i_{F_r} & = 0 \text{ (modulo 2)} \\ i_{A_r} & + & i_{B_r} & + & i_{C_r} & & & & & & & = 0 \text{ (modulo 2)} \\ & & & & i_{C_r} & + & i_{D_r} & + & i_{E_r} & & & = 0 \text{ (modulo 2)} \end{array}$$

and the treatment combinations that are to be applied to the columns by finding the solutions to the system of equations

$$\begin{array}{ccccccccl} a_c & & b_c & & c_c & & d_c & & e_c \\ \hline i_{A_c} & + & i_{B_c} & + & i_{C_c} & & & & = 0 \text{ (modulo 2)} \\ i_{A_c} & & & & & + & i_{D_c} & + \ i_{E_c} & = 0 \text{ (modulo 2).} \end{array}$$

□

Exercise 22.12: Treatment Combinations for an Additional Block. Find the systems of equations that must be solved to obtain the treatment combinations that are to be applied to rows and columns for a block constructed from cell 6 of Table 22.11. □

22.5 GENERAL COMMENTS ON STRIP-UNIT EXPERIMENTS

The choice of fractional factorial plans to use for the row and column strata is quite broad. All of our illustrations are from the 2^N system because these are the most useful. The theory extends directly to fractions from the p^n systems, where p is any prime number or power of a prime number. However, when p is equal to three or more, the plans that allow at least some two-factor interactions to be estimated cleanly are typically too large for practical use. If the experimenter finds that estimates of quadratic effects are really required, we suggest considering using four levels with pseudofactors. Pseudofactor techniques are explained in Chapter 20.

The statistical analysis of strip-unit experiments with fractional replication can be summarized Miller (1997) with the following six rules:

1. Row factors and interactions among row factors are estimable in the row stratum or are confounded with blocks.
2. Column factors and interactions among column factors are estimable in the column stratum or are confounded with blocks.
3. Generalized interactions between estimable row effects or interactions and nonestimable column effects or interactions are in the row stratum.
4. Generalized interactions between estimable column effects or interactions and nonestimable row effects or interactions are in the column stratum.
5. Any generalized interaction between a nonestimable row effect or interaction and a nonestimable column effect or interaction is either part of the defining contrast for the fraction or is in the block stratum.
6. Generalized interactions that involve at least one row factor and one column factor and are not covered in the cases above are in the row×column stratum.

We recommend that the experimenter make up some synthetic data and perform a test statistical analysis before investing too many resources in the experiment. If the experimenter has access to GLM in SAS® this construction and analysis is relatively simple. In addition, using the options in `proc GLM`, it is relatively easy to examine the possibilities for adding extra factors in the row×column stratum as illustrated in Section 22.4.5.

22.6 SPLIT-LOT DESIGNS

In this section we return to the strategy introduced in Section 22.4.3. The key requirement for applying the techniques to be illustrated in this section is that there is no a priori natural grouping of the experimental units. These designs are appropriate for situations in which it is convenient to apply treatments to batches (also called *lots*) of units in a sequence of steps and it is possible (advisable) to reallocate individual units to different batches for each step. Notice that this rules out the classical split-block field experiment in agriculture, where fields are divided into rows and columns of plots and the plots cannot be moved. We also assume that there are experimental errors associated with the application of treatments of individual lots. Examples of the types of applications that come to mind are:

1. Fabricating integrated circuits where wafers are treated in batches and there is a sequence of treatments. Wafers can be reassigned to different batches at each stage.
2. Education where students are taught in groups and also must be taught a sequence of skills in a particular order. Student groups can be reorganized.
3. Animals in a complex nutrition study where groups of animals are first fed early starter rations, then grower rations, and eventually, finishing rations. It may be possible to reorganize groups of animals.
4. An egg storage study in which batches of freshly laid eggs are washed in various solutions, batches of freshly washed eggs are dried by various means, batches of dried eggs are cooled, and then eventually packed for storage.

We focus on the common feature of all of these examples, which is that treatments are applied to batches of units and that units are assigned to different batches between stages of the experiment. We illustrate with a sequence of possible plans.

22.6.1 Motivating Example

As a motivating example for the techniques to be developed, we refer to an example in Mee and Bates (1998). The example concerns a silicon wafer production process. The process involves a sequence of three distinct treatments, each applied to lots of four wafers and in a specific sequence. Denote the treatments (each at two levels) by A_1, A_2, and A_3. For our purposes we identify the wafers as $w_1, w_2, \ldots, w_{16}$. Assign the 16 wafers to four batches $\{w_1, w_2, w_3, w_4\}$, $\{w_5, w_6, w_7, w_8\}$, $\{w_9, w_{10}, w_{11}, w_{12}\}$, and $\{w_{13}, w_{14}, w_{15}, w_{16}\}$. This defines the A_1 stratum. Apply the low level of A_1 to the first two batches and high level to the last two batches. Reallocate the wafers to batches $\{w_1, w_5, w_9, w_{13}\}$, $\{w_2, w_6, w_{10}, w_{14}\}$, $\{w_3, w_7, w_{11}, w_{15}\}$, and $\{w_4, w_8, w_{12}, w_{16}\}$. This defines the A_2 stratum. Apply treatment A_2, at the low level to batches 1 and 2, and at the high level to batches 3 and 4. Finally, reallocate the wafers to $\{w_1, w_6, w_{11}, w_{16}\}$, $\{w_3, w_8, w_9, w_{14}\}$, $\{w_2, w_5, w_{12}, w_{15}\}$, and $\{w_4, w_7, w_{10}, w_{13}\}$. This defines the A_3 stratum. The important features of this design are that (1) no two wafers appear together in the same batch in more than one stratum; and (2) two wafers receive treatments A_1, A_2, A_3 in each of the eight possible sequences LLL, LLH, ..., HHH.

Table 22.12 Analysis of Variance for the Silicon Wafer Example

Source	df	E[MS]
A_1	1	$\sigma_\epsilon^2 + 4\sigma_1^2 + 8\phi_{A_1}$
Error A_1-stratum	2	$\sigma_\epsilon^2 + 4\sigma_1^2$
A_2	1	$\sigma_\epsilon^2 + 4\sigma_2^2 + 8\phi_{A_2}$
Error A_2-stratum	2	$\sigma_\epsilon^2 + 4\sigma_2^2$
A_3	1	$\sigma_\epsilon^2 + 4\sigma_3^2 + 8\phi_{A_3}$
Error A_3-stratum	2	$\sigma_\epsilon^2 + 4\sigma_3^2$
$A_1 \times A_2$	1	$\sigma_\epsilon^2 + 4\phi_{A_1 \times A_2}$
$A_1 \times A_3$	1	$\sigma_\epsilon^2 + 4\phi_{A_1 \times A_3}$
$A_2 \times A_3$	1	$\sigma_\epsilon^2 + 4\phi_{A_2 \times A_3}$
$A_1 \times A_2 \times A_3$	1	$\sigma_\epsilon^2 + \phi_{A_1 \times A_2 \times A_3}$
Error	2	σ_ϵ^2
Total	$16 - 1$	

It is important to note the error structure. (i) There is an initial unique error associated with each wafer. (ii) Errors are imparted to the batches as the treatments are imposed. These errors have variances σ_1^2, σ_2^2, and σ_3^2, respectively. (iii) There is a unique error associated with the final evaluation of each wafer. Errors from sources (i) and (iii) combine to have variance σ_ϵ^2. The analysis of variance takes the form shown in Table 22.12. It is important to see that the stratum errors cancel out in the interaction contrasts.

22.6.2 Formal Structure

From the motivating example, we see that the key operation is to form the batches that constitute the strata. These strata must form orthogonal partitions of the contrasts (degrees of freedom) among the individual units. One way to accomplish such partitioning is to utilize a factorial structure and associate units with treatment combinations as is done in the lattice designs.

Let p be a prime number or a power of a prime. The p^n units can be partitioned into p^{m_1} batches of size p^{n-m_1} by selecting m_1 independent generators (effects or interactions) from the total set of $(p^n - 1)/(p - 1)$ effects and interactions. These p^{m_1} batches constitute a stratum. Levels of the first factor are assigned to the batches. A second orthogonal stratum of p^{m_2} batches of size p^{n-m_2} is formed by selecting another set of m_2 independent generators. The restriction is that the set of m_2 generators and all generalized interactions have no overlap with the initial set of m_1 generators and all generalized interactions.

Successive strata can be constructed as long as it is possible to find generators such that sets of generators and generalized interactions have no overlap. In the motivating example, $p = 2$, $n = 4$ and the three strata with $m_1 = m_2 = m_3 = 2$

Table 22.13 Analysis of Variance for Experiment with Four Strata

Source	df	E[MS]
A_1	1	$\sigma_\epsilon^2 + 8\sigma_1^2 + 16\phi_{A_1}$
Error A_1-stratum	2	$\sigma_\epsilon^2 + 8\sigma_1^2$
A_2	1	$\sigma_\epsilon^2 + 8\sigma_2^2 + 16\phi_{A_2}$
Error A_2-stratum	2	$\sigma_\epsilon^2 + 8\sigma_2^2$
A_3	1	$\sigma_\epsilon^2 + 8\sigma_3^2 + 16\phi_{A_3}$
Error A_3-stratum	2	$\sigma_\epsilon^2 + 8\sigma_3^2$
A_4	1	$\sigma_\epsilon^2 + 8\sigma_4^2 + 16\phi_{A_4}$
Error A_4-stratum	2	$\sigma_\epsilon^2 + 8\sigma_4^2$
$A_1 \times A_2$	1	$\sigma_\epsilon^2 + 8\phi_{A_1 \times A_2}$
$A_1 \times A_3$	1	$\sigma_\epsilon^2 + 8\phi_{A_1 \times A_3}$
$A_1 \times A_4$	1	$\sigma_\epsilon^2 + 8\phi_{A_1 \times A_4}$
$A_2 \times A_3$	1	$\sigma_\epsilon^2 + 8\phi_{A_2 \times A_3}$
$A_2 \times A_4$	1	$\sigma_\epsilon^2 + 8\phi_{A_2 \times A_4}$
$A_3 \times A_4$	1	$\sigma_\epsilon^2 + 8\phi_{A_3 \times A_4}$
Error	13	σ_ϵ^2
Total	31	

were defined by the sets of effects and interactions $\{A, B, AB\}$, $\{C, D, CD\}$, and $\{AC, BD, ABCD\}$. The levels of A_1 were assigned to batches according to the sign of A, the sign of C defined the levels of A_2, and the levels of A_3 were assigned to batches according to the sign of BD. It is important that the generalized interaction of A and C not be used to assign the levels of the third factor to batches. If AC were used to assign the levels of A_3, only half of the possible 2^3 distinct orderings would be assigned to wafers. It is possible to define two more orthogonal strata using $\{ABC, BCD, AD\}$ and $\{ACD, ABD, BC\}$. If treatments A_4 and A_5 each at two levels are assigned to these strata, the analysis of variance shows five strata, one treatment and two error degrees of freedom in each stratum and no degrees of freedom for interactions or residual error.

If p is a prime number, and $m_i > 1$, a factor with p levels can be assigned to the p^{m_i} batches in the ith stratum and there will be $p^{m_i} - p$ degrees of freedom for error in the stratum. Corresponding statements hold for p a power of a prime.

22.6.3 Split-Lot or Split-Batch Designs

Thirty-Two Units in Batches of Eight

Wu (1989) shows that it is possible to partition 31 effects and interactions in the 2^5 factorial into nine sets of three: $\{A, B, AB\}$, $\{D, E, ED\}$, $\{BDE, ABCE,$

$ADE\}$, $\{ABCD, ACE, BDE\}$, $\{ACD, AE, CDE\}$, $\{AD, DE, ABDE\}$, $\{ABD, BCE, ACDE\}$, $\{BD, CE, BCDE\}$, and $\{CD, ABE, ABCDE\}$. This shows that there exist designs with up to nine strata, each stratum consisting of four batches of eight units. Treatment combinations that correspond to units in the first lot of eight in the first stratum are obtained as the eight solutions to

$$\begin{array}{ccccc|l} a & b & c & d & e & \\ \hline i_A & & & & & = 0 \text{ (modulo 2)} \\ & i_B & & & & = 0 \text{ (modulo 2).} \end{array}$$

The remaining batches are obtained as in blocking. An experiment with four factors, each at two levels, would have the analysis of variance shown in Table 22.13.

Thirty-Two Units in Batches of Four

It is not possible to find two nonoverlapping sets of seven effects and interactions to provide two orthogonal groupings of eight lots of four. However, there are some possibilities. One possibility is one stratum consisting of eight batches of four and several strata with batches of eight. The batches of four are defined by $\{A, B, AB, C, AC, BC, ABC\}$. Six more strata with batches of eight are defined by $\{D, E, DE\}$, $\{CD, BE, BCDE\}$, $\{BD, AE, ABDE\}$, $\{AD, BDE, ABE\}$, $\{ABD, BCE, ACDE\}$, and $\{ACD, ADE, CE\}$.

An alternative that can be used if one really needs two strata, each consisting of batches of four (in order to increase degrees of freedom for testing main effects) is to construct two strata using $\{A, B, AB, C, AC, BC, ABC\}$ and $\{BD, AE, ABDE, ACD, ABC, CDE, BCE\}$. Notice that one contrast, ABC, is common to the two sets. Assign the two levels of treatments A_1 and A_2 to contrasts other than ABC in the two strata. Assign a dummy treatment to contrast ABC. Then in the analysis of variance, first remove the sum of squares attributable to the dummy treatment and then proceed with the two strata. The analysis will show five degrees of freedom for error in both strata.

Sixty-Four Units in Lots of Eight

Taguchi (1987a, p. 1152) gives the following nine sets of effects and interactions each that can be used to construct an experiment with up to nine stages, with eight batches of eight units used at each stage:

$\{A, B, AB, C, AC, BC, ABC\}$
$\{D, E, DE, F, DF, EF, DEF\}$
$\{AD, BE, ABDE, CF, BCEF, ACDF, ABCDEF\}$
$\{BD, CE, BCDE, ABF, ADF, ABCEF, ACDEF\}$
$\{ABD, BCE, ACDE, ABCF, CDF, AEF, BDEF\}$
$\{CD, ABE, ABCDE, BCF, BDF, ACEF, ADEF\}$
$\{ACD, AE, CDE, BF, ABCDF, ABEF, BCDEF\}$
$\{BCD, ABCE, ADE, ACF, ABDF, BEF, CDEF\}$
$\{ABCD, ACE, BDE, AF, BCDF, CEF, ABDEF\}$

Notice that one can have one factor at two levels and a six-degree-of-freedom error estimate, two factors each at two levels, their interaction and a four-degree-of-freedom estimate of error, or one factor at four levels and a three-degree-of-freedom error estimate at each stage. If one uses fewer than nine stages, it is also possible to estimate other interactions at the unit stratum. Taguchi (1987a) gives two alternative sets of groupings for this problem, and Mee and Bates (1998) give two constructions for such plans.

APPENDIX 22A: FRACTIONAL FACTORIAL PLANS FOR SPLIT-PLOT DESIGNS*

Catalog of Minimum Aberration 16-Run Fractional Factorial Plans for Split-Plot Designs

Plan	Defining Contrasts[a]	WLP[b]
(1 + 4) − (0 + 1)	*F*8	0 0 1 0
(2 + 3) − (0 + 1)	*F*8	0 0 1 0
(3 + 2) − (0 + 1)	*F*8	0 0 1 0
(1 + 5) − (0 + 2)	*E*8, *D*4	0 3 0 0
(2 + 4) − (0 + 2)	*E*8, *D*4	0 3 0 0
(2 + 4) − (0 + 2)	*E*8, *B*4	0 3 0 0
(3 + 3) − (0 + 2)	*D*8, *B*4	0 3 0 0
(3 + 3) − (1 + 1)	*E*0, 9*C*	1 1 1 0
(4 + 2) − (1 + 1)	*F*0, *CC*	0 3 0 0
(1 + 6) − (0 + 3)	*E*8, *D*4, *B*2	0 7 0 0
(2 + 5) − (0 + 3)	*E*8, *D*4, *B*2	0 7 0 0
(3 + 4) − (0 + 3)	*D*8, *B*4, 72	0 7 0 0
(3 + 4) − (1 + 2)	*E*0, 94, 5*A*	2 3 2 0
(4 + 3) − (1 + 2)	*F*0, *CC*, *AA*	0 7 0 0
(5 + 2) − (2 + 1)	*D*0, *A*8, 66	2 3 2 0
(1 + 7) − (0 + 4)	*E*8, *D*4, *B*2, 71	0 14 0 0
(2 + 6) − (0 + 4)	*E*8, *D*4, *B*2, 71	0 14 0 0
(3 + 5) − (0 + 4)	98, 54, 32, *F*1	3 7 4 0
(3 + 5) − (1 + 3)	*E*0, 94, 8*A*, 59	3 7 4 0
(4 + 4) − (1 + 3)	*F*0, *CC*, *AA*, 69	0 14 0 0
(5 + 3) − (2 + 2)	*D*0, *A*8, 86, 65	3 7 4 0
(6 + 2) − (3 + 1)	*D*0, *A*8, 64, *E*3	4 6 4 0
(1 + 8) − (0 + 5)	*C*80, *A*40, 710, *F*08 920,	4 14 8 0
(1 + 8) − (0 + 5)	*C*80, 640, *B*10, *F*08 520,	4 14 8 0
(2 + 7) − (0 + 5)	*A*80, 640, 310, *F*08 *D*20,	4 14 8 0
(3 + 6) − (0 + 5)	980, 540, 710, *F*08 *B*20,	6 9 9 6
(3 + 6) − (1 + 4)	*E*00, 940, 590, *D*88 8*A*0,	4 14 8 0
(4 + 5) − (1 + 4)	*F*00, 8*C*0, 290, *E*88 4*A*0,	4 14 8 0
(5 + 4) − (2 + 3)	*D*00, *A*80, 650, *E*48 860,	4 14 8 0
(6 + 3) − (3 + 2)	*D*00, *A*80, 830, 628 640,	6 10 8 4

(continued)

Catalog of Minimum Aberration 16-Run Fractional Factorial Plans for Split-Plot Designs *(continued)*

Plan	Defining Contrasts[a]	WLP[b]
(7 + 2) − (4 + 1)	*D*00, *A*80, *E*20, 818 640,	8 10 4 4
(1 + 9) − (0 + 6)	*C*80, *A*40, 620, 910, 708, *F*04	8 18 16 8
(1 + 9) − (0 + 6)	*C*80, *A*40, 620, 510, *B*08, *F*04	8 18 16 8
(2 + 8) − (0 + 6)	*A*80, 640, *E*20, 910, 508, *D*04	8 18 16 8
(2 + 8) − (0 + 6)	*A*80, 640, 920, 510, *B*08, 704	8 18 16 8
(2 + 8) − (0 + 6)	*A*80, 640, *E*20, *D*10, 308, *F*04	8 18 16 8
(3 + 7) − (0 + 6)	980, 540, *D*20, 310, *B*08, 704	9 16 15 12
(3 + 7) − (1 + 5)	*E*00, 940, 520, 890, 588, *D*84	8 18 16 8
(4 + 6) − (1 + 5)	*F*00, 8*C*0, 4*A*0, *C*90, 288, *E*84	8 18 16 8
(5 + 5) − (2 + 4)	*D*00, *A*80, 640, 830, 628, *E*24	8 18 16 8
(6 + 4) − (3 + 3)	*D*00, *A*80, 640, 830, 628, *E*24	8 18 16 8
(7 + 3) − (4 + 2)	*C*80, *A*40, 620, *E*10, 908, 504	10 16 12 12
(1 + 10) − (0 + 7)	*C*80, *A*40, 620, 910, 508, *B*04, 702	12 26 28 24
(1 + 10) − (0 + 7)	*C*80, *A*40, 620, 910, 508, 704, *F*02	12 26 28 24
(2 + 9) − (0 + 7)	*A*80, 640, *E*20, 910, 508, *D*04, 302	12 26 28 24
(2 + 9) − (0 + 7)	*A*80, 640, *E*20, 910, 508, *D*04, *F*02	12 26 28 24
(2 + 9) − (0 + 7)	*A*80, 640, 920, 510, *B*08, 704, *F*02	12 26 28 24
(3 + 8) − (0 + 7)	980, 540, *D*20, 310, *B*08, 704, *F*02	12 26 28 24
(3 + 8) − (1 + 6)	*E*00, 940, 520, 890, 488, 984, 582	12 26 28 24
(4 + 7) − (1 + 6)	*F*00, 8*C*0, 4*A*0, *C*90, 288, *A*84, 682	12 26 28 24
(5 + 6) − (2 + 5)	*D*00, *A*80, 860, 450, 248, 644, *E*42	12 26 28 24
(6 + 5) − (3 + 4)	*D*00, *A*80, 640, 830, 428, *A*24, 622	12 26 28 24
(7 + 4) − (4 + 3)	*D*00, *A*80, 640, *E*20, 818, 414, 212	13 25 25 27
(1 + 11) − (0 + 8)	*C*80, *A*40, 620, 910, 508, *B*04, 702, *F*01	16 39 48 48
(2 + 10) − (0 + 8)	*A*80, 640, *E*20, 910, 508, *D*04, 302, *F*01	16 39 48 48
(3 + 9) − (1 + 7)	*E*00, 940, 520, 890, 488, 984, 582, *D*81	16 39 48 48
(4 + 8) − (1 + 7)	*F*00, 8*C*0, 4*A*0, *C*90, 148, *A*84, 682, *E*81	16 39 48 48
(5 + 7) − (2 + 6)	*D*00, *A*80, 860, 450, *C*48, 244, *A*42, 641	17 38 44 52
(6 + 6) − (3 + 5)	*D*00, *A*80, 640, 830, 428, *A*24, 622, *E*21	16 39 48 48
(7 + 5) − (4 + 4)	*D*00, *A*80, 640, *E*20, 818, 414, *C*12, 211	17 38 44 52
(1 + 12) − (0 + 9)	*C*800, *A*400, 6200, *E*100, 9080, 5040, *D*020, 3010, *B*008	22 55 72 96
(1 + 12) − (0 + 9)	*C*800, *A*400, 6200, *E*100, 9080, 5040, *D*020, 3010, 7008	22 55 72 96
(2 + 11) − (0 + 9)	*A*800, 6400, *E*200, 9100, 5080, *D*040, 3020, *B*010, 7008	22 55 72 96

Catalog of Minimum Aberration 16-Run Fractional Factorial Plans for Split-Plot Designs *(continued)*

Plan	Defining Contrasts[a]	WLP[b]
(3 + 10) − (1 + 8)	*E*000, 9400, 5200, *D*100, 8880, 4840, *C*820, 1810, 9808	22 55 72 96
(3 + 10) − (1 + 8)	*E*000, 9400, 5200, *D*100, 8880, 4840, 1820, 9810, 5808	22 55 72 96
(5 + 8) − (2 + 7)	*D*000, *A*800, 8600, 4500, *C*480, 2440, *A*420, 6410, *E*408	22 55 72 96
(6 + 7) − (3 + 6)	*D*000, *A*800, 6400, 8300, 4280, *C*240, 2220, *A*210, 6208	22 55 72 96
(7 + 6) − (4 + 5)	*D*000, *A*800, 6400, *E*200, 8180, 4140, *C*120, 2110, *A*108	22 55 72 96
(1 + 13) − (0 + 10)	*C*800, *A*400, 6200, *E*100, 9080, 5040, *D*020, 3010, *B*008, 7004	28 77 112 168
(2 + 12) − (0 + 10)	*C*800, 6400, *E*200, 9100, 5080, *D*040, 3020, *B*010, 7008, *F*004	28 77 112 168
(3 + 11) − (1 + 9)	*E*000, 9400, 5200, *D*100, 8880, 4840, *C*820, 1810, 9808, 5804	28 77 112 168
(6 + 8) − (3 + 7)	*D*000, *A*800, 6400, 8300, 4280, *C*240, 2220, *A*210, 6208, *E*204	28 77 112 168
(7 + 7) − (4 + 6)	*D*000, *A*800, 6400, *E*200, 8180, 4140, *C*120, 2110, *A*108, 6104	28 77 112 168
(1 + 14) − (0 + 11)	*C*800, *A*400, 6200, *E*100, 9080, 5040, *D*020, 3010, *B*008, 7004, *F*002	35 105 168 280
(3 + 12) − (1 + 10)	*E*000, 9400, 5200, *D*100, 8880, 4840, *C*820, 1810, 9808, 5804, *D*802	35 105 168 280
(7 + 8) − (4 + 7)	*D*000, *A*800, 6400, *E*200, 8180, 4140, *C*120, 2110, *A*108, 6104, *E*102	35 105 168 280

[a]Defining contrasts are given in hexadecimal notation. To reconstruct the defining contrast, convert to binary notation and truncate to the number of factors. See page 577.
Example: F8 converts to 11111000 in binary, which is truncated to 11111 for five factors.
[b]$(\ell_3, \ell_4, \ell_5, \ell_6)$ Truncated word-length pattern.
*Adapted with permission from Bingham and Sitter (1999a).

CHAPTER 23

Orthogonal Arrays and Related Structures

23.1 INTRODUCTION

The object of this chapter is to introduce orthogonal arrays, asymmetrical orthogonal arrays, and nearly orthogonal arrays. These structures can be translated directly into fractional factorial plans. This will become important to potential users for two reasons. First, it provides access to a large collection of fractional factorial plans that are not derivable using the techniques presented up to this point, especially plans for factors with more than three levels and for asymmetrical factorials. Second, a growing body of the literature on fractional factorial experiments is couched in terms of orthogonal arrays. In particular, the L-arrays that figure so prominently in the work of Taguchi (1986, 1987a,b) are either orthogonal arrays or nearly orthogonal arrays. In a subsequent chapter we examine some of the experimental strategies based on the use of orthogonal arrays. In this chapter we focus on presenting a collection of orthogonal, asymmetrical orthogonal, and nearly orthogonal arrays and numerous derivations. These derivations suggest means to obtain arrays not given.

In 1942, Kishen introduced into the statistical literature mathematical structures which he called *hyper-Graeco–Latin* cubes and *hypercubes*. These structures are now known as *orthogonal arrays of strength* 2. A few years later, Rao (1946a) used finite geometries to construct combinatorial arrangements which were generalizations of the Kishen structures. He called these *hypercubes of strength d*. In a subsequent paper, Rao (1946b) systematized the construction of these hypercubes and used them as a device for constructing confounded plans for factorial experiments. Here was the early connection between orthogonal arrays or hypercubes and factorial experiments. Shortly thereafter came the Plackett and Burman (1946) paper, which can be interpreted as the complete solution to the problem of constructing and using arrays of strength 2 when each

Planning, Construction, and Statistical Analysis of Comparative Experiments,
by Francis G. Giesbrecht and Marcia L. Gumpertz
ISBN 0-471-21395-0

factor in the experiment has two levels. Further developments in the early construction and application of orthogonal arrays or hypercubes were given by Rao (1947a,b).

While the early papers by Rao clearly emphasize the connection with fractional factorial plans, there are a number of papers, beginning with Bose and Bush (1952), that regarded orthogonal arrays as generalizations of orthogonal Latin squares. In this book we emphasize and utilize the connection between orthogonal arrays and plans for fractional factorial experiments. Currently, the principal reference on orthogonal arrays is Hedayat et al. (1999), which also contains (in Chapter 12) a large collection of orthogonal and asymmetrical orthogonal arrays.

23.1.1 Motivating Example

Fractional factorial and Hadamard plans both fit into the framework of orthogonal arrays. Orthogonal arrays provide a different perspective, a somewhat more abstract approach, to the problem of constructing suitable fractional factorials. To provide some links with previous material, we begin with some constructions of orthogonal arrays from known factorial experiments. Once some familiarity has been achieved, we proceed to alternative constructions. However, before doing this we introduce a simple example of an orthogonal array that translates directly into a fraction of the 3^7 factorial that has immediate application but cannot be obtained via the projective geometry or Hadamard matrices used to this point. The example is an array with 18 rows and seven columns, taken from Bose and Bush (1952):

$$
\begin{array}{ccccccc}
0 & 0 & 0 & 0 & 0 & 0 & 0 \\
1 & 1 & 1 & 1 & 1 & 1 & 0 \\
2 & 2 & 2 & 2 & 2 & 2 & 0 \\
0 & 0 & 1 & 2 & 1 & 2 & 0 \\
1 & 1 & 2 & 0 & 2 & 0 & 0 \\
2 & 2 & 0 & 1 & 0 & 1 & 0 \\
0 & 1 & 0 & 2 & 2 & 1 & 1 \\
1 & 2 & 1 & 0 & 0 & 2 & 1 \\
2 & 0 & 2 & 1 & 1 & 0 & 1 \\
0 & 2 & 2 & 0 & 1 & 1 & 1 \\
1 & 0 & 0 & 1 & 2 & 2 & 1 \\
2 & 1 & 1 & 2 & 0 & 0 & 1 \\
0 & 1 & 2 & 1 & 0 & 2 & 2 \\
1 & 2 & 0 & 2 & 1 & 0 & 2 \\
2 & 0 & 1 & 0 & 2 & 1 & 2 \\
0 & 2 & 1 & 1 & 2 & 0 & 2 \\
1 & 0 & 2 & 2 & 0 & 1 & 2 \\
2 & 1 & 0 & 0 & 1 & 2 & 2
\end{array}
$$

If we interpret the columns of this array as factors, each at three levels, denoted by 0, 1, or 2, and the rows as runs, or treatment combinations, we have a fraction of the 3^7. In fact, if we look carefully we see that the factors are all mutually orthogonal. We look at any two columns and see that every level of the factor in one of the columns appears exactly twice with every level of the second factor. An estimate of, say, the quadratic effect of any factor is orthogonal to any contrast among levels of any other factor. This is a plan with much potential, but which is not part of the collection of 3^N fractions we have encountered up to this point.

Exercise 23.1: Orthogonality of Columns

(a) Verify that factors assigned to any two distinct columns in the array in Section 23.1.1 are orthogonal. Use the type of argument used in Chapters 10 and 11.
(b) Establish that a sufficient criterion for factors in two distinct columns to be orthogonal is that all distinct pairs of symbols, one from each column appear equally frequently.
(c) Verify that an eighth column $[0\ 0\ 0\ 1\ 1\ 1\ 0\ 0\ 0\ 1\ 1\ 1\ 0\ 0\ 0\ 1\ 1\ 1]^t$ is orthogonal to the seven columns in the array in Section 23.1.1. □

23.1.2 Examples of Fractional Factorial Experiments Based on Orthogonal Arrays

Our first example is part of a study described by Vrescak and Reed (1990) to improve a process of cleaning Kovar components. The basic procedure was an acid-cleaning process of the Kovar parts in order to enhance brazing to metallized ceramic washers. The cleaning involved immersion of the parts into a series of acid baths, the first referred to as descale and the second as brite-dip. The reaction rates for both are functions of temperature and strength of the acid. The purpose of the descale is to remove a surface coat of metallic oxides, and the brite-dip acid is to remove a target amount of Kovar metal to alter the surface to enhance brazing. After studying the problem it was decided that there were seven important factors, six at three levels and one at two levels. The factors are summarized as follows:

Descale acid exposure time	t_1	t_2	t_3
Descale acid strength	s_1	s_2	s_3
Descale acid temperature	c_1	c_2	c_3
Status of part at brite-dip	dry	wet	
Nitric to acetic acid ratio	r_1	r_2	r_3
Brite-dip acid concentration	a_1	a_2	a_3
Brite-dip acid temperature	b_1	b_2	b_3

All possible combinations of these factors ($3^6 \times 2 = 1458$) would be too large a project. Two times a 3^{6-3} requires 54 combinations and was still considered too large. Using six of the columns in the array on page 599 plus an extra column of three 0's and three 1's repeated six times gave a satisfactory plan that required 18 treatment combinations. This 18-run fractional factorial plan cannot be obtained using the standard geometric approach.

Our second example is adapted from a parameter design study by Hirai and Koga (1990) to determine stable metal oxide semiconductor (MOS) transistor manufacturing conditions. The actual project was a computer simulation study. The variable of interest was the performance (drain-to-source current) of simulated transistors. Engineering experience identified nine factors, each at three levels, that required examination. Simulating all 3^9 combinations was clearly impossible. A fraction was required. A possible candidate fraction would be the 3^{9-6}, which requires 27 runs. A difficulty with this plan is that it is a main-effects plan and all factors are aliased with potential two-factor interactions. Instead, the authors chose to use nine columns from a 36-run orthogonal array. We illustrate the construction of suitable arrays in Section 23.3.3. This choice requires 36 rather than just 27 runs. However, the extra runs provide advantages. First, they provide more degrees of freedom for error. In addition, they provide some protection against two-factor interactions. In the 27-run plan, all main effects are perfectly aliased with two-factor interactions. They cannot be separated. In the 36-run plan it is possible to use the techniques introduced in Sections 16.4 and 16.5 to look for evidence of two-factor interactions. Again we have a fractional factorial plan that cannot be obtained using the standard geometric approach.

Our third example is from a parachute design optimization problem. Parvey (1990) describes a 54-run experiment to improve the design of a special type of parachute. These parachutes consist of 28 lines that connect four fabric panels to each other and to an attachment plate. The experimental objective was to determine panel size and line dimensions for the parachute to meet performance specifications. A team of experts decided that they needed to try two panel sizes and three line lengths for each of 28 lines.

The smallest geometry-based fraction that accommodates the 28 factors is a 3^{28-24} fraction which requires 81 runs. This plan can be obtained by superimposing 28 factors on a subset of the 40 columns of an array obtained from a full 3^4 factorial. In addition, one column can be collapsed to two levels. The result is an 81-run plan. Alternatively, one could go to 162 runs and add the extra factor at two levels. Both were considered too large. However, there exists an orthogonal array, (23.6) on page 618, that will accommodate one factor at two levels and 25 factors, each at three levels in 54 runs. The investigators decided to compromise. The design of the parachutes allowed them to group four of the centerline lengths into one factor at three levels. This left them with exactly 25 three-level factors and one two-level factor for a fully saturated 54-run plan.

Exercise 23.2: Constructing a 3^{40-36} Fractional Factorial. Identify the 40 columns that can be obtained via the 3^4 factorial and used to construct a 3^{40-36} fractional factorial by superimposing factors. □

23.1.3 Definition of Orthogonal Array

An orthogonal array, normally written as OA[N;k;s;t], is a special $N \times k$ matrix of elements, selected from a set $\mathcal{S}$ that contains s distinct elements. The distinguishing feature of the matrix is that any $N \times t$ submatrix, i.e., any submatrix constructed by taking t distinct columns will have each of the s^t possible row vectors represented exactly λ times. This is an orthogonal array of strength t. The value k, the number of columns, is occasionally referred to as the *number of constraints*. We prefer to use the more suggestive notation OA[N;s^k;t], where the s^k reminds us of a factorial, k factors, each at s levels.

Occasionally, one encounters an author who will interchange rows and columns, i.e., talk about the transpose of this definition. Notice that we always have $\lambda = N/s^t$. The value λ is often called the *index of the array*.

23.2 ORTHOGONAL ARRAYS AND FRACTIONAL FACTORIALS

23.2.1 Fractional Factorials to Construct Orthogonal Arrays

We illustrate the definition of orthogonal arrays given above with a series of examples. For the first example, consider the 8×7 array, which we recognize as coming from the 2^{7-4} main effect plan, or equivalently, the eight-run Plackett–Burman plan.

$$\begin{array}{ccccccc}
+1 & +1 & +1 & +1 & +1 & +1 & +1 \\
+1 & +1 & +1 & -1 & -1 & -1 & -1 \\
+1 & -1 & -1 & +1 & +1 & -1 & -1 \\
+1 & -1 & -1 & -1 & -1 & +1 & +1 \\
-1 & +1 & -1 & +1 & -1 & +1 & -1 \\
-1 & +1 & -1 & -1 & +1 & -1 & +1 \\
-1 & -1 & +1 & +1 & -1 & -1 & +1 \\
-1 & -1 & +1 & -1 & +1 & +1 & -1
\end{array} \qquad (23.1)$$

We can easily check that (23.1) is an OA[8;2^7;2]. There are eight rows, corresponding to the eight runs and seven columns corresponding to seven treatments. The set $\mathcal{S}$ consists of the two elements $+1$ and -1 and $s = 2$. We could just as well have used the elements $+$ and $-$ or 0 and 1 or any other pair of distinct elements in the array. It is usual but not necessary to use the integers as elements of the array. Finally, we look at the strength of the array. Consider any pair of columns. Each pair of columns will contain the s^t, i.e., four pairs $\{+1, +1\}$, $\{+1, -1\}$, $\{-1, +1\}$, and $\{-1, -1\}$ an equal number of times, in this case exactly two times. This property gives rise to the term *orthogonal* in the name *orthogonal array*: the two columns are orthogonal to each other. This is an array with index 2. It is also easy to show that this array is not of strength 3 by showing that, for example, the first three columns do not contain the $2^3 = 8$ triplets $\{+1, +1, +1\}$, $\{+1, +1, -1\}$, $\{+1, -1, +1\}$, $\{+1, -1, -1\}$, $\{-1, +1, +1\}$, $\{-1, +1, -1\}$, $\{-1, -1, +1\}$, and $\{-1, -1, -1\}$ equally frequently. Dropping columns (constraints) from the array does not decrease the strength of the array.

***Exercise 23.3: Equivalent* 8 x 7 *Arrays*.** Show by interchanging the +1's and −1's and then rearranging rows and columns that this array is equivalent to the array shown in (16.1). □

For a second example of an orthogonal array, consider the 3^{3-1} fraction defined by $\mathcal{I} = ABC$. This gives us an OA[9;3^3;2]:

$$\begin{matrix} 0 & 0 & 0 \\ 0 & 1 & 2 \\ 0 & 2 & 1 \\ 1 & 0 & 2 \\ 1 & 1 & 1 \\ 1 & 2 & 0 \\ 2 & 0 & 1 \\ 2 & 1 & 0 \\ 2 & 2 & 2. \end{matrix} \tag{23.2}$$

Inspection shows that in this case the set $\mathcal{S}$ consists of the three elements, 0, 1, and 2 and that the array has strength 2, since any two columns contain all of the $3^2 = 9$ pairs {0, 0}, {0, 1}, {0, 2}, {1, 0}, {1, 1}, {1, 2}, {2, 0}, {2, 1}, and {2, 2} equally frequently.

It should be clear to the reader that any fractional factorial, resolution III plan defines an orthogonal array of strength 2.

***Exercise 23.4: Constructing an OA*[8;2^4;3].** Show by construction that an OA[8;2^4;3] can be constructed from the resolution IV plan defined by $\mathcal{I} = ABCD$. □

23.2.2 Orthogonal Arrays to Construct Fractional Factorials

From the proceeding constructions we can see that the process can be inverted; i.e., given an orthogonal array, one can readily construct an orthogonal fractional factorial plan. One simply associates columns with factors in the factorial, symbols in the columns with levels of the factors, and rows in the array with runs. An orthogonal array of strength 2 yields a main-effect (resolution III) plan. An orthogonal array of strength 3 yields a resolution IV plan, and so on. It is easy to see where the comment that orthogonal arrays are nothing but fractional factorials comes from.

The reader may well ask, "Why bother with orthogonal arrays?" or "What do we gain from orthogonal arrays?" The answer is that we gain considerable flexibility in the construction of fractional factorial plans that are not part of the systems obtained from the projective geometries and Hadamard matrices studied so far. A possible disadvantage may be that dealing with interactions becomes somewhat more troublesome. Potential interactions must not be ignored. They do not go away. They must be dealt with.

23.2.3 Summary of Some Classes of Orthogonal Arrays

In this section we give a brief review of some of the major classes of orthogonal arrays and recipes for their construction. The list is designed to cover the cases that an experimenter is likely to need. For a thorough and complete discussion of the theory, construction, and application of orthogonal arrays, the reader is urged to consult Hedayat et al. (1999). We note in passing that if OA[N;s^k;d] exists, then an OA[mN;$s^{k'}$;d'] exists for any $m \geq 1$, $k' \leq k$ and/or $d' \leq d$. The changes repeat the complete array or delete some columns.

Construction from Hadamard Matrices

It is well known that except for the two trivial cases, Hadamard matrices must be of order $4m \times 4m$, where m is an integer. Except for the five cases ($m =$ 107, 167, 179, 191, and 223), Hadamard matrices are known to exist for $m < 250$. It follows that there is an almost complete set of OA[$4m$;2^{4m-1};2] for at least $m < 250$. One constructs the Hadamard matrix and deletes the column with all elements equal to $+1$. (One may need to change signs on all elements.) The rows are then associated with runs and the columns with treatment factors. Each treatment factor is at two levels. These are main-effect plans. An additional sequence of OA[$8m$;2^{4m-1};3] is constructed by adding the foldover. These give the resolution IV plans, plans with main effects clear of two-factor interactions.

Construction Directly from the 2^N System

The first and simplest class here consists of OA[2^N;2^{2^N-1};2] for $N > 1$. These are the resolution III main effect plans. To construct them one writes out the full model for the 2^N factorial and then associates the $2^N - 1$ columns for effects and interactions with the new set of factors. These can all be obtained as a subset of the arrays derived from Hadamard matrices. We give them separately because of the ease of construction.

The OA[2^{N+1};2^{2^N-1};3] class is obtained via the foldover technique from the main-effect plans. These arrays yield resolution IV plans. However, we can do slightly better than this construction. If we proceed as in the construction of resolution III plans outlined above by writing out the full model and then assigning treatments to all the columns, but instead, assign treatments only to the columns corresponding to main effects, three-factor interactions, five-factor interactions, and so on, we obtain resolution IV plans with one extra treatment. The result is an OA[2^{N+1};2^{2^N};3] class. Just deleting a column from an OA[2^{N+1};2^{2^N};3] obtained in this way does not give us a member of the OA[2^{N+1};2^{2^N-1};3] class obtained via the foldover technique, even though they are the same size and have similar properties.

We are unaware of any easy trick to obtain the OA[2^N;2^k;4]. The necessary resolution V plans are difficult to obtain. One approach is to use the

Franklin–Bailey algorithm. It is easy to show that one can obtain the OA[2^4;2^5;4] from the fraction defined by $\mathcal{I} = ABCDE$ and the OA[2^5;2^6;4] from $\mathcal{I} = ABCDEF$. An OA[2^6;2^8;4] can be obtained from $\mathcal{I} = ADEFH = BCDEFG$, an OA[$2^7$;$2^{11}$;4] from $\mathcal{I} = ABCDEFGH = DEFGI = BCFGJ = ACEGK$, and an OA[$2^8$;$2^{17}$;4] from $\mathcal{I} = ABCDEFGHI = DEFGHJ = BCFGHK = ACEGHL = BDGHM = CEFHN = ADFHO = ABEHP = ABCDHQ$. These arrays are large in the sense that their use requires a large number of runs relative to the number of factors. However, they also permit the user to estimate two-factor interactions.

Construction Directly from the 3^N System

The simplest class here consists of OA[3^N;$3^{(3^N-1)/2}$;2] for $N > 1$. These are constructed from main-effect plans just as in the 2^N system. To construct them one writes out the full model for the 3^N factorial with all $(3^N - 1)/2$ two-degree-of-freedom effects and interactions. These columns, which contain 0's, 1's, and 2's, then form the columns of the array.

Strength 3 arrays are constructed in a similar manner, except that fewer of the columns are accepted. Assume that the columns are in standard order, i.e., A, B, AB, AB^2, C, AC, AC^2, BC, BC^2, D, and so on. To construct an OA[3^4;3^8;3] we begin with a full 3^4 factorial and keep only columns corresponding to A, B, C, D, ABC, ABD, ACD, and BCD. To construct an OA[3^5;3^{16};3], begin with a full 3^5 factorial and keep only columns corresponding to the five main effects, the 10 three-factor interactions (all letters to the unit power), and the one-five factor interaction with all letters to the unit power. The corresponding fractional factorial lets one estimate all 16 main effects free of two-factor interactions. This argument can be extended to produce, for example, the OA[3^6;3^{32};3]. However, the requirement for $3^6 = 729$ runs probably makes use of this array in experimental work rather unlikely. Plans in this class have their appeal to investigators who are studying many factors and are particularly interested in linear and quadratic effects.

Other Constructions Directly from the 2^N System

A class of OA[2^N;4^k;2] is constructed from the 2^N system by Wu (1989). In particular, if one writes out the full model for the 2^4 and retains the columns obtained by collapsing the columns associated with the pairs A and B, C and D, AC and BD, BC and ABD, and ABC and AD, one obtains the OA[2^4;4^5;2]. By collapsing we mean replacing elements $(-1, -1)$ by 0, $(-1, +1)$ by 1, $(+1, -1)$ by 2, and $(+1, +1)$ by 3. By retaining the columns obtained by collapsing column pairs (A, B), (D, E), $(BCD, ABCE)$, $(ABCD, ACE)$, (ACD, AE), (AD, BE), (ABD, BCE), (BD, CE), and (CD, ABE) in the complete 2^5, we obtain the OA[2^5;4^9;2]. Similarly, retaining the columns obtained by collapsing column pairs (A, B), (C, D), (AC, BD), (BC, ABD), (ABC, AD), (E, F), (AE, BF), (ABE, AF), (BE, ABF), (CE, DF), (CDE, CF), (DE, CDF), (ACE, BDF), $(ABCDE, ACF)$, $(BDE, ABCDF)$, $(ABCDE, ACF)$, $(BDE,$

$ABCDF$), ($BCE, ABDF$), ($ACDE, BCF$), ($ABDE, ACDF$), and ($ABCE$, ADF) in the complete 2^6, we obtain the OA[2^6;4^{21};2].

Classic Bose and Bush Construction

Bose and Bush (1952) gave a very general recipe for the construction of orthogonal arrays. We present a restricted version of their construction. Let q be a prime number. Let $\boldsymbol{T}$ be an $r \times k$ matrix with elements from $\{0, \ldots, q-1\}$, such that any d columns are linearly independent with arithmetic modulo q. Let $\boldsymbol{X}$ be a $q^r \times r$ matrix made up of all q^r r-tuples. Then the matrix product $\boldsymbol{XT}$, can be shown to be an OA[q^r;q^k;d].

As an example, consider $q = 2$, $r = 3$, $k = 4$,

$$\boldsymbol{T} = \begin{bmatrix} 1 & 0 & 0 & 1 \\ 0 & 1 & 0 & 1 \\ 0 & 0 & 1 & 1 \end{bmatrix}, \quad \text{and} \quad \boldsymbol{X} = \begin{bmatrix} 0 & 1 & 0 & 0 & 1 & 1 & 0 & 1 \\ 0 & 0 & 1 & 0 & 1 & 0 & 1 & 1 \\ 0 & 0 & 0 & 1 & 0 & 1 & 1 & 1 \end{bmatrix}^t.$$

The product $\boldsymbol{XT}$ is an OA[8;2^4;3].

23.3 OTHER CONSTRUCTION METHODS

At this point the reader may well question the real value of the orthogonal arrays since the process up to this point is essentially circular in nature. We used factorials to construct orthogonal arrays which we then use to construct fractional factorials. We now illustrate constructions that do not derive from factorial structures and illustrate some of the power of orthogonal arrays. For the purpose of these constructions we need some special mathematical tools.

23.3.1 Difference Matrix

The difference matrix turns out to be especially useful in the construction of orthogonal arrays. A difference matrix is an $r \times c$ matrix with elements from the set $\mathcal{S}$. This set $\mathcal{S}$ consists of s distinct elements and has operation addition (subtraction) defined. An example would be $\mathcal{S} = \{0, 1, 2\}$ with addition modulo 3. The defining characteristic of the difference matrix is that elements must be selected and arranged so that in the set of all differences between corresponding elements of any two selected columns, all elements of $\mathcal{S}$ appear equally frequently. In general, difference matrices are difficult to construct, although there are some general methods that lead to classes of difference matrices. A good reference is Beth et al. 1993. We will use the notation $\boldsymbol{D}_{r,c;s}$ to represent an $r \times c$ difference matrix with elements drawn from $\mathcal{S}$, which has s distinct elements and the arithmetic is the arithmetic associated with $\mathcal{S}$. We give a small collection of difference matrices that are useful in the construction of orthogonal arrays of reasonable size. A number of these are taken from Wang and Wu (1991). The first two are $\boldsymbol{D}_{3,3;3} = \begin{bmatrix} 0 & 0 & 0 \\ 0 & 1 & 2 \\ 0 & 2 & 1 \end{bmatrix}$, where $\mathcal{S} = \{0, 1, 2\}$ and addition is modulo 3 and

$$\boldsymbol{D}_{5,5;5} = \begin{bmatrix} 0 & 0 & 0 & 0 & 0 \\ 0 & 1 & 2 & 3 & 4 \\ 0 & 2 & 4 & 1 & 3 \\ 0 & 3 & 1 & 4 & 2 \\ 0 & 4 & 3 & 2 & 1 \end{bmatrix},$$

where $\mathcal{S} = \{0, 1, 2, 3, 4\}$ and addition is modulo 5. These two illustrate a simple method of construction. If s is a prime or a power of a prime number, the multiplication table associated with $\mathcal{S}$ but with addition(subtraction) as the arithmetic operation is a difference matrix. Note that permuting rows and columns of a difference matrix still leaves a difference matrix. Also adding a constant, modulo s to all elements in a row or a column leaves a difference matrix. Consequently, common practice is to write these matrices with first row and column all equal to zero.

A second construction is from Hadamard matrices by simply replacing -1's by 0's. Examples are

$$\boldsymbol{D}_{2,2;2} = \begin{bmatrix} 0 & 0 \\ 0 & 1 \end{bmatrix} \quad \text{and} \quad \boldsymbol{D}_{4,4;2} = \begin{bmatrix} 0 & 0 & 0 & 0 \\ 0 & 1 & 0 & 1 \\ 0 & 0 & 1 & 1 \\ 0 & 1 & 1 & 0 \end{bmatrix}.$$

Clearly, any Hadamard matrix can be used to construct a difference matrix.

For the next construction we need the Kronecker sum of two matrices. This operator is indicated by the $\oplus$ symbol. We write

$$\underset{r_a \times c_a}{\boldsymbol{A}} \oplus \underset{r_b \times c_b}{\boldsymbol{B}} = \underset{r_a r_b \times c_a c_b}{\boldsymbol{C}},$$

where $\boldsymbol{C}$ takes the form

$$\begin{bmatrix} a_{1,1}\boldsymbol{J} + \boldsymbol{B} & \cdots & a_{1,c_a}\boldsymbol{J} + \boldsymbol{B} \\ \vdots & & \vdots \\ a_{r_a,1}\boldsymbol{J} + \boldsymbol{B} & \cdots & a_{r_a,c_a}\boldsymbol{J} + \boldsymbol{B} \end{bmatrix}$$

and $\boldsymbol{J}$ is the $r_b \times c_b$ matrix with all elements equal to 1. Beth et al. 1993 show that $\boldsymbol{D}_{r_1,c_1;p} \oplus \boldsymbol{D}_{r_2,c_2;p}$ yields a difference matrix $\boldsymbol{D}_{r_1 r_2, c_1 c_2; p}$. In particular, this construction yields difference matrices of the form $\boldsymbol{D}_{p^n,p^n;p}$,where p is prime and $n > 1$.

Four other difference matrices constructed by assorted means are

$$\boldsymbol{D}_{6,6;3} = \begin{bmatrix} 0 & 0 & 0 & 0 & 0 & 0 \\ 0 & 1 & 2 & 0 & 1 & 2 \\ 0 & 2 & 1 & 1 & 0 & 2 \\ 0 & 0 & 2 & 1 & 2 & 1 \\ 0 & 2 & 0 & 2 & 1 & 1 \\ 0 & 1 & 1 & 2 & 2 & 0 \end{bmatrix},$$

$$\boldsymbol{D}_{10,10;5} = \begin{bmatrix} 0 & 0 & 0 & 0 & 0 & 0 & 0 & 0 & 0 & 0 \\ 0 & 1 & 2 & 3 & 4 & 1 & 2 & 3 & 4 & 0 \\ 0 & 2 & 4 & 1 & 3 & 0 & 2 & 4 & 1 & 3 \\ 0 & 3 & 1 & 4 & 2 & 2 & 0 & 3 & 1 & 4 \\ 0 & 4 & 3 & 2 & 1 & 2 & 1 & 0 & 4 & 3 \\ 0 & 1 & 0 & 2 & 2 & 3 & 4 & 4 & 3 & 1 \\ 0 & 2 & 2 & 0 & 1 & 4 & 3 & 1 & 3 & 4 \\ 0 & 3 & 4 & 3 & 0 & 4 & 1 & 2 & 2 & 1 \\ 0 & 4 & 1 & 1 & 4 & 3 & 3 & 2 & 0 & 2 \\ 0 & 0 & 3 & 4 & 3 & 1 & 4 & 1 & 2 & 2 \end{bmatrix},$$

$$\boldsymbol{D}_{12,12;3} = \begin{bmatrix} 0 & 0 & 0 & 1 & 1 & 0 & 0 & 1 & 0 & 2 & 2 & 0 \\ 0 & 0 & 0 & 0 & 2 & 0 & 2 & 0 & 2 & 0 & 0 & 1 \\ 0 & 0 & 1 & 0 & 0 & 2 & 1 & 2 & 0 & 0 & 1 & 0 \\ 0 & 0 & 2 & 2 & 0 & 1 & 0 & 0 & 1 & 1 & 0 & 0 \\ 0 & 1 & 2 & 2 & 0 & 0 & 1 & 1 & 2 & 0 & 2 & 2 \\ 0 & 1 & 2 & 1 & 2 & 1 & 2 & 2 & 2 & 2 & 1 & 0 \\ 0 & 1 & 0 & 0 & 2 & 2 & 0 & 2 & 1 & 1 & 2 & 2 \\ 0 & 1 & 1 & 2 & 1 & 2 & 2 & 0 & 0 & 2 & 0 & 2 \\ 0 & 2 & 1 & 2 & 1 & 0 & 0 & 2 & 2 & 1 & 1 & 1 \\ 0 & 2 & 1 & 0 & 0 & 1 & 2 & 1 & 1 & 2 & 2 & 1 \\ 0 & 2 & 2 & 1 & 2 & 2 & 1 & 1 & 0 & 1 & 0 & 1 \\ 0 & 2 & 0 & 1 & 1 & 1 & 1 & 0 & 1 & 0 & 1 & 2 \end{bmatrix},$$

$$\boldsymbol{D}_{12,6;6} = \begin{bmatrix} 0 & 0 & 0 & 0 & 0 & 0 \\ 0 & 1 & 3 & 2 & 4 & 0 \\ 0 & 2 & 0 & 1 & 5 & 2 \\ 0 & 3 & 1 & 5 & 4 & 2 \\ 0 & 4 & 3 & 5 & 2 & 1 \\ 0 & 5 & 5 & 3 & 1 & 1 \\ 0 & 0 & 2 & 3 & 2 & 3 \\ 0 & 1 & 2 & 4 & 0 & 5 \\ 0 & 2 & 5 & 2 & 3 & 4 \\ 0 & 3 & 4 & 1 & 1 & 4 \\ 0 & 4 & 1 & 0 & 3 & 5 \\ 0 & 5 & 4 & 4 & 5 & 3 \end{bmatrix}.$$

These four as well as a number of other difference matrices can be found in Wang and Wu (1991). At this stage we have recipes or examples of $k \times k$ difference matrices for $2 \leq k \leq 13$.

For future reference we also define the matrix

$$\boldsymbol{H}_{12} = \begin{bmatrix} 0 & 0 & 0 & 0 & 0 & 0 & 0 & 0 & 0 & 0 & 0 & 0 \\ 0 & 1 & 1 & 0 & 1 & 1 & 1 & 0 & 0 & 0 & 1 & 0 \\ 0 & 0 & 1 & 1 & 0 & 1 & 1 & 1 & 0 & 0 & 0 & 1 \\ 0 & 1 & 0 & 1 & 1 & 0 & 1 & 1 & 1 & 0 & 0 & 0 \\ 0 & 0 & 1 & 0 & 1 & 1 & 0 & 1 & 1 & 1 & 0 & 0 \\ 0 & 0 & 0 & 1 & 0 & 1 & 1 & 0 & 1 & 1 & 1 & 0 \\ 0 & 0 & 0 & 0 & 1 & 0 & 1 & 1 & 0 & 1 & 1 & 1 \\ 0 & 1 & 0 & 0 & 0 & 1 & 0 & 1 & 1 & 0 & 1 & 1 \\ 0 & 1 & 1 & 0 & 0 & 0 & 1 & 0 & 1 & 1 & 0 & 1 \\ 0 & 1 & 1 & 1 & 0 & 0 & 0 & 1 & 0 & 1 & 1 & 0 \\ 0 & 0 & 1 & 1 & 1 & 0 & 0 & 0 & 1 & 0 & 1 & 1 \\ 0 & 1 & 0 & 1 & 1 & 1 & 0 & 0 & 0 & 1 & 0 & 1 \end{bmatrix}, \qquad (23.3)$$

which can be obtained from the 12×12 Hadamard matrix by writing the row with all -1's first, replacing all -1's with 0's and replacing the first column with 0's.

Exercise 23.5: Verifying a Difference Matrix. Verify that $\boldsymbol{D}_{5,5;5}$ and $\boldsymbol{D}_{12,6;6}$ defined in Section 23.3.1 are difference matrices. □

23.3.2 Construction of Orthogonal Arrays Using Difference Matrices

Illustration of the Method

We now follow Wang and Wu (1991) to present a series of constructions for a collection of orthogonal arrays. The Kronecker sum utilizing difference matrices provides the major tool in these constructions. For example, it is simple to verify that

$$\begin{bmatrix} 0 \\ 1 \\ 2 \end{bmatrix} \oplus \boldsymbol{D}_{6,6;3} \qquad (23.4)$$

with the arithmetic modulo 3 is an OA[18;3^6;2]. This is an example of a small, very useful main-effect plan that cannot be obtained via projective geometries or Hadamard matrices. This array is of some historical interest, since it is one of the first orthogonal arrays that could not be obtained from projective geometries to be constructed. However, we again caution the potential user that even though interactions are not explicitly mentioned in the construction, the investigator must be aware of their potential existence. Existence or nonexistence of interactions is a property of experimental material and factors in the experiment and not of the design or experimental plan used.

Exercise 23.6: Resolvability in Orthogonal Arrays

(a) Verify that the 6×6 array used in (23.4) is a difference matrix.

(b) Construct the orthogonal array defined by expression (23.4) and establish the following:

1. That this really is an orthogonal array.
2. That one can set up the fractional factorial, a fraction of the 3^6 in six blocks of three if you put treatment combinations corresponding to lines 1, 7, and 13 in one block, 2, 8, and 14 in a second block, ..., 6, 12, and 18 in the last block . Each treatment has every level in each block an equal number of times. Every contrast among the levels of the first factor is orthogonal to every contrast among levels of the second factor.
3. Illustrate by example that one can take any column in the array and use it to construct blocks. Clearly, this reduces room for the number of factors in the experiment by one. □

Defining Asymmetrical Orthogonal Arrays

Before continuing on with the development of orthogonal arrays we digress to define asymmetrical orthogonal arrays. These have the same relationship to orthogonal array as asymmetrical factorials have to the usual factorials. We use the notation AOA$[N;s_1^{k_1}, \ldots ,s_r^{k_r};t]$ to represent an $N \times \sum k_i$ array of symbols with k_i columns having symbols from $\mathcal{S}_i$ which itself has s_i symbols for $i = 1, \ldots, r$ and t refers to the strength of the array. Recall that an array of strength t means that one can select an $N \times t$ submatrix from the array, and all possible row vectors of length t will be represented equally frequently. For example, an AOA$[N;2^{k_1},3^{k_2};2]$ is an array that corresponds to a factorial with k_1 factors each at two levels and k_2 factors each at three levels. Any pair of columns selected will have one of the sets of pairs $\{(0\ 0), (0\ 1), (1\ 0), (1\ 1)\}$, $\{(0\ 0), (0\ 1), (0\ 2), (1\ 0), (1\ 1), (1\ 2)\}$, $\{(0\ 0), (0\ 1), (1\ 0), (1\ 1), (2\ 0), (2\ 1)\}$, or $\{(0\ 0), (0\ 1), (0\ 2), (1\ 0), (1\ 1), (1\ 2), (2\ 0), (2\ 1), (2\ 2)\}$ represented equally frequently. This corresponds to a main-effect plan with some factors at two levels and other factors at three levels. Possibly the simplest example of such an asymmetric orthogonal array is

$$\begin{bmatrix} 0 & 0 & 0 & 1 & 1 & 1 \\ 1 & 1 & 2 & 0 & 1 & 2 \end{bmatrix}^t,$$

which in our notation is designated as AOA$[6;3^1,2^1;2]$.

Array (23.5) demonstrates construction of another asymmetrical orthogonal array.

$$\begin{bmatrix} \begin{bmatrix} 0 \\ 1 \\ 2 \end{bmatrix} \oplus \boldsymbol{D}_{6,6;3}, & \begin{bmatrix} 0 \\ 0 \\ 0 \end{bmatrix} \oplus \begin{bmatrix} 0 & 0 \\ 1 & 0 \\ 2 & 0 \\ 0 & 1 \\ 1 & 1 \\ 2 & 1 \end{bmatrix} \end{bmatrix} \tag{23.5}$$

It is simple to verify that array (23.5) is an AOA[18;3^7,2^1;2]. This construction illustrates several important features. First we notice that the second part of (23.5) is nothing more than stacking three copies of an orthogonal array. Additionally, putting two orthogonal arrays side by side often yields a new orthogonal array. In general, one must check to see that things match up correctly. In this case they do. These properties will be used repeatedly in subsequent sections.

It is also easy to verify that the two arrays

$$\begin{bmatrix} \begin{bmatrix} 0 \\ 1 \\ 2 \end{bmatrix} \oplus \boldsymbol{D}_{6,6;3}, & \begin{bmatrix} 0 \\ 0 \\ 0 \end{bmatrix} \oplus \begin{bmatrix} 0 \\ 1 \\ 2 \\ 3 \\ 4 \\ 5 \end{bmatrix} \end{bmatrix} \quad \text{and} \quad \begin{bmatrix} \begin{bmatrix} 0 \\ 1 \\ 2 \end{bmatrix} \oplus \boldsymbol{D}_{6,6;3}, & \begin{bmatrix} 0 \\ 0 \\ 0 \\ 0 \\ 0 \\ 0 \end{bmatrix} \oplus \begin{bmatrix} 0 \\ 1 \\ 2 \end{bmatrix} \end{bmatrix}$$

are examples of an AOA[18;3^6,6^1;2] and an OA[18;3^7;2], respectively. The latter is nothing more than an array obtained from (23.5) by deleting a column.

Formal Result

It is proved in Beth et al. 1993 that

$$\mathrm{OA}[as;\, s^k;\, 2] \oplus \boldsymbol{D}_{bs,c;s} \Rightarrow \mathrm{OA}[abs^2;\, s^{kc};\, 2]$$

where $\boldsymbol{D}_{bs,c;s}$ is a $bs \times c$ difference matrix with elements from the set $\mathcal{S} = \{0, 1, 2, \ldots, s-1\}$, s is a prime, and the arithmetic is modulo s. Note that $c \leq bs$. This result can actually be generalized in a number of ways. One is to let s be a power of a prime. With the tools developed to this point, we can construct a large number of very useful symmetric and asymmetric orthogonal arrays.

23.3.3 Series of Constructions

12-*Run Arrays*

We have already encountered the OA[12;2^{11};2] obtained from the 12×12 Hadamard matrix, which can be used as a main-effect plan with 11 factors.

In addition, we have the three arrays

$$
\begin{bmatrix}
0 & 0 & 0 \\
0 & 0 & 1 \\
0 & 0 & 2 \\
0 & 1 & 3 \\
0 & 1 & 4 \\
0 & 1 & 5 \\
1 & 1 & 0 \\
1 & 1 & 1 \\
1 & 1 & 2 \\
1 & 0 & 3 \\
1 & 0 & 4 \\
1 & 0 & 5
\end{bmatrix},
\quad
\begin{bmatrix}
0 & 0 \\
0 & 1 \\
0 & 2 \\
0 & 3 \\
1 & 0 \\
1 & 1 \\
1 & 2 \\
1 & 3 \\
2 & 0 \\
2 & 1 \\
2 & 2 \\
2 & 3
\end{bmatrix},
\quad
\begin{bmatrix}
0 & 0 & 0 & 0 & 0 \\
0 & 1 & 0 & 1 & 0 \\
1 & 0 & 1 & 1 & 0 \\
1 & 1 & 1 & 0 & 0 \\
0 & 0 & 1 & 1 & 1 \\
0 & 1 & 1 & 0 & 1 \\
1 & 0 & 0 & 1 & 1 \\
1 & 1 & 0 & 0 & 1 \\
0 & 0 & 1 & 0 & 2 \\
0 & 1 & 0 & 1 & 2 \\
1 & 0 & 0 & 0 & 2 \\
1 & 1 & 1 & 1 & 2
\end{bmatrix}.
$$

These three represent an AOA[12;2^2,6^1;2], an AOA[12;3^1,4^1;2], and an AOA[12;2^4,3^1;2], respectively. Each of these can be used as an experimental plan.

18-*Run Arrays*

We have already seen the three arrays constructed as

$$
\left[\begin{bmatrix} 0 \\ 1 \\ 2 \end{bmatrix} \oplus \boldsymbol{D}_{6,6;3}, \begin{bmatrix} 0 \\ 0 \\ 0 \end{bmatrix} \oplus \boldsymbol{A} \right],
$$

where $\boldsymbol{A}$ is one of

$$
\begin{bmatrix} 0 \\ 1 \\ 2 \\ 0 \\ 1 \\ 2 \end{bmatrix},
\quad
\begin{bmatrix} 0 & 0 \\ 1 & 0 \\ 2 & 0 \\ 0 & 1 \\ 1 & 1 \\ 2 & 1 \end{bmatrix},
\quad \text{or} \quad
\begin{bmatrix} 0 \\ 1 \\ 2 \\ 3 \\ 4 \\ 5 \end{bmatrix}.
$$

20-*Run Arrays*

In addition to the OA[20;2^{19};2] constructed by cyclically expanding the vector

$$
\begin{bmatrix} 1 & 1 & 0 & 0 & 1 & 1 & 1 & 1 & 0 & 1 & 0 & 1 & 0 & 0 & 0 & 0 & 1 & 1 & 0 \end{bmatrix}
$$

into a 19×19 matrix and then adding an initial row of 0's, and the trivial AOA[20;5^1,4^1;2], there are two other asymmetrical orthogonal arrays:

AOA[20; 2^2, 10^1; 2]

$$
= \begin{bmatrix}
0 & 0 & 0 & 0 & 0 & 0 & 0 & 0 & 0 & 0 & 1 & 1 & 1 & 1 & 1 & 1 & 1 & 1 & 1 & 1 \\
0 & 0 & 0 & 0 & 0 & 1 & 1 & 1 & 1 & 1 & 1 & 1 & 1 & 1 & 1 & 0 & 0 & 0 & 0 & 0 \\
0 & 1 & 2 & 3 & 4 & 5 & 6 & 7 & 8 & 9 & 0 & 1 & 2 & 3 & 4 & 5 & 6 & 7 & 8 & 9
\end{bmatrix}^t
$$

AOA[20; $2^8, 5^1$; 2]

$$
= \begin{bmatrix}
0 & 0 & 0 & 0 & 1 & 1 & 1 & 1 & 2 & 2 & 2 & 2 & 3 & 3 & 3 & 3 & 4 & 4 & 4 & 4 \\
0 & 0 & 1 & 1 & 0 & 0 & 1 & 1 & 0 & 0 & 1 & 1 & 0 & 0 & 1 & 1 & 0 & 0 & 1 & 1 \\
0 & 1 & 0 & 1 & 0 & 1 & 0 & 1 & 0 & 1 & 0 & 1 & 0 & 1 & 0 & 1 & 0 & 1 & 0 & 1 \\
0 & 1 & 1 & 0 & 0 & 0 & 1 & 1 & 1 & 1 & 0 & 0 & 0 & 1 & 0 & 1 & 1 & 0 & 1 & 0 \\
0 & 1 & 0 & 1 & 1 & 0 & 0 & 1 & 0 & 0 & 1 & 1 & 1 & 1 & 0 & 0 & 1 & 0 & 1 & 0 \\
0 & 1 & 0 & 1 & 1 & 0 & 1 & 0 & 1 & 0 & 0 & 1 & 0 & 0 & 1 & 1 & 1 & 1 & 0 & 0 \\
0 & 0 & 1 & 1 & 0 & 1 & 0 & 1 & 1 & 0 & 1 & 0 & 1 & 0 & 0 & 1 & 1 & 1 & 0 & 0 \\
0 & 1 & 1 & 0 & 1 & 1 & 0 & 0 & 1 & 0 & 0 & 1 & 1 & 0 & 0 & 1 & 0 & 0 & 1 & 1 \\
0 & 1 & 1 & 0 & 1 & 0 & 0 & 1 & 1 & 0 & 1 & 0 & 0 & 1 & 1 & 0 & 0 & 1 & 0 & 1
\end{bmatrix}^t .
$$

This last array can be used to construct a plan for a main effects experiment with eight factors, each at two levels in five blocks of four.

24-Run Arrays

Here we have three arrays constructed as

$$
\left[\begin{bmatrix} 0 \\ 1 \end{bmatrix} \oplus \boldsymbol{H}_{12}, \begin{bmatrix} 0 \\ 0 \end{bmatrix} \oplus \boldsymbol{A} \right],
$$

where $\boldsymbol{H}_{12}$ is as defined in (23.3). Notice that for columns 2 to 12 the $\begin{bmatrix} 0 \\ 1 \end{bmatrix} \oplus \boldsymbol{H}_{12}$ operation can be interpreted as adding a foldover to a 12-run Plackett–Burman plan. This frees all interactions among any factors assigned to those columns as in a resolution IV plan. Now for $\boldsymbol{A}$ one has the choice of $[0\ 1\ 2\ \cdots\ 10\ 11]^t$, AOA[12;$2^2$,$6^1$;2], AOA[12;$3^1$,$4^1$;2], or AOA[12;$2^4$,$3^1$;2]. The first gives us the AOA[24;2^{12},12^1;2]. This is a saturated plan with all potential two-factor interactions between pairs of factors (at two levels) aliased with contrasts among the levels of the factor at 12 levels. Also note that this array can be used as the basis to construct a 12×2 row–column design for 11 two-level factors. Examples of possible applications would be field experiments with two parallel rows of 12 plots each or greenhouse experiments.

Selecting AOA[12;2^2,6^1;2] for $\boldsymbol{A}$ as defined earlier in this section gives us the AOA[24;2^{14},6^1;2] directly. However, if we rearrange the columns of the $\boldsymbol{A}^t$ matrix as

$$
\begin{bmatrix}
0 & 1 & 0 & 1 & 0 & 0 & 0 & 1 & 1 & 1 & 0 & 1 \\
1 & 1 & 0 & 1 & 0 & 0 & 1 & 1 & 0 & 0 & 1 & 0 \\
5 & 0 & 0 & 1 & 1 & 2 & 3 & 2 & 3 & 4 & 4 & 5
\end{bmatrix},
$$

one can collapse columns 1, 2, and 13 into a column with four levels to obtain the AOA[24;6^1,4^1,2^{11};2]. Notice that any one of the three columns mentioned can be interpreted as the interaction of the remaining two columns. We perform the collapsing by taking any pair of columns and replacing 0 0 $\Rightarrow$ 0, 0 1 $\Rightarrow$ 1, 1 0 $\Rightarrow$ 2, and 1 1 $\Rightarrow$ 3, and deleting the third column. This array can be used as the basis for a 4×6 row–column design for 11 factors, each at two levels. Also, from the foldover nature, we know that potential two-factor interactions between pairs of the first 10 two-level factors are aliased with contrasts among levels of the four- and six-level factors or their interaction. Randomization in the row–column design is rather limited. One assigns treatment combinations to the cells in a 4×6 table and then randomizes rows and columns independently.

The last two choices for $\boldsymbol{A}$ yield AOA[$24;2^{12},3^1,4^1;2$] and AOA[$24;2^{16},3^1;2$] directly. Smaller arrays (fewer factors) can always be obtained by dropping columns.

Exercise 23.7: Twelve-Run AOA

(a) Using the recipe given, construct the AOA[$24;6^1,4^1,2^{11};2$].
(b) Indicate the exact treatment combinations assigned for the factors of the 2^{11} in a row–column design with four rows and six columns.
(c) Describe a possible randomization scheme.
(d) Give the lines and degrees of freedom in the analysis of variance.
(e) Show that within the set consisting of the first 10 two-level factors, main effects are free of two-factor interactions. □

36-*Run Arrays*

A collection of six different 36-run plans can be obtained from

$$\left[\begin{bmatrix}0\\1\\2\end{bmatrix} \oplus \boldsymbol{D}_{12,12;3}, \begin{bmatrix}0\\0\\0\end{bmatrix} \oplus \boldsymbol{A}\right],$$

where $\boldsymbol{A}$ is one of $[0\ 1\ 2\ 0\ 1\ 2\ 0\ 1\ 2\ 0\ 1\ 2]^t$, $[0\ 1\ \cdots\ 10\ 11]^t$, OA[$12;2^{11};2$], AOA[$12;2^2,6^1;2$], AOA[$12;3^1,4^1;2$], or AOA[$12;2^4,3^1;2$], which we prefer to write as AOA[$12;3^1,2^4;2$] because of the order of the factors implied or the matrix obtained by dropping the first column of $\boldsymbol{H}_{12}$ defined in (23.3). The results are OA[$36;3^{13};2$], AOA[$36;3^{12},12^1;2$], AOA[$36;3^{12},2^{11};2$], AOA[$36;3^{12},2^2,6^1;2$], AOA[$36;3^{13},4^1;2$], and AOA[$36;3^{13},2^4;2$].

One can use these arrays to construct plans consisting of three blocks of 12, 12 blocks of 3, two blocks of 12, six blocks of 6, or four blocks of 9. Additionally, one can construct the following row–column designs:

1. 3×3 with four treatment combinations per cell
2. 3×12 with one treatment combination per cell
3. 3×6 with two treatment combinations per cell
4. 2×6 with three treatment combinations per cell
5. 3×4 with three treatment combinations per cell
6. 3×2 with six treatment combinations per cell

Whenever possible, the assignments within cells should be randomized, as should row and column assignments.

If one treats the factors as qualitative, one can make no claim about potential two-factor interactions. If they should exist, they will be partially aliased with other main effects, somewhat in the nature of interactions in the 12-run Plackett–Burman plan. If one of these designs is used in an exploratory setting and one expects and finds few main effects to be active, one can resort to something like a stepwise regression procedure outlined in the chapter on supersaturated plans or the Box and Meyer (1993) procedure to check for possible two-factor interactions of active factors. If the factors are qualitative and equally spaced, some of the linear × linear interactions become estimable. However, they are not orthogonal to main effects. If the investigator anticipates such an analysis, the advice is to construct a set of fake data and test the anticipated statistical analysis to make sure that quantities that are expected to be important really are estimable before conducting the experiment.

Exercise 23.8: Three-Row, Four-Column Design. Using the AOA[36;3^{13}, 4^1;2] as a base, construct the row–column design with three rows and four columns and three entries per cell to accommodate a main-effect plan with up to 12 factors, each at three levels. □

40-*Run Arrays*

Here we have four arrays constructed as

$$\left[\begin{bmatrix}0\\1\end{bmatrix} \oplus \boldsymbol{H}_{20}, \begin{bmatrix}0\\0\end{bmatrix} \oplus \boldsymbol{A}\right],$$

where $\boldsymbol{A}$ is one of $[\ 0\ 1\ \cdots\ 18\ 19\ ]^t$, AOA[20;$2^2$,$10^1$;2], AOA[20;$2^8$,$5^1$;2], or OA[20;$2^{19}$;2].

The first choice for $\boldsymbol{A}$ yields AOA[40;2^{19},20^1;2]. Notice that the first ⊕ operation is comparable to adding a foldover set to a resolution III plan, and consequently, the first 20 columns function as a resolution IV plan; i.e., main effects are all free of two-factor interactions. The complete array can also be used to construct a design with 20 blocks of two, or a 20 × 2 row–column design.

The second choice for $\boldsymbol{A}$ yields an AOA[40;2^{22},10^1;2]. However, if one sorts the rows of the AOA[20;2^2,10^1;2] so that the first two columns duplicate the second and third columns of the $\boldsymbol{H}_{20}$ array, it is easy to see that columns 1, 2, and 21 of the complete construction form a set in the sense that one column is the generalized interaction of the remaining two. Consequently, these three can be collapsed to give one column with four levels. The result is an AOA[40;2^{21},4^1,10^1;2]. If necessary, this can be used to construct a 10 × 4 row–column design with a resolution IV fraction with 19 factors each at two levels.

The third choice for $\boldsymbol{A}$ yields an AOA[40;2^{27},5^1;2] directly with any row permutation of the AOA[20;2^8,5^1;2] earlier in this section. However, again, if the rows are sorted so that the first two columns duplicate columns 2 and 3 of H_{20}, columns 1, 2, and 21 of the constructed array can be collapsed into a column with four elements. The result is an AOA[40;2^{25},4^1,5^1;2]. Since the first 19 remaining two-factor columns have the resolution IV property, it is recommended that these be selected first if not all columns are to be used.

Finally, the last choice given for $\boldsymbol{A}$ yields the OA[40;2^{36};2]. This array can be collapsed to an AOA[40;2^{36},4^1;2] by combining columns 1, 2, and 21 of the constructed array.

48-*Run Arrays*

The base construction for one set of 48-run arrays is

$$\left[\begin{bmatrix} 0 & 0 & 0 \\ 0 & 1 & 1 \\ 1 & 0 & 1 \\ 1 & 1 & 0 \end{bmatrix} \oplus \boldsymbol{H}_{12}, \begin{bmatrix} 0 \\ 0 \\ 0 \\ 0 \end{bmatrix} \oplus \boldsymbol{A}\right],$$

where $\boldsymbol{H}_{12}$ is defined in (23.3).

Before looking at specific constructions, we first examine

$$\begin{bmatrix} 0 & 0 & 0 \\ 0 & 1 & 1 \\ 1 & 0 & 1 \\ 1 & 1 & 0 \end{bmatrix} \oplus \boldsymbol{H}_{12}$$

in some detail. Notice that this can be thought of as three distinct foldover constructions (combined with duplications). Also columns 1, 13, and 25 have very distinct patterns. The implication is that up to three suitably chosen sets of three columns can be combined into columns with four elements. Three suitable sets are {1, 16, 23}, {2, 14, 25}, and {3, 13, 27}. The foldover nature also throws potential interactions between pairs of columns (not involving columns 1, 13, or 25) onto columns orthogonal to the entire set of 36. The 33-column array (array of strength 3) can be used to good advantage in an exploratory study since there is room to look for potential two-factor interactions in the remaining columns using a stepwise regression search algorithm.

We now have the following:

1. Using an OA[12;2^{11};2] for $\boldsymbol{A}$ yields either an OA[48;2^{47};2] or an AOA[48;2^{47-3m},4^m;2], where $m = 1, 2,$ or 3.
2. Using an AOA[12;2^2,6^1;2] for $\boldsymbol{A}$ yields one of the AOA[48;2^{38-3m},4^m,6^1;2], where $m = 0, 1, 2,$ or 3.
3. Using an AOA[12;2^4,3^1;2] for $\boldsymbol{A}$ yields one of the AOA[48;2^{40-3m},4^m,3^1;2], where $m = 0, 1, 2,$ or 3.
4. Using an AOA[12;3^1,4^1;2] for $\boldsymbol{A}$ yields one of the AOA[48;2^{36-3m},4^{m+1},3^1;2], where $m = 0, 1, 2,$ or 3.

5. Using $[0\ 1\ \ldots\ 11]^t$ in place of $\boldsymbol{A}$ yields one of the AOA[48;2^{36-3m},4^m,12^1;2], where $m = 0, 1, 2,$ or 3.

It is also possible to modify the construction to provide a column for a factor with eight levels by sorting the rows of $\boldsymbol{A}$ so that the first column matches exactly with the second column (first nonzero column) of $\boldsymbol{H}_{12}$. If one uses an OA[12;2^{12};2] with rows properly sorted, columns 1, 13, 25, 2, 14, 26, and 37 can be collapsed into one column with eight levels. A suitable rule is to use columns 2, 14, and 26 and make the replacement

$$
\begin{array}{llll}
0\,0\,0 \Rightarrow 0 & & 1\,0\,0 \Rightarrow 4 \\
0\,0\,1 \Rightarrow 1 & & 1\,0\,1 \Rightarrow 5 \\
0\,1\,0 \Rightarrow 2 & & 1\,1\,0 \Rightarrow 6 \\
0\,1\,1 \Rightarrow 3 & & 1\,1\,1 \Rightarrow 7.
\end{array}
$$

The result is an AOA[48;2^{40},8^1;2]. Similar computations with an AOA[12;2^2,6^1;2] yields an AOA[48;2^{31},6^1,8^1;2] and with an AOA[12;2^4,3^1;2] in place of $\boldsymbol{A}$ yields an AOA[48;2^{33},3^1,8^1;2].

It is also possible to construct a number of orthogonal arrays with more factors at four levels if the problem requires the ability to estimate linear and quadratic effects of the factors. However, some of the preliminary computations are slightly more complex since 4 is not a prime number. A simple device is to represent the symbols for levels (elements of $\mathcal{S}$) by a pair of numbers, i.e., by (0,0), (0,1), (1,0), and (1,1). Addition in this new system by coordinate, i.e., $(a_1, a_2) + (b_1, b_2) = ((a_1 + b_1)$ (modulo 2), $(a_2 + b_2)$ (modulo 2)). For example, $(1, 0) + (1, 1) = (0, 1)$. The base calculation for our arrays now is

$$
\left[\begin{bmatrix} (0,0) \\ (0,1) \\ (1,0) \\ (1,1) \end{bmatrix} \oplus \boldsymbol{D}_{12,12;4}, \begin{bmatrix} 0 \\ 0 \\ 0 \\ 0 \end{bmatrix} \oplus \boldsymbol{A} \right],
$$

where $\boldsymbol{D}_{12,12;4}$ is

$$
\begin{bmatrix}
(0,0) & (0,0) & (0,0) & (0,0) & (0,0) & (0,0) & (0,0) & (0,0) & (0,0) & (0,0) & (0,0) & (0,0) \\
(0,0) & (0,0) & (0,0) & (0,1) & (0,1) & (0,1) & (1,1) & (1,1) & (1,1) & (1,0) & (1,0) & (1,0) \\
(0,0) & (0,0) & (0,0) & (1,1) & (1,1) & (1,1) & (1,0) & (1,0) & (1,0) & (0,1) & (0,1) & (0,1) \\
(0,0) & (1,1) & (0,1) & (1,0) & (0,1) & (1,1) & (0,1) & (1,0) & (0,0) & (1,1) & (0,0) & (1,0) \\
(0,0) & (1,1) & (0,1) & (1,1) & (1,0) & (0,1) & (0,0) & (0,1) & (1,0) & (1,0) & (1,1) & (0,0) \\
(0,0) & (1,1) & (0,1) & (0,1) & (1,1) & (1,0) & (1,0) & (0,0) & (0,1) & (0,0) & (1,0) & (1,1) \\
(0,0) & (0,1) & (1,0) & (1,1) & (0,0) & (1,0) & (0,1) & (0,0) & (1,1) & (0,1) & (1,1) & (1,0) \\
(0,0) & (0,1) & (1,0) & (1,0) & (1,1) & (0,0) & (1,1) & (0,1) & (0,0) & (1,0) & (0,1) & (1,1) \\
(0,0) & (0,1) & (1,0) & (0,0) & (1,0) & (1,1) & (0,0) & (1,1) & (0,1) & (1,1) & (1,0) & (0,1) \\
(0,0) & (1,0) & (1,1) & (0,1) & (1,0) & (0,0) & (0,1) & (1,1) & (1,0) & (0,1) & (0,0) & (1,1) \\
(0,0) & (1,0) & (1,1) & (0,0) & (0,1) & (1,0) & (1,0) & (0,1) & (1,1) & (1,1) & (0,1) & (0,0) \\
(0,0) & (1,0) & (1,1) & (1,0) & (0,0) & (0,1) & (1,1) & (1,0) & (0,1) & (0,0) & (1,1) & (0,1)
\end{bmatrix}.
$$

Once the complete array has been constructed, the experimenter assigns levels of the factors to the $(0, 0)$, $(0, 1)$, $(1, 0)$, and $(1, 1)$.

Without any need for collapsing columns the construction using:

1. An OA[12;2^{12};2] for $\boldsymbol{A}$ produces an AOA[48;4^{12},2^{11};2].
2. An AOA[12;2^2,6^1;2] for $\boldsymbol{A}$ produces an AOA[48;4^{12},6^1,2^2;2].
3. An AOA[12;3^1,2^4;2] for $\boldsymbol{A}$ produces an AOA[48;4^{12},3^1,2^4;2].
4. An AOA[12;4^1,3^1;2] for $\boldsymbol{A}$ produces an AOA[48;4^{13},3^1;2].
5. $[0\ 1\ \cdots\ 11]^t$ for $\boldsymbol{A}$ produces an AOA[48;4^{12},12^1;2].

50-*Run Arrays*

Two 50-run arrays, an AOA[50;5^{11},2^1;2] and an AOA[50;10^1,5^{10};2], for factors at five levels can be constructed from

$$\left[\begin{bmatrix}0\\1\\2\\3\\4\end{bmatrix} \oplus \boldsymbol{D}_{10,10;5},\ \begin{bmatrix}0\\0\\0\\0\\0\end{bmatrix} \oplus \boldsymbol{A}\right]$$

by using $\begin{bmatrix}0 & 0 & 0 & 0 & 0 & 1 & 1 & 1 & 1 & 1\\ 0 & 1 & 2 & 3 & 4 & 0 & 1 & 2 & 3 & 4\end{bmatrix}^t$ and $[0\ 1\ \cdots\ 9]^t$ for $\boldsymbol{A}$.

54-*Run Arrays*

If one requires an array for a large number of factors each at three levels (say, to estimate linear and quadratic effects), one can use

$$\left[\begin{bmatrix}0&0&0&0\\0&1&1&2\\0&2&2&1\\1&0&1&1\\1&1&2&0\\1&2&0&2\\2&0&2&2\\2&1&0&1\\2&2&1&0\end{bmatrix} \oplus \boldsymbol{D}_{6,6;3},\ \begin{bmatrix}0\\0\\0\\0\\0\\0\\0\\0\\0\end{bmatrix} \oplus \boldsymbol{A}\right], \tag{23.6}$$

where $\boldsymbol{A}$ is either AOA[6;3^1,2^1;2] or $[0\ 1\ \cdots\ 5]^t$. The resulting arrays are AOA[54;3^{25},2^1;2] and AOA[54;3^{24},6^1;2]. In addition, notice that the four columns in the initial array in (23.6) can be collapsed into one column with nine elements. In addition, since $\boldsymbol{D}_{6,6;3}$ contains an initial column with only 0's, these four columns get copied six times into the first four columns of the resulting array. Clearly, the first four columns of the resulting array can be collapsed into one column with nine elements. This can be done for both choices of $\boldsymbol{A}$, yielding AOA[54;3^{21},9^1,2^1;2] and AOA[54;3^{20},9^1,6^1;2].

96-*Run Arrays*

The last set of arrays we present in this section are 96-run arrays. Among other things, these arrays can be used as the basis for designs that fit on the 8×12 ELISA plates. The base construction here is

$$\left[\begin{bmatrix}(0,0)\\(0,1)\\(1,0)\\(1,1)\\(0,0)\\(0,1)\\(1,0)\\(1,1)\end{bmatrix} \oplus \boldsymbol{D}_{12,12;4}, \begin{bmatrix}0&0&0&0\\0&0&1&1\\0&1&0&1\\1&0&0&1\\0&1&1&0\\1&0&1&0\\1&1&0&0\\1&1&1&1\end{bmatrix} \oplus \boldsymbol{H}_{12}, \begin{bmatrix}0\\0\\0\\0\\0\\0\\0\\0\end{bmatrix} \oplus \boldsymbol{A}\right],$$

where $\boldsymbol{A}$ is an array with 12 rows. Suitable choices for $\boldsymbol{A}$ are AOA$[12;2^4,3^1;2]$, AOA$[12;2^2,6^1;2]$, AOA$[12;2^{11};2]$, and $[0\ 1\ \cdots\ 11]^t$. Direct construction without any collapsing of columns yields an AOA$[96;4^{12},2^{52},3^1;2]$, an AOA$[96;4^{12},2^{50},6^1;2]$, an AOA$[96;4^{12},2^{49};2]$, and an AOA$[96;4^{12},2^{48}, 12^1;2]$.

If instead of $\boldsymbol{H}_{12}$ one uses

$$\boldsymbol{H}_{12}^* = \begin{bmatrix}0&0&0&0&0&0&0&0&0&0&0&0\\0&0&1&0&1&1&1&1&1&0&0&0\\0&1&0&1&1&1&1&0&0&1&0&0\\0&1&1&1&0&1&0&1&0&0&1&0\\0&0&0&1&1&0&1&1&0&0&1&1\\0&0&1&1&0&0&1&0&1&1&1&0\\0&1&0&0&1&0&0&1&1&1&1&0\\0&1&1&0&0&0&1&1&0&1&0&1\\0&0&0&1&0&1&0&1&1&1&0&1\\0&0&1&0&1&1&0&0&0&1&1&1\\0&1&0&0&0&1&1&0&1&0&1&1\\0&1&1&1&1&0&0&0&1&0&0&1\end{bmatrix}$$

and $\boldsymbol{A}$ = AOA$[12;2^4,3^1;2]$, column sets {13, 14, and 61}, {25, 26, and 62}, {37, 38, and 63}, and {49, 50, and 64} can each be collapsed into columns with four elements. The result is an AOA$[96;4^{16},2^{40},3^1;2]$. Notice that $\boldsymbol{H}_{12}^*$ is obtained from a Hadamard matrix, but it is sorted so that the first four nonzero columns of $\boldsymbol{H}_{12}^*$ are duplicates of the first four columns of AOA$[12;2^4,3^1;2]$. In a similar manner we can obtain an AOA$[96;4^{14},2^{44},6^1;2]$ if we sort rows of $\boldsymbol{A}$ =AOA$[12;2^2,6^1;2]$ so that the first two columns duplicate $\boldsymbol{H}_{12}^*$ or an AOA$[96;4^{16},2^{47};2]$ if the $\boldsymbol{A}$ consists of the 11 nonzero columns of $\boldsymbol{H}_{12}^*$.

23.4 NEARLY ORTHOGONAL ARRAYS

23.4.1 Concepts and Definitions

There is no really satisfactory definition for a nearly orthogonal array, except that the array is nearly orthogonal. The problem is that there are many ways in which an array can deviate from orthogonality. The trick is to find arrays that can be used in cases where no orthogonal array exists, yet lead to experiments that can be interpreted reasonably well. Orthogonality in this case can be interpreted as uncorrelated estimates of effects. Before using any of these plans, the investigator should always evaluate the situation carefully and make sure that there is a reasonable chance that the results of the experiment will be worth the effort or whether the best alternative is simply to wait until more resources for an adequate orthogonal experiment are available. A good base reference on constructing and evaluating nonorthogonal arrays is Wang and Wu (1992).

We will use the notation NOA[N;$s_1^{k_1}$,$s_2^{k_2}$, ... ,$s_r^{k_r}$;2] for a nearly orthogonal array with k_1 factors with s_1 levels, k_2 factors with s_2 levels, ... , k_r factors with s_r levels, and which would be of strength 2 if it were orthogonal.

Three criteria that are often looked for in evaluating nonorthogonal plans are:

1. That the nonorthogonality be restricted to a subset of the factors.
2. That, in some sense, the maximum loss of information on each individual factor be minimized. A reasonable criterion used is

$$D_i = (\boldsymbol{x}_i^t\boldsymbol{x}_i - \boldsymbol{x}_i^t\boldsymbol{X}_{(i)}(\boldsymbol{X}_{(i)}^t\boldsymbol{X}_{(i)})^{-1}\boldsymbol{X}_{(i)}^t\boldsymbol{x}_i)/\boldsymbol{x}_i^t\boldsymbol{x}_i,$$

   which can be interpreted as the fraction of the information available on the ith factor relative to the information that would have been available had that factor been orthogonal to all other factors in the experiment. This criterion is based on the model

$$\boldsymbol{y} = \boldsymbol{X}\boldsymbol{\beta} + \boldsymbol{\epsilon},$$

   where Var($\boldsymbol{\epsilon}$) $= \boldsymbol{I}\sigma_\epsilon^2$, $\boldsymbol{x}_i$ is the column of $\boldsymbol{X}$ corresponding to factor i and $\boldsymbol{X}_{(i)}$ is the matrix obtained from $\boldsymbol{X}$ by deleting the column $\boldsymbol{x}_i$.
3. That the overall quantity

$$D = |\widetilde{\boldsymbol{X}}^t\widetilde{\boldsymbol{X}}|^{1/k}$$

   be maximized, with k the number of columns in $\boldsymbol{X}$ and $\widetilde{\boldsymbol{X}}$ the standardized form of $\boldsymbol{X}$,

$$\widetilde{\boldsymbol{X}} = \left(\frac{\boldsymbol{x}_1}{\|\boldsymbol{x}_1\|}, \frac{\boldsymbol{x}_2}{\|\boldsymbol{x}_2\|}, \ldots, \frac{\boldsymbol{x}_k}{\|\boldsymbol{x}_k\|}\right),$$

   where $|\,\boldsymbol{.}\,|$ denotes the determinant and $\|\boldsymbol{x}\|$ the $\sqrt{\boldsymbol{x}^t\boldsymbol{x}}$.

Statistical Analysis Warning

Before going into too much detail with nearly orthogonal arrays, we need to interject a note of warning. One of the big advantages of data generated from experiments based on any orthogonal plan, whether obtained via the classical geometric approach or manipulating orthogonal arrays, is that simply calculating treatment means is often a very adequate statistical analysis. This is especially true if interactions are negligible and the experimental error is small. This is not true in the case of nonorthogonal experiments. Here, simply looking at treatment means is not only inadequate, but may be quite misleading. A proper least squares (multiple regression) analysis is required to avoid the danger of misinterpretation, and even then there is a danger of ambiguous results. The investigator must always remember this as one of the shortcomings of experiments based on nearly orthogonal arrays. In fact, this is exactly the same problem that arises in the interpretation of results from many unplanned studies.

23.4.2 Collection of Small Nearly Orthogonal Arrays

10-*Run Arrays*

Ten experimental units presents a relatively awkward situation. Except for the trivial cases of one factor at five levels and/or one factor at two levels, nothing really fits. The following array allows up to seven two-level factors in a nonorthogonal arrangement:

$$\begin{bmatrix} 0 & 0 & 0 & 0 & 0 & 0 & 0 \\ 0 & 0 & 0 & 0 & 0 & 0 & 0 \\ 0 & 0 & 0 & 1 & 1 & 1 & 1 \\ 0 & 1 & 1 & 0 & 0 & 1 & 1 \\ 0 & 1 & 1 & 1 & 1 & 0 & 0 \\ 1 & 0 & 1 & 0 & 1 & 0 & 1 \\ 1 & 0 & 1 & 1 & 0 & 1 & 0 \\ 1 & 1 & 0 & 0 & 1 & 1 & 0 \\ 1 & 1 & 0 & 1 & 0 & 0 & 1 \\ 1 & 1 & 1 & 1 & 1 & 1 & 1 \end{bmatrix}.$$

If fewer than seven factors are present, columns are selected beginning from the left. The array was constructed by sequentially adding assignments closest to being orthogonal, using the D_i criterion. The D_i values all exceed .888.

12-*Run Arrays*

We have seen that if 12 experimental units are available, the investigator has a choice of a Plackett–Burman plan with up to 11 two-level factors or an asymmetrical orthogonal array with one three-level factor and up to four two-level factors. If the problem demands more than 11 two-level factors (and only two-level factors), the methods discussed in Chapter 21 are available. Wang and Wu

(1992) present several arrays suitable for one factor at three levels and up to six factors, each at two levels. Two NOA[12;3^1,2^6;2]'s are obtained from

$$\begin{bmatrix} 0 & 0 & 0 & 0 & 0 & 0 & 0 & 0 \\ 0 & 0 & 1 & 0 & 1 & 1 & 0 & 1 \\ 0 & 1 & 0 & 1 & 1 & 1 & 1 & 0 \\ 0 & 1 & 1 & 1 & 0 & 0 & 1 & 1 \\ 1 & 0 & 0 & 1 & 1 & 0 & 1 & 0 \\ 1 & 0 & 1 & 1 & 0 & 1 & 1 & 1 \\ 1 & 1 & 0 & 0 & 1 & 0 & 0 & 1 \\ 1 & 1 & 1 & 0 & 0 & 1 & 0 & 0 \\ 2 & 0 & 0 & 1 & 0 & 1 & 0 & 1 \\ 2 & 0 & 1 & 0 & 1 & 0 & 1 & 0 \\ 2 & 1 & 0 & 0 & 0 & 1 & 1 & 1 \\ 2 & 1 & 1 & 1 & 1 & 0 & 0 & 0 \end{bmatrix}$$

by selecting the first six columns and one of the last two. In both arrays, the two-level factor columns are all orthogonal to the three-level factor column. The first array has only the fourth and sixth and fifth and seventh columns nonorthogonal, while the second array has the fifth, sixth and seventh columns nonorthogonal. If the investigator feels that the experiment involves three factors (each at two levels) that are likely to be important, the second array is definitely to be recommended.

If the model is set up to estimate linear and quadratic effects for the first factor, the D_s values for the factors are:

A_L	A_Q	B	C	D	E	F	G
1	1	1	1	.89	.89	.89	.89
1	1	1	1	1	.83	.83	.83

An alternative nonorthogonal array that allows for one factor to measure linear and quadratic effects and up to nine two-level factors can be obtained from the 12-run Plackett–Burman plan by collapsing the first two columns:

$$\begin{array}{ll} 0\,0 \Rightarrow 0 & 1\,0 \Rightarrow 3 \\ 0\,1 \Rightarrow 1 & 1\,1 \Rightarrow 2. \end{array}$$

Only linear and quadratic effects can be estimated from the first column. The D_s values for this array are

A_L	A_Q	B	C	D	E	F	G	I	J	K
.8	1	.97	.97	.97	.97	.97	.97	.97	.97	.97

We note that the D_s values for the A_L and A_Q values are not directly comparable for the two arrays, since in the first two arrays the levels of the three-level factor are optimally spaced to estimate quadratic effects, while in the third they are not. The price is an increase in variance by a factor of 10/9.

14-*Run Arrays*

In animal research it occasionally happens that investigators find that in a building designed for experimental work with $2^4 = 16$ pens, the two end pens tend to be different, leaving 14 relatively homogeneous pens. Only one factor with seven levels and/or one factor with two levels fit properly. The array with 12 columns,

$$
\begin{bmatrix}
0 & 0 & 0 & 0 & 0 & 0 & 0 & 0 & 0 & 0 & 0 & 0 \\
0 & 0 & 0 & 0 & 0 & 0 & 0 & 0 & 0 & 1 & 1 & 1 \\
0 & 0 & 0 & 0 & 0 & 0 & 1 & 1 & 1 & 0 & 0 & 0 \\
0 & 0 & 0 & 1 & 1 & 1 & 0 & 0 & 0 & 0 & 0 & 0 \\
0 & 1 & 1 & 0 & 1 & 1 & 0 & 1 & 1 & 0 & 1 & 1 \\
0 & 1 & 1 & 1 & 0 & 1 & 1 & 0 & 1 & 1 & 0 & 1 \\
0 & 1 & 1 & 1 & 1 & 0 & 1 & 1 & 0 & 1 & 1 & 0 \\
1 & 0 & 1 & 0 & 1 & 1 & 1 & 1 & 0 & 1 & 0 & 1 \\
1 & 0 & 1 & 1 & 0 & 1 & 0 & 1 & 1 & 1 & 1 & 0 \\
1 & 0 & 1 & 1 & 1 & 0 & 1 & 0 & 1 & 0 & 1 & 1 \\
1 & 1 & 0 & 0 & 1 & 1 & 1 & 0 & 1 & 1 & 1 & 0 \\
1 & 1 & 0 & 1 & 0 & 1 & 1 & 1 & 0 & 0 & 1 & 1 \\
1 & 1 & 0 & 1 & 1 & 0 & 0 & 1 & 1 & 1 & 0 & 1 \\
1 & 1 & 1 & 0 & 0 & 0 & 0 & 0 & 0 & 0 & 0 & 0
\end{bmatrix},
$$

has been constructed by sequentially adding columns based on maximizing the D_s criterion for the incoming column at each step. In the final array, all D_s values are equal to .908. Adding a thirteenth column results in a marked decrease in D_s values. The fact that no column is orthogonal to all the remaining columns implies that no column can be used for blocking. The array is also of interest in that interactions between factors assigned to any two columns are partially aliased (not quit uniformly) with all other columns, somewhat in the nature of the 12-run Plackett–Burman plan.

23.4.3 General Method of Construction

The need for nonorthogonal plans tends to be much more acute for small experiments in the sense that in larger experiments it is usually simpler to adjust the number of experimental units and/or number of factors. However, the need still arises to make compromises. The general techniques for constructing asymmetrical orthogonal arrays can be extended to the nearly orthogonal arrays as well. In the construction of asymmetrical orthogonal arrays, one of the key techniques was to use Kronecker addition to combine orthogonal arrays with difference matrices and then add additional columns based on another orthogonal array. Wang and Wu (1992) extend the methods to nearly orthogonal arrays. One possible extension is to replace one of the orthogonal arrays with a nearly orthogonal array. As an example of this type of modification, we construct a NOA[18;3^7,2^3;2] by taking the expression in (23.5) and replacing A by the NOA[6;3^1,2^3;2]

$$\begin{bmatrix} 0 & 0 & 0 & 0 \\ 1 & 0 & 1 & 1 \\ 2 & 0 & 1 & 0 \\ 0 & 1 & 1 & 1 \\ 1 & 1 & 0 & 0 \\ 2 & 1 & 0 & 1 \end{bmatrix}.$$

A second possible extension is to replace the difference matrix with a nearly difference matrix. However, here we first need a definition.

Definition of Nearly Difference Matrix

Recall that an $r \times c$ matrix $\boldsymbol{D}$ is a difference matrix if the elements are all from a set $\mathcal{S}$ with the property that the set of all differences between corresponding elements of two selected columns consists of all elements of $\mathcal{S}$ represented equally often. Following this definition we say that $D^*_{r,c;s}$ is a nearly difference matrix if the set of differences between corresponding elements of distinct columns contains the s distinct elements of $\mathcal{S}$ as evenly as possible.

23.4.4 Collection of Nearly Difference Matrices

For purposes of constructing a collection of nearly orthogonal arrays, we present the following collection of nearly difference matrices given by Wang and Wu (1992).

$$\boldsymbol{D}^*_{3,3;2} = \begin{bmatrix} 0 & 0 & 0 \\ 0 & 1 & 1 \\ 0 & 1 & 0 \end{bmatrix}, \quad \boldsymbol{D}^*_{4,4;3} = \begin{bmatrix} 0 & 0 & 0 & 0 \\ 0 & 1 & 2 & 0 \\ 0 & 2 & 1 & 1 \\ 0 & 1 & 0 & 2 \end{bmatrix},$$

$$\boldsymbol{D}^*_{5,5;2} = \begin{bmatrix} 0 & 0 & 0 & 0 & 0 \\ 0 & 0 & 1 & 1 & 1 \\ 0 & 1 & 0 & 1 & 1 \\ 0 & 0 & 1 & 0 & 0 \\ 0 & 1 & 1 & 1 & 0 \end{bmatrix}, \quad \boldsymbol{D}^*_{5,5;3} = \begin{bmatrix} 0 & 0 & 0 & 0 & 0 \\ 0 & 0 & 1 & 2 & 2 \\ 0 & 1 & 2 & 2 & 1 \\ 0 & 2 & 2 & 1 & 0 \\ 0 & 2 & 1 & 0 & 1 \end{bmatrix},$$

$$\boldsymbol{D}^*_{8,8;3} = \begin{bmatrix} 0 & 0 & 0 & 0 & 0 & 0 & 0 & 0 \\ 0 & 1 & 2 & 0 & 1 & 2 & 0 & 1 \\ 0 & 2 & 1 & 0 & 2 & 1 & 0 & 2 \\ 0 & 0 & 0 & 1 & 1 & 1 & 2 & 2 \\ 0 & 1 & 2 & 1 & 2 & 0 & 2 & 0 \\ 0 & 2 & 1 & 1 & 0 & 2 & 2 & 1 \\ 0 & 0 & 0 & 2 & 2 & 2 & 1 & 1 \\ 0 & 1 & 2 & 2 & 0 & 1 & 1 & 2 \end{bmatrix}, \quad \boldsymbol{D}^*_{9,8;2} = \begin{bmatrix} 0 & 0 & 0 & 0 & 0 & 0 & 0 & 0 \\ 0 & 1 & 0 & 0 & 1 & 1 & 0 & 1 \\ 0 & 0 & 1 & 0 & 1 & 0 & 1 & 1 \\ 0 & 1 & 1 & 0 & 0 & 1 & 1 & 0 \\ 0 & 0 & 0 & 1 & 0 & 1 & 1 & 1 \\ 0 & 1 & 0 & 1 & 1 & 0 & 1 & 0 \\ 0 & 0 & 1 & 1 & 1 & 1 & 0 & 0 \\ 0 & 1 & 1 & 1 & 0 & 0 & 0 & 1 \\ 0 & 0 & 0 & 0 & 0 & 0 & 0 & 0 \end{bmatrix}.$$

Note that we can also obtain $D^*_{r,c';s}$ from $D^*_{r,c;s}$ for $c' < c$ by deleting $c - c'$ columns.

In addition, we will need the difference matrix

$$D^*_{8,8;4} = \begin{bmatrix} (0,0) & (0,0) & (0,0) & (0,0) & (0,0) & (0,0) & (0,0) & (0,0) \\ (0,0) & (0,1) & (1,0) & (0,0) & (0,1) & (1,1) & (1,1) & (1,0) \\ (0,0) & (1,0) & (0,0) & (0,1) & (1,1) & (1,1) & (1,0) & (0,1) \\ (0,0) & (0,0) & (0,1) & (1,1) & (1,1) & (1,0) & (0,1) & (1,0) \\ (0,0) & (0,1) & (1,1) & (1,1) & (1,0) & (0,1) & (1,0) & (0,0) \\ (0,0) & (1,1) & (1,1) & (1,0) & (0,1) & (1,0) & (0,0) & (0,1) \\ (0,0) & (1,1) & (1,0) & (0,1) & (1,0) & (0,0) & (0,1) & (1,1) \\ (0,0) & (1,0) & (0,1) & (1,0) & (0,0) & (0,1) & (1,1) & (1,1) \end{bmatrix}.$$

Sequence of Constructions

We now present a set of three related methods that illustrate how a wide variety of nearly orthogonal arrays can be constructed. The reader is urged to consult Wang and Wu (1992) for details about each of these methods. The first construction takes the form

$$[L_1 \oplus D^* \ , \ L_2 \oplus A]. \tag{23.7}$$

Table 23.1 gives the details for a series of useful arrays constructed using this recipe.

Table 23.1 Specific Constructions Using (23.7)

Array Constructed	Pieces Required			
	L_1	D^*	L_2	A
NOA[6;3[1],2[3];2]	$[01]^t$	$D^*_{3,3;2}$	$[0\ 0]^t$	$[0\ 1\ 2\ ]^t$
NOA[10;5[1],2[5];2]	$[01]^t$	$D^*_{5,5;2}$	$[0\ 0]^t$	$[0\ 1\ 2\ 3\ 4]^t$
NOA[12;4[1],3[4];2]	$[012]^t$	$D^*_{4,4;3}$	$[0\ 0\ 0]^t$	$[0\ 1\ 2\ 3]^t$
NOA[12;3[4],2[3];2]	$[012]^t$	$D^*_{5,5;2}$	$[0\ 0]^t$	OA[4;2[3];2]
NOA[12;6[1],2[5];2]	$[01]^t$	$D^*_{6,5;2}$	$[00]^t$	$[012345]^t$
NOA[12;3[1],2[9];2]	OA[4;2[3];2]	$D^*_{3,3;2}$	$[0000]^t$	$[0123]^t$
NOA[15;5[1],3[5];2]	$[012]^t$	$D^*_{5,5;3}$	$[000]^t$	$[01234]^t$
NOA[18;9[1],2[8];2]	$[01]^t$	$D^*_{9,8;2}$	$[00]^t$	$[012345678]^t$
NOA[18;3[4],2[8];2]	$[01]^t$	$D^*_{9,8;2}$	$[00]^t$	OA[9;3[4];2]
NOA[20;9[1],2[15];2]	OA[4;2[3];2]	$D^*_{5,5;2}$	$[0000]^t$	$[01234]^t$
NOA[24;8[1],3[8];2]	$[012]^t$	$D^*_{8,8;3}$	$[0000]^t$	$[01234567]^t$
NOA[24;3[8],2[7];2]	$[012]^t$	$D^*_{8,8;3}$	$[0000]^t$	OA[8;2[7];2]
NOA[24;4[1],3[8],2[4];2]	$[012]^t$	$D^*_{8,8;3}$	$[0000]^t$	OA[8;4[1],2[4];2]
NOA[24;3[1],2[21];2]	OA[8;2[7];2]	$D^*_{3,3;2}$	$[00000000]^t$	$[012]^t$
NOA[24;6[1],2[15];2]	OA[4;2[3];2]	$D^*_{6,5;2}$	$[000000]^t$	$[012345]^t$

Table 23.2 Four Arrays Constructed Using (23.8)

Array Constructed	Actual Construction
NOA[12;3^5,2^1;2]	$[01]^t \oplus \boldsymbol{D}_{6,6;3}$
NOA[24;3^{11},2^1;2]	$[01]^t \oplus \boldsymbol{D}_{12,12;3}$
NOA[24;4^7,3^1;2]	$[012]^t \oplus \boldsymbol{D}_{8,8;4}$
NOA[36;4^{11},3^1;2]	$[012]^t \oplus \boldsymbol{D}_{12,12;4}$

Table 23.3 Five Arrays Constructed Using (23.9)

Array Constructed	First Part	$\boldsymbol{L}_1$	$\boldsymbol{A}$
NOA[12;3^4,2^2;2]	$[01]^t \oplus \boldsymbol{D}_{6,4;3}$	$[00]^t$	OA[6;3^1,2^1;2]
NOA[24;4^1,3^9,2^1;2]	$[01]^t \oplus \boldsymbol{D}_{12,9;3}$	$[00]^t$	OA[12;4^1,3^1;2]
NOA[24;3^9,2^5;2]	$[01]^t \oplus \boldsymbol{D}_{12,9;3}$	$[00]^t$	OA[2;3^1,2^4;2]
NOA[24;4^6,3^1,2^3;2]	$[01]^t \oplus \boldsymbol{D}_{8,6;4}$	$[000]^t$	OA[8;4^1,2^3;2]
NOA[36;4^{10},3^2;2]	$[012]^t \oplus \boldsymbol{D}_{12,10;4}$	$[000]^t$	OA[12;4^1,3^1;2]

The second method of construction makes use of difference matrices and takes the form

$$[[012\ldots(s-2)]^t \oplus \boldsymbol{D}_{ns,c;s}]. \tag{23.8}$$

Table 23.2 gives the details for four useful arrays constructed using this method.

The third method of construction combines aspects of the first two. It takes the form

$$[[012\ldots(s-2)]^t \oplus \boldsymbol{D}_{ns,c;s}\ ,\ \boldsymbol{L}_1 \oplus \boldsymbol{A}]. \tag{23.9}$$

Table 23.3 gives the details for five useful arrays constructed using this method.

23.5 LARGE ORTHOGONAL ARRAYS

In this section we present constructions for a number of large orthogonal arrays. The reader thinking in terms of the conventional laboratory factorial experiment may want to skip this section until the need for these very large array becomes clear in Chapter 25. We present three distinct methods for constructing large arrays and give a number of examples.

23.5.1 OA[N;2^{N-1};2] Arrays

These arrays are easily constructed from Hadamard matrices by dropping the first column (the one with all elements alike). In practice, it is often convenient to replace the -1's with 0's and retain the $+1$'s. In Chapter 16 we present methods for constructing large Hadamard matrices. Recall that a necessary restriction was that N be a multiple of 4. Also, for some values of N the construction is easier than others. However, the investigator using the large arrays typically has sufficient flexibility that these restrictions matter little. Arrays with fewer columns can easily be obtained by dropping columns.

23.5.2 OA[s^n;$s^{(s^n-1)/(s-1)}$;2] Arrays

This sequence of orthogonal arrays is easily obtained by the superimposing method used to construct fractional factorial plans in Section 14.5. The construction is simple and is valid for s equal to either a prime number or the power of a prime number. We illustrate first with s a prime number and $n = 2$. Generate an $s^2 \times 2$ matrix whose rows consist of all possible pairs of elements from {0, 1, ..., $s - 1$}. Denote these two columns by $\boldsymbol{X}_1$ and $\boldsymbol{X}_2$. These represent the first two columns of the orthogonal array. Construct the remaining $s - 1$ columns as $\boldsymbol{X}_1 + i\boldsymbol{X}_2$ where $i = 1, 2, \ldots, s - 1$ and all arithmetic is done modulo s. For $n = 3$ begin with the $s^3 \times 3$ matrix consisting of all triples from {0, 1, ..., $s-1$}. Let the three columns be $\boldsymbol{X}_1$, $\boldsymbol{X}_2$, and $\boldsymbol{X}_3$. Then construct the remaining columns in the desired array as $\boldsymbol{X}_1+i\boldsymbol{X}_2$, $\boldsymbol{X}_1+i\boldsymbol{X}_3$, $\boldsymbol{X}_2+i\boldsymbol{X}_3$, and $\boldsymbol{X}_1+i\boldsymbol{X}_2+j\boldsymbol{X}_1+i\boldsymbol{X}_2$ for $i = 1, 2, \ldots, s - 1$ and $i = j, 2, \ldots, s - 1$ and all arithmetic is done modulo s. The pattern for $n > 3$ is clear.

It is informative to look at these constructions in more detail. Consider first the $n = 2$ and $s = 3$ case. Begin with the two columns $\boldsymbol{X}_1 = [0\ 0\ 0\ 1\ 1\ 1\ 2\ 2\ 2]^t$ and $\boldsymbol{X}_2 = [0\ 1\ 2\ 0\ 1\ 2\ 0\ 1\ 2]^t$. We then follow the recipe to construct the 9×4 array

$$\text{OA}[9; 3^4; 2] = \begin{bmatrix} 0 & 0 & 0 & 0 \\ 0 & 1 & 1 & 2 \\ 0 & 2 & 2 & 1 \\ 1 & 0 & 1 & 1 \\ 1 & 1 & 2 & 0 \\ 1 & 2 & 0 & 2 \\ 2 & 0 & 2 & 2 \\ 2 & 1 & 0 & 1 \\ 2 & 2 & 1 & 0 \end{bmatrix}. \tag{23.10}$$

As an alternative, consider the difference matrix $\boldsymbol{D}_{3,3;3} = \begin{bmatrix} 0 & 0 & 0 \\ 0 & 1 & 2 \\ 0 & 2 & 1 \end{bmatrix}$ and the Kronecker sum $\begin{bmatrix} 0 \\ 1 \\ 2 \end{bmatrix} \oplus \begin{bmatrix} 0 & 0 & 0 \\ 0 & 1 & 2 \\ 0 & 2 & 1 \end{bmatrix}$ using modulo 3 arithmetic. Notice that the difference matrix used is actually a copy of the multiplication table from modulo 3 arithmetic. When we perform these calculations, we generate the array (23.10) with the second column corresponding to $\boldsymbol{X}_2$ missing. We are, in fact, doing exactly the same arithmetic in both procedures. Since column order does not matter in orthogonal arrays, one can summarize the construction of OA[9:3^4;2] as $\begin{bmatrix} 0 \\ 1 \\ 2 \end{bmatrix} \oplus \begin{bmatrix} 0 & 0 & 0 \\ 0 & 1 & 2 \\ 0 & 2 & 1 \end{bmatrix}$ and then adding a column with elements 0, 1, 2, in that order, repeated three times. Notice that when the construction involved a column of the form $[0, 1, \ldots, s-1]^t$ and an $s \times s$ difference matrix, one can always append a column $[0, 1, \ldots, s-1]^t$ repeated s times to the resulting array.

The next construction for $n = 3$ and $s = 3$ is easily written as

$$\text{OA}[9; 3^4; 2] \oplus \begin{bmatrix} 0 & 0 & 0 \\ 0 & 1 & 2 \\ 0 & 2 & 1 \end{bmatrix}$$

and then adding the additional column consisting of 0, 1, 2 repeated nine times. The result is OA[9;3^{13};2].

This procedure also extends directly to larger values of n and s equal to higher prime numbers or powers of a prime number. The requisite difference matrix can always be obtained as a copy of the multiplication table for the corresponding finite field.

In practice, the first procedure is usually easier to implement. Also, if one has access to SAS®, either programming the procedure into a `DATA` step or using `PROC FACTEX` generates the necessary array. Some arrays from this series that can prove useful are OA[25;5^6;2], OA[49;7^8;2], OA[64;8^9;2], OA[81;9^{10};2], OA[121;11^{12};2], OA[169;13^{14};2], OA[256;16^{17};2], OA[289;17^{18};2], OA[125;5^{31};2], OA[343;7^{57};2], OA[512;8^{73};2], OA[729;9^{91};2], and OA[625;5^{156};2].

23.5.3 OA[$2s^n$;$s^{2(s^n-1)/(s-1)-1}$;2] Arrays

This sequence of arrays represents a special subset from a larger family. We present this sequence because the members tend to fill in some of the intermediate spots between arrays given in Section 23.5.2. The general construction is rather involved and we will only give several examples. For the general construction the reader is urged to consult Addelman and Kempthorne (1961), Mukhopadhyay (1981), and Hedayat et al. 1999.

Before we begin the construction, we need to digress to construct a $\boldsymbol{D}_{2s,2s;s}$ difference matrix. Hedayat et al. 1999 give the following recipe. Let $\alpha_0, \alpha_1, \ldots, \alpha_{s-1}$ be the elements of a finite field. Let these be such that $\alpha_0 = 0$, $\alpha_{s-1} = 1$, and the remaining elements ordered so that $\alpha_i = \alpha^i$ for $i = 1, \ldots, s-1$. This is always possible. A suitable α is known as a *primitive root* of the field. Now construct the four matrices, $\boldsymbol{A} = (a_{ij})$, $\boldsymbol{B} = (b_{ij})$, $\boldsymbol{C} = (c_{ij})$, and $\boldsymbol{D} = (d_{ij})$ for $0 \le i, j \le s-1$, with entries

$$a_{ij} = \alpha_i \alpha_j$$

$$b_{ij} = \alpha_i \alpha_j + \gamma \alpha_i^2$$

$$c_{ij} = \alpha_i \alpha_j + \beta \alpha_j^2$$
$$d_{ij} = \alpha \alpha_i \alpha_j + \epsilon \alpha_i^2 + \delta \alpha_j^2,$$

where β, γ, δ, and ϵ are $\beta = \frac{1}{2}$, $\gamma = (\alpha - 1)/2\alpha$, $\delta = \alpha/2$, and $\epsilon = (\alpha - 1)/2$. It then follows that the matrix

$$\boldsymbol{D}_{2s,2s;s} = \begin{bmatrix} \boldsymbol{A} & \boldsymbol{B} \\ \boldsymbol{C} & \boldsymbol{D} \end{bmatrix}$$

is a suitable difference matrix. As an example, we can take the finite field with five elements, 0, 2, 4, 3, 1, and arithmetic modulo 5. Note that we could also have chosen 3 as the primitive root. We have $\alpha = 2$, $\beta = 3$, $\gamma = 4$, $\delta = 1$, and $\epsilon = 3$. The four 5×5 matrices are

$$\begin{bmatrix} 0 & 0 & 0 & 0 & 0 \\ 0 & 4 & 3 & 1 & 2 \\ 0 & 3 & 1 & 2 & 4 \\ 0 & 1 & 2 & 4 & 3 \\ 0 & 2 & 4 & 3 & 1 \end{bmatrix}, \quad \begin{bmatrix} 0 & 0 & 0 & 0 & 0 \\ 1 & 0 & 4 & 2 & 3 \\ 4 & 2 & 0 & 1 & 3 \\ 1 & 2 & 3 & 0 & 4 \\ 4 & 1 & 3 & 2 & 0 \end{bmatrix},$$

$$\begin{bmatrix} 0 & 2 & 3 & 2 & 3 \\ 0 & 1 & 1 & 3 & 0 \\ 0 & 0 & 4 & 4 & 2 \\ 0 & 3 & 0 & 1 & 1 \\ 0 & 4 & 2 & 0 & 4 \end{bmatrix}, \quad \begin{bmatrix} 0 & 4 & 1 & 4 & 1 \\ 2 & 4 & 4 & 3 & 2 \\ 3 & 3 & 1 & 1 & 2 \\ 2 & 3 & 2 & 2 & 4 \\ 3 & 1 & 2 & 3 & 1 \end{bmatrix}.$$

Exercise 23.9: Verifying the Difference Matrix. Verify that the 10×10 matrix produced above is a difference matrix. □

We now compute

$$\text{OA}[50;\, 5^{10};\, 2] = \begin{bmatrix} 0 \\ 2 \\ 4 \\ 3 \\ 1 \end{bmatrix} \oplus \boldsymbol{D}_{10,10;5},$$

where the arithmetic is modulo 5. This can be expanded to an OA[50;5^{11};2] by adding an extra column consisting of the five elements 0, 2, 4, 3, 1 repeated 10 times in that order. Hedayat et al. (1999) show that this construction is general and can be utilized to construct the OA[$2s^2$;$s^{2(s^2-1)/(s-1)-1}$;2] for any s that is a power of an odd prime. This construction gives us OA[18;3^7;2], OA[50;5^{11};2], OA[98;7^{15};2], OA[162;9^{19};2], OA[242;11^{23};2], OA[338;13^{27};2], OA[578;17^{35};2], and so on. Of those listed, all but the one with 162 runs require only simple modular arithmetic for their construction.

Our final construction yields the OA[$2s^3$;$s^{2(s^3-1)/(s-1)-1}$;2] using the formula

$$\boldsymbol{D}_{2s^2,2s^2;s} \oplus \begin{bmatrix} a_0 \\ \vdots \\ a_{s-1} \end{bmatrix}, \quad \boldsymbol{D}_{2s,2s;s} \oplus \begin{bmatrix} a_0 \\ \vdots \\ a_{s-1} \end{bmatrix} \oplus \boldsymbol{L}_s, \begin{bmatrix} a_0 \\ \vdots \\ a_{s-1} \end{bmatrix} \oplus \boldsymbol{L}_{2s^2},$$

where $\boldsymbol{L}_s$ is the column with s 1's and $\boldsymbol{L}_{2s^2}$ the column with $2s^2$ 1's. If we restrict ourselves to odd prime numbers where the arithmetic is easy, we obtain the arrays OA[54;3^{25};2], OA[250;5^{61};2], OA[343;7^{113};2], and OA[2662;11^{265};2].

23.6 SUMMARY

To summarize the material in this chapter, we repeat that any orthogonal, asymmetrical orthogonal, or nearly orthogonal array can be interpreted directly as a fractional factorial plan. The advantage of the approach is that it greatly broadens the collection of fractional factorial plans available. These plans have their place in any investigator's or consulting statistician's collection of tools. The disadvantage is that the lack of stress on interactions may foster a cavalier attitude toward them and mislead unsuspecting investigators. We repeat: Factorial experiments do not cause interactions, and ignoring interactions does not cause them to go away.

Orthogonal arrays provide a very general tool for the design of experiments. However, we must give a warning. It is very easy to miss important design possibilities by relying too much on just looking at orthogonal arrays. For example, in an experiment that involves two factors, each at four levels and another three factors at two levels, it is easy to simply use columns in the OA[32;4^2,2^9;2] obtained directly from the OA[32;2^{15};2] by collapsing two sets of three columns and not realizing that there exists a resolution IV 2^{7-2} plan and that proper assignment of factors allows one to keep all main effects free of two-factor interactions.

CHAPTER 24

Factorial Plans Derived via Orthogonal Arrays

24.1 INTRODUCTION

Historically, the design topics that we cover in this chapter relate to the quality evaluation and product enhancement problems in industry. Our objective is to address and explore the design aspects and options. The definition of quality and production issues are beyond the scope of this book.

An underlying feature of all the designs and discussions in this chapter is the importance of interactions. In essence, all of the problems considered boil down to trying to locate and take advantage of interactions between factors that can be easily controlled and those that cannot. The object is to find combinations of factor levels that make the difficult-to-control factors inactive. In a quality sense, one wants to produce items that function as advertised regardless of the conditions encountered.

24.2 PRELIMINARIES

24.2.1 Motivating Example

To motivate the material in this chapter we consider an example given by Grize (1995) encountered in the production of a particular pigment. Traditional statistical process control techniques (Shewhart control charts) had revealed a high level of batch-to-batch variability in color strength. Color strength was measured as the percentage deviation from a standard, and consequently, zero was the ideal value. It was known that the most crucial phase of the process was kneading the pigment mass. At this stage there were six easily adjustable, potentially important variables. These six factors, which we refer to as $A, \ldots, F$, were the kneading

Planning, Construction, and Statistical Analysis of Comparative Experiments,
by Francis G. Giesbrecht and Marcia L. Gumpertz
ISBN 0-471-21395-0

Table 24.1 Factors and Levels for Pigment Kneading Example

Factor		Level	
Easy-to-control factors			
A	Temperature of mixture	60°C	80°C
B	Excess of salt	10%	20%
C	Kneading speed	Slow	Fast
D	Kneading time	2 hr	3 hr
E	Order of introduction of materials	A	B
F	Transfer time	Short	Long
Difficult-to-control factors			
M	Salt grain size	Small	Medium
N	Resin texture	Bulk	Powder
O	Temperature of cooling water	5°C	20°C

temperature, the excess of salt affecting the viscosity of the mass, the kneading speed, the kneading time, the order of introduction of the products in the kneader, and the transfer time. These factors were easy to control in the manufacturing process. In addition, prior knowledge suggested one or more of (1) the salt quality as quantified by grain size, (2) the texture of the resin, and (3) the temperature of the cooling water (coming from the river) could have an affect on color strength. These factors could be controlled for the sake of some testing or some experiments, but not during routine production. Cost of controlling them during routine production runs would be prohibitive. These factors we label as M, N, and O, respectively.

The problem in the experiment is to find combinations of the $A, \ldots, F$ factors that give good results, and possibly more important, to make the entire process robust to variation in the M, N, and O factors so that they can be safely ignored. The manufacturer would like to be able to ignore the grain size of the salt, the texture of the resin, and most important, the temperature of the cooling water that changed with the seasons.

The factors and levels for an experiment are summarized in Table 24.1. Before continuing with this example we digress and consider some underlying principles.

24.2.2 Control Factors, Noise Factors, Control Array, and Noise Array

The key thing that stands out in the pigment kneading example is that there are two distinct types of factors. The first set of factors consists of those that are under the direct control of the manufacturer or processor. We refer to these as *control factors*. The second set of factors consists of all sorts of things. Here we include factors that are expensive to control and the processor would rather not control (like the water temperature) to factors that can be controlled for the sake of an experiment or a special study, but cannot be controlled in general. In the latter category we could include factors such as variations in the structure of the material where the pigment is ultimately to be used. It is in the manufacturers' best interest to produce a product that performs satisfactorily in a wide range of situations. We call this second set of factors *noise factors*.

This distinction between the two types of factors was forcibly brought to the attention of applied scientists and engineers in a series of publications by Taguchi (1986, 1987a,b, 1992) and Taguchi and Wu (1985). The overall purpose in these publications was to present techniques for improving the quality of products and/or processes. The definition of quality in this context was to produce a product that performed in a consistent manner under a broad range of conditions. Hence the terms *control factors* and *noise factors*. The definition of quality as consistency also forces a recognition of variance as an important factor.

For simplicity we consider mostly cases where all factors are at two levels, although the principles developed extend directly to more levels.

Taguchi Approach

The standard Taguchi approach is to separate the two sets of factors, organize a fractional factorial plan with N_1 treatment combinations for the first set, a second independent fractional factorial plan with N_2 treatment combinations for the second set, and then examining all N_1N_2 combinations from the two plans. Standard terminology in the Taguchi literature is to refer to *L-arrays* (they are orthogonal arrays or, in some cases, nearly orthogonal arrays) rather than fractional factorials and inner and outer arrays rather than the more suggestive control and noise or error arrays. The usual terminology is to refer to the complete plan as a *product array*, the product of an *inner* and an *outer array*. Some have suggested that both sets of factors be combined into one set and one fractional factorial plan implemented. We return to that point later.

Broadly speaking, the Taguchi philosophy can be summarized under the following topics:

1. Two major components of the quality of a product or process as measured on any scale are the mean level and the variability. For example, it is not enough that on average the items be the correct size, but also that there be very little variability in the size of units with a specific size label. For example, the exact size of a size 12 shirt is not nearly as crucial as that all shirts from that manufacturer that have a size 12 label be the same size. The important feature is that a consumer can rely on getting exactly the same product on each purchase. This applies to clothing, drugs, bolts, screws, tires, records, and many other items.
2. The factors affecting the quality of a product or process can usefully be grouped into two sets. The first set, referred to as *design* or *control factors*, consists of those factors that can readily be controlled in the manufacturing process and the second set, referred to as *noise factors*, consisting of those

that can only be controlled in a special study or at great cost. Often, this set will include factors that are related to how the product is used (or misused) by the consumer. The motivating idea behind this classification is that one tries to find levels of the control or design factors that make the product or process robust to the noise factors and consequently, of better quality.

3. The use of fractional factorials (Taguchi called them L-arrays) in the construction of the experimental plans to study and improve the quality of the product or process. The control factors are assigned to one array (fractional factorial) and the noise factors to a second array. The actual plan for the study then involves the direct product of these two arrays; that is, for every combination of a setting of the control factors and a setting of the error factors an observation is taken.
4. Take into account both first-order (mean, main effects, interactions, etc.) and second-order (dispersion) effects in the statistical analysis of the data observed in the experiments.

Some aspects of the approach that Taguchi proposed were quite radical and new. The formal emphasis on reduction in the variance of the product produced and designing experiments specifically to accomplish this appeared to be new. The statistical analysis that focused on both mean level (signal) and variance (noise) was also new, although the use of signal-to-noise ratios (discussed later) has generated considerable controversy. Experience has shown that not only does the use of these ratios tend to be inefficient, but also often leads to misleading interpretations (Box 1988; Grize 1995). Formal separation of the design factors from the noise factors in the design was also quite new, although the basic idea of considering "noise" factors in experiments has been incorporated into good-quality agricultural and biological research work for many years. Two good references are Cox (1958) and John and Quenouille (1977).

24.2.3 Possible Applications

We consider several hypothetical examples of applications.

1. *Dye manufacturer.* There are many factors that affect the exact shade of color produced by a dye. Some of these are under the direct control of the manufacturer, and some are not. For example, the formula, amount of mixing, and temperature in the process can be controlled easily. However, the storage time and temperature of the dye cannot be controlled as easily, and probably not at all. The conditions under which the dye is used cannot be controlled at all. Also, the exact shade of color produced may not be as crucial as the consistency and durability of the color produced.
2. *Food processor.* The processor cannot easily control the variety, ripeness, amount of abuse, and so on, of the crop being processed. Also, the storage

time and storage conditions after processing cannot be controlled. However, the temperature, time, amount of additives, and so on, in the product are easily controlled. A consistent product regardless of the uncontrollable factors is desired.

3. *Auto manufacturer.* The manufacturer has control over the design and construction of the auto. However, there is little or no control over the use or abuse that the auto will receive at the hands of the eventual purchaser. The quality of the product is eventually evaluated on the basis of the performance in the hands of the purchaser after some use (or abuse).
4. *Furniture manufacturer.* The manufacturer can control the species of wood used, type of construction, and finish put on the piece of furniture. However, there is limited control over the temperature and humidity conditions under which the pieces are to be used. Quality is judged by the performance without regard to conditions.
5. *Premixed foods such as cake mixes.* The manufacturer has control over the ingredients that go into a cake mix as well as the instructions on the box. However, there is no way to control what gets added to the mix at the time it is used. There is no control over the time and temperature of baking. But it is the manufacturer's product that gets evaluated.

24.2.4 Motivating Example Revisited

The investigators in the pigment production example used a 2^{6-3} resolution III plan for the control array. Since they were concerned about the likelihood of a kneading speed × kneading time interaction, they assigned factors $A, \ldots, F$ to columns a, b, ab, c, ac, and bc in a 2^3, respectively. This is a good example of a case where a resolution III plan is used, but there is room to protect one two-factor interaction and it is used to protect the CD interaction. The noise factors were assigned to the columns in a 2^2 plan.

At the next stage some care is needed. There are two distinct possibilities for randomization. There are 32 treatment combinations in the product array. One can randomize the order of the complete set for a basic completely randomized design. Alternatively, one could randomize the order of the eight treatment combinations in the control array. Then, as each is selected, the order of the treatment combinations is randomized in the noise array. The basic plan is now a split-plot experiment. There is nothing wrong with running this as a completely randomized design or as a split plot, provided that the analysis honors the design used. It is not unusual to find this type of experiment performed as a split plot and analyzed as a completely randomized experiment. This sort of mistake results in experimenters declaring too many control effects significant and a failure to take advantage of extra sensitivity in the measure of noise effects. Grize (1995) makes no mention of the randomization used.

Several different statistical analyses are possible for data resulting from this experiment. Recall that this study was set up in response to complaints of too much variability in the product. No mention was made of a problem with the color being off target. This is really an experiment to find control factors that can be adjusted to reduce variability caused by uncontrolled noise factors. A straightforward analysis of variance for either the CRD or the split plot could be performed, but this seems somewhat inappropriate, since the investigator is probably convinced that salt grain size, resin texture, and water temperature have an effect. The aim is to find conditions where these factors are not active. An alternative statistical analysis that has considerable appeal would be to compute a mean and sample variance for the observations at each of the eight control factor combinations. One could then analyze the means and the logarithm of the variances. Since the control array is saturated, the only real options here are normal probability plotting as proposed by Daniel (1959, 1976) or the Lenth (1989) and Dong (1993) (Chapter 12) techniques. Note that estimated contrasts among control treatment combinations have a common variance regardless of the homogeneity of variance hypothesis.

Rather than giving actual data for the pigment study, Grize assumes a model, generates some data, and presents the analysis with comments. An informative way to present the plan, data, and preliminary analyses is illustrated in Table 24.2. The final statistical analysis is then based on $\overline{y}_{1\cdot}, \ldots, \overline{y}_{8\cdot}$ and on $\ln s_1^2, \ldots, \ln s_8^2$.

Grize uses the model

$$y = 3.5 + A - .75B + .75M - N + .75A * M + (4.5 - 3.5D)\epsilon, \qquad (24.1)$$

where ϵ is a random error from the $N(0, 1)$ distribution. Rather than repeat the Grize analysis, we used this model to simulate 1000 data sets and computed the means corresponding to $\overline{y}_{1\cdot}, \ldots, \overline{y}_{8\cdot}$ and $\ln s_1^2, \ldots, \ln s_8^2$ and estimates of

Table 24.2 Grize Pigment Study[a]

							Noise Array					
						M	−1[a]	+1	−1	+1		
Control Array						N	−1	−1	+1	+1		
A	B	C	D	E	F	O	+1	−1	−1	+1		
−1	−1	+1	−1	+1	+1		y_{11}	y_{12}	y_{13}	y_{14}	$\overline{y}_{1\cdot}$	s_1^2
+1	−1	−1	−1	−1	+1		y_{21}	y_{22}	y_{23}	y_{24}	$\overline{y}_{2\cdot}$	s_2^2
−1	+1	−1	−1	+1	−1		y_{31}	y_{32}	y_{33}	y_{34}	$\overline{y}_{3\cdot}$	s_3^2
+1	+1	+1	−1	−1	−1		y_{41}	y_{42}	y_{43}	y_{44}	$\overline{y}_{4\cdot}$	s_4^2
−1	−1	+1	+1	−1	−1		y_{51}	y_{52}	y_{53}	y_{54}	$\overline{y}_{5\cdot}$	s_5^2
+1	−1	−1	+1	+1	−1		y_{61}	y_{62}	y_{63}	y_{64}	$\overline{y}_{6\cdot}$	s_6^2
−1	+1	−1	+1	−1	+1		y_{71}	y_{72}	y_{73}	y_{74}	$\overline{y}_{7\cdot}$	s_7^2
+1	+1	+1	+1	+1	+1		y_{81}	y_{82}	y_{83}	y_{84}	$\overline{y}_{8\cdot}$	s_8^2

[a] +1 represents the high level and −1 the low level.

Table 24.3 Results of 1000 Simulations with Grize's Model

Analysis of Means		Analysis of $\ln s^2$	
$\widehat{\mu}$	3.57		
$\widehat{F}$	−.02	$\widehat{CD}$	.01
$\widehat{E}$	.09	$\widehat{C}$	−.01
$\widehat{C}$	−.09	$\widehat{B}$	.01
$\widehat{D}$	−.11	$\widehat{F}$	.02
$\widehat{CD}$	.12	$\widehat{E}$	.49
$\widehat{B}$	−1.48	$\widehat{A}$	.50
$\widehat{A}$	1.90	$\widehat{D}$	−2.77

$\widehat{A}, \widehat{B}, \ldots, \widehat{CD}$ for both these variables. The average estimates are in Table 24.3. The results from the simulation are as one would expect from the model. Since the data represent percent deviations from the standard (zero is the ideal), the two estimates $\widehat{B}$ and $\widehat{A}$ both indicate the correct adjustments required. The estimate of A indicates that increasing the temperature of the mixture increases the deviation from standard, and consequently, temperature should be reduced. Similarly, the sign on $\widehat{B}$ indicates that excess of salt should be increased. Also, analysis of the variances indicates that larger values of D should decrease the variance. This is exactly consistent with the model in equation (24.1). Note that there is no way to detect the temperature × grain size (AM) interaction.

The Taguchi references suggest that when the object of the project is to reduce the response, one should base the statistical analysis on the signal-to-noise ratio,

$$sn_i = \ln\left(\overline{y}_{i\cdot}^2/s_i^2\right),$$

with smaller values being better. We computed $\ln\left(\overline{y}_{i\cdot}^2/s_i^2\right)$ values and performed the analysis of variance for each of the 1000 data sets. The resulting average estimates were $\widehat{CD} = .15$, $\widehat{C} = .19$, $\widehat{E} = -.20$, $\widehat{F} = -.26$, $\widehat{A} = .49$, $\widehat{B} = -.77$, and $\widehat{D} = 2.83$. The large positive $\widehat{D}$ value suggests reducing D since smaller values of y are better. But D has values -1 and $+1$ in the model (24.1). Making D small will not affect the mean, but will increase the variance. This demonstration agrees with the Box (1988) observation that the signal-to-noise ratios provide an inadequate summary of the data, can be extremely inefficient, and can occasionally be quite misleading. Conventional statistical analyses based on means and logarithms of variances are much more reliable. Such analyses are much more likely to point the investigator in the right direction than is an analysis of signal-to-noise ratios.

Exercise 24.1: CRD or Split Plot. The experiment discussed in the Grize example could be conducted in either a CRD or a split-plot design. Explain in detail the appropriate randomization for the study in a CRD and as a split-plot experiment. □

24.2.5 Comments on Statistical Analyses

It was already pointed out in the introduction that there are still some unanswered questions about appropriate statistical analyses. First, the tacit assumption that the $\{s_i^2\}$ values computed in Table 24.2 are from a scaled χ^2 distribution is not really defensible since the $y_{i1}, \ldots, y_{i4}$ do not represent a random sample from a single population. However, the analysis is simple to perform, easy to understand, and simulations with various models show that it tends to lead to good interpretations. The signal-to-noise ratios that are often advocated also suffer from a lack of proper theoretical justification, tend to make inefficient use of the data, and occasionally (Box 1988; Grize 1995) lead to incorrect conclusions. A much more elaborate statistical analysis based on a similar model is found in Engel (1992). Analyses based on the $\{s_i^2\}$ are often referred to as *modeling loss statistics directly*. The term clearly reflects the early history of applications to quality production problems.

Welch et al. (1990) argue that a more efficient approach is to model the variables observed in a study directly as a function of both control and noise factors. This approach often leads to better insight into the interrelationships among factors. For this reason it is often a good idea to perform this sort of analysis in addition to the analyses based on loss functions. An advantage of this direct modeling approach is that it permits the designer of the study to combine the control array and the noise array into one comprehensive plan and get by with fewer observations. However, a word of warning is in order. If this approach is taken, the analysis must take the design of the study into account. If the treatment combinations are randomized as in a split-unit design, the statistical analysis must honor that protocol.

A third method of statistical analysis that has much to offer has been suggested by Box and Meyer (1986) and Bergman and Hynén (1997) under the name *dispersion analysis*. This analysis is discussed in Section 12.3.

24.3 PRODUCT ARRAY DESIGNS

The product array designs discussed so far were introduced into the scientific literature in a series of publications by Taguchi (1986, 1987a,b, 1992) and Taguchi and Wu (1985). However, the presentation was in terms of L-arrays. Most of these L-arrays are either symmetric or asymmetric orthogonal arrays of strength 2. The remainder are nearly orthogonal arrays, but again of strength 2. Whenever the reader encounters the term *L-array* in the literature, it is generally safe to think orthogonal or nearly orthogonal array or even fractional factorial. These arrays provide the basic building blocks for what he called *parameter design experiments*. A more evocative term that some have used is *robust design experiments*. Note that the latter term is particularly appropriate since Taguchi is generally concerned with improving the quality of a product or a process. The approach is to formally split the factors affecting the product or process into two sets. The first set is called the *control* or *design set*, and the second, the *noise set*. The distinction between the two sets is that the control set consists of those factors that are under the direct control of the manufacturing process. The noise factors are the factors that impinge on the performance or quality of the product or process but can generally not be controlled under general operating conditions. They may be controlled by a consumer. They can be controlled, possibly with some difficulty, for the purpose of a study, but beyond that they are uncontrolled. Two separate fractional factorial arrangements are set up. Taguchi often uses the less suggestive terms *inner* and *outer array* for the two.

However, this emphasis on fractional factorial plans for both arrays is probably too confining. The control array in particular can be replaced by any design plan. Consider, for example, a caramel-making example used by Kackar (1985). As he points out, caramel is a mixture of more than 10 ingredients, but mostly sugar. One of the performance characteristics of interest in the particular study was the plasticity of the product. The problem was that the plasticity of the product being produced was very sensitive to environmental temperature, with caramels with plasticity less sensitive to temperature considered to be of higher quality. In Kackar's example, the error array involved only one factor, temperature. In place of a control array consisting of a fractional factorial, one could use a response surface design, a split-unit design, a Box–Bhenkin design, or a sequence of fractions to be developed over a period of time.

A problem with the product array designs is that the number of runs, the product of the treatment combinations in the control array and in the noise array, becomes excessive. Notice that the investigator is really looking for interactions between control effects and noise effects. Success normally amounts to finding a level or combination of levels of control factors that render the noise factors inactive. In many cases it may well be a combination (interaction) of control factors that leads to reducing efficacy of noise factors. Keeping all two-factor interactions in the control array estimable means a large array. A nice feature in the product array approach is that all interactions between control factors and noise factors are preserved automatically. This is not the case with the single-array approach.

The selection of sequences of fractional factorial plans discussed in Chapter 19, particularly in Sections 19.7 and 19.9.4, combines very nicely with the product array strategy. The investigator begins with a control array (design) with a large number of factors. To make the study manageable, only main effects, each at two levels, are considered at this stage. For each control treatment combination the complete set of combinations of factors in the noise array is investigated. The statistical analysis is then used to remove control factors that turn out to have no effect. A subsequent set of control treatment factor combinations (an additional fraction) is then examined to clear up lingering uncertainties concerning interactions. Similarly, extra levels of factors can be inserted if the need arises. At each point a complete set of noise factors (noise array) is observed. This approach allows the investigator to evaluate a large number of control factors, yet eventually get to examine detailed interrelationships that appear.

24.3.1 More Formal Construction

It is useful to consider the structure of the Taguchi product array in a somewhat more formal setting. Assume that there are c control factors and n noise factors. The control array (Taguchi's inner array) consists of N_c rows (runs) and c columns (factors), and the noise array (Taguchi's outer array) consists if N_n rows (runs) and n columns (factors). Let $\boldsymbol{D}_c$ and $\boldsymbol{D}_n$ represent these two arrays. We can also think of them as design matrices specifying the design, the relationship between factors and runs.

Now the design matrix for the overall product array, the design that is actually performed, can be written as

$$\boldsymbol{D} = \boldsymbol{D}_c \otimes \boldsymbol{j}_{N_n} \| \boldsymbol{j}_{N_c} \otimes \boldsymbol{D}_n, \tag{24.2}$$

where $\boldsymbol{j}_N$ is the column of N 1's. This design requires $N_n N_c$ runs.

Exercise 24.2: Formal Construction of a Product Array Design. Let $\boldsymbol{D}_8$ be the 8×6 matrix of $+1$'s and -1's obtained by retaining columns corresponding to a, b, ab, c, ac, and bc in a 2^3 and let $\boldsymbol{D}_4$ be the 4×3 matrix of $+1$'s and -1's obtained by retaining columns corresponding to a, b, and ab of a 2^2 factorial. Show that the design matrix for the Grize example can be written as

$$\boldsymbol{D} = \boldsymbol{D}_8 \otimes \boldsymbol{j}_4 \| \boldsymbol{j}_8 \otimes \boldsymbol{D}_4. \qquad \square$$

In the literature cited earlier in the section, $\boldsymbol{D}_c$ and $\boldsymbol{D}_n$ are orthogonal arrays, typically of strength 2, or nearly orthogonal arrays. If they are strength 2 orthogonal arrays (resolution III factions), main effects are clear of two-factor interactions. Formally, this means that within the set of control factors and the set of noise factors, main effects are clear of two-factor interactions. If nearly orthogonal arrays are used, presumably these conditions are nearly satisfied. It is also clear from the cross structure of $\boldsymbol{D}$ that control factors are not confounded with either noise main effects or noise factor interactions.

These designs are well suited to accomplish two objectives. The first is to find levels of the control factors that achieve the desired mean level of the process. The second is to find ways to take advantage of interactions between control factors (or their interactions) and noise factors, to find settings for the control factors that make the noise factors inactive. A potential problem appears, however, if the control array is a resolution III plan and a two-factor (control) interaction is responsible for allowing some noise factors to be active. It will not be possible to identify the proper control two-factor interaction because it is confounded with a main effect. We come back to this point later.

A problem with this approach is that it becomes costly as the number of runs required, N_cN_n, becomes large. The remainder of this chapter is devoted to finding ways of reducing the number of runs, yet protecting the ability of the design to accomplish the twin objectives. A solution is to construct one large array for $\boldsymbol{D}$ that satisfies all estimability requirements and minimizes the number of runs. Such a $\boldsymbol{D}$ can be found using the Franklin–Bailey algorithm. The requirements set will include at least all of the control and noise main effects, possibly some of the control two-factor interactions, and at least the two-factor interactions between control and noise effects. Possibly some three-factor interactions, two control factors and a noise factor, should also be protected. The problem is that this is difficult. Simply selecting a resolution V or higher plan is possible, but will be wasteful. These plans require too many runs. We will examine smaller, simpler alternatives. An additional problem with the single-array approach is that it restricts the investigator randomizing the full set of treatment combinations at one time as one complete design.

24.4 BLOCK CROSSED ARRAYS

24.4.1 Construction

Fang and Zhou (1998) have developed a systematic, yet simple method for constructing plans, with all factors at two levels, with reasonable size and desirable estimation properties. They often require fewer runs than either the Taguchi plans or the single-array plans. The construction depends somewhat on the number of factors and is best explained by two examples. Recall that we have c control factors and n noise factors. The first construction designs can be fit into split-plot designs in the sense that one can assign control factors to whole-plots and noise factors to split-plots. Designs from the second construction cannot.

Their first construction requires a resolution IV plan or higher to form $\boldsymbol{D}_c$ the control array and a resolution III plan to form $\boldsymbol{D}_n$ the noise array. These are then combined as in a product array (24.2). However, there are restrictions on the two arrays. It is immediately clear that all main effects will be estimable, even if there are two-factor interactions present.

The second construction begins with two resolution III plans. Now the strategy is to partition $\boldsymbol{D}_c$ into two parts, one part with $c^{(1)}$ columns that constitute a resolution IV plan and the remaining $c^{(2)} = c - c^{(1)}$ columns. This requirement clearly restricts the $\boldsymbol{D}_c$ choice. Let the corresponding parts of $\boldsymbol{D}_c$ be $\boldsymbol{D}_c^{(1)}$ and $\boldsymbol{D}_c^{(2)}$. Note that $\boldsymbol{D}_c^{(1)}$ is a resolution IV plan. For the noise part of the design we need $\boldsymbol{D}_n$, a resolution III plan with room for $n + 1$ factors. Partition $\boldsymbol{D}_n$ into $\boldsymbol{D}_n^{(1)}$ with room for n factors and $\boldsymbol{D}_n^{(2)}$ the remaining column. Then construct the block-crossed design

$$\boldsymbol{D} = \boldsymbol{D}_c^{(1)} \otimes \boldsymbol{j}_{N_n} \| \boldsymbol{D}_c^{(2)} \otimes \boldsymbol{D}_n^{(2)} \| \boldsymbol{j}_{N_c} \otimes \boldsymbol{D}_n^{(1)}. \quad (24.3)$$

Notice the restriction on the resolution III plan used for $\boldsymbol{D}_c$. It must be possible to select a subset of columns that form a resolution IV plan. This rules out the possibility of using a Hadamard plan for the control array. There is no such restriction for the noise array.

To illustrate the second construction, consider a plan for six control factors and six noise factors. For $\boldsymbol{D}_c$ consider columns a, b, c, abc, ab, and ac from a 2^3 factorial. Notice, however, that using only columns a, b, c, and abc gives a resolution IV plan for four factors. These columns define $\boldsymbol{D}_c^{(1)}$. Columns ab and ac define $\boldsymbol{D}_c^{(2)}$. For $\boldsymbol{D}_n$ we need a resolution III plan with room for $6 + 1 = 7$ factors. An eight-run Plackett–Burman plan is good. Six columns form $\boldsymbol{D}_n^{(1)}$, and the seventh column forms $\boldsymbol{D}_n^{(2)}$. The 64-run design is

$$\boldsymbol{D} = \boldsymbol{D}_c^{(1)} \otimes \boldsymbol{j}_8 \| \boldsymbol{D}_c^{(2)} \otimes \boldsymbol{D}_n^{(2)} \| \boldsymbol{j}_8 \otimes \boldsymbol{D}_n^{(1)}. \tag{24.4}$$

24.4.2 Estimability Properties

We summarize the estimability properties of the block-crossed arrays of Section 24.4.1.

1. All main effects are mutually orthogonal. We get this from the properties of $\boldsymbol{D}_c$ and $\boldsymbol{D}_n$ and the nature of the cross structure defined in (24.2) and (24.3).
2. All control main effects are orthogonal to all control two-factor interactions. In the (24.2) construction we get this directly from the resolution of $\boldsymbol{D}_c$. In the (24.3) construction we also need to check the $\boldsymbol{D}_c^{(1)} \otimes \boldsymbol{j}_{N_n}$ and $\boldsymbol{D}_c^{(2)} \otimes \boldsymbol{D}_n^{(2)}$ columns. This involves verifying that any column in either of the matrices is orthogonal to the column obtained as the element-by-element product of any two columns in either of the matrices.
3. At least some control two-factor interactions may be mutually confounded. We get this from $\boldsymbol{D}_c$, which need only be resolution IV.
4. All two-factor control–noise interactions are orthogonal to control main effects and to noise main effects. This is verified by checking the orthogonality of elementwise products of control and noise columns and individual control columns and noise columns.
5. All two-factor control–noise interactions are orthogonal to all two-factor interactions. This comes from the nature of the cross structure and is checked by examining orthogonality of columns.
6. Noise main effects and noise two-factor interactions need not be orthogonal. The nature of $\boldsymbol{D}_n$ does not guarantee this orthogonality.
7. Three-factor and higher interactions involving two control factors may be confounded, but inheritance considerations may help identification. This traces back to the nature of $\boldsymbol{D}_c$.

Exercise 24.3: Details of Example. Verify that the example constructed in expression (24.4) can be written as

Factor:	A	B	C	D	E	F	G	H	K	L	M	N
	$+\boldsymbol{j}_8$	$+\boldsymbol{j}_8$	$+\boldsymbol{j}_8$	$+\boldsymbol{j}_8$	$+\boldsymbol{o}$	$+\boldsymbol{o}$	$\boldsymbol{g}$	$\boldsymbol{h}$	$\boldsymbol{k}$	$\boldsymbol{l}$	$\boldsymbol{m}$	$\boldsymbol{n}$
	$+\boldsymbol{j}_8$	$+\boldsymbol{j}_8$	$-\boldsymbol{j}_8$	$-\boldsymbol{j}_8$	$+\boldsymbol{o}$	$-\boldsymbol{o}$	$\boldsymbol{g}$	$\boldsymbol{h}$	$\boldsymbol{k}$	$\boldsymbol{l}$	$\boldsymbol{m}$	$\boldsymbol{n}$
	$+\boldsymbol{j}_8$	$-\boldsymbol{j}_8$	$+\boldsymbol{j}_8$	$-\boldsymbol{j}_8$	$-\boldsymbol{o}$	$+\boldsymbol{o}$	$\boldsymbol{g}$	$\boldsymbol{h}$	$\boldsymbol{k}$	$\boldsymbol{l}$	$\boldsymbol{m}$	$\boldsymbol{n}$
	$+\boldsymbol{j}_8$	$-\boldsymbol{j}_8$	$-\boldsymbol{j}_8$	$+\boldsymbol{j}_8$	$-\boldsymbol{o}$	$-\boldsymbol{o}$	$\boldsymbol{g}$	$\boldsymbol{h}$	$\boldsymbol{k}$	$\boldsymbol{l}$	$\boldsymbol{m}$	$\boldsymbol{n}$
	$-\boldsymbol{j}_8$	$+\boldsymbol{j}_8$	$+\boldsymbol{j}_8$	$-\boldsymbol{j}_8$	$-\boldsymbol{o}$	$-\boldsymbol{o}$	$\boldsymbol{g}$	$\boldsymbol{h}$	$\boldsymbol{k}$	$\boldsymbol{l}$	$\boldsymbol{m}$	$\boldsymbol{n}$
	$-\boldsymbol{j}_8$	$+\boldsymbol{j}_8$	$-\boldsymbol{j}_8$	$+\boldsymbol{j}_8$	$-\boldsymbol{o}$	$+\boldsymbol{o}$	$\boldsymbol{g}$	$\boldsymbol{h}$	$\boldsymbol{k}$	$\boldsymbol{l}$	$\boldsymbol{m}$	$\boldsymbol{n}$
	$-\boldsymbol{j}_8$	$-\boldsymbol{j}_8$	$+\boldsymbol{j}_8$	$+\boldsymbol{j}_8$	$+\boldsymbol{o}$	$-\boldsymbol{o}$	$\boldsymbol{g}$	$\boldsymbol{h}$	$\boldsymbol{k}$	$\boldsymbol{l}$	$\boldsymbol{m}$	$\boldsymbol{n}$
	$-\boldsymbol{j}_8$	$-\boldsymbol{j}_8$	$-\boldsymbol{j}_8$	$-\boldsymbol{j}_8$	$+\boldsymbol{o}$	$+\boldsymbol{o}$	$\boldsymbol{g}$	$\boldsymbol{h}$	$\boldsymbol{k}$	$\boldsymbol{l}$	$\boldsymbol{m}$	$\boldsymbol{n}$

where the symbols $\boldsymbol{g}$, $\boldsymbol{h}$, $\boldsymbol{k}$, $\boldsymbol{l}$, $\boldsymbol{m}$, $\boldsymbol{n}$, and $\boldsymbol{o}$ represent the seven columns in the eight-run Hadamard plan. By $+\boldsymbol{o}$ we mean the vector $\boldsymbol{o}$ from the Hadamard plan with signs intact and $-\boldsymbol{o}$ the vector with all signs reversed. □

Exercise 24.4: Properties of Cross-Product Arrays. Verify statistical properties 1 to 5 for the design in Exercise 24.3. □

24.4.3 Statistical Analysis of Data

We recommend that the experimenter consider at least three different statistical analyses on the data from one of these studies. Note, however, that the analyses and tests of significance from the multiple analyses are not independent. We justify this recommendation of multiple analyses with the view that the analyst wants insight and will probably require more experimental work before arriving at a final conclusion. The analyses consist of:

- A conventional analysis of variance of the data, ignoring the distinction between control and noise factors. However, this analysis must honor the design used. For example, if the control factors are randomized and then the noise factors are randomized separately for each control factor, i.e., a split-plot experiment, the analysis must follow the split-plot rules. If the experiment is saturated or too small to provide sufficient degrees of freedom for appropriate error terms, one of the graphical or semigraphical techniques discussed in Chapter 12 should be used.
- If the design is of the type obtained in the first construction, sort the data by control factor treatment combinations, compute sample variances for the control factor treatment combination subsets, and perform an analysis of variance on the logarithms of the sample variances. Again, a shortage of degrees of freedom may force a graphical technique. Such as a half-normal plot of the log (variance) values. Particular attention should be given to two-factor (control factor) interactions. Since the control design is resolution IV, these interactions will have aliases. However, if some interactions appear large, an attempt should be made to understand them, based on inheritance considerations. The conclusion may well be that

more experimental work is required. Some of the techniques discussed in Chapter 19 may prove appropriate.

- Perform an analysis for dispersion effects along the lines discussed in Chapter 12. The analysis may suggest some possible control factor interactions, interacting with noise factors, and the need for more experimental work.

24.5 COMPOUND ARRAYS

It is clear from the discussions of the dispersion analyses (Sections 24.2.5, 24.3, and 24.4.3), and quite contrary to what is written in some of the literature on robust parameter design, that interactions are of major importance. In particular, it is necessary that interactions between control and noise factors be free of interactions between control factors. In a series of papers, Rosenbaum (1994, 1996, 1999) developed a class of plans, called *compound arrays*. The objective in these plans was to maximize the number of estimable interactions, yet control the number of runs. Hedayat and Stufken (1999) formalized these structures and referred to them as *compound orthogonal arrays*.

To motivate the construction of these compound arrays, we recall the product arrays defined in expression (24.2). In the compound arrays we replace the noise array, $\boldsymbol{D}_n$, with a set of N_c orthogonal arrays. If we let $\boldsymbol{D}_n^{(1)}, \ldots, \boldsymbol{D}_n^{(N_c)}$ represent such a set of orthogonal arrays, each consisting of N_n rows and n columns (n noise factors), the compound array can be written as

$$\boldsymbol{D} = \boldsymbol{D}_c \otimes \boldsymbol{j}_{N_n} \left\| \begin{bmatrix} \boldsymbol{D}_n^{(1)} \\ \vdots \\ \boldsymbol{D}_n^{(N_c)} \end{bmatrix} \right. . \tag{24.5}$$

Note that a factorial plan that has been set up in N_c blocks provides a suitable set of N_c orthogonal arrays. So, instead of using just two fractional factorials, one for the control array and one for the noise array, the compound array uses one fraction for the control array, but splits a much larger fractional factorial into pieces (like blocks) and assigns them to the control treatment combinations.

Following the Hedayat and Stufken notation, $\boldsymbol{D}_c$ is an OA[N_c; 2^c; t_1], each $\boldsymbol{D}_n^{(i)}$ is an OA[N_n; 2^n; t_2], and $\boldsymbol{D}$ is a COA[N_cN_n; 2^{c+n}; t_1, t_2, t_3] (compound orthogonal array). The $\boldsymbol{D}_c$ array has strength t_1, the $\boldsymbol{D}_n^{(i)}$ arrays have strength t_2, and the combined array, $\boldsymbol{D}$ has strength t_3. The $\boldsymbol{D}_c$ and $\{\boldsymbol{D}_n^{(i)}\}$ arrays are selected so that they have appropriate strengths, i.e., values of t_1 and t_2. The strategy is to make t_3 as large as possible. Recall that in Section 24.2 we saw that three-factor interactions involving two control factors and one noise factor could easily be important. In fact, it is quite possible that four-factor interactions, (two from the control set and two from the noise set,) could be important. Hedayat and Stufken classify COA's as type(t_1, t_2, t_3) and argue that arrays with values for t_1, t_2, and t_3 as large as possible for given values of c, n, N_c, and N_n are most desirable.

They have examined the general problem of choices for $\boldsymbol{D}_c$ and $\{\boldsymbol{D}_n^{(i)}\}$ in the construction of two-level compound arrays. They discuss a series of four cases where $\boldsymbol{D}_c$ and the $\{\boldsymbol{D}_n^{(i)}\}$ are either full factorials or classical fractional factorials. The first is the simplest case where $\boldsymbol{D}_c$ is a full factorial and all $\boldsymbol{D}_n^{(i)} = \boldsymbol{D}_n$, with $\boldsymbol{D}_n$ also a full factorial. This implies that $\boldsymbol{D}$ is also a full factorial. These plans are simple, but the number of runs required tends to be too large. Next, there is the case of $\boldsymbol{D}_c$ a fractional factorial and one $\boldsymbol{D}_n$ as in the first case. This can be interpreted as standard construction of orthogonal arrays. The third case becomes more complex. Here $\boldsymbol{D}_c$ is a full factorial and the $\{\boldsymbol{D}_n^{(i)}\}$ consist of fractional factorials. Table 24.4 gives the construction of 40 compound arrays with desirable statistical properties.

The final case they consider is the one where $\boldsymbol{D}_c$ and the $\{\boldsymbol{D}_n^{(i)}\}$ are all fractions of full factorials. These can again be interpreted as orthogonal arrays. Table 24.5 gives the construction of 40 compound arrays in this class with desirable statistical properties.

These tables are very simple to use. Note that entries are just fractional factorials. Use the defining contrasts given to generate the fraction and then sort by the control factors to display the compound array structure. This sorting is also necessary if the experiment is to be randomized and performed as a split-unit study. If the study is to be performed as a completely randomized design, the sorting prior to randomization is not necessary. In many cases it will be desirable to introduce blocking into these experiments. Again, this introduces no difficulties. One simply selects a suitable subset of interactions and confounds them with blocks.

Example 24.1: Peanut Processor. A peanut butter manufacturer is studying his procedures with the aim of reducing cost by relaxing controls on the purchase of raw peanuts. In the manufacturing process nuts are roasted to a tightly controlled uniform color before grinding to a paste. After grinding, measured amounts of sugar, molasses, salt, and hydrogenated soybean oil are added to produce the desired product. However, there are a number of extraneous, difficult, or expensive factors to control that affect flavor.

We list five such factors, all of which can be controlled for experimental purposes.

- *Size of kernels.* Generally, larger kernels are more mature and have better flavor. Peanuts can be sorted and often are sold by size grade in the market.
- *Type of kernel.* Two major types are *runner* and *Virginia large seeded.* It is possible to ensure that one purchases only one type or the other. However, it would simplify matters if this control were not necessary.
- *Storage.* Peanuts change in storage. The rate of change depends in part on storage temperature. The manufacturer does not want the flavor of the product to be influenced by the aging of peanuts. For experimental

Table 24.4 First Table of Compound Arrays Construct as Fractional Factorial

Plan	N_c	N_n	c	n	Res_{tot}	Res_{noise}	Defining Contrasts
1	4	2	2	2	IV	II	$A_cB_cC_nD_n$
2	4	2	2	3	III	II	$A_cC_nD_n$, $B_cC_nE_n$
3	4	4	2	3	V	III	$A_cB_cC_nD_nE_n$
4	4	2	2	4	III	II	$A_cC_nD_n$, $B_cC_nE_n$, $A_cE_nF_n$
5	4	4	2	4	IV	II	$A_cB_cC_nD_n$, $C_nD_nE_nF_n$
6	4	8	2	4	VI	IV	$A_cB_cC_nD_nE_nF_n$
7	4	2	2	5	III	^{a}	$A_cB_cC_n$, $C_nD_nE_n$, $E_nF_nG_n$, $B_cD_nF_n$
8	4	4	2	5	IV	II	$A_cB_cC_nD_n$, $A_cB_cE_nF_n$, $A_cC_nE_nG_n$
9	4	8	2	5	IV	III	$A_cC_nD_nE_n$, $B_cC_nF_nG_n$
10	4	16	2	5	VII	V	$A_cB_cC_nD_nE_nF_n$
11	8	2	3	2	V	II	$A_cB_cC_cD_nE_n$
12	8	2	3	3	IV	II	$A_cB_cD_nE_n$, $A_cC_cE_nF_n$
13	8	4	3	3	VI	III	$A_cB_cC_cD_nE_nF_n$
14	8	2	3	4	IV	II	$A_cB_cD_nG_n$, $A_cB_cE_nF_n$, $A_cC_cE_nG_n$
15	8	4	3	4	IV	II	$A_cB_cD_nE_nF_n$, $A_cC_cF_nG_n$
16	8	8	3	4	VII	IV	$A_cB_cC_cD_nE_nF_nG_n$
17	8	2	3	5	IV	^{a}	$A_cB_cC_cD_n$, $C_cD_nE_nF_n$, $E_nF_nG_nH_n$, $A_cB_cG_nH_n$, $A_cC_cE_nG_n$
18	8	4	3	5	IV	II	$A_cC_cD_nE_n$, $A_cB_cE_nF_n$, $C_cD_nG_nH_n$
19	8	8	3	5	V	III	$A_cC_cD_nE_nF_n$, $A_cB_cF_nG_nH_n$
20	8	16	3	5	VIII	V	$A_cB_cC_cD_nE_nF_nG_nH_n$
21	16	2	4	2	VI	II	$A_cB_cC_cD_cE_nF_n$
22	16	2	4	3	IV	II	$A_cD_cE_nG_n$, $B_cC_cD_cE_nF_n$
23	16	4	4	3	VII	III	$A_cB_cC_cD_cE_nF_nG_n$
24	16	2	4	4	IV	II	$A_cC_cE_nG_n$, $A_cB_cE_nF_n$, $A_cD_cE_nH_n$
25	16	4	4	4	V	II	$A_cB_cF_nG_nH_n$, $A_cC_cD_cE_nF_n$
26	16	8	4	4	VIII	IV	$A_cB_cC_cD_cE_nF_nG_nH_n$
27	16	2	4	5	IV	II	$A_cB_cE_nF_n$, $A_cC_cE_nG_n$, $A_cD_cE_nH_n$, $B_cC_cE_nH_nI_n$
28	16	4	4	5	IV	II	$A_cB_cC_cD_cE_nF_n$, $A_cB_cE_nG_nH_n$, $A_cD_cE_nG_nI_n$
29	16	8	4	5	VI	III	$A_cB_cC_cE_nF_nG_n$, $A_cD_cE_nF_nH_nI_n$
30	16	16	4	5	IX	V	$A_cB_cC_cD_cE_nF_nG_nH_nI_n$
31	32	2	5	2	VII	II	$A_cB_cC_cD_cE_cF_nG_n$
32	32	2	5	3	IV	II	$A_cB_cC_cF_nG_n$, $A_cD_cE_cF_nH_n$
33	32	4	5	3	VIII	III	$A_cB_cC_cD_cE_cF_nG_nH_n$
34	32	2	5	4	IV	II	$A_cB_cC_cD_cF_nG_n$, $A_cB_cE_cG_nH_n$, $A_cD_cE_cG_nI_n$

Table 24.4 *(continued)*

Plan	N_c	N_n	c	n	Res_{tot}	Res_{noise}	Defining Contrasts
35	32	4	5	4	VI	II	$A_cB_cC_cD_cF_nG_n$, $A_cB_cE_cF_nH_nI_n$
36	32	8	5	4	IX	IV	$A_cB_cC_cD_cE_cF_nG_nH_nI_n$
37	32	2	5	5	IV	II	$A_cB_cF_nG_n$, $A_cC_cD_cE_cF_nH_n$, $A_cE_cF_nG_nH_nI_n$, $A_cD_cF_nG_nH_nJ_n$
38	32	4	5	5	V	II	$A_cB_cF_nG_nH_n$, $A_cC_cF_nI_nJ_n$, $A_cD_cE_cF_nG_nJ_n$
39	32	8	5	5	VI	III	$A_cB_cC_cD_cF_nG_nH_n$, $A_cB_cE_cF_nI_nJ_n$
40	32	16	5	5	X	V	$A_cB_cC_cD_cE_cF_nG_nH_nI_nJ_n$

[a]Lose at least one main effect.

Source: Adapted with permission from Hedayat and Stufken (1999).

purposes, it is possible to hold part of a lot of nuts under optimal conditions and accelerate the aging process of the remainder by increasing the temperature.

- *Alcohol level.* Poor handling and/or immature harvesting lead to peanuts with excessive alcohol content. This in turn leads to an off-flavor in peanut products. The manufacturer would like to have the process robust to the occasional lot with high alcohol level. For experimental purposes it is possible to induce high alcohol levels in some lots and to obtain other lots that do not have high alcohol levels.
- *Growing locations.* There are differences between peanuts produced in different growing regions. The manufacturer can insist that all peanuts be purchased from one or another region. However, it would simplify purchasing if geographic region of origin could be ignored.

The manufacturer wants a uniform product. In a very real sense, this is the definition of quality. The manufacturer easily can control roast color, amount of sugar, amount of molasses, amount of salt, and amount of hydrogenated soybean oil that is added. These are control factors. On the other hand, size, type of kernel, storage, alcohol level, and growing location are five factors that affect flavor but are difficult or expensive to control. They can however, be controlled for purposes of an experiment. These are noise factors.

Ideally, the manufacturer would like a set of control conditions that would yield a good product uniformly, regardless of the state of the noise factors. It may be that a moderately dark roast, a slight increase in molasses, and a slight decrease in sugar gives a recipe that is more robust to peanut storage and/or off-flavor due to alcohol level.

Table 24.5 Second Table of Compound Arrays Construct as Fractional Factorial

Plan	N_c	N_n	c	n	$\text{Res}_{control}$	Res_{tot}	Res_{noise}	Defining Contrasts
1	4	2	3	2	IV	III	II	$A_cB_cC_c$, $C_cD_nE_n$
2	4	2	3	3	III	III	II	$A_cB_cC_c$, $C_cD_nE_n$, $A_cC_cE_nF_n$
3	4	4	3	3	III	III	III	$A_cB_cC_c$, $A_cD_nE_nF_n$
4	4	2	3	4	III	III	II	$A_cB_cC_c$, $A_cB_cD_nE_n$, $A_cC_cD_nF_n$, $B_cC_cD_nG_n$
5	4	4	3	4	III	III	II	$A_cB_cC_c$, $A_cD_nE_n$, $B_cD_nF_nG_n$
6	4	8	3	4	III	III	IV	$A_cB_cC_c$, $A_cB_cD_nE_nF_n$
7	4	2	3	5	III	II	II	$A_cB_cC_c$, $A_cD_nE_n$, $B_cD_nF_n$, $C_cD_nG_n$, D_nH_n
8	4	4	3	5	III	III	II	$A_cB_cC_c$, $A_cB_cD_nE_n$, $A_cC_cD_nF_nG_n$, $B_cC_cD_nF_nH_n$
9	4	8	3	5	III	III	III	$A_cB_cC_c$, $A_cD_nE_nF_n$, $B_cD_nG_nH_n$
10	4	16	3	5	III	III	V	$A_cB_cC_c$, $A_cD_nE_nF_nG_nH_n$
11	8	2	4	2	IV	IV	II	$A_cB_cC_cD_c$, $A_cB_cE_nF_n$
12	8	2	4	3	IV	IV	II	$A_cB_cC_cD_c$, $A_cB_cE_nF_n$, $A_cC_cE_nG_n$
13	8	4	4	3	IV	IV	III	$A_cB_cC_cD_c$, $A_cB_cE_nF_nG_n$
14	8	2	4	4	IV	IV	II	$A_cB_cC_cD_c$, $A_cB_cE_nF_n$, $A_cC_cE_nG_n$, $A_cD_cE_nH_n$
15	8	4	4	4	IV	IV	II	$A_cB_cC_cD_c$, $A_cB_cE_nF_nG_n$, $A_cC_cE_nH_n$
16	8	8	4	4	V	IV	IV	$A_cB_cC_cD_c$, $A_cB_cE_nF_nG_nH_n$
17	8	2	4	5	V	III	II	$A_cB_cC_cD_c$, $A_cB_cE_nF_n$, $A_cC_cE_nG_n$, $A_cB_cG_nH_n$, $A_cB_cE_nG_nI_n$
18	8	4	4	5	V	IV	II	$A_cB_cC_cD_c$, $A_cB_cE_nF_n$, $A_cC_cE_nG_n$, $A_cD_cE_nH_nI_n$

Table 24.5 (*continued*)

Plan	N_c	N_n	c	n	$\text{Res}_{control}$	Res_{tot}	Res_{noise}	Defining Contrasts
19	8	8	4	5	V	IV	III	$A_cB_cC_cD_c$, $A_cB_cE_nF_nG_n$, $A_cC_cE_nH_nI_n$
20	8	16	4	5	V	IV	V	$A_cB_cC_cD_c$, $A_cB_cE_nF_nG_nH_nI_n$
21	8	2	5	2	III	III	II	$A_cB_cC_c$, $A_cD_cE_c$, $B_cD_cF_nG_n$
22	16	2	5	2	V	IV	II	$A_cB_cC_cD_cE_c$, $A_cB_cF_nG_n$
23	8	2	5	3	III	III	II	$A_cB_cC_c$, $A_cD_cE_c$, $B_cD_cF_nG_n$, $B_cE_cF_nH_n$
24	16	2	5	3	V	IV	II	$A_cB_cC_cD_cE_c$, $A_cB_cF_nG$, $A_cC_cF_nH_n$
25	8	4	5	3	III	III	III	$A_cB_cC_c$, $A_cD_cE_c$, $B_cD_cF_nG_nH_n$
26	16	4	5	3	V	V	III	$A_cB_cC_cD_cE_c$, $A_cB_cF_nG_nH_n$
27	8	2	5	4	III	III	II	$A_cB_cC_c$, $A_cD_cE_c$, $A_cF_nG_n$, $B_cF_nH_n$, $D_cH_nI_n$
28	16	2	5	4	V	IV	II	$A_cB_cC_cD_cE_c$, $A_cB_cF_nG_n$, $A_cC_cF_nH_n$, $A_cD_cF_nI_n$
29	8	4	5	4	III	III	II	$A_cB_cC_c$, $A_cD_cE_c$, $A_cF_nG_nH_n$, $B_cD_cF_nI_n$
30	16	4	5	4	V	IV	II	$A_cB_cC_cD_cE_c$, $A_cB_cF_nG_n$, $A_cC_cF_nH_n$, $A_cB_cD_cH_nI_n$
31	8	8	5	4	III	III	IV	$A_cB_cC_c$, $A_cD_cE_c$, $B_cD_cF_nG_nH_nI_n$
32	16	8	5	4	V	V	IV	$A_cB_cC_cD_cE_c$, $A_cB_cF_nG_nH_nI_n$
33	8	2	5	5	III	III	II	$A_cB_cC_c$, $A_cD_cE_c$, $A_cF_nG_n$, $A_cH_nI_n$, $B_cI_nJ_n$, $D_cF_nJ_n$
34	16	2	5	5	V	IV	II	$A_cB_cC_cD_cE_c$, $A_cB_cF_nG_n$, $A_cC_cF_nH_n$, $A_cD_cF_nI_n$, $A_cE_cF_nJ_n$

(*continued*)

Table 24.5 ***(continued)***

Plan	N_c	N_n	c	n	$\text{Res}_{control}$	Res_{tot}	Res_{noise}	Defining Contrasts
35	8	4	5	5	III	III	II	$A_cB_cC_c$, $A_cD_cE_c$, $B_cD_cF_nG_n$, $A_cF_nH_nI_n$, $A_cG_nH_nJ_n$
36	16	4	5	5	V	IV	II	$A_cB_cC_cD_cE_c$, $A_cB_cF_nG_n$, $C_cE_cI_nJ_n$, $A_cC_cF_nH_nI_n$
37	8	8	5	5	III	III	III	$A_cB_cC_c$, $A_cD_cE_c$, $B_cD_cF_nG_nH_n$, $B_cE_cF_nI_nJ_n$
38	16	8	5	5	V	V	III	$A_cB_cC_cD_cE_c$, $A_cB_cF_nG_nH_n$, $A_cC_cF_nI_nJ_n$
39	8	16	5	5	III	III	V	$A_cB_cC_c$, $A_cD_cE_c$, $B_cD_cF_nG_nH_nI_nJ_n$
40	16	16	5	5	V	V	V	$A_cB_cC_cD_cE_c$, $A_cB_cF_nG_nH_nI_nJ_n$

Source: Adapted with permission from Hedayat and Stufken (1999).

A possible experimental plan is number 38 in Table 24.5. The plan is a 128-run fractional factorial with defining contrast

$$\begin{aligned}\mathcal{I} = & \quad + ABCDE = +ABFGH = +CDEFGH \\ & = +ACFIJ = +BDEFIJ = +BCGHIJ = +ADEGHIJ.\end{aligned}$$

The five control factors are assigned to letters A to E and the five noise factors to letters F to J. In the laboratory it is necessary to make up 128 different peanut butters. These must be tested.

We now encounter an additional difficulty that does not occur in many areas of research. Flavor must be evaluated by a trained taste panel. There are two distinct strategies that can be adopted here. The simpler is to test each of these experimental peanut butters against a controlled standard. The test peanut butters are then given a positive score if they are "better" than the control and a negative score if they are "poorer." The alternative strategy is to evaluate the 128 peanut butters among themselves as a group. Now the problem is that there is a limit to the number of samples that even a well-trained taste panel can evaluate in a session. We will be conservative and assume that four samples is the limit. This means that we have to confound 31 interactions with panel sessions. Panel sessions are blocks.

A reasonable set of interactions to confound with blocks, found after some searching, is $\{DE, CD, AJ, GJ, BI\}$. As a consequence of this, two-factor interactions $\{CE, AG, FJ, AF, FG, CH, DH, EH\}$ and a number of higher-order interactions are also confounded with blocks. The remaining $45 - 13 = 32$ two-factor interactions are clear and estimable.

CHAPTER 25

Experiments on the Computer

25.1 INTRODUCTION

A standard occurrence in science is to encounter problems that are too large and too complex and often too poorly understood to allow for simple direct experimentation. Occasionally, cost and/or ethics of conducting experiments contribute reasons for a reluctance to experiment. One approach around these problems is to construct a theoretical model based on the best existing knowledge, to simulate the model, and to conduct experiments on the computer to obtain better understanding. These simulation models are often extremely complex and require large amounts of computer time. The special nature of these experiments presents a unique set of design problems. The object of this chapter is to examine some of these designs. A key feature that makes these problems different is that repeated runs with a simulation will give identical results. There is no need for replication. Replication, which was the hallmark of the designs studied up to this point, represents wasted effort in the computer experiment setting.

Generally speaking, the immediate objective of a simulation run on a computer is to evaluate some quantity. If the quantity of interest, y, is a function of random variables $(v_1, v_2, \ldots, v_k)$, and the objective is to evaluate the expected value of y, this sort of problem can be thought of as numerical integration, where the objective of the simulation is to evaluate the expectation

$$E\big[y(v_1 \cdots v_k)\big] = \int \cdots \int y(v_1, \ldots, v_k)\, dFv_1 \cdots dFv_k. \qquad (25.1)$$

A special case of this that occurs frequently in statistics is that one wants to evaluate the probability distribution of a statistic that happens to be a nasty (but known) function of a set of random variables with known distribution. This is

Planning, Construction, and Statistical Analysis of Comparative Experiments,
by Francis G. Giesbrecht and Marcia L. Gumpertz
ISBN 0-471-21395-0

equivalent to evaluating the expectation $E[g]$, where

$$g = \begin{cases} 0 & \text{for } y(v_1, \ldots, v_k) > \theta \\ 1 & \text{for } y(v_1, \ldots, v_k) \le \theta \end{cases} \tag{25.2}$$

for assorted values of θ.

This is essentially the problem addressed by Mckay et al. (1979). They point out that one classical approach to the problem expressed in equation (25.1) is to draw N random evaluations of the vector $(v_1, \ldots, v_k)$ from the appropriate distribution(s), evaluate $y(v_1, \ldots, v_k)$ for each, and compute the average. In the case of expression (25.2), one simply counts and evaluates the proportion less than or equal to the selected values of θ. The difficulty comes when the simulation is complex and the computer runs become expensive. The sampling literature provides an answer, stratified sampling. It is well known that stratified sampling leads to better estimates of the mean than does simple random sampling. The gain in efficiency depends on the nature of the stratification.

To provide some intuitive feeling for the orthogonal array–based Latin hypercube sampling that we are about to develop, and more important, to generate a suitable system of notation, we begin with the simple stratified sample.

25.2 STRATIFIED AND LATIN HYPERCUBE SAMPLING

25.2.1 Stratified Sampling

We think of the simulation program as a black box that returns a value in response to an input vector. Let $\boldsymbol{v}_i$ represent the row vector of k random variables that serve as input to the simulation program and $y(\boldsymbol{v}_i)$ the output. In general, v_{ij} represents a realization of a random variable from the distribution F_j for $j = 1, \ldots, k$. The $\{F_j\}$ do not have to be identical, and in fact, the $\{v_{ij}, \ldots, v_{ij}\}$ do not need to be independent, although our notation suggests otherwise. We also note that a sample of N computer runs is defined completely by the $N \times k$ matrix $\boldsymbol{V}$. Notice that we can write $\boldsymbol{V}^t = \boldsymbol{v}_1^t \| \cdots \| \boldsymbol{v}_N^t$. Simple random sampling to find an estimate of $E[y]$ can now be described as drawing a sample N realizations of $\boldsymbol{v}_i$, computing $y(\boldsymbol{v}_i)$, $i = 1, \ldots, N$, and then computing the average $\overline{y}$. An unbiased estimate of the variance of this estimate is computed as

$$\frac{1}{N} \sum_i \left(y(\boldsymbol{v}_i) - \overline{y}\right)^2 / (N - 1).$$

The procedure that comes to mind to increase the efficiency (reduce the variance of the estimate) is stratified sampling. A simplifying device is to think of the $\boldsymbol{v}_i$ coming from a multidimensional space denoted by S. The first step is to partition S into M disjoint strata $\{S_\ell\}$. Let $p_\ell = \text{Prob}(\boldsymbol{v} \in S_\ell)$ represent the size of stratum S_ℓ. The stratified sampling is to select n_ℓ samples from stratum S_ℓ

subject to the condition that $\sum n_\ell = N$. The simple random sample is the special case of this with only one stratum. The advantage of stratified sampling comes from the fact that it is possible to force all strata to be represented with prespecified frequencies. Care must be exercised to make sure that weights attached to cell means are appropriate to yield an unbiased estimate of $E[y]$.

Example 25.1: Simulation to Find the Expected Value of the F-Statistic. This example is somewhat artificial in the sense that the simulation is simple and that we know the exact answer from general theory. The problem is to find the expected value of the F-statistic computed in an analysis of variance for a data set described by the model

$$y_{ij} = \mu + a_i + e_{ij}$$

where $j = 1, 2, 3$, $i = 1, 2, \ldots, 7$, $a_i \sim N(0, 1)$, $e_{ij} \sim N(0, 1)$ and all random variables are mutually independent.

Simple Random Sampling. We can use simple random sampling to estimate this by generating seven $\{a_i\}$ values and 21 $\{e_{ij}\}$ values, constructing 21 $\{y_{ij}\}$ values and computing the F-statistic repeatedly. The unbiased estimate of $E[F]$ is the mean of the generated values. An estimate of the variance of the estimate is obtained directly from the simulated F-values.

25.2.2 Latin Hypercube

When we consider stratified sampling we immediately encounter the question of how to construct strata to yield the greatest benefit. One reasonable criterion is that the full range of values of each of the k random variables be represented. Hence the stratification is done on the probability scale rather than on the scale of the random variables. To simplify the notation for the remainder of this chapter, we assume that each of the k random variables has a uniform (0, 1) distribution and the $\boldsymbol{v}_i$ vectors are points in the k-dimensional hypercube. We can think of the transformation to any other distribution as being absorbed into the $y(v_1, \ldots, v_k)$ function, as demonstrated in the continuation of Example 25.1. Our requirement of appropriate representation of all values for all variables now translates into the requirement that the realizations for each variable be uniformly distributed on the (0, 1) interval.

The Latin hypercube sample is constructed in steps:

1. Construct the $N \times k$ matrix $\boldsymbol{X}$ in which the k columns consist of distinct permutations of the elements $0, 1, 2, \ldots, (N - 1)$.
2. Construct a matrix $\boldsymbol{V}$ by adding a uniform (0, 1) number to each element in $\boldsymbol{X}$.
3. Divide each element in $\boldsymbol{V}$ by N.

All elements in the resulting $\boldsymbol{V}$ matrix lie between 0 and 1. This matrix defines the Latin hypercube sample.

The origin of the term *Latin hypercube* can be understood if we think of the columns of the $\boldsymbol{X}$ matrix as defining sparsely populated $N \times N \times N \cdots$ hypercube, where the values of the first column select the rows of the square, those of the second select the columns, and so on. One can think of the $\boldsymbol{X}$ matrix as describing the strata to be selected and the $\boldsymbol{V}$ matrix the actual sample to be drawn. A big advantage of the Latin hypercube appears when the response $y(v_1, \ldots, v_k)$ happens to be dominated by only a few of the k components. This can be thought of as erasing columns from the $\boldsymbol{V}$ matrix. The values in the remaining columns are uniformly distributed across their full range. The method ensures that each important component is represented in a fully stratified manner.

Stein (1987) shows that as long as N is large compared to k, Latin hypercube sampling gives a smaller variance than simple random sampling for any $y(\cdot)$ having finite second moment, and that the mean based on Latin hypercube sampling is asymptotically normal as N increases.

We haven't yet discussed how to generate the $\boldsymbol{X}$ matrix. A suitable procedure given by Stein is to generate an $N \times k$ matrix $\boldsymbol{Z}$ in which the rows are independent, identically distributed vectors from the joint distribution of the k input variables. For example, if the input variables to the simulation represent independent normal variables, the rows of $\boldsymbol{Z}$ are sets of k independent normal variates. If the input variables are from a multivariate normal with known correlation matrix, the rows of $\boldsymbol{Z}$ are from the multivariate normal. The elements in the columns of $\boldsymbol{Z}$ are then ranked from 0 to $N - 1$. This represents the $\boldsymbol{X}$ matrix. The $\boldsymbol{X}$ matrix is then converted to the $\boldsymbol{V}$ matrix for input to the simulation.

Since, in general, the variance of the sample mean depends on the stratification, standard practice is to run, say, r sets of N simulations and then estimate the variance of the overall mean from the sample variance computed from the r subsample means.

Example 25.1 (continued): Simulation to Find the Expected Value of the F-Statistic—Latin Hypercube Sampling. In this example there are seven random a_i effects and 21 random errors ϵ_{ij}, where all are independent and follow standard normal distributions. If we want to use a Latin hypercube, we need to construct the $N \times 28$ matrix $\boldsymbol{X}$. Since the 28 random variables involved are all mutually independent, we first use our random number generator to produce an $N \times 28$ matrix of independent $N(0, 1)$ variates. This represents the $\boldsymbol{Z}$ matrix. In each of the columns, replace the variates by their ranks within the column, i.e., 0 to $N - 1$, in some order. This is the $\boldsymbol{X}$ matrix. Next add an independent uniform (0, 1) variate to each element in the matrix and divide by N. This produces the $\boldsymbol{V}$ matrix. Finally, replace each of the (0, 1) values by $\Phi^{-1}(\cdot)$ to obtain a_i or ϵ_{ij}. The rows of the resulting matrix are then used to construct the $\{y_{ij}\}$ values and subsequently the F-values. The estimate of $E[F]$ is again the mean of the F-values computed. A simple method of estimating the error of the estimate is to repeat the simulation of sets of N values a number of times and then compute the variance from the set of means.

25.3 USING ORTHOGONAL ARRAYS FOR COMPUTER SIMULATION STUDIES

We begin by noting that the $\boldsymbol{X}$ matrices that we have used can be interpreted as OA$[N;\ N^k;\ 1]$'s. All of the values 0 to $N-1$ appear once in each column in these arrays, ensuring that each of the marginal distributions is properly represented. The logical extension is to use OA$[n;\ N^k;\ 2]$'s which will spread the sampling strata over the bivariate distribution for each pair of random variables being simulated. To illustrate the construction of an $\boldsymbol{X}$ matrix for a Latin hypercube sample, we begin with an OA$[n;\ s^k;\ 2]$ and notice that n, the number of runs, must be some multiple of s^2, i.e., $n = \lambda s^2$. It follows that in each column, each of s elements occurs λs times. The construction now follows:

1. Replace the s symbols in the orthogonal array with a random permutation of the numbers, $0, 1, \ldots, s-1$.
2. In each column, 0 appears λs times, 1 appears λs times, and so on. In each column replace the 0's by a random permutation of the numbers $0, 1, \ldots, \lambda s - 1$, the 1's by a random permutation of the numbers $\lambda s + 0, \lambda s + 1, \ldots, 2\lambda s - 1$, and so on, until finally, the $s - 1$'s by a random permutation of the numbers $(s-1)\lambda s + 0, (s-1)\lambda s + 1, \ldots, \lambda s^2 - 1$. This is done independently for each of the k columns. The result is a $\lambda s^2 \times k$ matrix with columns consisting of independent permutations of the values from 0 to $\lambda s^2 - 1$, i.e., an $\boldsymbol{X}$ matrix defining a Latin hypercube sample of size $N = \lambda s^2$. The $\boldsymbol{X}$ matrix satisfies the additional constraint inherited from the orthogonal array.
3. Randomize the columns.
4. Convert to a $\boldsymbol{V}$ matrix.

To obtain an estimate of error for the final estimate, r distinct randomly generated samples are drawn and the variance is estimated from the sample means. Clearly, the construction extends to orthogonal arrays of strength d, which ensure that all d-variate distributions are represented properly. The disadvantage is that the arrays become excessively large. Methods for constructing orthogonal arrays with suitably large numbers of columns are given in Section 23.5. N. J. A. Sloan at AT&T maintains a Web site, *www.research.alt.com/~ njas/oadir*, which contains a large collection of orthogonal arrays. Hedayat et al. (1999) give a comprehensive review and discussion of the problem of constructing orthogonal arrays.

Example 25.1 (continued): Simulation to Find the Expected Value of the F-Statistic—Orthogonal Array Sampling. The first step is to select an orthogonal array with 28 columns. Recall that one can discard excess columns in an orthogonal array. Three possible choices would be subsets from OA$[125;\ 5^{31};\ 2]$, OA$[841;\ 29^{30};\ 2]$, or OA$[32;\ 2^{31};\ 2]$ obtained from the 32×32 Hadamard matrix. The first array is obtained as a main effect plan from the 5^3 factorial. The second array is even simpler to obtain since 29 is a prime number and columns are obtained directly as a main-effect plan by superimposing on a 29^2 factorial. We demonstrate the calculations first with the OA[125; 5^{31}; 2].

Since we only need an OA[125; 5^{28}; 2], we generate only 28 columns. Begin with the 5^3 factorial, construct the model, and select 28 columns. The 125 rows of the matrix formed by columns 1, 2, and 3 are $0\ 0\ 0, 0\ 0\ 1, 0\ 0\ 2, \ldots, 4\ 4\ 3, 4\ 4\ 4$. Denote the columns as $C1$, $C2$, and $C3$. Generate the remaining 25 columns as $C1 + C2, \ldots, C1 + 4C2, C1 + C3, \ldots, C1 + 4C3, C2 + C3, \ldots, C2 + 4C3, C1 + C2 + C3, C1 + C2 + 2C3, \ldots, C1 + 4C2 + C3$, where all of the arithmetic is done modulo 5. Use a random order to assemble these 28 columns into a 125×28 matrix. Note that each column consists of a set of 25 zeros, a set of 25 ones, ... and a set of 25 fours in some order. Within each column use a random permutation of $0, \ldots, 24$ to replace one of the sets, a random permutation of $25, \ldots, 49$ to replace another set, ..., and a random permutation of $100, \ldots, 124$ to replace the last set. For example, in one column the 25 zeros are replaced by a permutation of $25, \ldots, 49$, the 25 ones by a permutation of $0, \ldots, 24$, and so on. In a second column, the zeros are replaced by $100, \ldots, 124$, and so on. The sets are selected in random order. Essentially what we want is to randomize the order of the columns and the symbols in the orthogonal array. The result of these replacements is a matrix with 28 randomly generated permutations of the values, 0 to 124, i.e., a suitable $\boldsymbol{X}$ matrix. Now convert this to a $\boldsymbol{V}$ matrix by adding an independent uniform (0, 1) variate to each element and dividing by 125. This defines the sample. Finally, apply the $\Phi^{-1}(\cdot)$ function and compute the 125 F-values as in the simple random sample. The estimate of $E[F]$ is obtained as the average. To obtain an estimate of the variance of the estimate, use the orthogonal array to generate r sets of 125 F-values and compute the sample variance from among the r averages.

25.4 DEMONSTRATION SIMULATIONS

25.4.1 Comparison of Three Sampling Methods

To illustrate properties of the three sampling methods: simple random sampling, Latin hypercube sampling, and orthogonal array sampling, we present a simulation of Example 25.1. Recall that the problem is to find the expected value of the F-statistic computed in an analysis of variance for a data set described by the model

$$y_{ij} = \mu + a_i + e_{ij},$$

where $j = 1, 2, 3, i = 1, 2, \ldots, 7, a_i \sim N(0, 1), e_{ij} \sim N(0, 1)$, and all random variables mutually independent. Under this model, $\mathrm{E}[F] = 14/3$.

We simulated 1000 sets of 125 F-values using simple random sampling, using the OA[125; 5^{28}; 2], and using Latin hypercube sampling. The results are given in Table 25.1.

Table 25.1 Summary of Three Sampling Procedures, Each with 125 Monte Carlo Samples

Method	Mean	SD	Variance
Simple random	4.6636	.320	.102
Latin hypercube	4.6772	.260	.068
Orthogonal array	4.6635	.162	.026
True value	4.6667		

Table 25.2 Summary of Two Sampling Procedures, Each with 32 Monte Carlo Samples

Method	Mean	SD	Variance
Simple random	4.6147	.628	.394
Orthogonal array	4.6697	.423	.179
True value	4.6667		

It is clear that all three methods are unbiased. The big difference is as advertised, in the standard deviations. The orthogonal array, which forces bivariate marginals, leads to a 50% reduction in standard deviation.

It is also informative to construct a Latin hypercube sample based on the OA[32; 2^{28}; 2] obtained from the Hadamard matrix. The randomization is again in two steps. First, the columns are randomized. Then, within columns the 0's and 1's are replaced by random permutations of 0 to 15 and 16 to 31, or vice versa, at random. This defines the 32×28 $\boldsymbol{X}$ matrix. This is then converted to a $\boldsymbol{V}$ matrix and then to the input variables. This leads to samples of 32 F-values. We repeated this 1000 times and also for comparison generated 1000 sets of 32 F-values using simple random sampling. The results are given in Table 25.2

The standard deviations for the simple random samples are in the proportion $\sqrt{(125/32)}$. However, the orthogonal array sampling did not lead to as big a gain in efficiency. The reason is that the OA[32; 2^{28}; 2] split the sampling region into 2^{28} strata and selected a specially balanced sample. The OA[125; 5^{28}; 2] split the region into 5^{28} strata and selected a sample. The latter provided a much tighter control on the sampling and consequently, a much smaller variance. This suggests that the OA[841; 29^{28}; 2] might provide an even better sampling scheme. However, 100 sets consisting of samples of 841 based on OA[841; 29^{28}; 2] gave an overall mean of 4.6763 and a standard deviation among set means of .075. This would compare with a simple random sampling standard deviation equal to $\sqrt{(125/841)} \times .320 = .123$. There is an almost 50% reduction in standard error in response to using the stratification. It appears that the fact that such an extremely small fraction of strata is visited in the simulation tends to limit the gain in efficiency somewhat. However, notice that the standard deviation from the orthogonal array with 32 samples is nearly as small as from the simple random sample with 125 Monte Carlo runs.

Example 25.2: Simulation to Evaluate an Alpha Design. This example is more realistic than Example 25.1 in the sense that calculations based on general theory are mathematically intractable, so simulation is a good way to answer the research question.

Simulation is a very handy tool for evaluating experimental designs before doing an actual physical experiment. Suppose that an investigator is confronted with 30 different treatments, with the object to select the best from the set. We assume a worst-case scenario, 29 equal and one better by an amount Δ. The investigator proposes using an α-design with 18 blocks of five. After the experiment is completed, the two best treatments will be selected and the trial declared a success if the true best is one of the pair selected. The question is: What are the chances of success? Notice that pure chance gives odds of success of $2/30$.

As a simulation problem we have 18 random block effects and 90 random errors associated with experimental units. For this example we assume that both block variance and unit variance are equal to 1.0. Independent normal variates are assumed throughout.

Orthogonal Array Sampling. For this simulation we need an orthogonal array with 108 columns. A suitable choice is AO[1331;11^{108};2], which is easily generated using `Proc FACTEX` in SAS® or equivalently, working directly from the 11^3 factorial. We present the results for Δ, the difference between the desired treatment and the remainder equal to .1, .2, .5, and 1.0. We report the estimated probability of success for 25×1331 simulation runs and the observed variance among 25 estimates, each based on 1331 runs.

For comparison the simple random sampling simulations were also performed with the same model. The results are shown in Table 25.3. We notice that the simple random sampling variances are roughly three to four times the size of the orthogonal array sampling variances. The estimates of probability of success are equivalent. Both schemes are unbiased.

Table 25.3 Simulation of Success Probabilities, α-Design

Method	Δ	Estimate	Variance
Orthogonal array	.1	.089, 98	.000, 028
	.2	.111, 44	.000, 038
	.5	.217, 22	.000, 034
	1.0	.477, 93	.000, 056
Simple random	.1	.088, 42	.000, 081
	.2	.112, 85	.000, 042
	.5	.217, 28	.000, 191
	1.0	.477, 27	.000, 280

Summarizing Comments

In a simulation problem as small as this one, we typically would not consider using anything more complex than simple random sampling. The amount of work involved in generating the Latin hypercube or the orthogonal array samples is not justified. However, the example does illustrate the potential gains. At this point much further research is needed to provide guidance in selecting suitable orthogonal arrays. The small example certainly suggests that arrays based on Hadamard matrices are probably not good choices. For more details on Latin hypercube sampling, the reader is encouraged to consult Mckay et al. (1979), Iman and Conover (1980), Stein (1987), Owen (1992, 1994), and Tang (1993, 1994).

References

Abeyasekera, S. and R. N. Curnow (1984). The desirability of adjusting for residual effects in a crossover design. *Biometrics* *40*(4), 1071–1078.

Abraham, B., H. Chipman, and K. Vijayan (1999). Some risks in the construction and analysis of supersaturated designs. *Technometrics* *41*(2), 135–141.

Addelman, S. (1962, February). Symmetrical and asymmetrical fractional factorial plans. *Technometrics* *4*(1), 47–58.

Addelman, S. (1967, March). Equal and proportional frequency squares. *Journal of the American Statistical Association* *62*(1), 226–240.

Addelman, S. (1969, August). Sequences of two-level fractional factorial plans. *Technometrics* *11*(3), 477–509.

Addelman, S. and O. Kempthorne (1961, December). Some main effect plans and orthogonal arrays of strength two. *Annals of Mathematical Statistics* *32*(4), 1167–1176.

Afsarinejad, K. (1990). Repeated measurements designs—a review. *Communications in Statistics—Theory and Methods* *19*(11), 3985–4028.

Atkinson, A. and A. Donev (1992). *Optimum Experimental Designs*. Oxford Statistical Science Series. Oxford: Clarendon Press.

Bailey, R. A. (1992). Efficient semi- Latin squares. *Statistica Sinica* *2*, 413–437.

Bailey, R. A., C. S. Cheng, and P. Kipnis (1992, July). Construction of trend-resistant factorial designs. *Statistica Sinica* *2*(2), 383–411.

Balagtas, C. C., M. P. Becker, and J. B. Lang (1995). Marginal modeling of categorical data from crossover experiments. *Applied Statistics* *44*(1), 63–77.

Barnard, G. A. (1986). Causation. See Kotz and Johnson (1986), pp. 387–389.

Beal, S. L. (1989). Sample size determination for confidence intervals on the population mean and on the difference between two population means. *Biometrics* *45*, 969–977.

Bechhofer, R. E., T. J. Santner, and D. M. Goldsman (1995). *Design and Analysis of Experiments for Statistical Selection, Screening, and Multiple Comparisons*. Wiley Series in Probability and Statistics. New York: John Wiley & Sons.

Becker, M. P. and C. C. Balagtas (1993). Marginal modeling of binary cross-over data. *Biometrics* *49*(4), 997–1009.

Planning, Construction, and Statistical Analysis of Comparative Experiments,
by Francis G. Giesbrecht and Marcia L. Gumpertz
ISBN 0-471-21395-0

Bellavance, F., S. Tardif, and M. A. Stephens (1996). Tests for the analysis of variance of crossover designs with correlated errors. *Biometrics* *52*(2), 607–612.

Berger, R. L. and J. C. Hsu (1996, November). Bioequivalence trials, intersection–union tests and equivalence confidence sets. *Statistical Science* *11*(4), 283–302.

Bergman, B. and A. Hynén (1997, May). Dispersion effects from unreplicated designs in the 2^{k-p} series. *Technometrics* *39*(2), 191–198.

Beth, T., D. Jungnickel, and H. Lenz (1993). *Design Theory*. Cambridge: Cambridge University Press.

Bingham, D. (1998). Design and analysis of fractional factorial split-plot designs. Ph.D. dissertation, Simon Fraser University, Burnaby, B.C., Canada.

Bingham, D. and R. R. Sitter (1999a, February). Minimum-aberration two-level fractional factorial split-plot designs. *Technometrics* *41*(1), 62–70.

Bingham, D. R. and R. R. Sitter (1999b). Some theoretical results for fractional factorial split-plot designs. *Annals of Statistics* *27*(4), 1240–1255.

Bisgaard, S. (1994, October). Blocking generators for small 2^{k-p} designs. *Journal of Quality Technology* *26*(4), 288–296.

Bose, R. C. and K. A. Bush (1952). Orthogonal arrays of strength two and three. *Annals of Mathematical Statistics* *23*, 508–524.

Bose, R. C., S. S. Shrikhande, and E. T. Parker (1960). Further results on the construction of mutually orthogonal Latin squares and the falsity of Euler's conjecture. *Canadian Journal of Mathematics* *12*, 189–203.

Box, G. E. P. (1954a). Some theorems on quadratic forms applied in the study of analysis of variance problems, I. Effect of inequality of variance in the one-way classification. *Annals of Mathematical Statistics* *25*, 290–302.

Box, G. E. P. (1954b). Some theorems on quadratic forms applied in the study of analysis of variance problems, II. Effects of inequality of variance and of correlation between errors in the two-way classification. *Annals of Mathematical Statistics* *25*, 484–498.

Box, G. E. P. (1988, February). Signal-to-noise ratios, performance criteria, and transformations. *Technometrics* *30*(1), 1–17.

Box, G. E. P. and D. W. Behnken (1960, November). Some new three level designs for the study of quantitative variables. *Technometrics* *2*(4), 455–475.

Box, G. E. and N. R. Draper (1987). *Empirical Model-Building and Response Surfaces*. Wiley Series in Probability and Mathematical Statistics. New York: John Wiley & Sons.

Box, G. E. P. and J. S. Hunter (1957, March). Multi-factor experimental designs for exploring response surfaces. *Annals of Mathematical Statistics* *28*(1), 195–241.

Box, G. and S. Jones (1992). Split-plot designs for robust product experimentation. *Journal of Applied Statistics* *19*(1), 3–26.

Box, G. E. P. and R. D. Meyer (1985). Some new ideas in the analysis of screening designs. *Journal of Research of the National Bureau of Standards* *90*, 495–502.

Box, G. E. P. and R. D. Meyer (1986, February). Dispersion effects from fractional designs. *Technometrics* *28*(1), 19–27.

Box, G. E. P. and R. D. Meyer (1993, April). Finding the active factors in fractionated screening experiments. *Journal of Quality Technology* *25*(2), 94–105.

Box, G. and J. Tyssedal (1996). Projective properties of certain orthogonal arrays. *Biometrika* *83*(4), 950–955.

Box, G. E. P. and K. B. Wilson (1951). On the experimental attainment of optimum conditions. *Journal of the Royal Statistical Society, Series B* *13*(1), 1–45.

Bozivich, H., T. A. Bancroft, and H. O. Hartley (1956, December). Power of analysis of variance test procedures for certain incompletely specified models, I. *Annals of Mathematical Statistics* *27*(4), 1017–1043.

Bromwell, J., D. Witt, and M. White (1990). Process optimization of continuous silicone coating of label stock. See Wu (1990), pp. 47–61.

Brown, H. K. and R. A. Kempton (1994). The application of REML in clinical trials. *Statistics in Medicine* *13*, 1601–1617.

Brownie, C., D. Bowman, and J. Burton (1993). Estimating spatial variation in analysis of data from yield trials: A comparison of methods. *Agronomy Journal* *85*, 1244–1253.

Burns, J. C., K. R. Pond, D. S. Fisher, and J. M. Luginbuhl (1997). Changes in forage quality, ingestive mastication, and digesta kinetics resulting from switchgrass maturity. *Journal of Animal Science* *75*, 1368–1379.

Burrows, P. M., S. W. Scott, O. W. Barnett, and M. R. McLaughlin (1984, November). Use of experimental designs with quantitative ELISA. *Journal of Virological Methods* *8*, 207–216.

Carmichael, R. D. (1956). *Introduction to the Theory of Groups of Finite Order*. New York: Dover Publications.

Carriere, K. C. (1994). Crossover designs for clinical trials. *Statistics in Medicine* *13*, 1063–1069.

Chen, J. (1992). Some results on 2^{n-k} fractional factorial designs and search for minimum aberration designs. *Annals of Statistics* *20*(4), 2124–2141.

Chen, J. and D. K. J. Lin (1991). On the identity relationships of 2^{k-p} designs. *Journal of Statistical Planning and Inference* *28*, 95–98.

Chen, J. and C. F. J. Wu (1991). Some results on s^{n-k} fractional factorial design with minimum aberration or optimal moments. *Annals of Statistics* *19*(2), 1028–1041.

Chen, J., D. X. Sun, and C. F. J. Wu (1993). A catalogue of two-level and three-level fractional factorial designs with small runs. *International Statistical Review* *61*(1), 131–145.

Cheng, C.-S. (1990). Construction of run orders of factorial designs. In S. Ghosh (Ed.), *Statistical Design and Analysis of Industrial Experiments*, Chapter 14, pp. 423–439. New York: Marcel Dekker.

Cheng, C.-S. and M. Jacroux (1988, December). The construction of trend-free run orders of two-level factorial designs. *Journal of the American Statistical Association* *83*(404), 1152–1158.

Cheng, C.-S. and D. M. Steinberg (1991). Trend robust two-level factorial designs. *Biometrika* *78*(2), 325–36.

Cheng, C.-S., R. J. Martin, and B. Tang (1998). Two-level factorial designs with extreme numbers of level changes. *Annals of Statistics* *26*(4), 1522–1539.

Chi, S. and T. Chen (1992). Predicting optimum monosodium glutamate and sodium chloride concentrations in chicken broth as affected by spice addition. *Journal of Food Processing and Preservation* *16*, 313–326.

Christensen, R. (1996a). *Analysis of Variance, Design and Regression*. London: Chapman & Hall.

Christensen, R. (1996b). *Plane Answers to Complex Questions: The Theory of Linear Models*. New York: Springer-Verlag.

Clemmitt, M. (1991). Clinical researchers adapting to mandate for more diversity in study populations. *The Scientist* 5(18), 1, 8–9, 13.

Cochran, W. G. (1938). Recent advances in mathematical statistics: Recent work on the analysis of variance. *Journal of the Royal Statistical Society* *101*(2), 434–449.

Cochran, W. G. (1939). Long-term agricultural experiments. *Supplement to the Journal of the Royal Statistical Society* *6*(2), 104–148.

Cochran, W. G. (1957, September). Analysis of covariance: Its nature and uses. *Biometrics* *13*(3), 261–281.

Cochran, W. G. and G. M. Cox (1957). *Experimental Designs* (2nd ed.). A Wiley Publication in Applied Statistics. New York: John Wiley & Sons.

Cohen, J. (2001, April). Statistical analysis of vehicle test data from a crossover study. ICF Consulting, prepared for USEPA National Vehicle and Fuel Emissions Laboratory. Available at *http://www.epa.gov/otaq/regs /fuels/additive/iv-a-01.pdf*.

Colbourn, C. J. and J. H. Dinitz (Eds.) (1996). *The CRC Handbook of Combinatorial Designs*. Boca Raton, FL: CRC Press.

Conner, W. S. and S. Young (1961). *Fractional Factorial Designs for Experiments with Factors at Two and Three Levels*. Volume 58 of Applied Mathematics Series. Washington, DC: National Bureau of Standards.

Conner, W. S. and M. Zellin (1959). *Fractional Factorial Experiment Designs for Factors at Three Levels*. Volume 54 of Applied Mathematics Series. Washington, DC: National Bureau of Standards.

Cornell, J. A. (2002). *Experiments with Mixtures: Designs, Models, and the Analysis of Mixture Data* (3rd ed.). Wiley Series in Probability and Mathematical Statistics. New York: John Wiley & Sons.

Coster, D. C. and C.-S. Cheng (1988). Minimum cost trend-free run orders of fractional factorial designs. *Annals of Statistics* *16*(3), 1188–1205.

Cox, D. R. (1956, December). A note on weighted randomization. *Annals of Mathematical Statistics* *27*(4), 1144–1151.

Cox, D. R. (1957). The use of a concomitant variable in selecting an experimental design. *Biometrika* *44*(1–2), 150–158.

Cox, D. R. (1958). *Planning of Experiments*. A Wiley Publication in Applied Statistics. New York: John Wiley & Sons.

Czitrom, V. (1999, May). One-factor-at-a-time versus designed experiments. *The American Statistician* *53*, 126–131.

Daniel, C. (1959). Use of half-normal plots in interpreting factorial two-level experiments. *Technometrics* *1*, 311–341.

Daniel, C. (1962, June). Sequences of fractional replicates in the 2^{p-q} series. *Journal of the American Statistical Association* *57*(298), 403–429.

Daniel, C. (1973, June). One-at-a-time plans. *Journal of the American Statistical Association* *68*(342), 353–360.

Daniel, C. (1975, July). Calibration designs for machines with carry-over and drift. *Journal of Quality Technology* *7*(3), 103–108.

Daniel, C. (1976). *Application of Statistics to Industrial Experimentation*. New York: John Wiley & Sons.

Darby, L. A. and N. Gilbert (1958). The Trojan square. *Euphytica 7*, 183–188.

Dawid, A. P. (2000). Causal inference without counterfactuals and discussion. *Journal of the American Statistical Association 95*(450), 407–448.

Dean, A. and D. Voss (1999). *Design and Analysis of Experiments*. New York: Springer-Verlag.

DeMates, J. J. (1990). Dynamic analysis of injection molding using Taguchi methods. See Wu (1990), pp. 313–332.

Denbow, D. M. (1980). Involvement of cations in temperature regulation in chickens. Ph.D. dissertation, North Carolina State University, Raleigh NC.

Desu, M. M. and D. Raghavarao (1990). *Sample Size Methodology: Statistical Modeling and Decision Science*. San Diego, CA: Academic Press.

Dey, A. (1986). *Theory of Block Designs*. A Halstead Press Book. New York: John Wiley & Sons.

Diamond, N. T. (1995). Some properties of a foldover design. *Australian Journal of Statistics 37*(3), 345–352.

Diaz, E., D. Pazo, A. Esquifino, and B. Diaz (2000). Effects of ageing and exogenous melatonin on pituitary responsiveness to GnRH in rats. *Journal of Reproduction and Fertility 119*, 151–156.

Dobson, A. J. (1990). *An Introduction to Generalized Linear Models*. London: Chapman & Hall.

Dong, F. (1993). On the identification of active contrasts in unreplicated fractional factorials. *Statistica Sinica 3*, 209–217.

Dourleijn, C. (1993). On statistical selection in plant breeding. Ph.D. dissertation, Agricultural University, Wageningen, The Netherlands.

Driessen, S. (1992). Statistical selection: multiple comparison approach. Ph.D. dissertation, University of Technology, Eindhoven, The Netherlands.

Dunnett, C. W. (1989). Multivariate normal probability integrals with product correlation structure. *Applied Statistics 38*(3), 564–579.

Dunnett, C. W. (1993). Correction to algorithm AS 251: Multivariate normal probability integrals with product correlation structure. *Applied Statistics 42*(4), 709.

Edgington, E. S. (1986). Randomization tests. See Kotz and Johnson (1986), pp. 530–538.

Engel, J. (1992). Modeling variation in industrial experiments. *Applied Statistics 41*(3), 579–593.

EPA (2000). Comments on the use of data from the testing of human subjects. EPA-SAB-EC-00-017. September 2000, www.epa.gov/sab/pdf/ec0017.pdf.

Eun, J.-S., V. Fellner, and M. Gumpertz (2004). Methane production by mixed ruminal cultures incubated in dual-flow fermentors. *Journal of Dairy Science 87*, 112–121.

Fai, A. H. T. and P. L. Cornelius (1993). Approximate F-tests of multiple degree of freedom hypotheses in generalized least squares analyses of unbalanced split-plot experiments. Technical Report 342. University of Kentucky Department of Statistics, Lexington, KY.

Fang, Y. and B. Zhou (1998). Block-crossed arrays with applications to robust design. *Journal of Statistical Planning and Inference 74*, 169–175.

Farewell, V. T. and D. A. Sprott (1992, February). Some thoughts on randomization and causation. *Liaison* *6*(2), 6–10.

Federer, W. T. (1955). *Experimental Design, Theory and Application*. New York: Macmillan.

Federer, W. T. and L. N. Balaam (1972). *Bibliography on Experiment and Treatment Design pre-1968*. Edinburgh: Oliver & Boyd.

Federer, W. and M. Meredith (1992). Covariance analysis for split-plot and split-block designs. *The American Statistician* *46*, 155–162.

Federer, W. T., A. Hedayat, C. C. Lowe, and D. Raghavarao (1976). Application of statistical design theory to crop estimation with special reference to legumes and mixtures of cultivars. *Agronomy Journal* *68*, 914–919.

Felton, R. P. and D. W. Gaylor (1989). Multistrain experiments for screening toxic substances. *Journal of Toxicology and Environmental Health* *26*, 399–411.

Festing, M. F. W. and D. P. Lovell (1996). Reducing the use of laboratory animals in toxicological research and testing by better experimental design. *Journal of the Royal Statistical Society, Series B* *58*(1), 127–140.

Finney, D. J. (1945). Some orthogonal properties of the 4×4 and 6×6 Latin square. *Annals of Eugenics* *12*, 213–219.

Finney, D. J. (1946a). Orthogonal partitions of the 5×5 Latin squares. *Annals of Eugenics* *13*, 1–3.

Finney, D. J. (1946b). Orthogonal partitions of the 6×6 latin squares. *Annals of Eugenics* *13*, 184–196.

Finney, D. J. (1987). *Statistical Method in Biological Assay* (3rd ed.). London: Griffin.

Fisher, R. A. (1926). The arrangement of field experiments. *Journal of the Ministry of Agriculture* *33*, 503–513.

Fisher, R. A. (1935). *The Design of Experiments*. Edinburgh: Oliver & Boyd.

Franklin, M. F. (1984, August). Constructing tables of minimum aberration p^{n-m} designs. *Technometrics* *26*(3), 225–232.

Franklin, M. F. (1985, May). Selecting defining contrasts and confounded effects in p^{n-m} factorial experiments. *Technometrics* *27*(2), 165–172.

Franklin, M. F. and R. A. Bailey (1977). Selection of defining contrasts and confounded effects in two-level experiments. *Applied Statistics* *26*(3), 321–326.

Freeman, G. H. (1966). Some non-orthogonal partitions of 4×4, 5×5 and 6×6 Latin squares. *Annals of Mathematical Statistics* *37*, 666–681.

Freeman, G. H. (1979a). Complete Latin squares and related experimental designs. *Journal of the Royal Statistical Society, Series B* *40*, 253–262.

Freeman, G. H. (1979b). Some two-dimensional designs balanced for nearest neighbours. *Journal of the Royal Statistical Society, Series B* *41*, 88–95.

Freeman, G. H. (1981). Further results on quasi-complete Latin squares. *Journal of the Royal Statistical Society, Series B* *43*, 314–320.

Freeman, P. R. (1989). The performance of the two-stage analysis of two-treatment, two-period crossover trials. *Statistics in Medicine* *8*, 1421–1432.

Fries, A. and W. G. Hunter (1980, November). Minimum aberration 2^{k-p} designs. *Technometrics* *22*(4), 601–608.

Gibbons, J. D., I. Olkin, and M. Sobel (1977). *Selecting and Ordering Populations: A New Statistical Methodology*. A Wiley Publication in Applied Statistics. New York: John Wiley & Sons.

Giesbrecht, F. G. and J. C. Burns (1985). Two-stage analysis based on a mixed model: Large-sample asymptotic theory and small-sample simulation results. *Biometrics 41*, 477–486.

Goad, C. L. and D. E. Johnson (2000). Crossover experiments: A comparison of ANOVA tests and alternative analyses. *Journal of Agricultural, Biological, and Environmental Statistics 5*, 69–87.

Good, P. (2000). *Permutation Tests: A Practical Guide to Resampling Methods for Testing Hypotheses* (2nd ed.). New York: Springer-Verlag.

Graybill, F. (1961). *An Introduction to Linear Statistical Models*, Vol. 1. New York: McGraw-Hill.

Greenfield, A. A. (1976). Selection of defining contrasts in two-level experiments. *Applied Statistics 25*(1), 64–67.

Grize, Y. L. (1995). A review of robust process design approaches. *Journal of Chemometrics 9*, 239–262.

Grizzle, J. E. (1965, June). The two-period change-over design and its use in clinical trials. *Biometrics 21*(2), 467–480.

Grizzle, J. E. (1974). Corrigenda to Grizzle (1965). *Biometrics 30*, 727.

Gupta, S. (1956). On a decision rule for a problem in ranking means. Ph.D. dissertation, University of North Carolina, Chapel Hill, NC.

Gupta, S. (1965, May). On some multiple decision (selection and ranking) rules. *Technometrics 6*(2), 225–245.

Gupta, B. C. (1991). Some results on main effect plus one for 2^m factorials. *Communications in Statistics—Theory and Methods 20*(9), 2955–2963.

Haaland, P. D. and M. R. O'Connell (1995, February). Inference for effect-saturated fractional factorials. *Technometrics 37*(1), 82–93.

Hader, R. J. (1973, April). An improper method of randomization in experimental design. *The American Statistician 27*(2), 82–84.

Hald, A. (1952). *Statistical Theory with Engineering Applications*. Wiley Publications in Statistics. New York: John Wiley & Sons.

Hall, M. J. (1961). Hadamard matrix of order 16. Research Summary 36–10. Summary 1. Jet Propulsion Laboratory, Pasadena, CA.

Hall, W. B. and R. G. Jarrett (1981). Nonresolvable incomplete block designs with few replicates. *Biometrika, 68*(3), 617–627.

Hamada, M. and N. Balakrishnan (1998). Analyzing unreplicated factorial experiments: A review with some new proposals. *Statistica Sinica 8*, 1–41.

Hamada, M. and J. A. Nelder (1997). Generalized linear models for quality-improvement experiments. *Journal of Quality Technology 29*, 292–304.

Hamada, M. and C. F. J. Wu (1992, July). Analysis of designed experiments with complex aliasing. *Journal of Quality Technology 24*(3), 130–137.

Hardin, R. H. and N. J. A. Sloane (1993). A new approach to the construction of optimal designs. *Journal of Statistical Planning and Inference 37*, 339–369.

Harwit, M. and N. J. A. Sloan (1979). *Hadamard Transform Optics*. New York: Academic Press.

Hedayat, A. and K. Afsarinejad (1978). Repeated measurements designs, II. *Annals of Statistics* *6*, 619–628.

Hedayat, A. S. and J. Stufken (1999, February). Compound orthogonal arrays. *Technometrics* *41*(1), 57–61.

Hedayat, A. and W. D. Wallis (1978, November). Hadamard matrices and their applications. *Annals of Statistics* *6*(6), 1184–1238.

Hedayat, A. S., N. J. A. Sloan, and J. Stufken (1999). *Orthogonal Arrays Theory and Applications*. Springer Series in Statistics. New York: Springer-Verlag.

Hill, S. W. (1990). Design of a multiple pair data transmission cable. See Wu (1990), pp. 197–210.

Hinkelmann, K. and O. Kempthorne (1994). Design and Analysis of Experiments. Volume I of Wiley Series in Probability and Mathematical Statistics. New York: John Wiley & Sons.

Hirai, S. and M. Koga (1990). Robust design for transistors: Parameter design using simulation. See Wu (1990), pp. 179–196.

Hollander, M. and D. A. Wolf (1973). *Nonparametric Statistical Methods*. A Wiley Publication in Applied Statistics. New York: John Wiley & Sons.

Huitson, A., J. Poloniecki, R. Hews, and N. Barker (1982). A review of cross-over trials. *The Statistician* *31*(1), 71–80.

Hunter, G. B., F. S. Hodi, and T. W. Eager (1982). High-cycle fatigue of weld repaired cast Ti-6Al-4V. *Metallurgical Transactions* *13A*, 1589–1594.

Huynh, H. and L. S. Feldt (1970, December). Conditions under which mean square ratios in repeated measurements designs have exact F-distributions. *Journal of the American Statistical Association* *65*(332), 1582–1589.

Iman, R. L. and W. J. Conover (1980). Small sample sensitivity analysis techniques for computer models, with an application to risk assessment (with discussion). *Communications in Statistics—Theory and Methods* *9A*(17), 1749–1874.

Inkpen, K. M. (2001). Drag-and-drop versus point-and-click mouse interaction styles for children. *ACM Transactions on Computer–Human Interaction* *8*, 1–33.

Jackson, L. S., T. W. Joyce, J. J. A. Heitmann, and F. G. Giesbrecht (1996). Enzyme activity recovery from secondary fiber treated with cellulase and xylanase. *Journal of Biotechnology* *45*, 33–44.

Janky, D. G. (2000, November). Sometimes pooling for analysis of variance hypothesis tests: A review and study of a split-plot model. *The American Statistician* *54*(4), 269–279.

John, J. (1981). Efficient cyclic designs. *Journal of the Royal Statistical Society, series B* *43*, 76–80.

John, J. (1987). *Cyclic Designs*. London: Chapman & Hall.

John, J. A. and M. H. Quenouille (1977). *Experiments: Design and Analysis* (2nd ed.). New York: Macmillan.

John, J. and E. Williams (1995). *Cyclic and Computer Generated Designs* (2nd ed.). London: Chapman & Hall.

John, P. W. M. (1961). Three-quarter replicates of 2^4 and 2^5 designs. *Biometrics* *17*, 319–321.

John, P. W. M. (1962). Three quarter replicates of 2^n designs. *Biometrics* *18*, 172–184.

John, P. W. M. (1971). *Statistical Design and Analysis of Experiments*. New York: Macmillan.

John, P. W. M. (2000). Breaking alias chains in fractional factorials. *Communications in Statistics—Theory and Methods* *29*(9–10), 2143–2155.

Joiner, B. L. and C. Campbell (1976, August). Designing experiments when run order is important. *Technometrics* *18*(3), 249–259.

Jones, B. and M. G. Kenward (1989). *Design and Analysis of Cross-over Trials*. Monographs on Statistics and Applied Probability. London: Chapman & Hall.

Kackar, R. N. (1985). Off-line quality control, parameter design, and the Taguchi method. *Journal of Quality Technology* *17*(4), 176–209.

Kempthorne, O. (1952). *The Design and Analysis of Experiments*. Wiley Publications in Statistics. New York: John Wiley & Sons.

Kempton, R., J. Seraphin, and A. Sword (1994). Statistical analysis of two-dimensional variation in variety yield trials. *Journal of Agricultural Science* *122*, 335–342.

Kenward, M. G. and J. H. Roger (1997). Small sample inference for fixed effects from restricted maximum likelihood. *Biometrics* *53*, 983–997.

Keppel, G. (1982). *Design and Analysis: A Researcher's Handbook* (2nd ed.). Englewood Cliffs, NJ: Prentice Hall.

Kershner, R. P. and W. T. Federer (1981, September). Two-treatment crossover designs for estimating a variety of effects. *Journal of the American Statistical Association* *76*(3), 612–619.

Kettaneh-Wold, N. and D. K. J. Lin (1995). Response to letter to the editor. *Technometrics* *37*(3), 359.

Khuri, A. I. and J. A. Cornell (1996). *Response Surfaces: Designs and Analyses* (2nd ed.). Statistics: Textbooks and Monographs. New York: Marcel Dekker.

Kim, H., M. Lieffering, S. Miura, K. Kobayashi, and M. Okada (2001). Growth and nitrogen uptake of CO_2-enriched rice under field conditions. *New Phytologist* *150*, 223–229.

Kishen, K. (1942). On Latin and hyper- Graeco- Latin cubes and hyper-cubes. *Current Science* *11*(3), 98–99.

Kobayashi, K., M. Okada, and H. Y. Kim (1999). The free-air CO_2 enrichment (FACE) with rice in Japan. In proceedings of *International Symposium on World Food Security*, Kyoto, pp. 213–215.

Kotz, S. and N. L. Johnson (Eds.) (1986). *Encyclopedia of Statistical Sciences*. New York: John Wiley & Sons.

Kotz, S., N. L. Johnson, and C. B. Read (Eds.) (1988). *Encyclopedia of Statistical Sciences*, Vol. 8. New York: John Wiley & Sons.

Kunert, J. (1987). On variance estimation in crossover designs. *Biometrics* *43*(4), 833–845.

Kunert, J. and B. P. Utzig (1993). Estimation of variance in crossover designs. *Journal of the Royal Statistical Society, Series B* *55*(4), 919–927.

Lakatos, E. and D. Raghavarao (1987). Undiminished residual effects designs and their suggested applications. *Communications in Statistics—Theory and Methods* *16*(5), 1345–1359.

Lamacraft, R. and W. Hall (1982). Tables of incomplete block designs: $r = k$. *Australian Journal of Statistics* *24*, 350–360.

Lattemai, P., C. Ohlsson, and P. Lingvall (1996). Influence of molasses or molasses-formic acid treated red clover silage on feed intake and milk yield. *Swedish Journal of Agriculture* *26*, 91–100.

Laywine, C. F. and G. L. Mullen (1998). *Discrete Mathematics Using Latin Squares*. Wiley-Interscience Series in Discrete Mathematics and Optimization. New York: John Wiley & Sons.

Lenth, R. V. (1989, November). Quick and easy analysis of unreplicated factorials. *Technometrics* *31*(4), 469–473.

Lenth, R. V. (2001, August). Some practical guidelines for effective sample size determination. *The American Statistician* *55*(3), 187–193.

Li, W. and C. J. Nachtsheim (2000, November). Model-robust factorial designs. *Technometrics* *42*(4), 345–352.

Lin, D. K. J. (1995, May). Generating systematic supersaturated designs. *Technometrics* *37*(2), 213–225.

Lin, D. K. J. and N. R. Draper (1992, November). Projection properties of Plackett and Burman designs. *Technometrics* *34*(4), 423–428.

Lin, D. K. J. and N. R. Draper (1993). Generating alias relationships for two-level Plackett and Burman designs. *Computational Statistics and Data Analysis* *15*, 147–157.

Lucas, H. L. (1948). Design and analysis of feeding experiments with milking dairy cattle. Mimeo series 18, Institute of Statistics, University of North Carolina System.

Ma, R. H. and J. B. Harrington (1948). A study on field experiments of semi- Latin square design. *Scientific Agriculture* *29*, 241–251.

Mandl, R. (1985). Orthogonal Latin squares: An application of experimental design to compiler testing. *Communications of the ACM* *28*(10), 1054–1058.

Martin, B., D. Parker, and L. Zenick (1987). Minimize slugging by optimizing controllable factors on topaz windshield molding. In *Fifth Symposium on Taguchi Methods*, pp. 519–526. Dearborn, MI: American Supplier Institute.

Matthews, J. N. S. (1990). The analysis of data from crossover designs: The efficiency of ordinary least squares. *Biometrics* *46*, 689–696.

McGrath, R. and D. K. Lin (2001, November). Testing multiple dispersion effects in unreplicated fractional factorial designs. *Technometrics* *43*(4), 406–414.

Mckay, M. D., R. J. Beckman, and W. J. Conover (1979, May). A comparison of three methods for selecting values of input variables in the analysis of output from a computer code. *Technometrics* *21*(2), 239–245.

McLean, R. A. and V. L. Anderson (1984). *Applied Factorial and Fractional Designs*. Volume 55 of Statistics, Textbooks and Monographs. New York: Marcel Dekker.

McLean, R. A. and W. L. Sanders (1988). Approximating degrees of freedom for standard errors in mixed linear models. In *Proceedings of the Statistical Computing Section*, pp. 50–59. Washington, DC: American Statistical Association.

Mead, R. (1988). *The Design of Experiments: Statistical Principles for Practical Applications*. Cambridge: Cambridge University Press.

Mee, R. W. and R. L. Bates (1998, August). Split-lot designs: Experiments for multistage batch processes. *Technometrics* *40*(2), 127–140.

Miller, A. (1997, May). Strip-plot configurations of fractional factorials. *Technometrics 39*(2), 153–161.

Mukhopadhyay, A. C. (1981). Construction of some series of orthogonal arrays. *Sankhyā: Indian Journal of Statistics, Series B 43*(1), 81–92.

Myers, R. H. (1990). *Classical and Modern Regression with Applications* (2nd ed.). Duxbury Advanced Series in Statistics and Decision Sciences. Boston: PWS-Kent.

Myers, R. H. and D. C. Montgomery (1995). *Response Surface Methodology: Process and Product Optimization Using Designed Experiments*. Wiley Series in Probability and Statistics. New York: John Wiley & Sons.

National Bureau of Standards (1957). *Fractional Factorial Experiment Designs for Factors at Two Levels*. Volume 48 of Applied Mathematics Series. Washington DC: National Bureau of Standards.

Natrella, M. G. (1963). *Experimental Statistics*. Volume 91 of *National Bureau of Standards Handbook*. Washington DC: National Bureau of Standards.

NCSU Bulletin (2001, September 21). Program takes new approach to intro courses. North Carolina State University, Raleigh, NC.

Neter, J., W. Wasserman, and M. Kutner (1990). *Applied Linear Statistical Models: Regression, Analysis of Variance, and Experimental Design* (3rd ed.). Irwin Series in Statistics. Homewood, IL: Richard D. Irwin.

Nguyen, N.-K. (1996, February). An algorithmic approach to constructing supersaturated designs. *Technometrics 38*(1), 69–73.

Okada, M., M. Lieffering, H. Nakamura, M. Yoshimoto, H. Y. Kim, and K. Kobayashi. (2001) Free-air CO_2 enrichment (FACE) using pure CO_2 injection: system description. *New Phytologist* 150

Owen, A. B. (1992). Orthogonal arrays for computer experiments, integration and visualization. *Statistica Sinica 2*, 439–452.

Owen, A. B. (1994, December). Controlling correlations in Latin hypercube samples. *Journal of the American Statistical Association 89*(428), 1517–1522.

Paley, R. E. A. C. (1933). On orthogonal matrices. *Journal of Mathematics and Physics 12*, 311–320.

Pan, G. (1999, November). The impact of unidentified location effects on dispersion-effects identification from unreplicated factorial designs. *Technometrics 41*(4), 313–326.

Parvey, D. E. (1990). Application of Taguchi methods for parachute design optimization. See Wu (1990), pp. 85–92.

Paterson, L. and H. Patterson (1983). An algorithm for constructing α-lattice designs. *Ars Combinatoria 16-A*, 87–98.

Patterson, H. D. (1951). Change-over trials. *Journal of the Royal Statistical Society, Series B 13*(2), 256–271.

Patterson, H. D. and R. Thompson (1971). Recovery of inter-block information when block sizes are unequal. *Biometrika 58*(3), 545–554.

Patterson, H. and E. Williams (1976). A new class of resolvable incomplete block designs. *Biometrika 63*, 83–92.

Patterson, H., E. Williams, and E. Hunter (1978). Block designs for variety trials. *Journal of Agricultural Science, Cambridge 90*, 395–400.

Plackett, R. L. and J. P. Burman (1946). The design of optimum multifactorial experiments. *Biometrika* *33*(3), 305–325.

Raghavarao, D. (1971). *Constructions and Combinatorial Problems in Design of Experiments*. New York: Dover Publications.

Raghavarao, D. (1990). *Crossover Designs in Industry*. Volume 109 of *Statistics, Textbooks and Monographs*, Chapter 18, pp. 517–530. New York: Marcel Dekker.

Raghavarao, D. and W. Federer (1979). Block total response as an alternative to the randomized response method in surveys. *Journal of the Royal Statistical Society, Series B* *41*, 40–45.

Raktoe, B. L., A. Hedayat, and W. T. Federer (1981). *Factorial Designs*. New York: John Wiley & Sons.

Rao, C. R. (1946a). Difference sets and combinatorial arrangements derivable from finite geometries. *National Institute of Sciences of India* *12*, 123–135.

Rao, C. R. (1946b). Hypercubes of strength *d* leading to confounded designs in factorial experiments. *Bulletin of the Calcutta Mathematics Society* *38*, 67–78.

Rao, C. R. (1947a). Factorial experiments derivable from combinatorial arrangements of arrays. *Journal of the Royal Statistical Society, Supplement* *9*, 128–139.

Rao, C. R. (1947b). On a class of arrangements. *Edinburg Mathematical Society Proceedings, Series 2* *8*, 119–125.

Rao, C. (1971). Estimation of variance and covariance components—MINQUE theory. *Journal of Multivariate Analysis* *1*, 257–275.

Robillard, P. (1968). Combinatorial problems in the theory of factorial designs and error correcting codes. Ph.D. dissertation, University of North Carolina, Chapel Hill, NC.

Robinson, D. L., C. Kershaw, and R. Ellis (1988). An investigation of two-dimensional yield variability in breeders' small plot barley trials. *Journal of Agricultural Science, Cambridge* *111*, 419–426.

Rojas, B. and R. F. White (1957). The modified Latin square. *Journal of the Royal Statistical Society, Series B* *19*(2), 890–317.

Rosenbaum, P. R. (1994). Dispersion effects from fractional factorials in Taguchi's method of quality design. *Journal of the Royal Statistical Society, Series B* *56*(4), 641–652.

Rosenbaum, P. R. (1996, November). Some useful compound dispersion experiments in quality design. *Technometrics* *38*(4), 354–364.

Rosenbaum, P. R. (1999, May). Blocking in compound dispersion experiments. *Technometrics* *41*(2), 125–134.

Rosier, C. L. (2002). Factors affecting rooting of Frasier fir and Virginia pine cuttings. Master's thesis, NCSU Department of Forestry, North Carolina State University, Raleigh, NC.

Russell, K. (1991). The construction of good change-over designs when there are fewer units than treatments. *Biometrika* *78*(2), 305–313.

Ryser, H. J. (1963). *Combinatorial Mathematics*. Carus Mathematical Monographs. Washington, DC: Mathematical Association of America.

Sarker, A., M. Singh, and W. Erskine (2001). Efficiency of spatial methods in yield trials in lentil (*Lens culinaris* ssp. *culinaris*). *Journal of Agricultural Science, Cambridge* *137*, 427–438.

SAS® (1999). *SAS®, Version 8*. Cary, NC: SAS Institute.

Satterthwaite, F. (1946). An approximate distribution of estimates of variance components. *Biometrics Bulletin 2*, 110–114.

Saunders, I. W. and J. A. Eccleston (1992). Experimental design for continuous processes. *Australian Journal of Statistics 34*(1), 77–89.

Saunders, I. W., J. A. Eccleston, and R. J. Martin (1995). An algorithm for the design of 2^p factorial experiments on continuous processes. *Australian Journal of Statistics 37*(3), 353–365.

Scheffé, H. (1956). *The Analysis of Variance*. A Wiley Publication in Mathematical Statistics. New York: John Wiley & Sons.

Searle, S. (1971). *Linear Models*. A Wiley Publication in Mathematical Statistics. York, New York: John Wiley & Sons.

Sharma, V. K. (1977). Change-over designs with complete balance for first and second residual effects. *Canadian Journal of Statistics 5*(1), 121–132.

Sharma, V. K. (1981). A class of experimental designs balanced for first residuals. *Australian Journal of Statistics 23*(3), 365–370.

Shellabarger, C. J., J. P. Stone, and S. Holtzman (1978). Rat differences in mammary tumor induction with estrogen and neutron radiation. *Journal of the National Cancer Institute 61*, 1505–1508.

Shrikhande, S. S. and N. K. Singh (1962). On a method of constructing symmetrical balanced incomplete block designs. *Sankhyā A 24*, 25–32.

Sitter, R. R., J. Chen, and M. Feder (1997, November). Fractional resolution and minimum aberration in blocked 2^{n-k} designs. *Technometrics 39*(4), 382–390.

Skillings, J. H. and G. A. Mack (1981). On the use of a Friedman-type statistic in balanced and unbalanced block designs. *Technometrics 23*(2), 171–177.

Smith, C. A. B. and H. O. Hartley (1948). The construction of Youden squares. *Journal of the Royal Statistical Society, Series B 10*, 262–264.

Snedecor, G. W. (1946). *Statistical Methods* (4th ed.). Ames, IA: Iowa State College Press.

Snedecor, G. W. and W. G. Cochran (1989). *Statistical Methods* (8th ed.). Ames, IA: Iowa State University Press.

Sprott, D. A. (2000). *Statistical Inference in Science*. Springer Series in Statistics. New York: Springer-Verlag.

Sprott, D. A. and V. T. Farewell (1993). Randomization in experimental science. *Statistical Papers 34*, 89–94.

Srivastava, J. and R. Hveberg (1992). Sequential factorial probing designs for identifying and estimating non-negligible factorial effects for the 2^m experiment under the tree structure. *Journal of Statistical Planning and Inference 30*, 141–162.

Steel, R. and J. Torrie (1980). *Principles and Procedures of Statistics: A Biometrical Approach (2nd ed.)*. New York: McGraw-Hill.

Steel, R. G. D., J. H. Torrie, and D. A. Dickey (1997). *Principles and Procedures of Statistics: A Biometrical Approach* (3rd ed.). New York: McGraw-Hill.

Stein, M. (1987, May). Large sample properties of simulations using Latin hypercube sampling. *Technometrics 29*(2), 143–151.

Steinberg, D. M. (1988). Factorial experiments with time trends. *Technometrics* *30*(3), 259–269.

Stoltz, M., T. Arumugham, C. Lippert, D. Yu, V. Bhargava, M. Eller, and S. Weir (1997). Effect of food on the bioavailability of fexofenadine hydrochloride (MDL 16455A). *Biopharmaceutics and Drug Disposition* *18*, 645–648.

Street, A. P. and D. J. Street (1987). *Combinatorics of Experimental Design*. Oxford: Clarendon Press.

Taguchi, G. (1986). *Introduction to Quality Engineering: Designing Quality into Products and Processes*. White Plains, NY: Krause International Publications.

Taguchi, G. (1987a). *System of Experimental Design: Engineering Methods to Optimize Quality and Minimize Costs*, Vol. 1. Dearborn, MI: UNIPUB/Krause International Publications; American Supplier Institute.

Taguchi, G. (1987b). *System of Experimental Design: Engineering Methods to Optimize Quality and Minimize Costs*, Vol. 2. Dearborn, MI: UNIPUB/Krause International Publications; American Supplier Institute.

Taguchi, G. (1992). *Methods of Research and Development*. Dearborn, MI: American Supplier Institute.

Taguchi, G. and Y. Wu (1985). *Introduction to Off-Line Quality Control*. Nagaya, Japan: Central Japan Quality Control Association.

Tang, B. (1993, December). Orthogonal array-based Latin hypercubes. *Journal of the American Statistical Association* *88*(424), 1392–1397.

Tang, B. (1994). A theorem for selecting OA-based Latin hypercubes using a distance criterion. *Communications in Statistics—Theory and Methods* *23*(7), 2047–2058.

Vrescak, W. D. and T. L. Reed (1990). Variation reduction and linearization of Kovar material removal during chemical cleaning using an $l_{18} \times l_4 \times m_3$ dynamic experimental design. See Wu (1990), pp. 135–156.

Wallis, W. D. (1988). *Combinatorial Designs*. Monographs and Textbooks in Pure and Applied Mathematics. New York: Marcel Dekker.

Wang, P. C. (1990, June). On the trend-free run orders in orthogonal plans. In M. T. Chao and P. E. Cheng (Eds.), *Proceedings of the 1990 Taipei Symposium in Statistics*, pp. 605–613. Taipei, Taiwan: Institute of Statistical Science, Academia Sinica.

Wang, S.-J. and H. M. J. Hung (1997). Use of two-stage test statistic in the two-period crossover trials. *Biometrics* *53*(3), 1081–1091.

Wang, P. C. and H. W. Jan (1995). Designing two-level factorial experiments using orthogonal arrays when the run order is important. *The Statistician* *44*(3), 379–388.

Wang, J. C. and C. F. J. Wu (1991, June). An approach to the construction of asymmetrical orthogonal arrays. *Journal of the American Statistical Association* *86*(414), 450–456.

Wang, J. C. and C. F. J. Wu (1992, November). Nearly orthogonal arrays with mixed levels and small runs. *Technometrics* *34*(4), 409–422.

Warrick, J. (2000, June 7). U.S. rejects pesticide tests on humans. *Washington Post*.

Welch, W. J., T.-K. Yu, S. M. Kang, and J. Sacks (1990). Computer experiments for quality control by parameter design. *Journal of Quality Technology* *22*(1), 15–22.

Welham, S. J. and R. Thompson (1997). Likelihood ratio tests for fixed model terms using residual maximum likelihood. *Journal of the Royal Statistical Society, Series B* *59*, 701–714.

Wilk, M. and R. Gnanadesikan (1968). Probability plotting methods for the analysis of data. *Biometrika* *55*(1), 1–18.

Williams, E. (1949). Experimental designs balanced for the estimation of residual effects of treatments. *Australian Journal of Scientific Research, Series A* *2*, 149–168.

Wu, C. F. J. (1989, December). Constructions of 2^m4^n designs via a grouping scheme. *Annals of Statistics* *17*(4), 1880–1885.

Wu, Y. (Ed.) (1990). *Eighth Symposium on Taguchi Methods*. Dearborn MI: American Supplier Institute.

Yamada, S. and D. K. L. Lin (1997). Supersaturated design including an orthogonal base. *Canadian Journal of Statistics* *25*(2), 203–213.

Yang, C. H. (1966). Some designs for maximal (+1, −1)-determinant of order $n \equiv 2 \pmod 4$. *Mathematics of Computation* *20*, 147–148.

Yang, C. H. (1968). On designs of maximal (+1, −1)-matrices of order $n \equiv 2 \pmod 4$. *Mathematics of Computation* *22*, 174–180.

Yates, F. W. (1933). The formation of Latin squares for use in field experiments. *Empire Journal of Experimental Agriculture* *1*(3), 235–244.

Yates, F. W. (1935). Complex experiments. *Journal of the Royal Statistical Society, Supplement* *2*, 181–247.

Yau, S. (1997). Efficiency of alpha-lattice designs in international variety yield trials of barley and wheat. *Journal of Agricultural Science, Cambridge* *128*, 5–9.

Zahn, D. A. (1975a, May). An empirical study of the half-normal plot. *Technometrics* *17*(2), 201–211.

Zahn, D. A. (1975b, May). Modifications of and revised critical values for the half-normal plot. *Technometrics* *17*(2), 189–200.

Zucker, D. M., O. Lieberman, and O. Manor (2000). Improved small sample inference in the mixed linear model: Bartlett correction and adjusted likelihood. *Journal of the Royal Statistical Society, Series B* *62*, 827–838.

Index

Planning, Construction, and Statistical Analysis of Comparative Experiments, by Francis G. Giesbrecht and Marcia L. Gumpertz
ISBN 0-471-21395-0

Yu-Ming Albert Shen

Jan. 2011

WILEY SERIES IN PROBABILITY AND STATISTICS

ESTABLISHED BY WALTER A. SHEWHART AND SAMUEL S. WILKS

The ***Wiley Series in Probability and Statistics*** is well established and authoritative. It covers many topics of current research interest in both pure and applied statistics and probability theory. Written by leading statisticians and institutions, the titles span both state-of-the-art developments in the field and classical methods.

Reflecting the wide range of current research in statistics, the series encompasses applied, methodological and theoretical statistics, ranging from applications and new techniques made possible by advances in computerized practice to rigorous treatment of theoretical approaches.

This series provides essential and invaluable reading for all statisticians, whether in academia, industry, government, or research.

ABRAHAM and LEDOLTER · Statistical Methods for Forecasting
AGRESTI · Analysis of Ordinal Categorical Data
AGRESTI · An Introduction to Categorical Data Analysis
AGRESTI · Categorical Data Analysis, *Second Edition*
ALTMAN, GILL, and McDONALD · Numerical Issues in Statistical Computing for the Social Scientist
AMARATUNGA and CABRERA · Exploration and Analysis of DNA Microarray and Protein Array Data
ANDĚL · Mathematics of Chance
ANDERSON · An Introduction to Multivariate Statistical Analysis, *Third Edition*
*ANDERSON · The Statistical Analysis of Time Series
ANDERSON, AUQUIER, HAUCK, OAKES, VANDAELE, and WEISBERG · Statistical Methods for Comparative Studies
ANDERSON and LOYNES · The Teaching of Practical Statistics
ARMITAGE and DAVID (editors) · Advances in Biometry
ARNOLD, BALAKRISHNAN, and NAGARAJA · Records
*ARTHANARI and DODGE · Mathematical Programming in Statistics
*BAILEY · The Elements of Stochastic Processes with Applications to the Natural Sciences
BALAKRISHNAN and KOUTRAS · Runs and Scans with Applications
BARNETT · Comparative Statistical Inference, *Third Edition*
BARNETT and LEWIS · Outliers in Statistical Data, *Third Edition*
BARTOSZYNSKI and NIEWIADOMSKA-BUGAJ · Probability and Statistical Inference
BASILEVSKY · Statistical Factor Analysis and Related Methods: Theory and Applications
BASU and RIGDON · Statistical Methods for the Reliability of Repairable Systems
BATES and WATTS · Nonlinear Regression Analysis and Its Applications
BECHHOFER, SANTNER, and GOLDSMAN · Design and Analysis of Experiments for Statistical Selection, Screening, and Multiple Comparisons
BELSLEY · Conditioning Diagnostics: Collinearity and Weak Data in Regression

*Now available in a lower priced paperback edition in the Wiley Classics Library.

BELSLEY, KUH, and WELSCH · Regression Diagnostics: Identifying Influential Data and Sources of Collinearity
BENDAT and PIERSOL · Random Data: Analysis and Measurement Procedures, *Third Edition*
BERRY, CHALONER, and GEWEKE · Bayesian Analysis in Statistics and Econometrics: Essays in Honor of Arnold Zellner
BERNARDO and SMITH · Bayesian Theory
BHAT and MILLER · Elements of Applied Stochastic Processes, *Third Edition*
BHATTACHARYA and JOHNSON · Statistical Concepts and Methods
BHATTACHARYA and WAYMIRE · Stochastic Processes with Applications
BILLINGSLEY · Convergence of Probability Measures, *Second Edition*
BILLINGSLEY · Probability and Measure, *Third Edition*
BIRKES and DODGE · Alternative Methods of Regression
BLISCHKE AND MURTHY (editors) · Case Studies in Reliability and Maintenance
BLISCHKE AND MURTHY · Reliability: Modeling, Prediction, and Optimization
BLOOMFIELD · Fourier Analysis of Time Series: An Introduction, *Second Edition*
BOLLEN · Structural Equations with Latent Variables
BOROVKOV · Ergodicity and Stability of Stochastic Processes
BOULEAU · Numerical Methods for Stochastic Processes
BOX · Bayesian Inference in Statistical Analysis
BOX · R. A. Fisher, the Life of a Scientist
BOX and DRAPER · Empirical Model-Building and Response Surfaces
*BOX and DRAPER · Evolutionary Operation: A Statistical Method for Process Improvement
BOX, HUNTER, and HUNTER · Statistics for Experimenters: An Introduction to Design, Data Analysis, and Model Building
BOX and LUCEÑO · Statistical Control by Monitoring and Feedback Adjustment
BRANDIMARTE · Numerical Methods in Finance: A MATLAB-Based Introduction
BROWN and HOLLANDER · Statistics: A Biomedical Introduction
BRUNNER, DOMHOF, and LANGER · Nonparametric Analysis of Longitudinal Data in Factorial Experiments
BUCKLEW · Large Deviation Techniques in Decision, Simulation, and Estimation
CAIROLI and DALANG · Sequential Stochastic Optimization
CHAN · Time Series: Applications to Finance
CHATTERJEE and HADI · Sensitivity Analysis in Linear Regression
CHATTERJEE and PRICE · Regression Analysis by Example, *Third Edition*
CHERNICK · Bootstrap Methods: A Practitioner's Guide
CHERNICK and FRIIS · Introductory Biostatistics for the Health Sciences
CHILÈS and DELFINER · Geostatistics: Modeling Spatial Uncertainty
CHOW and LIU · Design and Analysis of Clinical Trials: Concepts and Methodologies, *Second Edition*
CLARKE and DISNEY · Probability and Random Processes: A First Course with Applications, *Second Edition*
*COCHRAN and COX · Experimental Designs, *Second Edition*
CONGDON · Applied Bayesian Modelling
CONGDON · Bayesian Statistical Modelling
CONOVER · Practical Nonparametric Statistics, *Second Edition*
COOK · Regression Graphics
COOK and WEISBERG · Applied Regression Including Computing and Graphics
COOK and WEISBERG · An Introduction to Regression Graphics
CORNELL · Experiments with Mixtures, Designs, Models, and the Analysis of Mixture Data, *Third Edition*
COVER and THOMAS · Elements of Information Theory

*Now available in a lower priced paperback edition in the Wiley Classics Library.

COX · A Handbook of Introductory Statistical Methods
*COX · Planning of Experiments
CRESSIE · Statistics for Spatial Data, *Revised Edition*
CSÖRGŐ and HORVÁTH · Limit Theorems in Change Point Analysis
DANIEL · Applications of Statistics to Industrial Experimentation
DANIEL · Biostatistics: A Foundation for Analysis in the Health Sciences, *Sixth Edition*
*DANIEL · Fitting Equations to Data: Computer Analysis of Multifactor Data, *Second Edition*
DASU and JOHNSON · Exploratory Data Mining and Data Cleaning
DAVID and NAGARAJA · Order Statistics, *Third Edition*
*DEGROOT, FIENBERG, and KADANE · Statistics and the Law
DEL CASTILLO · Statistical Process Adjustment for Quality Control
DENISON, HOLMES, MALLICK and SMITH · Bayesian Methods for Nonlinear Classification and Regression
DETTE and STUDDEN · The Theory of Canonical Moments with Applications in Statistics, Probability, and Analysis
DEY and MUKERJEE · Fractional Factorial Plans
DILLON and GOLDSTEIN · Multivariate Analysis: Methods and Applications
DODGE · Alternative Methods of Regression
*DODGE and ROMIG · Sampling Inspection Tables, *Second Edition*
*DOOB · Stochastic Processes
DOWDY, WEARDEN, and CHILKO · Statistics for Research, *Third Edition*
DRAPER and SMITH · Applied Regression Analysis, *Third Edition*
DRYDEN and MARDIA · Statistical Shape Analysis
DUDEWICZ and MISHRA · Modern Mathematical Statistics
DUNN and CLARK · Basic Statistics: A Primer for the Biomedical Sciences, *Third Edition*
DUPUIS and ELLIS · A Weak Convergence Approach to the Theory of Large Deviations
*ELANDT-JOHNSON and JOHNSON · Survival Models and Data Analysis
ENDERS · Applied Econometric Time Series
ETHIER and KURTZ · Markov Processes: Characterization and Convergence
EVANS, HASTINGS, and PEACOCK · Statistical Distributions, *Third Edition*
FELLER · An Introduction to Probability Theory and Its Applications, Volume I, *Third Edition,* Revised; Volume II, *Second Edition*
FISHER and VAN BELLE · Biostatistics: A Methodology for the Health Sciences
*FLEISS · The Design and Analysis of Clinical Experiments
FLEISS · Statistical Methods for Rates and Proportions, *Third Edition*
FLEMING and HARRINGTON · Counting Processes and Survival Analysis
FULLER · Introduction to Statistical Time Series, *Second Edition*
FULLER · Measurement Error Models
GALLANT · Nonlinear Statistical Models
GHOSH, MUKHOPADHYAY, and SEN · Sequential Estimation
GIESBRECHT and GUMPERTZ · Planning, Construction, and Statistical Analysis of Comparative Experiments
GIFI · Nonlinear Multivariate Analysis
GLASSERMAN and YAO · Monotone Structure in Discrete-Event Systems
GNANADESIKAN · Methods for Statistical Data Analysis of Multivariate Observations, *Second Edition*
GOLDSTEIN and LEWIS · Assessment: Problems, Development, and Statistical Issues
GREENWOOD and NIKULIN · A Guide to Chi-Squared Testing
GROSS and HARRIS · Fundamentals of Queueing Theory, *Third Edition*
*HAHN and SHAPIRO · Statistical Models in Engineering
HAHN and MEEKER · Statistical Intervals: A Guide for Practitioners

*Now available in a lower priced paperback edition in the Wiley Classics Library.

HALD · A History of Probability and Statistics and their Applications Before 1750
HALD · A History of Mathematical Statistics from 1750 to 1930
HAMPEL · Robust Statistics: The Approach Based on Influence Functions
HANNAN and DEISTLER · The Statistical Theory of Linear Systems
HEIBERGER · Computation for the Analysis of Designed Experiments
HEDAYAT and SINHA · Design and Inference in Finite Population Sampling
HELLER · MACSYMA for Statisticians
HINKELMAN and KEMPTHORNE: · Design and Analysis of Experiments, Volume 1: Introduction to Experimental Design
HOAGLIN, MOSTELLER, and TUKEY · Exploratory Approach to Analysis of Variance
HOAGLIN, MOSTELLER, and TUKEY · Exploring Data Tables, Trends and Shapes
*HOAGLIN, MOSTELLER, and TUKEY · Understanding Robust and Exploratory Data Analysis
HOCHBERG and TAMHANE · Multiple Comparison Procedures
HOCKING · Methods and Applications of Linear Models: Regression and the Analysis of Variance, *Second Edition*
HOEL · Introduction to Mathematical Statistics, *Fifth Edition*
HOGG and KLUGMAN · Loss Distributions
HOLLANDER and WOLFE · Nonparametric Statistical Methods, *Second Edition*
HOSMER and LEMESHOW · Applied Logistic Regression, *Second Edition*
HOSMER and LEMESHOW · Applied Survival Analysis: Regression Modeling of Time to Event Data
HUBER · Robust Statistics
HUBERTY · Applied Discriminant Analysis
HUNT and KENNEDY · Financial Derivatives in Theory and Practice
HUSKOVA, BERAN, and DUPAC · Collected Works of Jaroslav Hajek—with Commentary
IMAN and CONOVER · A Modern Approach to Statistics
JACKSON · A User's Guide to Principle Components
JOHN · Statistical Methods in Engineering and Quality Assurance
JOHNSON · Multivariate Statistical Simulation
JOHNSON and BALAKRISHNAN · Advances in the Theory and Practice of Statistics: A Volume in Honor of Samuel Kotz
JUDGE, GRIFFITHS, HILL, LÜTKEPOHL, and LEE · The Theory and Practice of Econometrics, *Second Edition*
JOHNSON and KOTZ · Distributions in Statistics
JOHNSON and KOTZ (editors) · Leading Personalities in Statistical Sciences: From the Seventeenth Century to the Present
JOHNSON, KOTZ, and BALAKRISHNAN · Continuous Univariate Distributions, Volume 1, *Second Edition*
JOHNSON, KOTZ, and BALAKRISHNAN · Continuous Univariate Distributions, Volume 2, *Second Edition*
JOHNSON, KOTZ, and BALAKRISHNAN · Discrete Multivariate Distributions
JOHNSON, KOTZ, and KEMP · Univariate Discrete Distributions, *Second Edition*
JUREČKOVÁ and SEN · Robust Statistical Procedures: Aymptotics and Interrelations
JUREK and MASON · Operator-Limit Distributions in Probability Theory
KADANE · Bayesian Methods and Ethics in a Clinical Trial Design
KADANE AND SCHUM · A Probabilistic Analysis of the Sacco and Vanzetti Evidence
KALBFLEISCH and PRENTICE · The Statistical Analysis of Failure Time Data, *Second Edition*
KASS and VOS · Geometrical Foundations of Asymptotic Inference
KAUFMAN and ROUSSEEUW · Finding Groups in Data: An Introduction to Cluster Analysis

*Now available in a lower priced paperback edition in the Wiley Classics Library.

KEDEM and FOKIANOS · Regression Models for Time Series Analysis
KENDALL, BARDEN, CARNE, and LE · Shape and Shape Theory
KHURI · Advanced Calculus with Applications in Statistics, *Second Edition*
KHURI, MATHEW, and SINHA · Statistical Tests for Mixed Linear Models
KLEIBER and KOTZ · Statistical Size Distributions in Economics and Actuarial Sciences
KLUGMAN, PANJER, and WILLMOT · Loss Models: From Data to Decisions
KLUGMAN, PANJER, and WILLMOT · Solutions Manual to Accompany Loss Models: From Data to Decisions
KOTZ, BALAKRISHNAN, and JOHNSON · Continuous Multivariate Distributions, Volume 1, *Second Edition*
KOTZ and JOHNSON (editors) · Encyclopedia of Statistical Sciences: Volumes 1 to 9 with Index
KOTZ and JOHNSON (editors) · Encyclopedia of Statistical Sciences: Supplement Volume
KOTZ, READ, and BANKS (editors) · Encyclopedia of Statistical Sciences: Update Volume 1
KOTZ, READ, and BANKS (editors) · Encyclopedia of Statistical Sciences: Update Volume 2
KOVALENKO, KUZNETZOV, and PEGG · Mathematical Theory of Reliability of Time-Dependent Systems with Practical Applications
LACHIN · Biostatistical Methods: The Assessment of Relative Risks
LAD · Operational Subjective Statistical Methods: A Mathematical, Philosophical, and Historical Introduction
LAMPERTI · Probability: A Survey of the Mathematical Theory, *Second Edition*
LANGE, RYAN, BILLARD, BRILLINGER, CONQUEST, and GREENHOUSE · Case Studies in Biometry
LARSON · Introduction to Probability Theory and Statistical Inference, *Third Edition*
LAWLESS · Statistical Models and Methods for Lifetime Data, *Second Edition*
LAWSON · Statistical Methods in Spatial Epidemiology
LE · Applied Categorical Data Analysis
LE · Applied Survival Analysis
LEE and WANG · Statistical Methods for Survival Data Analysis, *Third Edition*
LEPAGE and BILLARD · Exploring the Limits of Bootstrap
LEYLAND and GOLDSTEIN (editors) · Multilevel Modelling of Health Statistics
LIAO · Statistical Group Comparison
LINDVALL · Lectures on the Coupling Method
LINHART and ZUCCHINI · Model Selection
LITTLE and RUBIN · Statistical Analysis with Missing Data, *Second Edition*
LLOYD · The Statistical Analysis of Categorical Data
MAGNUS and NEUDECKER · Matrix Differential Calculus with Applications in Statistics and Econometrics, *Revised Edition*
MALLER and ZHOU · Survival Analysis with Long Term Survivors
MALLOWS · Design, Data, and Analysis by Some Friends of Cuthbert Daniel
MANN, SCHAFER, and SINGPURWALLA · Methods for Statistical Analysis of Reliability and Life Data
MANTON, WOODBURY, and TOLLEY · Statistical Applications Using Fuzzy Sets
MARCHETTE · Random Graphs for Statistical Pattern Recognition
MARDIA and JUPP · Directional Statistics
MASON, GUNST, and HESS · Statistical Design and Analysis of Experiments with Applications to Engineering and Science, *Second Edition*
McCULLOCH and SEARLE · Generalized, Linear, and Mixed Models
McFADDEN · Management of Data in Clinical Trials
McLACHLAN · Discriminant Analysis and Statistical Pattern Recognition
McLACHLAN and KRISHNAN · The EM Algorithm and Extensions

*Now available in a lower priced paperback edition in the Wiley Classics Library.

McLACHLAN and PEEL · Finite Mixture Models
McNEIL · Epidemiological Research Methods
MEEKER and ESCOBAR · Statistical Methods for Reliability Data
MEERSCHAERT and SCHEFFLER · Limit Distributions for Sums of Independent Random Vectors: Heavy Tails in Theory and Practice
MICKEY, DUNN, and CLARK · Applied Statistics: Analysis of Variance and Regression, *Third Edition*
*MILLER · Survival Analysis, *Second Edition*
MONTGOMERY, PECK, and VINING · Introduction to Linear Regression Analysis, *Third Edition*
MORGENTHALER and TUKEY · Configural Polysampling: A Route to Practical Robustness
MUIRHEAD · Aspects of Multivariate Statistical Theory
MULLER and STOYAN · Comparison Methods for Stochastic Models and Risks
MURRAY · X-STAT 2.0 Statistical Experimentation, Design Data Analysis, and Nonlinear Optimization
MURTHY, XIE, and JIANG · Weibull Models
MYERS and MONTGOMERY · Response Surface Methodology: Process and Product Optimization Using Designed Experiments, *Second Edition*
MYERS, MONTGOMERY, and VINING · Generalized Linear Models. With Applications in Engineering and the Sciences
NELSON · Accelerated Testing, Statistical Models, Test Plans, and Data Analyses
NELSON · Applied Life Data Analysis
NEWMAN · Biostatistical Methods in Epidemiology
OCHI · Applied Probability and Stochastic Processes in Engineering and Physical Sciences
OKABE, BOOTS, SUGIHARA, and CHIU · Spatial Tesselations: Concepts and Applications of Voronoi Diagrams, *Second Edition*
OLIVER and SMITH · Influence Diagrams, Belief Nets and Decision Analysis
PALTA · Quantitative Methods in Population Health: Extensions of Ordinary Regressions
PANKRATZ · Forecasting with Dynamic Regression Models
PANKRATZ · Forecasting with Univariate Box-Jenkins Models: Concepts and Cases
*PARZEN · Modern Probability Theory and Its Applications
PEÑA, TIAO, and TSAY · A Course in Time Series Analysis
PIANTADOSI · Clinical Trials: A Methodologic Perspective
PORT · Theoretical Probability for Applications
POURAHMADI · Foundations of Time Series Analysis and Prediction Theory
PRESS · Bayesian Statistics: Principles, Models, and Applications
PRESS · Subjective and Objective Bayesian Statistics, *Second Edition*
PRESS and TANUR · The Subjectivity of Scientists and the Bayesian Approach
PUKELSHEIM · Optimal Experimental Design
PURI, VILAPLANA, and WERTZ · New Perspectives in Theoretical and Applied Statistics
PUTERMAN · Markov Decision Processes: Discrete Stochastic Dynamic Programming
*RAO · Linear Statistical Inference and Its Applications, *Second Edition*
RAUSAND and HØYLAND · System Reliability Theory: Models, Statistical Methods, and Applications, *Second Edition*
RENCHER · Linear Models in Statistics
RENCHER · Methods of Multivariate Analysis, *Second Edition*
RENCHER · Multivariate Statistical Inference with Applications
RIPLEY · Spatial Statistics
RIPLEY · Stochastic Simulation
ROBINSON · Practical Strategies for Experimenting
ROHATGI and SALEH · An Introduction to Probability and Statistics, *Second Edition*

*Now available in a lower priced paperback edition in the Wiley Classics Library.

ROLSKI, SCHMIDLI, SCHMIDT, and TEUGELS · Stochastic Processes for Insurance and Finance
ROSENBERGER and LACHIN · Randomization in Clinical Trials: Theory and Practice
ROSS · Introduction to Probability and Statistics for Engineers and Scientists
ROUSSEEUW and LEROY · Robust Regression and Outlier Detection
RUBIN · Multiple Imputation for Nonresponse in Surveys
RUBINSTEIN · Simulation and the Monte Carlo Method
RUBINSTEIN and MELAMED · Modern Simulation and Modeling
RYAN · Modern Regression Methods
RYAN · Statistical Methods for Quality Improvement, *Second Edition*
SALTELLI, CHAN, and SCOTT (editors) · Sensitivity Analysis
*SCHEFFE · The Analysis of Variance
SCHIMEK · Smoothing and Regression: Approaches, Computation, and Application
SCHOTT · Matrix Analysis for Statistics
SCHOUTENS · Levy Processes in Finance: Pricing Financial Derivatives
SCHUSS · Theory and Applications of Stochastic Differential Equations
SCOTT · Multivariate Density Estimation: Theory, Practice, and Visualization
*SEARLE · Linear Models
SEARLE · Linear Models for Unbalanced Data
SEARLE · Matrix Algebra Useful for Statistics
SEARLE, CASELLA, and McCULLOCH · Variance Components
SEARLE and WILLETT · Matrix Algebra for Applied Economics
SEBER and LEE · Linear Regression Analysis, *Second Edition*
SEBER · Multivariate Observations
SEBER and WILD · Nonlinear Regression
SENNOTT · Stochastic Dynamic Programming and the Control of Queueing Systems
*SERFLING · Approximation Theorems of Mathematical Statistics
SHAFER and VOVK · Probability and Finance: It's Only a Game!
SMALL and McLEISH · Hilbert Space Methods in Probability and Statistical Inference
SRIVASTAVA · Methods of Multivariate Statistics
STAPLETON · Linear Statistical Models
STAUDTE and SHEATHER · Robust Estimation and Testing
STOYAN, KENDALL, and MECKE · Stochastic Geometry and Its Applications, *Second Edition*
STOYAN and STOYAN · Fractals, Random Shapes and Point Fields: Methods of Geometrical Statistics
STYAN · The Collected Papers of T. W. Anderson: 1943–1985
SUTTON, ABRAMS, JONES, SHELDON, and SONG · Methods for Meta-Analysis in Medical Research
TANAKA · Time Series Analysis: Nonstationary and Noninvertible Distribution Theory
THOMPSON · Empirical Model Building
THOMPSON · Sampling, *Second Edition*
THOMPSON · Simulation: A Modeler's Approach
THOMPSON and SEBER · Adaptive Sampling
THOMPSON, WILLIAMS, and FINDLAY · Models for Investors in Real World Markets
TIAO, BISGAARD, HILL, PEÑA, and STIGLER (editors) · Box on Quality and Discovery: with Design, Control, and Robustness
TIERNEY · LISP-STAT: An Object-Oriented Environment for Statistical Computing and Dynamic Graphics
TSAY · Analysis of Financial Time Series
UPTON and FINGLETON · Spatial Data Analysis by Example, Volume II: Categorical and Directional Data
VAN BELLE · Statistical Rules of Thumb
VAN BELLE, FISHER, HEAGERTY, and LUMLEY · Biostatistics: A Methodology for the Health Sciences, *Second Edition*

*Now available in a lower priced paperback edition in the Wiley Classics Library.

VESTRUP · The Theory of Measures and Integration
VIDAKOVIC · Statistical Modeling by Wavelets
WEISBERG · Applied Linear Regression, *Second Edition*
WELSH · Aspects of Statistical Inference
WESTFALL and YOUNG · Resampling-Based Multiple Testing: Examples and Methods for p-Value Adjustment
WHITTAKER · Graphical Models in Applied Multivariate Statistics
WINKER · Optimization Heuristics in Economics: Applications of Threshold Accepting
WONNACOTT and WONNACOTT · Econometrics, *Second Edition*
WOODING · Planning Pharmaceutical Clinical Trials: Basic Statistical Principles
WOOLSON and CLARKE · Statistical Methods for the Analysis of Biomedical Data, *Second Edition*
WU and HAMADA · Experiments: Planning, Analysis, and Parameter Design Optimization
YANG · The Construction Theory of Denumerable Markov Processes
*ZELLNER · An Introduction to Bayesian Inference in Econometrics
ZHOU, OBUCHOWSKI, and McCLISH · Statistical Methods in Diagnostic Medicine

*Now available in a lower priced paperback edition in the Wiley Classics Library.